Stochastic Mechanics
Random Media
Signal Processing
and Image Synthesis
Mathematical Economics and Finance
Stochastic Optimization
Stochastic Control
Stochastic Models in Life Sciences

Stochastic Modelling and Applied Probability

(Formerly:
Applications of Mathematics)

69

For further volumes:
http://www.springer.com/series/602

Etienne Pardoux • Aurel Răşcanu

Stochastic Differential Equations, Backward SDEs, Partial Differential Equations

Springer

Etienne Pardoux
Institut de Mathématiques de Marseille
Université d'Aix–Marseille
Marseille, France

Aurel Răşcanu
Faculty of Mathematics,
Alexandru Ioan Cuza University
Iaşi, Romania

ISSN 0172-4568 Stochastic Modelling and Applied Probability
ISSN 2197-439X (electronic)
ISBN 978-3-319-05713-2 ISBN 978-3-319-05714-9 (eBook)
DOI 10.1007/978-3-319-05714-9
Springer Cham Heidelberg New York Dordrecht London

Library of Congress Control Number: 2014941443

Mathematics Subject Classification (2010): 60H05, 60H10, 60J60, 35D40

Printed on acid-free paper

Springer is part of Springer Science+Business Media (www.springer.com)

Contents

Introduction

The main goal of this monograph is to present the theories of stochastic differential equations (in short SDEs), backward stochastic differential equations (in short BSDEs), and their connections with linear and semilinear second order partial differential equations (in short PDEs) both of elliptic and parabolic type, with various types of boundary conditions. In particular, we shall present an original version of the celebrated Feynman–Kac formula. But one of our main goals in this monograph is to present an extension of that formula to semilinear PDEs, with the help of a coupled system of forward and backward SDEs.

There has been in the past at least three ways of extending the Feynman–Kac formula to nonlinear equations. One is to replace the diffusion $\{X_t : t \geq 0\}$ by a controlled diffusion (see Fleming and Soner [31]), the second is to replace it by a branching-diffusion process (or a "superprocess", see e.g. Dynkin [25]), the third is to replace it by a "nonlinear Markov process" in the sense that the evolution of X_t depends not only on X_t but also on its probability law, see e.g. McKean [45]. What we shall present here is a fourth such nonlinear generalization of the Feynman–Kac formula, based on BSDEs.

The book starts with a preliminary chapter presenting several tools from probability theory and stochastic processes. Chapter 2 gives a complete introduction to Itô's stochastic calculus, including the Itô integral, Itô's change of variable formula, a martingale representation theorem (which we discuss in great generality), and Girsanov's theorem. Of course, almost all of this material is presented in existing textbooks. However, some of our proofs and even a few results are new. Moreover, we introduce here some specific techniques of proofs which are used repeatedly later.

Chapter 3 gives a complete treatment of the theory of strong solutions of very general SDEs driven by a Brownian motion, whose coefficients may be random. We treat both the classical case of Lipschitz coefficients and the case where the drift satisfies a monotonicity type condition (which is a type of one-sided Lipschitz condition). Apparently this type of condition first appeared in the context of nonlinear parabolic PDEs. It was then transferred to stochastic PDEs, before Jacod [40] and Krylov–Gyöngy [36] formulated the condition for strong solutions of SDEs. We consider both global and local monotonicity conditions. Some of the results presented here are new. We next consider SDEs with deterministic coefficients and establish the Markov property of its solution. We finally discuss the connection with partial differential equations of second order, both parabolic and elliptic, in the whole space as well as with Dirichlet boundary condition. We thus establish a new version of the Feynman–Kac formula, which says that a certain functional of the solution of an SDE is a viscosity solution of a PDE.

Chapter 4 starts with SDEs with a multivalued drift, which can be, for instance, the subdifferential of a convex function. In the case of the subdifferential of the indicator of a convex set, we obtain one way to construct an SDE reflected at the boundary of a convex set. We extend the same methodology to the study of SDEs reflected at the boundary of a (reasonably smooth) arbitrary domain. We consider both normal and oblique reflection. We then study the Markov property of the solution of reflected SDEs and establish the Feynman–Kac formula for parabolic and elliptic PDEs with Neumann boundary condition.

Chapter 5 studies BSDEs. Again we consider those equations with a Lipschitz coefficient, as well as in the case of a monotonicity type condition. Note that unlike in the case of forward SDEs, there is no known general existence and uniqueness theory for the case of locally Lipschitz or locally monotone coefficients. We consider BSDEs with a coefficient which is the subdifferential of a convex function. This allows us to study BSDEs reflected at the boundary of a convex set. Note that the corresponding problem for BSDEs reflected at the boundary of a non-convex domain is presently still open. BSDEs, coupled with a forward SDE, are then used to give a probabilistic representation of solutions of semilinear parabolic and elliptic PDEs with various boundary conditions. Note that BSDEs with a deterministic final time are associated with parabolic PDEs, while the probabilistic representation of semilinear elliptic PDEs requires BSDEs with random final time.

Finally the last chapter contains various technical results which are used in the book. Most of them are known. In particular several uniqueness results for viscosity solutions of second order PDEs and systems of PDEs are discussed. Note that most of the probabilistic representation results for PDEs are given in terms of viscosity solutions of PDEs. This is a specific aspect of our treatment, which allows minimal assumptions on the coefficients and permits us to avoid any non-degeneracy assumptions on the matrix of second order coefficients. We do not presuppose the existence of a solution to our PDEs. Existence is provided by the probabilistic representation formula. On the other hand, uniqueness results for the same PDEs are established by purely analytic arguments.

We claim that we present in this monograph a rather complete treatment of the connections between SDEs and BSDEs on one side and linear and semilinear PDEs on the other. We regret that we have not covered the connection between fully coupled forward–backward stochastic differential equations (in short FBSDEs) and quasilinear PDEs. This addition would have probably made the book too long. We refer the reader to the papers of Delarue [22] and Delarue and Guatteri [23] for recent results on FBSDEs. See also Ma and Yong [44].

This book is intended as a reference manual and requires from its reader a good knowledge of analysis, measure, integration and probability theory. The reader who has a good knowledge of stochastic processes and Itô calculus can of course skip the first two chapters. He/she will be referred to them for specific technical results which are used in further chapters.

Marseille, France
Iaşi, Romania
April 2013

Etienne Pardoux
Aurel Răşcanu

Notations

- Abbreviations:
  r.v. $\stackrel{def}{=}$ Random variable(s);
  s.p. $\stackrel{def}{=}$ Stochastic process(es);
  BM $\stackrel{def}{=}$ Brownian motion;
  a.s. $\stackrel{def}{=}$ Almost sure;
  a.e. $\stackrel{def}{=}$ Almost everywhere;
  $\mathcal{P}$-m.s.p. $\stackrel{def}{=}$ Progressively measurable stochastic process(es);
  $\mathcal{P}$-m.i.s.p. $\stackrel{def}{=}$ Progressively measurable increasing stochastic process(es);
  ($\{X_t : t \geq 0\}$ is increasing if $t < s$ implies $X_t \leq X_s$);
  $\mathcal{P}$-m.b-v.s.p. $\stackrel{def}{=}$ Progressively measurable bounded-variation stochastic process(es);
  $\mathcal{P}$-m.c.s.p. $\stackrel{def}{=}$ Progressively measurable continuous stochastic process(es);
  $\mathcal{P}$-m.i.c.s.p. $\stackrel{def}{=}$ Progressively measurable increasing continuous stochastic process(es);
  $\mathcal{P}$-m.b-v.c.s.p. $\stackrel{def}{=}$ Progressively measurable bounded-variation continuous stochastic process(es);
  SDE $\stackrel{def}{=}$ Stochastic differential equation(s);
  BSDE $\stackrel{def}{=}$ Backward stochastic differential equation(s);
  PDE $\stackrel{def}{=}$ Partial differential equation(s)
- $\mathbb{N} = \{0, 1, 2, \cdots\}$ denotes the set of natural number; $\mathbb{N}^* = \mathbb{N} \setminus \{0\}$; if $k, n \in \mathbb{N}$, $k \leq n$, then

$$\overline{k,n} \stackrel{def}{=} \{i \in \mathbb{N} : k \leq i \leq n\}.$$

- $\mathbb{R}$ is the set of real numbers; $\mathbb{Q}$ is the set of rational numbers; $\mathbb{C}$ is the set of complex numbers; $\mathbb{R}^* = \mathbb{R} \setminus \{0\}$; $\mathbb{Q}^* = \mathbb{Q} \setminus \{0\}$;

$\mathbb{R}_+ = [0, \infty[\,, \quad \mathbb{R}_+^* = \mathbb{R}_+ \setminus \{0\},$
$\overline{\mathbb{R}} = \mathbb{R} \cup \{-\infty, +\infty\}, \quad \overline{\mathbb{R}}_+ = [0, +\infty],$
$\mathbb{R}^d = \left\{ (x_i)_{d\times 1} : x_i \in \mathbb{R},\ i \in \overline{1,d} \right\};$
$\mathbb{R}^{d\times k} = \left\{ \left(x_{i,j}\right)_{d\times k} : x_{i,j} \in \mathbb{R}, i \in \overline{1,d},\ j \in \overline{1,k} \right\}.$

- $f : \mathbb{R} \to \mathbb{R}$ is an increasing function (or equivalently nondecreasing) if for all $x < y$ it follows that $f(x) \le f(y)$.
- $f : \mathbb{R} \to \mathbb{R}$ is a strictly increasing function if for all $x < y$ it follows that $f(x) < f(y)$.
- $f : \mathbb{R} \to \mathbb{R}$ is a decreasing function (or equivalently nonincreasing) if for all $x < y$ it follows that $f(x) \ge f(y)$.
- $f : \mathbb{R} \to \mathbb{R}$ is a strictly decreasing function if for all $x < y$ it follows that $f(x) > f(y)$.
- $\mathbb{S}^d \subset \mathbb{R}^{d\times d}$ is the set of symmetric matrices.
- $\mathbb{X}$, $\mathbb{Y}$ are metric spaces; $\mathbb{H}$ is a Hilbert space.
- $2^{\mathbb{X}} \overset{def}{=} \{S : S \subset \mathbb{X}\}$.
- $A : \mathbb{X} \rightrightarrows \mathbb{Y}$ means $A : \mathbb{X} \to 2^{\mathbb{Y}}$ (a multivalued operator from $\mathbb{X}$ to $\mathbb{Y}$).
- $\inf \emptyset = +\infty$.
- If $a, b \in \mathbb{R}$, then

$$a \vee b = \max\{a,b\}, \quad a \wedge b = \min\{a,b\},$$
$$a^+ = \max\{0,a\}, \quad a^- = \max\{0,-a\}.$$

- $\lceil x \rceil$ denotes the smallest integer larger than or equal to $x \in \mathbb{R}$.
- $\lfloor x \rfloor$ denotes the largest integer smaller than or equal to $x \in \mathbb{R}$.
- If $x \in \mathbb{R}^d$ and $r \in \mathbb{R}_+$, then

$$B(x,r) \overset{def}{=} \left\{ y \in \mathbb{R}^d : |y - x| < r \right\},$$
$$\bar{B}(x,r) = \overline{B(x,r)} \overset{def}{=} \left\{ y \in \mathbb{R}^d : |y - x| \le r \right\}.$$

- $\mathrm{cl}(D) = \overline{D} \overset{def}{=}$ the closure of the set D.
- $\mathrm{int}(G) \overset{def}{=} \{x \in G : \exists\, r > 0,\ B(x,r) \subset G\}$.
- $\partial D = \mathrm{Bd}(D) \overset{def}{=} \overline{D} \setminus \mathrm{int}(G)$.
- If $x = (x_i)_{d\times 1} \in \mathbb{R}^d$, then

$$|x| \overset{def}{=} \left(x_1^2 + \cdots + x_d^2\right)^{1/2}, \qquad x^+ \overset{def}{=} \left(x_i^+\right)_{d\times 1}.$$

- If x is a vector or a matrix, then x^* denotes its transposed, and

$$\mathbf{1} \overset{def}{=} (1, \ldots, 1)^* \in \mathbb{R}^d.$$

- The indicator function of the set A is the function $\mathbf{1}_A$ defined by

$$\mathbf{1}_A(s) \overset{def}{=} \begin{cases} 1, \text{ if } s \in A, \\ 0, \text{ if } s \notin A. \end{cases}$$

- The convex indicator function of the set A is the function $\mathbf{I}_A$ defined by

$$\mathbf{I}_A(s) \overset{def}{=} \begin{cases} 0, & \text{if } s \in A, \\ +\infty, & \text{if } s \notin A. \end{cases}$$

- $(\Omega, \mathcal{F}, \mathbb{P})$ is a complete probability space.
- $(\Omega, \mathcal{F}, \mathbb{P}, \{\mathcal{F}_t\}_{t\geq 0})$ is a stochastic basis (see Sect. 1.1.2).
- $\mathcal{P} = \mathcal{P}(\mathcal{F}_t)$ is the σ-algebra of progressively measurable subsets of $\Omega \times \mathbb{R}_+$ associated to the filtration $\{\mathcal{F}_t : t \geq 0\}$ (see Sect. 1.1.2).
- $\mathcal{F} \vee \mathcal{G}$ denotes the smallest σ-algebra containing $\mathcal{F} \cup \mathcal{G}$, where $\mathcal{F}$ and $\mathcal{G}$ are two σ-algebras of subsets of Ω.
- $\mathcal{B}_d$ denotes the Borel σ-algebra over $\mathbb{R}^d$.
- $\{X \in B\} \overset{def}{=} \{\omega \in \Omega : X(\omega) \in B\}$.
- $\mathbb{P}(X \in B) = \mathbb{P}(\{X \in B\})$.
- $\mathbb{E}X$ is the expectation of the random variable X.
- $\mathbb{E}(X; A) = \mathbb{E}(X\,\mathbf{1}_A)$.
- $\mathbb{E}|X|^p = \mathbb{E}(|X|^p)$.
- $L^0(\Omega, \mathcal{F}, \mathbb{P}; \mathbb{R}^d)$ is the space of (equivalence classes of) random variables $X : \Omega \to \mathbb{R}^d$.
- $L^p(\Omega, \mathcal{F}, \mathbb{P}; \mathbb{R}^d)$, $p > 0$, is the space of (equivalence classes of) random variables $X : \Omega \to \mathbb{R}^d$ such that $\mathbb{E}|X|^p < \infty$ (see Sect. 1.1.1).
- $L^p(\Omega, \mathcal{F}, \mathbb{P}) \overset{def}{=} L^p(\Omega, \mathcal{F}, \mathbb{P}; \mathbb{R})$.
- If $x : [a, b] \to \mathbb{R}^d$, then we define

$$x(t-) = x(t-0) \overset{def}{=} \lim_{\varepsilon \searrow 0} x(t-\varepsilon), \quad x(t+) = x(t+0) \overset{def}{=} \lim_{\varepsilon \searrow 0} x(t+\varepsilon)$$

and

$$\|x\|_{[a,b]} \overset{def}{=} \sup\{|x(s)| : a \leq s \leq b\}, \quad \|x\|_T = \|x\|_{[0,T]}.$$

- If $H : \Omega \times [0, \infty[\to \mathbb{R}^d$, then we define

$$H(\omega, t) = H_t(\omega), \quad H(\cdot, t) = H_t,$$

$$H_{t-} = H_{t-0} \overset{def}{=} \lim_{\varepsilon \searrow 0} H_{t-\varepsilon}, \quad H_{t+} = H_{t+0} \overset{def}{=} \lim_{\varepsilon \searrow 0} H_{t+\varepsilon}$$

and

$$\|H\|_t = \sup\{|H_s| : 0 \le s \le t\}.$$

- If $H : \Omega \times [0, \infty[\times \mathbb{R}^d \to \mathbb{R}^{d \times k}$, then we define

$$H^{\#}_{u,R}(t) = \sup\{|H(t, u + x)| : |x| \le R\},$$
$$H^{\#}_{R}(t) = \sup\{|H(t, x)| : |x| \le R\}.$$

- $\updownarrow g \updownarrow_{[a,b]} \overset{def}{=} \sup\left\{\sum_{i=0}^{n-1} |g(t_{i+1}) - g(t_i)| : n \in \mathbb{N}^*, a = t_0 < \cdots < t_n = b\right\}$ is the total variation of g on $[a, b]$.
- $\updownarrow g \updownarrow_T \overset{def}{=} \updownarrow g \updownarrow_{[0,T]}$.
- $C([0, T]; \mathbb{R}^d)$ is the space of continuous functions $x : [0, T] \to \mathbb{R}^d$; equipped with the norm $\|x\|_T$, $C([0, T]; \mathbb{R}^d)$ is a Banach space.
- $C([0, \infty[; \mathbb{R}^d) = C(\mathbb{R}_+; \mathbb{R}^d)$ is the space of continuous functions $x : [0, \infty[\to \mathbb{R}^d$.
- $BV([0, T]; \mathbb{R}^d)$ is the space of all functions $g : [0, T] \to \mathbb{R}^d$ such that $\updownarrow g \updownarrow_T < \infty$; equipped with the norm

$$||g||_{BV([0,T];\mathbb{R}^d)} \overset{def}{=} |g(0)| + \updownarrow g \updownarrow_T.$$

  $BV\left([0, T]; \mathbb{R}^d\right)$ is a Banach space. We identify it with the dual of $C([0, T]; \mathbb{R}^d)$.
- If $\psi : \mathbb{R}^d \to \mathbb{R}$ is a continuous function of class C^2, we define

$$\nabla_x \psi(x) = \psi'_x(x) = \left(\frac{\partial \psi(x)}{\partial x_i}\right)_{d \times 1} \in \mathbb{R}^d$$

  the gradient of ψ with respect to x, and

$$D^2_{xx}\psi(x) = \psi''_{xx}(x) = \left(\frac{\partial^2 \psi(x)}{\partial x_i \partial x_j}\right)_{d \times d} \in \mathbb{R}^{d \times d}$$

  the Hessian matrix of ψ with respect to x.
- Given a function $f : \mathbb{R}^d \to \mathbb{R}$, its support is defined as

$$\text{supp}(f) \overset{def}{=} \overline{\{x \in \mathbb{R}^d : f(x) \neq 0\}}.$$

- $C_c^\infty\left(\mathbb{R}^d\right)$ is the space of functions $f : \mathbb{R}^d \to \mathbb{R}$ of class C^∞ with compact support.
- $C_c\left(\mathbb{R}^d\right)$ is the space of continuous functions $f : \mathbb{R}^d \to \mathbb{R}$ with compact support.

- $C_0(\mathbb{R}^d)$ is the space of continuous functions $f:\mathbb{R}^d \to \mathbb{R}$ satisfying $\lim_{|x|\to\infty} |f(x)| = 0$.
- $C_b(\mathbb{R}^d)$ is the space of bounded continuous functions $f : \mathbb{R}^d \to \mathbb{R}$.
- $B_b(\mathbb{R}^d)$ is the space of bounded Borel measurable functions $f : \mathbb{R}^d \to \mathbb{R}$.
- $S_d^p[0,T]$ is the space of (equivalence classes of) $\mathcal{P}$-measurable continuous stochastic processes $X : \Omega \times [0,T] \to \mathbb{R}^d$ such that:

$$\mathbb{E} \sup_{t\in[0,T]} |X_t|^p < +\infty, \quad \text{if } p > 0;$$

two processes X, Y are equivalent if $(X_t = Y_t,\ \forall t \in [0,T])$ $\mathbb{P}$-a.s.
- S_d^p is the space of (equivalence classes of) $\mathcal{P}$-measurable continuous stochastic processes $X : \Omega \times [0,+\infty[\to \mathbb{R}^d$ such that for all $T > 0$ the restriction $X_{|[0,T]}$ of X to $[0,T]$ belongs to $S_d^p[0,T]$.
- $S_d^p([0,T];V) \overset{def}{=} \{X \in S_d^0[0,T] : VX \in S_d^p[0,T]\}$; $V:\Omega\times[0,+\infty) \to (0,\infty)$ is a $\mathcal{P}$-measurable continuous stochastic process.
- $\Lambda_d^p(0,T)$ is the space of (equivalent classes of) $\mathcal{P}$-measurable processes $X:\Omega\times\,]0,T[\, \to \mathbb{R}^d$ such that

$$\int_0^T |X_t|^2\,dt < +\infty, \quad \mathbb{P}\text{-}a.s.\ \omega \in \Omega, \quad \text{if } p = 0,$$

and

$$\mathbb{E}\left(\int_0^T |X_t|^2\,dt\right)^{p/2} < +\infty, \quad \text{if } p > 0;$$

two processes X, Y are equivalent if $(X_t = Y_t$ a.e. $t \in\,]0,T[)$ $\mathbb{P}$-a.s. $\omega \in \Omega$.
- Λ_d^p is the space of (equivalence classes of) $\mathcal{P}$-measurable processes $X : \Omega \times\,]0,+\infty[\, \to \mathbb{R}^d$ such that for all $T > 0$ the restriction $X_{|]0,T[}$ of X to $]0,T[$ belongs to $\Lambda_d^p(0,T)$.
- $\Lambda_d^p(0,T;V) \overset{def}{=} \{X \in \Lambda_d^0(0,T) : VX \in \Lambda_d^p(0,T)\}$; $V : \Omega\times[0,+\infty) \to (0,\infty)$ is a $\mathcal{P}$-measurable continuous stochastic process.
- $\mathcal{M}_d^p[0,T]$, $\mathcal{M}_d^p$, are spaces of d-dimensional continuous p-martingales.
- $\mathcal{M}_d^0[0,T]$, $\mathcal{M}_d^0$, are spaces of d-dimensional continuous local martingales.
- In the case $d = 1$ the subscript d in the notations $S_d^p[0,T]$, S_d^p, $\Lambda_d^p(0,T)$, Λ_d^p, $\mathcal{M}_d^p[0,T]$, $\mathcal{M}_d^0[0,T]$, $\mathcal{M}_d^0$ will usually be omitted.
- $\{<M>_t; t \ge 0\}$ is the scalar increasing continuous stochastic process and $\{<<M>>_t : t \ge 0\}$ is the symmetric matrix increasing continuous stochastic process associated to a d-dimensional continuous (local) martingale $\{M_t; t \ge 0\}$ (Theorem 1.69 and Proposition 1.70).
- The same letter C will be used repeatedly to denote various constants; the notation C_a will be used to insist upon the fact that this constant depends only upon the parameter a and nothing else.

Chapter 1
Background of Stochastic Analysis

The goal of this chapter is to introduce several tools from the theory of probability theory and stochastic processes, which will be useful throughout this book. After presenting several basic results from the theory of stochastic processes, we will discuss continuous martingales, and finally we introduce Brownian motion, which will be central to the text.

Although the material in this chapter is mostly standard, our presentation has some originality. We present both a criterion for tightness and a criterion for uniform integrability which are not so well-known, and at least one inequality in Proposition 1.56 seems to be new. But the main reason for including this material here is to introduce some specific tools which will be used systematically in later chapters, notably the notion of a "basic partition", see Definition 1.54.

1.1 Preliminaries

1.1.1 Preliminaries of Probability Theory

We assume given a probability space $(\Omega, \mathcal{F}, \mathbb{P})$. Let

$$\mathcal{N} = \{A : A \subset \Omega,\ \exists\ N \in \mathcal{F}, \mathbb{P}(N) = 0 \text{ and } A \subset N\}$$

be the collection of null probability sets. The probability space $(\Omega, \mathcal{F}, \mathbb{P})$ is complete if $\mathcal{N} \subset \mathcal{F}$. In this book the probability space $(\Omega, \mathcal{F}, \mathbb{P})$ will always be considered as completed by the collection of null probability sets.

If $\mathbb{X}$ is a topological space, then $\mathcal{B}_{\mathbb{X}}$ denotes the Borel σ-algebra over $\mathbb{X}$, that is the σ-algebra generated by the family of open subsets of $\mathbb{X}$; in particular $\mathcal{B}_d \stackrel{def}{=} \mathcal{B}_{\mathbb{R}^d}$ and $\mathcal{B} \stackrel{def}{=} \mathcal{B}_1$.

E. Pardoux and A. Răşcanu, *Stochastic Differential Equations, Backward SDEs, Partial Differential Equations*, Stochastic Modelling and Applied Probability 69,
DOI 10.1007/978-3-319-05714-9_1,

A mapping $X : \Omega \to \mathbb{X}$ is an ($\mathbb{X}$-valued) random variable if for all $B \in \mathcal{B}_{\mathbb{X}}$

$$\{X \in B\} \overset{def}{=} \{\omega \in \Omega : X(\omega) \in B\} \in \mathcal{F}.$$

If $\mathbb{X} = \mathbb{R}$ then X will be called a (real) random variable (or scalar random variable). If $\mathbb{X} = \mathbb{R}^d$ then X will be called a d-dimensional random vector (or a d-dimensional random variable).

If $A \subset \Omega$ and $X : \Omega \to \mathbb{R}$ is given by

$$X(\omega) = \mathbf{1}_A(\omega) = \begin{cases} 1, & \text{if } \omega \in A, \\ 0, & \text{if } \omega \in \Omega \setminus A, \end{cases}$$

then X is random variable iff $A \in \mathcal{F}$.

The probability measure $\mathcal{L}(X) = \mathbb{P}_X : \mathcal{B}_{\mathbb{X}} \to [0,1]$ defined by $\mathbb{P}_X(B) = \mathbb{P}(X \in B)$ is called the (probability) law of X. We shall write $X \sim Q$ if $\mathbb{P}_X = Q$.

Let $X : (\Omega; \mathcal{F}, \mathbb{P}) \to \mathbb{X}$ be a random vector. Then

$$\sigma(X) = \{X^{-1}(B) : B \in \mathcal{B}_{\mathbb{X}}\}$$

is a σ-algebra of subsets of Ω (called the σ-algebra generated by X). Since X is a random variable, $\sigma(X) \subset \mathcal{F}$, and $\sigma(X)$ is the smallest σ-algebra which makes X measurable. It is the class of events for which one knows whether or not they are realized, once $X(\omega)$ is observed. In this sense $\sigma(X)$ represents the information carried by X. We also define

$$\mathcal{F}^X = \sigma(X) \vee \mathcal{N}$$

to be the smallest σ-algebra which contains both $\sigma(X)$ and $\mathcal{N}$. In probabilistic terms, the σ-algebra $\mathcal{F}^X$ can be interpreted as containing all *relevant information* about the random variable X (see [61], Section 1.2, Proposition 3):

Lemma 1.1. *Let X (resp. Y) be a d (resp. k)-dimensional random vector defined on the probability space $(\Omega, \mathcal{F}, \mathbb{P})$. The following two conditions are equivalent:*

a) there exists a Borel measurable function $f : \mathbb{R}^d \to \mathbb{R}^k$ such that $Y = f(X)$;
b) $\sigma(Y) \subset \sigma(X)$.

Recall that the σ-algebras $\mathcal{G}_1, \ldots, \mathcal{G}_n$ are said to be independent whenever

$$\mathbb{P}\left(\bigcap_{i=1}^n G_i\right) = \prod_{i=1}^n \mathbb{P}(G_i), \quad \forall\, G_i \in \mathcal{G}_i,\ 1 \le i \le n.$$

To an arbitrary collection $\{X_i,\ i \in I\}$ of random variables, we can associate the σ-algebra $\sigma(X_i\,;\ i \in I)$, which is the smallest σ-algebra containing $\bigcup_{i \in I} \sigma(X_i)$.

We denote by $\mathcal{L}^0(\Omega, \mathcal{F}, \mathbb{P}; \mathbb{X})$ the space of random variables $X : \Omega \to \mathbb{X}$. If $X, Y \in \mathcal{L}^0(\Omega, \mathcal{F}, \mathbb{P}; \mathbb{X})$ are random variables, we say $X = Y$ a.s. (almost surely) if $\mathbb{P}(X = Y) = 1$. This is an equivalence relation. We can partition the set of random variables into equivalence classes with respect to this relation. The space of equivalence classes will be denoted by $L^0(\Omega, \mathcal{F}, \mathbb{P}; \mathbb{X})$ and we shall usually not distinguish a random variable X from its equivalence class.

If $X : \Omega \to \mathbb{R}$ is of the form

$$X = \sum_{k=1}^{n} a_k \mathbf{1}_{A_k}, \quad a_k \in \mathbb{R}, \; A_k \in \mathcal{F}, \; k \in \overline{1, n},$$

we say that $X \in \mathcal{S}$ and the expectation of X is defined by

$$\mathbb{E}X \stackrel{def}{=} \sum_{k=1}^{n} a_k \mathbb{P}(A_k).$$

If $X : \Omega \to [0, \infty]$ is a random variable then

$$\mathbb{E}X \stackrel{def}{=} \sup_{Y \in \mathcal{S}, 0 \leq Y \leq X} \mathbb{E}Y.$$

If $X : \Omega \to \mathbb{R}$ is a random variable such that $\mathbb{E}X^+ < \infty$ or $\mathbb{E}X^- < \infty$, then the expectation $\mathbb{E}X$ exists and

$$\mathbb{E}X \stackrel{def}{=} \mathbb{E}X^+ - \mathbb{E}X^-.$$

If $X = (X_1, \cdots, X_d)^* : \Omega \to \mathbb{R}^d$ is a random variable such that $\mathbb{E}|X| < \infty$, then we say that X is integrable and

$$\mathbb{E}X \stackrel{def}{=} (\mathbb{E}X_1, \cdots, \mathbb{E}X_d)^*;$$

in particular $\mathbb{E}(X_1 + iX_2) \stackrel{def}{=} \mathbb{E}X_1 + i\mathbb{E}X_2$.

In other words the expectation of a random variable is the integral

$$\mathbb{E}X = \int_{\Omega} X(\omega)\, d\mathbb{P}(\omega)$$

(we shall also use the notation $\mathbb{P}(d\omega)$ for $d\mathbb{P}(\omega)$). For $A \in \mathcal{F}$ we define

$$\mathbb{E}(X; A) = \mathbb{E}(X\mathbf{1}_A) = \int_A X(\omega)\, d\mathbb{P}(\omega).$$

Denote by $L^p\left(\Omega, \mathcal{F}, \mathbb{P}; \mathbb{R}^d\right)$, $p > 0$, the linear space of random variables (equivalence classes) $X : \Omega \to \mathbb{R}^d$, such that $\mathbb{E}|X|^p < \infty$.

Theorem 1.2. *$L^p\left(\Omega,\mathcal{F},\mathbb{P};\mathbb{R}^d\right)$ is a complete metric linear space with respect to the metric*

$$\rho(X,Y)=\begin{cases}\mathbb{E}\left(1\wedge|X-Y|\right), & \text{if } p=0\\ \mathbb{E}\ |X-Y|^p, & \text{if } 0<p<1,\end{cases}$$

and it is a Banach space with respect to the norm

$$\|X\|=\begin{cases}\left(\mathbb{E}\ |X|^p\right)^{1/p}, & \text{if } 1\le p<\infty,\\ \inf\{M: |X|\le M,\ a.s.\}, & \text{if } p=\infty.\end{cases}$$

Proposition 1.3 (Markov–Chebyshev). *Let $X:\Omega\to\mathbb{R}$ be a random variable and $g:\mathbb{R}\to[0,\infty[$ be a Borel measurable function such that, for all $0\le x<y$, it follows that $g(x)\le g(y)$ and $g(y)>0$. Then*

$$\mathbb{P}(X\ge\varepsilon)\le\frac{\mathbb{E}g(X)}{g(\varepsilon)},\quad\forall\,\varepsilon>0. \tag{MC1}$$

If moreover there exists an $M>0$ such that $g(|X|)\le M$ a.s., then

$$\frac{\mathbb{E}g(|X|)-g(\varepsilon)}{M}\le\mathbb{P}(|X|\ge\varepsilon),\quad\forall\,\varepsilon>0. \tag{MC2}$$

Proof. The inequalities follow by taking the expectation in

$$g(\varepsilon)\,\mathbf{1}_{X\ge\varepsilon}\le g(X)\quad\text{and}\quad g(|X|)\le g(\varepsilon)+M\,\mathbf{1}_{|X|\ge\varepsilon}.$$

■

Corollary 1.4. *Let $X:\Omega\to\mathbb{R}$ be a random variable. Then for all $\varepsilon,\delta>0$*

$$\mathbb{P}(|X|\ge\varepsilon)\le\frac{\mathbb{E}\left(|X|^r\right)}{\varepsilon^r},$$

and

$$\frac{\mathbb{E}(\delta\wedge|X|)-\delta\wedge\varepsilon}{\delta}\le\mathbb{P}(|X|\ge\varepsilon)\le\frac{\mathbb{E}(\delta\wedge|X|)}{\delta\wedge\varepsilon}.$$

Let $X_n,\ X:(\Omega;\mathcal{F},\mathbb{P})\to\mathbb{R}^d$, $n\in\mathbb{N}$, be random variables. The following types of convergence will be used in this book:

- $X_n\to X$ a.s. if there exists an $\Omega_0\in\mathcal{F}$, $\mathbb{P}(\Omega_0)=1$ such that

$$\lim_{n\to\infty}X_n(\omega)=X(\omega),\ \text{for all } \omega\in\Omega_0\ ;$$

- $X_n \to X$ *in probability* if for all $\varepsilon > 0$:

$$\lim_{n\to\infty} \mathbb{P}(|X_n - X| \geq \varepsilon) = 0,$$

or equivalently

$$\lim_{n\to\infty} \mathbb{E}(1 \wedge |X_n - X|) = 0,$$

or equivalently

$$\lim_{n\to\infty} \mathbb{E}g(|X_n - X|) = 0,$$

where $g : [0,\infty[\to [0,\infty[$ is any bounded increasing continuous function such that $0 = g(0) < g(x)$ for all $x > 0$.
We shall use this equivalence for the following functions

$$g(x) = \delta \wedge x, \quad 0 \leq g(x) \leq \delta,$$
$$g(x) = \frac{x^r}{\delta + x^r}, \quad 0 \leq g(x) \leq 1,$$
$$g(x) = \frac{x}{\sqrt{1+\delta x^2}}, \quad 0 \leq g(x) \leq \frac{1}{\delta},$$

where $r, \delta > 0$ are arbitrary constants;
- $X_n \to X$ *in* L^p, $p > 0$, if $X_n, X \in L^p$ and

$$\lim_{n\to\infty} \mathbb{E}|X_n - X|^p = 0\,;$$

- $X_n \to X$ *in law* if for any bounded continuous function $g : \mathbb{R}^d \to \mathbb{R}$

$$\lim_{n\to\infty} \mathbb{E}g(X_n) = \mathbb{E}g(X)\,.$$

Remark 1.5. The various notions of convergence of random variables with values in a metric space are defined analogously.

We recall some classical convergence results (sketching some of the proofs). First, we remark that

$$X_n \xrightarrow{a.s.} X \iff \sum_{n=1}^{\infty} \mathbf{1}_{|X_n - X|\geq\varepsilon} < \infty, \ \mathbb{P}\text{-}a.s. \ \forall\varepsilon > 0$$

and we have:

Lemma 1.6 (Borel–Cantelli). *If* $\{A_n : n \in \mathbb{N}^*\} \subset \mathcal{F}$ *and*

$$\sum_{n\in\mathbb{N}^*} \mathbb{P}(A_n) < \infty,$$

then

$$\sum_{n\in\mathbb{N}^*} \mathbf{1}_{A_n} < \infty, \ \mathbb{P}\text{-}a.s., \quad \text{or equivalently} \quad \mathbb{P}\left(\limsup_{n\to+\infty} A_n\right) = 0.$$

Proof. For all $m \in \mathbb{N}^*$,

$$\mathbb{P}\left(\sum_{n\in\mathbb{N}^*} \mathbf{1}_{A_n} = \infty\right) = \mathbb{P}\left(\bigcap\nolimits_{n\geq 1}\bigcup\nolimits_{k\geq n} A_k\right) \leq \mathbb{P}\left(\bigcup\nolimits_{k\geq m} A_k\right) \leq \sum_{k\geq m} \mathbb{P}(A_k).$$

■

Proposition 1.7. *Let* δ_n *be a positive sequence such that* $\delta_n \to 0$. *Let* $g : \mathbb{R}_+ \to \mathbb{R}_+$ *be a continuous function such that* $0 = g(0) < g(x) \leq g(y)$ *for all* $0 < x < y$. *If one of the following conditions is satisfied*

$$(a) \quad \sum_{n=1}^{\infty} \mathbb{P}(|X_n - X| \geq \varepsilon) < \infty, \ \forall\, \varepsilon > 0,$$

$$(b) \quad \sum_{n=1}^{\infty} \mathbb{E}g(|X_n - X|) < \infty,$$

$$(c) \quad \sum_{n=1}^{\infty} \mathbb{P}(|X_n - X| \geq \delta_n) < \infty,$$

then $X_n \xrightarrow{a.s.} X$. *Moreover, the condition* (c) *and* $\sum_{n=1}^{\infty} \delta_n < \infty$ *yield*

$$\sum_{n=1}^{\infty} |X_n - X| < \infty, \ a.s.$$

Proof. The results follow by setting in the Borel–Cantelli Lemma $A_n = \{|X_n - X| \geq \varepsilon\}$ for (a) and, respectively, $A_n = \{|X_n - X| \geq \delta_n\}$ for (c). The result from (b) follows by the Markov–Chebyshev inequality:

$$\mathbb{P}(|X_n(\omega) - X(\omega)| \geq \varepsilon) \leq \frac{1}{g(\varepsilon)} \mathbb{E}g(|X_n - X|).$$

■

Corollary 1.8. *We have*

$$\sum_{n=1}^{\infty} \mathbb{P}(|X_n| < R) < \infty, \ \forall \ R > 0, \quad \Longrightarrow \quad |X_n| \xrightarrow{a.s.} \infty.$$

It is well-known that if $X_n \xrightarrow{prob.} X$, then $g(X_n) \xrightarrow{prob.} g(X)$ whenever $g \in C(\mathbb{R}^d)$, and there exists a subsequence $X_{n_k} \xrightarrow{a.s.} X$. Also, if $X_n \xrightarrow{prob.} X$ then $X_n \xrightarrow{law} X$ and conversely if $X_n \xrightarrow{law} a \in \mathbb{R}^d$, then $X_n \xrightarrow{prob.} a$.

Theorem 1.9 (Monotone Convergence; Beppo Levi). *Let Y, Y_n be random variables, $n \in \mathbb{N}^*$. If*

$$(i) \ \ 0 \le Y_1 \le \cdots \le Y_n \le \cdots \le Y, \ \ \mathbb{P}\text{-}a.s.$$
$$(ii) \ \ Y_n \xrightarrow{a.s.} Y,$$

then

$$\lim_{n\to\infty} \mathbb{E}Y_n = \mathbb{E}Y.$$

Theorem 1.10 (Fatou's Lemma). *Let Y, Y_n be random variables and $Y_n \ge 0$, a.s., for all $n \in \mathbb{N}^*$. Then*

$$\mathbb{E}\left(\liminf_{n\to+\infty} Y_n\right) \le \liminf_{n\to+\infty} \mathbb{E}Y_n. \tag{1.1}$$

If moreover $Y_n \xrightarrow{law} Y$, then $Y \ge 0$ a.s. and

$$\mathbb{E}Y \le \liminf_{n\to+\infty} \mathbb{E}Y_n. \tag{1.2}$$

Definition 1.11. A family of d-dimensional random variables $\{X_i : i \in I\}$ is said to be:

- tight if

$$\lim_{N\to\infty}\left[\sup_{i\in I} \mathbb{P}(|X_i| \ge N)\right] = 0;$$

- uniformly integrable if

$$\lim_{N\to\infty}\left[\sup_{i\in I} \mathbb{E}\left(|X_i|\,\mathbf{1}_{|X_i|\ge N}\right)\right] = 0.$$

Clearly if $\{X_i : i \in I\}$ is a uniformly integrable family of random variables, then

$$\sup_{i\in I} \mathbb{E}\left(|X_i|\right) < \infty.$$

Indeed, let $N_1 > 0$ be such that

$$\sup_{i\in I} \mathbb{E}\left(|X_i| \mathbf{1}_{|X_i|\geq N_1}\right) \leq 1;$$

then for all $i \in I$,

$$\begin{aligned}\mathbb{E}|X_i| &= \mathbb{E}\left(|X_i| \mathbf{1}_{|X_i|\geq N_1}\right) + \mathbb{E}\left(|X_i| \mathbf{1}_{|X_i|<N_1}\right) \\ &\leq 1 + N_1,\end{aligned}$$

and $\{X_i : i \in I\}$ is tight, since

$$\sup_{i\in I} \mathbb{P}\left(|X_i| \geq N\right) \leq \sup_{i\in I} \frac{\mathbb{E}|X_i|}{N} \leq \frac{1+N_1}{N} \to 0, \quad \text{as } N \to \infty.$$

A first criterion of uniformly integrability is:

Lemma 1.12. *Let $X_i : (\Omega; \mathcal{F}, \mathbb{P}) \to \mathbb{R}^d$, $i \in I$, be random variables. If*

- *there exists a $Y \in L^1(\Omega, \mathcal{F}, \mathbb{P})$ such that for all $i \in I$*

$$|X_i| \leq Y, \quad a.s.$$

or

- *there exists a $p > 1$ such that*

$$\sup_{i\in I} \mathbb{E}|X_i|^p < \infty,$$

then $\{X_i : i \in I\}$ is uniformly integrable.

Proof. The uniformly integrability of $\{X_i : i \in I\}$ follows, in the first case, from the inequality

$$|X_i| \mathbf{1}_{|X_i|\geq N} \leq Y \mathbf{1}_{|Y|\geq N}$$

and, in the second case, from

$$|X_i| \mathbf{1}_{|X_i|\geq N} \leq \frac{1}{N^{p-1}} |X_i|^p.$$

∎

We have the following general criterion:

Proposition 1.13. *Let $X_i : (\Omega; \mathcal{F}, \mathbb{P}) \to \mathbb{R}^d$, $i \in I$, be random variables.*

(i) The collection $\{X_i : i \in I\}$ is tight if and only if there exists a Borel measurable function $G : [0, \infty[\to [0, \infty[$ such that

$$(T) \quad \begin{cases} (a) & \lim_{r\to\infty} G(r) = \infty, \\ (b) & \sup_{i\in I} \mathbb{E}\, G(|X_i|) < \infty. \end{cases}$$

(ii) The collection $\{X_i : i \in I\}$ is uniformly integrable if and only if there exists a Borel measurable function $H : [0, \infty[\to [0, \infty[$ such that

$$(UI) \quad \begin{cases} (a) & \lim_{r\to\infty} \dfrac{H(r)}{r} = \infty, \\ (b) & \sup_{i\in I} \mathbb{E}\, H(|X_i|) < \infty. \end{cases}$$

Moreover one can choose G to be an increasing continuous function and the function H to be an increasing continuous convex function such that $G(0) = H(0) = 0$.

Proof. Since the conditions $(T - a)$ and $(UI - a)$ imply that for all $\varepsilon > 0$ there exists an $N_\varepsilon > 0$ such that

$$\mathbf{1}_{r\ge N_\varepsilon} \le \varepsilon G(r) \quad \text{and} \quad r\mathbf{1}_{r\ge N_\varepsilon} \le \varepsilon H(r), \quad \forall\, r \ge 0,$$

it is clear that (T) yields the tightness and (UI) yields the uniformly integrability of the family $\{X_i : i \in I\}$.

To each strictly increasing sequence $\{k_n : n \in \mathbb{N}\} \subset \mathbb{N}$ we associate the functions

$$G(r) = \sum_{n=0}^{\infty} \left(n + \frac{r - k_n}{k_{n+1} - k_n} \right) \mathbf{1}_{[k_n, k_{n+1}[}(r)$$

and

$$H(r) = \int_0^r G(s)\, ds$$

from $\mathbb{R}_+$ into $\mathbb{R}_+$. G is continuous increasing and H is moreover convex. If $\{X_i : i \in I\}$ is tight, we choose $\{k_n : n \in \mathbb{N}\}$ such that

$$\sup_{i\in I} \mathbb{P}(|X_i| \ge k_n) \le \frac{1}{(n+1)^3},$$

and then G satisfies (T).

If $\{X_i : i \in I\}$ is uniformly integrable, we choose $\{k_n : n \in \mathbb{N}\}$ such that

$$\sup_{i\in I} \mathbb{E}\left(|X_i|\,\mathbf{1}_{|X_i|\geq k_n}\right) \leq \frac{1}{(n+1)^3},$$

and then H satisfies (UI). ■

Proposition 1.14. *Let $Z_n, Y_n : (\Omega;\mathcal{F},\mathbb{P}) \to \mathbb{R}^d$, $n \in \mathbb{N}$, be random variables. If $\{Y_n : n \in \mathbb{N}\}$ is tight and $Z_n \xrightarrow{prob.} 0$, then*

$$Y_n Z_n \xrightarrow{prob.} 0.$$

Proof. The result follows easily from the inequality

$$\mathbf{1}_{[\varepsilon,\infty[}\left(|yz|\right) \leq \mathbf{1}_{[\frac{\varepsilon}{N},\infty[}\left(|z|\right) + \mathbf{1}_{[N,\infty[}\left(|y|\right),$$

for all $\varepsilon, N > 0$ and $y, z \in \mathbb{R}$. ■

Theorem 1.15. *Let X, X_n be d-dimensional random variables, $n \in \mathbb{N}^*$. If*

$$\begin{cases} (i) & X_n \xrightarrow{law} X, \\ (ii) & \{X_n : n \in \mathbb{N}^*\} \text{ is uniformly integrable,} \end{cases} \tag{1.3}$$

then

$$\mathbb{E}\,|X| < \infty \qquad \text{and} \qquad \lim_{n\to\infty} \mathbb{E}X_n = \mathbb{E}X.$$

Proof. Since $|X_n| \xrightarrow{law} |X|$, from Fatou's Lemma we deduce that

$$\mathbb{E}\,|X| \leq \liminf_{n\to+\infty} \mathbb{E}\,|X_n| \leq \sup_{n\in\mathbb{N}^*} \mathbb{E}\,|X_n| < \infty.$$

Let $\alpha_N : \mathbb{R}_+ \to [0,1]$, $\alpha_N(r) = \mathbf{1}_{[0,N]}(r) + (N+1-r)\,\mathbf{1}_{(N,N+1]}(r)$; $r \longmapsto r\alpha_N(r)$ is a bounded continuous function and

$$\begin{aligned} |\mathbb{E}X_n - \mathbb{E}X| &\leq |\mathbb{E}X_n\alpha_N(|X_n|) - \mathbb{E}X\alpha_N(|X|)| \\ &\quad + \sup_{k\in\mathbb{N}^*} \mathbb{E}\left(|X_k|\,\mathbf{1}_{|X_k|\geq N}\right) + \mathbb{E}\left(|X|\,\mathbf{1}_{|X|\geq N}\right). \end{aligned}$$

By the convergence in law of X_n to X we have

$$\limsup_{n\to+\infty} |\mathbb{E}X_n - \mathbb{E}X| \leq \sup_{k\in\mathbb{N}^*} \mathbb{E}\left(|X_k|\,\mathbf{1}_{|X_k|\geq N}\right) + \mathbb{E}\left(|X|\,\mathbf{1}_{|X|\geq N}\right)$$

for all $N \geq 1$. The result follows from (ii) by letting $N \to \infty$. ■

From Lemma 1.12 and Theorem 1.15 we immediately have:

Corollary 1.16. *Let X, X_n be d-dimensional random variables, $n \in \mathbb{N}^*$. If $X_n \xrightarrow{law} X$ and*

- *(**Dominated Convergence Theorem; Lebesgue Theorem**) there exists a positive random variable $Y \in L^1(\Omega, \mathcal{F}, \mathbb{P})$ such that for all $n \in \mathbb{N}^*$: $|X_n| \leq Y$, a.s., or*
- *there exists a $p > 1$ such that*

$$\sup_{n \in \mathbb{N}^*} \mathbb{E}|X_n|^p < \infty,$$

then

$$\mathbb{E}|X| < \infty \qquad \text{and} \qquad \lim_{n\to\infty} \mathbb{E}X_n = \mathbb{E}X.$$

Now we present a generalization of Fatou's Lemma.

Proposition 1.17. *Let $(\mathbb{X}, \rho)$ be a separable metric space. Let $\varphi : \mathbb{X} \to]-\infty, +\infty]$ be a lower semicontinuous function. If X, X_n are $\mathbb{X}$-valued r.v., $n \in \mathbb{N}^*$, such that*

$$(i)\ X_n \xrightarrow{law} X, \text{ as } n \to \infty,$$

and there exists a continuous function $\alpha : \mathbb{X} \to \mathbb{R}$ such that

$$(ii)\ \alpha(x) \leq \varphi(x), \quad \forall\, x \in \mathbb{X}, \quad \text{and}$$
$$(iii)\ \{\alpha(X_n) : n \in \mathbb{N}^*\} \text{ is a uniformly integrable family,}$$

then the expectations $\mathbb{E}\varphi(X)$ and $\mathbb{E}\varphi(X_n)$ exist for all $n \in \mathbb{N}$, and

$$-\infty < \mathbb{E}\varphi(X) \leq \liminf_{n\to+\infty} \mathbb{E}\varphi(X_n).$$

Proof. From Proposition 6.26 (Annex B) there exists a sequence of continuous functions $\varphi_k : \mathbb{X} \to \mathbb{R}$, $k \in \mathbb{N}^*$, such that for all $x \in \mathbb{X}$

$$\alpha(x) \leq \varphi_1(x) \leq \ldots \leq \varphi_k(x) \leq \ldots \leq \varphi(x) \quad \text{and} \quad \lim_{k\to\infty} \varphi_k(x) = \varphi(x).$$

Note that for every $k \in \mathbb{N}^*$:

$$\varphi_k(X_n) - \alpha(X_n) \xrightarrow[n\to\infty]{law} \varphi_k(X) - \alpha(X).$$

Then by Fatou's Lemma 1.10, we have

$$\begin{aligned} 0 &\leq \mathbb{E}\left[\varphi_k(X) - \alpha(X)\right] \\ &\leq \liminf_{n\to+\infty} \mathbb{E}\left[\varphi_k(X_n) - \alpha(X_n)\right] \\ &\leq \liminf_{n\to+\infty} \mathbb{E}\left[\varphi(X_n) - \alpha(X_n)\right]. \end{aligned}$$

By Beppo Levi's Theorem we can take the limit as $k \nearrow \infty$ in the above; hence

$$0 \leq \mathbb{E}\left[\varphi(X) - \alpha(X)\right] \leq \liminf_{n\to+\infty} \mathbb{E}\left[\varphi(X_n) - \alpha(X_n)\right].$$

Now by *(iii)* and Theorem 1.15

$$\mathbb{E}\left|\alpha(X)\right| < \infty, \qquad \mathbb{E}\alpha(X) = \lim_{n\to\infty} \mathbb{E}\alpha(X_n).$$

Then the expectations $\mathbb{E}\varphi(X)$, $\mathbb{E}\varphi(X_n)$ exist in $]-\infty, +\infty]$, and

$$\mathbb{E}\alpha(X) \leq \mathbb{E}\varphi(X) \leq \liminf_{n\to+\infty} \mathbb{E}\varphi(X_n).$$

■

If $\mathbb{X} = C\left([0,T];\mathbb{R}^d\right)$, then an $\mathbb{X}$-valued random variable is called a $\mathbb{R}^d$-valued continuous stochastic process and we write $X_t(\omega) = X(\omega,t)$. We denote the total variation of X on $[s,t]$ by $\updownarrow X \updownarrow_{[s,t]}$, that is

$$\updownarrow X \updownarrow_{[s,t]} = \sup\left\{ \sum_{i=0}^{n-1} \left|X_{t_{i+1}} - X_{t_i}\right| : n \in \mathbb{N}^*, s = t_0 < t_1 < \cdots < t_n = t \right\}.$$

We also use $\updownarrow X \updownarrow_T \stackrel{def}{=} \updownarrow X \updownarrow_{[0,T]}$.

Corollary 1.18. *Let* $0 \leq s \leq t \leq T$. *If* X, V, X^n, V^n, $n \in \mathbb{N}^*$, *are random variables with values in* $\mathbb{X} = C\left([0,T];\mathbb{R}^d\right)$, *such that*

$$(X^n, V^n) \xrightarrow{law} (X, V), \text{ as } n \to \infty,$$

and $g : C\left([0,T];\mathbb{R}^d\right) \to \mathbb{R}_+$ *is a continuous function, then the following implications hold:*

(a) $X_t^n \in F$, *a.s.* $\Rightarrow$ $X_t \in F$, *a.s., whenever* F *is closed;*
(b) $\updownarrow X^n \updownarrow_{[s,t]} \leq g(V^n)$, *a.s.* $\Rightarrow$ $\updownarrow X \updownarrow_{[s,t]} \leq g(V)$, *a.s.;*
(c) if $d = 1$ *and* $X_s^n \leq X_t^n$, *a.s., then* $X_s \leq X_t$, *a.s.*

Proof. In Proposition 1.17 one successively sets $\varphi(x) = \text{dist}(x(t); F)$,

$$\varphi(x,y) = \left(\sum_{i=0}^{N-1} |x(t_{i+1}) - x(t_i)| - g(y) \right)^+ ,$$

where $s = t_0 < t_1 < \ldots < t_N = t$ is an arbitrary partition of $[s,t]$, and $\varphi(x) = (x(s) - x(t))^+$. ■

Proposition 1.19. *Let* (X,K,V), (X^n,K^n,V^n), $n \in \mathbb{N}$, *be* $\mathbb{X} = C\left([0,T];\mathbb{R}^d\right)^2 \times C([0,T];\mathbb{R})$*-valued random variables such that*

$$(X^n,K^n,V^n) \xrightarrow[n\to\infty]{law} (X,K,V)$$

and for all $0 \le s < t$, *and* $n \in \mathbb{N}^*$,

$$\updownarrow K^n \updownarrow_t - \updownarrow K^n \updownarrow_s \le V_t^n - V_s^n \quad a.s.$$

If $\varphi : \mathbb{R}^d \to]-\infty,+\infty]$ *is a l.s.c. function and*

$$\int_s^t \varphi\left(X_r^n\right) dr \le \int_s^t \langle X_r^n, dK_r^n \rangle, \quad a.s. \quad \text{for all } n \in \mathbb{N}^*,$$

then

$$\updownarrow K \updownarrow_t - \updownarrow K \updownarrow_s \le V_t - V_s \quad a.s.$$

and

$$\int_s^t \varphi(X_r)\, dr \le \int_s^t \langle X_r, dK_r \rangle, \quad a.s.$$

Proof. Define the partition

$$\Delta_N : s = r_0 < r_1 < \ldots < r_N = t, \quad r_{i+1} - r_i = \frac{t-s}{N}.$$

For $\delta > 0$, $x,k \in C\left([0,T];\mathbb{R}^d\right)$ we define

$$\Phi_N(k) = \sum_{i=0}^{N-1} |k(r_{i+1}) - k(r_i)|,$$

$$S_N(x,k) = \sum_{i=0}^{N-1} \langle x(r_i), k(r_{i+1}) - k(r_i) \rangle \quad \text{and}$$

$$\mathbf{m}(\delta,x) = \sup\{|x(u) - x(r)| : u,r \in [0,T],\ |u-r| \le \delta\}.$$

Note that

$$\Phi_N(k) \leq \updownarrow k \updownarrow_{[s,t]}$$

and

$$\int_s^t \langle x(r), dk(r)\rangle \leq S_N(x,k) + \mathbf{m}\left(\frac{1}{N}, x\right) \updownarrow k \updownarrow_{[s,t]}.$$

Since

$$|\mathbf{m}(\delta, x) - \mathbf{m}(\delta, y)| \leq 2\|x - y\|_T$$

and $\varphi : \mathbb{R}^d \to]-\infty, +\infty]$ is bounded below on bounded sets of $\mathbb{R}^d$, we deduce that $g : \mathbb{X} \to [0, 1]$ given by

$$\begin{aligned} g(x,k,v) &= (\Phi_N(k) - v(t) + v(s))^+ \wedge 1 \\ &\quad + \left[\int_s^t \varphi(x(r))\,dr - S_N(x,k) - \mathbf{m}\left(\frac{1}{N}, x\right)(v(t) - v(s))\right]^+ \wedge 1 \end{aligned}$$

is a l.s.c. function. Hence by Proposition 1.17

$$0 \leq \mathbb{E}g(X, K, V) \leq \liminf_{n\to+\infty} \mathbb{E}g(X^n, K^n, V^n) = 0.$$

Consequently, $\mathbb{P}$-*a.s.*

$$\Phi_N(K) \leq V_t - V_s$$

and

$$\int_s^t \varphi(X_\tau)\,d\tau \leq S_N(X, K) + \mathbf{m}\left(\frac{1}{N}, X\right)(V_t - V_s).$$

Passing to the limit as $N \to \infty$ the result follows. ■

Proposition 1.20. *Let (X, K), (X^n, K^n), $n \in \mathbb{N}$, be $C\left([0,T];\mathbb{R}^d\right)^2$-valued random variables. Assume*

$$(i)\quad \exists\, p > 0 \text{ such that } L \stackrel{def}{=} \sup_{n\in\mathbb{N}^*} \mathbb{E} \updownarrow K^n \updownarrow_T^p < \infty \quad \text{and}$$

$$(ii)\quad \|X^n - X\|_T + \|K^n - K\|_T \xrightarrow{prob.} 0, \text{ as } n \to \infty.$$

Then for all $0 \leq s \leq t \leq T$:

$$\int_s^t \langle X_r^n, dK_r^n\rangle \xrightarrow{prob.} \int_s^t \langle X_r, dK_r\rangle, \text{ as } n \to \infty, \tag{1.4}$$

and moreover

$$\mathbb{E}\ \updownarrow K \updownarrow_T^p \leq \liminf_{n\to+\infty} \mathbb{E}\ \updownarrow K^n \updownarrow_T^p. \tag{1.5}$$

Proof. The inequality (1.5) is a consequence of Proposition 1.17 since the function $\varphi : C\left([0,T];\mathbb{R}^d\right) \to [0,+\infty]$, $\varphi(k) = \updownarrow k \updownarrow_T^p$, is l.s.c.

Using the notation from Proposition 1.19 we have

$$\begin{aligned}
&\left|\int_s^t \langle X_r^n, dK_r^n\rangle - \int_s^t \langle X_r, dK_r\rangle\right| \\
&\quad \leq \left|\int_s^t \langle X_r^n - X_r, dK_r^n\rangle\right| + \left|\int_s^t \langle X_r, dK_r^n - dK_r\rangle - S_N\left(X, K^n - K\right)\right| \\
&\qquad\qquad + \left|S_N\left(X, K^n - K\right)\right| \\
&\quad \leq \|X^n - X\|_T\ \updownarrow K^n \updownarrow_T + \mathbf{m}\left(\frac{1}{N}, X\right)\left[\updownarrow K^n \updownarrow_T + \updownarrow K \updownarrow_T\right] + \left|S_N\left(X, K^n - K\right)\right|.
\end{aligned}$$

Let $A > 0$ be arbitrary. Since for all $x, y, z, w, u \geq 0$,

$$\begin{aligned}
1 \wedge (x + y + z + w) &\leq 1 \wedge x + 1 \wedge y + 1 \wedge z + 1 \wedge w \quad \text{and} \\
1 \wedge (x\ u) &\leq 1 \wedge (x\ A) + \frac{u^p}{A^p},
\end{aligned}$$

and using the assumptions of our Proposition, we deduce that

$$\begin{aligned}
&\limsup_{n\to+\infty} \mathbb{E}\left[1 \wedge \left|\int_s^t \langle X_r^n, dK_r^n\rangle - \int_s^t \langle X_r, dK_r\rangle\right|\right] \\
&\quad \leq 2\,\mathbb{E}\left[1 \wedge \left(A\,\mathbf{m}\left(\frac{1}{N}, X\right)\right)\right] + \frac{3\,L}{A^p}.
\end{aligned}$$

Passing here to limit, first for $N \to \infty$ and then for $A \to \infty$, we complete the proof. ∎

Remark 1.21. It is clear from the proof that in the above Proposition we can replace the convergences in probability by a.s. convergences.

Now from Proposition 1.20 and Proposition 1.17 we clearly deduce:

Corollary 1.22. *Let the assumptions of Proposition 1.20 be satisfied. If $A : \mathbb{R}^d \rightrightarrows \mathbb{R}^d$ is a (multivalued) maximal monotone operator, then the following implication holds $\mathbb{P}$-a.s. $\omega \in \Omega$:*

$$dK_t^n \in A\left(X_t^n\right)(dt) \ \text{ on } \ [0,T] \ \Rightarrow dK_t \in A\left(X_t\right)(dt) \ \text{ on } \ [0,T].$$

In particular if $\varphi : \mathbb{R}^d \to]-\infty, +\infty]$ *is a proper convex l.s.c. function then*

$$dK_t^n \in \partial\varphi\left(X_t^n\right)(dt) \;\; on\, [0,T] \;\Rightarrow dK_t \in \partial\varphi\left(X_t\right)(dt) \;\; on\, [0,T].$$

Proof. To prove the Corollary it suffices to recall from Annex B (Remark 6.18) the definitions

(A) $dk(t) \in A(x(t))(dt)$ if

- (a_1) $x \in C\left(\mathbb{R}_+; \overline{\mathrm{Dom}(A)}\right)$,
- (a_2) $k \in C\left(\mathbb{R}_+; \mathbb{R}^d\right) \bigcap BV_{loc}\left(\mathbb{R}_+; \mathbb{R}^d\right)$, $k(0) = 0$,
- (a_3) $\langle x(t) - u, dk(t) - \hat{u}\, dt\rangle \geq 0, \quad \forall\, (u, \hat{u}) \in A$.

In the case of $A = \partial\varphi$ we have an equivalent definition.

(B) $dk(t) \in \partial\varphi(x(t))(dt)$ if

- (b_1) $x \in C\left(\mathbb{R}_+; \overline{\mathrm{Dom}(\varphi)}\right)$,
- (b_2) $k \in C\left(\mathbb{R}_+; \mathbb{R}^d\right) \bigcap BV_{loc}\left(\mathbb{R}_+; \mathbb{R}^d\right)$, $k(0) = 0$,
- (b_3) for all $0 \leq s \leq t \leq T, u \in \mathbb{R}^d$

$$\int_s^t \langle u - x(r), dk(r)\rangle + \int_s^t \varphi(x(r))dr \leq (t-s)\,\varphi(u).$$

■

We close this section with two statements about convergence in probability.

Proposition 1.23. *Let* $X, X_n : (\Omega; \mathcal{F}, \mathbb{P}) \to \mathbb{R}^d$, $n \in \mathbb{N}$, *be random variables and* $0 < p < \infty$. *The following are equivalent:*

(i) $X_n, X \in L^p\left(\Omega, \mathcal{F}, \mathbb{P}; \mathbb{R}^d\right)$ *and* $X_n \to X$ *in* L^p;
(ii) $X_n \to X$ *in probability and* $\{|X_n|^p : n \in \mathbb{N}\}$ *is uniformly integrable.*

Proof. $(ii) \Longrightarrow (i)$: This implication is a consequence of Theorem 1.15.
$(i) \Longrightarrow (ii)$: By the Markov inequality,

$$\mathbb{P}\left(|X_n - X| \geq \varepsilon\right) \leq \frac{\mathbb{E}\,|X_n - X|^p}{\varepsilon^p}, \quad \varepsilon > 0,$$

we clearly have that $X_n \to X$ in probability, as $n \to \infty$.
Let $C_p = 1 \vee 2^{p-1}$. We have for all $N, A > 0$:

$$|X_n|^p\, \mathbf{1}_{|X_n|^p > N} \leq C_p\left[|X_n - X|^p + |X|^p\, \mathbf{1}_{|X|^p > A} + \frac{|X_n|^p}{N} A\right]. \tag{1.6}$$

Since $X_n \to X$ in L^p as $n \to \infty$, there exists a $B > 0$ such that

$$\mathbb{E}\,|X_n|^p \leq B, \quad \forall\, n \in \mathbb{N}^*,$$

and for all $\varepsilon > 0$ there exists a $k_\varepsilon \in \mathbb{N}^*$ such that

$$\mathbb{E}\,|X_n - X|^p < \varepsilon, \quad \forall\, n > k_\varepsilon.$$

Let $Y_\varepsilon = \max\left\{|X_n| : n \in \overline{1, k_\varepsilon}\right\}$. Then from (1.6) we have

$$\begin{aligned}\sup_{n\geq 1} \mathbb{E}\left(|X_n|^p \mathbf{1}_{|X_n|^p > N}\right) &\leq \mathbb{E}\left(Y_\varepsilon^p \mathbf{1}_{|Y_\varepsilon|^p > N}\right) + \sup_{n > k_\varepsilon} \mathbb{E}\left(|X_n|^p \mathbf{1}_{|X_n|^p > N}\right) \\ &\leq \mathbb{E}\left(Y_\varepsilon^p \mathbf{1}_{|Y_\varepsilon|^p > N}\right) + C_p\left[\varepsilon + \mathbb{E}\left(|X|^p \mathbf{1}_{|X|^p > A}\right) + \frac{B}{N} A\right],\end{aligned}$$

which yields

$$\limsup_{N\to+\infty}\left[\sup_{n\geq 1} \mathbb{E}\left(|X_n|^p \mathbf{1}_{|X_n|^p > N}\right)\right] \leq C_p\left[\varepsilon + \mathbb{E}\left(|X|^p \mathbf{1}_{|X|^p > A}\right)\right],$$

for all $\varepsilon, A > 0$. Letting, in the last inequality, $\varepsilon \searrow 0$ and $A \nearrow \infty$, the uniform integrability follows. This completes the proof. ■

Proposition 1.24. *Let Y, Y_n be random variables and $Y_n \geq 0$, a.s., for all $n \in \mathbb{N}^*$. If $\mathbb{E}\,|Y_n| + \mathbb{E}\,|Y| < \infty$, $Y_n \xrightarrow{prob.} Y$ and $\mathbb{E}Y_n \to \mathbb{E}Y$, then*

$$\mathbb{E}\,|Y_n - Y| \to 0.$$

Proof. Since

$$(Y - Y_n)^+ \leq Y^+ \quad \text{and} \quad |Y - Y_n| = 2(Y - Y_n)^+ - (Y - Y_n)$$

the result follows from Lebesgue's dominated convergence theorem. ■

Let $X : \Omega \to \mathbb{R}^d$ be a random variable. Define

$$\|X\|_{L^p} = \begin{cases} \left(\mathbb{E}\,|X|^p\right)^{1/p}, & \text{if } 1 \leq p < \infty, \\ \inf\{a > 0 : \mathbb{P}\left(|X| > a\right) = 0\}, & \text{if } p = \infty, \\ \mathbb{E}\,|X|^p, & \text{if } 0 < p < 1, \\ \mathbb{E}\left(1 \wedge |X|\right), & \text{if } p = 0. \end{cases}$$

Note that $\|X\|_{L^0} < \infty$ for any random variable X.

Denote by $L^p\left(\Omega, \mathcal{F}, \mathbb{P}; \mathbb{R}^d\right)$, $0 \leq p \leq \infty$, the space of random variables $X : \Omega \to \mathbb{R}^d$ such that $\|X\|_{L^p} < \infty$ (we shall usually not distinguish a random variable X from its equivalence class with respect to a.s. equivalence). The space $L^p\left(\Omega, \mathcal{F}, \mathbb{P}; \mathbb{R}^d\right)$, $1 \leq p \leq \infty$, is a Banach space with respect to the norm $\|\cdot\|_{L^p}$. If $0 \leq p < 1$ then $L^p\left(\Omega, \mathcal{F}, \mathbb{P}; \mathbb{R}^d\right)$ is a complete metric linear space with the metric $d_p\left(X, Y\right) = \|X - Y\|_{L^p}$. The convergence in the metric d_0 is convergence in probability.

The space $L^2\left(\Omega, \mathcal{F}, \mathbb{P}; \mathbb{R}^d\right)$ is a Hilbert space with respect to the scalar product $\langle X, Y\rangle_{L^2} = \mathbb{E}\langle X, Y\rangle$.

1.1.2 Filtrations

We say that $(\Omega, \mathcal{F}, \mathbb{P}, \{\mathcal{F}_t\}_{t\geq 0})$ is a stochastic basis if:

(i) $(\Omega, \mathcal{F}, \mathbb{P})$ is a complete probability space;
(ii) $\{\mathcal{F}_t\}_{t\geq 0}$ is a filtration i.e. an increasing collection of sub-σ-algebras of $\mathcal{F}$ satisfying the condition:

  (a) $\mathcal{F}_t$ is a sub-σ-algebra of $\mathcal{F}$ which contains all $\mathbb{P}$-null sets of $\mathcal{F}$;
  (b) $t \to \mathcal{F}_t$ is right continuous, that is $\mathcal{F}_t = \mathcal{F}_{t+}$, where $\mathcal{F}_{t+} \overset{def}{=} \bigcap_{s>t} \mathcal{F}_s$.

The σ-algebra $\mathcal{F}$ can be thought of as the set of observable events and the σ-algebra $\mathcal{F}_t$ can be thought of as the set of observable events before time t.

To such a filtration, we associate the σ-algebra $\mathcal{P}$ of progressively measurable subsets of $\Omega \times \mathbb{R}_+$, defined as follows:

Definition 1.25. $\mathcal{P} = \mathcal{P}(\mathcal{F}_t)$ is the σ-algebra of the sets $A \subset \Omega \times \mathbb{R}_+$ such that *for all* $t \geq 0$,

$$A \cap (\Omega \times [0, t]) \in \mathcal{F}_t \otimes \mathcal{B}_{[0,t]} \, .$$

1.1.3 Conditional Expectation

In this section, we present the notions of conditional expectation and conditional probability, together with their main properties.

All random variables will be assumed to be defined on a probability space $(\Omega, \mathcal{F}, \mathbb{P})$. $\mathcal{G}$, $\mathcal{H}$ will denote sub-σ-algebras of $\mathcal{F}$, and we will assume (in order to simplify slightly the beginning of our exposition) that each of them contains the collection $\mathcal{N}$ of all $\mathbb{P}$-null-sets of $\mathcal{F}$.

Thanks to the hypothesis we have just formulated, $L^2(\Omega, \mathcal{G}, \mathbb{P})$ is a sub-Hilbert space (hence in particular a closed sub-vector space) of $L^2(\Omega, \mathcal{F}, \mathbb{P})$. Therefore, we can state the following:

Definition 1.26. We will call the conditional expectation with respect to $\mathcal{G}$ the orthogonal projection operator from $L^2(\Omega, \mathcal{F}, \mathbb{P})$ onto $L^2(\Omega, \mathcal{G}, \mathbb{P})$.

Given a square integrable random variable X, we can associate to X its equivalence class, which is an element of $L^2(\Omega, \mathcal{F}, \mathbb{P})$, which by an abuse of notation we shall also call X. We denote by

$$\mathbb{E}(X|\mathcal{G}) \quad \text{or} \quad \mathbb{E}^{\mathcal{G}}(X)$$

its orthogonal projection on $L^2(\Omega, \mathcal{G}, \mathbb{P})$. In practice, $\mathbb{E}(X|\mathcal{G})$ will rather denote an (arbitrary!) element in this equivalence class, i.e. $\mathbb{E}(X|\mathcal{G})$ will be for us a random variable. It is usually unimportant, but sometimes crucial, to remember that the choice of the particular element in the equivalence class is arbitrary.

$\mathbb{E}(X|\mathcal{G})$ is characterized as the unique (equivalence class of) random variables such that:

(i) $\mathbb{E}(X|\mathcal{G})$ is $\mathcal{G}$-measurable;
(ii) $\mathbb{E}(YX) = \mathbb{E}[Y\mathbb{E}(X|\mathcal{G})]$, $\forall Y \in L^2(\Omega, \mathcal{G}, \mathbb{P})$.

If we choose in particular $Y = \mathbf{1}$ in (ii), we obtain

(iii) $\mathbb{E}[\mathbb{E}(X|\mathcal{G})] = \mathbb{E}X$.

If we choose $Y = \mathbf{1}_{\{\mathbb{E}(X|\mathcal{G})<0\}}$ in (ii), we deduce that

$$\text{whenever } X \geq 0 \text{ a.s., then } \mathbb{E}(X|\mathcal{G}) \geq 0 \text{ a.s.}$$

Now, for any square integrable X,

$$X = X^+ - X^-$$

then by linearity of the orthogonal projection,

$$\begin{aligned}\mathbb{E}(X|\mathcal{G}) &= \mathbb{E}(X^+|\mathcal{G}) - \mathbb{E}(X^-|\mathcal{G})\\ |\mathbb{E}(X|\mathcal{G})| &\leq \mathbb{E}(X^+|\mathcal{G}) + \mathbb{E}(X^-|\mathcal{G})\\ &= \mathbb{E}(|X||\mathcal{G}),\end{aligned}$$

and we then deduce from (iii)

$$\mathbb{E}[|\mathbb{E}(X|\mathcal{G})|] \leq \mathbb{E}|X|.$$

We have just established that the conditional expectation is continuous with respect to the norm in the space $L^1(\Omega, \mathcal{F}, \mathbb{P})$. Since $L^2(\Omega, \mathcal{F}, \mathbb{P})$ is dense in $L^1(\Omega, \mathcal{F}, \mathbb{P})$, we easily deduce that the conditional expectation given $\mathcal{G}$ can be extended as a continuous linear operator from $L^1(\Omega, \mathcal{F}, \mathbb{P})$ onto $L^1(\Omega, \mathcal{G}, \mathbb{P})$.

Given an integrable random variable X (i.e. such that $\mathbb{E}|X| < \infty$), its conditional expectation given $\mathcal{G}$ is the unique (equivalence class of) random variable(s) $\mathbb{E}(X|\mathcal{G})$ such that:

(i) $\mathbb{E}(X|\mathcal{G})$ is $\mathcal{G}$-measurable;
(ii′) $\mathbb{E}(YX) = \mathbb{E}(Y\mathbb{E}(X|\mathcal{G})$ for all $\mathcal{G}$-measurable and bounded random variables Y.

(ii′) can be replaced by:

(ii″) $\mathbb{E}(X; A) = \mathbb{E}(\mathbb{E}(X|\mathcal{G}); A)\ \forall A \in \mathcal{G}$.

If $X = (X_1, \cdots, X_d)^* : (\Omega; \mathcal{F}, \mathbb{P}) \to \mathbb{R}^d$ is a random vector such that $\mathbb{E}|X| < \infty$, then the conditional expectation of X is defined by

$$\mathbb{E}^{\mathcal{G}} X \stackrel{def}{=} \left(\mathbb{E}^{\mathcal{G}} X_1, \cdots, \mathbb{E}^{\mathcal{G}} X_d\right)^*.$$

Example 1.27. Suppose that $\mathcal{G}$ is generated by a finite measurable partition $\{G_1, \cdots, G_n\}$ of Ω. Condition (i) forces $\mathbb{E}(X|\mathcal{G})$ to be constant on each G_i, and we then deduce from (ii) the formula

$$\mathbb{E}(X|\mathcal{G}) = \sum_{i=1}^{n} \frac{\mathbb{E}(X\mathbf{1}_{G_i})}{\mathbb{P}(G_i)} \mathbf{1}_{G_i}.$$

If in particular $\mathcal{G} = \{\varnothing, \Omega\}$, then $\mathbb{E}(X|\mathcal{G}) = \mathbb{E}X$.

The main properties of the conditional expectation are stated in the next proposition (we recall that $\mathcal{G} \vee \mathcal{H}$ is the smallest σ-algebra which contains both $\mathcal{G}$ and $\mathcal{H}$):

Proposition 1.28. *Let X, $Y \in L^1\left(\Omega, \mathcal{F}, \mathbb{P}; \mathbb{R}^d\right)$, $Z \in L^1(\Omega, \mathcal{F}, \mathbb{P}; \mathbb{R})$ and a, b be real numbers. Then:*

(a) $\mathbb{E}(aX + bY|\mathcal{G}) = a\mathbb{E}(X|\mathcal{G}) + b\mathbb{E}(Y|\mathcal{G})$.
(b) $Z \geq 0$ *a.s.* $\Rightarrow \mathbb{E}(Z|\mathcal{G}) \geq 0$ *a.s.*
(c) If $\mathcal{H}$ and $\mathcal{G} \vee \sigma(X)$ are independent, then $\mathbb{E}(X|\mathcal{G} \vee \mathcal{H}) = \mathbb{E}(X|\mathcal{G})$. In particular, if $\mathcal{G}$ and X are independent, then $\mathbb{E}(X|\mathcal{G}) = \mathbb{E}X$.
(d) If $\mathcal{G} \subset \mathcal{H}$, then

$$\mathbb{E}^{\mathcal{G}}\left[\mathbb{E}^{\mathcal{H}} X\right] = \mathbb{E}^{\mathcal{G}}(X),$$

and in particular, $\mathbb{E}\left[\mathbb{E}^{\mathcal{G}}(X)\right] = \mathbb{E}X$.
(e) If X is $\mathcal{G}$-measurable and $\langle X, Y \rangle$ is integrable, then

$$\mathbb{E}^{\mathcal{G}} \langle X, Y \rangle = \left\langle X, \mathbb{E}^{\mathcal{G}} Y \right\rangle.$$

(f) (Jensen's inequality) If $\varphi : \mathbb{R}^d \to \mathbb{R}$ is a convex function such that $\mathbb{E}\varphi^+(X) < \infty$, then $\mathbb{E}|\varphi(X)| < \infty$ and

$$\varphi\left(\mathbb{E}^{\mathcal{G}}(X)\right) \leq \mathbb{E}^{\mathcal{G}}(\varphi(X)).$$

(g) If

$$\mathbb{E}(\xi|\mathcal{G}) = \mathbb{E}(\xi|\mathcal{H})$$

for all bounded $\mathcal{G} \vee \mathcal{H}$-measurable random variables ξ, then $\mathcal{G} = \mathcal{H}$ (recall that we assumed that $\mathcal{N} \subset \mathcal{G}, \mathcal{H}$).

From the property (f) with $\varphi(x) = |x|^p$, $p \geq 1$ and (iii), we easily deduce the following:

Proposition 1.29. *If $p \geq 1$, then $\mathbb{E}^{\mathcal{G}}$ is a linear continuous operator from $L^p\left(\Omega, \mathcal{F}, \mathbb{P}; \mathbb{R}^d\right)$ into $L^p\left(\Omega, \mathcal{G}, \mathbb{P}; \mathbb{R}^d\right)$.*

Remark 1.30. If X is a positive random variable, $\mathbb{E}^{\mathcal{G}} X$ is the class of random variables with values in $[0, \infty]$ defined by

$$\mathbb{E}^{\mathcal{G}} X \stackrel{def}{=} \lim_{n \to \infty} \mathbb{E}^{\mathcal{G}}\left(X \wedge n\right), \ \mathbb{P}\text{-}a.s.$$

We shall define

$$\mathbb{P}(X \in B | \mathcal{F}_s) = \mathbb{E}\left(\mathbf{1}_B\left(X\right) | \mathcal{F}_s\right).$$

Since the mapping $X \mapsto \mathbb{E}\left(X | \mathcal{G}\right)$ is linear and preserves non-negativity, one might hope that for each $\omega \in \Omega$ there is a probability measure $\mathbb{P}^{\omega}$ on $(\Omega, \mathcal{F})$ such that

$$\mathbb{E}\left(X | \mathcal{G}\right)(\omega) = \int_{\Omega} X\left(\omega'\right) \mathbb{P}^{\omega}\left(d\omega'\right).$$

Unfortunately, this hope does not hold in the general case. However, it is fulfilled when Ω is a Polish space (i.e. Ω is a topological space which admits a complete separable metrization) and $\mathcal{F} = \mathcal{B}_{\Omega}$ is the Borel σ-algebra on Ω.

Denote by $\mathcal{M}_1\left(\Omega\right)$ the set of probability measures $\mathbb{P} : \mathcal{B}_{\Omega} \to [0, 1]$.

Theorem 1.31. *Let Ω be a Polish space and $\mathcal{G} \subset \mathcal{B}_{\Omega}$. For every $\mathbb{P} \in \mathcal{M}_1\left(\Omega\right)$, there is a map $\omega \mapsto \mathbb{P}^{\omega} : \Omega \to \mathcal{M}_1\left(\Omega\right)$, uniquely determined up to a $\mathbb{P}$-null set $N \in \mathcal{G}$, such that:*

(a) $\omega \mapsto \mathbb{P}^{\omega}\left(B\right)$ *is $\mathcal{G}$-measurable for all $B \subset \mathcal{B}_{\Omega}$;*

(b) $\mathbb{P}\left(A \cap B\right) = \int_A \mathbb{P}^{\omega}\left(B\right) \mathbb{P}\left(d\omega\right)$, *for all $A \in \mathcal{G}$ and $B \in \mathcal{B}_{\Omega}$.*

Moreover if $\mathcal{G}$ is countably generated, then $\omega \mapsto \mathbb{P}^{\omega}$ can be chosen so that

$$\mathbb{P}^{\omega}\left(A\right) = \mathbf{1}_A\left(\omega\right), \quad \forall\, \omega \in \Omega \text{ and } A \in \mathcal{G}.$$

The interested reader can find a proof of this theorem in Stroock [68] (Theorem 1.2).

Definition 1.32. The map $\omega \mapsto \mathbb{P}^{\omega}$ from Theorem 1.31 is called the conditional probability distribution of $\mathbb{P}$ given $\mathcal{G}$. If, moreover,

$$\mathbb{P}^{\omega}\left(A\right) = \mathbf{1}_A\left(\omega\right), \quad \forall\, \omega \in \Omega \text{ and } A \in \mathcal{G},$$

then $\omega \mapsto \mathbb{P}^{\omega}$ is called a regular conditional probability distribution of $\mathbb{P}$ given $\mathcal{G}$.

If $\Omega = W^d \stackrel{def}{=} C\left(\mathbb{R}_+;\mathbb{R}^d\right)$ is the space of continuous functions $\omega : [0,\infty[\to \mathbb{R}^d$ with the topology of uniform convergence on compact sets, then Ω is a Polish space with the metric

$$\rho\left(\omega,\omega'\right) = \sum_{n=1}^{\infty} \frac{1}{2^n} \frac{\|\omega-\omega'\|_n}{1+\|\omega-\omega'\|_n},$$

where

$$\left\|\omega-\omega'\right\|_n = \sup_{t\in[0,n]} \left|\omega(t)-\omega'(t)\right|.$$

For every $s \geq 0$ the map $\xi_s : (\Omega;\mathcal{B}_\Omega) \to \mathbb{R}^d$, $\xi_s(\omega) = \omega(s)$, is a random variable. Define

$$\mathcal{B}_t(W^d) = \sigma\left\{\xi_s : 0 \leq s \leq t\right\}.$$

Then from Theorem 1.31, for every $\mathbb{P} \in \mathcal{M}_1(\Omega)$, there exists a regular conditional probability distribution $\omega \mapsto \mathbb{P}_t^\omega$ of $\mathbb{P}$ given $\mathcal{B}_t(W^d)$, which is uniquely determined up to a $\mathbb{P}$-null set $N \in \mathcal{B}_t(W^d)$. Hence $\omega \mapsto \mathbb{P}_t^\omega(B)$ is $\mathcal{F}_t$-measurable and for all $B \in \mathcal{B}_\Omega$,

$$\mathbb{P}(A\cap B) = \int_A \mathbb{P}_t^\omega(B)\,\mathbb{P}(d\omega), \ \forall\, A \in \mathcal{B}_t(W^d),$$

$$\mathbb{P}_t^\omega(A) = \mathbf{1}_A(\omega), \ \ \forall\, \omega \in \Omega \text{ and } A \in \mathcal{B}_t(W^d).$$

We now want to define the conditional expectation, given a random vector. Recall that to a d-dimensional random vector X we associate the σ*-algebra*

$$\mathcal{F}^X \stackrel{def}{=} \{X^{-1}(B) : \ B \in \mathcal{B}_d\} \vee \mathcal{N},$$

where $\mathcal{B}_d$ denotes the Borel σ-field of $\mathbb{R}^d$ and $\mathcal{N}$ denotes the collection of all $\mathbb{P}$-null-sets of $\mathcal{F}$.

We can now state the following definition:

Definition 1.33. The conditional expectation of a random variable X, given the random vector Y, is the random variable

$$\mathbb{E}(X|Y) \stackrel{def}{=} \mathbb{E}(X|\mathcal{F}^Y).$$

We also introduce the notation

$$\mathbb{P}(X \in B|Y) = \mathbb{E}\left(\mathbf{1}_B(X)\,|\mathcal{F}^Y\right).$$

Proposition 1.34. *Let X and Y be respectively a d-dimensional and a k-dimensional random vector, and let $\psi : \mathbb{R}^{d+k} \to \mathbb{R}$ be Borel measurable such that $\psi(X,Y) \geq 0$ a.s., or $\mathbb{E}|\psi(X,Y)| < \infty$. Let $\mathcal{G}$ be a sub-σ-algebra of $\mathcal{F}$ such that:*

(i) X is $\mathcal{G}$-measurable;
(ii) Y and $\mathcal{G}$ are independent.

Then

$$\mathbb{E}[\psi(X,Y)|\mathcal{G}] = \int_{\mathbb{R}^k} \psi(X,y)\mathbb{P}_Y(dy) = \mathbb{E}[\psi(X,Y)|X].$$

Proof. If $\psi(x,y) = g(x)h(y)$ then the result follows from Proposition 1.28(e). The general case follows via approximation by linear combinations of such functions. ■

Let us prove three auxiliary results, which will be used in Sects. 1.3.3 and 2.4, respectively.

Lemma 1.35. *Let $X \in L^0(\Omega,\mathcal{F},\mathbb{P};\mathbb{R}^d)$ and $\mathcal{G}$ be a sub-σ-algebra of $\mathcal{F}$. If $\varphi(u) = \mathbb{E}\left(e^{i\langle u,X\rangle}|\mathcal{G}\right)$, $u \in \mathbb{R}^d$, is a deterministic function then:*

(j) φ is the characteristic function of X;
(jj) X is independent of $\mathcal{G}$.

Proof. Since $\varphi(u) = \mathbb{E}\varphi(u) = \mathbb{E}e^{i\langle u,X\rangle}$ it follows that φ is the characteristic function of X.

Let $Y \in L^0(\Omega,\mathcal{G},\mathbb{P};\mathbb{R}^d)$ be arbitrary. Then the characteristic function of (X,Y) is

$$\begin{aligned}\varphi_{(X,Y)}(u,v) &= \mathbb{E}\left[e^{i\langle u,X\rangle+i\langle v,Y\rangle}\right] \\ &= \mathbb{E}\left[e^{i\langle v,Y\rangle}\mathbb{E}\left(e^{i\langle u,X\rangle}|\mathcal{G}\right)\right] \\ &= \varphi(u)\,\mathbb{E}e^{i\langle v,Y\rangle} \\ &= \varphi_X(u)\,\varphi_Y(v),\end{aligned}$$

for all $u, v \in \mathbb{R}^d$ and consequently X and Y are independent. ■

Lemma 1.36. *Let $U \in L^1(\Omega,\mathcal{F},\mathbb{P})$ and $Y \in L^0(\Omega,\mathcal{F},\mathbb{P};\mathbb{R}^d)$. Then*

$$\mathbb{E}(U|Y) = 0 \iff \mathbb{E}(Ue^{i<\mu,Y>}) = 0, \ \forall \mu \in \mathbb{R}^d.$$

Proof. ($\Rightarrow$): If $\mathbb{E}(U|Y) = 0$ then

$$\mathbb{E}(Ue^{i<\mu,Y>}) = \mathbb{E}(e^{i<\mu,Y>}\mathbb{E}(U|Y)) = 0.$$

($\Leftarrow$): We know that there exists a Borel measurable function $g : \mathbb{R}^d \to \mathbb{R}$ such that

$$\mathbb{E}(U|Y) = g(Y).$$

Then

$$\begin{aligned}\mathbb{E}(Ue^{i<\mu,Y>}) = 0 &\Leftrightarrow \mathbb{E}(g(Y)e^{i<\mu,Y>}) = 0\\ &\Leftrightarrow \mathbb{E}(g^+(Y)e^{i<\mu,Y>}) = \mathbb{E}(g^-(Y)e^{i<\mu,Y>})\\ &\Leftrightarrow \int_{\mathbb{R}^d} g^+(y)e^{i<\mu,y>}\mathbb{P}_Y(dy) = \int_{\mathbb{R}^d} g^-(y)e^{i<\mu,y>}\mathbb{P}_Y(dy).\end{aligned}$$

Choosing $\mu = 0$ we deduce that $\mathbb{E}g^+(Y) = \mathbb{E}g^-(Y) = c$.

If $c = 0$ then $g^+(Y) = g^-(Y) = 0$ a.s. and hence $\mathbb{E}(U|Y) = 0$.

If $c > 0$ then the probability measures

$$Q^+(dy) = \frac{1}{c}g^+(y)\mathbb{P}_Y(dy), \quad Q^-(dy) = \frac{1}{c}g^-(y)\mathbb{P}_Y(dy)$$

on $(\mathbb{R}^d, \mathcal{B}_d)$ have the same characteristic functions (Fourier transforms). Hence $Q^+ = Q^-$, which implies that $g^+(Y) = g^-(Y)$ $\mathbb{P}_Y$ - a.s., that is $\mathbb{E}(U|Y) = 0$. ■

Lemma 1.37. *Let $\varphi \in C_b(\mathbb{R}^d)$, $\mathcal{G}$ be a sub-σ-algebra of $\mathcal{F}$ and $X_n, X \in L^0(\Omega, \mathcal{F}, \mathbb{P}; \mathbb{R}^d)$. If $X_n \xrightarrow{prob.} X$ as $n \to \infty$, then $\mathbb{E}^{\mathcal{G}}\varphi(X_n) \to \mathbb{E}^{\mathcal{G}}\varphi(X)$ in $L^p(\Omega, \mathcal{F}, \mathbb{P})$ as $n \to \infty$ for any $p \geq 1$.*

Proof. Clearly, by the Lebesgue dominated convergence theorem $\varphi(X_n) \to \varphi(X)$ in $L^p(\Omega, \mathcal{F}, \mathbb{P})$, for every $p \geq 1$, and the result follows from the continuity on $L^p(\Omega, \mathcal{F}, \mathbb{P})$ of the conditional expectation. ■

1.1.4 Stochastic Processes

Let $\mathbb{X}$ be a topological space and $\mathbb{T} \subset \mathbb{R}^d$ be a Borel set. A mapping $X : \Omega \times \mathbb{T} \to \mathbb{X}$ is an $\mathbb{X}$-valued stochastic process if for all $t \in \mathbb{T}$

$$X(\cdot, t) \text{ is an } \mathbb{X}\text{-valued random variable}$$

(i.e. $X(\cdot, t)$ is $(\mathcal{F}, \mathcal{B}_{\mathbb{X}})$-measurable). We shall write $X_t = X(\cdot, t)$. The mappings $t \to X(\omega, t)$, $\omega \in \Omega$, are called the sample paths (or trajectories) of the stochastic process X.

An $\mathbb{R}^d$-valued stochastic process will be called a d-dimensional stochastic process. By a stochastic process we will usually mean a one-dimensional stochastic process.

A stochastic process $\{Y_t : t \in \mathbb{T}\}$ is a modification of $\{X_t : t \in \mathbb{T}\}$ if

$$\mathbb{P}(X_t = Y_t) = 1, \quad \text{for all } t \in \mathbb{T}.$$

The stochastic processes X and Y are indistinguishable if

$$\mathbb{P}\left(\bigcup_{t\in\mathbb{T}} (X_t \neq Y_t)\right) = 0$$

or equivalently

$$\mathbb{P}\left(\bigcap_{t\in\mathbb{T}} (X_t = Y_t)\right) = 1.$$

If the mapping $X : \Omega \times \mathbb{T} \to \mathbb{X}$ is $(\mathcal{F} \otimes \mathcal{B}_{\mathbb{T}}, \mathcal{B}_{\mathbb{X}})$-measurable then we shall say that X is an $\mathbb{X}$-valued measurable stochastic process.

Remark 1.38. In this book, all stochastic processes will be understood to be measurable. We will usually have $\mathbb{T} = [0, \infty[$, or $\mathbb{T} = [0, T]$.

We shall say that the stochastic process $X : \Omega \times \mathbb{T} \to \mathbb{X}$ is continuous (abbreviated c.s.p.) (resp. right continuous, left continuous, increasing for $\mathbb{X} = \mathbb{R}$) if the trajectories (the paths) $X(\omega, .) : \mathbb{T} \to \mathbb{X}$ are continuous (resp. right continuous, left continuous, increasing) $\mathbb{P}$-$a.s.$

It easy to prove that if $\{Y_t : t \in \mathbb{T}\}$ is a modification of $\{X_t : t \in \mathbb{T}\}$ and both X and Y are right (or left) continuous stochastic processes then X and Y are indistinguishable. Hence:

Remark 1.39. If both X and Y are right (or left) continuous stochastic processes on an interval $\mathbb{T} \subset \mathbb{R}$, then

$$(X_t = Y_t,\ \forall t \in \mathbb{T})\ \ \mathbb{P}\text{-}a.s. \quad \Longleftrightarrow \quad (X_t = Y_t,\ \mathbb{P}\text{-}a.s.)\ \ \forall t \in \mathbb{T}\,.$$

The natural filtration associated to a stochastic process $X : \Omega \times \mathbb{R}_+ \to \mathbb{X}$ is defined by

$$\mathcal{F}_t^X = \sigma\{X_s : s \leq t\} \vee \mathcal{N},$$

where $\mathcal{N}$ denotes the collection of all $\mathbb{P}$-null-sets of $\mathcal{F}$. $\mathcal{F}_t^X$ is called the *history* of the process X until (and including) time $t \geq 0$. The right continuous version of the natural filtration is by definition

$$\mathcal{F}_{t+}^X = \bigcap_{\varepsilon>0} \mathcal{F}_{t+\varepsilon}^X = \left[\bigcap_{\varepsilon>0} \sigma\{X_s : s \leq t+\varepsilon\}\right] \vee \mathcal{N}\,.$$

A stochastic process $X : \Omega \times \mathbb{R}_+ \to \mathbb{X}$ is *progressively measurable* (abbreviated $\mathcal{P}$-m.s.p.), if X is $(\mathcal{P}, \mathcal{B}_{\mathbb{X}})$-measurable or equivalently if for all $t \geq 0$,

$$(\omega, s) \longmapsto X(\omega, s) : \Omega \times [0, t] \to \mathbb{X}$$

is $(\mathcal{F}_t \otimes \mathcal{B}_{[0,t]}, \mathcal{B}_{\mathbb{X}})$-measurable; we shall say that X is *$\mathcal{P}$-measurable.*

A stochastic process $X : \Omega \times \mathbb{R}_+ \to \mathbb{X}$ is adapted to the filtration $\{\mathcal{F}_t : t \geq 0\}$, if

$$\omega \longmapsto X_t(\omega) : \Omega \to \mathbb{X}$$

is $(\mathcal{F}_t, \mathcal{B}_{\mathbb{X}})$-measurable for all $t \geq 0$. It is easy to prove that:

- every progressively measurable stochastic process is adapted;
- if X is an adapted right (or left) continuous stochastic process, then X is progressively measurable.

We now recall the very well known Kolmogorov's criterion for the existence of a continuous version of a process.

Theorem 1.40 (Kolmogorov's Criterion). *Let* $(\mathbb{X}, \|\cdot\|)$ *be a Banach space and* $\{X_v\, ;\, v \in \mathbb{R}_+^k\}$ *be an* $\mathbb{X}$*-valued stochastic process, for which there exists three strictly positive constants* M, a, b *such that*

$$\mathbb{E}(\|X_u - X_v\|^a) \leq M|u - v|^{k+b}, \quad u, v \in [0, R]^k.$$

Then there exists a process $\{\tilde{X}_v,\ v \in [0, R]^k\}$ *which is a modification of* X *such that for all* $0 < \delta \leq b/a$, $\mathbb{P}$*-a.s.* $\omega \in \Omega$*:*

$$\|\tilde{X}_u(\omega) - \tilde{X}_v(\omega)\| \leq \xi_\delta(\omega)\ \ |u - v|^{\frac{b}{a} - \delta}, \ \ \textit{for all } u, v \in [0, R]^k,$$

where the random variable ξ_δ *satisfies:*

$$\mathbb{E}\xi_\delta^a \leq \frac{M\ k\ R^k 2^{a+b}}{\left(2^\delta - 1\right)^a \wedge \left(2^{a\delta} - 1\right)}.$$

Let $p \in [0, \infty[$ and $0 < T < +\infty$. We define the spaces of stochastic processes:

- $S_d^p[0, T]$: the space of (equivalence classes of) $\mathcal{P}$-measurable and continuous stochastic processes (abbreviated $\mathcal{P}$-m.c.s.p.) $X : \Omega \times [0, T] \to \mathbb{R}^d$ such that

  $$\mathbb{E} \sup_{t \in [0,T]} |X_t|^p < +\infty, \quad \text{if } p > 0;$$

  two processes X, Y are equivalent if $(X_t = Y_t$ for all $t \in [0, T])$ $\mathbb{P}$-a.s. $\omega \in \Omega$.
- S_d^p: the space of (equivalence classes of) $\mathcal{P}$-measurable and continuous processes $X : \Omega \times [0, +\infty[\ \to \mathbb{R}^d$ such that for all $T > 0$ the restriction of X to $\Omega \times [0, T]$ belongs to $S_d^p[0, T]$.

One easily sees that for every $p \in [1, \infty[$, the space $S_d^p[0,T]$ is a Banach space, when equipped with the norm

$$\|X\|_{S_d^p[0,T]} = \left(\mathbb{E}\,\|X\|_T^p\right)^{1/p} = \left(\mathbb{E}\sup_{t\in[0,T]}|X_t|^p\right)^{1/p}.$$

For $0 \le p < 1$ the space $S_d^p[0,T]$ is a complete metric linear space with respect to the metric

$$\Delta_p(X,Y) = \begin{cases} \mathbb{E}\left(\|X-Y\|_T^p\right), & \text{if } 0 < p < 1,\\ \mathbb{E}\left(1 \wedge \|X-Y\|_T\right), & \text{if } p = 0. \end{cases}$$

$\Delta_0(X,Y)$ is the metric of convergence in probability uniformly in $t \in [0,T]$. On $S_d^0[0,T]$ we shall also use the topological equivalent metrics

$$\rho_1(X,Y) = \mathbb{E}\frac{|X-Y|_T}{1+|X-Y|_T} \quad \text{and} \quad \rho_a(X,Y) = \left(\mathbb{E}\frac{|X-Y|_T^a}{\left(1+\delta|X-Y|_T^2\right)^{a/2}}\right)^{1/a},$$

where $a \ge 1$ and $\delta > 0$. Note that in the case $d = 1$ the subscript d in S_d^p and $S_d^p[0,T]$ will usually be omitted.

1.1.5 Complements on Tightness

Let $(\mathbb{X}, \rho)$ be a metric space. Denote by $\mathcal{B}_{\mathbb{X}}$ the σ-algebra of Borel subsets of $\mathbb{X}$ and by $\mathcal{M}_1(\mathbb{X})$ the space of probability measures $Q : \mathcal{B}_{\mathbb{X}} \to [0,1]$. If $g : (\mathbb{X}; \mathcal{B}_{\mathbb{X}}, Q) \to \mathbb{R}$ is a random variable (i.e. a $\mathcal{B}_{\mathbb{X}}$-measurable function), then we write

$$Q(g) = \mathbb{E}_Q(g) = \int_{\mathbb{X}} g(x)\,dQ(x),$$

whenever at least $Q\left(g^+\right) < \infty$ or $Q\left(g^-\right) < \infty$. Hence $Q(\mathbf{1}_B) = Q(B)$.

Definition 1.41. We say that Q_n converges weakly to Q, denoted $Q_n \Longrightarrow Q$, whenever $Q_n(g) \to Q(g)$, for all $g \in C_b(\mathbb{X})$.

Let $(\Omega, \mathcal{F}, \mathbb{P})$ be a probability space and $Y : \Omega \to \mathbb{X}$ be a random variable; then the probability measure $\mathcal{L}(Y) = \mathbb{P}_Y : \mathcal{B}_{\mathbb{X}} \to [0,1]$ defined by

$$\mathbb{P}_Y(B) = \mathbb{P}(Y \in B)$$

is called the (probability) law of Y.

Note that for all $Q \in \mathcal{M}_1(\mathbb{X})$ there exists a probability space $(\Omega, \mathcal{F}, \mathbb{P})$ and a random variable $Y : \Omega \to \mathbb{X}$ such that $\mathcal{L}(Y) = Q$. In this case

$$Q(g) \overset{def}{=} \int_{\mathbb{X}} g(x)\, dQ(x) = \int_{\Omega} g(Y(\omega))\, d\mathbb{P}(\omega) \overset{def}{=} \mathbb{E}(g(Y)).$$

Let $X_n : \left(\Omega^{(n)}; \mathcal{F}^{(n)}, \mathbb{P}^{(n)}\right) \to \mathbb{X}$, $n \in \mathbb{N}^*$, and $X : (\Omega; \mathcal{F}, \mathbb{P}) \to \mathbb{X}$ be random variables and

$$Q_n = \mathcal{L}(X_n), \quad Q = \mathcal{L}(X).$$

Definition 1.42. $X_n \overset{law}{\longrightarrow} X$, as $n \to \infty$, if $Q_n \Longrightarrow Q$, or equivalently

$$\lim_{n\to\infty} \mathbb{E}_{\mathbb{P}^{(n)}}(g(X_n)) = \mathbb{E}_{\mathbb{P}}(g(X))$$

for all $g \in C_b(\mathbb{X})$.

Definition 1.43. (a) A family $X^{(i)} : \left(\Omega^{(i)}; \mathcal{F}^{(i)}, \mathbb{P}^{(i)}\right) \to \mathbb{X}$, $i \in I$, of random variables is tight if for every $\varepsilon > 0$ there exists a compact set $K_\varepsilon \subset \mathbb{X}$ (abbreviated $K_\varepsilon \subset\subset \mathbb{X}$), such that for all $i \in I$:

$$\mathbb{P}^{(i)}\left(X^{(i)} \notin K_\varepsilon\right) < \varepsilon.$$

(b) The family $\left\{X^{(i)} : i \in I\right\}$ is relatively compact in law if every sequence $\{X_n : n \in \mathbb{N}^*\} \subset \left\{X^{(i)} : i \in I\right\}$ contains a subsequence $\{X_{n_k} : k \in \mathbb{N}^*\}$ convergent in law, i.e. there exists a probability space $(\Omega, \mathcal{F}, \mathbb{P})$ and a random variable $X : \Omega \to \mathbb{X}$ such that $X_{n_k} \overset{law}{\longrightarrow} X$.

A famous result due to Prohorov and Varadarajan gives an equivalence between the compactness in law and the tightness property.

Theorem 1.44 (Prohorov–Varadarajan). *Let* $(\mathbb{X}, \rho)$ *be a metric space and* $\left\{X^{(i)} : i \in I\right\}$ *be a family of* $\mathbb{X}$*-valued random variables.*

(A) *If* $\left\{X^{(i)} : i \in I\right\}$ *is tight then it is relatively compact in law.*
(B) *Suppose that* $(\mathbb{X}, \rho)$ *is a Polish space. If* $\left\{X^{(i)} : i \in I\right\}$ *is relatively compact in law then it is tight.*

The interested reader can find a proof of this theorem in Billingsley [11] or Stroock [68] (Theorem 2.6).

Concerning the convergence in law of sequences of random variables we have another famous result due to Skorohod.

Theorem 1.45 (Skorohod). *Let* $\mathbb{X}$ *be a Polish space (i.e.* $\mathbb{X}$ *is a topological space which admits a complete separable metric* ρ*),* $X_n : \left(\Omega^{(n)}; \mathcal{F}^{(n)}, \mathbb{P}^{(n)}\right) \to \mathbb{X}$, $n \in \mathbb{N}^*$, *and* $X : (\Omega; \mathcal{F}, \mathbb{P}) \to \mathbb{X}$ *be random variables. If*

$$X_n \xrightarrow{law} X$$

then there exist a probability space $\left(\hat{\Omega}, \hat{\mathcal{F}}, \hat{\mathbb{P}}\right)$ *on which we can define a sequence of random variables* $Y, Y_n : \hat{\Omega} \to \mathbb{X}, \quad n \in \mathbb{N}^*$, *such that*

$$\begin{aligned} &(c)\ \mathbb{P}_Y = \mathbb{P}_X, \quad \text{and } \mathbb{P}_{Y_n} = \mathbb{P}_{X_n} \text{ for all } n \in \mathbb{N}^*, \\ &(cc)\ Y_n \xrightarrow{a.s.} Y, \quad \text{as } n \to \infty. \end{aligned}$$

Moreover one can choose $\left(\hat{\Omega}, \hat{\mathcal{F}}, \hat{\mathbb{P}}\right) \equiv \left([0,1]\,; \mathcal{B}_{[0,1]}, \mu\right)$, *where* μ *is the Lebesgue measure on* $[0,1]$.

If $x \in C(\mathbb{R}_+; \mathbb{R}^d)$ we write

$$\|x\|_T = \sup_{t \in [0,T]} |x(t)|, \quad \text{and}$$

$$\mathbf{m}_x(\varepsilon; [0,T]) = \sup\{|x(t) - x(s)| : t, s \in [0,T],\ |t-s| \le \varepsilon\}.$$

Recall that, when equipped with uniform convergence on compact sets, $C(\mathbb{R}_+; \mathbb{R}^d)$ is a Polish space. A metric for this topology is given by

$$\rho(x,y) = \sum_{n=1}^{\infty} \frac{1}{2^n} \frac{\|x-y\|_n}{1 + \|x-y\|_n}.$$

If $\{X_t^n : t \ge 0\}$, $n \in \mathbb{N}^*$, is a family of continuous stochastic processes such that Q_n is the law of the random variable $\omega \mapsto X_\cdot^n : \left(\Omega^{(n)}; \mathcal{F}^{(n)}, \mathbb{P}^{(n)}\right) \to C(\mathbb{R}_+; \mathbb{R}^d)$, then the following result is a consequence of the Arzelà–Ascoli theorem (see Billingsley [11], Theorem 7.3).

Theorem 1.46. *$\{X^n : n \in \mathbb{N}^*\}$ is tight on $C(\mathbb{R}_+; \mathbb{R}^d)$ if and only if for every $T \ge 0$:*

$$\begin{aligned} &(i)\ \lim_{N \nearrow \infty} \left[\sup_{n \ge 1} \mathbb{P}^{(n)}\left(\left|X_0^n\right| \ge N\right)\right] = 0, \\ &(ii)\ \lim_{\varepsilon \searrow 0} \left[\sup_{n \ge 1} \mathbb{P}^{(n)}\left(\mathbf{m}_{X^n}(\varepsilon; [0,T]) \ge a\right)\right] = 0, \quad \forall\, a > 0. \end{aligned}$$

Moreover, tightness yields that for all $T > 0$:

$$\lim_{N \nearrow \infty} \left[\sup_{n \ge 1} \mathbb{P}^{(n)}\left(\|X^n\|_T \ge N\right)\right] = 0.$$

Without using the above Theorem, we establish a criterion for tightness which is well adapted to our needs.

Proposition 1.47. *Let* $\{X_t^n : t \geq 0\}$, $n \in \mathbb{N}^*$, *be a family of* $\mathbb{R}^d$*-valued continuous stochastic processes defined on a probability space* $(\Omega, \mathcal{F}, \mathbb{P})$. *Suppose that for every* $T \geq 0$, *there exist* $\alpha = \alpha_T > 0$ *and* $b = b_T \in C(\mathbb{R}_+)$ *satisfying* $b(0) = 0$, *such that:*

$$(j)\ \lim_{N\to\infty}\left[\sup_{n\in\mathbb{N}^*}\mathbb{P}(\{\left|X_0^n\right| \geq N\})\right] = 0,$$

$$(jj)\ \mathbb{E}\left[1 \wedge \sup_{0\leq s\leq\varepsilon}\left|X_{t+s}^n - X_t^n\right|^\alpha\right] \leq \varepsilon \cdot b(\varepsilon),\ \forall\ \varepsilon > 0, n \geq 1,\ t \in [0, T].$$

Then $\{X^n : n \in \mathbb{N}^*\}$ *is tight in* $C(\mathbb{R}_+; \mathbb{R}^d)$.

Proof. We fix $\varepsilon, T > 0$. From (j), there exists an $M = M_\varepsilon \geq 1$ such that

$$\sup_{n\in\mathbb{N}^*}\mathbb{P}(\{|X_0^n| \geq M\}) < \frac{\varepsilon}{2}.$$

Let $\gamma_k = \dfrac{1}{2^{(k-1)/\alpha}}$ and $\varepsilon_k \searrow 0$ be such that $b(\varepsilon_k) \leq \dfrac{\varepsilon}{4^k\, T}$. Let $N_k = \lfloor \dfrac{T}{\varepsilon_k} \rfloor$, and $t_i = \dfrac{(i-1)T}{N_k}$. By Corollary 6.12 from Annex B the set

$$\mathcal{K}_\varepsilon = \{z \in C([0, T]; \mathbb{R}^d) : |z(0)| \leq M,$$
$$\sup_{1\leq i\leq N_k}\ \sup_{0<s\leq\varepsilon_k}|z(t_i + s) - z(t_i)| \leq \gamma_k, \forall k \in \mathbb{N}^*\}$$

is compact in $C([0, T]; \mathbb{R}^d)$.

From Markov's inequality and (jj)

$$\begin{aligned}\mathbb{P}(X^n \notin \mathcal{K}_\varepsilon) &\leq \mathbb{P}(\{|X_0^n| > M\}) + \sum_{k\in\mathbb{N}^*}\sum_{i=1}^{N_k}\mathbb{P}(\{\sup_{0\leq s\leq\varepsilon_k}\left|X_{t_i+s}^n - X_{t_i}^n\right| > \gamma_k\}) \\ &< \frac{\varepsilon}{2} + \sum_{k\in\mathbb{N}^*}\sum_{i=1}^{N_k}\frac{\varepsilon_k \times b(\varepsilon_k)}{\gamma_k^\alpha} \\ &= \frac{\varepsilon}{2} + \sum_{k\in\mathbb{N}^*}\frac{N_k \times \varepsilon_k \times b(\varepsilon_k)}{\gamma_k^\alpha} \\ &\leq \varepsilon.\end{aligned}$$

The proof is complete. ■

Proposition 1.48. *Let* $g : \mathbb{R}_+ \to \mathbb{R}_+$ *be a continuous function satisfying* $g(0) = 0$ *and* $G : C(\mathbb{R}_+; \mathbb{R}^d) \to \mathbb{R}_+$ *be a mapping which is bounded on compact subsets of*

$C\left(\mathbb{R}_+;\mathbb{R}^d\right)$. Let X^n, Y^n, $n \in \mathbb{N}^*$, be random variables with values in $C\left(\mathbb{R}_+;\mathbb{R}^d\right)$. If $\{Y^n : n \in \mathbb{N}^*\}$ is tight and for all $n \in \mathbb{N}^*$

$$\begin{aligned} &(i) \quad \left|X_0^n\right| \leq G\left(Y^n\right), \; a.s. \\ &(ii) \quad \mathbf{m}_{X^n}\left(\varepsilon;[0,T]\right) \leq G\left(Y^n\right) g\left(\mathbf{m}_{Y^n}\left(\varepsilon;[0,T]\right)\right), \; a.s., \quad \forall\, \varepsilon, T > 0, \end{aligned}$$

then $\{X^n : n \in \mathbb{N}^*\}$ is tight.

Proof. Let $\delta > 0$ be arbitrary. Then there exists a compact set $K_\delta \subset C\left([0,\infty[\,;\mathbb{R}^d\right)$ such that for all $n \in \mathbb{N}^*$

$$\mathbb{P}\left(Y^n \notin K_\delta\right) < \delta.$$

Define $N_\delta = \sup\limits_{x \in K_\delta} G\left(x\right)$. Then

$$\mathbb{P}\left(\left|X_0^n\right| > N_\delta\right) < \delta.$$

Let $a > 0$ be arbitrary. There exists an $\varepsilon_0 > 0$ such that

$$\sup_{x \in K_\delta}\left[g\left(\mathbf{m}_x\left(\varepsilon;[0,T]\right)\right)\right] < \frac{a}{N_\delta}, \quad \forall\, 0 < \varepsilon < \varepsilon_0.$$

Consequently for all $n \in \mathbb{N}^*$, $0 < \varepsilon < \varepsilon_0$,

$$\begin{aligned} &\mathbb{P}\left(\mathbf{m}_{X^n}\left(\varepsilon;[0,T]\right) \geq a\right) \\ &\leq \mathbb{P}\left[g\left(\mathbf{m}_{Y^n}\left(\varepsilon;[0,T]\right)\right) \geq \frac{a}{N_\delta},\; Y^n \in K_\delta\right] + \mathbb{P}\left(Y^n \notin K_\delta\right) \\ &\leq \delta. \end{aligned}$$

The result follows. ■

1.1.6 Stopping Times

Fix a stochastic basis $(\Omega, \mathcal{F}, \mathbb{P}, \{\mathcal{F}_t\}_{t\geq 0})$. A random variable $\tau : \Omega \to [0,\infty]$ is a stopping time if

$$\{\tau \leq t\} \overset{def}{=} \{\omega \in \Omega : \tau\left(\omega\right) \leq t\} \in \mathcal{F}_t, \quad \forall\, t \in [0,\infty].$$

Since $\mathcal{F}_t = \mathcal{F}_{t+}$ (the filtration is right continuous) we have

$$\tau \text{ is a stopping time} \quad \Leftrightarrow \quad \{\tau < t\} \in \mathcal{F}_t, \ \forall \, t \in [0, \infty] .$$

Proposition 1.49. *Let $X : \Omega \times [0, \infty[\to \mathbb{R}^d$ be a $\mathcal{P}$-measurable continuous stochastic process and D be a closed (or open) subset of $\mathbb{R}^d$. The hitting time of D by X (the exit time of X from $D^c = \mathbb{R}^d \setminus D$) defined by*

$$\tau(\omega) = \begin{cases} \inf\{t \geq 0 : X_t(\omega) \in D\}, & \text{if } \{t \geq 0 : X_t(\omega) \in D\} \neq \emptyset, \\ +\infty, & \text{if } \{t \geq 0 : X_t(\omega) \in D\} = \emptyset, \end{cases}$$

is a stopping time.

Proof. If D is an open set, then

$$\{\tau < t\} = \bigcup\nolimits_{r \in [0,t) \cap \mathbb{Q}} \{X_r \in D\}$$

and in the case of a closed set D

$$\{\tau \leq t\} = \bigcap\nolimits_{n \in \mathbb{N}^*} \bigcup\nolimits_{r \in [0,t] \cap \mathbb{Q}} \{X_r \in D_n\},$$

where $D_n = \left\{x \in \mathbb{R}^d : d(x, D) < \dfrac{1}{n}\right\}$ and $d(x, D)$ is the distance from the point x to D. ■

We define the σ-algebra $\mathcal{F}_\tau$ of events *prior* to the stopping time τ by

$$\mathcal{F}_\tau = \{A \in \mathcal{F} : A \cap \{\tau \leq t\} \in \mathcal{F}_t, \ \forall \, t \geq 0\} .$$

As an exercise for the reader we give the following proposition:

Proposition 1.50. *If τ, θ, σ are stopping times and a is a real number such that $a \geq 1$, then:*

(a) $\tau \vee \theta$, $\tau \wedge \theta$, $\tau + \theta$, $a\tau$ are stopping times;
(b) if $\tau \leq \theta$ a.s., then $\mathcal{F}_\tau \subset \mathcal{F}_\theta$;
(c) if $\tau = t$ a.s., then $\mathcal{F}_\tau = \mathcal{F}_t$;
(d) $\mathcal{F}_\tau \cap \mathcal{F}_\theta = \mathcal{F}_{\tau \wedge \theta}$ and the sets $\{\tau < \theta\}$, $\{\tau = \theta\}$, $\{\tau \leq \theta\}$ are in $\mathcal{F}_{\tau \wedge \theta}$;
(e) $\{\tau \leq \sigma < \theta\} \in \mathcal{F}_\sigma$;
(f) the stochastic processes $\left\{\mathbf{1}_{[0,\tau[}(t) : t \geq 0\right\}$ and $\left\{\mathbf{1}_{[0,\tau]}(t) : t \geq 0\right\}$ are $\mathcal{P}$-measurable.

Hint for (d). If $A \in \mathcal{F}_\tau \cap \mathcal{F}_\theta$ then

$$A \bigcap \{\tau \wedge \theta \leq t\} = (A \cap \{\tau \leq t\}) \bigcup (A \cap \{\theta \leq t\}) \in \mathcal{F}_t .$$

Also

$$\{\tau < \theta\} \bigcap \{\tau \wedge \theta \le t\}$$
$$= (\{\tau \le t\} \cap \{\theta > t\}) \bigcup \bigcup_{r \in \mathbb{Q} \cap]-\infty, t]} (\{\tau \le r\} \cap \{r < \theta\} \cap \{\theta \le t\}) \in \mathcal{F}_t.$$

Proposition 1.51. *Let* $\{\tau_n\}_{n\in\mathbb{N}}$ *be a sequence of stopping times. Then*

(a) $\sup_n \tau_n$ *is a stopping time;*
and moreover, by the right continuity of filtration $\{\mathcal{F}_t : t \ge 0\}$,
(b) $\inf_n \tau_n$, $\liminf_{n\to+\infty} \tau_n$, $\limsup_{n\to+\infty} \tau_n$ *are stopping times and if* $\tau_n \downarrow \tau$ *then*

$$\mathcal{F}_\tau = \bigcap_n \mathcal{F}_{\tau_n}.$$

Proof. We have

$$\left\{\sup_n \tau_n \le t\right\} = \bigcap_{n\in\mathbb{N}} \{\tau_n \le t\}$$

and

$$\left\{\inf_n \tau_n < t\right\} = \bigcup_{n\in\mathbb{N}} \{\tau_n < t\}.$$

■

If X is a stochastic process and $\theta : \Omega \to \mathbb{R}_+$ is a random variable we denote by X_θ the random variable $\omega \longrightarrow X_{\theta(\omega)}(\omega)$ and by $\{X_{t\wedge\theta} : t \ge 0\}$ the process X stopped at θ, that is $X_{t\wedge\theta}(\omega) = X(\omega, t \wedge \theta(\omega))$ for all $t \ge 0$.

Proposition 1.52. *Let* $\{X_t : t \ge 0\}$ *be a* $\mathcal{P}$*-measurable stochastic process and* $\tau : \Omega \to \mathbb{R}_+$ *be a stopping time. Then:*

(a) X_τ *is* $\mathcal{F}_\tau$*-measurable;*
(b) $\{X_{t\wedge\tau} : t \ge 0\}$ *is progressively measurable with respect to the filtration* $\{\mathcal{F}_{t\wedge\tau} : t \ge 0\}$*, that is for all* $t \ge 0$ *and* $B \in \mathcal{B}_1$*:*

$$\left\{(\omega, s) \in \Omega \times [0, t] : X_{s\wedge\tau(\omega)}(\omega) \in B\right\} \in \mathcal{F}_{t\wedge\tau} \otimes \mathcal{B}_{[0,t]}.$$

Proof. We sketch the proof of the first part only. Let $t \ge 0$, $B \in \mathcal{B}_1$ and

$$F = \{(\omega, s) \in \Omega \times [0, t] : X_s(\omega) = X(\omega, s) \in B\}.$$

From the $\mathcal{P}$-measurability of X, $F \in \mathcal{F}_t \otimes \mathcal{B}_{[0,t]}$. The mapping $\alpha(\omega) = (\omega, \tau(\omega))$ is measurable from $(\{\tau \le t\}; \{\tau \le t\} \cap \mathcal{F}_t)$ to $\left(\Omega \times [0, t]; \mathcal{F}_t \otimes \mathcal{B}_{[0,t]}\right)$.

Hence

$$\{\omega : X(\omega, \tau(\omega)) \in B\} = \alpha^{-1}(F) \in \mathcal{F}_t.$$

■

Proposition 1.53. *If $\tau : \Omega \to \mathbb{R}_+$ is a stopping time, then*

$$\mathcal{F}_\tau = \sigma\{X_\tau : X \text{ is a } \mathcal{P}\text{-measurable right continuous stochastic process}\}. \tag{1.7}$$

Proof. Denote by $\mathcal{G}$ the right-hand side of (1.7).

If $A \in \mathcal{F}_\tau$, then $X_t(\omega) = \mathbf{1}_A(\omega)\mathbf{1}_{[0,t]}(\tau(\omega))$ is a $\mathcal{P}$-measurable right continuous stochastic process. Since $X_\tau = \mathbf{1}_A$ it follows that $A \in \mathcal{G}$.

To show that $\mathcal{G} \subset \mathcal{F}_\tau$ it is sufficient to prove that X_τ is $\mathcal{F}_\tau$-measurable. For $t \geq 0$, $B \in \mathcal{B}_1$, let

$$F = \{(\omega, s) \in \Omega \times [0,t] : X_s(\omega) \in B\}.$$

From the $\mathcal{P}$-measurability of X, $F \in \mathcal{F}_t \otimes \mathcal{B}_{[0,t]}$. The mapping $\alpha(\omega) = (\omega, \tau(\omega))$ is measurable from $(\{\tau \leq t\}; \{\tau \leq t\} \cap \mathcal{F}_t)$ to $(\Omega \times [0,t]; \mathcal{F}_t \otimes \mathcal{B}_{[0,t]})$.

Hence

$$\{\omega : X(\omega, \tau(\omega)) \in B\} = \alpha^{-1}(F) \in \mathcal{F}_t.$$

■

Given a d-dimensional $\mathcal{P}$-measurable continuous process $\{X_t : t \geq 0\}$, two stopping times σ, τ, such that $0 \leq \sigma \leq \tau$ and a real sequence $\delta_n \searrow 0$, we construct a "basic partition":

Definition 1.54 (Basic Partition). A sequence $\{(\theta_i^n, k_n) : i, n \in \mathbb{N}\}$, where θ_i^n are stopping times and $k_n \in \mathbb{N}$, $k_n \nearrow \infty$, is called a "basic partition" of $[\sigma, \tau]$ associated to $\{(X_t, \delta_n) : t \geq 0, n \in \mathbb{N}\}$ if

$$\begin{aligned}
&(i) && \sigma = \theta_0^n \leq \theta_1^n \leq \cdots \leq \theta_i^n \leq \cdots \leq \tau,\\
&(ii) && \exists N_n(\omega) \in \mathbb{N}^* \text{ s.t. } \theta_i^n(\omega) = \tau(\omega), \quad \text{for } i \geq N_n(\omega),\\
&(iii) && 0 \leq \theta_{i+1}^n - \theta_i^n \leq \delta_n,\\
&(iv) && \sup_{\theta_i^n \leq s \leq t \leq \theta_{i+1}^n} |X_t - X_s| \leq \delta_n, \text{ for all } i \in \mathbb{N},\\
&(v) && \{\theta_i^n : i \in \mathbb{N}\} \subset \{\theta_i^{n+1} : i \in \mathbb{N}\}, \ \forall\, n \in \mathbb{N}
\end{aligned} \tag{1.8}$$

and

$$\begin{aligned}
&(vi) && \sigma = \theta_{k_0}^0 \leq \theta_{k_1}^1 \leq \cdots \leq \theta_{k_n}^n \leq \cdots \leq \tau,\\
&(vii) && \mathbb{P}\left(\theta_{k_n}^n < \tau\right) \leq \delta_n.
\end{aligned} \tag{1.9}$$

Such a partition can be defined as follows:

$$\theta_0^0 = \sigma, \ \theta_i^0 = \tau, \ \text{for } i \geq 1,$$

and given $\left(\theta_i^{n-1}\right)_{i\in\mathbb{N}}$, the interval $\left[\theta_j^{n-1}, \theta_{j+1}^{n-1}\right]$ is partitioned by the sequence:

$$\theta_\ell^n = \theta_j^{n-1},$$

$$\theta_{\ell+1}^n = \theta_{j+1}^{n-1} \wedge \left(\theta_\ell^n + \delta_n\right) \wedge \inf\left\{ t \geq \theta_\ell^n : \sup_{\theta_\ell^n \leq s \leq t} \left|X_s - X_{\theta_\ell^n}\right| \geq \frac{\delta_n}{2} \right\}.$$

For each ω, there exists an $L = L_j^n(\omega) \in \mathbb{N}^*$ such that

$$\theta_j^{n-1}(\omega) = \theta_\ell^n(\omega) < \theta_{\ell+1}^n(\omega) < \cdots < \theta_{\ell+L}^n(\omega) = \theta_{j+1}^{n-1}(\omega).$$

Since $\left\{\theta_i^n\right\}_{i\in\mathbb{N}} \subset \left\{\theta_i^{n+1}\right\}_{i\in\mathbb{N}}$, $\theta_i^n \nearrow \tau$ and $\theta_i^{n+1} \nearrow \tau$ as $i \to \infty$, there exists an increasing sequence of natural numbers $\{k_n : n \in \mathbb{N}\}$ such that (1.9) holds.

1.1.7 Fundamental Inequalities

Definition 1.55. a) A positive stochastic process $\{X_t : t \geq 0\}$ is $\mathcal{F}_t$-dominated on $[0, T]$ by a positive random variable U if for all $t \in [0, T]$:

$$\mathbb{E}\left(X_t|\mathcal{F}_t\right) \leq \mathbb{E}\left(U|\mathcal{F}_t\right), \ \mathbb{P}\text{-}a.s.$$

b) A positive stochastic process $\{X_t : t \geq 0\}$ is dominated on $[0, T]$ by a positive increasing stochastic process $\{A_t : t \geq 0\}$ if for any stopping time $\theta, 0 \leq \theta \leq T$:

$$\mathbb{E}\left(X_\theta\right) \leq \mathbb{E}\left(A_\theta\right).$$

Proposition 1.56. *Let $\{X_t : t \geq 0\}$ be a positive continuous $\mathcal{P}$-measurable stochastic process.*

A. *If $\{X_t : t \geq 0\}$ is $\mathcal{F}_t$-dominated on $[0, T]$ by a positive random variable U, then:*

$$(A_1) \quad \mathbb{P}\Big(\sup_{t\in[0,T]} X_t \geq \varepsilon\Big) \leq 1 \wedge \left[\frac{1}{\varepsilon}\mathbb{E}\Big(U\,;\ \sup_{t\in[0,T]} X_t \geq \varepsilon\Big)\right], \ \forall\, \varepsilon > 0;$$

$$(A_2) \quad \mathbb{E}\sup_{t\in[0,T]} X_t^p \leq \left(\frac{p}{p-1}\right)^p \mathbb{E}U^p, \ \ \forall\, p > 1;$$

$$(A_3) \quad \mathbb{E}\sup_{t\in[0,T]} X_t^p \leq \frac{1}{1-p}\left(\mathbb{E}U\right)^p, \ \ \forall\, 0 < p < 1.$$

B. *If* $\{X_t : t \geq 0\}$ *is dominated on* $[0, T]$ *by a positive increasing* $\mathcal{P}$*-measurable continuous stochastic process* $\{A_t : t \geq 0\}$, $X_0 = A_0 = 0$, *then for all* $\varepsilon, \delta > 0$*:*

$$(B_1) \quad \mathbb{P}\Big(\sup_{t\in[0,T]} X_t \geq \varepsilon,\ A_T \leq \delta \Big) \leq \frac{1}{\varepsilon}\mathbb{E}\left(A_T \wedge \delta\right),$$
$$(B_2) \quad \mathbb{P}\Big(\sup_{t\in[0,T]} X_t \geq \varepsilon \Big) \leq \left(\frac{1}{\varepsilon} + \frac{1}{\delta}\right)\mathbb{E}\left(A_T \wedge \delta\right)$$

and

$$(B_3) \quad \mathbb{E} \sup_{t\in[0,T]} X_t^p \leq \frac{2-p}{1-p}\mathbb{E}A_T^p, \quad \forall\, 0 < p < 1,$$
$$(B_4) \quad \mathbb{E}\Big(\sup_{t\in[0,T]} \sqrt{X_t} \wedge 1 \Big) \leq 3\left[\mathbb{E}\left(A_T \wedge 1\right)\right]^{1/3}. \tag{1.10}$$

Proof. A. Let $\tilde{X}_i = X_{\frac{iT}{N}}$, $i \in \overline{0, N}$, $N = 2^n$, and

$$F_0 = \left(\tilde{X}_0 \geq \varepsilon\right),$$
$$F_i = \left(\tilde{X}_0 < \varepsilon\right) \cap \cdots \cap \left(\tilde{X}_{i-1} < \varepsilon\right) \cap \left(\tilde{X}_i \geq \varepsilon\right), \quad \text{for } i \in \overline{1, N},$$
$$F = \bigcup_{i=0}^{N} F_i.$$

We have $\left\{\max_{i\in\overline{0,N}} \tilde{X}_i \geq \varepsilon\right\} = F$ and

$$\mathbb{P}(F) = \sum_{i=0}^{N} \mathbb{P}(F_i) \leq \sum_{i=0}^{N} \frac{1}{\varepsilon}\mathbb{E}\left(\tilde{X}_i \mathbf{1}_{F_i}\right) \leq \sum_{i=0}^{N} \frac{1}{\varepsilon}\mathbb{E}\left(U\mathbf{1}_{F_i}\right) = \frac{1}{\varepsilon}\mathbb{E}\left(U\mathbf{1}_{F}\right).$$

Hence

$$\mathbb{P}\left(\max_{i\in\overline{0,N}} \tilde{X}_i \geq \varepsilon \right) \leq \frac{1}{\varepsilon}\mathbb{E}\left(U;\ \sup_{t\in[0,T]} X_t \geq \varepsilon \right).$$

Since $\left\{ \max_{i\in\overline{0,N}} \tilde{X}_i \geq \varepsilon \right\} \nearrow \left\{ \sup_{t\in[0,T]} X_t \geq \varepsilon \right\}$ as $N = 2^n \nearrow \infty$ the inequality (A_1) follows.

To prove (A_2) let $Y = \sup_{t\in[0,T]} X_t$. We can assume that $0 < \mathbb{E}U < \infty$ and $\mathbb{E}U^p < \infty$, since otherwise (A_2) is clearly satisfied. Let $n \in \mathbb{N}^*$. By Fubini's theorem and (A_1) we have

$$\begin{aligned}
\mathbb{E}\left[(Y\wedge n)^p\right] &= \mathbb{E}\int_0^{Y\wedge n} p\lambda^{p-1}d\lambda \\
&= \mathbb{E}\int_0^n \mathbf{1}_{\{Y\geq\lambda\}}p\lambda^{p-1}d\lambda \\
&= \int_0^n \mathbb{P}\left(Y\geq\lambda\right)p\lambda^{p-1}d\lambda \\
&\leq \int_0^n \left\{1\wedge\left[\frac{1}{\lambda}\mathbb{E}\left(U\,;Y\geq\lambda\right)\right]\right\}p\lambda^{p-1}d\lambda \stackrel{def}{=} I_n.
\end{aligned}$$

If $p>1$, then

$$\begin{aligned}
I_n &\leq \mathbb{E}U\int_0^{Y\wedge n} p\lambda^{p-2}d\lambda \\
&= \frac{p}{p-1}\mathbb{E}\left[U\left(Y\wedge n\right)^{p-1}\right] \\
&\leq \frac{p}{p-1}\left(\mathbb{E}U^p\right)^{1/p}\left\{\mathbb{E}\left(Y\wedge n\right)^p\right\}^{(p-1)/p}.
\end{aligned}$$

Hence

$$\left\{\mathbb{E}\left(Y\wedge n\right)^p\right\}^{1/p}\leq\frac{p}{p-1}\left(\mathbb{E}U^p\right)^{1/p}$$

and (A_2) follows by passing to the limit as $n\to\infty$.

If $0<p<1$, then

$$\begin{aligned}
\mathbb{E}\left[(Y\wedge n)^p\right] &\leq I_n \\
&\leq p\int_0^{\mathbb{E}U}\lambda^{p-1}d\lambda+\int_{\mathbb{E}U}^{\infty}\frac{1}{\lambda}\mathbb{E}\left(U\;;Y\geq\lambda\right)p\lambda^{p-1}d\lambda \\
&\leq \left(\mathbb{E}U\right)^p+p\left(\mathbb{E}U\right)\int_{\mathbb{E}U}^{\infty}\lambda^{p-2}d\lambda \\
&= \left(\mathbb{E}U\right)^p+\frac{p}{1-p}\left(\mathbb{E}U\right)^p \\
&= \frac{1}{1-p}\left(\mathbb{E}U\right)^p
\end{aligned}$$

and (A_3) follows letting $n\to\infty$.

B. Define the stopping times

$$\begin{aligned}
\tau\left(\omega\right) &= \inf\left\{t\geq 0: X_t\left(\omega\right)\geq\varepsilon\right\}, \\
\theta\left(\omega\right) &= \inf\left\{t\geq 0: A_t\left(\omega\right)\geq\delta\right\}.
\end{aligned}$$

Then $\{A_T < \delta\} \subset \{\theta > T\}$ and

$$\begin{aligned}\mathbb{P}\left(\sup_{t\in[0,T]} X_t \geq \varepsilon,\ A_T < \delta\right) &\leq \mathbb{P}\left(X_{\tau\wedge T} \geq \varepsilon,\ \theta > T\right)\\ &\leq \mathbb{P}\left(X_{\tau\wedge\theta\wedge T} \geq \varepsilon\right)\\ &\leq \frac{1}{\varepsilon}\mathbb{E}\left(X_{\tau\wedge\theta\wedge T}\right)\\ &\leq \frac{1}{\varepsilon}\mathbb{E}\left(A_{\tau\wedge\theta\wedge T}\right)\\ &\leq \frac{1}{\varepsilon}\mathbb{E}\left(A_T \wedge \delta\right),\end{aligned}$$

since A_t is increasing and $A_0 = 0$. The inequality (B_1) follows easily.

(B_2) follows from

$$\begin{aligned}\mathbb{P}\left(\sup_{t\in[0,T]} X_t \geq \varepsilon\right) &\leq \mathbb{P}\left(\sup_{t\in[0,T]} X_t \geq \varepsilon,\ A_T \leq \delta\right) + \mathbb{P}\left(A_T \geq \delta\right)\\ &\leq \frac{1}{\varepsilon}\mathbb{E}\left(A_T \wedge \delta\right) + \frac{1}{\delta}\mathbb{E}\left(A_T \wedge \delta\right).\end{aligned}$$

Let us prove (B_3). Let $0 < p < 1$ and $Y = \sup_{t\in[0,T]} X_t$. Then

$$\begin{aligned}\mathbb{E}\sup\nolimits_{t\in[0,T]} X_t^p &= \mathbb{E}\int_0^Y p\lambda^{p-1}d\lambda\\ &= \int_0^\infty \mathbb{P}\left(Y \geq \lambda\right) p\lambda^{p-1}d\lambda\\ &= \int_0^\infty \left[\mathbb{P}\left(Y \geq \lambda,\ A_T \leq \lambda\right) + \mathbb{P}\left(A_T > \lambda\right)\right] p\lambda^{p-1}d\lambda\\ &\leq \int_0^\infty \left[\frac{1}{\lambda}\mathbb{E}\left(A_T \wedge \lambda\right) + \mathbb{P}\left(A_T > \lambda\right)\right] p\lambda^{p-1}d\lambda\\ &= \int_0^\infty \left(\frac{1}{\lambda}\mathbb{E}\left(A_T \mathbf{1}_{A_T\leq\lambda}\right) + 2\mathbb{P}\left(A_T > \lambda\right)\right) p\lambda^{p-1}d\lambda\\ &\leq \mathbb{E}\left(A_T\int_{A_T}^\infty p\lambda^{p-2}d\lambda\right) + 2\mathbb{E}\int_0^{A_T} p\lambda^{p-1}d\lambda\\ &= \frac{p}{1-p}\mathbb{E}A_T^p + 2\mathbb{E}A_T^p\\ &= \frac{2-p}{1-p}\mathbb{E}A_T^p.\end{aligned}$$

The inequality (B_4) is obtained as follows. Using successively the inequality

$$r \wedge 1 \leq \mathbf{1}_{[\varepsilon,+\infty[}(r) + \varepsilon, \quad \text{for } r \geq 0 \text{ and } \varepsilon > 0,$$

and (B_2) with $\delta = 1$, we obtain

$$\begin{aligned} \mathbb{E}\left(\sup_{t\in[0,T]} \sqrt{X_t} \wedge 1 \right) &\leq \mathbb{P}\left(\sup_{t\in[0,T]} \sqrt{X_t} \geq \varepsilon \right) + \varepsilon \\ &\leq \mathbb{P}\left(\sup_{t\in[0,T]} X_t \geq \varepsilon^2 \right) + \varepsilon \\ &\leq \left(\frac{1}{\varepsilon^2} + 1 \right) \mathbb{E}\left[A_T \wedge 1\right] + \varepsilon. \end{aligned}$$

Clearly (B_4) holds if $\mathbb{E}\left[A_T \wedge 1\right] = 0$. We then assume that $\mathbb{E}\left[A_T \wedge 1\right] > 0$ and set $\varepsilon = \left[\mathbb{E}\left(A_T \wedge 1\right)\right]^{1/3}$. Since

$$\mathbb{E}\left(A_T \wedge 1\right) \leq \left[\mathbb{E}\left(A_T \wedge 1\right)\right]^{1/3} \leq 1,$$

it follows that

$$\begin{aligned} \mathbb{E}[\sup_{t\in[0,T]} \sqrt{X_t} \wedge 1] &\leq \frac{1}{\varepsilon^2}\mathbb{E}\left[A_T \wedge 1\right] + \mathbb{E}\left[A_T \wedge 1\right] + \varepsilon \\ &\leq 3\left[\mathbb{E}\left(A_T \wedge 1\right)\right]^{1/3}. \end{aligned}$$

That is (B_4). ■

1.2 Continuous Martingales

In this section $(\Omega, \mathcal{F}, \mathbb{P}, \{\mathcal{F}_t\}_{t\geq 0})$ is a given stochastic basis and $\mathcal{P}$ is the associated σ-algebra of progressively measurable subsets of $\Omega \times \mathbb{R}_+$.

Definition 1.57. a) A $\mathcal{P}$-measurable stochastic process $\{M_t : t \geq 0\}$ is an $\mathcal{F}_t$-submartingale (resp. $\mathcal{F}_t$-supermartingale) if

m_1) $\mathbb{E}\,|M_t| < \infty$, for all $t \geq 0$,
m_2) $\mathbb{E}\,(M_t|\mathcal{F}_s) \geq$ (resp. $\leq$) M_s, $\mathbb{P}$-a.s., for all $s \leq t$.

b) A $\mathcal{P}$-measurable d-dimensional stochastic process $\{M_t : t \geq 0\}$ is an $\mathcal{F}_t$-martingale if

m_1) $\mathbb{E}\,|M_t| < \infty$, for all $t \geq 0$,
m_2) $\mathbb{E}\,(M_t|\mathcal{F}_s) = M_s$, $\mathbb{P}$-a.s., for all $s \leq t$.

c) A $\mathcal{P}$-measurable d-dimensional stochastic process $\{M_t : t \geq 0\}$ is an $\mathcal{F}_t$-local-martingale if there exists an increasing sequence $\{\tau_n\}$ of stopping times such that $\tau_n \to \infty$ a.s., and for every $n \in \mathbb{N}^*$, $M_{\cdot \wedge \tau_n}$ is a martingale.

Remark 1.58. In the sequel, the stochastic basis will be fixed, and we will say martingale (resp. submartingale, supermartingale local-martingale) instead of $\mathcal{F}_t$-martingale (resp. $\mathcal{F}_t$-submartingale, $\mathcal{F}_t$-supermartingale, $\mathcal{F}_t$-local-martingale).

By Jensen's inequality for conditional expectation (see Proposition 1.28), we have:

Proposition 1.59. *If $\{M_t : t \geq 0\}$ is a d-dimensional martingale and $\varphi : \mathbb{R}^d \to \mathbb{R}$ is a convex function such that*

$$\mathbb{E}\varphi^+(M_t) < \infty \quad \textit{for all } t \geq 0,$$

then $\{\varphi(M_t) : t \geq 0\}$ is a submartingale.

Let $p \geq 1$ and $0 < T < +\infty$. We denote by $\mathcal{M}_d^p[0,T]$, (resp. $\mathcal{M}_d^p$), the linear space of continuous d-dimensional martingales $\{M_t : t \in [0,T]\}$, (resp. $\{M_t : t \geq 0\}$), satisfying

$$\mathbb{E}\,|M_t|^p < \infty, \quad \text{for all } t \geq 0.$$

An element of $\mathcal{M}_d^p[0,T]$ is called a d-dimensional *continuous p-martingale*. The space of local martingales will be denoted by $\mathcal{M}_d^0[0,T]$, (resp. $\mathcal{M}_d^0$).

1.2.1 Basic Results

As a consequence of the fundamental inequalities (Proposition 1.56) we have:

Theorem 1.60 (Doob's Inequality). *If $M \in \mathcal{M}_d^1[0,T]$, or M is a continuous positive submartingale, then for all $\varepsilon > 0$,*

$$\begin{aligned}
&(A_1)\ \mathbb{P}\left(\sup_{t\in[0,T]}|M_t| \geq \varepsilon\right) \leq \frac{1}{\varepsilon}\mathbb{E}\left(|M_T|\ ;\ \sup_{t\in[0,T]}|M_t| \geq \varepsilon\right),\\
&(A_2)\ \mathbb{E}\sup_{t\in[0,T]}|M_t|^p \leq \left(\frac{p}{p-1}\right)^p \mathbb{E}\,|M_T|^p,\ \textit{if } p > 1,\\
&(A_3)\ \mathbb{E}\sup_{t\in[0,T]}|M_t|^p \leq \frac{1}{1-p}\left(\mathbb{E}|M_T|\right)^p,\ \textit{if } 0 < p < 1.
\end{aligned} \tag{1.11}$$

Proof. Since for $0 \leq t \leq T$

$$|M_t| \leq \mathbb{E}^{\mathcal{F}_t}(|M_T|),$$

we have that $\{M_t : 0 \le t \le T\}$ is $\mathcal{F}_t$-dominated by $U = |M_T|$ and the result follows from part (A) of Proposition 1.56. ■

We note that from (A_2), if $p > 1$, $\mathcal{M}_d^p[0,T]$ is a closed linear subspace of $S_d^p[0,T]$ and on $\mathcal{M}_d^p[0,T]$ we have the equivalent norm

$$\|M\|_{\mathcal{M}_d^p[0,T]} = \left(\mathbb{E}\,|M_T|^p\right)^{1/p}.$$

Moreover $\mathcal{M}_d^2[0,T]$ is a Hilbert space with the inner product

$$\langle M, N\rangle_{\mathcal{M}_d^2[0,T]} = \mathbb{E}\,\langle M(T), N(T)\rangle.$$

The reader can find the proofs of the next two theorems in many textbooks, for example in [10, 64, 68].

Theorem 1.61 (Martingale Convergence). *Let $(M_t)_{t\ge 0}$ be a one-dimensional martingale. The next three conditions are equivalent.*

(i) M_t converges in $L^1(\Omega)$, as $t \to \infty$.
(ii) There exists an $M_\infty \in L^1(\Omega)$ such that $M_t = \mathbb{E}[M_\infty|\mathcal{F}_t]$, $t \ge 0$.
(iii) The collection of random variables $(M_t)_{t\ge 0}$ is uniformly integrable.

If these conditions hold, then $M_t \to M_\infty$ a.s.
If moreover,

$$\sup_{t\ge 0} \mathbb{E}[|M_t|^p] < \infty,$$

for some $p > 1$, then the above equivalent conditions are satisfied, and M_t converges to M_∞ in $L^p(\Omega)$.

We now state Doob's celebrated optional stopping theorem, which essentially says that the martingale property is true at stopping times

$$M_\tau = \mathbb{E}\,(M_\theta|\mathcal{F}_\tau)$$

and the *stopped stochastic process* $\{M_{t\wedge\theta} : t \ge 0\}$ is a martingale whenever M is a martingale.

Theorem 1.62 (Optional Stopping; Doob). *Let $T > 0$, $\{M_t : t \ge 0\}$ be a right continuous $\mathcal{F}_t$-martingale and θ, τ, σ be stopping times such that $0 \le \tau \le \theta \le T$ a.s. Then*

$$M_\tau = \mathbb{E}\,(M_\theta|\mathcal{F}_\tau) \tag{1.12}$$

and $\{M_{t\wedge\sigma} : t \ge 0\}$ is an $\mathcal{F}_t$-martingale.

Combining this result with Jensen's inequality we deduce the following:

Corollary 1.63. *If* $\{M_t : t \geq 0\}$ *is a* d*-dimensional right-continuous martingale and* $\varphi : \mathbb{R}^d \to \mathbb{R}$ *is a convex function such that* $\mathbb{E}\varphi^+ (M_t) < \infty$ *for all* $t > 0$, *then*

$$\varphi(M_\tau) \leq \mathbb{E}\left[\varphi(M_\theta) \,|\mathcal{F}_\tau\right], \tag{1.13}$$

for all bounded stopping times τ, θ, *such that* $0 \leq \tau \leq \theta$.

Another consequence of Doob's optional stopping Theorem 1.62 is the following:

Corollary 1.64. *Let* $\{M_t : t \geq 0\}$ *be a* d*-dimensional* $\mathcal{P}$*-measurable continuous stochastic process and* $\{\theta_k\}$ *the sequence of stopping times*

$$\theta_k = \inf\{t \geq 0 : |M_t| \geq k\}, \quad k \in \mathbb{N}^*.$$

Then $\{M_t : t \geq 0\}$ *is a local martingale if and only if* $\{M_{t\wedge\theta_k} : t \geq 0\}$ *is a martingale for every* $k \in \mathbb{N}^*$.

Proof. We need only prove the necessity of the assertion. Let $\tau_n \nearrow \infty$ be a sequence of stopping times such that $\{M_{t\wedge\tau_n} : t \geq 0\}$ is a martingale for every $n \in \mathbb{N}^*$. By Doob's optional stopping Theorem 1.62 $\{M_{t\wedge\tau_n\wedge\theta_k} : t \geq 0\}$ are martingales for all $n, k \in \mathbb{N}^*$; then for all $0 \leq s \leq t$ and $F \in \mathcal{F}_s$

$$\mathbb{E}\left(M_{t\wedge\tau_n\wedge\theta_k}; F\right) = \mathbb{E}\left(M_{s\wedge\tau_n\wedge\theta_k}; F\right).$$

It remains to take the limit in this identity as $n \to \infty$, which is possible due to the definition of θ_k and Lebesgue's dominated convergence theorem. ■

1.2.2 Martingales and Bounded Variation Processes

The next goal in this section is to show that $\mathbb{P}$-almost surely the trajectories of a continuous martingale have unbounded variation on any time interval.

Proposition 1.65. *Let* $M \in \mathcal{M}_d^1, M_0 = 0$ *and* $V : \Omega \times \mathbb{R}_+ \to \mathbb{R}$, $V_0 = 0$, *be a* $\mathcal{P}$*-measurable continuous stochastic process such that* $V_\cdot(\omega) \in BV_{loc}(\mathbb{R}_+)$, $\mathbb{P}$*-a.s.* $\omega \in \Omega$. *If*

$$\mathbb{E}\left(\|M\|_T \updownarrow V \updownarrow_T\right) < \infty, \ \forall\, T > 0,$$

then $\left\{M_t V_t - \int_0^t M_s dV_s : t \geq 0\right\}$ *is a* d*-dimensional continuous martingale.*

Proof. I. Let $0 \le s = t_0 < t_1 < \cdots < t_n = t$ and

$$\delta_n = \max\{t_{i+1} - t_i : 0 \le i < n\} \to 0.$$

Let $M_i = M_{t_i}$, $V_i = V_{t_i}$ and $\Delta_i V = V_{i+1} - V_i$. Let $A \in \mathcal{F}_s$. Then for all $n \in \mathbb{N}^*$:

$$\begin{aligned}\mathbb{E}\left[(M_t V_t - M_s V_s)\mathbf{1}_A\right] &= \sum_{i=0}^{n-1} \mathbb{E}\left[(M_{i+1} V_{i+1} - M_i V_i)\mathbf{1}_A\right] \\ &= \mathbb{E}\sum_{i=0}^{n-1} \mathbf{1}_A M_{i+1}(V_{i+1} - V_i) \\ &\to \mathbb{E}\mathbf{1}_A \int_s^t M_s dV_s.\end{aligned}$$

This concludes the proof. ■

Corollary 1.66. *If M is a continuous martingale, $M_0 = 0$, and*

$$\theta = \sup\{t \ge 0 : \updownarrow M \updownarrow_t < \infty\},$$

then

$$M_{t\wedge\theta} = 0, \ a.s.$$

In particular if

$$\updownarrow M \updownarrow_{[a,b]}(\omega) < \infty, \ \ a.s. \ \ on \ A \in \mathcal{F},$$

then

$$M_t(\omega) = M_a(\omega), \ \forall\, t \in [a,b], \ a.s. \ \ on \ A.$$

Proof. Let

$$\theta_n = \sup\{t \ge 0 : \updownarrow M \updownarrow_t < n\}.$$

Then θ_n is a stopping time and $\theta_n \nearrow \theta$. The integration by parts formula yields

$$\int_0^{t\wedge\theta_n} M_s dM_s = \frac{1}{2} M_{t\wedge\theta_n}^2.$$

On the other hand, by the above Proposition 1.65

$$M_{t\wedge\theta_n}^2 - \int_0^{t\wedge\theta_n} M_s dM_s$$

is a martingale and therefore $M_{t\wedge\theta_n}^2$ is a martingale. Hence $\mathbb{E}M_{t\wedge\theta_n}^2 = 0$, and so $M_{t\wedge\theta_n} = 0$ a.s. Letting $n \to \infty$ we deduce that $M_{t\wedge\theta} = 0$ a.s.

Applying the result to $N_t = M_t - M_{t\wedge a}$ and

$$\theta(\omega) = \sup\{t \geq 0 : \updownarrow N_{\cdot}(\omega)\updownarrow_t < \infty\},$$

we have $N_{t\wedge\theta} = 0$ a.s.

On the set A, $\theta(\omega) \geq b$ a.s. and consequently for all $t \in [a,b]$

$$M_t(\omega) - M_a(\omega) = N_{t\wedge\theta(\omega)}(\omega) = 0.$$

■

Definition 1.67. A stochastic process $\{X_t : t \geq 0\}$ is called a continuous semi-martingale if X is of the form $X = M + V$, where M is a continuous local martingale with $M_0 = 0$ and V is a $\mathcal{P}$-measurable continuous stochastic processes such that

$$V_{\cdot}(\omega) \in BV_{loc}(\mathbb{R}_+), \ \mathbb{P}\text{-}a.s. \ \omega \in \Omega.$$

Corollary 1.68. *Let $X = M + V$ and $X' = M' + V'$ be two semimartingales. Then $X = X'$ if and only if $M = M'$ and $V = V'$.*

Proof. We need only prove that $X = X'$ implies $M = M'$ and $V = V'$.

Let $X = X'$ and define the sequence of stopping times

$$\theta_n = \inf\{t \geq 0 : |M_t| + |M_t'| \geq n\}.$$

Then

$$V_{\cdot\wedge\theta_n} - V'_{\cdot\wedge\theta_n} = M'_{\cdot\wedge\theta_n} - M_{\cdot\wedge\theta_n}$$

is a martingale starting from 0 and

$$\updownarrow V_{\cdot\wedge\theta_n} - V'_{\cdot\wedge\theta_n}\updownarrow_T < \infty, \ \forall\, T > 0, \ \mathbb{P}\text{-a.s.}$$

Hence $V_{t\wedge\theta_n} - V'_{t\wedge\theta_n} = 0$ for all $t \geq 0$, $\mathbb{P}$-a.s. and the result follows letting $n \to \infty$. ■

We recall now another celebrated result: the Doob–Meyer decomposition. For the proof we recommend to the reader Stroock [68].

Theorem 1.69 (Doob–Meyer). *If M is a d-dimensional continuous local martingale, then there exists a unique progressively measurable increasing continuous stochastic process (abbreviated $\mathcal{P}$-m.i.c.s.p.) $\{<M>_t; t \geq 0\}$ such that:*

$$a)\ <M>_0 = 0,\ a.s.$$
$$b)\ |M|^2 - <M>\ \text{is a continuous local martingale.}$$

In particular if $M \in \mathcal{M}_d^2$, then $|M|^2 - <M> \ \in \mathcal{M}^1 \cap S^1$.

Denote by $\mathbb{S}^d$ the set of symmetric matrices $Q, P \in \mathbb{R}^{d\times d}$. If $Q, P \in \mathbb{S}^d$, we shall say that $Q \leq P$ if $\langle Qx, x\rangle \leq \langle Px, x\rangle$, for all $x \in \mathbb{R}^d$; Q is semipositive definite if $Q \geq 0$.

It is easy to extend Theorem 1.69 to the vector case.

Proposition 1.70. *If M is a d-dimensional continuous local martingale then there exists a unique $\mathbb{S}^d$-valued $\mathcal{P}$-m.i.c.s.p. $\{<<M>>_t : t \geq 0\}$ such that $\ll M \gg_0 = 0$, a.s. and*

$$M \otimes M - \ll M \gg \ \text{is a continuous local martingale.}$$

Moreover

◊ $<M> \ = \mathbf{Tr} \ll M \gg$;
◊ *for any stopping time τ:*

$$\ll M_{\cdot\wedge\tau} \gg \ = \ \ll M \gg_{\cdot\wedge\tau} \quad \text{and} \quad <M_{\cdot\wedge\tau}> \ = \ <M>_{\cdot\wedge\tau};$$

and

◊ *if $M \in \mathcal{M}_d^2$, then*

$$M \otimes M - \ll M \gg \ \in \mathcal{M}^1_{d\times d} \cap S^1_{d\times d}.$$

As a particular case of this proposition we have:

Remark 1.71. If M and N are two scalar continuous local $(\mathcal{F}_t$-)martingale, then there exists a unique $\mathcal{P}$-m.b-v.c.s.p. usually denoted $<M, N>$ such that $MN - <M, N>$ is a continuous local martingale. We have

$$<M, N> \ = \frac{1}{2} <M + N> - \frac{1}{2} <M> - \frac{1}{2} <N>.$$

If M is a d-dimensional continuous local martingale and $u, v \in \mathbb{R}^d$, then $u^* << M >>_t v =< u^*M, v^*M >_t$, from which we deduce that for any $i, j \in \overline{1, d}$, the (i, j) term of the matrix $<< M >>_t$ coincides with $< M^i, M^j >_t$, which is $< M^i >_t$ in the case $i = j$.

To avoid confusion with the inner product, in this book we shall not use the notation $< M, N >$.

We complete the inequalities (1.11) with the following:

Proposition 1.72. *Let* $M \in \mathcal{M}_d^2[0,T]$ *and* $M_0 = 0$, *then for all* $\varepsilon, \delta > 0$

$$
\begin{aligned}
&(B_1) \quad \mathbb{P}\left(\sup_{t\in[0,T]} |M_t| \geq \sqrt{\varepsilon},\ \langle M\rangle_T \leq \delta \right) \leq \frac{1}{\varepsilon}\mathbb{E}\left[<M>_T \wedge\delta\right], \\
&(B_2) \quad \mathbb{P}\left(\sup_{t\in[0,T]} |M_t| \geq \sqrt{\varepsilon} \right) \leq \left(\frac{1}{\varepsilon}+\frac{1}{\delta}\right)\mathbb{E}\left[<M>_T \wedge\delta\right],
\end{aligned}
$$

and

$$
\begin{aligned}
&(B_3) \quad \mathbb{E} \sup_{t\in[0,T]} |M_t|^p \leq C_p\, \mathbb{E} <M>_T^{p/2}, \quad \forall\, 0 < p \leq 2, \\
&(B_4) \quad \mathbb{E}[\sup_{t\in[0,T]} |M_t| \wedge 1] \leq 3\left[\mathbb{E}\left(<M>_T \wedge 1\right)\right]^{1/3},
\end{aligned}
$$

where

$$
C_p = \begin{cases} \dfrac{4-p}{2-p}, & \text{if } 0 < p < 2, \\ 4, & \text{if } p = 2. \end{cases}
$$

Proof. (B_1), (B_2), (B_3 with $0 < p < 2$) and (B_4) follow from the corresponding statements in Proposition 1.56 and the fact that $X_t = |M_t|^2$ is dominated by $A_t =< M>_t$; (B_3 with $p = 2$) follows from $(1.11{-}A_2)$ and $\mathbb{E}|M_T|^2 = \mathbb{E} <M>_T$. ∎

We want to extend the inequality (B_3) to the case $p > 2$. First we have:

Lemma 1.73. *Let* $M \in \mathcal{M}_d^1$, $M_0 = 0$. *If*

$$
\begin{aligned}
&(i) \quad \varphi \in C^{1,2}\left(\mathbb{R}_+ \times \mathbb{R}^d; \mathbb{R}\right), \\
&(ii) \quad \mathbb{E}\left[\|\varphi(\cdot, M)\|_T + \|\varphi_t'(\cdot, M)\|_T\right] < \infty,\ \forall\, T > 0, \\
&(iii) \quad \mathbb{E}\left[\|\varphi_{xx}''(\cdot, M)\|_T <M>_T\right] < \infty,\ \forall\, T > 0,
\end{aligned}
$$

then

$$
U_t \stackrel{def}{=} \varphi(t, M_t) - \int_0^t \varphi_t'(r, M_r)\, dr - \frac{1}{2}\mathbf{Tr}\int_0^t \varphi_{xx}''(r, M_r)\, d \ll M \gg_r \tag{1.14}
$$

is a continuous martingale. Moreover

$$\mathbb{E}^{\mathcal{F}_\sigma}\varphi(\theta, M_\theta) \le \varphi(\sigma, M_\sigma) + \mathbb{E}^{\mathcal{F}_\sigma}\int_\sigma^\theta \varphi_t'(s, M_s)\, ds + \frac{1}{2}\mathbb{E}^{\mathcal{F}_\sigma}\int_\sigma^\theta |\varphi_{xx}''(s, M_s)|\, d <M>_s, \ a.s., \tag{1.15}$$

for all bounded stopping times $0 \le \sigma \le \theta$, a.s.

Proof. Let $0 \le \sigma \le \theta \le T$ and $R > 0$ be arbitrary. Define

$$\Psi(r, x) = \left(x, \varphi(r, x), \varphi_r'(r, x), \varphi_x'(r, x), \varphi_{xx}''(r, x)\right)$$

and the modulus of continuity

$$\mathbf{m}_\Psi\left(\tfrac{1}{n}\right) = \sup\left\{|\Psi(r, u) - \Psi(r', u')| : 0 \le r, r' \le T, \ |u| \le R, |u'| \le R, \ |r - r'| + |u - u'| \le \frac{1}{n}\right\}.$$

Let

$$C_R = \sup\{|\Psi(r, x)| : r \in [0, T], \ |x| \le R\}$$

and the stopping time

$$\tau_R = \inf\{r \ge 0 : |M_r| \ge R\}.$$

Let $\{(\theta_i^n, k_n) : i, n \in \mathbb{N}\}$ be a *basic partition* (see Definition 1.54) of the interval $[\sigma \wedge \tau_R, \theta \wedge \tau_R]$, associated to $\left\{\left(M_t, \frac{1}{n}\right) : t \ge 0, n \in \mathbb{N}^*\right\}$.

Let $\theta_i = \theta_i^n$, $\Delta_i\theta = \theta_{i+1} - \theta_i$, $Y_i = Y_{\theta_i^n}$, $\Delta_i Y = Y_{i+1} - Y_i$ for any arbitrary stochastic process Y.

Note that

$$\sum_{i=0}^{k_n-1} \mathbb{E}|\Delta_i M|^2 = \mathbb{E}\left|M_{\theta_{k_n}^n}\right|^2 \le R^2.$$

We have

$$\varphi(\theta \wedge \tau_R, M_{\theta\wedge\tau_R}) - \varphi(\sigma \wedge \tau_R, M_{\theta\wedge\tau_R})$$
$$= \sum_{i=0}^{k_n-1} [\varphi(\theta_{i+1}, M_{i+1}) - \varphi(\theta_i, M_i)] + \left[\varphi(\theta \wedge \tau_R, M_{\theta\wedge\tau_R}) - \varphi\left(\theta_{k_n}^n, M_{\theta_{k_n}^n}\right)\right].$$

Then for $A \in \mathcal{F}_\sigma$ and some $\theta_i', \theta_i'' \in \left[\theta_i^n, \theta_{i+1}^n\right]$,

$$\begin{aligned}
&\mathbb{E}\left[\varphi\left(\theta \wedge \tau_R, M_{\theta\wedge\tau_R}\right) - \varphi\left(\sigma \wedge \tau_R, M_{\sigma\wedge\tau_R}\right); A\right] \\
&= \mathbb{E}\left[\sum_{i=0}^{k_n-1} \varphi_t'\left(\theta_i', M_{i+1}\right) \Delta_i\theta; A\right] + \sum_{i=0}^{k_n-1} \mathbb{E}\left[\left\langle \varphi_x'\left(\theta_i, M_i\right), \Delta_i M\right\rangle; A\right] \\
&\quad + \frac{1}{2}\sum_{i=0}^{k_n-1} \mathbb{E}\left[\left\langle \varphi_{xx}''\left(\theta_i, M_i\right) \Delta_i M, \Delta_i M\right\rangle; A\right] + R_n,
\end{aligned} \tag{1.16}$$

where

$$\begin{aligned}
R_n &= \frac{1}{2}\sum_{i=0}^{k_n-1} \mathbb{E}\left[\left\langle \varphi_{xx}''\left(\theta_i, M_{\theta_i''}\right) - \varphi_{xx}''\left(\theta_i, M_i\right) \Delta_i M, \Delta_i M\right\rangle; A\right] \\
&\quad + \mathbb{E}\left[\left(\varphi\left(\theta \wedge \tau_R, M_{\theta\wedge\tau_R}\right) - \varphi\left(\theta_{k_n}^n, M_{\theta_{k_n}^n}\right)\right); A\right]
\end{aligned}$$

satisfies

$$\begin{aligned}
|R_n| &\le \frac{1}{2}\mathbf{m}_\Psi\left(\tfrac{1}{n}\right) \times \sum_{i=0}^{k_n-1} \mathbb{E}\left(|\Delta_i M|^2; A\right) + 2C_R \times \mathbb{P}\left(\theta_{k_n}^n < \theta \wedge \tau_R\right) \\
&\le C_R' \times \left(\mathbf{m}_\Psi\left(\tfrac{1}{n}\right) + \frac{1}{n}\right).
\end{aligned}$$

Since $\mathbf{Tr}\left[H \times (x \otimes y)\right] = \langle Hx, y\rangle$,

$$\begin{aligned}
&\frac{1}{2}\sum_{i=0}^{k_n-1} \mathbb{E}\left[\mathbf{Tr}\left(\varphi_{xx}''\left(\theta_i, M_i\right) \Delta_i \lt\!\!\lt M \gt\!\!\gt\right); A\right] \\
&\qquad = \frac{1}{2}\sum_{i=0}^{k_n-1} \mathbb{E}\left[\left\langle \varphi_{xx}''\left(\theta_i, M_i\right) \Delta_i M, \Delta_i M\right\rangle; A\right] \\
&\qquad \le \frac{1}{2}\sum_{i=0}^{k_n-1} \mathbb{E}\left[\left|\varphi_{xx}''\left(\theta_i, M_i\right)\right| |\Delta_i M|^2; A\right] \\
&\qquad = \frac{1}{2}\sum_{i=0}^{k_n-1} \mathbb{E}\left[\left|\varphi_{xx}''\left(\theta_i, M_i\right)\right| \Delta_i \lt M \gt; A\right].
\end{aligned}$$

Note also that

$$\mathbb{E}\left[\left\langle \varphi_x'\left(\theta_i, M_i\right), \Delta_i M\right\rangle; A\right] = 0.$$

Passing to the limit in (1.16) as $n \to \infty$ we obtain

$$\begin{aligned}
&\mathbb{E}\left[\varphi\left(\theta \wedge \tau_R, M_{\theta\wedge\tau_R}\right) - \varphi\left(\sigma \wedge \tau_R, M_{\sigma\wedge\tau_R}\right); A\right] \\
&= \mathbb{E}\left[\int_{\sigma\wedge\tau_R}^{\theta\wedge\tau_R} \varphi_t'\left(r, M_r\right) dr; A\right] + \frac{1}{2}\mathbb{E}\left[\mathbf{Tr}\int_{\sigma\wedge\tau_R}^{\theta\wedge\tau_R} \varphi_{xx}''\left(r, M_r\right) d \ll M \gg_r; A\right] \\
&\leq \mathbb{E}\left[\int_{\sigma\wedge\tau_R}^{\theta\wedge\tau_R} \varphi_t'\left(r, M_r\right) dr; A\right] + \frac{1}{2}\mathbb{E}\left[\mathbf{Tr}\int_{\sigma\wedge\tau_R}^{\theta\wedge\tau_R} \left|\varphi_{xx}''\left(r, M_r\right)\right| d <M>_r; A\right].
\end{aligned}$$

Letting $R \to \infty$, we deduce that $\{U_t : t \geq 0\}$ is a martingale, and the inequality (1.15) holds. ∎

Corollary 1.74. *Let the assumptions of Lemma 1.73 be satisfied. If $\varphi(t,x) \geq 0$ for all $(t,x) \in \mathbb{R}_+ \times \mathbb{R}^d$, then we have*

$$\mathbb{E}\varphi\left(\theta, M_\theta\right) \leq \varphi(0,0) + \mathbb{E}\int_0^\theta \left|\varphi_t'\left(s, M_s\right)\right| ds + \frac{1}{2}\mathbb{E}\int_0^\theta \left|\varphi_{xx}''\left(s, M_s\right)\right| d <M>_s \tag{1.17}$$

for all a.s. finite stopping times θ.

Proof. We write the inequality (1.15) for the bounded stopping times $\sigma := 0$ and $\theta := N \wedge \theta$ and we pass to $\liminf_{N\to+\infty}$. ∎

Corollary 1.75. *If M is a bounded continuous martingale and $M_0 = 0$, then the increasing process associated to the martingale $|M_t|^2 - <M>_t$ is $4\,\mathbf{Tr}\int_0^t M_s \otimes M_s d \ll M \gg_s$.*

Proof. By Lemma 1.73 applied to $\varphi(t,x) = |x|^4$

$$|M_t|^4 - \frac{1}{2}\mathbf{Tr}\int_0^t \left(8M_s \otimes M_s + 4\left|M_s\right|^2 I_{d\times d}\right) d \ll M \gg_s$$

is a continuous martingale.

Also by Proposition 1.65

$$\left(|M_t|^2 - <M>_t\right) <M>_t - \int_0^t \left(|M_s|^2 - <M>_s\right) d <M>_s$$

is a continuous martingale.

Since

$$\left(|M_t|^2 - <M>_t\right)^2 = |M_t|^4 - 2\left(|M_t|^2 - <M>_t\right) <M>_t - <M>_t^2,$$

we deduce that

$$\left(|M_t|^2 - <M>_t\right)^2 - 4\mathbf{Tr}\int_0^t M_s \otimes M_s d \ll M \gg_s$$

is a continuous martingale. ■

Theorem 1.76 (Burkholder–Davis–Gundy (BDG) Inequality). *Let M be a continuous local martingale and $M_0 = 0$. Then, for all $p > 0$, there exist two constants $c_p > 0$ and $C_p > 0$ such that for all $T > 0$:*

$$c_p \mathbb{E} <M>_T^{p/2} \leq \mathbb{E} \sup_{t\in[0,T]} |M_t|^p \leq C_p \, \mathbb{E} <M>_T^{p/2} . \tag{1.18}$$

Moreover

$$C_p = \begin{cases} \dfrac{4-p}{2-p} \leq 3, & \text{if } 0 < p \leq 1, \\ 4, & \text{if } p = 2, \\ \leq \left(3p^3\right)^p, & \text{if } p > 1. \end{cases}$$

Proof. It is sufficient to treat the case where M is a bounded continuous martingale. Indeed the result would then apply to $M_{\cdot\wedge\tau_R}$, where

$$\tau_R = \inf\{t \geq 0 : |M_t| \geq R\} .$$

But, since $<M_{\cdot\wedge\tau_R}> = <M>_{\cdot\wedge\tau_R}$, the inequality (1.18) for M follows by letting $R \to \infty$.

Second Inequality. For $0 < p \leq 2$ the inequality was proved in Proposition 1.72. Let $q \geq 2$. The inequality (1.17) tells us that

$$|M_t|^q \text{ is dominated by } A_t = \frac{q\,(q-1)}{2}\int_0^t |M_s|^{q-2} d <M>_s .$$

Let $0 < r < 1$. By the inequality (B_3) from Proposition 1.56 we obtain

$$\begin{aligned} \mathbb{E} \sup_{t\in[0,T]} |M_t|^{qr} &\leq \frac{2-r}{1-r}\mathbb{E}A_T^r \\ &\leq \frac{2-r}{1-r}\left(\frac{q\,(q-1)}{2}\right)^r \mathbb{E} \sup_{s\in[0,T]} |M_s|^{(q-2)r} <M>_T^r \\ &\leq \frac{2-r}{1-r}\left(\frac{q\,(q-1)}{2}\right)^r \left(\mathbb{E} \sup_{s\in[0,T]} |M_s|^{qr}\right)^{\frac{q-2}{q}} \left(\mathbb{E} <M>_T^{rq/2}\right)^{2/q} , \end{aligned}$$

which yields

$$\mathbb{E} \sup_{t\in[0,T]} |M_t|^{qr} \le C_{r,q} \left(\mathbb{E} <M>_T^{rq/2}\right)$$

with

$$C_{r,q} = \left(\frac{2-r}{1-r}\right)^{q/2} \left(\frac{q\,(q-1)}{2}\right)^{rq/2}.$$

Let $p \geq 1$, $q = p+1$, $r = p/\,(p+1)$. Then the second inequality in (1.18) holds with

$$\begin{aligned} C_p &= (p+2)^{(p+1)/2} \left(\tfrac{(p+1)p}{2}\right)^{p/2} \\ &\le \left(\tfrac{p(p+1)(p+2)}{2}\right)^{(p+1)/2} \\ &\le \left(3p^3\right)^p. \end{aligned}$$

First Inequality. Since $4\mathbf{Tr}\int_0^t M_s \otimes M_s d \ll M \gg_s$ is the increasing process associated to the martingale $|M_t|^2 - <M>_t$, from the second part of the inequality (1.18), which we have just proved, we obtain that for all $p > 0$:

$$\begin{aligned} \mathbb{E} \sup_{t\in[0,T]} \left| |M_t|^2 - <M>_t \right|^{p/2} &\le C_p' \,\mathbb{E} \left(\int_0^T |M_s|^2 \, d <M>_s \right)^{p/4} \\ &\le C_p' \,\mathbb{E} \left(\sup_{t\in[0,T]} |M_t|^{p/2} <M>_T^{p/4} \right) \end{aligned}$$

and

$$\begin{aligned} \mathbb{E} <M>_T^{p/2} &\le a_p \,\mathbb{E} \left| <M>_T - |M_T|^2 \right|^{p/2} + a_p \,\mathbb{E}\, |M_T|^p \\ &\le b_p \,\mathbb{E} \left(\sup_{t\in[0,T]} |M_t|^{p/2} <M>_T^{p/4} \right) + a_p \,\mathbb{E}\, |M_T|^p \\ &\le C_p'' \,\mathbb{E} \sup_{t\in[0,T]} |M_t|^p + \frac{1}{2} \mathbb{E} <M>_T^{p/2}, \end{aligned}$$

which yields the first inequality in (1.18). ■

1.3 Brownian Motion

1.3.1 Gaussian Spaces

Recall that a random variable X is said to be Gaussian if its law either has a density of the form

$$g_{\mu,\sigma}(x) = \frac{1}{\sigma\sqrt{2\pi}} \exp\left[-\frac{(x-\mu)^2}{2\sigma^2}\right],$$

for some $\mu \in \mathbb{R}, \sigma > 0$, in which case $\mathbb{E}X = \mu$ and $\mathrm{Var}(X) = \sigma^2$, or else is a Dirac mass at μ, in which case $\mathbb{E}X = \mu$, $\mathrm{Var}(X) = 0$. The corresponding probability law on $\mathbb{R}$ is denoted by $N(\mu, \sigma^2)$, and we write $X \sim N(\mu, \sigma^2)$.

If X is Gaussian, then for all $a \in \mathbb{C}$,

$$\mathbb{E}\exp(aX) = \exp\left[a\,\mathbb{E}X + \frac{a^2}{2}\mathrm{Var}(X)\right],$$

and conversely whenever the above formula holds either for all real a or for all imaginary a, then X is a Gaussian random variable.

A d-dimensional vector $X = (X_1, \cdots, X_d)^*$ is said to be Gaussian if for any $a \in \mathbb{R}^d$,

$$\sum_1^d a_i X_i \quad \text{is a Gaussian random variable,}$$

or equivalently for any $a \in \mathbb{R}^d$,

$$\mathbb{E}\exp\langle a, X\rangle = \exp\left[\langle a, \mathbb{E}X\rangle + \frac{1}{2}\langle C_X a, a\rangle\right],$$

where $C_X = \left[\mathbb{E}\left(X_i X_j\right) - \mathbb{E}X_i \mathbb{E}X_j\right]_{d\times d}$.

Note that if X is a Gaussian random vector, its coordinates are Gaussian random variables, which are independent iff the covariance matrix of X is diagonal. If $X_1, \ldots, X_d$ are Gaussian random variables, then the vector $X = (X_1, \ldots, X_d)^*$ need not be Gaussian. As an example if $U \sim N(0,1)$, then $V = U\mathbf{1}_{|U|\le 1} - U\mathbf{1}_{|U|>1} \sim N(0,1)$ and $X = (U, V)^*$ is not Gaussian since $U + V = 2U\mathbf{1}_{|U|\le 1}$ is not a Gaussian random variable. However, if $X_1, \ldots, X_d$ are independent Gaussian random variables then $X = (X_1, \ldots, X_d)^*$ is a Gaussian random vector.

Definition 1.77. A d-dimensional stochastic process $\{X_t, t \in \mathbb{T}\}$ is said to be *Gaussian* if for any $k \in \mathbb{N}$, $t_1, \ldots, t_k \in \mathbb{T}$, the $d \times k$-dimensional random vector $(X_{t_1}, \ldots, X_{t_k})$ is a Gaussian vector.

Definition 1.78. A closed sub-vector space H of $L^2(\Omega, \mathcal{F}, \mathbb{P}; \mathbb{R})$ is called a *Gaussian space* if all its elements are zero mean Gaussian random variables. The Gaussian space $H[X]$ associated to the Gaussian stochastic process $\{X_t, \, t \in \mathbb{T}\}$ is the closed vector subspace of $L^2(\Omega, \mathcal{F}, \mathbb{P}; \mathbb{R})$ spanned by:

$$\{\langle a, \, X_t - \mathbb{E}(X_t)\rangle \, , \, a \in \mathbb{R}^d \, ; \, t \in \mathbb{T}\}.$$

1.3.2 Definition and Main Properties

Definition 1.79. A (one-dimensional) *Brownian motion* (abbreviated Bm or BM) is a continuous stochastic process $B : \Omega \times [0, \infty[\to \mathbb{R}$ such that:

(i) $B_0 = 0$;
(ii) $B_t - B_s \sim N(0, t - s)$, for any $0 \leq s < t$;
(iii) $B_{t_1} - B_{t_0}, B_{t_2} - B_{t_1}, \cdots, B_{t_k} - B_{t_{k-1}}$ are independent random variables for each $k \geq 2$ and $0 = t_0 < t_1 < \cdots < t_k$.

As a consequence of (i): for all $p > 0$:

$$\mathbb{E}\,|B_t - B_s|^p = \frac{1}{\sqrt{\pi}}\,(2\,|t-s|)^{p/2}\,\Gamma\left(\tfrac{p+1}{2}\right), \tag{1.19}$$

where $\Gamma :]0, \infty[\to]0, \infty[$ is defined by

$$\Gamma(x) \stackrel{def}{=} \int_0^\infty t^{x-1} e^{-t}\, dt.$$

Note that

$$\Gamma(x+1) = x\Gamma(x), \quad \Gamma(1) = 1, \quad \Gamma\left(\tfrac{1}{2}\right) = \sqrt{\pi},$$

and in particular

$$\mathbb{E}\,|B_t - B_s| = \sqrt{\frac{2}{\pi}\,|t-s|},$$
$$\mathbb{E}\,|B_t - B_s|^2 = |t-s|\,,$$
$$\mathbb{E}\,|B_t - B_s|^4 = 3\,(t-s)^2.$$

Also

$$\mathbb{E}\exp\left[z\,(B_t - B_s)\right] = \exp\left[\frac{1}{2}z^2\,(t-s)\right], \quad \forall\, z \in \mathbb{C}.$$

The importance of Brownian motion follows from the universality of the central limit theorem. Let $(X_n, n = 1,2\ldots)$ be a sequence of independent identically distributed random variables such that $\mathbb{E}(X_1) = 0$, $\mathbb{E}(X_1^2) = \sigma^2$. Define

$$S_n = X_1 + \cdots + X_n,\ n = 1,2,\ldots$$

Then $S_n/\sigma\sqrt{n}$ converges in law to $N(0,1)$ as $n \to \infty$. Now define for $t \geq 0$, $n \in \mathbb{N}^*$,

$$B_t^n = \frac{S_{[nt]}}{\sigma\sqrt{n}}.$$

Clearly, the limit in law of $B_t^n - B_s^n (0 \leq s < t)$ is $N(0, t-s)$, the increments are independent and the limiting process is likely to be continuous. Indeed, one can show that if $\{\tilde{B}_t^n : t \geq 0\}$ is the piecewise linear process which coincides with $\{B_t^n : t \geq 0\}$ at all times $t = \dfrac{k}{n}$, $k \in \mathbb{N}$, then $\tilde{B}_t^n \to B$ in law in $C(\mathbb{R}_+)$.

Note that so far we have not proved the existence of a continuous process having the properties listed in Definition 1.79, i.e. of the Brownian motion.

If $\{e_n : n \in \mathbb{N}^*\}$ is an orthonormal basis in $L^2(\mathbb{R}_+)$ and $\{\xi_n : n \in \mathbb{N}^*\}$ are independent normal $\mathcal{N}(0,1)$ random variables then the stochastic process $\{B_t : t \geq 0\}$ defined by

$$B_t = \sum_{n=1}^{\infty} \xi_n \int_0^t e_n(r)\,dr = \sum_{n=1}^{\infty} \left\langle \mathbf{1}_{[0,t]}, e_n \right\rangle_{L^2} \xi_n \tag{1.20}$$

is a Gaussian stochastic process such that

$$\mathbb{E}B_t = 0 \quad \text{and} \quad \mathbb{E}(B_t B_s) = t \wedge s.$$

Choosing as orthonormal basis of $L^2(\mathbb{R}_+)$ the Haar basis $\left\{\mathbf{1}_{(k,k+1]},\ e_{n,k},\right\}_{n,k\in\mathbb{N}}$, where $e_{n,k}(t) = 2^{n/2}\left[\mathbf{1}_{(2k,2k+1]}\left(2^{n+1}t\right) - \mathbf{1}_{(2k+1,2k+2]}\left(2^{n+1}t\right)\right]$, one can show that the trajectories of $\{B_t : t \geq 0\}$ are a.s. continuous; hence $\{B_t : t \geq 0\}$ is a Brownian motion (this is the Lévy–Ciesielski's construction of Brownian motion).

We now give some elementary properties of Brownian motion.

Proposition 1.80. *(a) A stochastic process $\{X_t, t \geq 0\}$ is a Brownian motion iff it is a continuous centered Gaussian process whose covariance is given by: $\mathbb{E}(X_s X_t) = s \wedge t;\ s,t \geq 0$.*

(b) If $\{B_t : t \geq 0\}$ is a Brownian motion, $h, c > 0$ and

$$W_t = -B_t,\quad H_t = B_{t+h} - B_h,\quad U_t = cB_{\frac{t}{c^2}},\quad V_t = tB_{\frac{1}{t}} \text{ and } V_0 = 0,$$

then $\{W_t : t \geq 0\}$, $\{H_t : t \geq 0\}$, $\{U_t : t \geq 0\}$ and $\{V_t : t \geq 0\}$ are Brownian motions.

Proof.

(a) Let $\{B_t, t \geq 0\}$ be a Brownian motion, and let $k \in \mathbb{N}^*$, $0 = t_0 < t_1 < \cdots < t_k$. $(B(t_1), B(t_2), \ldots, B(t_k))$ is the image under a linear mapping of the Gaussian random vector $(B(t_1), B(t_2)-B(t_1), \ldots, B(t_k)-B(t_{k-1}))$, hence it is Gaussian. Moreover, if $s \leq t$,

$$\begin{aligned}\mathbb{E}(B_s B_t) &= \mathbb{E}B_s^2 + \mathbb{E}(B_s(B_t - B_s)) \\ &= s + 0,\end{aligned}$$

since B_s and $B_t - B_s$ are independent.
Conversely, if $\{X_t\}$ has the properties of the statement, then $X_0 = 0$ a.s., $X_t - X_s$ has the law $N(0, t-s)$, and for $0 < t_1 < \cdots < t_k$, $(X_{t_1}, X_{t_2} - X_{t_1}, \ldots, X_{t_n} - X_{t_{n-1}})$ is Gaussian vector whose covariance matrix is diagonal, hence the sequence is independent.

(b) This follows easily from a). The continuity at $t = 0$ of V_t follows from Proposition 1.83 (see below), which shows in particular that

$$\lim_{t\to 0} |V_t| = \lim_{t\to\infty} \frac{|B_t|}{t} = 0, \quad \mathbb{P}\text{-}a.s.$$

■

Definition 1.81 (*d*-**Dimensional Brownian Motion).** A stochastic process $\{B_t, t \geq 0\}$ with values in $\mathbb{R}^d$ is called a d-dimensional Brownian motion if one of the following three equivalent properties (*I*)–(*III*)holds:

(*I*) its components $\{B_t^1, t \geq 0\}, \ldots, \{B_t^d, t \geq 0\}$ are mutually independent scalar Brownian motions;

(*II*) (*i*) $B_0 = 0$;
(*ii*) $B_t - B_s \sim N(0, (t-s)\, I_{d\times d})$ for any $0 \leq s < t$;
(*iii*) $B_{t_1} - B_{t_0}, B_{t_2} - B_{t_1}, \cdots, B_{t_k} - B_{t_{k-1}}$ are independent random vectors for each $k \geq 2$ and $0 = t_0 < t_1 < \cdots < t_k$.

(*III*) $\{B_t, t \geq 0\}$ is a d-dimensional continuous centered Gaussian process whose covariance function is given by

$$\mathbb{E}\left(B_t B_s^*\right) = (t \wedge s)\, I_{d\times d}, \quad t, s \geq 0.$$

The equivalences stated in the above definition are proposed as an exercise (Exercise 1.5).

Let $t \geq 0$ and

$$\mathcal{F}_t^B = \sigma\left(\{B_s : 0 \leq s \leq t\}\right) \vee \mathcal{N}$$

be the natural filtration associated to the Brownian motion B.

Proposition 1.82. *Let $\{B_t : t \geq 0\}$ be a Brownian motion. Then $\{B_t : t \geq 0\}$ and $\{B_t^2 - t : t \geq 0\}$ are $\mathcal{F}_t^B$-martingales. Consequently $<B>_t = t$ and*

$$\mathbb{E}B_\theta^2 = \mathbb{E}\theta \tag{1.21}$$

for any bounded stopping time θ.

Proof. Since $\sigma\{B_r : 0 \leq r \leq s\} = \sigma\{B_r - B_0, B_s - B_r : 0 \leq r \leq s\}$ it follows that $B_t - B_s$ is independent of $\mathcal{F}_s^B$ and hence

$$\begin{aligned}\mathbb{E}\left(B_t|\mathcal{F}_s^B\right) &= \mathbb{E}\left(B_t - B_s|\mathcal{F}_s^B\right) + B_s \\ &= B_s\end{aligned}$$

and

$$\begin{aligned}\mathbb{E}\left(B_t^2 - t|\mathcal{F}_s^B\right) &= \mathbb{E}\left(B_t^2 - B_s^2|\mathcal{F}_s^B\right) - t + B_s^2 \\ &= \mathbb{E}\left[(B_t - B_s)^2 - 2B_s\left(B_t - B_s\right)|\mathcal{F}_s^B\right] - t + B_s^2 \\ &= \mathbb{E}\left(B_t - B_s\right)^2 - 2B_s\mathbb{E}\left(B_t - B_s|\mathcal{F}_s^B\right) - t + B_s^2 \\ &= B_s^2 - s.\end{aligned}$$

It follows that $< B >_t = t$. The Doob optional stopping Theorem 1.62 now yields (1.21). ■

Proposition 1.83. *a. For each $p > 0$, there exists a constant C_p such that*

$$\mathbb{E}\sup_{r\in[s,t]}|B_r - B_s|^p \leq C_p\,(t-s)^{p/2}$$

for all $0 \leq s \leq t$.

b. For each $\alpha > 0$ and $p > 0$, there exists a constant $C_{\alpha,p}$ such that

$$\mathbb{E}\sup_{s\geq t}\left|\frac{B_s}{s^{\frac{1}{2}+\alpha}}\right|^p \leq \frac{C_{\alpha,p}}{t^{\alpha p}}, \quad \forall\, t > 0.$$

In particular, again for each $\alpha > 0$

$$\lim_{t\to\infty}\left(\sup_{s\geq t}\frac{|B_s|}{s^{\frac{1}{2}+\alpha}}\right) = 0, \quad \mathbb{P}\text{-}a.s.$$

Proof. a. Since $M_t = B_t - B_{t\wedge s}$ is an $\mathcal{F}_t^B$-martingale with $<M>_t = t - t\wedge s$, the result follows from the Burkholder–Davis–Gundy inequality (1.18).

b. We have

$$\begin{aligned}
\mathbb{E}\sup_{s\geq t}\left|\frac{B_s}{s^{\frac{1}{2}+\alpha}}\right|^p &\leq \sum_{n=0}^{\infty}\mathbb{E}\sup_{2^n t\leq s\leq 2^{n+1}t}\left|\frac{B_s}{s^{\frac{1}{2}+\alpha}}\right|^p \\
&\leq \sum_{n=0}^{\infty}\frac{1}{(2^n t)^{\frac{p}{2}+p\alpha}}\mathbb{E}\sup_{0\leq s\leq 2^{n+1}t}|B_s|^p \\
&\leq \sum_{n=0}^{\infty}\frac{1}{(2^n t)^{\frac{p}{2}+p\alpha}}C_p\left|2^{n+1}t\right|^{p/2} \\
&= \frac{C_p 2^{p/2}}{1-2^{-p\alpha}}\times\frac{1}{t^{p\alpha}}.
\end{aligned}$$

The $\mathbb{P}$-a.s. convergence follows by the decreasing monotonicity of

$$\xi_t = \sup_{s\geq t}\frac{|B_s|}{s^{\frac{1}{2}+\alpha}}.$$

■

In fact we can show that $\mathbb{P}$-$a.s.$ the trajectories $t \longrightarrow B_t(\omega)$ are Hölder-continuous of exponent $\frac{1}{2}-\varepsilon$, $0<\varepsilon<\frac{1}{2}$.

Indeed let $0<\varepsilon<\frac{1}{2}$ and $p>\frac{2}{\varepsilon}$. Using Kolmogorov's Criterion (Theorem 1.40) with $k=1$, $a=p$, $b=\frac{p}{2}-1$, $\delta=\varepsilon-\frac{1}{p}>\frac{1}{p}$, we obtain:

Proposition 1.84 (Hölder-Continuity). *Let $\{B_t : t\geq 0\}$ be a Brownian motion. Then for every $0<\varepsilon<1/2$ and $T\geq 0$, there exists a positive random variable $\xi_{\varepsilon,T}$ such that for all $t,s\in[0,T]$:*

$$|B_t(\omega)-B_s(\omega)|\leq \xi_{\varepsilon,T}(\omega)\,|t-s|^{\frac{1}{2}-\varepsilon},\quad \mathbb{P}\text{-}a.s.\ \omega\in\Omega, \tag{1.22}$$

where

$$\mathbb{E}\left(\xi^p_{\varepsilon,T}\right) < C_{p,\varepsilon}T,\quad \forall\, p>\frac{2}{\varepsilon}.$$

Let $t\geq 0$. The random variable $\frac{B_{t+h}-B_t}{h}$ has the same law $N(0,h^{-1})$ as $\frac{B_1}{\sqrt{h}}$. Hence it would be unreasonable to expect that $h^{-1}(B_{t+h}-B_t)$ converges in any sense to a finite limit as $h\to 0$. We recall from Revuz and Yor [64] (Exercise 2.9) or Karatzas and Shreve [42] (Chapter 2, Section 2.9, Theorem 9.18) the non-differentiability property of the paths of the Brownian motion:

Proposition 1.85 (Paley-Wiener-Zygmund). *Almost any trajectory of* $\{B_t, t \geq 0\}$ *is nowhere differentiable on* $\mathbb{R}_+$.

We shall show now that, $\mathbb{P}$-a.s. the Brownian motion has unbounded variation on every interval $[s,t]$. Let

$$s = t_0 < t_1 < \ldots < t_n = t,$$
$$\delta_n = \max\left\{t_{i+1} - t_i : i \in \overline{0, n-1}\right\}.$$

Proposition 1.86. *If* $S_n^{(2)} = \sum_{i=0}^{n-1} \left(B_{t_{i+1}} - B_{t_i}\right)^2$, *then* $S_n^{(2)} \xrightarrow{prob.} (t-s)$ *as* $\delta_n \to 0$. *Moreover:*

1. $\mathbb{E}\left|S_n^{(2)} - (t-s)\right|^2 \leq 2(t-s)\,\delta_n$;
2. *if* $\sum_{n=1}^{\infty} \delta_n < \infty$, *then* $S_n^{(2)} \xrightarrow{a.s.} t - s$;
3. $\updownarrow B_\bullet(\omega) \updownarrow_{[s,t]} = \infty$, $\mathbb{P}$-*a.s.* $\omega \in \Omega$.

Proof. Since

$$\begin{aligned}
\mathbb{E}\left|S_n^{(2)} - (t-s)\right|^2 &= \operatorname{Var}\left(S_n^{(2)}\right) \\
&= \sum_{i=0}^{n-1} \operatorname{Var}\left(\left(B_{t_{i+1}} - B_{t_i}\right)^2\right) \\
&= 2\sum_{i=0}^{n-1} (t_{i+1} - t_i)^2 \\
&\leq 2(t-s)\,\delta_n,
\end{aligned}$$

we see that $S_n^{(2)} \to (t-s)$ in $L^2(\Omega, \mathcal{F}, \mathbb{P})$.

The last assertion follows from the inequality

$$S_n^{(2)}(\omega) \leq \updownarrow B_\bullet(\omega) \updownarrow_{[s,t]} \mathbf{m}_{B_\bullet(\omega)}(\delta_n),$$

(with $\delta_n = 2^{-n}$), where

$$\mathbf{m}_{B_\bullet(\omega)}(\delta) = \sup\left\{|B_u(\omega) - B_v(\omega)| : u, v \in [s,t],\ |u - v| \leq \delta\right\}$$

is the modulus of continuity of $\{B_r(\omega) : r \in [s,t]\}$. ∎

Hence the path of the Brownian motion is a.s. of unbounded variation on any interval of nonzero length. Nevertheless, in the next chapter, we shall define an integral of the type

$$\int_0^t X_r dB_r.$$

1.3.3 $\mathcal{F}_t$-Brownian Motion

In many cases, we have together with a Brownian motion $\{B_t,\ t \geq 0\}$ other random variables or processes, so that the past information at time t is richer than that carried by $\mathcal{F}_t^B$. This motivates the following definition.

Let $(\Omega, \mathcal{F}, \mathbb{P}, \{\mathcal{F}_t\}_{t\geq 0})$ be a given stochastic basis and $\mathcal{P}$ the corresponding σ-algebra of progressively measurable subsets of $\Omega \times \mathbb{R}_+$.

Definition 1.87. A d-dimensional $\mathcal{P}$-m.c.s.p. $\{B_t; t \geq 0\}$ is called a d-dimensional $\mathcal{F}_t$-Brownian motion if $B_0 = 0$ and

(a) for all $0 \leq s < t$,

$$\begin{aligned} &(i)\ B_t - B_s \ \text{ is independent of } \mathcal{F}_s, \\ &(ii)\ B_t - B_s \ \sim\ N(0,\ (t-s)I_{d\times d}), \end{aligned}$$

or equivalently (by Lemma 1.35)

(b) for all $0 \leq s < t$ and $u \in \mathbb{R}$,

$$\mathbb{E}^{\mathcal{F}_s} \exp i\ \langle u, B_t - B_s\rangle = \exp\left[-\frac{|u|^2}{2}(t-s)\right].$$

By the Burkholder–Davis–Gundy inequality (1.18) we then have:

Proposition 1.88. *If $\{B_t\ ;\ t \geq 0\}$ is a d-dimensional $\mathcal{F}_t$-Brownian motion, then for all $p > 0$ there exist some constants $c_p > 0$ and $C_p > 0$, such that*

$$c_p d^{p/2} |t-s|^{p/2} \leq \mathbb{E} \sup_{r\in[s,t]} |B_r - B_s|^p \leq C_p d^{p/2} |t-s|^{p/2}.$$

Let $\{B_t\ ;\ t \geq 0\}$ be a d-dimensional Brownian motion and

$$\mathcal{F}_{t+}^B \stackrel{def}{=} \bigcap\nolimits_{r>t} \mathcal{F}_r^B = \bigcap\nolimits_{n\in\mathbb{N}^*} \mathcal{F}_{t+1/n}^B,$$

where

$$\mathcal{F}_t^B \stackrel{def}{=} \sigma\{B_s : 0 \leq s \leq t\} \vee \mathcal{N}$$

is the natural filtration associated to a stochastic process $\{B_t\ ;\ t \geq 0\}$.

Proposition 1.89. *The natural filtration of the d-dimensional Brownian motion $\{B_t\ ;\ t \geq 0\}$ is right continuous: $\mathcal{F}_t^B = \mathcal{F}_{t+}^B$, for all $t \geq 0$.*

Proof. Step 1: $\ B_t - B_s$ *is independent of $\mathcal{F}_{s+}^B$ for all $0 \leq s < t$.*
Let $n \in \mathbb{N}^*$ such that $s < s + 1/n < t$. Since the increment $B_t - B_{s+1/n}$ is independent of $\mathcal{F}_{s+1/n}^B \supset \mathcal{F}_{s+}^B$ it follows that for all $u \in \mathbb{R}^d$:

$$\mathbb{E}^{\mathcal{F}^B_{s+}} \exp i \left\langle u, B_t - B_{s+1/n} \right\rangle = \exp\left[-\frac{|u|^2}{2}(t-s-1/n)\right]$$

and passing to the limit as $n \to \infty$ we obtain

$$\mathbb{E}^{\mathcal{F}^B_{s+}} \exp i \left\langle u, B_t - B_s \right\rangle = \exp\left[-\frac{|u|^2}{2}(t-s)\right]$$

and $B_t - B_s$ is independent of $\mathcal{F}^B_{s+}$.

Step 2: *For all* $t \geq 0$, $0 = t_0 < t_1 < t_2 < \cdots < t_k$ *and* $u_1, \ldots, u_k \in \mathbb{R}^d$:

$$\mathbb{E}^{\mathcal{F}^B_{t+}} \exp i \left(\langle u_1, B_{t_1}\rangle + \cdots + \langle u_k, B_{t_k}\rangle\right) = \mathbb{E}^{\mathcal{F}^B_t} \exp i \left(\langle u_1, B_{t_1}\rangle + \cdots + \langle u_k, B_{t_k}\rangle\right).$$

Indeed the equality is clear if $t \geq t_k$.

If $t_{j-1} \leq t < t_j$, then

$$\begin{aligned}\langle u_1, B_{t_1}\rangle + \cdots + \langle u_k, B_{t_k}\rangle &= \langle u_1, B_{t_1}\rangle + \cdots + \left\langle u_{j-1}, B_{t_{j-1}}\right\rangle + \left\langle v_j, B_t\right\rangle \\ &\quad + \left\langle v_j, B_{t_j} - B_t\right\rangle + \left\langle v_{j+1}, B_{t_{j+1}} - B_{t_j}\right\rangle \\ &\quad + \cdots + \left\langle v_{k-1}, B_{t_{k-1}} - B_{t_{k-2}}\right\rangle + \left\langle v_k, B_{t_k} - B_{t_{k-1}}\right\rangle\end{aligned}$$

with $v_k = u_k, v_{k-1} = v_k + u_{k-1}, \ldots, v_j = v_{j+1} + u_j$. Writing

$$\mathbb{E}^{\mathcal{F}^B_{t+}} = \mathbb{E}^{\mathcal{F}^B_{t+}} \mathbb{E}^{\mathcal{F}^B_{t_j}} \ldots \mathbb{E}^{\mathcal{F}^B_{t_k}},$$

and using step 1, we deduce that

$$\begin{aligned}&\mathbb{E}^{\mathcal{F}^B_{t+}} \exp i \left(\langle u_1, B_{t_1}\rangle + \cdots + \langle u_k, B_{t_k}\rangle\right) \\ &= \exp i \left(\langle u_1, B_{t_1}\rangle + \cdots + \left\langle u_{j-1}, B_{t_{j-1}}\right\rangle + \left\langle v_j, B_t\right\rangle\right) \\ &\times \exp\left[-\frac{\left|v_j\right|^2}{2}\left(t_j - t\right) - \frac{\left|v_{j+1}\right|^2}{2}\left(t_{j+1} - t_j\right) - \cdots - \frac{|v_k|^2}{2}(t_k - t_{k-1})\right] \\ &= \mathbb{E}^{\mathcal{F}^B_t} \exp i \left(\langle u_1, B_{t_1}\rangle + \cdots + \langle u_k, B_{t_k}\rangle\right).\end{aligned}$$

Step 3: The equality proved in Step 2 yields that

$$\mathbb{E}\left(\xi|\mathcal{F}^B_t\right) = \mathbb{E}\left(\xi|\mathcal{F}^B_{t+}\right),$$

for all bounded $\sigma\left\{\mathcal{F}^B_r : 0 \leq r < \infty\right\}$-measurable random variables ξ. Setting $\xi = \mathbf{1}_F$, $F \in \mathcal{F}^B_{t+}$, we obtain that

$$\mathbf{1}_F = \mathbb{E}\left(\mathbf{1}_F|\mathcal{F}^B_{t+}\right) = \mathbb{E}\left(\mathbf{1}_F|\mathcal{F}^B_t\right)$$

is $\mathcal{F}_t^B$-measurable. Hence $\mathcal{F}_{t+}^B \subset \mathcal{F}_t^B \subset \mathcal{F}_{t+}^B$. ■

Hence we have:

Remark 1.90.

▲ A d-dimensional $\mathcal{F}_t$-Brownian motion is a d-dimensional Brownian motion.

▲ A d-dimensional Brownian motion is a d-dimensional $\mathcal{F}_t^B$-Brownian motion, where $\mathcal{F}_t^B = \sigma(\{B_s : 0 \leq s \leq t\}) \vee \mathcal{N}$ is the natural filtration (associated to the Brownian motion B, which is right continuous by the above Proposition).

We now prove a well known and useful characterization of Brownian motion.

Theorem 1.91 (Paul Lévy). *Let $\{B_t\,;\, t \geq 0\}$, $B_0 = 0$, be a d-dimensional $\mathcal{P}$-m.c.s.p. Then the following statements are equivalent:*

(I) $\{B_t\,;\, t \geq 0\}$ is a d-dimensional $\mathcal{F}_t$-Brownian motion.
(II) B and $\{B_t \otimes B_t - t I_{d\times d} : t \geq 0\}$ are continuous $\mathcal{F}_t$-martingales.

Remark 1.92. The condition *(II)* is clearly equivalent to

(II′) *(j)* $\mathbb{E}\,|B_t|^2 < \infty$, for all $t \geq 0$,
(jj) $\mathbb{E}\,(B_t|\mathcal{F}_s) = B_s$, for each $0 \leq s \leq t$,
(jjj) $\mathbb{E}\,[(B_t - B_s) \otimes (B_t - B_s)\,|\mathcal{F}_s] = (t-s)\, I_{d\times d}$, for each $0 \leq s \leq t$;
and implies that $\{|B_t|^2 - td : t \geq 0\}$ is an $\mathcal{F}_t$-martingale.

of Paul Lévy's Theorem. $(I) \Rightarrow (II)$: Let $0 \leq s \leq t$. We have

$$\begin{aligned}\mathbb{E}\,(B_t - B_s|\mathcal{F}_s) &= \mathbb{E}\,(B_t - B_s)\\ &= 0\end{aligned}$$

and

$$\begin{aligned}\mathbb{E}\,[(B_t - B_s) \otimes (B_t - B_s)\,|\mathcal{F}_s] &= \mathbb{E}\,[(B_t - B_s) \otimes (B_t - B_s)]\\ &= (t-s)\, I_{d\times d}.\end{aligned}$$

(II) $\Rightarrow$ (I): For $u \in \mathbb{R}^d$ fixed, let

$$\varphi(t,x) = \exp\left(i\,\langle u, x\rangle + \frac{|u|^2}{2} t\right), \quad t \geq 0,\ x \in \mathbb{R}^d.$$

We have

$$\varphi_t'(t,x) = \frac{|u|^2}{2}\varphi(t,x),\, \varphi_{xx}''(t,x) = -\varphi(t,x)\, u \otimes u.$$

Then from Lemma 1.73, $\{\varphi(t, B_t) : t \geq 0\}$ is an $\mathcal{F}_t$-martingale.
Consequently,

$$\mathbb{E}^{\mathcal{F}_s} \exp i \langle u, B_t - B_s \rangle = \exp\left[-\frac{|u|^2}{2}(t-s)\right], \; 0 \le s < t, \; u \in \mathbb{R}^d.$$

Then, by Lemma 1.35, for each $0 \le s < t$, the increment $B_t - B_s$ is independent of $\mathcal{F}_s$, and its law is $N(0, (t-s)I_{d\times d})$. ■

The following corollary is immediate.

Corollary 1.93. *A d-dimensional $\mathcal{P}$-m.c.s.p. $\{B_t; t \ge 0\}, B_0 = 0$, is a d-dimensional $\mathcal{F}_t$-Brownian motion iff for all $u \in \mathbb{R}^d$:*

$$M_t = \exp\left(i\langle u, B_t\rangle + \frac{|u|^2}{2}t\right), t \ge 0, \;\; \textit{is an } \mathcal{F}_t\textit{-martingale}. \tag{1.23}$$

We now establish the strong Markov property of Brownian motion (Brownian motion renews itself at stopping times).

Proposition 1.94. *Let $\{B_t\,;\, t \ge 0\}$ be a d-dimensional $\mathcal{F}_t$-Brownian motion and τ an $\mathcal{F}_t$-stopping time such that $\tau < \infty$, $\mathbb{P}$-a.s. Let $\tilde{B}_t \overset{def}{=} B_{\tau+t} - B_\tau$ and $\tilde{\mathcal{F}}_t = \mathcal{F}_{\tau+t}$. Then $\{\tilde{B}_t : t \ge 0\}$ is an $\tilde{\mathcal{F}}_t$-Brownian motion which is independent of $\mathcal{F}_\tau$.*

Proof. Assume first that τ is bounded. We know that

$$M_t = \exp\left(i\langle u, B_t\rangle + \frac{|u|^2}{2}t\right), \quad t \ge 0,$$

is an $\mathcal{F}_t$-martingale for all $u \in \mathbb{R}^d$.

Let $0 \le s < t$, $A \in \tilde{\mathcal{F}}_s = \mathcal{F}_{\tau+s}$ and

$$\tilde{M}_t = \exp\left(i\left\langle u, \tilde{B}_t\right\rangle + \frac{|u|^2}{2}t\right) = \tilde{M}_s \frac{M_{\tau+t}}{M_{\tau+s}}.$$

Clearly $\tilde{M}_t$ is $\tilde{\mathcal{F}}_t$-measurable and by the Doob optional stopping theorem (Theorem 1.62)

$$\begin{aligned}\mathbb{E}\left(\mathbf{1}_A \tilde{M}_t\right) &= \mathbb{E}\left(\mathbf{1}_A \tilde{M}_s \frac{M_{\tau+t}}{M_{\tau+s}}\right) \\ &= \mathbb{E}\left(1_A \tilde{M}_s \frac{1}{M_{\tau+s}} \mathbb{E}^{\mathcal{F}_{\tau+s}} M_{\tau+t}\right) \\ &= \mathbb{E}\left(1_A \tilde{M}_s\right).\end{aligned}$$

Hence, by Proposition 1.93, $\tilde{B}_t$ is an $\tilde{\mathcal{F}}_t$-Brownian motion.

Since

$$\mathbb{E}^{\mathcal{F}_\tau} \exp i \left\langle u, \tilde{B}_t \right\rangle = \left(\mathbb{E}^{\tilde{\mathcal{F}}_0} \tilde{M}_t\right) \exp\left(-\frac{|u|^2}{2} t\right)$$

$$= \tilde{M}_0 \exp\left(-\frac{|u|^2}{2} t\right)$$

$$= \exp\left(-\frac{|u|^2}{2} t\right),$$

we conclude by Lemma 1.35 that $\tilde{B}_t$ is independent of $\mathcal{F}_\tau$.

If τ is not bounded we replace τ by $\tau \wedge n$ and A by $A_n = A \cap \{\tau \leq n\}$. Taking limits and using the Lebesgue dominated convergence theorem, we obtain the desired result. ∎

Finally we give a result which will be useful for proving the convergence of stochastic integrals in the next chapter.

Let $B : \Omega \times [0, \infty[\longrightarrow \mathbb{R}^k$ and $X : \Omega \times [0, \infty[\longrightarrow \mathbb{R}^l$ be two stochastic processes. Let

$$\mathcal{F}_t^{B,X} \overset{def}{=} \sigma\{B_s, X_s;\ s \leq t\} \vee \mathcal{N}_{\mathbb{P}}, \quad t \geq 0,$$

be the natural filtration generated jointly by B and X and

$$\mathcal{F}_{t+}^{B,X} \overset{def}{=} \bigcap_{\varepsilon > 0} \mathcal{F}_{t+\varepsilon}^{B,X}$$

its right continuous version. If $X \in L^0\left(\Omega; L^1_{loc}\left(\mathbb{R}_+; \mathbb{R}^d\right)\right)$, we shall also consider the filtration

$$\hat{\mathcal{F}}_t^{B,X} = \sigma\{B_s, \frac{1}{\varepsilon}\int_{0 \vee (s-\varepsilon)}^{s} X_r dr;\ \varepsilon > 0,\ s \in [0, t]\} \vee \mathcal{N}_{\mathbb{P}}, \quad t \geq 0,$$

and $\hat{\mathcal{F}}_{t+}^{B,X}$ its right continuous version. It is clear that $\hat{\mathcal{F}}_{t+}^{B,X} \subset \mathcal{F}_{t+}^{B,X}$, and if X is a continuous stochastic process then $\hat{\mathcal{F}}_{t+}^{B,X} = \mathcal{F}_{t+}^{B,X}$.

Proposition 1.95. *Let $B, \tilde{B} \in L^0\left(\Omega; C\left(\mathbb{R}_+; \mathbb{R}^k\right)\right)$ and $X, \tilde{X} \in L^0\left(\Omega; \mathbb{X}\right)$ with $\mathbb{X} = L^p_{loc}\left(\mathbb{R}_+; \mathbb{R}^l\right)$, $p \geq 1$, (or $\mathbb{X} = C\left(\mathbb{R}_+; \mathbb{R}^l\right)$) be stochastic processes such that:*

(i) $\tilde{B}$ is an $\hat{\mathcal{F}}_{t+}^{\tilde{B},\tilde{X}}$-Brownian motion; and
(ii) $\mathcal{L}(B, X) = \mathcal{L}(\tilde{B}, \tilde{X})$ on $C(\mathbb{R}_+, \mathbb{R}^k) \times \mathbb{X}$ ((B, X) and $(\tilde{B}, \tilde{X})$ have the same law on $C(\mathbb{R}_+, \mathbb{R}^k)) \times \mathbb{X}$).

Then B is an $\hat{\mathcal{F}}_{t+}^{B,X}$-Brownian motion.

Proof. Let $m \in \mathbb{N}^*$ and choose an arbitrary bounded continuous function $\Phi : (\mathbb{R}^k)^m \times (\mathbb{R}^l)^m \to \mathbb{R}$. Let $0 \le t_1 \le t_2 \le \ldots \le t_m = t < t + \delta$, $u \in \mathbb{R}^k$, and $k_1, k_2, \ldots, k_m \ge 1$. In virtue of (i) and (ii) we get, using the notations $F_{t_i}^{k_i} = k_i \int_{t_i - \frac{1}{k_i}}^{t_i} X_r dr$, $\tilde{F}_{t_i}^{k_i} = k_i \int_{t_i - \frac{1}{k_i}}^{t_i} \tilde{X} dr$, that

$$\begin{aligned}
&\mathbb{E}\Big[\exp\{i(\langle B_{t+\delta}, u\rangle - \langle B_t, u\rangle)\}\,\Phi\Big(B_{t_1}, \ldots, B_{t_m}, F_{t_1}^{k_1}, F_{t_2}^{k_2}, \ldots, F_{t_m}^{k_m}\Big)\Big] \\
&= \mathbb{E}\Big[\exp\Big\{i\Big(\big\langle \tilde{B}_{t+\delta}, u\big\rangle - \big\langle \tilde{B}_t, u\big\rangle\Big)\Big\}\,\Phi\Big(\tilde{B}_{t_1}, \ldots, \tilde{B}_{t_m}, \tilde{F}_{t_1}^{k_1}, \tilde{F}_{t_2}^{k_2}, \ldots, \tilde{F}_{t_m}^{k_m}\Big)\Big] \\
&= \exp\left\{-\frac{\delta}{2}|u|^2\right\}\mathbb{E}\Big[\Phi\Big(\tilde{B}_{t_1}, \ldots, \tilde{B}_{t_m}, \tilde{F}_{t_1}^{k_1}, \tilde{F}_{t_2}^{k_2}, \ldots, \tilde{F}_{t_m}^{k_m}\Big)\Big] \\
&= \exp\left\{-\frac{\delta}{2}|u|^2\right\}\mathbb{E}\Big[\Phi\Big(B_{t_1}, \ldots, B_{t_m}, F_{t_1}^{k_1}, F_{t_2}^{k_2}, \ldots, F_{t_m}^{k_m}\Big)\Big].
\end{aligned}$$

Consequently,

$$\mathbb{E}\Big[\exp\{i(\langle B_{t+\delta}, u\rangle - \langle B_t, u\rangle)\}\,|\hat{\mathcal{F}}_{t+}^{B,X}\Big] = \exp\left\{-\frac{\delta}{2}|u|^2\right\}$$

for all $u \in \mathbb{R}^k$, and $t, \delta > 0$ such that $t + \delta \le T$, and the result follows. ■

We now give a convergence result for $(B^n, \mathcal{F}_t^n)$ Brownian motions.

Corollary 1.96. *Let $B, B^n, \tilde{B}^n \in L^0(\Omega; C(\mathbb{R}_+; \mathbb{R}^k))$ and $X, X^n, \tilde{X}^n \in L^0(\Omega; \mathbb{X})$ with $\mathbb{X} = L^p_{loc}(\mathbb{R}_+; \mathbb{R}^l)$, $p \ge 1$, (or $\mathbb{X} = C(\mathbb{R}_+; \mathbb{R}^l)$) be stochastic processes such that:*

- (i) *$\tilde{B}^n$ is an $\hat{\mathcal{F}}_{t+}^{\tilde{B}^n, \tilde{X}^n}$-Brownian motion $\forall\, n \ge 1$;*
- (ii) *$\mathcal{L}(B^n, X^n) = \mathcal{L}(\tilde{B}^n, \tilde{X}^n)$ on $C(\mathbb{R}_+, \mathbb{R}^k) \times \mathbb{X}$ ((B^n, X^n) and $(\tilde{B}^n, \tilde{X}^n)$ have the same law on $C(\mathbb{R}_+, \mathbb{R}^k)) \times \mathbb{X}$);*
- (iii) *$|B_t^n - B_t| \longrightarrow 0$ in probability, as $n \to \infty$, $\forall t \ge 0$;*
- (iv) *as $n \to \infty$, for all $t \ge 0$,*

  (a) *if $\mathbb{X} = L^p_{loc}(\mathbb{R}_+; \mathbb{R}^l)$ then*

  $$\int_0^t |X_s^n - X_s|^p\, ds \longrightarrow 0 \quad \textit{in probability},$$

  (b) *if $\mathbb{X} = C(\mathbb{R}_+; \mathbb{R}^l)$ then*

  $$|X_t^n - X_t| \longrightarrow 0 \quad \textit{in probability.}$$

Then $\Big(B^n, \{\hat{\mathcal{F}}_{t+}^{B^n, X^n}\}\Big)$, $n \ge 1$, and $\Big(B, \{\hat{\mathcal{F}}_{t+}^{B,X}\}\Big)$ are Brownian motions.

Proof. By Proposition 1.95 B^n is an $\hat{\mathcal{F}}_{t+}^{B^n,X^n}$-Brownian motion. Hence for all $m \in \mathbb{N}^*$, $\Phi : (\mathbb{R}^k)^m \times (\mathbb{R}^{d\times k})^m \to \mathbb{R}$ an arbitrary bounded continuous function, $0 \le t_1 \le t_2 \le \ldots \le t_m = t < t+\delta$, $u \in \mathbb{R}^k$, and $k_1, k_2, \ldots, k_m \ge 1$, we have

$$\begin{aligned}&\mathbb{E}\left[\exp\left\{i\left(\langle B^n_{t+\delta}, u\rangle - \langle B^n_t, u\rangle\right)\right\}\Phi\left(B^n_{t_1},\ldots,B^n_{t_m},F^{k_1,n}_{t_1},F^{k_2,n}_{t_2},\ldots,F^{k_m,n}_{t_m}\right)\right]\\ &= \exp\left\{-\frac{\delta}{2}|u|^2\right\}\mathbb{E}\left[\Phi\left(B^n_{t_1},\ldots,B^n_{t_m},F^{k_1,n}_{t_1},F^{k_2,n}_{t_2},\ldots,F^{k_m,n}_{t_m}\right)\right],\end{aligned}$$

where $F^{k_i,n}_{t_i} = k_i\int_{t_i-\frac{1}{k_i}}^{t_i} X^n_r dr$, $\tilde{F}^{k_i,n}_{t_i} = k_i\int_{t_i-\frac{1}{k_i}}^{t_i} \tilde{X}^n dr$ and $F^{k_i}_{t_i} = k_i\int_{t_i-\frac{1}{k_i}}^{t_i} X_r dr$.

Passing to the limit as $n \to \infty$, we obtain

$$\begin{aligned}&\mathbb{E}\left[\exp\left\{i\left(\langle B_{t+\delta}, u\rangle - \langle B_t, u\rangle\right)\right\}\Phi\left(B_{t_1},\ldots,B_{t_m},F^{k_1}_{t_1},F^{k_2}_{t_2},\ldots,F^{k_m}_{t_m}\right)\right]\\ &= \exp\left\{-\frac{\delta}{2}|u|^2\right\}\mathbb{E}\left[\Phi\left(B_{t_1},\ldots,B_{t_m},F^{k_1}_{t_1},F^{k_2}_{t_2},\ldots,F^{k_m}_{t_m}\right)\right]\end{aligned}$$

and then

$$\mathbb{E}\left[\exp\left\{i\left(\langle B_{t+\delta}, u\rangle - \langle B_t, u\rangle\right)\right\}|\hat{\mathcal{F}}^{B,X}_{t+}\right] = \exp\left\{-\frac{\delta}{2}|u|^2\right\},$$

that is, B is an $\hat{\mathcal{F}}^{B,X}_{t+}$-Brownian motion. ■

1.4 Exercises

Exercise 1.1 (Carathéodory Functions). Let $(\Omega,\mathcal{F})$ be a measurable space and $(\mathbb{X},\rho)$ be a separable metric space. A mapping $F : \Omega\times\mathbb{X}\to\mathbb{R}^d$ is an $(\mathcal{F},\mathbb{X})$-Carathéodory function if

$$\begin{aligned}&c_1)\ F(\cdot,x) \quad \text{is } (\mathcal{F},\mathcal{B}_d)\text{-measurable } \forall x\in\mathbb{X},\\ &c_2)\ F(\omega,\cdot) \quad \text{is continuous } \forall\omega\in\Omega.\end{aligned}$$

Show that $F(\cdot,\cdot)$ is $(\mathcal{F}\otimes\mathcal{B}_{\mathbb{X}},\mathcal{B}_d)$-measurable ($\mathcal{B}_d$ denotes the Borel σ-algebra on $\mathbb{R}^d$ and $\mathcal{B}_{\mathbb{X}}$ the Borel σ-algebra on $\mathbb{X}$).

Exercise 1.2. If $(\mathbb{S},\mathcal{S},\mu)$ is a measurable space, $(\Omega,\mathcal{F},\mathbb{P})$ is a probability space and $f : \Omega\times S\to\mathbb{R}$ is an $\mathcal{F}\otimes\mathcal{S}$-measurable function, then for all $p\in[1,\infty[$:

$$\left\|\int_{\mathbb{S}} f(s)\,d\mu(s)\right\|_{L^p(\Omega,\mathcal{F},\mathbb{P})} \le \int_{\mathbb{S}} \|f(s)\|_{L^p(\Omega,\mathcal{F},\mathbb{P})}\,d\mu(s) \qquad (1.24)$$

(Minkowski's inequality).

Exercise 1.3. Let $\{S_t : t \in [0,T]\}$ be a real right-continuous sub-martingale. Show that for all stopping times $0 \le \tau \le \theta \le T$:

$$S_\tau \le \mathbb{E}\left(S_\theta | \mathcal{F}_\tau\right), \text{ a.s.}$$

Exercise 1.4. Let $\tau < \infty$ a.s. be a stopping time and

$$\tau_n = \frac{\lceil 2^n \tau \rceil}{2^n},$$

where $\lceil x \rceil$ denotes the smallest integer greater than or equal to x. Show that $\{\tau_n\}$ is a decreasing sequence of stopping times such that

$$0 \le \tau_n - \tau \le \frac{1}{2^n}, \text{ a.s.}$$

and if τ is bounded then τ_n takes a finite number of values.

Hint: Note that

$$\tau_n = \sum_{k=1}^{\infty} \frac{k}{2^n} \mathbf{1}_{\frac{k-1}{2^n} < \tau \le \frac{k}{2^n}}.$$

Exercise 1.5. Prove the equivalences stated in Definition 1.81.

Exercise 1.6. Let $\{B_t : t \ge 0\}$ be a real Brownian motion. Show that $\{B_t^2 - t, t \ge 0\}$, $\{B_t^4 - 6tB_t^2 + 3t^2, t \ge 0\}$ and $\{\exp(\lambda B_t - \lambda^2 t/2), t \ge 0\}$ (where $\lambda \in \mathbb{R}$ is arbitrary) are martingales.

Exercise 1.7. Let $(\Omega, \mathcal{F}, \mathbb{P}, \{\mathcal{F}_t\}_{t\ge 0})$ be a stochastic basis and $\{B_t ; t \ge 0\}$ is a scalar $\mathcal{F}_t$-Brownian motion. Let $a \in \mathbb{R}$ and $g : \mathbb{R} \to \mathbb{R}$ be a bounded Borel measurable function. Show that for all $0 \le t < T$:

$$(a)\ \mathbb{E}\left(g\left(B_T\right)|\mathcal{F}_t\right) = \int_{\mathbb{R}} g\left(x\sqrt{T-t} + B_t\right)\rho\left(x\right)dx, \quad \mathbb{P}\text{-}a.s.,$$

$$(b)\ \mathbb{E}\left(\mathbf{1}_{B_T \le a}|\mathcal{F}_t\right) = \Phi\left(\frac{a - B_t}{\sqrt{T-t}}\right), \mathbb{P}\text{-}a.s.,$$

where $\rho(x) = \frac{1}{\sqrt{2\pi}}\exp\left(-\frac{x^2}{2}\right)$ and $\Phi(x) = \int_{-\infty}^{x} \rho(u)\,du$.

Exercise 1.8. Let $\{B_t : t \ge 0\}$ be a real Brownian motion and for $a \in R$, let $T_a = \inf\{t \ge 0,\ B_t = a\}$.

1. Show that T_a is a stopping time.
2. For $a < 0 < b$, compute $\mathbb{P}(T_a < T_b)$ and $\mathbb{P}(T_b < T_a)$.

3. Exploiting the fact that $\{B_t^2 - t,\ t \geq 0\}$ is a martingale, compute $\mathbb{E}(T_a \wedge T_b)$, and then $\mathbb{E}T_a$. Note that if we would have $\mathbb{E}T_a < \infty$, then there would be two contradictory ways of computing $\mathbb{E}B_{T_a}$.

Exercise 1.9. Let $\{B_t : t \geq 0\}$ be a real Brownian motion. For $a > 0$, let $S_a = \inf\{t \geq 0,\ |B_t| = a\}$.

1. Show that S_a is stopping time.
2. Compute $\mathbb{E}S_a$, $\mathbb{E}S_a^2$ and $\mathbb{E}\exp(-\lambda S_a)$ for $\lambda > 0$.

Exercise 1.10. Let $X_t = B_t + \mu t$, for $t \geq 0$, where $\mu \in \mathbb{R}$, and $T_a = \inf\{t,\ X_t = a\}$, for $a \in \mathbb{R}$.

1. Show that for any $\sigma \in \mathbb{R}$, the following is a martingale:

$$Z_t = \exp\left(\sigma X_t - (\frac{\sigma^2}{2} + \mu\sigma)t\right).$$

2. Show that for any $t > 0$, $\mathbb{E}(Z_{t\wedge T_a}) = 1$.
3. Deduce that if $a > 0$ and $\sigma > (-2\mu)^+$, or else $a < 0$ and $\sigma < -2\mu^+$,

$$\mathbb{E}\left[\mathbf{1}_{T_a<\infty}\exp\left(-(\frac{\sigma^2}{2} + \mu\sigma)T_a\right)\right] = \exp(-\sigma a).$$

4. Show that $\mathbb{P}(T_a < \infty) = 1 \wedge e^{2\mu a}$.

Exercise 1.11. Let $\{B_t : t \geq 0\}$ be a scalar Brownian motion defined on a probability space $(\Omega, \mathcal{F}, \mathbb{P})$. We define the random process

$$Z_t = B_t - tB_1,\ \ 0 \leq t \leq 1.$$

1. Show that $\{Z_t : 0 \leq t \leq 1\}$ is a centered Gaussian process which is independent of B_1.
2. Let $\bar{Z}_t = Z_{1-t}, 0 \leq t \leq 1$. Show that $\{\bar{Z}_t : 0 \leq t \leq 1\}$ has the same law as $\{Z_t : 0 \leq t \leq 1\}$.
3. Let $Y_t = (1-t)B_{\frac{t}{1-t}}, 0 \leq t < 1$. Show that Y_t converges a.s. to 0 as $t \to 1$. We define $Y_1 = 0$. Show that $\{Y_t : 0 \leq t \leq 1\}$ and $\{Z_t : 0 \leq t \leq 1\}$ have the same law.
4. The last questions aim at identifying the law of $\sup_{0\leq t\leq 1} Z_t$. We shall compute

$$F(a) = \mathbb{P}\left(\sup_{0\leq t\leq 1} Z_t \leq a\right).$$

Let $\tau_a = \inf\{t \geq 0, B_t - ta = a\}$. Show that for any $a > 0$, τ_a is a stopping time with respect to the Brownian filtration, and that $F(a) = 1 - \mathbb{P}(\tau_a < \infty)$.

6. Show that for all $t \geq 0$

$$\mathbb{E}\left(\exp\left[2a(B_{t\wedge\tau_a} - at \wedge \tau_a)\right]\right) = 1.$$

7. Deduce that $\mathbb{E}[\exp(2a(B_{\tau_a} - a\tau_a))\mathbf{1}_{\tau_a<\infty}] = 1$ and compute $\mathbb{P}(\tau_a < \infty)$.

Exercise 1.12. For $\alpha > 0$, let

$$X_n^\alpha = n^{-\alpha} \int_0^n B_s \, ds.$$

For which values of α does the sequence $\{X_n^\alpha;\, n \geq 1\}$ converge in $L^2(\Omega)$ as $n \to \infty$? Do we have a.s. convergence?

Exercise 1.13. Let $g : [0, \infty[\to [0, \infty[$, $g(0) = 0$, be an increasing right continuous function and

$$\gamma(t) = \begin{cases} \inf\{s \geq 0 : g(s) > t\}, & \text{if } \{s \geq 0 : g(s) > t\} \neq \emptyset, \\ +\infty, & \text{otherwise,} \end{cases}$$

and by convention let $\gamma(0_-) = 0$. Show that:

- γ is an increasing right continuous function;
- $\gamma(t_-) = \inf\{s \geq 0 : g(s) \geq t\}$;
- $g(\gamma(t)) \geq t$;
- $g(t) = \inf\{s \geq 0 : \gamma(s) > t\}$;
- $\gamma(t) < \infty$ iff $\lim_{s\to\infty} g(s) > t$;
- for all Borel functions $f : [0, \infty[\to [0, \infty[$,

$$\int_{[0,\infty[} f(t)\, dg(t) = \int_0^\infty f(\gamma(t))\, \mathbf{1}_{[0,\infty[}(\gamma(t))\, dt;$$

- if g is strictly increasing then γ is continuous;
- if $\{M_t; t \geq 0\}$ is a d-dimensional continuous local martingale and

$$\Gamma_t = \inf\{s \geq 0 : s + \langle M\rangle_s > t\},$$

then $\{\Gamma_t : t \geq 0\}$ is a $\mathcal{P}$-m.i.c.s.p. and $0 = \Gamma_0 \leq \Gamma_t \leq t$.

Exercise 1.14. Let $f : \mathbb{R}_+ \to \mathbb{R}$ be a locally square integrable function. For $t \geq 0$, we let $X_t = \displaystyle\int_0^t f(s) dB_s$, and define

$$a(t) = \int_0^t f^2(s) ds; \quad c(t) = \inf\{u \geq 0;\ a(u) > t\} \quad \text{and } W_t = X_{c(t)}.$$

1. Show that the process

$$\left\{W_t : 0 \le t < \int_0^\infty f^2(s)ds\right\}$$

is a Brownian motion. Show that for all $t \ge 0$, $X_t = W_{a(t)}$ a.s.
2. Consider the process $\{Z_t;\ 0 \le t < 1\}$ defined by

$$Z_t = \int_0^t \frac{1-t}{1-s} dB_s.$$

Show that $Z_t \to 0$, $\mathbb{P}$-$a.s.$, as $t \to 1$.

Exercise 1.15. Let $\{B_t : t \ge 0\}$ be a d-dimensional Brownian motion. Prove that

$$\limsup_{\varepsilon \searrow 0} \frac{|B_{t+\varepsilon} - B_t|}{\varepsilon} = \infty \text{ a.s.}$$

Exercise 1.16. Let $0 \le s < t$ and $\mathcal{D}_{[s,t]}$ be the set of partitions

$$\Delta : s = t_0 < t_1 < \cdots < t_n = t, \quad n = n_\Delta \in \mathbb{N}^*.$$

Denote by $\|\Delta\| = \sup\left\{t_i - t_{i-1} : i \in \overline{1,n}\right\}$ the norm of the partition Δ. Let $p \ge 1$, $X : \Omega \times \mathbb{R}_+ \to \mathbb{R}^d$ be a stochastic process and

$$S_\Delta^{(p)}(X;[s,t]) = \sum_{i=1}^{n} |X_{t_i} - X_{t_{i-1}}|^p.$$

The p-variation of X on $[s,t]$ is defined as a limit in probability by

$$\mathrm{Var}^{(p)}(X;[s,t]) \stackrel{def}{=} \lim_{\|\Delta_n\| \to 0} S_{\Delta_n}^{(p)}(X;[s,t]),$$

if the limit exists and is independent of the sequence $\Delta_n \in D_{[s,t]}$, $\|\Delta_n\| \to 0$. The length of the trajectory $\{(r, X_r(\omega)) : r \in [s,t]\}$ is defined as

$$\mathbf{L}(X_\cdot(\omega);[s,t]) \stackrel{def}{=} \sup_{\Delta \in \mathcal{D}_{[a,b]}} \sum_{i=1}^{n_\Delta} \sqrt{(t_i - t_{i-1})^2 + |X_{t_i}(\omega) - X_{t_{i-1}}(\omega)|^2}.$$

Show that:

1. (Quadratic Variation Process) If $\{M_t : t \ge 0\}$ is a continuous local martingale, then for all $t \ge 0$:

$$\mathrm{Var}^{(2)}(M;[0,t]) = \langle M \rangle_t,\ a.s.$$

2. If X has C^1-trajectories, then

$$\mathrm{Var}^{(p)}(X;[s,t]) = \begin{cases} \int_s^t \left|\frac{dX_r}{dr}\right| dr, & \text{if } p=1, \\ 0, & \text{if } p>1. \end{cases}$$

3. If $\{B_t : t \geq 0\}$ is a Brownian motion, $p \geq 1$ and $0 \leq s < t$, then $\mathbb{P}$-$a.s.$:

$$\mathrm{Var}^{(p)}(B;[s,t]) = \begin{cases} +\infty, & \text{if } 1 \leq p < 2, \\ t-s, & \text{if } p = 2, \\ 0, & \text{if } p > 2, \end{cases}$$

and

$$\mathbf{L}(B;[s,t]) = \infty.$$

Moreover if $\sum_{n=1}^{\infty} \|\Delta_n\| < \infty$, then

$$\text{a.s-}\lim_{\|\Delta_n\|\to 0} S_{\Delta_n}^{(p)}(B;[s,t]) = \mathrm{Var}^{(p)}(B;[s,t]).$$

Exercise 1.17 (Hölder-Continuity of BM). Prove the Hölder-continuity of Brownian motion as a consequence of the following Garcia–Rademich–Ramsey inequality:

For $T \geq 0$, $p \geq 1$ and $\alpha > \frac{1}{p}$, there exists a constant $C_{\alpha,p} > 0$ such that for any function $f \in C([0,T];\mathbb{R})$, and for all $t, s \in [0,T]$, one has:

$$|f(t) - f(s)| \leq C_{\alpha,p} |t-s|^{\alpha - \frac{1}{p}} \left(\int_0^T \int_0^T \frac{|f(\theta) - f(\tau)|^p}{|\theta - \tau|^{p\alpha+1}} d\theta d\tau \right)^{1/p} \tag{1.25}$$

(with the convention $0/0 = 0$), see e.g. Stroock [68].

Exercise 1.18. Let $\{B_t : t \geq 0\}$ be a real Brownian motion. Show that for all $a \geq 0$, $c > 0$, $0 < s < t$:

$$\begin{aligned} &j)\ \mathbb{P}\big(B_t < c-a,\ \sup_{s\in[0,t]} B_s \geq c\big) = 2\mathbb{P}(B_t > a+c), \\ &jj)\ \mathbb{P}\big(\sup_{s\in[0,t]} B_s > c\big) = 2\mathbb{P}(B_t > c), \\ &jjj)\ \mathbb{P}\big(\sup_{u\in]s,t]} B_u > 0,\ B_s < 0\big) = 2\mathbb{P}(B_t > 0,\ B_s < 0). \end{aligned}$$

Exercise 1.19 (Tightness Criterion). Let $\{X_t^n,\ 0 \leq t \leq T\}_{n\geq 1}$ be a sequence of continuous $\mathbb{R}^d$-valued stochastic processes. Assume that:

(*i*) the sequence of r.v.'s $\{X_0^n,\ n \geq 1\}$ is tight;
(*ii*) there exist $a, b, M > 0$ such that for all $n \geq 1$,

$$\mathbb{E}\ |X_t^n - X_s^n|^a \leq M|t-s|^{1+b}.$$

Show that the sequence $\{(X^n)_{n\geq 1}\}$ is tight as a family of $C([0,T];\mathbb{R}^d)$-valued random variables.

Exercise 1.20. Let $T > 0$ and $\lambda \in \mathbb{R}$. Let $(M_t)_{t\in[0,T]}$ be a scalar local continuous martingale starting from $M_0 = 0$. Let $Z_t = e^{M_t - \frac{1}{2}<M>_t}$,

$$Z_t^{(\lambda)} = e^{\lambda M_t - \frac{\lambda^2}{2}<M>_t} \quad \text{and} \quad U_t^{(\lambda)} = e^{\lambda M_t} - \frac{\lambda^2}{2}\int_0^t e^{\lambda M_s} d <M>_s .$$

Show that:

1.

$$Z_t^{(\lambda)} = U_t^{(\lambda)} e^{-\frac{\lambda^2}{2}<M>_t} - \int_0^t U_s^{(\lambda)} d\left(e^{-\frac{\lambda^2}{2}<M>_s}\right), \qquad t \geq 0;$$

2. if $(M_t)_{t\in[0,T]}$ is a bounded continuous martingale, then $U^{(\lambda)}, Z^{(\lambda)} \in \mathcal{M}^1[0,T]$, $Z_0^{(\lambda)} = U_0^{(\lambda)} = 1$;
3. $(Z_t^{(\lambda)})_{t\in[0,T]}$ is a super-martingale, that is $\mathbb{E}^{\mathcal{F}_s} Z_t^{(\lambda)} \leq Z_s^{(\lambda)}$, $\mathbb{P}$-*a.s.*, for all $0 \leq s \leq t \leq T$; moreover

$$0 < \mathbb{E}\ Z_T^{(\lambda)} \leq \mathbb{E}\ Z_t^{(\lambda)} \leq \mathbb{E}\ Z_s^{(\lambda)} \leq \mathbb{E}\ Z_0^{(\lambda)} = 1;$$

4. $\mathbb{E}\ Z_T^{(\lambda)} = 1$ if and only if $Z^{(\lambda)} \in \mathcal{M}^1[0,T]$ ($(Z_t^{(\lambda)})_{t\in[0,T]}$ is a martingale);
5. if $0 < a < b \leq c$, $A \in \mathcal{F}$, and $\mathbb{E}\left(e^{cM_T}\right) < \infty$, then for any stopping time θ

$$\mathbb{E}\left(\mathbf{1}_A \sup_{t\in[0,T]} e^{aM_{t\wedge\theta}}\right) \leq C_{\alpha,\beta,\lambda}\left(\mathbb{P}(A)\right)^{b/(b-a)}\left(\mathbb{E}\ e^{cM_{T\wedge\theta}}\right)^{a/c};$$

6. if $0 < \lambda < 1$, $\theta_n = \inf\{t \geq 0 : |M_t| + <M>_t \geq n\}$, then for all $A \in \mathcal{F}$

$$\mathbb{E}\left(\mathbf{1}_A Z_{T\wedge\theta_n}^{(\lambda)}\right) \leq \left(\mathbb{E}Z_{T\wedge\theta_n}\right)^{\lambda^2}\left(\mathbb{E}\ e^{\frac{\lambda}{1+\lambda}M_{T\wedge\theta_n}}\mathbf{1}_A\right)^{1-\lambda^2}$$

and if $\mathbb{E}\left(e^{\frac{1}{2}M_T}\right) < \infty$ deduce that $Z_{T\wedge\theta_n}^{(\lambda)} \to Z_T^{(\lambda)}$ in $L^1(\Omega,\mathcal{F},\mathbb{P})$, as $n \to \infty$; consequently $\mathbb{E}\ Z_T^{(\lambda)} = 1$;
7. (Kazamaki) if $\mathbb{E}\left(e^{\frac{1}{2}M_T}\right) < \infty$, then $\mathbb{E}\ Z_T = 1$;
8. (Novikov) if $\mathbb{E}\left(e^{\frac{1}{2}<M>_T}\right) < \infty$, then $\mathbb{E}\ Z_T = 1$.

Chapter 2
Itô's Stochastic Calculus

In this chapter we construct Itô's stochastic integral (first introduced in [39]), and prove the famous Itô formula. We also establish several not quite standard versions of that formula, in particular for certain functions which do not satisfy the regularity assumptions of the basic result. In particular, we prove a d-dimensional version of the famous Tanaka formula, see Proposition 2.26 and the corollaries which follow. Those refined results will be useful later in the book. We also discuss in great detail in Sect. 2.4 a martingale representation theorem which will play an essential role in the study of BSDEs. Finally we present Girsanov's theorem.

Throughout this chapter $(\Omega, \mathcal{F}, \mathbb{P}, \{\mathcal{F}_t\}_{t\geq 0})$ will denote a given stochastic basis, $\mathcal{P}$ the corresponding σ-algebra of progressively measurable subsets of $\Omega \times \mathbb{R}_+$ and

$$\{B_t : t \geq 0\} \text{ a } k\text{-dimensional } \mathcal{F}_t\text{-Brownian motion.}$$

2.1 Notations: Preliminaries

Let $0 \leq p < \infty$ and $0 < T < \infty$. We introduce the notations:

◇ $\Lambda_d^p(0, T)$: the space of (equivalence classes of) $\mathcal{P}$-measurable processes $X : \Omega \times [0, T] \to \mathbb{R}^d$ such that

$$\int_0^T |X_t|^2 \, dt < +\infty, \quad \mathbb{P}\text{-}a.s.\ \omega \in \Omega, \quad \text{if } p = 0$$

and

$$\mathbb{E}\left(\int_0^T |X_t|^2 \, dt\right)^{p/2} < +\infty, \quad \text{if } p > 0;$$

two processes X, Y are equivalent if $(X_t = Y_t \text{ a.e. } t \in [0, T])$ $\mathbb{P}$-a.s. $\omega \in \Omega$.

E. Pardoux and A. Răşcanu, *Stochastic Differential Equations, Backward SDEs, Partial Differential Equations*, Stochastic Modelling and Applied Probability 69,
DOI 10.1007/978-3-319-05714-9_2,

◊ Λ_d^p : the space of (equivalence classes of) $\mathcal{P}$-measurable processes $X : \Omega \times [0, +\infty[\to \mathbb{R}^d$ such that for all $T > 0$ the restriction $X^{(T)}$ of X to $[0, T]$ belongs to $\Lambda_d^p(0, T)$.

Note that the property of progressive measurability is independent of the choice of an element in an equivalence class X, and for every $p \geq 0$,

$$\Lambda_d^p(0,T) \subset L^p\left(\Omega, \mathcal{F}, \mathbb{P}; L^2(0,T;\mathbb{R}^d)\right),$$

as a closed linear subspace.

Hence, for all $p \in [1, \infty[$, the space $\Lambda_d^p(0, T)$ is a Banach space with respect to the norm

$$\|X\|_{\Lambda_d^p(0,T)} = \left(\mathbb{E}\left(\int_0^T |X_t|^2\, dt\right)^{p/2}\right)^{1/p}.$$

Moreover the space $\Lambda_d^2(0, T)$ is a Hilbert space.

For $0 \leq p < 1$ the space $\Lambda_d^p(0, T)$ is a complete metric linear space with the metric

$$d_p(X,Y) = \begin{cases} \mathbb{E}\left(\|X - Y\|^p_{L^2(0,T;\mathbb{R}^d)}\right), & \text{if } 0 < p < 1, \\ \mathbb{E}\left(1 \wedge \|X - Y\|_{L^2(0,T;\mathbb{R}^d)}\right), & \text{if } p = 0. \end{cases}$$

$d_0(X, Y)$ is the metric of convergence in probability of the $L^2\left(0, T; \mathbb{R}^d\right)$-valued random variables.

The definitions for the case $T = +\infty$ are similar.

2.2 Definition of Itô's Stochastic Integral

Define $\mathcal{E}_{d\times k}$ to be the linear space of stochastic processes of the form

$$X_t(\omega) = \sum_{i=0}^{n-1} X_i(\omega)\, \mathbf{1}_{[t_i, t_{i+1}[}(t), \quad t \geq 0, \tag{2.1}$$

with $n \in \mathbb{N}^*$, $0 \leq t_0 < t_1 < \ldots < t_n$ and for $0 \leq i \leq n-1$, $X_i : \Omega \to \mathbb{R}^{d\times k}$ is an $\mathcal{F}_{t_i}$-measurable bounded random variable.

Denote by $\mathcal{E}_{d\times k}(0, T)$ the same space with the restriction $t_n \leq T$.

Proposition 2.1. *Let $T > 0$ and $p \in [0, \infty[$. Then $\mathcal{E}_{d\times k}(0, T)$ is a dense linear subspace of $\Lambda_{d\times k}^p(0, T)$.*

Proof. Let $X \in \Lambda^p_{d\times k}(0,T)$. Extend X by $X_t = 0$ for $t > T$. To approximate X by $X^n \in \mathcal{E}_{d\times k}(0,T)$ we proceed in three steps.

◇ First we approximate X in $\Lambda^p_{d\times k}(0,T)$ by the bounded progressively measurable stochastic processes $X^n = X_t \mathbf{1}_{[0,n]}(|X_t|)$, $n \in \mathbb{N}^*$.

◇ Second, let X be a bounded progressively measurable stochastic process. Then

$$X_t^n = n \int_{(t-1/n)\vee 0}^{t} X_s ds$$

defines a continuous bounded progressively measurable stochastic process X^n, which by the Lebesgue dominated convergence theorem converges to X in $\Lambda^p_{d\times k}(0,T)$ for all $p \geq 0$.

◇ Finally if X is a bounded continuous progressively measurable stochastic process and $X^n(t,\omega) = X\left(\frac{[nt]}{n},\omega\right)$, then $X^n \in \mathcal{E}_{d\times k}(0,T)$ and $X^n \to X$ in $\Lambda^p_{d\times k}(0,T)$.

■

For $X \in \mathcal{E}_{d\times k}$ of the form

$$X_t(\omega) = \sum_{i=0}^{n-1} X_i(\omega)\, \mathbf{1}_{[t_i, t_{i+1}[}(t),$$

we define the stochastic Itô integral

$$\mathbb{B}_t(X) = \int_0^t X_r dB_r, \quad t \geq 0,$$

by

$$\int_0^t X_r dB_r \overset{def}{=} \sum_{i=0}^{n-1} X_i \left(B_{t\wedge t_{i+1}} - B_{t\wedge t_i}\right).$$

Moreover we let

$$\int_s^t X_r dB_r \overset{def}{=} \int_0^t X_r dB_r - \int_0^s X_r dB_r, \text{ for } 0 \leq s \leq t.$$

Since $X_i\left(B_{\cdot\wedge t_{i+1}} - B_{\cdot\wedge t_i}\right) \in \mathcal{M}^p_d$, $\forall p \geq 1$, it clearly follows that

$$\mathbb{B} : \mathcal{E}_{d\times k} \to \mathcal{M}^p_d \text{ is a linear operator.}$$

Recall from Annex A that, if $x, y \in \mathbb{R}^d$, then

$$x \otimes y \overset{def}{=} \left(x_i y_j\right)_{d\times d} = xy^*$$

and

$$\mathbf{Tr}\,(x \otimes y) = \langle x, y \rangle\,.$$

Lemma 2.2. *Let $X, Y \in \mathcal{E}_{d\times k}$ and $0 \le s \le t$. Then:*

$$j)\ \mathbb{E}^{\mathcal{F}_s} \int_s^t X_r dB_r = 0,$$

$$jj)\ \mathbb{E}^{\mathcal{F}_s} \left[\int_s^t X_r dB_r \otimes \int_s^t Y_r dB_r \right] = \mathbb{E}^{\mathcal{F}_s} \int_s^t X_r Y_r^* \, dr,$$

$$jjj)\ \mathbb{E}^{\mathcal{F}_s} \left| \int_s^t X_r dB_r \right|^2 = \mathbb{E}^{\mathcal{F}_s} \int_s^t |X_r|^2 \, dr.$$

In particular the following are continuous p-martingales for all $p \ge 1$:

$$M_t = |\mathbb{B}_t(X)|^2 - \int_0^t |X_r|^2 \, dr \text{ and } N_t = \mathbb{B}_t(X) \otimes \mathbb{B}_t(Y) - \int_0^t X_r Y_r^* \, dr.$$

Proof. Let $X, Y \in \mathcal{E}_{d\times k}$ and $0 \le s < t$ be fixed. There exist n and $s = t_0 < t_1 < \cdots < t_n = t < t_{n+1}$ such that for $s \le r \le t$:

$$X_r(\omega) = \sum_{i=0}^{n} X_i(\omega) \mathbf{1}_{[t_i, t_{i+1}[}(r),\ Y_r(\omega) = \sum_{i=0}^{n} Y_i(\omega) \mathbf{1}_{[t_i, t_{i+1}[}(r),$$

where X_i, Y_i are $\mathcal{F}_{t_i}$-measurable and bounded.

We shall use the notations

$$\Delta_i t = t_{i+1} - t_i, \quad \Delta_i B = B_{t_{i+1}} - B_{t_i},$$
$$\Delta_i \mathbb{B}(X) = \mathbb{B}_{t_{i+1}}(X) - \mathbb{B}_{t_i}(X) \,(= X_i \Delta_i B)\,.$$

Then

$$\mathbb{E}^{\mathcal{F}_s} \int_s^t X_r dB_r = \sum_{i=0}^{n-1} \mathbb{E}^{\mathcal{F}_s} \left[X_i \mathbb{E}^{\mathcal{F}_{t_i}} \Delta_i B \right]$$
$$= 0.$$

Now

$$\mathbb{E}^{\mathcal{F}_s} \left[\int_s^t X_r dB_r \otimes \int_s^t Y_r dB_r \right]$$
$$= \mathbb{E}^{\mathcal{F}_s} \sum_{i=0}^{n-1} X_i \Delta_i B \,(\Delta_i B)^* \, Y_i^* + \mathbb{E}^{\mathcal{F}_s} \sum_{0 \le i < j < n} X_i \Delta_i B \left(\Delta_j B\right)^* Y_j^*$$
$$+ \mathbb{E}^{\mathcal{F}_s} \sum_{0 \le j < i < n} X_i \Delta_i B \left(\Delta_j B\right)^* Y_j^*$$

$$
\begin{aligned}
&= \sum_{i=0}^{n-1} \mathbb{E}^{\mathcal{F}_s} \left\{ X_i \mathbb{E}^{\mathcal{F}_{t_i}} \left[\Delta_i B \left(\Delta_i B \right)^* \right] Y_i^* \right\} \\
&\quad + \sum_{0 \le i < j < n} \mathbb{E}^{\mathcal{F}_s} \left\{ X_i \Delta_i B \left[\mathbb{E}^{\mathcal{F}_{t_j}} \left(\Delta_j B \right)^* \right] Y_j^* \right\} \\
&\quad + \sum_{0 \le j < i < n} \mathbb{E}^{\mathcal{F}_s} \left\{ X_i \left[\mathbb{E}^{\mathcal{F}_{t_i}} \Delta_i B \right] \left(\Delta_j B \right)^* Y_j^* \right\} \\
&= \sum_{i=0}^{n-1} \mathbb{E}^{\mathcal{F}_s} X_i Y_i^* \Delta_i t \\
&= \mathbb{E}^{\mathcal{F}_s} \int_s^t X_r Y_r^* dr.
\end{aligned}
$$

The equality *jjj*) follows by taking the **trace** in *jj*) with $Y = X$. In particular $N \in \mathcal{M}^p_{d\times d}$ and $M \in \mathcal{M}^p$, since

$$
\begin{aligned}
N_t = N_s &+ \int_s^t X_r dB_r \otimes \int_s^t Y_r dB_r - \int_s^t X_r Y_r^* dr \\
&+ \int_0^s X_r dB_r \otimes \int_s^t Y_r dB_r + \int_s^t X_r dB_r \otimes \int_0^s Y_r dB_r
\end{aligned}
$$

and

$$
M_t = \mathbf{Tr}\left(N_t\right).
$$

■

Corollary 2.3. *If* $X, Y \in \mathcal{E}_{d\times k}$, *then* $\mathbb{B}(X) \in \mathcal{M}^p_d$ *for all* $p \ge 1$,

$$
\ll\mathbb{B}(X)\gg_t \;=\; \int_0^t X_r X_r^* dr \quad \text{and} \quad <\mathbb{B}(X)>_t = \int_0^t |X_r|^2 dr.
$$

Hence by the Burkholder–Davis–Gundy Theorem 1.76 and by the inequality (B_4) from Proposition 1.72 we have:

Proposition 2.4. *Let* $p > 0$. *Then there exist two constants* $c_p > 0$, $C_p > 0$ *such that for all* $X, Y \in \mathcal{E}_{d\times k}$ *and* $T > 0$,

(i) Burkholder–Davis–Gundy's (BDG) inequality:

$$
c_p \mathbb{E} \left\| X \right\|^p_{L^2\left(0,T;\mathbb{R}^{d\times k}\right)} \le \mathbb{E} \sup_{t\in[0,T]} \left| \mathbb{B}_t(X) \right|^p \le C_p \mathbb{E} \left\| X \right\|^p_{L^2\left(0,T;\mathbb{R}^{d\times k}\right)}, \tag{2.2}
$$

(ii) continuity in probability $\mathbb{B} : \mathcal{E}_{d\times k} \subset \Lambda^0_{d\times k}(0,T) \longrightarrow S^0_d[0,T]$:

$$\begin{aligned}\mathbb{E}\left[1 \wedge \|\mathbb{B}(X) - \mathbb{B}(Y)\|_T\right] &\leq 3\left[\mathbb{E}\left(1 \wedge \|X - Y\|^2_{L^2(0,T;\mathbb{R}^{d\times k})}\right)\right]^{1/3} \\ &\leq 3\left[\mathbb{E}\left(1 \wedge \|X - Y\|_{L^2(0,T;\mathbb{R}^{d\times k})}\right)\right]^{1/3}.\end{aligned} \tag{2.3}$$

Let $0 < T < \infty$. Since $\mathcal{E}_{d\times k}(0,T)$ is a dense linear subspace of $\Lambda^p_{d\times k}(0,T)$ for each $p \in [0,\infty[$, the inequalities (2.2) and (2.3) allow us to extend by continuity the operator

$$\mathbb{B} : \mathcal{E}_{d\times k}(0,T) \subset \Lambda^p_{d\times k}(0,T) \to S^p_d[0,T]$$

to a linear continuous operator

$$\mathbb{B} : \Lambda^p_{d\times k}(0,T) \to S^p_d[0,T],$$

called *Itô's stochastic integral*, which still satisfies the inequalities (2.2) and (2.3).

If X is a deterministic process then the integral

$$\mathbb{B}_t(X) = \int_0^t X_r dB_r$$

is called a *Wiener integral.*

Remark 2.5. Let $0 \leq t_0 < t_1 < \ldots < t_n$, $X_i \in L^0\left(\Omega, \mathcal{F}_{t_i}, \mathbb{P}; \mathbb{R}^{d\times k}\right)$, $\forall i \in \overline{0, n-1}$ and

$$X_t(\omega) = \sum_{i=0}^{n-1} X_i(\omega)\mathbf{1}_{[t_i,t_{i+1}[}(t).$$

Define for $N \geq 1$, $X_t^{(N)} = \sum_{i=0}^{n-1} X_i \mathbf{1}_{[0,N]}(|X_i|)\mathbf{1}_{[t_i,t_{i+1}]}(t)$. Since $X^{(N)} \in \mathcal{E}_{d\times k}(0,T)$ and $X^{(N)} \to X$ in $\Lambda^0_{d\times k}(0,T)$ as $N \to \infty$, it follows that

$$\int_0^t X_s dB_s = \sum_{i=0}^{n-1} X_i\left(B_{t\wedge t_{i+1}} - B_{t\wedge t_i}\right).$$

By continuity of the expectation operator $\mathbb{E} : L^p\left(\Omega, \mathcal{F}, P; \mathbb{R}^d\right) \to \mathbb{R}^d$, $p \geq 1$, and of the conditional expectation operator $\mathbb{E}^{\mathcal{F}_s} : L^p\left(\Omega, \mathcal{F}, P; \mathbb{R}^d\right) \to L^p\left(\Omega, \mathcal{F}_s, P; \mathbb{R}^d\right)$, $p \geq 1$, we deduce easily, from Lemma 2.2 and Proposition 2.4, that the stochastic Itô integral introduced in this section has the following properties:

Theorem 2.6. *Consider the stochastic Itô integral*

$$\mathbb{B}_t(X) = \int_0^t X_r dB_r, \quad t \geq 0.$$

We have the following:

i_1) $\mathbb{B} : \Lambda^p_{d\times k}(0,T) \to S^p_d[0,T]$ *is a linear continuous operator, for all* $T > 0$, $p \geq 0$;

i_2) $\mathbb{B}(X) \in \mathcal{M}^p_d$ *for all* $X \in \Lambda^p_{d\times k}$ *and* $p \in [1,\infty[$;

i_3) $\mathbb{B} : \Lambda^2_{d\times k}(0,T) \to \mathcal{M}^2_d[0,T]$, $0 < T < \infty$, *is an isometry, that is a linear continuous operator such that*

$$\|\mathbb{B}(X)\|_{\mathcal{M}^2_d[0,T]} = \|X\|;$$

i_4) *for all* $X \in \Lambda^1_{d\times k}$, $Y \in \Lambda^1_{d\times k}$, $0 \leq s \leq t$ *such that* $\mathbb{E}\int_s^t \left|X_r Y_r^*\right| dr < +\infty$:

$$c) \quad \mathbb{E}^{\mathcal{F}_s} \int_s^t X_r dB_r = 0,$$

$$cc)\ \mathbb{E}^{\mathcal{F}_s}\left[\int_s^t X_r dB_r \otimes \int_s^t Y_r dB_r\right] = \mathbb{E}^{\mathcal{F}_s}\int_s^t X_r Y_r^* dr;$$

i_5) *for all* $X, Y \in \Lambda^2_{d\times k}$

$$M_\cdot = \left|\int_0^\cdot X_r dB_r\right|^2 - \int_0^\cdot |X_r|^2 dr \in \mathcal{M}^1_d,$$

$$N_\cdot = \int_0^\cdot X_r dB_r \otimes \int_0^\cdot Y_r dB_r - \int_0^\cdot X_r Y_r^* dr \in \mathcal{M}^1_{d\times d};$$

i_6) *for all* $X \subset \Lambda^0_{d\times k}$ *and* $\eta \in L^0\left(\Omega, \mathcal{F}_s, \mathbb{P}; \mathbb{R}^{m\times d}\right)$,

$$\int_s^t \eta X_r dB_r = \eta \int_s^t X_r dB_r, \quad a.s.;$$

i_6) *(Burkholder–Davis–Gundy inequality, or for short, BDG inequality) for every* $p > 0$, *there exist two constants* $c_p > 0$ *and* $C_p > 0$ *such that for all* $X \in \Lambda^p_{d\times k}(0,T)$:

$$\begin{aligned} c_p\mathbb{E}\Big(\int_0^T |X_r|^2 dr\Big)^{p/2} &\leq \mathbb{E}\sup_{t\in[0,T]}\left|\int_0^t X_r dB_r\right|^p \\ &\leq C_p\mathbb{E}\Big(\int_0^T |X_r|^2 dr\Big)^{p/2}; \end{aligned} \tag{2.4}$$

moreover

$$C_p = \begin{cases} \dfrac{4-p}{2-p} \le 3, & \text{if } 0 < p \le 1, \\ 4, & \text{if } p = 2, \\ \le (3p^3)^p & \text{if } p > 1; \end{cases}$$

i_7) *continuity in probability* ($\mathbb{B} : \Lambda^0 \longrightarrow S^0$)*: for all* $X, Y \in \Lambda^0_{d\times k}(0,T)$

$$\begin{aligned} \mathbb{E}\left[1 \wedge \|\mathbb{B}(X) - \mathbb{B}(Y)\|_T\right] &\le 3\left[\mathbb{E}\left(1 \wedge \int_0^T |X_r - Y_r|^2 dr\right)\right]^{1/3} \\ &\le 3\left[\mathbb{E}\left(1 \wedge \|X - Y\|_{L^2(0,T;\mathbb{R}^{d\times k})}\right)\right]^{1/3}. \end{aligned} \tag{2.5}$$

The monotonicity with respect to T in (2.4) and (2.5) shows that, for $X \in \Lambda^p_{d\times k}(0,\infty)$, we can extend the stochastic integral as follows.

$$\int_0^\infty X_t dB_t \overset{def}{=} L^p - \lim_{T\to\infty} \int_0^T X_t dB_t.$$

Replacing T by ∞ the properties (2.4) and (2.5) are conserved. Also if $\mathbb{E}\int_0^\infty |X_t|^2 dt < \infty$, then the isometry property holds:

$$\mathbb{E}\left|\int_0^\infty X_t dB_t\right|^2 = \mathbb{E}\int_0^\infty |X_t|^2 dt.$$

If $X \in \Lambda^p_{d\times k}(0,\infty)$ then for every $p > 0$ there exist two constants $\hat{c}_p, \hat{C}_p > 0$ such that

$$\hat{c}_p \mathbb{E}\left(\int_t^\infty |X_r|^2 dr\right)^{p/2} \le \mathbb{E}\sup_{s\ge t}\left|\int_s^\infty X_r dB_r\right|^p \le \hat{C}_p \mathbb{E}\left(\int_t^\infty |X_r|^2 dr\right)^{p/2} \tag{2.6}$$

(*backward Burkholder–Davis–Gundy inequality, or backward BDG inequality for short*).

Indeed

$$\begin{aligned} c_p \mathbb{E}\left(\int_t^\infty |X_r|^2 dr\right)^{p/2} &\le \mathbb{E}\sup_{s\ge t}\left|\int_t^s X_r dB_r\right|^p \\ &= \mathbb{E}\sup_{s\ge t}\left|\int_t^\infty X_r dB_r - \int_s^\infty X_r dB_r\right|^p \end{aligned}$$

$$\leq 2^p \mathbb{E} \sup_{s \geq t} \left| \int_s^\infty X_r dB_r \right|^p$$

$$\leq C_p \left[\mathbb{E} \left| \int_t^\infty X_r dB_r \right|^p + \mathbb{E} \sup_{s \geq t} \left| - \int_t^s X_r dB_r \right|^p \right]$$

$$\leq C_p' \mathbb{E} \left(\int_t^\infty |X_r|^2 dr \right)^{p/2}.$$

Remark 2.7. From i_4) we deduce that

$$\mathbb{E}^{\mathcal{F}_s} \left| \int_s^t X_r dB_r \right|^2 = \mathbb{E}^{\mathcal{F}_s} \int_s^t |X_r|^2 dr, \ \forall X \in \Lambda^2_{d \times k},$$

and for all $s, t \geq 0$, $X, Y \in \Lambda^1_{d \times k}$ such that $\mathbb{E} \int_0^t \left| X_r Y_r^* \right| dr < +\infty$:

$$a) \ \mathbb{E} \left[\int_0^s X_r dB_r \otimes \int_0^t Y_r dB_r \right] = \mathbb{E} \int_0^{s \wedge t} X_r Y_r^* dr,$$

$$b) \ \mathbb{E} \langle \int_0^s X_r dB_r, \int_0^t Y_r dB_r \rangle = \mathbb{E} \int_0^{s \wedge t} \mathbf{Tr} \left(X_r Y_r^* \right) dr.$$

The following proposition allows us to extend the stochastic Itô integral to random intervals.

Lemma 2.8. *Let $X \in \Lambda^p_{d \times k}$, $p \geq 0$, and τ be a stopping time. Then $\mathbf{1}_{[0,\tau]} X \in \Lambda^p_{d \times k}$ and a.s.*

$$\int_0^{t \wedge \tau} X_s \, dB_s = \int_0^t \mathbf{1}_{[0,\tau]}(s) X_s \, dB_s, \quad t \geq 0. \tag{2.7}$$

Proof. The first part of the statement is obvious since $\mathbf{1}_{[0,\tau]}$ is a $\mathcal{P}$-measurable process.

Since for all $p \geq 0$

$$X \longrightarrow \mathbb{B}_\cdot \left(\mathbf{1}_{[0,\tau]} X \right) : \Lambda^p_{d \times k} \to S^p_d \, [0, T]$$

is continuous and $\Lambda^q_{d \times k} \subset \Lambda^2_{d \times k} \subset \Lambda^r_{d \times k}$ (with density) for all $0 \leq r < 2 < q$, it is clear that it suffices to prove (2.7) in the case $p = 2$.

Note that by the Doob optional stopping Theorem 1.62, for all $Y \in \Lambda^2_{d \times k}$:

$$\begin{aligned}
&\{\mathbb{B}_{t \wedge \tau} (Y)\}_{t \geq 0} \in \mathcal{M}^2_d, \\
&\mathbb{E} \left\langle \mathbb{B}_{t \wedge \tau} (Y), \mathbb{B}_t (Y) - \mathbb{B}_{t \wedge \tau} (Y) \right\rangle = 0, \quad \text{and} \\
&\{|\mathbb{B}_{t \wedge \tau} (Y)|^2 - \textstyle\int_0^{t \wedge \tau} |Y_r|^2 dr\}_{t \geq 0} \in \mathcal{M}^1_d.
\end{aligned}$$

Next we have that $\mathbb{B}_{t\wedge\tau}(X) = \mathbb{B}_{t\wedge\tau}\left(\mathbf{1}_{[0,\tau]}X\right)$. Indeed

$$\begin{aligned}\mathbb{E}\left|\mathbb{B}_{t\wedge\tau}(X) - \mathbb{B}_{t\wedge\tau}\left(\mathbf{1}_{[0,\tau]}X\right)\right|^2 &= \mathbb{E}\left|\mathbb{B}_{t\wedge\tau}\left(\left(1-\mathbf{1}_{[0,\tau]}\right)X\right)\right|^2 \\ &= \mathbb{E}\int_0^{t\wedge\tau}\left|\left(1-\mathbf{1}_{[0,\tau]}(r)\right)X_r\right|^2 dr \\ &= 0.\end{aligned}$$

Second $\mathbb{B}_{t\wedge\tau}\left(\mathbf{1}_{[0,\tau]}X\right) = \mathbb{B}_t\left(\mathbf{1}_{[0,\tau]}X\right)$, since

$$\begin{aligned}&\mathbb{E}\left|\mathbb{B}_t\left(\mathbf{1}_{[0,\tau]}X\right) - \mathbb{B}_{t\wedge\tau}\left(\mathbf{1}_{[0,\tau]}X\right)\right|^2 \\ &\quad= \mathbb{E}\left|\mathbb{B}_t\left(\mathbf{1}_{[0,\tau]}X\right)\right|^2 - \mathbb{E}\left|\mathbb{B}_{t\wedge\tau}\left(\mathbf{1}_{[0,\tau]}X\right)\right|^2 \\ &\quad= \mathbb{E}\int_0^t \mathbf{1}_{[0,\tau]}(r)\left|X_r\right|^2 dr - \mathbb{E}\int_0^{t\wedge\tau}\mathbf{1}_{[0,\tau]}(r)\left|X_r\right|^2 dr \\ &\quad= 0.\end{aligned}$$

■

From (2.4) we infer:

Corollary 2.9 (Burkholder–Davis–Gundy Inequality). *For every $p > 0$, there exist two constants $c_p > 0$ and $C_p > 0$, $(C_1 = 3,\ C_2 = 4)$, such that for all $X \in \Lambda^p_{d\times k}$ and for all bounded stopping times $0 \le \tau \le \sigma$ a.s.:*

$$\begin{aligned}c_p\mathbb{E}^{\mathcal{F}_\tau}\Big(\int_\tau^\sigma |X_r|^2\,dr\Big)^{p/2} &\le \mathbb{E}^{\mathcal{F}_\tau}\sup_{t\in[\tau,\sigma]}\left|\int_\tau^t X_r dB_r\right|^p \\ &\le C_p\mathbb{E}^{\mathcal{F}_\tau}\Big(\int_\tau^\sigma |X_r|^2\,dr\Big)^{p/2},\quad a.s.\end{aligned}\tag{2.8}$$

Proof. We replace in (2.4) X_r by $X_r\mathbf{1}_A\mathbf{1}_{\tau\le r\le\sigma}$, where $A \in \mathcal{F}_\tau$ and (2.8) follows in view of the definition of the conditional expectation. ■

We can now establish the following:

Proposition 2.10. *Let $0 < T < \infty$, $p \ge 0$ and $X \in \Lambda^0_{d\times k}$. Then*

(*i*) *$\mathbb{B}(X)$ is a continuous local martingale with*

$$\ll\!\mathbb{B}(X)\!\gg_t \;=\; \int_0^t X_r X_r^*\,dr \quad and \quad \langle\mathbb{B}(X)\rangle_t = \int_0^t |X_r|^2\,dr;$$

(*ii*) *$\mathbb{B}(X) \in S^p_d[0,T]$ if and only if $X \in \Lambda^p_{d\times k}(0,T)$;*
(*iii*) *$\mathbb{B} : \Lambda^p_{d\times k}(0,T) \to S^p_d[0,T]$ is a linear continuous injective operator.*

Proof. (*i*): For each $n \in \mathbb{N}^*$, define the stopping time

$$\tau_n(\omega) = \inf\left\{t \ge 0 : \int_0^t |X_r(\omega)|^2\,dr \ge n\right\}.$$

Then $\mathbb{B}_{t\wedge\tau_n}(X) = \mathbb{B}_t\left(\mathbf{1}_{[0,\tau_n]}X\right)$,

$$\begin{aligned}\mathbb{B}_{t\wedge\tau_n}(X) \otimes \mathbb{B}_{t\wedge\tau_n}(X) - \int_0^{t\wedge\tau_n} X_r X_r^* \, dr \\ = \mathbb{B}_t\left(\mathbf{1}_{[0,\tau_n]}X\right) \otimes \mathbb{B}_t\left(\mathbf{1}_{[0,\tau_n]}X\right) - \int_0^t \mathbf{1}_{[0,\tau_n]} X_r X_r^* \, dr\end{aligned}$$

and

$$\left|\mathbb{B}_{t\wedge\tau_n}(X)\right|^2 - \int_0^{t\wedge\tau_n} |X_r|^2 \, dr = \left|\mathbb{B}_t\left(\mathbf{1}_{[0,\tau_n]}X\right)\right|^2 - \int_0^t \left|\mathbf{1}_{[0,\tau_n]}X_r\right|^2 dr$$

define continuous martingales, since for all $T > 0$, $\mathbf{1}_{[0,\tau_n]}X \in \Lambda^2_{d\times k}(0,T)$. Indeed $\mathbf{1}_{[0,\tau_n]}X$ is a progressively measurable process and

$$\mathbb{E}\int_0^T \left|\mathbf{1}_{[0,\tau_n]}X_r\right|^2 dr = \mathbb{E}\int_0^{\tau_n\wedge T} |X_r|^2 \, dr \leq n < +\infty.$$

(ii): By the definition of the Itô integral, if $X \in \Lambda^p_{d\times k}(0,T)$, then $\mathbb{B}(X) \in S^p_d[0,T]$. Conversely if $\mathbb{B}(X) \in S^p_d[0,T]$, then from the Burkholder–Davis–Gundy inequality (2.4) for $X^n = \mathbf{1}_{[0,\tau_n]}X$ with τ_n as above we have

$$\begin{aligned}c_p\mathbb{E}\left(\int_0^{T\wedge\tau_n} |X_r|^2 \, dr\right)^{p/2} &\leq \mathbb{E}\sup_{t\in[0,T]} \left|\mathbb{B}_t(X^n)\right|^p \\ &= \mathbb{E}\sup_{t\in[0,T]} \left|\mathbb{B}_{t\wedge\tau_n}(X)\right|^p \\ &\leq \mathbb{E}\sup_{t\in[0,T]} \left|\mathbb{B}_t(X)\right|^p.\end{aligned} \tag{2.9}$$

$X \in \Lambda^p_{d\times k}(0,T)$ follows by taking the limit as $n \to \infty$.

(iii): The Burkholder–Davis–Gundy inequality shows that for all $p > 0$:

$$\mathbb{B} : \Lambda^p_{d\times k}(0,T) \to S^p_d[0,T]$$

is a linear continuous injective operator.

In the case $p = 0$, from the definition of the stochastic Itô integral, $\mathbb{B} : \Lambda^0_{d\times k}(0,T) \to S^0_d[0,T]$ is a linear continuous operator. Moreover $\mathbb{B}$ is an injective operator since if $X \in \Lambda^0_{d\times k}(0,T)$ and $\mathbb{B}_\cdot(X) = 0$, then by (2.9)

$$\mathbb{E}\left(\int_0^{T\wedge\tau_n} |X_r|^2 \, dr\right)^{p/2} = 0$$

for all $n \in \mathbb{N}^*$; hence $X = 0$. ■

Remark 2.11. When $\mathcal{F}_t = \mathcal{F}_t^B$ and $p > 1$ we shall show below (see the martingale representation Theorem 2.42) that

$$\mathbb{B} : \Lambda^p_{d\times k}(0,T) \to \mathcal{M}^p_d[0,T]$$

is also a surjective operator.

Remark 2.12. By linearity the stochastic Itô integral can be extended to $\mathbb{C}$-valued stochastic processes of the form $X_t = U_t + iV_t$, where $U, V \in \Lambda^p_{d\times k}$, $p \geq 0$.

Lemma 2.13. *Let $X \in S^0_l[0,T]$ and $G : \Omega \times [0,T] \times \mathbb{R}^l \to \mathbb{R}^{d\times k}$ be such that*

$$G(\cdot,\cdot,x) \text{ is } \mathcal{P}\text{-measurable } \forall\, x \in \mathbb{R}^l, \text{ and}$$
$$x \mapsto G(\omega,t,x) \text{ is continuous } d\mathbb{P}\otimes dt\text{-a.e.}$$

Suppose we are given a sequence of partitions $\Delta_n : 0 = t_0^n < t_1^n < \ldots < t_{k_n}^n = T$ with $\delta_n = \max\left\{t_{i+1}^n - t_i^n : i \in \overline{0, n-1}\right\}$ and $\delta_n \to 0$ as $n \to \infty$. If (here we write t_i for t_i^n)

$$\Gamma_t^n = \sum_{i=1}^{k_n-1} \frac{1}{t_i - t_{i-1}} \left[\int_{t_{i-1}}^{t_i} G(s, X_{t_i})\, ds\right] \left(B_{t\wedge t_{i+1}} - B_{t\wedge t_i}\right),$$

then

$$\Gamma_\cdot^n \xrightarrow[n\to\infty]{S_d^0[0,T]} \int_0^\cdot G(s, X_s)\, dB_s.$$

Proof. Let

$$\begin{aligned}
G_t^n(\omega) &= \sum_{i=1}^{k_n-1} \frac{1}{t_i - t_{i-1}} \left[\int_{t_{i-1}}^{t_i} G(s, Y_{t_i})\, ds\right] \mathbf{1}_{[t_i, t_{i+1}[}(t) \\
&= \sum_{i=1}^{k_n-1} \left[\int_0^1 G(t_{i-1} + (t_i - t_{i-1})r, Y_{t_i})\, dr\right] \mathbf{1}_{[t_i, t_{i+1}[}(t).
\end{aligned}$$

Since

$$G^n \xrightarrow[n\to\infty]{\Lambda^0_{d\times k}(0,T)} G(\cdot, Y_\cdot)$$

and

$$\Gamma_t^n = \mathbb{B}_t(G^n) = \int_0^t G_s^n dB_s$$

the result follows. ■

If $X \in S^0_{d\times k}$ and B is a $\mathbb{R}^k$-Brownian motion then the stochastic process $\{(X_t, B_t) : t \geq 0\}$ can be seen as a random variable with values in $C\left(\mathbb{R}_+, \mathbb{R}^{d\times k}\right) \times C\left(\mathbb{R}_+, \mathbb{R}^k\right)$. The law of this random variable will be denoted $\mathcal{L}(X, B)$. From the above Lemma 2.13 we easily deduce:

Corollary 2.14. *Let $X, \hat{X} \in S^0_d[0,T]$, $B, \hat{B}$ be two $\mathbb{R}^k$-Brownian motions and $g : \mathbb{R}_+ \times \mathbb{R}^d \to \mathbb{R}^{d\times k}$ be a function satisfying*

$$\begin{aligned}&g(\cdot, y) \text{ is measurable } \forall\, y \in \mathbb{R}^d, \text{ and}\\&y \mapsto g(t,y) \text{ is continuous } dt\text{-a.e.}\end{aligned}$$

If

$$\mathcal{L}(X, B) = \mathcal{L}(\hat{X}, \hat{B}) \text{ on } C\left(\mathbb{R}_+, \mathbb{R}^{d+k}\right),$$

then on $C\left(\mathbb{R}_+, \mathbb{R}^{d+k+d}\right)$,

$$\mathcal{L}\left(X, B, \int_0^{\cdot} g(s, X_s)\, dB_s\right) = \mathcal{L}\left(\hat{X}, \hat{B}, \int_0^{\cdot} g\left(s, \hat{X}_s\right) d\hat{B}_s\right).$$

Finally we present a continuity property of the mapping

$$(X, B) \longrightarrow \int_0^T X_s dB_s.$$

Given two stochastic processes $B, X : \Omega \times [0,\infty[\longrightarrow \mathbb{R}^k, \mathbb{R}^{d\times k}$, let $\mathcal{F}_t^{B,X}$ be the natural filtration generated jointly by B and X and $\mathcal{F}_{t+}^{B,X}$ its right continuous version. If $X \in L^0\left(\Omega; L^1(0,T;\mathbb{R}^d)\right)$, we shall also consider the filtration

$$\hat{\mathcal{F}}_t^{B,X} = \sigma\{B_s, \frac{1}{\varepsilon}\int_{0\vee(s-\varepsilon)}^{s} X_r dr;\ \varepsilon > 0,\ s \in [0,t]\} \vee \mathcal{N}_{\mathbb{P}}, \quad t \geq 0,$$

and $\hat{\mathcal{F}}_{t+}^{B,X}$ its right continuous version. Clearly $\hat{\mathcal{F}}_{t+}^{B,X} \subset \mathcal{F}_{t+}^{B,X}$, and if X is a continuous stochastic process then $\hat{\mathcal{F}}_{t+}^{B,X} = \mathcal{F}_{t+}^{B,X}$.

Proposition 2.15. *Let $B, B^n, \tilde{B}^n : \Omega \times [0,\infty[\to \mathbb{R}^k$ and $X, X^n, \tilde{X}^n : \Omega \times [0,\infty[\to \mathbb{R}^{d\times k}$, be stochastic processes such that:*

(i) $\tilde{B}^n$ *is an* $\hat{\mathcal{F}}_t^{\tilde{B}^n,\tilde{X}^n}$*-Brownian motion* $\forall\, n \geq 1$;
(ii) $\mathcal{L}(\tilde{B}^n, \tilde{X}^n) = \mathcal{L}(B^n, X^n)$ *on* $C(\mathbb{R}_+, \mathbb{R}^k) \times L^2_{loc}\left(\mathbb{R}_+; \mathbb{R}^{d\times k}\right)$, *for all* $n \geq 1$;
(iii) $\int_0^T |X_s^n - X_s|^2 ds + \sup_{t\in[0,T]} |B_t^n - B_t| \to 0$, *in probability, as* $n \to \infty$, *for all* $T > 0$.

Then $\left(B^n, \{\hat{\mathcal{F}}_t^{B^n,X^n}\}\right), n \geq 1$, *and* $\left(B, \{\hat{\mathcal{F}}_t^{B,X}\}\right)$ *are Brownian motions and as* $n \to \infty$

$$\sup_{t\in[0,T]} \left| \int_0^t X_s^n dB_s^n \longrightarrow \int_0^t X_s dB_s \right| \longrightarrow 0 \quad \textit{in probability.} \tag{2.10}$$

Moreover if in addition to i), ii) and iii), for some $p > 1$,

(*iv*) $\{X^n : n \geq 1\}$ *is bounded in* $L^p(\Omega; L^2(0,T;\mathbb{R}^{d\times k}))$,

then for all $q \in [1, p[$

$$\mathbb{E}\left[\sup_{t\in[0,T]} \left| \int_0^t X_s^n dB_s^n \longrightarrow \int_0^t X_s dB_s \right|^q \right] \longrightarrow 0, \quad \textit{as } n \to \infty. \tag{2.11}$$

Proof. The ideas of the proof are borrowed from Bensoussan [7] and Buckdahn and Răşcanu [16].

By Corollary 1.96 we deduce that for each $n \in \mathbb{N}^*$, $\left(B^n, \{\hat{\mathcal{F}}_t^{B^n,X^n}\}\right)$ and $\left(B, \{\hat{\mathcal{F}}_t^{B,X}\}\right)$ are Brownian motions.

We remark that

$$\mathbb{E}\left\{ \sup_{t\in[0,T]} |B_t^n|^{2a} + \sup_{t\in[0,T]} |B_t|^{2a} \right\} \leq C_a T^a, \quad \text{for all } a \geq 1,$$

and thus, by *iv*)

$$\lim_{n\to\infty} \mathbb{E} \sup_{t\in[0,T]} |B_t^n - B_t|^a = 0, \quad \text{for all } a \geq 1. \tag{2.12}$$

Given any $Y \in L^0\left(\Omega; L^2(0,T;\mathbb{R}^{d\times k})\right)$ progressively measurable with respect to the filtration $\left\{\hat{\mathcal{F}}_t^{B^n,X^n}\right\}_{t\in[0,T]}$ $\left(\left\{\hat{\mathcal{F}}_t^{B,X}\right\}_{t\in[0,T]}\right.$, respectively), we put

$$\mathbb{B}_t^n(Y) = \int_0^t Y_s dB_s^n \quad \text{and} \quad \mathbb{B}_t(Y) = \int_0^t Y_s dB_s, \text{ respectively,}$$

and

$$\begin{aligned} Y_t^\varepsilon &= \frac{1}{\varepsilon}\int_0^t Y_s \exp\{-\frac{t-s}{\varepsilon}\}ds \\ &= e^{-t/\varepsilon}\int_0^t Y_s\, d\left(e^{s/\varepsilon}\right), \quad t \geq 0,\ \varepsilon > 0. \end{aligned}$$

Then

$$\int_0^T |Y_t^\varepsilon|^2\,dt \le \int_0^T |Y_s|^2\,ds, \quad \forall\,\varepsilon > 0 \quad \text{and}$$

$$Y^\varepsilon \longrightarrow Y, \quad \text{in } L^0(\Omega; L^2(0,T;\mathbb{R}^{d\times k})), \quad \text{as } \varepsilon \to 0.$$

Moreover, if $Y \in L^r(\Omega; L^2(0,T;\mathbb{R}^{d\times k})),\ r \ge 1$, then

$$Y^\varepsilon \longrightarrow Y, \quad \text{in } L^r(\Omega; L^2(0,T;\mathbb{R}^{d\times k})), \quad \text{as } \varepsilon \to 0.$$

Since

$$\begin{aligned}
&\mathbb{B}_t^n(X^{n,\varepsilon}) - \mathbb{B}_t(X^\varepsilon)\\
&= \int_0^t X_s^{n,\varepsilon} dB_s^n - \int_0^t X_s^\varepsilon dB_s\\
&= \frac{1}{\varepsilon}\left[\int_0^t (X_s^n - X_s)\exp\{-\frac{t-s}{\varepsilon}\}ds\right] B_t^n\\
&- \frac{1}{\varepsilon}\int_0^t \left(X_s^n - X_s - \frac{1}{\varepsilon}\int_0^s (X_\tau^n - X_\tau)\exp\{-\frac{s-\tau}{\varepsilon}\}d\tau\right) B_s^n ds\\
&+ \frac{1}{\varepsilon}\left(\int_0^t X_s \exp\{-\frac{t-s}{\varepsilon}\}ds\right)(B_t^n - B_t)\\
&- \frac{1}{\varepsilon}\int_0^t \left(X_s - \frac{1}{\varepsilon}\int_0^s X_\tau \exp\{-\frac{s-\tau}{\varepsilon}\}d\tau\right)(B_s^n - B_s)\,ds,
\end{aligned}$$

and thanks to the assumptions $i) - v)$ and (2.12) we can deduce that

$$\sup_{t\in[0,T]} |\mathbb{B}_t^n(X^{n,\varepsilon}) - \mathbb{B}_t(X^\varepsilon)| \longrightarrow 0 \quad \text{in probability, as } n \to \infty, \text{ for every } \varepsilon > 0.$$

Now we have for all $n \ge 1$ and for every $\varepsilon > 0$:

$$\begin{aligned}
&\mathbb{E}\,[1 \wedge \|\mathbb{B}^n(X^n) - \mathbb{B}(X)\|_T]\\
&\le \mathbb{E}\,[1 \wedge \|\mathbb{B}^n(X^n - X^{n,\varepsilon})\|_T] + \mathbb{E}\,[1 \wedge \|\mathbb{B}^n(X^{n,\varepsilon}) - \mathbb{B}(X^\varepsilon)\|_T]\\
&+ \mathbb{E}\,[1 \wedge \|\mathbb{B}(X^\varepsilon - X)\|_T]\\
&\le 3\left[\mathbb{E}\left(1 \wedge \int_0^T |X_r^n - X^{n,\varepsilon}|^2\,dr\right)\right]^{1/3} + \mathbb{E}\,[1 \wedge \|\mathbb{B}^n(X^{n,\varepsilon}) - \mathbb{B}(X^\varepsilon)\|_T]\\
&+ 3\left[\mathbb{E}\left(1 \wedge \int_0^T |X_r^\varepsilon - X|^2\,dr\right)\right]^{1/3}.
\end{aligned}$$

Hence for all $\varepsilon > 0$,

$$\limsup_{n\to+\infty} \mathbb{E}\left[1 \wedge \|\mathbb{B}^n(X^n) - \mathbb{B}(X)\|_T\right] \le 6\left[\mathbb{E}\left(1 \wedge \int_0^T |X_r^\varepsilon - X|^2 dr\right)\right]^{1/3}$$

and consequently

$$\lim_{n\to\infty} \mathbb{E}\left[1 \wedge \|\mathbb{B}^n(X^n) - \mathbb{B}(X)\|_T\right] = 0$$

that is (2.10).

Let $1 \le q < p$. Then

$$\begin{aligned}
&\mathbb{E}\left\{\sup_{t\in[0,T]} |\mathbb{B}_t^n(X^n) - \mathbb{B}_t(X)|^q\right\} \\
&\le 3^{q-1}\mathbb{E}\left\{\sup_{t\in[0,T]} |\mathbb{B}_t^n(X^n - X^{n,\varepsilon})|^q\right\} + 3^{q-1}\mathbb{E}\left\{\sup_{t\in[0,T]} |\mathbb{B}_t^n(X^{n,\varepsilon}) - \mathbb{B}_t(X^\varepsilon)|^q\right\} \\
&+ 3^{q-1}\mathbb{E}\left\{\sup_{t\in[0,T]} |\mathbb{B}_t(X^\varepsilon - X)|^q\right\} \\
&\le C_q\mathbb{E}\left\{\left(\int_0^T |X^n - X^{n,\varepsilon}|^2 ds\right)^{\frac{q}{2}}\right\} + 3^{q-1}\mathbb{E}\left\{\sup_{t\in[0,T]} |\mathbb{B}_t^n(X^{n,\varepsilon}) - \mathbb{B}_t(X^\varepsilon)|^q\right\} \\
&+ C_q\mathbb{E}\left\{\left(\int_0^T |X^\varepsilon - X|^2 ds\right)^{\frac{q}{2}}\right\}.
\end{aligned}$$

Hence, in virtue of Lebesgue's theorem 1.15

$$\limsup_{n\to+\infty} \mathbb{E}\left\{\sup_{t\in[0,T]} |\mathbb{B}_t^n(X^n) - \mathbb{B}_t(X)|^q\right\} \le 2C_q\mathbb{E}\left\{\left(\int_0^T |X^\varepsilon - X|^2 ds\right)^{\frac{q}{2}}\right\}$$

for all $\varepsilon > 0$, and (2.11) follows. ■

2.3 Itô's Formula

Let $X \in S_d^0$ be of the form

$$X_t = X_0 + \int_0^t F_s ds + \int_0^t G_s dB_s, \quad \forall t \ge 0, \ a.s., \tag{2.13}$$

where $\{B_t, t \geq 0\}$ is a k-dimensional Brownian motion with respect to a fixed stochastic basis $(\Omega, \mathcal{F}, \mathbb{P}, \{\mathcal{F}_t\}_{t\geq 0})$ and

$$F : \Omega \times [0, \infty[\to \mathbb{R}^d, \qquad G : \Omega \times [0, \infty[\to \mathbb{R}^{d\times k}$$

are progressively measurable stochastic processes such that

$$F \in L^1_{loc}\left(\mathbb{R}_+; \mathbb{R}^d\right) \text{ a.s., and } \quad G \in \Lambda^0_{d\times k}.$$

Definition 2.16. A stochastic process $X \in S^0_d$ of the form (2.13) will be called an Itô process. F is the drift, and GG^* the matrix of diffusion coefficients of X.

Formally we shall write

$$dX_t = F_t dt + G_t dB_t.$$

If $\psi : \mathbb{R}^d \to \mathbb{R}$ is a function of class C^2 we denote by

$$\nabla_x \psi(x) = \psi'_x(x) = \left(\frac{\partial \psi(x)}{\partial x_i}\right)_{d\times 1} \in \mathbb{R}^d$$

the gradient of ψ with respect to x, and

$$D^2_{xx}\psi(x) = \psi''_{xx}(x) = \left(\frac{\partial^2 \psi(x)}{\partial x_i \partial x_j}\right)_{d\times d} \in \mathbb{R}^{d\times d}$$

the Hessian matrix of ψ with respect to x.

Theorem 2.17 (Itô's Formula). *Let $\varphi \in C^{1,2}\left(\mathbb{R}_+ \times \mathbb{R}^d\right)$ and*

$$\mathcal{A}\varphi(t,x) \stackrel{def}{=} \langle F_t, \varphi'_x(t,x)\rangle + \frac{1}{2}\mathbf{Tr}\left[G_t G_t^* \varphi''_{xx}(t,x)\right].$$

Then for all $t \geq 0$:

$$\begin{aligned}\varphi(t, X_t) = \varphi(0, X_0) &+ \int_0^t \left[\frac{\partial \varphi}{\partial t}(r, X_r) + \mathcal{A}\varphi(r, X_r)\right] dr \\ &+ \int_0^t \langle \varphi'_x(r, X_r), G_r dB_r\rangle, \ \mathbb{P}\text{-a.s.}\end{aligned} \tag{2.14}$$

From (2.14) with $\varphi(t,x) = |x|^2$ and the identity

$$\langle x, y\rangle = \frac{1}{2}|x+y|^2 - \frac{1}{2}|x|^2 - \frac{1}{2}|y|^2, \quad \forall\, x, y \in \mathbb{R}^d,$$

we deduce that:

Corollary 2.18. *If* $X, Y \in S_d^0$ *are Itô processes of the form*

$$X_t = X_0 + \int_0^t F_s ds + \int_0^t G_s dB_s, \quad t \geq 0, \quad \text{and}$$
$$Y_t = Y_0 + \int_0^t E_s ds + \int_0^t H_s dB_s, \quad t \geq 0,$$

then for all $t \geq 0$*:*

$$|X_t|^2 = |X_0|^2 + \int_0^t \left(2\langle X_r, F_r\rangle + |G_r|^2\right) dr + 2\int_0^t \langle X_r, G_r dB_r\rangle, \ \mathbb{P}\text{-a.s.} \tag{2.15}$$

and

$$\begin{aligned}\langle X_t, Y_t\rangle = \langle X_0, Y_0\rangle &+ \int_0^t \left[\langle F_s, Y_s\rangle + \langle X_s, E_s\rangle + \mathbf{Tr}\left(G_s H_s^*\right)\right] ds \\ &+ \int_0^t \left(Y_s^* G_s + X_s^* H_s\right) dB_s.\end{aligned} \tag{2.16}$$

The Itô formula (2.14) is a particular case of the following slightly more general result.

Proposition 2.19. *Let* $G \in \Lambda_{d\times k}^0$*,*

$$M_t = \int_0^t G_s dB_s$$

and $V \in S_m^0$ *be such that*

$$V_\cdot(\omega) \in BV_{loc}(\mathbb{R}_+; \mathbb{R}^m) \ \ \mathbb{P}\text{-a.s. } \omega \in \Omega.$$

If $(v, x) \longrightarrow \varphi(v, x)$ *is a function from* $C^{1,2}\left(\mathbb{R}^m \times \mathbb{R}^d\right)$*, then* $\forall\, t \geq 0,\ \mathbb{P}$*-a.s.:*

$$\begin{aligned}\varphi(V_t, M_t) = \varphi(V_0, 0) &+ \int_0^t \left[\langle \varphi_v'(V_s, M_s), dV_s\rangle + \langle \varphi_x'(V_s, M_s), G_s dB_s\rangle\right] \\ &+ \frac{1}{2}\int_0^t \mathbf{Tr}\left[G_s G_s^* \varphi_{xx}''(V_s, M_s)\right] ds.\end{aligned} \tag{2.17}$$

Proof. We mimic the proof of Lemma 1.73, with some minor changes. If

$$M_s = \int_0^s G_r dB_r, \quad s \geq 0,$$

then

$$\ll M \gg_s = \int_0^s G_r G_r^* dr \quad \text{and} \quad <M>_s = \int_0^s |G_r|^2 dr.$$

Let $0 \le t \le T$ and $R > 0$ be arbitrary.

Define

$$U_r = (M_r, V_r, \updownarrow V \updownarrow_r, <M>_r)$$

and the stopping time

$$\tau = \tau_R = \inf\{r \ge 0 : |U_r| \ge R\}.$$

Let

$$\Phi(u, v, x) = \left(u, \varphi(v, x), \varphi'_v(v, x), \varphi'_x(v, x), \varphi''_{xx}(v, x)\right),$$

and

$$\mathbf{m}_{n,R} = \sup \Big\{ |\Phi(u_1, v_1, x_1) - \Phi(u_2, v_2, x_2)| : |u_1|, |u_2|, |v_1|, |v_2| \le R$$
$$|x_1|, |x_2| \le R,\ |u_1 - u_2| + |v_1 - v_2| + |x_1 - x_2| \le \frac{1}{n} \Big\}.$$

Clearly $\mathbf{m}_{n,R} \to 0$ as $n \to \infty$ for each fixed $R > 0$, and

$$0 \le \mathbf{m}_{n,R} \le 2C_R \overset{def}{=} 2 \sup\{|\Phi(u, v, x)| : |u|, |v|, |x| \le R\}.$$

Let $\left\{\left(\theta_i^n, k_n\right) : i, n \in \mathbb{N}\right\}$ be a *basic partition* (see Definition 1.54) of the interval $[0, t \wedge \tau]$, associated to $\{(U_t, 1/n) : t \ge 0, \in \mathbb{N}^*\}$.

Denote $\theta_i = \theta_i^n$, $\Delta_i \theta = \theta_{i+1} - \theta_i$, $Y_i = Y_{\theta_i^n}$, $\Delta_i Y = Y_{i+1} - Y_i$ for any arbitrary stochastic process Y.

We have

$$\varphi(V_{t\wedge\tau}, M_{t\wedge\tau}) - \varphi(V_0, 0) = \sum_{i=0}^{k_n - 1} [\varphi(V_{i+1}, M_{i+1}) - \varphi(V_i, M_i)]$$
$$+ \left[\varphi(V_{t\wedge\tau}, M_{t\wedge\tau}) - \varphi\left(V_{\theta_{k_n}^n}, M_{\theta_{k_n}^n}\right)\right].$$

Since $\langle Hx, y\rangle = \mathbf{Tr}[H \times (x \otimes y)]$, we deduce by Taylor's formula that there exist some $\sigma_i, \tau_i \in \left[\theta_i^n, \theta_{i+1}^n\right]$ such that

$$\varphi(V_{t\wedge\tau}, M_{t\wedge\tau}) - \varphi(V_0, 0) = \sum_{i=0}^{k_n-1} \varphi'_v(V_i, M_i)\,\Delta_i V + \sum_{i=0}^{k_n-1} \langle \varphi'_x(V_i, M_i), \Delta_i M\rangle$$
$$+ \frac{1}{2}\sum_{i=0}^{k_n-1} \mathbf{Tr}\left[\varphi''_{xx}(V_i, X_i)\,\Delta_i \ll M \gg\right] + R_n, \tag{2.18}$$

where $R_n = R_n^{(1)} + R_n^{(2)} + R_n^{(3)} + R_n^{(4)}$ with

$$R_n^{(1)} = \sum_{i=0}^{k_n-1} \left[\varphi'_v(V_{\sigma_i}, M_{i+1}) - \varphi'_v(V_i, M_i)\right]\Delta_i V,$$

$$R_n^{(2)} = \frac{1}{2}\sum_{i=0}^{k_n-1} \left\langle \left[\varphi''_{xx}(V_i, M_{\tau_i}) - \varphi''_{xx}(V_i, M_i)\right]\Delta_i M, \Delta_i M\right\rangle,$$

$$R_n^{(3)} = \frac{1}{2}\sum_{i=0}^{k_n-1} \mathbf{Tr}\left[\varphi''_{xx}(V_i, M_i)\left(\Delta_i M \otimes \Delta_i M - \Delta_i \ll M \gg\right)\right],$$

$$R_n^{(4)} = \varphi(V_{t\wedge\tau}, M_{t\wedge\tau}) - \varphi\left(V_{\theta^n_{k_n}}, M_{\theta^n_{k_n}}\right).$$

We have

$$\mathbb{E}\,|R_n| \le C'_R \times \left(\frac{1}{\sqrt{n}} + \mathbf{m}_n\right) \to 0 \ \text{ as } n \to \infty,$$

since

$$\left|R_n^{(1)}\right| \le \mathbf{m}_n \times R,$$

$$\mathbb{E}\left|R_n^{(2)}\right| \le \frac{1}{2}\mathbf{m}_n \times \sum_{i=0}^{k_n-1} \mathbb{E}\,|\Delta_i M|^2 \le \frac{1}{2}\mathbf{m}_n \times R^2,$$

$$\mathbb{E}\left|R_n^{(4)}\right| \le 2C_R \times \mathbb{P}\left(\theta^n_{k_n} < t \wedge \tau\right) \le 2C_R \times \frac{1}{n},$$

and by mutually orthogonality in $L^2\left(\Omega, \mathcal{F}, \mathbb{P}; \mathbb{R}^d\right)$

$$\mathbb{E}\left|R_n^{(3)}\right|^2 = \sum_{i=0}^{k_n-1} \mathbb{E}\left|\frac{1}{2}\mathbf{Tr}\left[\varphi''_{xx}(V_i, X_i)\left(\Delta_i M \otimes \Delta_i M - \Delta_i \ll M \gg\right)\right]\right|^2$$
$$\le \left(\frac{C_R}{2}\right)^2 2\sum_{i=0}^{k_n-1} \mathbb{E}\left[|\Delta_i M|^4 + |\Delta_i < M >|^2\right]$$

$$\le \left(\frac{C_R}{2}\right)^2 \frac{2}{n} \sum_{i=0}^{k_n-1} \mathbb{E}\left[|\Delta_i M|^2 + \Delta_i <M>\right]$$

$$= \left(\frac{C_R}{2}\right)^2 \frac{4}{n} <M>_{\theta^n_{k_n}}$$

$$\le C_R^2 \frac{R}{n}.$$

Note also that if we define

$$\Phi^n_s = G^*_s \sum_{i=0}^{k_n-1} \varphi'_x (V_i, M_i) \mathbf{1}_{[\theta^n_i, \theta^n_{i+1}[} (s)$$

then

$$\int_0^{t\wedge\tau} \left|\Phi^n_s - G^*_s \varphi'_x (V_s, M_s)\right|^2 ds \xrightarrow{a.s.} 0, \quad \text{as } n \to \infty$$

and hence

$$\sum_{i=0}^{k_n-1} \left\langle \varphi'_x (V_i, M_i), \Delta_i M \right\rangle = \int_0^{t\wedge\tau} \langle \Phi^n_s, dB_s \rangle$$

$$\xrightarrow{prob.} \int_0^{t\wedge\tau} \left\langle \varphi'_x (V_s, M_s), G_s dB_s \right\rangle.$$

Consequently, again using the definition of the Riemann–Stieltjes integral, we can pass to the limit in (2.18) as $n \to \infty$ and we obtain (2.17) with t replaced by $t \wedge \tau_R$. It remains to let $R \to \infty$ and (2.17) follows. ■

Corollary 2.20. *If* $\varphi \in C^{1,2}\left(\mathbb{R}_+ \times \mathbb{R}^k; \mathbb{C}\right)$ *then*

$$\begin{aligned} \varphi(t, B_t) = \varphi(0,0) &+ \int_0^t \left[\varphi'_t (r, B_r) + \frac{1}{2}\Delta_x \varphi (r, B_r)\right] dr \\ &+ \int_0^t \langle \nabla_x \varphi (r, B_r), dB_r \rangle, \ \mathbb{P}\text{-}a.s., \end{aligned} \tag{2.19}$$

where $\Delta_x = \sum_{j=1}^k \partial^2/\partial^2 x_j$.

In particular

$$\int_0^t \langle B_r, dB_r \rangle = \frac{1}{2} |B_t|^2 - \frac{1}{2} t, \ \mathbb{P}\text{-}a.s.$$

Remark 2.21. By linearity the Itô formula (2.14) or (2.17) holds for a $\mathbb{C}$-valued function φ.

Remark 2.22. If we use an intuitive *symbolic linear calculus* based on

$$dt \otimes dt = 0, \qquad dt \otimes dB_t = 0 \quad \text{and} \quad dB_t \otimes dB_t = I_{k\times k}\, dt,$$

then Itô's formula can be rewritten formally as

$$d\varphi(t, X_t) = \varphi_t'(t, X_t)\, dt + \langle \varphi_x'(t, X_t), dX_t \rangle + \frac{1}{2}\mathbf{Tr}\left[\varphi_{xx}''(t, X_t)(dX_t \otimes dX_t)\right].$$

2.3.1 Applications of Itô's Formula

Let

$$U_t = U_0 + \int_0^t E_s ds + \int_0^t \langle G_s, dB_s\rangle,$$
$$V_t = V_0 + \int_0^t F_s ds + \int_0^t \langle H_s, dB_s\rangle$$

and

$$\begin{aligned} Y_t &= \exp\left[(U_0 + iV_0) + \int_0^t (E_s + iF_s)\, ds + \int_0^t \langle G_s + iH_s, dB_s\rangle\right] \\ &= \exp U_t \exp i V_t. \end{aligned}$$

By the Itô formula for both $\varphi(x) = \exp x$, $x \in \mathbb{R}$, and $\psi(x) = \exp ix$, $x \in \mathbb{R}$, we have

$$d(\exp U_t) = (\exp U_t)\left(E_t + \frac{1}{2}|G_t|^2\right) dt + (\exp U_t)\langle G_t, dB_t\rangle$$

and

$$d(\exp iV_t) = (\exp iV_t)\left(iF_t - \frac{1}{2}|H_t|^2\right) dt + i(\exp iV_t)\langle H_t, dB_t\rangle.$$

Hence:

Corollary 2.23. *Let E, F, G, H be progressively measurable stochastic processes such that*

$$E, F \in L^1_{loc}(\mathbb{R}_+; \mathbb{R}) \text{ a.s., and } \quad G, H \in \Lambda^0_k,$$

and U_0, V_0 be $\mathcal{F}_0$-measurable random variables. Let

$$Y_t = \exp\left[(U_0 + iV_0) + \int_0^t (E_s + iF_s)\,ds + \int_0^t \langle G_s + iH_s, dB_s\rangle\right].$$

Then

$$Y_t = \exp(U_0 + iV_0) + \int_0^t Y_s\left(E_s + iF_s + \frac{1}{2}|G_s + iH_s|^2\right)ds$$
$$+ \int_0^t Y_s \langle G_s + iH_s, dB_s\rangle$$

for all $t \geq 0$, $\mathbb{P}$-a.s.

Using Itô's formula we derive some exponential estimates.

Lemma 2.24. *Let U_0, E and G be as in Corollary 2.23 and*

$$U_t = U_0 + \int_0^t E_s ds + \int_0^t \langle G_s, dB_s\rangle\,,\ t \geq 0.$$

Assume there exist three constants r_0,a, b such that $|U_0| \leq r_0$ and $|E_s| \leq a$, $|G_s| \leq b$, $d\mathbb{P} \otimes dt$-a.e. Then for all $\lambda \in \mathbb{R}$,

$$\mathbb{E}e^{\lambda U_t} \leq \exp\left\{r_0\,|\lambda| + \left(a\,|\lambda| + \frac{\lambda^2 b^2}{2}\right)t\right\}.$$

Proof. Let $\tau_n = \inf\{n \in \mathbb{N} : |U_t| \geq n\}$. Then

$$\mathbb{E}e^{\lambda U_{t\wedge\tau_n}} = \mathbb{E}e^{\lambda U_0} + \mathbb{E}\int_0^{t\wedge\tau_n} e^{\lambda U_r}\left(\lambda E_r + \frac{\lambda^2}{2}|G_r|^2\right)dr$$
$$\leq e^{|\lambda| r_0} + \left(a\,|\lambda| + \frac{\lambda^2 b^2}{2}\right)\int_0^t \mathbb{E}e^{\lambda U_{r\wedge\tau_n}}dr$$

and by the Gronwall inequality

$$\mathbb{E}e^{\lambda U_{t\wedge\tau_n}} \leq e^{|\lambda| r_0} e^{t(a|\lambda| + \frac{\lambda^2 b^2}{2})}.$$

Passing to $\liminf_{n\to+\infty}$ the result follows. ∎

We also have:

Lemma 2.25. *Let $T > 0$, $\lambda \in \mathbb{R}$ and*

$$M_t = \int_0^t \langle G_s, dB_s\rangle\,,\ t \geq 0,$$

$$\mathcal{M}_t^\lambda = \exp\left(\lambda M_t - \frac{\lambda^2}{2}\int_0^t |G_s|^2\,ds\right),\ t \geq 0,$$

where $\{G_t : t \geq 0\}$ *is progressively measurable and satisfies* $|G_t| \leq b$, *a.s., for all* $t \geq 0$. *Then for all* $\lambda \in \mathbb{R}$, $\{\mathcal{M}_t^\lambda : t \geq 0\}$ *is a continuous* p*-martingale for all* $p \geq 1$ *and*

$$\mathbb{P}\left(\sup_{0\leq t\leq T} |M_t| \geq \delta\right) \leq 2\exp\left(-\frac{\delta^2}{2b^2T}\right),\quad \forall\,\delta > 0. \tag{2.20}$$

Proof. By Lemma 2.24 clearly $\mathbb{E}\left|\mathcal{M}_t^\lambda\right|^p < \infty$ and by Itô's formula

$$\mathcal{M}_t^\lambda = 1 + \lambda\int_0^t \mathcal{M}_s^\lambda\,\langle G_s, dB_s\rangle$$

and clearly $\{\mathcal{M}_t^\lambda : t \geq 0\}$ is a martingale.

We have

$$\mathbb{P}(\sup_{0\leq t\leq T} |M_t| \geq \delta) = \mathbb{P}(\sup_{0\leq t\leq T} M_t \geq \delta) + \mathbb{P}(\sup_{0\leq t\leq T} (-M_t) \geq \delta).$$

We estimate the first term on the right; the second one is bounded by the same quantity. Since $\{\mathcal{M}_t^\lambda,\ t \geq 0\}$ is a martingale we infer, by Doob's inequality (1.11-A_1), that

$$\begin{aligned}\mathbb{P}(\sup_{0\leq t\leq T} M_t \geq \delta) &\leq \mathbb{P}(\sup_{0\leq t\leq T} \mathcal{M}_t^\lambda \geq \exp(\lambda\delta - \lambda^2 b^2 T/2))\\ &\leq \exp(\lambda^2 b^2 T/2 - \lambda\delta).\end{aligned}$$

Setting $\lambda = \dfrac{\delta}{b^2T}$, we deduce that

$$\mathbb{P}(\sup_{0\leq t\leq T} M_t \geq \delta) \leq \exp\left(-\frac{\delta^2}{2b^2T}\right),$$

from which the result follows. ∎

We remark that if in Lemma 2.25 we replace G_s by $\mathbf{1}_{[\tau,\theta]}(s)\,G_s$, where τ and θ are two stopping times such that $0 \leq \tau \leq \theta \leq T$, then the inequality (2.20) becomes

$$\mathbb{P}\left(\sup_{\tau\leq t\leq\theta}\left|\int_\tau^t \langle G_s, dB_s\rangle\right| \geq \delta\right) \leq 2\exp\left(-\frac{\delta^2}{2b^2T}\right),\quad \text{for all } \delta > 0. \tag{2.21}$$

Finally we give a useful formula by applying Itô's formula to $\varphi^p(X_t)$, where $\varphi = \varphi_{\delta,\varepsilon} : \mathbb{R}^d \to]0,\infty[$

$$\varphi_{\delta,\varepsilon}(x) = \left(\frac{|x|^2}{1+\delta |x|^2} + \varepsilon \right)^{1/2},$$

$\delta \geq 0$, $0 < \varepsilon \leq 1$, and we shall take the limit as $\varepsilon \searrow 0$.

Proposition 2.26. *Let $p \geq 1$, $\delta \geq 0$ and $X \in S_d^0$ be a local semimartingale of the form*

$$X_t = X_0 + \int_0^t dK_s + \int_0^t G_s dB_s, \quad t \geq 0, \tag{2.22}$$

where $G \in \Lambda_{d\times k}^0$, $K \in S_d^0$, $K_\cdot(\omega) \in BV_{loc}\left(\mathbb{R}^d;\mathbb{R}\right)$ $\mathbb{P}$-a.s. $\omega \in \Omega$.
Then $\mathbb{P}$-a.s., for all $t \geq 0$:

$$\begin{aligned}
&\frac{|X_t|^p}{\left(1+\delta |X_t|^2\right)^{p/2}} + \frac{p}{2}\int_0^t Q_s^{(p,\delta)} ds = \frac{|X_0|^p}{\left(1+\delta |X_0|^2\right)^{p/2}} + p\int_0^t \left\langle U_s^{(p,\delta)}, dK_s \right\rangle \\
&+p\int_0^t \left\langle U_s^{(p,\delta)}, G_s dB_s \right\rangle + \frac{p}{2}\int_0^t R_s^{(p,\delta)} ds + \frac{1}{2} L_t^{(1,\delta)}\, \mathbf{1}_{p=1},
\end{aligned} \tag{2.23}$$

where

(j) $U_s^{(p,\delta)} = 0$, $R_s^{(p,\delta)} = Q_s^{(p,\delta)} = 0$, *if* $X_s = 0$, *and*

$$U_s^{(p,\delta)} = \frac{|X_s|^{p-2}}{\left(1+\delta |X_s|^2\right)^{(p+2)/2}} X_s,$$

$$\begin{aligned}
R_s^{(p,\delta)} = &\frac{|X_s|^{p-4}}{\left(1+\delta |X_s|^2\right)^{(p+2)/2}} \Big[\left(|G_s|^2 |X_s|^2 - |G_s^* X_s|^2 \right) \\
&+ \frac{p-1}{1+\delta |X_s|^2} |G_s^* X_s|^2 \Big],
\end{aligned}$$

$$Q_s^{(p,\delta)} = \frac{3\delta |X_s|^{p-2}}{\left(1+\delta |X_s|^2\right)^{(p+4)/2}} |G_s^* X_s|^2,$$

if $X_s \neq 0$;

(*jj*) $R_{\cdot}^{(p,\delta)} \geq 0$, $Q_{\cdot}^{(p,\delta)} \geq 0$ *and for all* $T > 0$, $\mathbb{P}$-*a.s.:*

$$\left\| U^{(p,\delta)} \right\|_T + \int_0^T R_s^{(p,\delta)} ds + \int_0^T Q_s^{(p,\delta)} ds < \infty;$$

(*jjj*) $\left\{ L_t^{(1,\delta)} : t \geq 0 \right\}$ *is an increasing continuous stochastic process such that* $L_0^{(1,\delta)} = 0$ *a.s. Moreover* $L_t^{(1,\delta)} = L_s^{(1,\delta)}$ *a.s., for every interval* $[s,t] \subset \{r \geq 0 : X_r(\omega) = 0\}$ *or* $[s,t] \subset int\{r \geq 0 : X_r(\omega) \neq 0\}$*;*

(*jv*) *if there exists a positive stochastic process* γ *such that* $\mathbb{P}$-*a.s.* $\gamma_{\cdot} \in L_{loc}^2(\mathbb{R}_+)$ *and*

$$|G_s| \leq \gamma_s \sqrt{|X_s|}, \quad d\mathbb{P} \otimes dt\text{-}a.e.,$$

then $L_t^{(1,\delta)} = 0$ *for all* $t \geq 0$.

Proof. Let $\delta \geq 0$ and $0 < \varepsilon \leq 1$. Let $\varphi = \varphi_{\delta,\varepsilon} : \mathbb{R}^d \to]0, \infty[$

$$\varphi(x) = \varphi_{\delta,\varepsilon}(x) = \left(\frac{|x|^2}{1 + \delta |x|^2} + \varepsilon \right)^{1/2}.$$

Note that

$$\nabla_x \varphi(x) = \varphi^{-1}(x) \frac{x}{\left(1 + \delta |x|^2\right)^2}$$

and

$$\begin{aligned}
\mathbf{Tr}\left[D_{xx}^2 \varphi(x) GG^*\right] &= -\varphi^{-3}(x) \frac{1}{\left(1 + \delta |x|^2\right)^4} |G^* x|^2 \\
&\quad - \varphi^{-1}(x) \frac{4\delta}{\left(1 + \delta |x|^2\right)^3} |G^* x|^2 + \varphi^{-1}(x) \frac{1}{\left(1 + \delta |x|^2\right)^2} |G|^2 \\
&= \varphi^{-3}(x) \frac{1}{\left(1 + \delta |x|^2\right)^3} \left(|G|^2 |x|^2 - |G^* x|^2 \right) + \\
&\quad + \varphi^{-3}(x) \frac{\varepsilon}{\left(1 + \delta |x|^2\right)^3} \left[|G|^2 + \delta \left(|G|^2 |x|^2 - |G^* x|^2 \right) \right] \\
&\quad - \varphi^{-1}(x) \frac{3\delta}{\left(1 + \delta |x|^2\right)^3} |G^* x|^2.
\end{aligned}$$

By Itô's formula for $\varphi^p(X_t)$, $p \geq 1$:

$$\varphi^p(X_t) = \varphi^p(X_0) + \int_0^t \langle \nabla_x \varphi^p(X_s), F_s \rangle ds + \int_0^t \langle \nabla_x \varphi^p(X_s), G_s dB_s \rangle$$
$$+ \frac{1}{2} \int_0^t \mathbf{Tr}\left(D_{xx}^2 \varphi^p(X_s) G_s G_s^*\right) ds.$$

Since

$$\nabla_x \varphi^p(x) = p \varphi^{p-1}(x) \nabla_x \varphi(x)$$

and

$$\begin{aligned} \mathbf{Tr}\left[D_{xx}^2 \varphi^p(x) GG^*\right] &= p(p-1) \varphi^{p-2}(x) \left|G^* \nabla_x \varphi(x)\right|^2 \\ &\quad + p \varphi^{p-1}(x) \mathbf{Tr}\left[D_{xx}^2 \varphi(x) GG^*\right], \end{aligned}$$

we have that

$$\begin{aligned} \varphi_{\delta,\varepsilon}^p(X_t) &= \varphi_{\delta,\varepsilon}^p(X_0) + p \int_0^t \left\langle U_s^{(p,\delta,\varepsilon)}, dK_s \right\rangle + p \int_0^t \left\langle U_s^{(p,\delta,\varepsilon)}, G_s dB_s \right\rangle \\ &\quad + \frac{p}{2} \int_0^t R_s^{(p,\delta,\varepsilon)} ds - \frac{p}{2} \int_0^t Q_s^{(p,\delta,\varepsilon)} ds + \frac{p}{2} L_t^{(p,\delta,\varepsilon)}, \end{aligned} \tag{2.24}$$

where

$$U_s^{(p,\delta,\varepsilon)} = \varphi_{\delta,\varepsilon}^{p-2}(X_s) \frac{1}{\left(1 + \delta |X_s|^2\right)^2} X_s,$$

$$\begin{aligned} R_s^{(p,\delta,\varepsilon)} &= (p-1) \varphi_{\delta,\varepsilon}^{p-4}(X_s) \frac{1}{\left(1 + \delta |X_s|^2\right)^4} |G_s^* X_s|^2 \\ &\quad + \varphi_{\delta,\varepsilon}^{p-4}(X_s) \frac{1}{\left(1 + \delta |X_s|^2\right)^3} \left(|G_s|^2 |X_s|^2 - |G_s^* X_s|^2\right), \end{aligned}$$

$$Q_s^{(p,\delta,\varepsilon)} = \varphi_{\delta,\varepsilon}^{p-2}(X_s) \frac{3\delta}{\left(1 + \delta |X_s|^2\right)^3} |G_s^* X_s|^2,$$

and

$$L_t^{(p,\delta,\varepsilon)} = \varepsilon \int_0^t \varphi_{\delta,\varepsilon}^{p-4}(X_s) \frac{1}{\left(1 + \delta |X_s|^2\right)^3} \left[|G_s|^2 + \delta\left(|G_s|^2 |X_s|^2 - |G_s^* X_s|^2\right)\right] ds.$$

Note that as $\varepsilon \to 0_+$

$$\Gamma_s^{(p,\delta,\varepsilon)} = \Gamma_s^{(p,\delta,\varepsilon)} \mathbf{1}_{X_s \neq 0} \to \Gamma_s^{(p,\delta)},$$

for $\Gamma_s^{(p,\delta,\varepsilon)} = U_s^{(p,\delta,\varepsilon)}$, $Q_s^{(p,\delta,\varepsilon)}$, or $R_s^{(p,\delta,\varepsilon)}$.

Since for all $\delta \geq 0$ and $0 < \varepsilon \leq 1$:

$$\left| U_s^{(p,\delta,\varepsilon)} \right| \leq (|X_s| + 1)^{p-1} \quad \text{and} \quad 0 \leq Q_s^{(p,\delta,\varepsilon)} \leq 3\delta \, (|X_s| + 1)^p \, |G_s|^2 ,$$

it follows, by the Lebesgue dominated convergence theorem, that for all $t \geq 0$:

$$\lim_{\varepsilon \to 0_+} \int_0^t \left\langle U_s^{(p,\delta,\varepsilon)}, dK_s \right\rangle = \int_0^t \left\langle U_s^{(p,\delta)}, dK_s \right\rangle, \quad \mathbb{P}\text{-a.s.},$$

and

$$\lim_{\varepsilon \to 0_+} \int_0^t \left| G_s^* U_s^{(p,\delta,\varepsilon)} - G_s^* U_s^{(p,\delta)} \right|^2 ds = 0, \quad \mathbb{P}\text{-a.s.},$$

which yields moreover, as $\varepsilon \to 0_+$,

$$\int_0^t \left\langle U_s^{(p,\delta,\varepsilon)}, G_s dB_s \right\rangle \xrightarrow{prob.} \int_0^t \left\langle U_s^{(p,\delta)}, G_s dB_s \right\rangle.$$

Also, by Lebesgue's dominated convergence theorem for $p \geq 4$, and the monotone convergence theorem for $1 \leq p < 4$:

$$\lim_{\varepsilon \to 0_+} \int_0^t R_s^{(p,\delta,\varepsilon)} ds = \int_0^t R_s^{(p,\delta)} ds, \quad \mathbb{P}\text{-a.s.}, \ \forall\, t \geq 0$$

and, since $R_t^{(p,\delta,\varepsilon)} \geq 0$, we get, from (2.24), that

$$0 \leq \int_0^t R_s^{(p,\delta)} ds < \infty, \quad \mathbb{P}\text{-}a.s., \ \forall\, t \geq 0.$$

Hence, once again from (2.24) and the definition of $L_t^{(p,\delta,\varepsilon)}$, for all $p \geq 1$, there exists a progressively measurable continuous increasing stochastic process $L^{(p,\delta)}$ such that for all $t \geq 0$, as $\varepsilon \to 0_+$:

$$L_t^{(p,\delta,\varepsilon)} \xrightarrow{prob.} L_t^{(p,\delta)}.$$

Letting $\varepsilon \to 0_+$ in (2.24), the equality (2.23) follows, as well as (j) and (jj).

If $p \geq 4$, by the Lebesgue dominated convergence theorem, we infer that

$$L_t^{(p,\delta)} = 0, \ \mathbb{P}\text{-a.s.}, \ \forall\, t \geq 0.$$

Let $1 < p < 4$ and $\beta = (4-p)/3$; then $0 < \beta < 1$.

If we define

$$H_s^{(\delta)} = \frac{1}{\left(1+\delta\left|X_s\right|^2\right)^3}\left[\left|G_s\right|^2 + \delta\left(\left|G_s\right|^2\left|X_s\right|^2 - \left|G_s^* X_s\right|^2\right)\right],$$

then

$$L_t^{(1,\delta,\varepsilon)} = \varepsilon \int_0^t \varphi_{\delta,\varepsilon}^{-3}\left(X_s\right) H_s^{(\delta)} ds.$$

We get using Hölder's inequality

$$\begin{aligned} 0 \leq L_t^{(p,\delta,\varepsilon)} &= \varepsilon \int_0^t \varphi_{\delta,\varepsilon}^{p-4}\left(X_s\right) H_s^{(\delta)} ds \\ &= \int_0^t \left(\varepsilon H_s^{(\delta)}\right)^{1-\beta}\left[\varepsilon \varphi_{\delta,\varepsilon}^{-3}\left(X_s\right) H_s^{(\delta)}\right]^\beta ds \\ &\leq \varepsilon^{1-\beta}\left(\int_0^t H_s^{(\delta)} ds\right)^{1-\beta}\left(L_t^{(1,\delta,\varepsilon)}\right)^\beta. \end{aligned}$$

Consequently for all $p > 1$ and $\delta \geq 0$:

$$L_t^{(p,\delta)} = 0, \ \mathbb{P}\text{-a.s.}, \ \forall\, t \geq 0.$$

We now study $L_t^{(1,\delta)}$.

Since

$$\begin{aligned} L_t^{(1,\delta,\varepsilon)} &= \varepsilon \int_0^t \frac{\left|G_s\right|^2 + \delta\left(\left|G_s\right|^2\left|X_s\right|^2 - \left|G_s^* X_s\right|^2\right)}{\left(1+\delta\left|X_s\right|^2\right)^{3/2}\left(\left|X_s\right|^2 + \varepsilon + \varepsilon\delta\left|X_s\right|^2\right)^{3/2}} ds \\ &\geq \frac{1}{(1+\delta\varepsilon)^{3/2}} \frac{1}{(2+\delta\varepsilon)^{3/2}} \frac{1}{\sqrt{\varepsilon}} \int_0^t \left|G_s\right|^2 \mathbf{1}_{\left|X_s\right| \leq \sqrt{\varepsilon}} ds, \end{aligned}$$

we obtain that

$$\frac{1}{2^{3/2}} \limsup_{\varepsilon \to 0_+} \frac{1}{\sqrt{\varepsilon}} \int_0^t \left|G_s\right|^2 \mathbf{1}_{\left|X_s\right| \leq \sqrt{\varepsilon}} ds \leq L_t^{(1,\delta)} \tag{2.25}$$

and, consequently,

$$\int_0^t \mathbf{1}_{X_s=0} |G_s|^2 \, ds = 0. \tag{2.26}$$

Let $0 \le s \le t$. By (2.26),

$$\begin{aligned} 0 &\le L_t^{(1,\delta,\varepsilon)}(\omega) - L_s^{(1,\delta,\varepsilon)}(\omega) \\ &\le \varepsilon \int_s^t \left(|X_u(\omega)|^2 + \varepsilon \right)^{-3/2} \mathbf{1}_{X_u(\omega)\neq 0} |G_u(\omega)|^2 \, du. \end{aligned}$$

◊ If $[s,t] \subset \{r \ge 0 : X_r(\omega) = 0\}$ then, clearly,

$$L_t^{(1,\delta)}(\omega) = L_s^{(1,\delta)}(\omega).$$

◊ If $[s,t] \subset \operatorname{int}\{r \ge 0 : X_r(\omega) \neq 0\}$, then there exists a $\delta(\omega) > 0$ such that

$$|X_r(\omega)| \ge \delta(\omega), \text{ for all } r \in [s,t]$$

and

$$0 \le L_t^{(1,\delta,\varepsilon)}(\omega) - L_s^{(1,\delta,\varepsilon)}(\omega) \le 2\varepsilon \left(\delta^2(\omega) + \varepsilon \right)^{-3/2} \int_s^t |G_u(\omega)|^2 \, du.$$

Setting $\varepsilon \to 0_+$ we obtain

$$L_t^{(1,\delta)}(\omega) = L_s^{(1,\delta)}(\omega).$$

Let us now prove (jv). We have

$$0 \le L_t^{(1,\delta,\varepsilon)} \le \int_0^t \gamma_s^2 \left(\Psi_s^\varepsilon \right)^{3/2} ds,$$

where

$$\Psi_s^\varepsilon = \frac{|X_s|^{2/3} \left(\varepsilon + \varepsilon\delta |X_s|^2 \right)^{2/3}}{|X_s|^2 + \varepsilon + \varepsilon\delta |X_s|^2}.$$

Note that $\lim_{\varepsilon \to 0_+} \Psi_s^\varepsilon = 0$ and, by Hölder's inequality,

$$0 \le \Psi_s^\varepsilon \le \frac{\frac{1}{3} |X_s|^2 + \frac{2}{3} \left(\varepsilon + \varepsilon\delta |X_s|^2 \right)}{|X_s|^2 + \varepsilon + \varepsilon\delta |X_s|^2} \le 1.$$

Hence

$$L_t^{(1,\delta)} = \lim_{\varepsilon\to 0_+} L_t^{(1,\delta,\varepsilon)} = 0.$$

This completes the proof. ∎

Remark 2.27. Let $y \in \mathbb{R}^d$. In Proposition 2.26 we can replace X_t by $X_t - y$ (and consequently X_0 by $X_0 - y$). Now from (2.25) it clearly follows that if $X \in S_d^0$ is a local semimartingale of the form (2.22), then for all $b > 0$ and $t \geq 0$:

$$\begin{aligned} &(\alpha)\ \limsup_{\varepsilon\to 0_+} \frac{1}{\varepsilon^{1-b}} \int_0^t \mathbf{1}_{|X_s - y|\leq\varepsilon} |G_s|^2\, ds = 0,\ \mathbb{P}\text{-a.s.},\\ &(\beta)\ \int_0^t \mathbf{1}_{X_s = y} |G_s|^2\, ds = 0,\ \mathbb{P}\text{-a.s.} \end{aligned} \tag{2.27}$$

We now deduce from Proposition 2.26 the following:

Corollary 2.28. *Let $X \in S_d^0$ be a local semimartingale of the form*

$$X_t = X_0 + \int_0^t dK_r + \int_0^t G_r dB_r, \quad t \geq 0,$$

where $G \in \Lambda^0_{d\times k}$, $K \in S_d^0$, $K_\cdot(\omega) \in BV_{loc}\left(\mathbb{R}^d;\mathbb{R}\right)$ $\mathbb{P}$-a.s. $\omega \in \Omega$.
Let $p \geq 1$, $\delta \geq 0$, $m_p = 1 \vee (p-1)$,

$$J_r^{(p,\delta)} = \frac{|X_r|^{p-2}\mathbf{1}_{X_r\neq 0}}{\left(1+\delta|X_r|^2\right)^{(p+2)/2}}.$$

Then for all $t \leq s$, $\mathbb{P}$-a.s.:

$$\begin{aligned} &\frac{|X_s|^p}{\left(1+\delta|X_s|^2\right)^{p/2}} \leq \frac{|X_t|^p}{\left(1+\delta|X_t|^2\right)^{p/2}}\\ &+p\int_t^s J_r^{(p,\delta)}\left[\langle X_r, dK_r\rangle + \frac{1}{2}m_p|G_r|^2 dr\right] + p\int_t^s J_r^{(p,\delta)}\langle X_r, G_r dB_r\rangle. \end{aligned} \tag{2.28}$$

Proof. First, let $p > 1$. From (2.23) and

$$\begin{aligned} &\left(|G_r|^2|X_r|^2 - |G_r^* X_r|^2\right) + \frac{p-1}{1+\delta|X_r|^2}|G_r^* X_r|^2\\ &\leq \left(|G_r|^2|X_r|^2 - |G_r^* X_r|^2\right) + (p-1)|G_r^* X_r|^2\\ &\leq m_p|G_r|^2|X_r|^2, \end{aligned}$$

the inequality (2.28) follows.

Now let $p = 1$.

To prove (2.28) note first that $m_1 = 1$. From the proof of Proposition 2.26 we have for $0 \le t \le s$:

$$\begin{aligned} &\frac{1}{2}\int_t^s \left(R_r^{(1,\delta,\varepsilon)} - Q_r^{(1,\delta,\varepsilon)}\right) dr + L_s^{(1,\delta,\varepsilon)} - L_t^{(1,\delta,\varepsilon)} \\ &= \frac{1}{2}\int_t^s \mathbf{Tr}\left[G_r G_r^* \, D_{xx}^2 \varphi_{\delta,\varepsilon}(X_r)\right] dr \\ &\le \frac{1}{2}\int_t^s \frac{1}{\varphi_{\delta,\varepsilon}(x)} \frac{1}{\left(1+\delta|x|^2\right)^2} |G_r|^2 dr \\ &\le \frac{1}{2}\int_t^s J_r^{(1,\delta)} \mathbf{1}_{X_r \neq 0} |G_r|^2 dr, \end{aligned}$$

which yields (2.28) passing to the limit as $\varepsilon \to 0_+$. ■

We remark that in the proof of Proposition 2.26, the relation (2.24) is also true for all $p \in \mathbb{R}$. Therefore we have for $\delta = 0$:

Corollary 2.29. *Let $p \in \mathbb{R}$ and $\varepsilon > 0$. If $X \in S_d^0$ is a local semimartingale of the form*

$$X_t = X_0 + \int_0^t dK_s + \int_0^t G_s dB_s, \quad \forall t \ge 0, \ a.s.;$$

with $G \in \Lambda_{d\times k}^0$, $K \in S_d^0$, $K_\cdot(\omega) \in BV_{loc}\left(\mathbb{R}^d;\mathbb{R}\right)$ $\mathbb{P}$-a.s. $\omega \in \Omega$, then $\mathbb{P}$-a.s., for all $t \ge 0$:

$$\begin{aligned} \left(|X_t|^2+\varepsilon\right)^{p/2} = \left(|X_0|^2+\varepsilon\right)^{p/2} + p\int_0^t \left\langle U_s^{(p,\varepsilon)}, dK_s\right\rangle + p\int_0^t \left\langle U_s^{(p,\varepsilon)}, G_s dB_s\right\rangle \\ + \frac{p}{2}\int_0^t R_s^{(p,\varepsilon)} ds + \frac{p}{2} L_t^{(p,\varepsilon)}, \end{aligned} \tag{2.29}$$

where

$$\begin{aligned} &(j) \quad U_s^{(p,\varepsilon)} = \left(|X_s|^2+\varepsilon\right)^{(p-2)/2} X_s, \\ &(jj) \quad R_s^{(p,\varepsilon)} = \left[|G_s|^2|X_s|^2 + (p-2)\left|G_s^* X_s\right|^2\right]\left(|X_s|^2+\varepsilon\right)^{(p-4)/2}, \\ &(jjj) \quad L_t^{(p,\varepsilon)} = \varepsilon\int_0^t |G_s|^2 \left(|X_s|^2+\varepsilon\right)^{(p-4)/2} ds \end{aligned}$$

and

$$n_p |G_s|^2 |X_s|^2 \left(|X_s|^2 + \varepsilon\right)^{(p-4)/2} \leq R_s^{(p,\varepsilon)} \leq m_p |G_s|^2 |X_s|^2 \left(|X_s|^2 + \varepsilon\right)^{(p-4)/2}$$

where $n_p = 1 \wedge (p-1)$ *and* $m_p = 1 \vee (p-1)$.

Choosing $\delta = 0$ in Proposition 2.26, or passing to the limit in (2.29) as $\varepsilon \searrow 0$, we infer:

Corollary 2.30. *Let* $p \geq 1$. *If* $X \in S_d^0$ *is a local semimartingale of the form*

$$X_t = X_0 + \int_0^t dK_s + \int_0^t G_s dB_s, \quad \forall t \geq 0, \ a.s.;$$

with $G \in \Lambda^0_{d\times k}$, $K \in S_d^0$, $K_\cdot(\omega) \in BV_{loc}\left(\mathbb{R}^d; \mathbb{R}\right)$ $\mathbb{P}$*-a.s.* $\omega \in \Omega$, *then* $\mathbb{P}$*-a.s., for all* $t \geq 0$*:*

$$\begin{aligned} |X_t|^p = |X_0|^p + p\int_0^t \left\langle U_s^{(p)}, dK_s\right\rangle + p\int_0^t \left\langle U_s^{(p)}, G_s dB_s\right\rangle \\ + \frac{p}{2}\int_0^t R_s^{(p)} ds + \frac{1}{2} L_t \, \mathbf{1}_{p=1}, \end{aligned} \tag{2.30}$$

where

(*j*) $U_s^{(p)} = 0$, $R_s^{(p)} = 0$, *when* $X_s = 0$ *and if* $X_s \neq 0$ *then*

$$\begin{aligned} U_s^{(p)} &= |X_s|^{p-2} X_s, \\ R_s^{(p)} &= |X_s|^{p-4} \left[\left(|G_s|^2 |X_s|^2 - |G_s^* X_s|^2\right) + (p-1) |G_s^* X_s|^2\right]; \end{aligned}$$

(*jj*) *setting* $n_p = 1 \wedge (p-1)$ *and* $m_p = 1 \vee (p-1)$ *we have*

$$n_p |X_s|^{p-2} \mathbf{1}_{X_s \neq 0} |G_s|^2 \leq R_s^{(p)} \leq m_p |X_s|^{p-2} \mathbf{1}_{X_s \neq 0} |G_s|^2$$

and

$$0 \leq \int_0^t R_s^{(p)} ds < \infty, \quad \mathbb{P}\text{-}a.s., \ \forall\, t \geq 0;$$

(*jjj*) $\{L_t : t \geq 0\}$ *is an increasing continuous stochastic process such that for all* $t \geq 0$:

$$L_t = \lim_{\varepsilon \to 0+} \int_0^t \frac{\varepsilon |G_s|^2}{\left(|X_s|^2 + \varepsilon\right)^{3/2}} ds \ \text{(convergence in probability)}$$

and $L_t(\omega) = L_s(\omega)$, $\mathbb{P}$*-a.s., for every interval* $[s,t] \subset \{r \geq 0 : X_r(\omega) = 0\}$, *or* $[s,t] \subset int\{r \geq 0 : X_r(\omega) \neq 0\}$;

(*jv*) *if there exists a positive stochastic process* γ *such that* $\mathbb{P}$*-a.s.* $\gamma_\cdot \in L^2_{loc}(\mathbb{R}_+)$ *and*

$$|G_s| \leq \gamma_s \sqrt{|X_s|}, \quad d\mathbb{P} \otimes dt\text{-}a.e.,$$

then $L_t = 0$ *for all* $t \geq 0$.

Setting $p = 1$ in (2.30) we obtain the d-dimensional generalization of the celebrated Tanaka formula.

Corollary 2.31. *If* $X \in S^0_d$ *is an Itô process of the form*

$$X_t = X_0 + \int_0^t F_s ds + \int_0^t G_s dB_s, \quad \forall t \geq 0, \; a.s.,$$

then $\mathbb{P}$*-a.s., for all* $t \geq 0$:

$$|X_t| = |X_0| + \int_0^t \langle sgn(X_s), F_s \rangle \, ds + \int_0^t \langle sgn(X_s), G_s dB_s \rangle + \int_0^t R_s ds + \frac{1}{2} L_t, \tag{2.31}$$

where

$$\text{sgn} : \mathbb{R}^d \to \mathbb{R}^d, \quad sgn(x) = \begin{cases} 0, & \text{if } x = 0, \\ \dfrac{x}{|x|}, & \text{if } x \neq 0, \end{cases}$$

$$R_s = \begin{cases} 0, & \text{if } X_s = 0, \\ \dfrac{1}{2|X_s|} \left(|G_s|^2 - |G_s^* \, \text{sgn}(X_s)|^2 \right), & \text{if } X_s \neq 0, \end{cases}$$

and $\{L_t : t \geq 0\}$ *is an increasing continuous stochastic process such that* $\mathbb{P}$*-a.s.* $\omega \in \Omega$*:*

$$L_0(\omega) = 0 \quad \text{and} \quad L_t(\omega) = L_s(\omega),$$

if $[s,t] \subset \{r \geq 0 : X_r(\omega) = 0\}$, *or* $[s,t] \subset \text{int}\{r \geq 0 : X_r(\omega) \neq 0\}$.

Remark 2.32. Comparing formula (2.31) with Tanaka's formula for the continuous scalar semimartingale $|X_t|$ (see [64], Ch. VI, Th 1.2) we deduce that L_t coincides with the local time at 0 of $|X_t|$.

We now derive in the case $d = 1$ the stochastic differential dX_t^+. Let

$$\rho(u) = \frac{1}{\sqrt{2\pi}} \exp\left(-\frac{u^2}{2}\right).$$

Proposition 2.33. *Let $X \in S_1^0$ be a local semimartingale of the form*

$$X_t = X_0 + \int_0^t dK_s + \int_0^t \langle G_s, dB_s \rangle, \quad t \geq 0,$$

where $K \in S_1^0$, $K_\cdot(\omega) \in BV_{loc}(\mathbb{R};\mathbb{R})$ $\mathbb{P}$-a.s. $\omega \in \Omega$ and $G \in \Lambda_{1\times k}^0$.
Then $\mathbb{P}$-a.s., for all $t \geq 0$:

$$X_t^+ = X_0^+ + \int_0^t \theta(X_s)\, dK_s + \int_0^t \langle \mathbf{1}_{X_s>0} G_s, dB_s \rangle + \frac{1}{2} P_t \tag{2.32}$$

where $\theta : \mathbb{R} \to \mathbb{R}$,

$$\theta(x) = \begin{cases} 0, & \text{if } x < 0, \\ \frac{1}{2}, & \text{if } x = 0, \\ 1, & \text{if } x > 0, \end{cases}$$

and $\{P_t : t \geq 0\}$, $P_0 = 0$, is an increasing continuous stochastic process such that for all $t \geq 0$:

$$\begin{aligned} P_t &= \lim_{\varepsilon \to 0+} \frac{1}{\varepsilon} \int_0^t \rho\left(\frac{X_s}{\varepsilon}\right) |G_s|^2 ds \\ &= \lim_{\varepsilon \to 0+} \frac{1}{\varepsilon} \int_0^t \rho\left(\frac{X_s}{\varepsilon}\right) \mathbf{1}_{0<|X_s|<\sqrt{\varepsilon}}\ |G_s|^2 ds \ \ \textit{(convergence in probability)}, \end{aligned} \tag{2.33}$$

with the properties:

(j) *for every interval $[s,t] \subset \{r \geq 0 : X_r(\omega) = 0\}$, or $[s,t] \subset \text{int}\{r \geq 0 : X_r(\omega) \neq 0\}$,*

$$P_t(\omega) = P_s(\omega), \quad \mathbb{P}\text{-a.s.};$$

(jj) *if there exists a positive stochastic process γ such that $\int_0^T \gamma_s^2 ds < \infty$, for all $T > 0$, and*

$$|G_s| \leq \gamma_s \sqrt{|X_s|}, \quad d\mathbb{P} \otimes dt\text{-a.e.},$$

then $P_t = 0$ for all $t \geq 0$.

Proof. Let $\varepsilon > 0$ and the function $\varphi_\varepsilon : \mathbb{R} \to [0, \infty[$ of class C^∞ be given by

$$\begin{aligned}\varphi_\varepsilon(x) &= \int_{\mathbb{R}} (x - \varepsilon u)^+ \rho(u)\, du \\ &= x \int_{-\infty}^{x/\varepsilon} \rho(u)\, du - \varepsilon \int_{-\infty}^{x/\varepsilon} u\rho(u)\, du.\end{aligned}$$

Then for all $x \in \mathbb{R}$

$$\begin{aligned}&0 < \varphi'_\varepsilon(x) = \int_{-\infty}^{x/\varepsilon} \rho(u)\, du < 1 \quad \text{and} \\ &0 < \varphi''_\varepsilon(x) = \frac{1}{\varepsilon}\rho\left(\frac{x}{\varepsilon}\right) < \frac{1}{\sqrt{2\pi}} \frac{2^p\, p!\, \varepsilon^{2p-1}}{2^p\, p! \varepsilon^{2p} + x^{2p}}, \quad \forall\ p \in \mathbb{N}^*.\end{aligned}$$

Since $\int_{\mathbb{R}} \rho(u)\, du = 1$ and $\left|(x+y)^+ - x^+\right| \le |y|$, we deduce that

$$\left|\varphi_\varepsilon(x) - x^+\right| \le \varepsilon.$$

Moreover

$$\lim_{\varepsilon \to 0} \varphi'_\varepsilon(x) = \begin{cases} 0, & \text{if } x < 0, \\ \frac{1}{2}, & \text{if } x = 0, \\ 1, & \text{if } x > 0, \end{cases}$$

and

$$\begin{aligned}0 \le \varphi''_\varepsilon(x) - \frac{1}{\varepsilon}\rho\left(\frac{x}{\varepsilon}\right) \mathbf{1}_{0<|x|<\sqrt{\varepsilon}} &= \frac{1}{\varepsilon}\rho(0)\, \mathbf{1}_{x=0} + \frac{1}{\varepsilon}\rho\left(\frac{x}{\varepsilon}\right) \mathbf{1}_{|x|\ge\sqrt{\varepsilon}} \\ &= \frac{1}{\varepsilon}\rho(0)\, \mathbf{1}_{x=0} + \frac{2^p p!}{\sqrt{2\pi}} \varepsilon^{p-1}.\end{aligned} \tag{2.34}$$

By Itô's formula we have

$$\begin{aligned}\varphi_\varepsilon(X_t) &= \varphi_\varepsilon(X_0) + \int_0^t \varphi'_\varepsilon(X_s)\, dK_s \\ &\quad + \int_0^t \varphi'_\varepsilon(X_s) \langle G_s, dB_s \rangle + \frac{1}{2} \int_0^t \varphi''_\varepsilon(X_s) |G_s|^2\, ds.\end{aligned}$$

Since

$$\lim_{\varepsilon \to 0_+} \varphi_\varepsilon(X_t) = X_t^+, \quad \lim_{\varepsilon \to 0_+} \varphi'_\varepsilon(X_s) = \theta(X_s) \quad \text{and } 0 \le \varphi'_\varepsilon(X_s) \le 1,$$

it follows that

$$\lim_{\varepsilon\to 0+}\int_0^t \varphi_\varepsilon'(X_s)\,dK_s = \int_0^t \theta(X_s)\,dK_s \quad \text{and}$$

$$\lim_{\varepsilon\to 0+}\int_0^t \varphi_\varepsilon'(X_s)\,\langle G_s, dB_s\rangle = \int_0^t \theta(X_s)\,\langle G_s, dB_s\rangle .$$

Consequently the limit

$$\lim_{\varepsilon\to 0+}\int_0^t \varphi_\varepsilon''(X_s)\,|G_s|^2\,ds \left(\stackrel{def}{=} P_t\right)$$

exists and by (2.27-β) and (2.34) we infer (2.33).

The properties of $(P_t)_{t\geq 0}$ easily follow. The last assertion on $(P_t)_{t\geq 0}$ is a consequence of

$$0 \leq \varphi_\varepsilon''(X_s)\,|G_s|^2 \leq \frac{1}{\sqrt{2\pi}}\frac{2\,\varepsilon}{2\varepsilon^2+X_s^2}\,|X_s| \leq 1$$

and

$$\lim_{\varepsilon\to 0+}\frac{2\,\varepsilon}{2\varepsilon^2+X_s^2}\,|X_s| = 0.$$

Finally observe that by (2.27)

$$\int_0^t \theta(X_s)\,\langle G_s, dB_s\rangle = \int_0^t \langle \mathbf{1}_{X_s>0} G_s, dB_s\rangle .$$

■

Remark 2.34. If we define

$$L_t = \int_0^t \mathbf{1}_{X_s=0}\,dK_s + P_t,$$

formula (2.32) becomes

$$X_t^+ = X_0^+ + \int_0^t \mathbf{1}_{X_s>0}\,dX_s + \frac{1}{2}L_t, \tag{2.35}$$

which is known as Tanaka's formula, where L_t is the local time of the continuous semimartingale X at level 0 and time t (see [64], Chapter VI, Theorem 1.2).

We finish the section with a slight generalization of Itô's formula.

Proposition 2.35. *Let $X \in S_d^0$ be an Itô process of the form*

$$X_t = X_0 + \int_0^t F_s ds + \int_0^t G_s dB_s, \quad t \geq 0,$$

and $\varphi \in C^1(\mathbb{R}^d)$ be such that $\nabla_x \varphi$ is differentiable on $\mathbb{R}^d \setminus \{0\}$ and for some positive constant L

$$|\nabla_x \varphi(x) - \nabla_x \varphi(y)| \leq L|x-y|, \quad \text{for all } x, y \in \mathbb{R}^d.$$

Then for all $0 \leq s \leq t$, $\mathbb{P}$-a.s.:

$$\varphi(X_t) = \varphi(X_s) + \int_s^t \langle \nabla_x \varphi(X_r), dX_r \rangle + \frac{1}{2}\int_s^t \mathbf{Tr}\left[G_r G_r^* D_{xx}^2 \varphi(X_r)\right] \mathbf{1}_{X_r \neq 0} dr. \tag{2.36}$$

Proof. Let

$$\rho(u) = \frac{1}{(2\pi)^{d/2}} \exp\left(-\frac{|u|^2}{2}\right).$$

We write the Itô formula (2.14) for the smooth function $\varphi_\varepsilon \in C^\infty(\mathbb{R}^d)$,

$$\varphi_\varepsilon(x) \stackrel{def}{=} \int_{\mathbb{R}^d} \varphi(x - \varepsilon u)\rho(u)\,du,$$

and taking into account (2.27-β) we have

$$\begin{aligned}\varphi_\varepsilon(X_t) = \varphi_\varepsilon(X_s) + \int_s^t \langle F_r, \nabla_x \varphi_\varepsilon(X_r)\rangle dr + \int_s^t \langle \nabla_x \varphi_\varepsilon(X_r), G_r dB_r\rangle \\ + \frac{1}{2}\int_s^t \mathbf{Tr}\left[G_r G_r^* D_{xx}^2 \varphi_\varepsilon(X_r)\right] \mathbf{1}_{X_r \neq 0} dr.\end{aligned} \tag{2.37}$$

Since

$$\begin{aligned}\nabla_x \varphi_\varepsilon(x) &= \int_{\mathbb{R}^d} \nabla_x \varphi(x - \varepsilon u)\rho(u)\,du \\ &= \int_{\mathbb{R}^d} \nabla_x \varphi(v)\rho\left(\frac{x-v}{\varepsilon}\right)\frac{dv}{\varepsilon^d},\end{aligned}$$

and for $x \neq 0$,

$$D_{xx}^2 \varphi_\varepsilon(x) = -\frac{1}{\varepsilon}\int_{\mathbb{R}^d} [\nabla_x \varphi(x - \varepsilon u)] \otimes u\,\rho(u)\,du$$

$$= \int_{\mathbb{R}^d} \frac{\nabla_x \varphi(x - \varepsilon u) - \nabla_x \varphi(x)}{-\varepsilon} \otimes u\, \rho(u)\, du,$$

we can pass to the limit in (2.37) as $\varepsilon \to 0_+$; the result follows. ∎

Remark 2.36. From the same proof we infer that if $\varphi \in C^1(\mathbb{R}^d)$ and

$$|\nabla_x \varphi(x+y) - \nabla_x \varphi(x)| \le L\,|y|, \quad \forall\, x, y \in \mathbb{R}^d,$$

then for all $0 \le s \le t$, $\mathbb{P}$-a.s.:

$$\varphi(X_t) \le \varphi(X_s) + \int_s^t \langle \nabla_x \varphi(X_r), dX_r \rangle + \int_s^t \alpha(X_r)\, |G_r|^2\, dr, \tag{2.38}$$

where

$$\alpha(r,x) \overset{def}{=} \limsup_{y \to 0} \frac{|\nabla_x \varphi(x+y) - \nabla_x \varphi(x)|}{|y|}.$$

If in (2.36) we put $d = 1$ and $\varphi(x) = (x^+)^2$, then we infer that for all $0 \le s \le t$:

$$(X_t^+)^2 = (X_s^+)^2 + 2\int_s^t X_r^+ dX_r + \int_s^t \mathbf{1}_{X_r > 0}\, |G_r|^2\, dr. \tag{2.39}$$

2.3.2 A Stochastic Subdifferential Inequality

We establish a stochastic subdifferential inequality which is very useful for estimates in the study of stochastic variational inequalities.

Lemma 2.37 (Stochastic Subdifferential Inequality). *Let $\psi : \Omega \times \mathbb{R}_+ \times \mathbb{R}^d \to \mathbb{R}$ be a function such that:*

- *$\psi(\cdot,\cdot,x)$ is $\mathcal{P}$-m.s.p. for all $x \in \mathbb{R}^d$;*
- *$\mathbb{P}$-a.s. $\omega \in \Omega$, $\psi(\omega,\cdot,\cdot)$ is of class $C^{1,1}(\mathbb{R}_+ \times \mathbb{R}^d)$; and*
- *for all $t \ge 0$, $\psi(\omega,t,\cdot) : \mathbb{R}^d \to \mathbb{R}$ is a convex function.*

Let $X \in S_d^0$ be a local semimartingale of the form

$$X_t = X_0 + \int_0^t dK_s + \int_0^t G_s dB_s, \quad t \ge 0,$$

($G \in \Lambda^0_{d \times k}$, K a $\mathcal{P}$-m.c.s.p., $K \in BV_{loc}(\mathbb{R}_+;\mathbb{R}^d)$, a.s.). Then $\mathbb{P}$-a.s. for all $s \le t$:

$$\psi(s, X_s) + \int_s^t \left[\frac{\partial \psi(r, X_r)}{\partial t} dr + \langle \nabla_x \psi(r, X_r), dX_r \rangle \right] \leq \psi(t, X_t). \quad (2.40)$$

Proof. Let $s = t_0 < t_1 < t_2 < \cdots < t_n = t$ such that $t_{i+1} - t_i = (t - s)/n$. For each $0 \leq i < n$, from the definition of the subdifferential operator, it follows that

$$\psi(t_i, X_{t_i}) + \left[\psi(t_{i+1}, X_{t_{i+1}}) - \psi(t_i, X_{t_{i+1}})\right] + \left\langle \nabla_x \psi(t_i, X_{t_i}), X_{t_{i+1}} - X_{t_i} \right\rangle \leq \psi(t_{i+1}, X_{t_{i+1}}).$$

The result follows by summing over i and by taking the limit $n \to \infty$. ■

From the proof it also follows that:

Lemma 2.38. *Let $\psi : \Omega \times \mathbb{R}_+ \times \mathbb{R}^d \to \mathbb{R}$ be a function such that $\psi(\cdot, \cdot, x)$ is $\mathcal{P}$-m.s.p. for all $x \in \mathbb{R}^d$ and $\mathbb{P}$-a.s. $\omega \in \Omega$,*

(*i*) $\psi(\omega, s, x) \leq \psi(\omega, t, x)$ *for all* $0 \leq s \leq t,\ x \in \mathbb{R}^d$,
(*ii*) $\psi(\omega, t, \cdot) : \mathbb{R}^d \to \mathbb{R}$ *is a convex function of class* C^1,
(*iii*) $(t, x) \longmapsto \nabla_x \psi(\omega, t, x) : \mathbb{R}_+ \times \mathbb{R}^d \to \mathbb{R}^d$ *is continuous.*

If $X \in S_d^0$ is a local semimartingale as in Lemma 2.37, then

$$\psi(t, X_t) + \int_t^T \langle \nabla_x \psi(r, X_r), dX_r \rangle \leq \psi(T, X_T), \quad \textit{for all } t \leq T, \quad \mathbb{P}\textit{-a.s.} \quad (2.41)$$

Remark 2.39. (*a*) Writing $X \in S_d^0$ (from Lemma 2.37) in the form

$$X_t = X_T + \int_t^T dU_r - \int_t^T Z_r dB_r$$

then (2.41) becomes

$$\psi(t, X_t) \leq \psi(T, X_T) + \int_t^T \langle \nabla_x \psi(r, X_r), dU_r \rangle - \int_t^T \langle \nabla_x \psi(r, X_r), Z_r dB_r \rangle, \quad (2.42)$$

$\mathbb{P}$-a.s., for $0 \leq t \leq T$.

(*b*) Setting $d = 1$ in (2.42),

$$\psi(t, x) = \varphi_\varepsilon(x) = \frac{1}{\sqrt{2\pi}} \int_{\mathbb{R}} (x - \varepsilon u)^+ e^{-u^2/2} du$$

and passing to the limit for $\varepsilon \to 0_+$ (as in the proof of Proposition 2.33) we have $\mathbb{P}$-a.s., for all $0 \leq t \leq T$:

$$X_t^+ \leq X_T^+ + \int_t^T \theta(X_r)\, dU_r - \int_t^T \theta(X_r) \langle Z_r, dB_r \rangle\,, \tag{2.43}$$

where

$$\theta(x) = \begin{cases} 0, & \text{if } x < 0, \\ \frac{1}{2}, & \text{if } x = 0, \\ 1, & \text{if } x > 0. \end{cases}$$

2.4 Martingale Representation Theorems

We have seen that Brownian motion and stochastic integrals of elements of Λ^p, $p \geq 1$, are martingales. In this section, we want to discuss conditions under which a martingale is a stochastic integral with respect to a given Brownian motion.

In this section, we again assume given a stochastic basis $(\Omega, \mathcal{F}, \mathbb{P}, \{\mathcal{F}_t\}_{t\geq 0})$ and that $\{B_t\,;\, t \geq 0\}$ is a k-dimensional Brownian motion. The main assumption is

(A_0) *the filtration $\{\mathcal{F}_t : t \geq 0\}$ is the natural filtration of $\{B_t : t \geq 0\}$, i.e. for all $t \geq 0$:*

$$\mathcal{F}_t = \mathcal{F}_t^B \overset{def}{=} \sigma(\{B_s : 0 \leq s \leq t\}) \vee \mathcal{N}.$$

Let $\mathcal{F}_\infty = \sigma(\{B_s : s \geq 0\}) \vee \mathcal{N} = \sigma\{\mathcal{F}_t : t \geq 0\}$.

Theorem 2.40 **(Λ^2-Martingale Representation Theorem).**

(i) If $0 < T \leq \infty$ and $\xi \in L^2(\Omega, \mathcal{F}_T, \mathbb{P}; \mathbb{R}^d)$, then there exists a unique $Z \in \Lambda^2_{d\times k}(0,T)$ such that

$$\xi = \mathbb{E}\xi + \int_0^T Z_s dB_s. \tag{2.44}$$

(ii) If $M \in \mathcal{M}_d^2$, then there exists a unique $Z \in \Lambda^2_{d\times k}$ such that

$$M_t = M_0 + \int_0^t Z_s dB_s,\ t \geq 0. \tag{2.45}$$

Proof. (ii): The representation result (2.45) follows from (2.44) applied to $\xi = M_T$, $T > 0$ arbitrary, by taking the conditional expectation $\mathbb{E}(\cdot|\mathcal{F}_t)$ for $0 \leq t \leq T$; it follows from (A_0) that $\mathbb{E}\xi = \mathbb{E}M_T = \mathbb{E}M_0 = M_0$.

(i) : *Uniqueness.* If Y, Z satisfy (2.44) then by the isometry property of Itô's integral (property (i_3) from Theorem 2.6)

$$\|Y - Z\|_{\Lambda^2_{d\times k}(0,T)} = \mathbb{E}\left|\int_0^T Y_s dB_s - \int_0^T Z_s dB_s\right|^2 = 0,$$

which yields $Y = Z$.

Existence. This will follow from the fact that the set

$$\mathcal{H} = \{h + \int_0^T Z_s dB_s \ : \ h \in \mathbb{R}^d, \ Z \in \Lambda^2_{d\times k}\}$$

coincides with $L^2(\Omega, \mathcal{F}_T, \mathbb{P}; \mathbb{R}^d)$. Indeed $\mathcal{H}$ is both closed and dense in $L^2(\Omega, \mathcal{F}_T, \mathbb{P}; \mathbb{R}^d)$. Clearly $\mathcal{H}$ is a linear subspace of $L^2(\Omega, \mathcal{F}_T, \mathbb{P}; \mathbb{R}^d)$.

a) *$\mathcal{H}$ is closed.*
Let

$$\xi_n = h_n + \int_0^T Z_s^n dB_s \ \in \mathcal{H},$$

and $\xi_n \longrightarrow \xi$ in $L^2(\Omega, \mathcal{F}_T, \mathbb{P}; \mathbb{R}^d)$. Then

$$h_n = \mathbb{E}\xi_n \longrightarrow \mathbb{E}\xi \ \text{ as } n \to \infty$$

and

$$\begin{aligned}\|Z_n - Z_m\|_{\Lambda^2_{d\times k}(0,T)} &= \mathbb{E}\left|\int_0^T Z_s^n dB_s - \int_0^T Z_s^m dB_s\right|^2 \\ &= \mathbb{E}|\xi_n - h_n - \xi_m + h_m|^2 \\ &\longrightarrow 0, \ \text{ as } n, m \longrightarrow \infty.\end{aligned}$$

Hence $\exists\, Z \in \Lambda^2_{d\times k}\,(0,T)$ such that $Z_n \to Z$ in $\Lambda^2_{d\times k}\,(0,T)$ and

$$\xi = E\xi + \int_0^T Z_s dB_s,$$

that is $\xi \in \mathcal{H}$.

b) *$\mathcal{H}$ is dense in $L^2(\Omega, \mathcal{F}_T, \mathbb{P}; \mathbb{R}^d)$.*
Let $\mathcal{R}$ denote the set of functions ρ of the form

$$\rho(t) = \sum_{j=0}^{n-1} \lambda_j \mathbf{1}_{[t_j, t_{j+1}[}(t),$$

where $n \in \mathbb{N}^*$, $0 = t_0 < t_1 < \cdots < t_n \leq T$ and $\lambda = (\lambda_0, \ldots, \lambda_{n-1}) \in \left(\mathbb{R}^k\right)^n$. Define

$$X_t^\rho = \int_0^t \langle \rho(s), dB_s \rangle .$$

Let $\varphi(t,x) = \exp\left(ix + \frac{1}{2}\int_0^t |\rho(s)|^2 ds\right)$ and $Y_t^\rho = \varphi\left(t, X_t^\rho\right)$. By the Itô formula

$$Y_t^\rho = 1 + i \int_0^t Y_s^\rho \langle \rho(s), dB_s \rangle .$$

Since $\mathbb{E}\int_0^T |Y_t^\rho \rho(t)|^2 dt = \int_0^T |\rho(t)|^2 \exp\left(\int_0^t |\rho(s)|^2 ds\right) dt < \infty$ and

$$\operatorname{Re} Y_T^\rho = 1 - \int_0^T \left(\operatorname{Im} Y_s^\rho\right) \rho^*(s)\, dB_s, \quad \operatorname{Im} Y_T^\rho = \int_0^T \left(\operatorname{Re} Y_s^\rho\right) \rho^*(s)\, dB_s,$$

it follows that for each $h \in \mathbb{R}^d$, $\left(\operatorname{Re} Y_T^\rho\right) h$ and $\left(\operatorname{Im} Y_T^\rho\right) h$ belong to $\mathcal{H}$.

The density of $\mathcal{H}$ in the space $L^2(\Omega, \mathcal{F}_T, \mathbb{P}; \mathbb{R}^d)$ will follow from the fact that for all $U \in L^2(\Omega, \mathcal{F}_T, \mathbb{P})$:

$$\mathbb{E}(U Y_T^\rho) = 0, \ \forall \rho \in \mathcal{R} \quad \Longrightarrow \quad U = 0.$$

But $\mathbb{E}(U Y_T^\rho) = 0, \ \forall \rho \in \mathcal{R}$, is equivalent to

$$\mathbb{E}\left[U \exp\left(i \sum_{j=0}^{n-1} \langle \lambda_j, B_{t_{j+1}} - B_{t_j} \rangle\right)\right] = 0, \ \forall \lambda \in \left(\mathbb{R}^k\right)^n,$$

for all $n \in \mathbb{N}^*$, $0 = t_0 < t_1 < \cdots < t_n < \infty$, $t_n \leq T$ and $\lambda = (\lambda_0, \ldots, \lambda_{n-1}) \in \left(\mathbb{R}^k\right)^n$, that is

$$\mathbb{E}[U \exp i \langle \mu, Y \rangle] = 0, \ \forall \mu = (\mu_1, \cdots, \mu_n) \in \left(\mathbb{R}^k\right)^n,$$

where $Y = (B_{t_1}, \cdots, B_{t_n})$ and

$$\langle \mu, Y \rangle \stackrel{def}{=} \langle \mu_1, B_{t_1} \rangle + \cdots + \langle \mu_n, B_{t_n} \rangle .$$

By Lemma 1.36, the last assertion is equivalent to

$$\mathbb{E}[U | Y] = 0.$$

Let $\mathcal{A} = \bigcup_{\{n, 0<t_1<\cdots<t_n<\infty,\ t_n \leq T\}} \sigma(B_{t_1}, B_{t_2}, \ldots, B_{t_n})$. We have that for any $A \in \mathcal{A}$,

$$\int_A U\, d\mathbb{P} = 0.$$

Hence the same is true for any $A \in \sigma(\mathcal{A}) \vee \mathcal{N} = \mathcal{F}_T$, and $U = 0$ a.s. follows. ■

Corollary 2.41. *Let $0 < T < \infty$, $\xi \in L^2(\Omega, \mathcal{F}_T, \mathbb{P}; \mathbb{R}^d)$ and $S \in S_d^2[0,T]$. Then there exists a unique pair $(Y, Z) \in S_d^2[0,T] \times \Lambda_{d\times k}^2(0,T)$ such that for all $t \in [0,T]$:*

$$Y_t = \xi + S_T - S_t - \int_t^T Z_s dB_s, \quad a.s. \tag{2.46}$$

Proof. *Uniqueness.* If (Y, Z) and $(\tilde{Y}, \tilde{Z})$ are two pairs satisfying (2.46), then

$$\begin{aligned}
\mathbb{E}\left|Y_t - \tilde{Y}_t\right|^2 + \mathbb{E}\int_t^T \left|Z_s - \tilde{Z}_s\right|^2 ds &= \mathbb{E}\left|Y_t - \tilde{Y}_t\right|^2 + \mathbb{E}\left|\int_t^T \left(Z_s - \tilde{Z}_s\right) dB_s\right|^2 \\
&= \mathbb{E}\left|Y_t - \tilde{Y}_t + \int_t^T \left(Z_s - \tilde{Z}_s\right) dB_s\right|^2 \\
&= 0
\end{aligned}$$

and the uniqueness follows.

Existence. By the Λ^2-martingale representation theorem, there exists a $Z \in \Lambda_{d\times k}^2(0,T)$ such that

$$\xi + S_T = \mathbb{E}(\xi + S_T) + \int_0^T Z_s dB_s.$$

Define

$$Y_t = \mathbb{E}(\xi + S_T) - S_t + \int_0^t Z_s dB_s.$$

Then the pair (Y, Z) solves (2.46). ■

We now give two extensions of Theorem 2.40.

Theorem 2.42 (**Λ^p-Martingale Representation Theorem**).

(i) If $0 < T \leq \infty$, $p > 1$ and $\xi \in L^p(\Omega, \mathcal{F}_T, \mathbb{P}; \mathbb{R}^d)$, then there exists a unique $Z \in \Lambda_{d\times k}^p(0,T)$ such that

$$\xi = \mathbb{E}\xi + \int_0^T Z_s dB_s. \tag{2.47}$$

(ii) If $p > 1$ and $M \in \mathcal{M}_d^p$, then there exists a unique $Z \in \Lambda_{d\times k}^p$ such that

$$M_t = M_0 + \int_0^t Z_s dB_s. \tag{2.48}$$

Proof. As in the proof of Theorem 2.40, it suffices to prove (i).

Uniqueness. If Z satisfies (2.47), then

$$\|Z\|^p_{\Lambda^p_{d\times k}(0,T)} \le C'_p\, \mathbb{E}(|\xi - \mathbb{E}\xi|^p). \tag{2.49}$$

Indeed, if $M_t = \int_0^t Z_s dB_s$, then by the Burkholder–Davis–Gundy inequality (1.18),

$$\|Z\|^p_{\Lambda^p_{d\times k}(0,T)} \le \frac{1}{c_p}\mathbb{E}\left(\sup_{0\le t\le T} |M_t|^p\right)$$

and by Doob's inequality (1.11), we have

$$\mathbb{E}\left(\sup_{0\le t\le T} |M_t|^p\right) \le \left(\frac{p}{p-1}\right)^p \mathbb{E}(|\xi - \mathbb{E}\xi|^p).$$

Uniqueness of Z follows from (2.49).

Existence. In the case $p \ge 2$ the existence of the solution follows from Theorem 2.40 and the inequality (2.49). Suppose now that $1 < p < 2$. Let $\xi_n = \xi \mathbf{1}_{[0,n]}(|\xi|)$, $n \in \mathbb{N}^*$. Then $\xi_n \in L^2(\Omega, \mathcal{F}_T, \mathbb{P}; \mathbb{R}^d)$ and

$$\xi_n \to \xi \text{ in } L^p(\Omega, \mathcal{F}_T, \mathbb{P}; \mathbb{R}^d), \text{ as } n \to \infty.$$

To each ξ_n we associate $Z^n \in \Lambda^2_{d\times k}(0,T)$ such that

$$\xi_n = \mathbb{E}\xi_n + \int_0^T Z_t^n dB_t. \tag{2.50}$$

From the inequality (2.49),

$$\|Z^n - Z^m\|_{\Lambda^p_{d\times k}(0,T)} \le \overline{c}_p \mathbb{E}\left[|\xi_n - \xi_m - \mathbb{E}(\xi_n - \xi_m)|^p\right]^{1/p} \to 0,$$

as $n, m \to \infty$. Hence there exists a $Z \in \Lambda^p_{d\times k}$ such that $Z^n \to Z$ in $\Lambda^p_{d\times k}$. The result follows by taking the limit in (2.50). ■

The next step is to extend the representations formula on random intervals. With this aim in mind we first prove the following:

Lemma 2.43. *Let $\tau:\Omega\to[0,\infty]$ be a stopping time and $Z\in\Lambda^1_{d\times k}(0,\infty)$. If $\int_0^\infty Z_s dB_s$ is $\mathcal{F}_\tau$-measurable then $Z_{t\vee\tau}=0$, a.s., a.e. $t\geq 0$.*

Proof. We have

$$\begin{aligned}\int_0^\infty \mathbf{1}_{[\tau,\infty[}(s)\,Z_s dB_s &= \int_0^\infty Z_s dB_s-\int_0^\tau Z_s dB_s\\ &=\mathbb{E}^{\mathcal{F}_\tau}\left(\int_0^\infty Z_s dB_s-\int_0^\tau Z_s dB_s\right)\\ &=\mathbb{E}^{\mathcal{F}_\tau}\left[\int_\tau^\infty Z_s dB_s\right]\\ &=0\end{aligned}$$

and the result follows since, by Proposition 2.10, the stochastic Itô integral is a linear injective operator from $\Lambda^0_{d\times k}(0,T)$ into $S^0_d[0,T]$. ■

Now from Theorem 2.42 we easily deduce:

Corollary 2.44. *Let $p>1$. If $\tau:\Omega\to[0,\infty]$ is a stopping time and $\xi\in L^p(\Omega,\mathcal{F}_\tau,\mathbb{P};\mathbb{R}^d)$, then:*

(A) there exists a unique $Z\in\Lambda^p_{d\times k}(0,\infty)$ such that $Z_{t\vee\tau}=0$ for all $t\geq 0$, and

$$\xi=\mathbb{E}\xi+\int_0^\tau Z_s dB_s,$$

or equivalently

(B) there exists a unique pair $(Y,Z)\in S^p_d\times\Lambda^p_{d\times k}(0,\infty)$ such that

$$\begin{aligned}&(a)\ Y_t=\xi-\int_{t\wedge\tau}^\tau Z_s dB_s,\ \forall\, t\geq 0,\ \ a.s.;\\ &(b)\ Y_{t\vee\tau}=\xi\quad and\quad Z_{t\vee\tau}=0,\ \forall\, t\geq 0,\quad a.s.\end{aligned}$$

or equivalently

(C) there exists a unique pair $(Y,Z)\in S^p_d\times\Lambda^p_{d\times k}(0,\infty)$ such that

$$\begin{aligned}&(a)\ Y_t=Y_{T\wedge\tau}-\int_{t\wedge\tau}^{T\wedge\tau} Z_s dB_s,\ \forall\, 0\leq t\leq T,\ \ a.s.;\\ &(b)\ \lim_{T\to\infty}\mathbb{E}\,|Y_{T\wedge\tau}-\xi|^p=0;\\ &(c)\ Y_{t\vee\tau}=\xi\quad and\quad Z_{t\vee\tau}=0,\ \forall\, t\geq 0,\quad a.s.\end{aligned}$$

Moreover, in (B) and (C), $Y_t=\mathbb{E}^{\mathcal{F}_{t\wedge\tau}}\xi$ and by the backward Burkholder–Davis–Gundy inequality (2.6), for all $t\geq 0$,

$$c_p\mathbb{E}\left(\int_{t\wedge\tau}^{\tau}|Z_r|^2\,dr\right)^{p/2}\leq\mathbb{E}\sup_{s\geq t}\left|\xi-\mathbb{E}^{\mathcal{F}_{s\wedge\tau}}\xi\right|^p\leq C_p\mathbb{E}\left(\int_{t\wedge\tau}^{\tau}|Z_r|^2\,dr\right)^{p/2}. \tag{2.51}$$

Corollary 2.45. *Let $0<T<\infty$, $p>1$, $\xi\in L^p(\Omega,\mathcal{F}_T,\mathbb{P};\mathbb{R}^d)$ and $S\in S_d^p[0,T]$. Then there exists a unique pair $(Y,Z)\in S_d^p[0,T]\times\Lambda_{d\times k}^p(0,T)$ satisfying $\mathbb{P}$-a.s., for all $t\in[0,T]$:*

$$Y_t=\xi+S_T-S_t-\int_t^T Z_s dB_s. \tag{2.52}$$

Moreover there exists a constant C_p such that if A is a $\mathcal{P}$-m.i.c.s.p., $A_0=0$, then

$$\mathbb{E}\sup_{t\in[0,T]}e^{pA_t}|Y_t|^p\leq C_p\,\mathbb{E}\left[e^{pA_T}|\xi|^p+\sup_{t\in[0,T]}e^{pA_t}|S_T-S_t|^p\right] \tag{2.53}$$

and

$$\begin{aligned}\mathbb{E}\sup_{t\in[0,T]}e^{pA_t}|Y_t|^p+\mathbb{E}\left(\int_0^T e^{2A_t}|Z_t|^2\,dt\right)^{p/2}\\ \leq C_p\,\mathbb{E}\left[e^{pA_T}|\xi|^p+\sup_{t\in[0,T]}e^{pA_t}|S_t|^p\right].\end{aligned} \tag{2.54}$$

Proof. By Theorem 2.42 there exists a unique $Z\in\Lambda_{d\times k}^p(0,T)$ such that

$$\xi+S_T=\mathbb{E}(\xi+S_T)+\int_0^T Z_s dB_s,$$

and the stochastic process $Y\in S_d^p[0,T]$ is uniquely defined by

$$Y_t=\mathbb{E}(\xi+S_T)-S_t+\int_0^t Z_s dB_s.$$

To prove (2.53) it suffices to consider the case where the right-hand side of the inequality is finite. The stochastic process $e^{A_t}|Y_t|$ is $\mathcal{F}_t$-dominated on $[0,T]$ by the positive random variable $e^{A_T}|\xi|+\sup\limits_{t\in[0,T]}e^{A_t}|S_T-S_t|$. Indeed

$$\begin{aligned}\mathbb{E}^{\mathcal{F}_t}\left(e^{A_t}|Y_t|\right)&=\mathbb{E}^{\mathcal{F}_t}\left(e^{A_t}\left|\mathbb{E}^{\mathcal{F}_t}(\xi+S_T-S_t)\right|\right)\\ &\leq\mathbb{E}^{\mathcal{F}_t}\left[e^{A_t}|\xi|+e^{A_t}|S_T-S_t|\right]\\ &\leq\mathbb{E}^{\mathcal{F}_t}\left[e^{A_T}|\xi|+\sup_{t\in[0,T]}e^{A_t}|S_T-S_t|\right].\end{aligned}$$

Then by Proposition 1.56 we have

$$\mathbb{E} \sup_{t\in[0,T]} e^{pA_t} |Y_t|^p \leq C_p\, \mathbb{E}\left(e^{pA_T} |\xi|^p + \sup_{t\in[0,T]} e^{pA_t} |S_T - S_t|^p \right).$$

To prove (2.54), again it suffices to consider the case where the right-hand side of the inequality is finite. From Proposition 6.80 for

$$(Y_t + S_t) = (\xi + S_T) + \int_t^T dK_s - \int_t^T Z_s dB_s$$

with $K = 0$, $\lambda = 0$, $V_t = A_t$, $R_t = N_t = 0$ and

$$dD_t = |Y_t + S_t|^2 \, dA_t,$$

we obtain

$$\mathbb{E}\left[\sup_{t\in[0,T]} \left|e^{A_t} (Y_t + S_t)\right|^p + \Big(\int_0^T e^{2A_t} |Z_t|^2 \, dt\Big)^{p/2} \right] \leq C_p\, \mathbb{E}\left[e^{pA_T} |Y_T + S_T|^p \right],$$

which completes the proof. ∎

Theorem 2.46 (Λ^1-Martingale Representation Theorem).

(i) If $0 < T \leq \infty$ and $\xi \in L^1(\Omega, \mathcal{F}_T, \mathbb{P}; \mathbb{R}^d)$, then there exists a unique $Z \in \bigcap_{0<q<1} \Lambda^q_{d\times k}(0,T)$ such that

$$\begin{aligned} &(a)\;\; \xi = \mathbb{E}\xi + \textstyle\int_0^T Z_r dB_r, \\ &(b)\;\; M_\cdot = \textstyle\int_0^\cdot Z_r dB_r \in \mathcal{M}^1_d[0,T]. \end{aligned} \tag{2.55}$$

(ii) If $M \in \mathcal{M}^1_d$, then there exists a unique $Z \in \bigcap_{0<q<1} \Lambda^q_{d\times k}$ such that

$$M_t = M_0 + \int_0^t Z_s dB_s. \tag{2.56}$$

Proof. Again (ii) follows from (i).

(i): *Uniqueness.* Let $0 < q < 1$. If Z and $\tilde{Z}$ are the integrands corresponding respectively to ξ and $\tilde{\xi}$, then by the Burkholder–Davis–Gundy inequality (2.4) and the inequality (1.11-A_3) in Theorem 1.60,

$$\begin{aligned} \|Z - \tilde{Z}\|^q_{\Lambda^q_{d\times k}(0,T)} &\leq \frac{1}{c_q} \mathbb{E} \sup_{t\in[0,T]} \left| \int_0^t (Z_r - \tilde{Z}_r)\, dB_r \right|^q \\ &\leq \frac{1}{c_q(1-q)} \left(\mathbb{E} \left| \int_0^T (Z_r - \tilde{Z}_r)\, dB_r \right| \right)^q \end{aligned}$$

$$= \frac{1}{c_q (1-q)} \left(\mathbb{E} \left| \xi - \tilde{\xi} - \mathbb{E}\xi + \mathbb{E}\tilde{\xi} \right| \right)^q$$
$$\leq \frac{2}{c_q (1-q)} \left(\mathbb{E} \left| \xi - \tilde{\xi} \right| \right)^q .$$

Hence the uniqueness follows.

Existence. Let $n \in \mathbb{N}^*$, $\xi_n = \xi \mathbf{1}_{[0,n]}(|\xi|) \in L^2(\Omega, \mathcal{F}_T, \mathbb{P}; \mathbb{R}^d)$ and $Z^n \in \Lambda^2_{d \times k}(0,T)$ be such that

$$\xi_n = \mathbb{E}\xi_n + \int_0^T Z_r^n dB_r. \tag{2.57}$$

Then for all $t \in [0,T]$:

$$\mathbb{E}^{\mathcal{F}_t}(\xi_n) = \mathbb{E}\xi_n + \int_0^t Z_t^n dB_t. \tag{2.58}$$

Let $0 < q < 1$. Since

$$\xi_n \to \xi \ \text{ in } \ L^1(\Omega, \mathcal{F}_T, \mathbb{P}; \mathbb{R}^d)$$

and

$$\|Z^n - Z^m\|^q_{\Lambda^q_{d \times k}(0,T)} \leq \frac{2}{c_q (1-q)} \left(\mathbb{E} |\xi_n - \xi_m| \right)^q ,$$

we deduce that there exists a $Z \in \Lambda^q_{d \times k}(0,T)$ such that $Z^n \to Z$ in $\Lambda^q_{d \times k}(0,T)$. The existence result follows by taking the limit in (2.57) and (2.58). ■

It is now clear that we have the following extension of Corollary 2.41:

Corollary 2.47. *Let $0 < T < \infty$, $\xi \in L^1(\Omega, \mathcal{F}_T, \mathbb{P}; \mathbb{R}^d)$ and $S \in S^1_d[0,T]$. Then there exists a unique pair $(Y, Z) \in S^0_d[0,T] \times \Lambda^0_{d \times k}(0,T)$, such that:*

$$\begin{aligned}
&(a)\ Y_t = \xi + S_T - S_t - \int_t^T Z_s dB_s, \quad a.s., \ \forall\, t \in [0,T]; \\
&(b)\ \sup_{t \in [0,T]} \mathbb{E}|Y_t| + \mathbb{E} \sup_{t \in [0,T]} |Y_t|^q + \mathbb{E}\left(\int_0^T |Z_t|^2 dt \right)^{q/2} < \infty, \ \forall\, 0 < q < 1; \\
&(c)\ M_\cdot = \int_0^\cdot Z_r dB_r \in \mathcal{M}^1_d[0,T].
\end{aligned} \tag{2.59}$$

Proof. Uniqueness. Let (Y, Z) and $(\tilde{Y}, \tilde{Z})$ be two pairs satisfying (2.59). Since Y_0, $\tilde{Y}_0$, S_0 are $\mathcal{F}_0$-measurable, they are deterministic quantities and

$$Y_0 = \mathbb{E}(\xi + S_T) - S_0 = \tilde{Y}_0.$$

By the Burkholder–Davis–Gundy inequality (1.18) and Doob's inequality (1.11-A_3), we have for $0 < q < 1$:

$$\begin{aligned}
\mathbb{E}\left[\left(\int_0^T \left|Z_t - \tilde{Z}_t\right|^2 dt\right)^{q/2}\right] &\leq \frac{1}{c_q}\mathbb{E}\left(\sup_{0\leq t\leq T}\left|\int_0^t \left(Z_s - \tilde{Z}_s\right) dB_s\right|^q\right) \\
&\leq \frac{1}{c_q\,(1-q)}\left[\mathbb{E}\left|\int_0^T \left(Z_s - \tilde{Z}_s\right) dB_s\right|\right]^q \\
&\leq \frac{1}{c_q\,(1-q)}\left[\mathbb{E}\left|Y_0 - \tilde{Y}_0\right|\right]^q \\
&= 0.
\end{aligned}$$

Existence. The pair (Y, Z) is well defined by

$$\xi + S_T = \mathbb{E}\,(\xi + S_T) + \int_0^T Z_s dB_s$$

and

$$Y_t = \mathbb{E}\,(\xi + S_T) - S_t + \int_0^t Z_s dB_s.$$

■

With arguments almost identical to those of Theorems 2.40 and 2.42 one can prove the following:

Theorem 2.48. *Let $\mathcal{F}_\infty = \sigma\left(\bigcup_{t>0}\mathcal{F}_t\right)$. If $p > 1$ and $\xi \in L^p(\Omega, \mathcal{F}_\infty, \mathbb{P}; \mathbb{R}^d)$, then there exists a unique $Z \in \Lambda^p_{d\times k}(\mathbb{R}_+)$ such that*

$$\xi = \mathbb{E}\xi + \int_0^\infty Z_t dB_t.$$

2.5 Girsanov's Theorem

In this section, we assume given a stochastic basis $(\Omega, \mathcal{F}, \mathbb{P}, \{\mathcal{F}_t\}_{t\geq 0})$ such that $\mathcal{F} = \sigma(\bigcup_{t>0}\mathcal{F}_t)$ and $\{B_t\,;\, t \geq 0\}$ is a k-dimensional Brownian motion.

Let $X \in \Lambda^0_k$, and define the process:

$$Z_t = \exp\left[\int_0^t \langle X_s, dB_s\rangle - \frac{1}{2}\int_0^t |X_s|^2 ds\right],\ t \geq 0.$$

$\{Z_t, t \geq 0\}$ is a positive continuous stochastic process and, furthermore, a *local martingale* since by the Itô formula

$$Z_t = 1 + \int_0^t Z_s \langle X_s, dB_s \rangle . \tag{2.60}$$

Define for each $n \in \mathbb{N}$ the stopping time

$$\tau_n = \inf \left\{ t \geq 0 : Z_t + \int_0^t |X_r|^2 \, dr \geq n \right\}$$

and let

$$X_r^n = X_r \mathbf{1}_{[0,\tau_n]}(r),$$
$$Z_t^n = Z_{t \wedge \tau_n} = \exp \left[\int_0^t \langle X_r^n, dB_r \rangle - \frac{1}{2} \int_0^t |X_r^n|^2 \, dr \right], \ t \geq 0.$$

Then it is clear that

- $Z_0^n = 1$,
- $\dfrac{Z_t^n}{Z_s^n} = 1 + \displaystyle\int_s^t \frac{Z_r^n}{Z_s^n} \langle X_r^n, dB_r \rangle$, for all $0 \leq s < t$,
- $\mathbb{E}^{\mathcal{F}_s} \left(\dfrac{Z_t^n}{Z_s^n} \right) = 1$ for all $0 \leq s < t$,
- $\mathbb{E} Z_t^n = 1$ and $\{Z_t^n, t \geq 0\}$ is a martingale,
- $Z_t^n \to Z_t$ and $\dfrac{Z_t^n}{Z_s^n} \to \dfrac{Z_t}{Z_s}$, $\mathbb{P}$-a.s., for each $0 \leq s < t$.

By Fatou's lemma for all $0 \leq s < t$ and $A \in \mathcal{F}_s$,

$$\mathbb{E} \left(\mathbf{1}_A \frac{Z_t}{Z_s} \right) \leq \liminf_{n \to +\infty} \mathbb{E} \left(\mathbf{1}_A \frac{Z_t^n}{Z_s^n} \right) = \mathbb{E} \left(\mathbf{1}_A \right),$$

which yields

$$\mathbb{E}^{\mathcal{F}_s} \left(\frac{Z_t}{Z_s} \right) \leq 1, \text{ that is } \{Z_t, t \geq 0\} \text{ is a supermartingale.}$$

In particular

$$\mathbb{E} Z_t \leq \mathbb{E} Z_s \leq \mathbb{E} Z_0 = 1, \quad \text{for all } 0 \leq s < t. \tag{2.61}$$

Lemma 2.49. *Let $T > 0$. Then $\mathbb{E} Z_T = 1$ if and only if $Z \in \mathcal{M}^1 [0, T]$ (i.e. $\{Z_t ; 0 \leq t \leq T\}$ is a continuous martingale).*

Proof. Since for all $0 \le s < t \le T$, $Z_s - \mathbb{E}^{\mathcal{F}_s}(Z_t) \ge 0$, $\mathbb{P}$-a.s., it follows that $Z_s = \mathbb{E}^{\mathcal{F}_s}(Z_t)$, $\mathbb{P}$-a.s., if and only if $\mathbb{E}\left[Z_s - \mathbb{E}^{\mathcal{F}_s}(Z_t)\right] = 0$, or equivalently, by (2.61) $\mathbb{E}Z_t = 1$, for all $t \in [0, T]$. ∎

Also we remark that by the Cauchy–Bunyakovski–Schwarz inequality and (2.61), for all $a, b \in \mathbb{R}$ and $0 \le s < t$

$$
\begin{aligned}
&\mathbb{E}\exp\left(a\int_s^t \langle X_r, dB_r\rangle - \frac{b}{2}\int_s^t |X_r|^2\, d\boldsymbol{r}\right)\\
&\quad = \mathbb{E}\exp\left(\int_s^t \langle aX_r, dB_r\rangle - \int_s^t |aX_r|^2\, d\boldsymbol{r}\right)\exp\left(\frac{2a^2-b}{2}\int_s^t |X_r|^2\, d\boldsymbol{r}\right)\\
&\quad \le \left[\mathbb{E}\exp\left(\int_s^t \langle 2aX_r, dB_r\rangle - \frac{1}{2}\int_s^t |2aX_r|^2\, d\boldsymbol{r}\right)\right]^{1/2} \times\\
&\qquad\qquad \times\left[\mathbb{E}\exp\left(\left(2a^2-b\right)\int_s^t |X_r|^2\, d\boldsymbol{r}\right)\right]^{1/2}\\
&\quad \le \left[\mathbb{E}\exp\left(\left(2a^2-b\right)\int_s^t |X_r|^2\, d\boldsymbol{r}\right)\right]^{1/2}\\
&\quad \le \left(\int_s^t \mathbb{E}\exp\left[(t-s)\left(2a^2-b\right)|X_r|^2\right]\frac{d\boldsymbol{r}}{t-s}\right)^{1/2},
\end{aligned}
$$

where for the last inequality we have used Jensen's inequality for the convex function $x \to \exp x$, and the probability measure $\dfrac{dr}{t-s}$ on the interval $[s, t]$.

Hence if $X \in \Lambda_k^0$, for all $0 \le s < t$ and $a \in \mathbb{R}$:

$$
(a)\ \mathbb{E}\left(\frac{Z_t}{Z_s}\right) = \mathbb{E}\exp\left[\int_s^t \langle X_r, dB_r\rangle - \frac{1}{2}\int_s^t |X_r|^2\, d\boldsymbol{r}\right] \le 1, \tag{2.62}
$$

$$
(b)\ \mathbb{E}\exp\left(a\int_s^t \langle X_r, dB_r\rangle\right) \le \left[\mathbb{E}\exp\left(2a^2\int_s^t |X_r|^2\, d\boldsymbol{r}\right)\right]^{1/2},
$$

$$
\begin{aligned}
(c)\ \left[\mathbb{E}\left(\frac{Z_t}{Z_s}\right)^a\right]^2 &\le \mathbb{E}\exp\left(\left(2a^2-a\right)\int_s^t |X_r|^2\, d\boldsymbol{r}\right)\\
&\le \frac{1}{t-s}\int_s^t \mathbb{E}\exp\left((t-s)\left(2a^2-a\right)|X_r|^2\right) dr.
\end{aligned}
$$

We set some sufficient conditions which yield $\mathbb{E}\left(Z_T\right) = 1$.

Proposition 2.50. *Let $X \in \Lambda_k^0$ and $T > 0$. Suppose that one of the following assumptions is satisfied:*

(*i*) *There exists a partition $0 = t_0 < t_1 < \cdots < t_N = T$ and for each $k \in \overline{1, N}$ there exists a $\delta_k > 0$ such that*

$$\mathbb{E}\exp\left((1+\delta_k)\int_{t_{k-1}}^{t_k}|X_t|^2\,dt\right)<\infty,\quad \textit{for all } k\in\overline{1,N}. \tag{2.63}$$

(*ii*) *There exist two constants $c,\gamma>0$ such that*

$$\mathbb{E}\exp\left(\gamma\,|X_t|^2\right)\leq c,\quad a.e.\ t\in[0,T]. \tag{2.64}$$

(*iii*) *(Kazamaki) $X\in\Lambda_d^1[0,T]$ and*

$$\mathbb{E}\exp\left(\frac{1}{2}\int_0^T\langle X_t,dB_t\rangle\right)<\infty. \tag{2.65}$$

(*iv*) *(Novikov)*

$$\mathbb{E}\exp\left(\frac{1}{2}\int_0^T|X_t|^2\,dt\right)<\infty. \tag{2.66}$$

Then $\mathbb{E}\,Z_t=1$ for all $t\in[0,T]$ and (by Lemma 2.49) $Z\in\mathcal{M}^1[0,T]$.

Proof. (*i*) Let $k\in\overline{1,N}$. From (2.62-c) with X replaced by X^n and $a=a_k=\frac{1}{4}\left(1+\sqrt{9+8\delta_k}\right)$ in (2.62-c) we obtain

$$\begin{aligned}\mathbb{E}\left(\frac{Z_{t_k}^n}{Z_{t_{k-1}}^n}\right)^{a_k}&\leq\left[\mathbb{E}\exp\left((2a_k^2-a_k)\int_{t_{k-1}}^{t_k}|X_r^n|^2\,dr\right)\right]^{1/2}\\&\leq\left[\mathbb{E}\exp\left((1+\delta_k)\int_{t_{k-1}}^{t_k}|X_r|^2\,dr\right)\right]^{1/2}.\end{aligned}$$

Hence

$$\frac{Z_{t_k}^n}{Z_{t_{k-1}}^n}\to\frac{Z_{t_k}}{Z_{t_{k-1}}}\quad a.s.\ \text{ and }\ \left\{\frac{Z_{t_k}^n}{Z_{t_{k-1}}^n}\right\}_{n\in\mathbb{N}}\ \text{ is uniformly integrable}$$

and consequently $\dfrac{Z_{t_k}^n}{Z_{t_{k-1}}^n}\to\dfrac{Z_{t_k}}{Z_{t_{k-1}}}$ in $L^1(\Omega)$; in particular

$$\mathbb{E}^{\mathcal{F}_{t_{k-1}}}\frac{Z_{t_k}}{Z_{t_{k-1}}}=\lim_{n\to\infty}\mathbb{E}^{\mathcal{F}_{t_{k-1}}}\frac{Z_{t_k}^n}{Z_{t_{k-1}}^n}=1.$$

For each $1\leq k\leq n$, $\dfrac{Z_{t_k}}{Z_{t_{k-1}}}$ is $\mathcal{F}_{t_k}$-measurable, and $\mathbb{E}^{\mathcal{F}_{t_{k-1}}}\left(\dfrac{Z_{t_k}}{Z_{t_{k-1}}}\right)=1$. The result then follows easily from the formula

$$Z_T = \prod_{k=1}^{n} \frac{Z_{t_k}}{Z_{t_{k-1}}}.$$

(*ii*) Via the inequality (2.62-c) with $a = 2$, the condition (2.64) implies (2.63) for $t_k = \frac{kT}{N}$, $N \geq \frac{6T}{\gamma}$ and $\delta_k = 5$.
(*iii*) See Exercise 1.20 and its hint in Annex E.
(*iv*) Via the inequality (2.62-b) with $a = \frac{1}{2}$, the condition (2.66) implies (2.65).

■

We note that the criterion (2.64) from (ii) is often the easiest to apply, because of the freedom of choice of the constant γ. It applies in particular in the case where $\{X_s\,;\,0 \leq s \leq t\}$ is a Gaussian process whose mean and variance are bounded on $[0, t]$.

The main result of this section is the following:

Theorem 2.51 (Girsanov). *Suppose that $\mathbb{E}Z_t = 1$ for all $t \geq 0$. Then there exists a unique probability Q on $(\Omega, \mathcal{F})$, $\mathcal{F} = \sigma\left(\bigcup_{t\geq 0} \mathcal{F}_t\right)$, such that for each $t \geq 0$, $A \in \mathcal{F}_t$,*

$$Q(A) = \int_A Z_t \, d\mathbb{P} \quad (i.e.\ \frac{dQ|_{\mathcal{F}_t}}{d\mathbb{P}|_{\mathcal{F}_t}} = Z_t),$$

and the stochastic process $\{\overline{B}_t\,;\,t \geq 0\}$ defined as

$$\overline{B}_t = B_t - \int_0^t X_s \, ds, \ t \geq 0,$$

is an $\mathbb{R}^k$-valued $\mathcal{F}_t$-Brownian motion under Q.

The next Proposition follows from a classical result of Daniell and Kolmogorov (see Karatzas and Shreve [42], Sections 2.2 and 3.5).

Proposition 2.52. *Let $\mathcal{F}_t = \mathcal{F}_t^B$ and $\mathcal{F} = \sigma\left(\bigcup_{t\geq 0} \mathcal{F}_t\right)$. Suppose that for each $t > 0$ we are given a probability measure R_t on $(\Omega, \mathcal{F}_t)$, with the property that for each $0 \leq s < t$,*

$$R_t|_{\mathcal{F}_s} = R_s.$$

Then there exists a unique probability measure R on $(\Omega, \mathcal{F})$ such that

$$R|_{\mathcal{F}_t} = R_t, \quad t \geq 0.$$

Proof (Sketch). Let $\{t_1, t_2, \ldots, t_n\}$, $n \in \mathbb{N}^*$, be a finite set of positive distinct numbers and $t = \max\{t_1, t_2, \ldots, t_n\}$. Define for $D \in \mathcal{B}_n = \mathcal{B}(\mathbb{R}^n)$,

$$Q_{t_1,t_2,\ldots,t_n}(D) = R_t\left(\{\omega \in \Omega : (B_{t_1}(\omega), \ldots, B_{t_n}(\omega)) \in D\}\right).$$

Then $\{Q_{t_1,t_2,\ldots,t_n} : t_1, t_2, \ldots, t_n \geq 0,\ t_i \neq t_j,\ n \in \mathbb{N}^*\}$ is a consistent family of finite-dimensional distributions and by the Daniell–Kolmogorov consistency theorem there is a probability Q on $\left(\mathbb{R}^{\mathbb{R}_+}, \mathcal{B}\left(\mathbb{R}^{\mathbb{R}_+}\right)\right)$, $\mathbb{R}_+ = [0, \infty)$, such that

$$Q_{t_1,t_2,\ldots,t_n}(D) = Q\left(\left\{w \in \mathbb{R}^{\mathbb{R}_+} : (w(t_1), \ldots, w(t_n)) \in D\right\}\right), \quad D \in \mathcal{B}_n,$$

and the probability R on $\mathcal{F}$ is given by

$$R(\{\omega \in \Omega : B.(\omega) \in \Gamma\}) = Q(\Gamma), \quad \Gamma \in \mathcal{B}\left(\mathbb{R}^{\mathbb{R}_+}\right).$$

■

We now proceed to the:

Proof of Theorem 2.51. Note that under the condition of the theorem, $Q_t = Z_t \cdot \mathbb{P}$, $t > 0$, is a probability measure, and provided $\{Z_t\}$ is a martingale, the compatibility condition of Proposition 2.52 is satisfied, since for $s < t$, $A \in \mathcal{F}_s$,

$$Q_t(A) = \int_A Z_t \, d\mathbb{P} = \int_A Z_s \, d\mathbb{P} = Q_s(A).$$

Hence, since our assumption implies that $\{Z_t\,;\, t \geq 0\}$ is a martingale, in view of Proposition 2.52, Theorem 2.51 will be proved once we establish that for each $t > 0$, $\{\overline{B}_s,\ 0 \leq s \leq t\}$ is a Brownian motion under $Q_t \stackrel{def}{=} Z_t \cdot \mathbb{P}$.

It suffices to prove that for any $p \in \mathbb{N}$, $0 < t_1 < \cdots < t_p \leq t$, $u_1, \ldots, u_p \in \mathbb{R}^k$,

$$\mathbb{E}_{Q_t} \exp\left(i \sum_{j=1}^{p} \langle u_j, \overline{B}_{t_j}\rangle\right) = \exp\left(-\frac{1}{2} \sum_{j,\ell=1}^{p} \langle u_j, u_\ell\rangle t_j \wedge t_\ell\right).$$

Let $p \in \mathbb{N}$, $0 < t_1 < \cdots < t_p \leq t$, $u_1, \ldots, u_p \in \mathbb{R}^k$ be fixed for the rest of the proof. Let τ_n, X^n and Z^n be given as at the beginning of this section. We also introduce the notations

$$Y_s^n = X_s^n + i \sum_{j=1}^{p} u_j \mathbf{1}_{[0,t_j]}(s) \quad \text{and} \quad U_t^n = \exp\left(\int_0^t \langle Y_s^n, dB_s\rangle - \frac{1}{2}\int_0^t \langle Y_s^n, Y_s^n\rangle\, ds\right).$$

Clearly

$$U_t^n = Z_t^n \exp\left(i \sum_{j=1}^{p} \langle u_j, \overline{B}_{t_j}\rangle + \frac{1}{2} \sum_{j,\ell=1}^{p} \langle u_j, u_\ell\rangle (t_j \wedge t_\ell)\right).$$

Since $\int_0^t |Y_s^n|^2\, ds \leq C_n$, $\mathbb{P}$-a.s., it follows, by an easy extension of Proposition 2.50-(i) to complex X's, that $\mathbb{E} U_t^n = 1$, i.e.

$$\mathbb{E}\left[Z_t^n \exp\left(i\sum_{j=1}^{p}\langle u_j, \overline{B}_{t_j}\rangle\right)\right] = \exp\left[-\frac{1}{2}\sum_{j,\ell=1}^{p}\langle u_j, u_\ell\rangle\left(t_j \wedge t_\ell\right)\right].$$

Passing to the limit as $n \to \infty$ we have that $Z_t^n \to Z_t$ in $L^1(\Omega)$, thanks to Lemma 1.24, and the desired result follows. ∎

2.6 Exercises

Exercise 2.1. Let $\Delta_n : 0 = t_0^n < t_1^n < \ldots < t_{k_n}^n = T$ be a fixed sequence of partitions of $[0, T]$ such that $\|\Delta_n\| = \max\left\{t_{i+1}^n - t_i^n : i \in \overline{0, k_n - 1}\right\} \to 0$ as $n \to \infty$. Let $g : [0, T] \to R$ be a continuous function.

Show that the sums

$$S_n(f) = \sum_{i=0}^{k_n-1} f(t_i^n)\left[g(t_{i+1}^n) - g(t_i^n)\right]$$

converge for all continuous function $f : [0, T] \to \mathbb{R}$ iff $g \in BV[0, T]$ (g is of finite variation on $[0, T]$).

Remark. Given $f \in C[0, T]$ and

$$f_n(t) = \sum_{i=0}^{n-1} f(t_i^n)\mathbf{1}_{\left[t_i^n, t_{i+1}^n\right)}(t),$$

then $f_n \to f$ in $L^2(0, T)$ as $n \to \infty$,

$$\mathbb{B}_T(f_n) = \sum_{i=0}^{k_n-1} f(t_i^n)\left[B_{t_{i+1}^n} - B_{t_i^n}\right] \to \mathbb{B}_T(f) \text{ in } L^2(\Omega), \text{ as } n \to \infty,$$

and as a consequence there exists a subsequence

$$\mathbb{B}_T\left(f_{n_k}\right) \to \mathbb{B}_T(f), \ \mathbb{P}\text{-a.s. } \omega \in \Omega,$$

while the paths $t \longrightarrow B_t(\omega)$ have a.s. unbounded variation. Is this a contradiction?

Exercise 2.2. Let $\tau, \theta : \Omega \to [0, \infty]$ be stopping times such that $\tau(\omega) \leq \theta(\omega)$, $\mathbb{P}$-a.s. $\omega \in \Omega$. Using

$$\int_\tau^\theta X_r dB_r = \int_0^\infty \mathbf{1}_{[\tau,\theta[}(r)\, X_r dB_r, \text{ for } X \in \Lambda_{d\times k}^0(\mathbb{R}_+),$$

prove that

1. $\mathbb{E}^{\mathcal{F}_\tau} \int_\tau^\theta X_r dB_r = 0,\ a.s.,\ \forall\ X \in \Lambda^1_{d\times k}(\mathbb{R}_+)$,
2. $\mathbb{E}^{\mathcal{F}_\tau} \left| \int_\tau^\theta X_r dB_r \right|^2 = \mathbb{E}^{\mathcal{F}_\tau} \int_\tau^\theta |X_r|^2 dr,\ a.s., \forall X \in \Lambda^2_{d\times k}(\mathbb{R}_+)$,
3. $\int_\tau^\theta \eta X_r dB_r = \eta \int_\tau^\theta X_r dB_r,\ a.s., \forall \eta \in L^0(\Omega, \mathcal{F}_\tau, \mathbb{P}; \mathbb{R}^{m\times d}),\ X \in \Lambda^0_{d\times k}(\mathbb{R}_+)$.

Exercise 2.3. Let $f \in L^2(\mathbb{R}_+)$ and

$$\mathbb{B}(f) = \int_0^\infty f(t)\, dB_t.$$

Show that the linear subspace $\{\mathbb{B}(f);\ f \in L^2(\mathbb{R}_+)\}$ of $L^2(\Omega, \mathcal{F}, \mathbb{P})$ coincides with the Gaussian space $H[B]$, and the (class of the) random variable $\mathbb{B}(f)$ is characterized by the two properties:

$$\begin{aligned} &(i)\ \ \mathbb{B}(f) \in H[B],\\ &(ii)\ \ \mathbb{E}[B_t \mathbb{B}(f)] = \textstyle\int_0^t f(s) ds,\quad \forall\, t > 0. \end{aligned}$$

Exercise 2.4. Let $f \in C^1(\mathbb{R}_+)$, $T > 0$. Show that

$$\int_0^T f(t)\, dB_t = f(T)B(T) - \int_0^T f'(t) B_t\, dt,$$

and if moreover $f(t) \to 0$ as $t \to \infty$, and

$$\int_0^\infty |f'(t)| \sqrt{t}\, dt < \infty,$$

then

$$\int_0^\infty f(t)\, dB_t = -\int_0^\infty f'(t)\, B_t\, dt.$$

Exercise 2.5. Let $X \in S_1^0$ be a local semimartingale of the form

$$X_t = X_0 + \int_0^t dK_s + \int_0^t \langle G_s, dB_s \rangle, \quad t \geq 0,$$

where $K \in S_1^0$, $K_\cdot(\omega) \in BV_{loc}(\mathbb{R};\mathbb{R})$ $\mathbb{P}$-a.s. $\omega \in \Omega$, $G \in \Lambda^0_{1\times k}$ and there exists a positive stochastic process γ such that P-a.s. $\gamma_\cdot \in L^2_{loc}(\mathbb{R}_+)$ and

$$|G_s| \leq \gamma_s \sqrt{|X_s|}, \quad d\mathbb{P} \otimes dt\text{-}a.e.$$

Let $g : \mathbb{R} \to \mathbb{R}$ be given by

$$g(x) = 6(x-1)^5 - 15(x-1)^4 + 10(x-1)^3$$

and

$$\varphi_\varepsilon(x) = \begin{cases} 0, & \text{if } x \le \varepsilon, \\ x\, g\left(\dfrac{x}{\varepsilon}\right), & \text{if } \varepsilon < x < 2\varepsilon, \\ x, & \text{if } x \ge 2\varepsilon. \end{cases}$$

Show that:

1. $\varphi_\varepsilon \in C^2(\mathbb{R})$ and $0 \le x^+ - \varphi_\varepsilon(x) \le 2\varepsilon$;
2. $|\theta(x) - \varphi_\varepsilon'(x)| \le 5 \times 1_{]0,2\varepsilon[}(x)$, where

$$\theta(r) = \begin{cases} 0, \text{ if } r \le 0, \\ 1, \text{ if } r > 0; \end{cases}$$

3. $|\varphi_\varepsilon''(x)| \le \dfrac{34}{\varepsilon} 1_{]\varepsilon,2\varepsilon[}(x)$;
4. $\mathbb{P}$-a.s., for all $t \ge 0$

$$X_t^+ = X_0^+ + \int_0^t \theta(X_s)\, dK_s + \int_0^t \theta(X_s) \langle G_s, dB_s \rangle. \tag{2.67}$$

Exercise 2.6. We let $S = \inf\{t \in [0,1],\ t + B_t^2 = 1\}$ and

$$X_t = \begin{cases} -2(1-t)^{-2} B_t \mathbf{1}_{t \le S}, & \text{if } \ 0 \le t < 1, \\ 0, \ \ \text{if } \ t = 1, \end{cases}$$

where $\{B_t : t \ge 0\}$ is a one-dimensional Brownian motion.

1. Show that S is a stopping time, that $\mathbb{P}(S < 1) = 1$, and that $\int_0^1 X_t^2 dt < \infty$, $\mathbb{P}$-a.s.
2. Deduce from Itô's formula applied to $(1-t)^{-2} B_t$ that

$$\int_0^1 X_s dB_s - \frac{1}{2} \int_0^1 X_s^2 ds = -1 - 2 \int_0^S t(1-t)^{-4} B_t^2 dt \le -1.$$

3. Deduce that for the above choice of X, the supermartingale

$$Z_t(X) = \exp\left(\int_0^t X_s dB_s - \frac{1}{2} \int_0^t X_s^2 ds \right), \quad 0 \le t \le 1,$$

is not a martingale.

Exercise 2.7. Let $X \in \Lambda^0_{d\times k}(\mathbb{R}_+)$ and $\{B_t : t \geq 0\}$ be a k-dimensional Brownian motion. Using an argument similar to that used in the proof of Proposition 1.56-$\mathbf{B}_1$, show that for all $T \geq 0$, $\varepsilon, \delta > 0$,

$$\mathbb{P}\left(\sup_{t\in[0,T]}\left|\int_0^t X_s dB_s\right| > \varepsilon\right) \leq \mathbb{P}\left(\int_0^T |X_s|^2 ds > \delta\right) + \frac{\delta}{\varepsilon^2}.$$

Exercise 2.8. Let $\{B_t : t \geq 0\}$ be a scalar Brownian motion. Show that the stochastic Langevin equation

$$dV_t + bV_t\,dt = \sigma\,dB_t, \quad t \geq 0,$$

has for each initial condition $V_0 \in R$ a unique solution given by

$$V_t = e^{-tb}V_0 + \int_0^t e^{-(t-s)b}\sigma\,dB_s, \quad t \geq 0,$$

called the Orsntein–Uhlenbeck process. If V_0 is a Gaussian random variable independent of $\{B_t, t \geq 0\}$, then $\{V_t, t \geq 0\}$ is a Gaussian process with mean and covariance

$$\begin{aligned}
&\mathbb{E}V_t = e^{-tb}\mathbb{E}V_0 \\
&\mathrm{Cov}(V_s, V_t) = e^{-sb}\mathrm{Cov}(V_0)e^{-tb} + \int_0^s e^{-(s-u)b}\sigma^2 e^{-(t-u)b)}\,du.
\end{aligned}$$

If moreover $b > 0$, $EV_0 = 0$ and $2b\mathrm{Cov}(V_0) = \sigma^2$, then $\{V_t, t \geq 0\}$ is a centered stationary Gaussian process whose covariance is given by

$$\mathrm{Cov}(V_s, V_t) = \mathrm{Cov}(V_0)e^{-(t-s)b}.$$

More generally, again if $b > 0$, the mean and covariance of $\{V(T+t), t \geq 0\}$ tend to those of the stationary process, as $T \to \infty$.

Exercise 2.9. Let M be a k-dimensional continuous square integrable martingale ($M \in \mathcal{M}^2_k$). Denote by $\Lambda^p_{d\times k}(0, T; <M>)$, $p \geq 0$, the space of progressively measurable stochastic processes $X : [0, T] \times \Omega \to \mathbb{R}^{d\times k}$ such that:

$$\begin{aligned}
\textstyle\int_0^T |X_t|^2\,d <M>_t &< \infty, \quad \text{a.s. for } p = 0, \text{ and} \\
\mathbb{E}\left(\textstyle\int_0^T |X_t|^2\,d <M>_t\right)^{p/2} &< \infty, \quad \text{for } p > 0.
\end{aligned}$$

The linear space $\Lambda^p_d(0, T; <M>)$ with the natural metric

$$d_p(X,Y) = \begin{cases} \left[\mathbb{E}\left(\int_0^T |X_t - Y_t|^2 d <M>_t\right)^{p/2}\right]^{(1/p)\wedge 1}, & \text{if } 0 < p < \infty, \\ \mathbb{E}\left[1 \wedge \left(\int_0^T |X_t - Y_t|^2 d <M>_t\right)^{1/2}\right], & \text{if } p = 0, \end{cases}$$

is a Polish space. If $p \geq 1$, $\Lambda_d^p(0,T;<M>)$ is a Banach space. Let $\mathcal{E}_{d\times k}(0,T)$ be the linear space of stochastic processes of the form

$$X_t(\omega) = \sum_{i=0}^{n-1} X_i(\omega) \mathbf{1}_{[t_i,t_{i+1}[}(t), \quad t \geq 0,$$

with $n \in \mathbb{N}^*, 0 \leq t_0 < t_1 < \ldots < t_n \leq T$ and for $0 \leq i \leq n-1$, let $X_i : \Omega \to \mathbb{R}^{d\times k}$ be an $\mathcal{F}_{t_i}$-measurable bounded random variable. Show that:

1. $\mathcal{E}_{d\times k}(0,T)$ is a dense linear subspace of $\Lambda_d^p(0,T;<M>)$ (see also Proposition 2.1);
2. the linear operator $\mathbb{M} : \mathcal{E}_{d\times k} \to \mathcal{M}_d^2$ given by

$$\mathbb{M}_t(X) = \int_0^t X_r dM_r \overset{def}{=} \sum_{i=0}^{N-1} X_i \left(M_{t\wedge t_{i+1}} - M_{t\wedge t_i}\right)$$

satisfies

$$\mathbb{E} \sup_{t\in[0,T]} \left|\int_0^t X_r dM_r\right|^p \leq C_p \mathbb{E}\left(\int_0^T |X_r|^2 d <M>_r\right)^{p/2} \tag{2.68}$$

and

$$\mathbb{E}[1 \wedge \sup_{t\in[0,T]} |\int_0^t X_r dM_r|] \leq 3\left[\mathbb{E}[1 \wedge \int_0^T |X_t|^2 d <M>_t]\right]^{1/3}; \tag{2.69}$$

Definition: If $X \in \Lambda_{d\times k}^p(0,T)$ then the integral $\displaystyle\int_0^t X_r dM_r \overset{def}{=} \mathbb{M}_t(X)$, where $\mathbb{M} : \Lambda_{d\times k}^p(0,T) \to S_d^p[0,T]$ is the unique extension by continuity (preserving the inequalities (2.68) and (2.69)) of the linear operator

$$\mathbb{M} : \mathcal{E}_{d\times k}(0,T) \subset \Lambda_{d\times k}^p(0,T;<M>) \to S_d^p[0,T];$$

3. if $p \geq 1$ then $\mathbb{M}(X) \in M_d^p[0,T]$;
4. if $\varphi \in C^{1,2}\left(\mathbb{R}^m \times \mathbb{R}^k\right)$, and $\{V_t \,;\, t \in [0,T]\}$ is an m-dimensional $\mathcal{P}$-m.b-v.c.s.p. then $\mathbb{P}$-a.s., for all $t \in [0,T]$:

$$
\begin{aligned}
\varphi\left(V_t, M_t\right) = \varphi\left(V_0, M_0\right) + \int_0^t \left\langle \varphi_v'\left(V_s, M_s\right), dV_s\right\rangle \\
+ \int_0^t \left\langle \varphi_x'\left(V_s, M_s\right), dM_s\right\rangle + \frac{1}{2}\mathbf{Tr}\int_0^t \varphi_{xx}''\left(V_s, M_s\right) d \ll M \gg_s .
\end{aligned}
\tag{2.70}
$$

Exercise 2.10. Let $0 = t_0 < t_1 < \ldots < t_n = t$ and

$$\delta_n = \max\left\{t_{i+1} - t_i : i \in \overline{0, n-1}\right\} \to 0.$$

Consider a scalar Brownian motion $\{B_t : t \geq 0\}$ and the Riemann–Stieltjes sum

$$S_n^\lambda = \sum_{i=0}^{n-1} B_{r_i}\left(B_{t_{i+1}} - B_{t_i}\right),$$

where $r_i = t_i + \lambda\left(t_{i+1} - t_i\right)$ and $\lambda \in [0, 1]$. Let

$$\mathbf{I}_t^\lambda = \frac{1}{2}B_t^2 + \left(\lambda - \frac{1}{2}\right)t.$$

Show that:

1. $S_n^\lambda = \frac{1}{2}\sum_{i=0}^{n-1}\left[\left|B_{r_i} - B_{t_i}\right|^2 - \left|B_{t_{i+1}} - B_{r_i}\right|^2\right] + \frac{1}{2}\left|B_t\right|^2;$
2. $\mathbb{E}\left(S_n^\lambda - \mathbf{I}_t^\lambda\right) = 0$ and $\mathbb{E}\left|S_n^\lambda - \mathbf{I}_t^\lambda\right|^2 \leq \frac{t}{2}\delta_n;$
3. $\left(L^2-\right)\lim_{n\to\infty}\sum_{i=0}^{n-1} B_{(1-\lambda)t_i + \lambda t_{i+1}}\left(B_{t_{i+1}} - B_{t_i}\right) = \int_0^t B_s\, dB_s + \lambda t;$ in particular deduce that
$$\int_0^t B_s\, dB_s = \frac{1}{2}B_t^2 - \frac{1}{2}t;$$
4. (Stratonovich integral) if $g \in C^1(\mathbb{R})$ then
$$
\begin{aligned}
\int_0^t g(B_s) \circ dB_s &\overset{def}{=} \left(L^2-\right)\lim_{n\to\infty}\sum_{i=0}^{n-1} g\left(B_{\frac{t_i + t_{i+1}}{2}}\right)\left(B_{t_{i+1}} - B_{t_i}\right) \\
&= \int_0^t g(B_s)\, dB_s + \frac{1}{2}\int_0^t g'(B_s)\, ds;
\end{aligned}
$$
in particular deduce that
$$\int_0^t B_s \circ dB_s = \frac{1}{2}B_t^2.$$

Chapter 3
Stochastic Differential Equations

3.1 Introduction

Let $\{B_t, t \geq 0\}$ be a k-dimensional Brownian motion with respect to the given stochastic basis $(\Omega, \mathcal{F}, \mathbb{P}, \{\mathcal{F}_t\}_{t\geq 0})$. Our goal in this chapter is to study stochastic differential equations (abbreviated SDE) of the form

$$\begin{cases} dX_t = F(t, X_t)dt + G(t, X_t)dB_t, & t \geq 0, \\ X_0 = \xi, \end{cases} \tag{3.1}$$

where $\xi : \Omega \to \mathbb{R}^d$ is the initial condition and the coefficients are given functions

$$F : \Omega \times [0, \infty[\times \mathbb{R}^d \to \mathbb{R}^d, \qquad G : \Omega \times [0, \infty[\times \mathbb{R}^d \to \mathbb{R}^{d\times k}.$$

$F(t, X_t)dt$ is called the drift of X and $G(t, X_t)G^*(t, X_t)$ is the matrix of diffusion coefficients of X.

We shall always assume that:

(*i*) *ξ is an $\mathcal{F}_0$-measurable random vector;*
(*ii*) *the functions F and G are $\mathcal{P} \otimes B_d$-measurable.*

In fact F and G will be $(\mathcal{P}, \mathbb{R}^d)$-Carathéodory functions, that is:

(c_1) $F(\cdot, \cdot, x)$ and $G(\cdot, \cdot, x)$ are $\mathcal{P}$-measurable for every $x \in \mathbb{R}^d$; and
(c_2) $F(\omega, t, \cdot)$ and $G(\omega, t, \cdot)$, are $d\mathbb{P} \otimes dt - a.e.$ continuous functions,

and then F and G are $\mathcal{P} \otimes \mathcal{B}_d$-measurable (see Exercise 1.1).

E. Pardoux and A. Răşcanu, *Stochastic Differential Equations, Backward SDEs, Partial Differential Equations*, Stochastic Modelling and Applied Probability 69,
DOI 10.1007/978-3-319-05714-9_3,

We state the following definition:

Definition 3.1. A stochastic process $X \in S_d^0$ is a (strong) solution of (3.1) if for all $T \geq 0$

$$\int_0^T |F(s, X_s)|\, ds + \int_0^T |G(s, X_s)|^2\, ds < \infty, \ a.s.,$$

and

$$X_t = \xi + \int_0^t F(s, X_s) ds + \int_0^t G(s, X_s) dB_s, \quad \forall\, t \geq 0, \ a.s. \tag{3.2}$$

Our discussion in this chapter is almost exclusively restricted to the theory of strong solutions of SDEs. We start with the classical results, essentially due to Itô, under Lipschitz assumptions upon the coefficients, see e.g. [32, 35, 42, 49, 64]. We then extend these results to the case of so-called monotone drift, following the original ideas of Jacod [40] and Krylov–Gyöngy [36]. We also discuss the particular case where the solution is a Markov diffusion process (i.e. the case where the coefficients are deterministic; the randomness of the solution is due solely to the driving Brownian motion). We then discuss the connection between diffusion processes and solutions of second order partial differential equations, via the Feynman–Kac formula. We establish in particular a result which proves that certain expectations of functionals of the solution of an SDE are viscosity solutions of linear second order PDEs, without any assumption concerning ellipticity. Finally we discuss weak solutions of SDEs (this is not an essential subject for us, but this tool will be needed in the next chapter).

3.2 A Basic Inequality

For convenience we recall a basic inequality from Annex C. First we introduce a notation used in this chapter.

Notation 3.2. *For $p \geq 1$ we define*

$$m_p \overset{def}{=} 1 \vee (p - 1)$$

and we recall the notations

$$\|U\|_{[t,s]} = \sup_{r \in [t,s]} |U_r| \quad \text{and} \quad \|U\|_T = \|U\|_{[0,T]} \cdot$$

Let $X \in S_d^0$ be a local semimartingale of the form

$$X_t = X_0 + K_t + \int_0^t G_s dB_s, \ t \geq 0, \quad \mathbb{P}\text{-a.s.}, \tag{3.3}$$

where

◊ $K \in S_d^0$; $K_\cdot \in BV_{loc}\left([0,\infty[\,;\mathbb{R}^d\right)$, $K_0 = 0$, $\mathbb{P}$-a.s.;
◊ $G \in \Lambda^0_{d\times k}$.

Concerning the triple (X, K, G) we assume:

$$\textbf{(SDE-FB)}: \tag{3.4}$$

Given $p \geq 1$ *and* $\lambda \geq 0$, *there exist three* $\mathcal{P}$*-m.i.c.s.p.* D, R, N *and a* $\mathcal{P}$*-m.b-v.c.s.p.* V, such *that* $D_0 = R_0 = N_0 = V_0 = 0$ and *as signed measures on* $[0,\infty[$:

$$dD_t + \langle X_t, dK_t\rangle + \left(\frac{1}{2}m_p + 9p\lambda\right)|G_t|^2\,dt \leq \mathbf{1}_{p\geq 2}dR_t + |X_t|dN_t + |X_t|^2 dV_t.$$

□

Propositions 6.71 and 6.74 are reformulated as:

Proposition 3.3. *Let* $p \geq 1$ *and the assumptions* (**SDE-FB**) *be satisfied.*

(*I*) *If the inequality (3.4) is satisfied for* $\lambda = 0$ *and* $R = N = 0$, *then:* $\mathbb{P}$*-a.s., with all* $0 \leq t \leq s$,

$$\mathbb{E}^{\mathcal{F}_t} e^{-pV_s}|X_s|^p + p\,\mathbb{E}^{\mathcal{F}_t}\int_t^s e^{-pV_r}|X_r|^{p-2}\,dD_r \leq e^{-pV_t}|X_t|^p \tag{3.5}$$

and moreover for all $\delta \geq 0$

$$\mathbb{E}^{\mathcal{F}_t}\frac{e^{-pV_s}|X_s|^p}{\left(1+\delta e^{-2V_s}|X_s|^2\right)^{p/2}} + p\,\mathbb{E}^{\mathcal{F}_t}\int_t^s \frac{e^{-pV_r}|X_r|^{p-2}}{\left(1+\delta\left|e^{-V_r}X_r\right|^2\right)^{(p+2)/2}}dD_r \leq \frac{e^{-pV_t}|X_t|^p}{\left(1+\delta e^{-2V_t}|X_t|^2\right)^{p/2}}. \tag{3.6}$$

(*II*) *For every* $p \geq 1$ *and* $\lambda > 1$ *satisfying (3.4) there exists a constant* $C_{p,\lambda}$ *such that:* $\mathbb{P}$*-a.s., for all* $0 \leq t \leq s$,

$$\begin{aligned}\mathbb{E}^{\mathcal{F}_t}\left\|e^{-V}X\right\|^p_{[t,s]} &+ \mathbb{E}^{\mathcal{F}_t}\int_t^s e^{-pV_r}|X_r|^{p-2}\,dD_r \\ &+\mathbb{E}^{\mathcal{F}_t}\left(\int_t^s e^{-2V_r}\left(dD_r + |G_r|^2\,dr\right)\right)^{p/2} \\ \leq C_{p,\lambda}\Bigg[e^{-pV_t}|X_t|^p &+ \mathbb{E}^{\mathcal{F}_t}\left(\int_t^s e^{-2V_r}\mathbf{1}_{p\geq 2}dR_r\right)^{p/2} \\ &+\mathbb{E}^{\mathcal{F}_t}\left(\int_t^s e^{-V_r}dN_r\right)^p\Bigg]\end{aligned} \tag{3.7}$$

and moreover for all $\delta \geq 0$

$$\begin{aligned}
&\mathbb{E}^{\mathcal{F}_t}\frac{\|e^{-V}X\|^p_{[t,s]}}{\left(1+\delta\|e^{-V}X\|^2_{[t,s]}\right)^{p/2}}+\mathbb{E}^{\mathcal{F}_t}\int_t^s\frac{e^{-pV_r}|X_r|^{p-2}}{\left(1+\delta|e^{-V_r}X_r|^2\right)^{(p+2)/2}}dD_r\\
&\quad+\mathbb{E}^{\mathcal{F}_t}\left(\int_t^s\frac{e^{-2V_r}}{\left(1+\delta|e^{-V_r}X_r|^2\right)^2}\left(dD_r+|G_r|^2dr\right)\right)^{p/2}\\
&\le C_{p,\lambda}\left[\frac{e^{-pV_t}|X_t|^p}{\left(1+\delta e^{-2V_t}|X_t|^2\right)^{p/2}}+\mathbb{E}^{\mathcal{F}_t}\left(\int_t^s e^{-2V_r}\mathbf{1}_{p\ge 2}dR_r\right)^{p/2}\right.\\
&\quad\left.+\mathbb{E}^{\mathcal{F}_t}\left(\int_t^s e^{-V_r}dN_r\right)^p\right].
\end{aligned}\tag{3.8}$$

3.3 Estimates, Uniqueness and Comparison Results

3.3.1 Classical SDE

We now formulate the main assumptions for the study of our SDE:

$$X_t=\xi+\int_0^t F(s,X_s)\,ds+\int_0^t G(s,X_s)\,dB_s,\ \mathbb{P}\text{-a.s., }\forall t\ge 0,\tag{3.9}$$

where $\{B_t,t\ge 0\}$ is a k-dimensional Brownian motion with respect to the given stochastic basis $(\Omega,\mathcal{F},\mathbb{P},\{\mathcal{F}_t\}_{t\ge0})$ and $\xi:\Omega\to\mathbb{R}^d$ is an $\mathcal{F}_0$-measurable random vector and for all $x\in\mathbb{R}^d$ the functions

$$F(\cdot,\cdot,x):\Omega\times[0,\infty[\,\to\mathbb{R}^d,\qquad G(\cdot,\cdot,x):\Omega\times[0,\infty[\,\to\mathbb{R}^{d\times k}$$

are $\mathcal{P}$-m.s.p.

We introduce the notation, for all $\rho\ge 0$:

$$F^{\#}_{\rho}(t)\overset{def}{=}\sup_{x\le\rho}|F(t,x)|\,.$$

The general assumptions on F and G under which we shall study the SDE (3.9) are the following:

$$(\textbf{SDE-H}_F):\tag{3.10}$$

♦ $\exists\ \mu:\Omega\times[0,\infty[\,\to\mathbb{R}$, *$\mathcal{P}$-m.s.p., such that,* $\forall T\ge 0,\ \int_0^T|\mu_t|\,dt<\infty$, *a.s., and*

($\mathbf{C}_F$) *Continuity:*

$$x \longrightarrow F(t,x) : \mathbb{R}^d \to \mathbb{R}^d \text{ is continuous, } d\mathbb{P} \otimes dt\text{-}a.e.;$$

($\mathbf{M}_F$) *Monotonicity condition:*

$$\langle x - y, F(t,x) - F(t,y) \rangle \leq \mu_t |x - y|^2, \quad d\mathbb{P} \otimes dt\text{-}a.e., \quad \forall\, x, y \in \mathbb{R}^d;$$

($\mathbf{B}_F$) *Boundedness condition:*

$$\int_0^T F_\rho^{\#}(s)\, ds < \infty, \quad a.s., \ \forall\, T, \rho \geq 0.$$

$$\textbf{(SDE-H}_G\textbf{)}: \tag{3.11}$$

♦ $\exists\ \ell : \Omega \times [0,\infty[\to \mathbb{R}_+$, *$\mathcal{P}$-m.s.p., such that,* $\forall T \geq 0$, $\int_0^T (\ell_t)^2\, dt < \infty$, *a.s., and*

($\mathbf{L}_G$) *Lipschitz condition:*

$$|G(t,x) - G(t,y)| \leq \ell_t |x - y|, \quad d\mathbb{P} \otimes dt\text{-}a.e. \quad \forall\, x, y \in \mathbb{R}^d;$$

($\mathbf{B}_G$) *Boundedness condition:*

$$\int_0^T |G(t,0)|^2 dt < \infty, \quad a.s.\ \forall\, T \geq 0.$$

□

The proof of the first lemma is left as an exercise.

Lemma 3.4. *Let the assumptions (**SDE-H**$_F$) and (**SDE-H**$_G$) be satisfied. If* $U \in S_d^0[0,T]$, *then*

$$\int_0^T |F(t,U_t)|\, dt + \int_0^T |G(t,U_t)|^2\, dt < \infty, \quad \mathbb{P}\text{-}a.s.$$

and the mapping

$$U \longrightarrow \left(\int_0^{\cdot} F(s,U_s) ds, \int_0^{\cdot} G(s,U_s) dB_s \right)$$

is continuous from $S_d^0[0,T]$ *to* $S_d^0[0,T] \times S_d^0[0,T]$.

We shall show that the above assumptions on F and G yield, for the solutions $\{X_t : t \geq 0\}$ of the SDE (3.2), inequalities of the form (3.7) and (3.8).

Lemma 3.5. *Let the assumption* $(\boldsymbol{M}_F)$ *from* (**SDE-H**$_F$) *be satisfied. Then for all* $r_0 \geq 0$, $x \in \mathbb{R}^d$, $t \geq 0$, $\mathbb{P}$-*a.s.*

$$\begin{aligned} r_0 \left|F(t,x)\right| + \langle F(t,x), x\rangle \leq{} & r_0 \left[F^{\#}_{r_0}(t) + r_0 \mu_t^+\right] \\ & + \left[F^{\#}_{r_0}(t) + 2r_0 \left|\mu_t\right|\right]|x| + \mu_t |x|^2 . \end{aligned} \tag{3.12}$$

Proof. Let $r_0 \geq 0$. The monotonicity property of F implies that for all $|u| \leq 1$:

$$\langle F(t, r_0 u) - F(t,x), r_0 u - x\rangle \leq \mu_t \left|r_0 u - x\right|^2,$$

and, consequently, $\forall\ |u| \leq 1$:

$$\begin{aligned} & r_0 \langle F(t,x), -u\rangle + \langle F(t,x), x\rangle \\ & \leq \mu_t \left|r_0 u - x\right|^2 + \left|F(t, r_0 u)\right| \left|x - r_0 u\right| \\ & \leq \mu_t \left(|x|^2 - 2r_0 \langle u, x\rangle + r_0^2 |u|^2\right) + F^{\#}_{r_0}(t)\left(|x| + r_0\right) \\ & \leq r_0 \left[F^{\#}_{r_0}(t) + r_0 \mu_t^+\right] + \left[F^{\#}_{r_0}(t) + 2r_0 \left|\mu_t\right|\right]|x| + \mu_t |x|^2 . \end{aligned}$$

(3.12) follows by taking the sup of the left-hand side over all vectors u such that $|u| \leq 1$. ■

Since for all $u, v \in \mathbb{R}^d$ and $\lambda > 1$

$$|u|^2 \leq \frac{\lambda}{\lambda - 1} |v|^2 + \lambda |u - v|^2 ,$$

we obtain, from (3.11-$\mathbf{L}_G$), that

$$|G(t,x)|^2 \leq \frac{\lambda}{\lambda - 1} |G(t,0)|^2 + \lambda \left(\ell_t\right)^2 |x|^2 . \tag{3.13}$$

Writing now the SDE (3.9) in the form

$$X_t = \xi + K_t + \int_0^t G_r dB_r ,$$

where

$$K_t = \int_0^t F(r, X_r)\, dr \quad \text{and} \quad G_r = G(r, X_r),$$

it follows from (3.12) and (3.13) that for all $p \geq 2$ and $\lambda > 1$

$$\begin{aligned} dD_r^{(r_0)} + \langle X_r, dK_r \rangle + \left(\frac{m_p}{2} + 9p\lambda\right) |G_r|^2 \, dr \\ \leq dR_r^{(r_0)} + |X_r| dN_r^{(r_0)} + |X_r|^2 dV_r, \end{aligned}$$

where $m_p = 1 \vee (p-1) = p-1$, $c_{p,\lambda} = \frac{\lambda}{\lambda - 1}(9p\lambda + (p-1)/2)$,

$$\begin{aligned} D_t^{(r_0)} &= r_0 \int_0^t |F(r, X_r)| \, dr, \\ R_t^{(r_0)} &= r_0 \int_0^t \left[F_{r_0}^{\#}(r) + r_0 \mu_r^+\right] dr + c_{p,\lambda} \int_0^t |G(r,0)|^2 \, dr, \\ N_t^{(r_0)} &= \int_0^t \left[F_{r_0}^{\#}(r) + 2r_0 |\mu_r|\right] dr, \\ V_t &= \int_0^t \left[\mu_r + \lambda \left(\frac{p-1}{2} + 9p\lambda\right) (\ell_r)^2\right] dr. \end{aligned} \tag{3.14}$$

We deduce from Proposition 3.3, first with $r_0 = 0$ and second with $r_0 > 0$, the following:

Proposition 3.6. *Let the assumptions (3.10-**SDE-H**$_F$) and (3.11-**SDE-H**$_G$) be satisfied and let $X \in S_d^0$ be a solution of the SDE (3.9). Then for every $p \geq 2$ and $\lambda > 1$ there exists a constant $C_{p,\lambda}$ such that (with V defined as in (3.14)), $\mathbb{P}$-a.s., for all $0 \leq t \leq s$:*

$$\begin{aligned} \mathbb{E}^{\mathcal{F}_t} \left\|e^{-V} X\right\|_{[t,s]}^p \leq C_{p,\lambda} \Big[e^{-pV_t} |X_t|^p + \mathbb{E}^{\mathcal{F}_t} \Big(\int_t^s e^{-V_r} |F(r,0)| \, dr \Big)^p \\ + \mathbb{E}^{\mathcal{F}_t} \Big(\int_t^s e^{-2V_r} |G(r,0)|^2 \, dr \Big)^{p/2} \Big], \end{aligned} \tag{3.15}$$

and for every $r_0 > 0$, there exists a constant C_{p,λ,r_0} such that for all $0 \leq t \leq s$:

$$\begin{aligned} &\mathbb{E} \int_t^s e^{-pV_r} |X_r|^{p-2} |F(r, X_r)| \, dr + \mathbb{E} \Big(\int_t^s e^{-2V_r} |F(r, X_r)| \, dr \Big)^{p/2} \\ &\leq C_{p,\lambda,r_0} \Big[\mathbb{E} e^{-pV_t} |X_t|^p + \mathbb{E} \Big(\int_t^s e^{-2V_r} \left[\mu_r^+ + \left|F_{r_0}^{\#}(r)\right| + |G(r,0)|^2\right] dr \Big)^{p/2} \\ &\qquad + \mathbb{E} \Big(\int_t^s e^{-V_r} \left[F_{r_0}^{\#}(r) + |\mu_r|\right] dr \Big)^p \Big]. \end{aligned}$$

Using Corollary 6.76 from Annex C we can derive other estimates. Let $p \geq 2$. Assume there exists $b > 0$ such that

$$\mu_t + (p-1)(\ell_t)^2 \leq b \quad \text{and} \quad (p-1)|G(t,0)|^2 + |F(t,0)| \leq b, \text{ a.e.}$$

Since

$$\langle X_t, dK_t\rangle + \frac{p-1}{2}\left|G(t, X_t)\right|^2 dt \le b\,dt + b|X_t|dt + b|X_t|^2 dt$$

and using Corollary 6.76 we have for all $t \ge 0$, $\rho > 0$:

$$\begin{aligned} &(a)\ \mathbb{E}\ |X_t|^p\, e^{-3(p-1)bt} \le \mathbb{E}\ |X_0|^p + \frac{1}{p-1}, \\ &(b)\ \int_0^\infty e^{-3(p-1)bt-\rho t}\ \mathbb{E}\ |X_t|^p\, dt \le \frac{1}{\rho}\left(\mathbb{E}\ |X_0|^p + \frac{1}{p-1}\right). \end{aligned} \tag{3.16}$$

The natural question now is to extend the inequality (3.15) to the case $1 \le p < 2$.

Proposition 3.7. *Let $X \in S_d^0$ be a solution of the SDE (3.9) and the assumptions (3.10-**SDE-H**$_F$) and (3.11-**SDE-H**$_G$) be satisfied. Assume there exists a $U \in S_d^0$ such that*

$$U_t = U_0 + \int_0^t G(r, U_r)\, dB_r.$$

Let $p \ge 1$, $\lambda > 1$ and

$$V_t = \int_0^t \left[\mu_r + \left(\frac{1}{2}m_p + 9p\lambda\right)(\ell_r)^2\right] dr.$$

Then there exists a constant $C_{p,\lambda}$ such that for all $\delta \ge 0$, $0 \le t \le s$:

$$\begin{aligned} \mathbb{E}^{\mathcal{F}_t} \left\|e^{-V}(X-U)\right\|_{[t,s]}^p \le C_{p,\lambda}\Big[& e^{-pV_t}\,|X_t - U_t|^p \\ & + \mathbb{E}^{\mathcal{F}_t}\left(\int_t^s e^{-V_r}\,|F(r, U_r)|\, dr\right)^p\Big],\ a.s. \end{aligned} \tag{3.17}$$

Proof. The SDE (3.9) can be written in the form

$$(X_t - U_t) = (\xi - U_0) + K_t + \int_0^t G_r dB_r$$

with

$$K_t = \int_0^t F(r, X_r)\, dr \quad \text{and} \quad G_r = G(r, X_r) - G(r, U_r).$$

Note that

$$\langle X_r - U_r, dK_r \rangle + \left(\frac{1}{2}m_p + 9p\lambda\right) |G_r|^2 \, dr$$
$$\leq |X_r - U_r| \, |F(r, U_r)| \, dr + |X_r - U_r|^2 \, dV_r .$$

Hence, by Proposition 3.3 with X_r replaced by $X_r - U_r$, the inequality (3.17) follows. ■

Theorem 3.8 (Uniqueness). *Let the assumptions (3.10-**SDE-H**$_F$) and (3.11-**SDE-H**$_G$) be satisfied. Let $X, \hat{X} \in S_d^0[0,T]$ be two solutions of the SDE (3.9) corresponding respectively to the initial data ξ and $\hat{\xi}$.*
Let $p \geq 1$ and $\lambda > 1$ be arbitrary.

(I) If

$$V_t = \int_0^t \left(\mu_r + \frac{1}{2}m_p (\ell_r)^2\right) dr,$$

then for all $\delta \geq 0$, $0 \leq t \leq s$, $\mathbb{P}$-a.s.:

$$\mathbb{E}^{\mathcal{F}_t} \frac{e^{-pV_s}|X_s - \hat{X}_s|^p}{\left(1 + \delta e^{-2V_s}|X_s - \hat{X}_s|^2\right)^{p/2}} \leq \frac{e^{-pV_t}|X_t - \hat{X}_t|^p}{\left(1 + \delta e^{-2V_t}|X_t - \hat{X}_t|^2\right)^{p/2}} . \tag{3.18}$$

(II) If

$$V_t = \int_0^t \left[\mu_r + \left(\frac{1}{2}m_p + 9p\lambda\right)(\ell_r)^2\right] dr,$$

then there exists a constant $C_{p,\lambda}$ such that for all $\delta \geq 0$, $0 \leq t \leq s$:

$$\mathbb{E}^{\mathcal{F}_t} \frac{\|e^{-V}(X - \hat{X})\|_{[t,s]}^p}{\left(1 + \delta \|e^{-V}(X - \hat{X})\|_{[t,s]}^2\right)^{p/2}} \leq C_{p,\lambda} \frac{|e^{-V_t}(X_t - \hat{X}_t)|^p}{\left(1 + \delta |e^{-V_t}(X_t - \hat{X}_t)|^2\right)^{p/2}}, \quad a.s. \tag{3.19}$$

The uniqueness in S_d^0 follows by choosing $t = 0$ and $\delta > 0$.

Proof. We have

$$X_t - \hat{X}_t = (\xi - \hat{\xi}) + \int_0^t d(K_r - \hat{K}_r) + \int_0^t \left[G(r, X_r) - G(r, \hat{X}_r)\right] dB_r,$$

where

$$K_t - \hat{K}_t = \int_0^t \left[F(r, X_r) - F(r, \hat{X}_r)\right] dr.$$

In view of the assumptions (3.10-**SDE-H**$_F$) and (3.11-**SDE-H**$_G$), for all $p \geq 1$ and $\gamma \geq 0$:

$$\begin{aligned}\langle X_r - \hat{X}_r, d(K_r - \hat{K}_r)\rangle + \left(\frac{1}{2}m_p + 9p\gamma\right)\left|G(r, X_r) - G(r, \hat{X}_r)\right|^2 dr \\ \leq \left|X_r - \hat{X}_r\right|^2 \left[\mu_r dr + \left(\frac{1}{2}m_p + 9p\gamma\right)(\ell_r)^2 dr\right].\end{aligned}$$

Hence, by Proposition 3.3 (or by Corollary 6.77 from Annex C) the inequalities (3.18) and (3.19) follow. ∎

3.3.2 SDEs with Stieltjes Integrals

We shall conclude this section with some remarks on the estimates for a more general equation

$$X_t = \xi + \int_0^t \Phi(s, X_s)\, dQ_s + \int_0^t G(s, X_s)\, dB_s, \tag{3.20}$$

where $\xi : \Omega \to \mathbb{R}^d$ is an $\mathcal{F}_0$-measurable random vector, G satisfies the assumption (3.11-**SDE-H**$_G$), $Q : \Omega \times [0, \infty[\to \mathbb{R}_+$ is such that

$$(\textbf{SDE-H}_Q): \qquad Q \textit{ is } \mathcal{P}\textit{-m.i.c.s.p.}, \ Q_0 = 0 \tag{3.21}$$

and $\Phi : \Omega \times [0, \infty[\times \mathbb{R}^d \to \mathbb{R}^d$.

Note that an equation of the form

$$X_t = \xi + \sum_{i=1}^{m} \int_0^t \Phi_i(s, X_s)\, dQ_s^{(i)} + \int_0^t G(s, X_s)\, dB_s,$$

can be transformed in the form (3.20) setting

$$Q_s = \sum_{i=1}^{m} Q_s^{(i)}, \qquad \Phi(s, x) = \sum_{i=1}^{m} \Phi_i(s, x)\, \alpha_s^{(i)},$$

where $\alpha^{(i)}$, $i \in \overline{1, m}$, are $\mathcal{P}$-m.s.p. given by the Radon–Nikodym theorem $dQ_s^{(i)} = \alpha_s^{(i)} dQ_s$.

For all $x \in \mathbb{R}^d$, we assume that

$$\Phi(\cdot, \cdot, x) : \Omega \times [0, \infty[\to \mathbb{R}^d \text{ and } G(\cdot, \cdot, x) : \Omega \times [0, \infty[\to \mathbb{R}^{d \times k}$$

are $\mathcal{P}$-m.s.p.

We introduce the notation, for all $\rho \geq 0$

$$\Phi^{\#}_{\rho}(t) \stackrel{def}{=} \sup_{|x|\leq\rho} |\Phi(t,x)|.$$

We shall assume:

$$\textbf{(SDE-H}_{\Phi}\textbf{)}: \tag{3.22}$$

there exists a $\mathcal{P}$-m.s.p. $\mu : \Omega \times [0,\infty[\to \mathbb{R}$ *such that* $\forall\ T \geq 0$,

$$\int_0^T |\mu_t|\, dQ_t < \infty,\ a.s.$$

and

$$\begin{array}{lll} (C_{\Phi}) & \textit{Continuity:} & \\ & x \longrightarrow \Phi(t,x) : \mathbb{R}^d \to \mathbb{R}^d \text{ are continuous, } d\mathbb{P}\otimes dt\text{-a.e.}; & \\ (M_{\Phi}) & \textit{Monotonicity condition:} & \\ & \langle x-y, \Phi(t,x)-\Phi(t,y)\rangle \leq \mu_t|x-y|^2, \quad d\mathbb{P}\otimes dt\text{-a.e.}, & (3.23) \\ & \qquad\qquad\qquad\qquad\qquad\qquad \forall\ x,y\in\mathbb{R}^d; & \\ (B_{\Phi}) & \textit{Boundedness condition:} & \\ & \displaystyle\int_0^T \left|\Phi^{\#}_{\rho}(t)\right| dQ_t < \infty, \quad a.s.,\ \forall\ T,\rho\geq 0. & \end{array}$$

□

With very similar proofs (as in Propositions 3.6, 3.7 and Theorem 3.8) we can establish the following results:

Proposition 3.9. *Let the assumptions (3.11-**SDE-H**$_G$), (3.22-**SDE-H**$_\Phi$) and (3.21-**SDE-H**$_Q$) be satisfied.*

(I) Then for every $p \geq 2$, $\lambda > 1$ there exists a constant $C_{p,\lambda}$ such that for every solution $X \in S^0_d$ of the SDE (3.20) and

$$V_t \stackrel{def}{=} \int_0^t \mu_r dQ_r + \lambda\left(\frac{p-1}{2} + 9p\lambda\right)\int_0^t (\ell_r)^2\, dr$$

the following inequality holds $\mathbb{P}$-a.s., for all $0 \leq t \leq s$:

$$\begin{aligned} \mathbb{E}^{\mathcal{F}_t}\left\|e^{-V}X\right\|^p_{[t,s]} \leq C_{p,\lambda}\Big[\left|e^{-V_t}X_t\right|^p + \mathbb{E}^{\mathcal{F}_t}\left(\int_t^s e^{-V_r}\left|\Phi(t,0)\right| dQ_r\right)^p \\ + \mathbb{E}^{\mathcal{F}_t}\left(\int_t^s e^{-2V_r}\left|G(r,0)\right|^2 dr\right)^{p/2}\Big]. \end{aligned} \tag{3.24}$$

(*II*) *Assuming there exists a* $U \in S_d^0$ *such that*

$$U_t = U_0 + \int_0^t G(r, U_r)\, dB_r,$$

then for every $p \geq 1$, $\lambda > 1$ *there exists a constant* $C_{p,\lambda}$ *such that if* $X \in S_d^0$ *is a solution of the SDE (3.20) and*

$$V_t \overset{def}{=} \int_0^t \mu_r dQ_r + \left(\frac{m_p}{2} + 9p\lambda\right) \int_0^t (\ell_r)^2\, dr,$$

the following inequality holds $\mathbb{P}$*-a.s., for all* $0 \leq t \leq s$,

$$\begin{aligned} \mathbb{E}^{\mathcal{F}_t} \left\| e^{-V}(X - U) \right\|_{[t,s]}^p \leq C_{p,\lambda} \Big[& e^{-pV_t} |X_t - U_t|^p \\ & + \mathbb{E}^{\mathcal{F}_t} \left(\int_t^s e^{-V_r} |\Phi(r, U_r)|\, dQ_r \right)^p \Big]. \end{aligned} \tag{3.25}$$

Finally we give the corresponding generalization of Theorem 3.8, which comes from the general formula from Corollary 6.77 (Annex C).

Theorem 3.10 (Uniqueness). *Let the assumptions (3.11-**SDE-H**$_G$), (3.22-**SDE-H**$_\Phi$) and (3.21-**SDE-H**$_Q$) be satisfied and* $X, \hat{X} \in S_d^0$ *be solutions of the SDE (3.20) corresponding to the initial data* ξ *and respectively* $\hat{\xi}$. *Let* $p \geq 1$, $m_p = 1 \vee (p - 1)$, $\lambda \geq 0$ *and*

$$V_t \overset{def}{=} \int_0^t \left[\mu_r dQ_r + \left(\frac{1}{2} m_p + 9p\lambda\right) (\ell_r)^2\, dr \right].$$

(*I*) *If* $\lambda = 0$, *then for all* $\delta \geq 0$, $0 \leq t \leq s$:

$$\mathbb{E}^{\mathcal{F}_t} \frac{e^{-pV_s} |X_s - \hat{X}_s|^p}{\left(1 + \delta e^{-2V_s} |X_s - \hat{X}_s|^2\right)^{p/2}} \leq \frac{e^{-pV_t} |X_t - \hat{X}_t|^p}{\left(1 + \delta e^{-2V_t} |X_t - \hat{X}_t|^2\right)^{p/2}}, \quad \mathbb{P}\text{-a.s.} \tag{3.26}$$

(*II*) *For every* $p \geq 1$ *and* $\lambda > 1$ *there exists a constant* $C_{p,\lambda}$ *such that for all* $\delta \geq 0$, $0 \leq t \leq s$:

$$\mathbb{E}^{\mathcal{F}_t} \frac{\left\| e^{-V}(X - \hat{X}) \right\|_{[t,s]}^p}{\left(1 + \delta \left\| e^{-V}(X - \hat{X}) \right\|_{[t,s]}^2\right)^{p/2}} \leq C_{p,\lambda} \frac{\left| e^{-V_t}(X_t - \hat{X}_t) \right|^p}{\left(1 + \delta \left| e^{-V_t}(X_t - \hat{X}_t) \right|^2\right)^{p/2}}, \quad a.s. \tag{3.27}$$

The uniqueness in S_d^0 *follows by choosing* $t = 0$ *and* $\delta > 0$.

3.3.3 Stochastic Linear Equations

Let $d = 1$ and consider the stochastic differential equation

$$\begin{cases} dX_t = (a_t X_t + b_t)\, dQ_t + \langle c_t X_t + e_t, dB_t \rangle, & t > 0, \\ X_0 = \xi \in L^0(\Omega, \mathcal{F}_0, \mathbb{P}; \mathbb{R}), \end{cases} \tag{3.28}$$

where $(a_t)_{t\geq 0}$, $(b_t)_{t\geq 0}$ are $\mathbb{R}$-valued $\mathcal{P}$-m.s.p., and $(c_t)_{t\geq 0}$, $(e_t)_{t\geq 0}$ are $\mathbb{R}^k$-valued $\mathcal{P}$-m.s.p. such that

$$\int_0^T (|a_t| + |b_t|)\, dQ_t + \int_0^T \left(|c_t|^2 + |e_t|^2\right) dt < \infty, \quad a.s. \tag{3.29}$$

Proposition 3.11. *Under the assumption (3.29) the SDE (3.28) has a unique solution $X \in S_1^0$ which is given by*

$$X_t = \Gamma_t \left[\xi + \int_0^t \Gamma_s^{-1} (b_s dQ_s - \langle c_s, e_s \rangle\, ds) + \int_0^t \Gamma_s^{-1} \langle e_s, dB_s \rangle\right], \tag{3.30}$$

where

$$\Gamma_t = \exp\left[\int_0^t \left(a_r dQ_r - \frac{1}{2}|c_r|^2 dr\right) + \int_0^t \langle c_r, dB_r \rangle\right].$$

Proof. We first verify that

$$Y_t = \Gamma_t \left[\xi + \int_0^t \Gamma_s^{-1} (b_s - \langle c_s, e_s \rangle)\, ds + \int_0^t \Gamma_s^{-1} \langle e_s, dB_s \rangle\right]$$

is a solution of (3.28). Indeed $Y \in S_1^0$, $Y_0 = \xi$ and

$$dY_t = (a_t Y_t + b_t)\, dt + \langle c_t Y_t + e_t, dB_t \rangle$$

follows from Itô's formula.

Uniqueness follows from Theorem 3.10, but we prove it independently in this particular case.

Let $X \in S_1^0$ be an arbitrary solution of (3.28). Since

$$d\Gamma_t^{-1} = \Gamma_t^{-1}\left(-a_t dQ_t + |c_t|^2 dt\right) - \Gamma_t^{-1} \langle c_t, dB_t \rangle$$

we deduce that

$$\begin{aligned} d\left(\Gamma_t^{-1} X_t\right) = & \left[\Gamma_t^{-1}\left(-a_t dQ_t + |c_t|^2 dt\right) - \Gamma_t^{-1} \langle c_t, dB_t \rangle\right] X_t \\ & + \Gamma_t^{-1} \left[(a_t X_t + b_t)\, dQ_t + \langle c_t X_t + e_t, dB_t \rangle\right] \end{aligned}$$

$$\begin{aligned}&+\Gamma_t^{-1}\left(X_t\,|c_t|^2-\langle c_t,e_t\rangle\right)dt\\&=\Gamma_t^{-1}\left[b_t dQ_t-\langle c_t,e_t\rangle\,dt+\langle e_t,dB_t\rangle\right]\\&=d\left(\Gamma_t^{-1}Y_t\right).\end{aligned}$$

Hence

$$\Gamma_t^{-1}X_t-\xi=\Gamma_t^{-1}Y_t-\xi,\quad\text{for all } t\geq 0,$$

which yields $X=Y$ and uniqueness. ∎

3.3.4 Comparison Results

In the case $d=1$ we also deduce uniqueness from a comparison result which is important by itself. Let $X\in S^0$ be a solution of the SDE

$$X_t=\xi+\int_0^t\Phi\left(s,X_s\right)dQ_s+\int_0^t\left\langle G\left(s,X_s\right),dB_s\right\rangle,\quad t\geq 0,\ a.s.,\tag{3.31}$$

and $\tilde{X}\in S^0$ a solution of the SDE

$$\tilde{X}_t=\tilde{\xi}+\int_0^t\tilde{\Phi}\left(s,\tilde{X}_s\right)dQ_s+\int_0^t\left\langle G\left(s,\tilde{X}_s\right),dB_s\right\rangle,\quad t\geq 0,\ a.s.\tag{3.32}$$

Assume that the functions $\Phi,\tilde{\Phi}:\Omega\times[0,\infty[\times\mathbb{R}\to\mathbb{R}$ and $G:\Omega\times[0,+\infty[\times\mathbb{R}\to\mathbb{R}^k$ are $(\mathcal{P},\mathbb{R})$-Carathéodory functions ($\mathcal{P}$-m.s.p. with respect to (ω,t) and continuous with respect to $x\in\mathbb{R}$) such that for all $T\geq 0$,

$$\begin{aligned}&(i)\ \int_0^T\left|G\left(t,X_t\right)\right|^2dt+\int_0^T\left|G\left(t,\tilde{X}_t\right)\right|^2dt<\infty,\quad a.s.,\\&(ii)\ \int_0^T\left|\Phi\left(t,X_t\right)\right|dQ_t+\int_0^T\left|\tilde{\Phi}\left(t,\tilde{X}_t\right)\right|dQ_t<\infty,\quad a.s.\end{aligned}\tag{3.33}$$

Also assume that there exist $\alpha\in[\frac{1}{2},1]$ and *a* $\mathcal{P}$-m.s.p. $\ell:\Omega\times[0,\infty[\to\mathbb{R}_+$ such that

$$\int_0^T\left(\ell_t\right)^2dt<\infty,\quad\mathbb{P}\text{-a.s.},\quad\forall\,T>0,\tag{3.34}$$

and $d\mathbb{P}\otimes dt$-*a.e.*,

$$\left|G(t,X_t)-G(t,\tilde{X}_t)\right|\leq\ell_t\left|X_t-\tilde{X}_t\right|^\alpha.\tag{3.35}$$

3.3.4.1 Lipschitz Case

We first give a comparison result when one of the functions Φ and $\tilde{\Phi}$ satisfies a Lipschitz condition. Without loss of generality we assume that there exists a $\mathcal{P}$-m.s.p. $L : \Omega \times [0, \infty[\to \mathbb{R}_+$ such that

$$\int_0^T L_t dQ_t < \infty, \quad \mathbb{P}\text{-a.s.}, \quad \forall\, T > 0, \tag{3.36}$$

and $d\mathbb{P} \otimes dt$-$a.e.$,

$$|\Phi(t,x) - \Phi(t,y)| \leq L_t\, |x - y|\,, \quad \forall\, x, y \in \mathbb{R}. \tag{3.37}$$

Proposition 3.12. *Let the assumptions (3.33), (3.34), (3.35 with $\alpha = 1$), (3.36) and (3.37) be satisfied and*

$$\begin{aligned} &(i)\ \xi \geq \tilde{\xi}, \quad \mathbb{P}\text{-}a.s.\ \ and \\ &(ii)\ \Phi(t, \tilde{X}_t) \geq \tilde{\Phi}(t, \tilde{X}_t),\ d\mathbb{P} \otimes dQ_t\text{-}a.e.\ on\ \Omega \times \mathbb{R}_+. \end{aligned}$$

(a) Then $\mathbb{P}$-a.s. $\omega \in \Omega$, $X_t(\omega) \geq \tilde{X}_t(\omega)$, for all $t \geq 0$; in particular the strong uniqueness for the SDE (3.31) holds.

(b) If moreover there exist $A \in \mathcal{F}$ and a stopping time $\tau > 0$ such that for all $\omega \in A$:

$$\begin{aligned} &(j)\ \xi(\omega) > \tilde{\xi}(\omega)\,, \quad or \\ &(jj)\ \int_0^{\tau(\omega)} \left[\Phi(\omega,t,\tilde{X}_t) - \tilde{\Phi}(\omega,t,\tilde{X}_t)\right] dQ_t(\omega) > 0, \end{aligned}$$

then $X_{\tau(\omega)}(\omega) > \tilde{X}_{\tau(\omega)}(\omega)$, for all $\omega \in A$; in particular if $\xi(\omega) > \tilde{\xi}(\omega)$ for $\omega \in A$, then $X_t(\omega) > \tilde{X}_t(\omega)$ for all $(\omega, t) \in A \times [0, \infty[$.

Proof. Let $U_t = X_t - \tilde{X}_t$. Then $\Phi(t, X_t) - \tilde{\Phi}(t, \tilde{X}_t) = b_t + a_t U_t$ and $G(t, X_t) - G(t, \tilde{X}_t) = U_t c_t$ with $b_t = \Phi(t, \tilde{X}_t) - \tilde{\Phi}(t, \tilde{X}_t)$,

$$a_t = \begin{cases} \dfrac{1}{U_t}\left[\Phi(t, X_t) - \Phi(t, \tilde{X}_t)\right], & \text{if } U_t \neq 0, \\ 0, \ \ \text{if } U_t = 0 \end{cases}$$

and

$$c_t = \begin{cases} \dfrac{1}{U_t}\left[G(t, X_t) - G(t, \tilde{X}_t)\right], & \text{if } U_t \neq 0, \\ 0, \ \ \text{if } U_t = 0. \end{cases}$$

Then

$$U_t = U_0 + \int_0^t (a_s U_s + b_s)\, dQ_s + \int_0^t U_s \langle c_s, dB_s\rangle .$$

By Proposition 3.11

$$U_t = \Gamma_t \left[\left(\xi - \tilde{\xi}\right) + \int_0^t \Gamma_s^{-1} b_s dQ_s \right]$$

where

$$\Gamma_t = \exp\left[\int_0^t \left(a_s dQ_s - \frac{1}{2} |c_s|^2 ds \right) ds + \int_0^t \langle c_s, dB_s\rangle \right],$$

and the results follow. ■

Remark 3.13. With the same proof the comparison result from Proposition 3.12 can be generalized to the following case.
Let $\Phi, b : \Omega \times \mathbb{R}_+ \to \mathbb{R}$, $G : \Omega \times \mathbb{R}_+ \to \mathbb{R}^k$ and $L, \ell : \Omega \times \mathbb{R}_+ \to \mathbb{R}_+$ be progressively measurable stochastic processes such that for all $T \geq 0$:

$$\int_0^T (|\Phi_t| + |b_t| + L_t)\, dQ_t + \int_0^T \left(|G_t|^2 + (\ell_t)^2 \right) dt < \infty, \quad a.s.$$

Let $X \in S^0$ be given by

$$X_t = \xi + \int_0^t \Phi_s dQ_s + \int_0^t \langle G_s, dB_s\rangle , \; t \geq 0,$$

and the following conditions are satisfied

$$\begin{aligned} &(i)\ |\Phi_t - b_t| \leq L_t |X_t|, \quad d\mathbb{P} \otimes dQ_t\text{-}a.e., \\ &(ii)\ |G_t| \leq \ell_t |X_t|, \quad d\mathbb{P} \otimes dt\text{-}a.e. \end{aligned} \tag{3.38}$$

1. If $\xi \geq 0$, $\mathbb{P}$-a.s. and $b_t \geq 0$, $\mathbb{P} \otimes dQ_t\text{-}a.e.$ on $\Omega \times \mathbb{R}_+$, then $\mathbb{P}$-a.s. $\omega \in \Omega$, $X_t(\omega) \geq 0$, for all $t \geq 0$.
2. If moreover there exist $A \in \mathcal{F}$ and a stopping time $\tau > 0$ such that for all $\omega \in A$,

$$\xi(\omega) > 0, \quad \text{or} \quad \int_0^{\tau(\omega)} b_t(\omega)\, dQ_t(\omega) > 0,$$

then $X_{\tau(\omega)}(\omega) > 0$, for all $\omega \in A$; in particular if $\xi(\omega) > 0$ for $\omega \in A$, then $X_t(\omega) > 0$ for all $(\omega, t) \in A \times [0, \infty[$.

3.3.4.2 Monotone Case

We now generalize the comparison result to the case where one of the functions Φ and $\tilde{\Phi}$ satisfies a monotonicity condition. Without loss of generality we assume that there exists a $\mathcal{P}$-m.s.p. $\mu : \Omega \times [0, \infty[\to \mathbb{R}$ such that

$$\int_0^T |\mu_t| \, dQ_t < \infty, \quad \mathbb{P}\text{-a.s.}, \qquad \forall\, T > 0, \tag{3.39}$$

and $d\mathbb{P} \otimes dQ_t$-$a.e.$,

$$\left(\Phi(t, \tilde{X}_t) - \Phi(t, X_t)\right)\left(\tilde{X}_t - X_t\right) \leq \mu_t \left(\tilde{X}_t - X_t\right)^2. \tag{3.40}$$

This last condition implies that

$$\left[\Phi\left(t, \tilde{X}_t\right) - \Phi\left(t, X_t\right)\right] \mathbf{1}_{\tilde{X}_t - X_t > 0} \leq \mu_t \left(\tilde{X}_t - X_t\right)^+.$$

Proposition 3.14. *Let the assumptions (3.33), (3.34), (3.35), (3.39) and (3.40, where μ is a deterministic process and $dQ_t = dt$, when $\frac{1}{2} \leq \alpha < 1$) be satisfied and*

$$\begin{aligned} &(i)\ \xi \geq \tilde{\xi}, \quad \mathbb{P}\text{-}a.s. \text{ and} \\ &(ii)\ \Phi(t, \tilde{X}_t) \geq \tilde{\Phi}(t, \tilde{X}_t),\ d\mathbb{P} \otimes dQ_t\text{-}a.e. \text{ on } \Omega \times \mathbb{R}_+. \end{aligned}$$

Then $\mathbb{P}$-a.s. $\omega \in \Omega$, $X_t(\omega) \geq \tilde{X}_t(\omega)$, for all $t \geq 0$. In particular the strong uniqueness for the SDE (3.31) holds.

Proof. We have

$$\begin{aligned} \tilde{X}_t - X_t = \tilde{\xi} - \xi &+ \int_0^t \left[\tilde{\Phi}\left(s, \tilde{X}_s\right) - \Phi\left(s, X_s\right)\right] dQ_s \\ &+ \int_0^t \left\langle G\left(s, \tilde{X}_s\right) - G\left(s, X_s\right), dB_s\right\rangle. \end{aligned}$$

Since

$$\left|G\left(s, \tilde{X}_s\right) - G\left(s, X_s\right)\right| \leq \ell_s \left|\tilde{X}_s - X_s\right|^{\alpha - 1/2} \sqrt{\left|\tilde{X}_s - X_s\right|},$$

we have, by Proposition 2.33, that

$$\begin{aligned} \left(\tilde{X}_t - X_t\right)^+ = \left(\tilde{\xi} - \xi\right)^+ &+ \int_0^t \left[\tilde{\Phi}\left(s, \tilde{X}_s\right) - \Phi\left(s, X_s\right)\right] \theta\left(\tilde{X}_s - X_s\right) dQ_s \\ &+ \int_0^t \left\langle \left[G\left(s, \tilde{X}_s\right) - G\left(s, X_s\right)\right] \mathbf{1}_{\tilde{X}_s - X_s > 0}, dB_s\right\rangle, \end{aligned} \tag{3.41}$$

where

$$\theta(x) = \begin{cases} 0, & \text{if } x < 0, \\ \frac{1}{2}, & \text{if } x = 0, \\ 1, & \text{if } x > 0. \end{cases}$$

Case $\alpha = 1$.
Let $p \geq 1, \lambda > 1$,

$$\begin{aligned} K_t &= \int_0^t \left[\tilde{\Phi}\left(s, \tilde{X}_s\right) - \Phi\left(s, X_s\right)\right] \theta\left(\tilde{X}_s - X_s\right) dQ_s \quad \text{and} \\ G_t &= \left[G\left(t, \tilde{X}_t\right) - G\left(t, X_t\right)\right] \mathbf{1}_{\tilde{X}_t - X_t > 0}. \end{aligned}$$

Since

$$\begin{aligned} \left(\tilde{X}_t - X_t\right)^+ dK_t &\leq \left(\tilde{X}_t - X_t\right)^+ \left[\tilde{\Phi}\left(t, \tilde{X}_t\right) - \Phi\left(t, \tilde{X}_t\right)\right] \mathbf{1}_{\tilde{X}_t - X_t > 0} dQ_t \\ &\quad + \left(\tilde{X}_t - X_t\right)^+ \left[\Phi\left(t, \tilde{X}_t\right) - \Phi\left(t, X_t\right)\right] \mathbf{1}_{\tilde{X}_t - X_t > 0} dQ_t \\ &\leq \mu_t^+ \left[\left(\tilde{X}_t - X_t\right)^+\right]^2 dQ_t \end{aligned}$$

and

$$|G_t| \leq \ell_t \left(\tilde{X}_t - X_t\right)^+,$$

it follows that

$$\left(\tilde{X}_t - X_t\right)^+ dK_t + \left(\frac{1}{2} m_p + 9p\lambda\right) |G_t|^2 \leq \left[\left(\tilde{X}_t - X_t\right)^+\right]^2 dV_t,$$

where

$$dV_t = \mu_t^+ dQ_t + \left(\frac{1}{2} m_p + 9p\lambda\right) (\ell_t)^2.$$

Using Proposition 3.3 (the inequality (3.8)) we deduce for all $t \geq 0$,

$$\mathbb{E} \frac{\left\| e^{-V} \left(\tilde{X} - X\right)^+ \right\|_t^p}{\left(1 + \delta \left\| e^{-V} \left(\tilde{X} - X\right)^+ \right\|_t^2\right)^{p/2}} \leq C_{p,\lambda}\, \mathbb{E} \frac{\left| \left(\tilde{\xi} - \xi\right)^+ \right|^p}{\left(1 + \delta \left| \left(\tilde{\xi} - \xi\right)^+ \right|^2\right)^{p/2}} = 0.$$

Hence $\mathbb{P}$-a.s. : $X_t \geq \tilde{X}_t$ for all $t \geq 0$.
Case $\frac{1}{2} \leq \alpha < 1$.

Consider again (3.41). Define the increasing sequence of stopping times

$$\tau_n = \inf\left\{t \geq 0 : \left(\tilde{X}_t - X_t\right)^+ + \int_0^t \ell_s^2 ds \geq n\right\}.$$

It is clear that $\tau_n \nearrow +\infty$, a.s. for $n \to \infty$.

Since

$$\mathbb{E}\int_0^{t\wedge\tau_n} \left|G\left(s, \tilde{X}_s\right) - G\left(s, X_s\right)\right|^2 \mathbf{1}_{\tilde{X}_s - X_s > 0} ds \leq \mathbb{E}\int_0^{t\wedge\tau_n} \ell_s^2 \left[\left(\tilde{X}_s - X_s\right)^+\right]^{2\alpha} ds$$
$$\leq n^{2\alpha+1},$$

we obtain, taking the expectation in (3.41), that

$$\begin{aligned}
\mathbb{E}\left(\tilde{X}_{t\wedge\tau_n} - X_{t\wedge\tau_n}\right)^+ &= \mathbb{E}\int_0^{t\wedge\tau_n} \left[\tilde{\Phi}\left(s, \tilde{X}_s\right) - \Phi\left(s, X_s\right)\right] \mathbf{1}_{\tilde{X}_s - X_s > 0} ds \\
&= \int_0^t \mathbb{E}\left[1_{[0,\tau_n]}(s)\left[\tilde{\Phi}\left(s, \tilde{X}_s\right) - \Phi\left(s, X_s\right)\right] \mathbf{1}_{\tilde{X}_s - X_s > 0}\right] ds \\
&\leq \int_0^t \mu^+(s)\, \mathbb{E}\left[1_{[0,\tau_n]}(s)\left(\tilde{X}_s - X_s\right)^+\right] ds \\
&\leq \int_0^t \mu^+(s)\, \mathbb{E}\left[\left(\tilde{X}_{s\wedge\tau_n} - X_{s\wedge\tau_n}\right)^+\right] ds,
\end{aligned}$$

and, by the Gronwall inequality, we deduce that $\mathbb{E}\left(\tilde{X}_{t\wedge\tau_n} - X_{t\wedge\tau_n}\right)^+ = 0$ for all $t \geq 0$. Hence for every $t \geq 0$, $n \geq 1$, $\left(\tilde{X}_{t\wedge\tau_n} - X_{t\wedge\tau_n}\right)^+ = 0$, $\mathbb{P}$-a.s. The result follows. ■

Remark 3.15. From Proposition 3.14 we infer that the SDE

$$X_t = \xi + \int_0^t F\left(s, X_s\right) ds + \int_0^t \left\langle G\left(s, X_s\right), dB_s\right\rangle, \quad t \geq 0, \ a.s., \tag{3.42}$$

with a monotone drift F and a Hölder continuous diffusion coefficient G has at most one solution $X \in S_1^0$. To be clear the coefficients F and G satisfy

- $F : \Omega \times \mathbb{R}_+ \times \mathbb{R} \to \mathbb{R}$ and $G : \Omega \times \mathbb{R}_+ \times \mathbb{R} \to \mathbb{R}^k$ are $\mathcal{P}$-m.s.p. with respect to $(\omega, t) \in \Omega \times \mathbb{R}_+$ and continuous with respect to $x \in \mathbb{R}$;
- F satisfies (3.39) and (3.40, with μ a deterministic process);
- $G : \Omega \times \mathbb{R}_+ \times \mathbb{R} \to \mathbb{R}^k$ satisfies (3.34) and (3.35) with $\frac{1}{2} \leq \alpha \leq 1$.

We now give a strict comparison result. Let $X \in S^0$ be a solution of the SDE

$$X_t = \xi + \int_0^t F\left(s, X_s\right) ds + \int_0^t \left\langle G\left(s, X_s\right), dB_s\right\rangle, \quad t \geq 0, \ a.s., \tag{3.43}$$

and $\hat{X} \in S^0$ a solution of the SDE

$$\tilde{X}_t = \tilde{\xi} + \int_0^t \tilde{F}\left(s, \tilde{X}_s\right) ds + \int_0^t \left\langle G\left(s, \tilde{X}_s\right), dB_s \right\rangle, \quad t \geq 0, \ a.s. \tag{3.44}$$

Assume that

- the function $G : \Omega \times [0, +\infty[\times \mathbb{R} \to \mathbb{R}^k$ is $\mathcal{P}$-m.s.p. with respect to (ω, t) and continuous with respect to $x \in \mathbb{R}$ such that for all $T \geq 0$

$$\int_0^T |G(t, X_t)|^2 dt + \int_0^T \left|G\left(t, \tilde{X}_t\right)\right|^2 dt < \infty, \quad a.s., \tag{3.45}$$

  and there exists *a* $\mathcal{P}$-m.s.p. $\ell : \Omega \times [0, \infty[\to \mathbb{R}_+$ with

$$\int_0^T (\ell_t)^2 dt < \infty, \quad \mathbb{P}\text{-a.s.}, \quad \forall\, T > 0, \tag{3.46}$$

  and $d\mathbb{P} \otimes dt$-*a.e.*,

$$\left|G(t, X_t) - G(t, \tilde{X}_t)\right| \leq \ell_t \left|X_t - \tilde{X}_t\right|. \tag{3.47}$$

- $F, \tilde{F} : \Omega \times [0, \infty[\times \mathbb{R} \to \mathbb{R}$ are $\mathcal{P}$-m.s.p. with respect to (ω, t) and continuous with respect to $(t, x) \in \mathbb{R}_+ \times \mathbb{R}$ and there exists a $\mathcal{P}$-m.s.p. $\mu : \Omega \times [0, \infty[\to \mathbb{R}$ such that

$$\int_0^T |\mu_t|\, dQ_t < \infty, \quad \mathbb{P}\text{-a.s.}, \quad \forall\, T > 0, \tag{3.48}$$

  and $d\mathbb{P} \otimes dQ_t$-*a.e.*,

$$\left(F(t, \tilde{X}_t) - F(t, X_t)\right)\left(\tilde{X}_t - X_t\right) \leq \mu_t \left(\tilde{X}_t - X_t\right)^2. \tag{3.49}$$

Proposition 3.16. *Let the assumptions (3.45), (3.46), (3.47), (3.48) and (3.49) be satisfied. If* $\mathbb{P}$*-a.s.*

$$\xi \geq \tilde{\xi} \quad and \quad F(\omega, t, x) > \tilde{F}(\omega, t, x), \quad for\ all\ (t, x) \in \mathbb{R}_+ \times \mathbb{R},$$

then $\mathbb{P}$*-a.s.,* $X_t(\omega) > \tilde{X}_t(\omega)$*, for all* $t > 0$.

Proof. Let

$$c_t = \begin{cases} \dfrac{1}{X_t - \tilde{X}_t}\left[G(t, X_t) - G(t, \tilde{X}_t)\right], & \text{if } X_t - \tilde{X}_t \neq 0, \\ 0, \ \text{if } X_t - \tilde{X}_t = 0. \end{cases}$$

Then

$$d\left(X_t - \tilde{X}_t\right) = \left[F\left(t, X_t\right) - \tilde{F}\left(t, \tilde{X}_t\right)\right] dt + \left(X_t - \tilde{X}_t\right) \langle c_t, dB_t \rangle$$

and by Proposition 3.11, $\mathbb{P}$-a.s.

$$\Gamma_t^{-1}\left(X_t - \tilde{X}_t\right) = \left(X_s - \tilde{X}_s\right) + \int_s^t \Gamma_r^{-1}\left[F\left(r, X_r\right) - \tilde{F}\left(r, \tilde{X}_r\right)\right] dr,$$

for all $0 \leq s \leq t$, where

$$\Gamma_t = \exp\left[\int_0^t \langle c_r, dB_r \rangle - \frac{1}{2}\int_0^t |c_r|^2 \, dr\right].$$

By Proposition 3.14 we have $X_s \geq \tilde{X}_s$, for all $s \geq 0$, $\mathbb{P}$-a.s. Hence multiplying by $\mathbf{1}_{X_t = \tilde{X}_t}$ we obtain, $\mathbb{P}$-a.s.:

$$0 \geq \mathbf{1}_{X_t = \tilde{X}_t} \frac{1}{\varepsilon} \int_{t-\varepsilon}^t \Gamma_r^{-1}\left[F\left(r, X_r\right) - \tilde{F}\left(r, \tilde{X}_r\right)\right] dr, \quad \text{for all } 0 < t-\varepsilon < t. \tag{3.50}$$

Since $\mathbb{P}$-a.s. the mapping $r \mapsto \Gamma_r^{-1}\left[F\left(r, X_r\right) - \tilde{F}\left(r, \tilde{X}_r\right)\right]$ is continuous on $\mathbb{R}_+$ we deduce, letting $\varepsilon \searrow 0$, that $\mathbb{P}$-a.s.,

$$0 \geq \mathbf{1}_{X_t = \tilde{X}_t} \Gamma_t^{-1}\left[F\left(t, X_t\right) - \tilde{F}\left(t, X_t\right)\right], \quad \text{for all } t > 0.$$

But $\Gamma_t^{-1}\left[F\left(t, X_t\right) - \tilde{F}\left(t, X_t\right)\right] > 0$ for all $t \geq 0$. Hence $\mathbb{P}\left(X_t = \tilde{X}_t\right) = 0$, for all $t > 0$ and consequently (using the continuity of X and $\tilde{X}$)

$$\mathbb{P}\left(X_t > \tilde{X}_t, \quad \text{for all } t > 0\right) = 1.$$

The proof is complete. ■

We remark that in the Lipschitz case $\xi > \tilde{\xi}$ and $\Phi \geq \tilde{\Phi}$ yield $X_t > \tilde{X}_t$ for all $t \geq 0$; in the monotone case the condition

$$\Phi(x) > \tilde{\Phi}(x), \quad \text{for all } x \in \mathbb{R}$$

is essential as we can see from the following example. Let $\Phi, \tilde{\Phi}, G : \mathbb{R} \to \mathbb{R}$, $\Phi(x) = -2\sqrt{x^+}$, $\tilde{\Phi}(x) = -4\sqrt{x^+}$ and $G(x) = 0$. The functions Φ and $\tilde{\Phi}$ are continuous monotone decreasing functions on $\mathbb{R}$: $(\Phi(x) - \Phi(y))(x - y) \leq 0$ and similar for $\tilde{\Phi}$.

Clearly $X_t = \left[(1-t)^+\right]^2$, $t \geq 0$, is the unique solution of the SDE (in fact an ordinary differential equation)

$$X_t = 1 - 2\int_0^t \sqrt{X_s^+} ds$$

and $\tilde{X}_t = 4\left[(\frac{1}{4} - t)^+\right]^2$, $t \geq 0$, is the unique solution of

$$\tilde{X}_t = \frac{1}{4} - 4\int_0^t \sqrt{\tilde{X}_s^+} ds.$$

We have $X_0 = 1 > \frac{1}{4} = \tilde{X}_0$, $\Phi(x) \geq \tilde{\Phi}(x)$ for all $x \in \mathbb{R}$ but we do not have $X_t > \tilde{X}_t$ for all $t \geq 0$ (in fact $X_t > \tilde{X}_t$ for $t \in [0, 1)$ and $X_t = \tilde{X}_t$ for $t \geq 1$).

3.4 Lipschitz Coefficients

3.4.1 Classical SDEs

We consider a slightly generalized version of Eq. (3.2). We shall obtain under some Lipschitz conditions the existence and uniqueness of the solution using the Banach fixed point theorem.

Consider the SDE

$$X_t = S_t + \int_0^t F(s, X) ds + \int_0^t G(s, X) dB_s, \quad t \geq 0, \ \mathbb{P}\text{-a.s.}, \tag{3.51}$$

where

◇ $S : \Omega \times [0, \infty[\to \mathbb{R}^d$ *is a* $\mathcal{P}$-m.c.s.p.,
◇ *the functions* $F(\cdot, \cdot, \varphi) : \Omega \times [0, +\infty[\to \mathbb{R}^d$ *and* $G(\cdot, \cdot, \varphi) : \Omega \times [0, +\infty[\to \mathbb{R}^{d\times k}$ *are* $\mathcal{P}$*-m.s.p. for every* continuous function $\varphi : \mathbb{R}_+ \to \mathbb{R}^d$,
◇ *there exist* $L \in L^1_{loc}(0, \infty; \mathbb{R}_+)$ *and* $\ell \in L^2_{loc}(0, \infty; \mathbb{R}_+)$,

such that $d\mathbb{P} \otimes dt$*-a.e.*, for all $\varphi, \psi \in C(\mathbb{R}_+, \mathbb{R}^d)$:

$$(\mathbf{LB}_{F,G}) \begin{cases} & \textit{Lipschitz condition:} \\ (\mathbf{L}_F) & |F(t,\varphi) - F(t,\psi)| \leq L(t) \|\varphi - \psi\|_t, \\ (\mathbf{L}_G) & |G(t,\varphi) - G(t,\psi)| \leq \ell(t) \|\varphi - \psi\|_t; \\ & \textit{Boundedness condition:} \\ (\mathbf{B}_F) & \displaystyle\int_0^T |F(t,0)| dt < \infty, \quad a.s., \ \forall T \geq 0, \\ (\mathbf{B}_G) & \displaystyle\int_0^T |G(t,0)|^2 dt < \infty, \quad a.s., \ \forall T \geq 0, \end{cases} \tag{3.52}$$

where for any $\alpha \in C(\mathbb{R}_+, \mathbb{R}^d)$, $\|\alpha\|_t \overset{def}{=} \sup\{|\alpha(s)| : 0 \leq s \leq t\}$.

Theorem 3.17. *If $S \in S_d^0$ and the assumptions* ($\mathbf{LB}_{FG}$) *are satisfied then the SDE (3.51) has a unique solution $X \in S_d^0$. Moreover if there exists a $p > 0$ such that for all $T \geq 0$:*

$$\mathbb{E} \sup_{t \in [0,T]} |S_t|^p + \mathbb{E}\left(\int_0^T |F(t,0)|\, dt\right)^p + \mathbb{E}\left(\int_0^T |G(t,0)|^2 dt\right)^{p/2} < +\infty,$$

then $X \in S_d^p$.

Proof. Uniqueness. Also we could use here Theorem 3.8, we prefer to present the classical uniqueness argument under Lipschitz conditions. Assume $X, Y \in S_d^0$ are two solutions. Define the increasing sequence of stopping times $\tau_n = \inf\{t \geq 0 : \|X - Y\|_t \geq n\}$, $n \in \mathbb{N}^*$. Then $\tau_n \nearrow +\infty$ a.s. and

$$\begin{aligned}
&\mathbb{E}\,\|X - Y\|_{t\wedge\tau_n} \\
&\leq \mathbb{E}\int_0^{t\wedge\tau_n} L(r)\,\|X - Y\|_r\, dr + 3\mathbb{E}\left[\left(\int_0^{t\wedge\tau_n} \ell^2(r)\,\|X - Y\|_r^2\, dr\right)^{1/2}\right] \\
&\leq \mathbb{E}\int_0^{t\wedge\tau_n} L(r)\,\|X - Y\|_r\, dr + \frac{1}{2}\mathbb{E}\,\|X - Y\|_{t\wedge\tau_n} + \frac{9}{2}\mathbb{E}\int_0^{t\wedge\tau_n} \ell^2(r)\,\|X - Y\|_r\, dr
\end{aligned}$$

which yields

$$\begin{aligned}
\mathbb{E}\left(1_{[0,\tau_n]}(t)\,\|X - Y\|_t\right) &\leq \mathbb{E}\,\|X - Y\|_{t\wedge\tau_n} \\
&\leq \mathbb{E}\int_0^{t\wedge\tau_n} \left[2L(r) + 9\ell^2(r)\right] \|X - Y\|_r\, dr \\
&= \int_0^t \left[2L(r) + 9\ell^2(r)\right] \mathbb{E}\left(1_{[0,\tau_n]}(r)\,\|X - Y\|_r\right) dr
\end{aligned}$$

and by Gronwall's inequality we get

$$\mathbb{E}\left(1_{[0,\tau_n]}(t)\,\|X - Y\|_t\right) = 0, \quad \text{for all } t \geq 0.$$

Hence $1_{[0,\tau_n]}(t)\,\|X - Y\|_t = 0$ for all $t \geq 0$, $\mathbb{P}$-a.s. and passing to the limit as $n \to \infty$, it follows that $X = Y$ in S_d^0.

Thanks to uniqueness, the existence of a solution on $\mathbb{R}_+$ will follow from the existence of a solution on an arbitrary interval $[0, T]$.

Existence (I) *Case* $p > 0$.

Let $M \in \mathbb{N}^*$ and $0 = T_0 < T_1 < \cdots < T_M = T$, with $T_i = \dfrac{iT}{M}$. Then

$$\boldsymbol{\alpha}\left(\tfrac{T}{M}\right) \stackrel{def}{=} \sup_{0 < s - t < \frac{T}{M}} \left(\int_t^s L(r)\, dr\right)^p + \left(\int_t^s \ell^2(r)\, dr\right)^{p/2} \to 0, \quad \text{as } M \to \infty.$$

The solution of the SDE (3.51) on $[0, T_1]$ is a fixed point of the mapping $\Gamma : S_d^p[0, T_1] \to S_d^p[0, T_1]$ defined by

$$\Gamma(U)_t = S_t + \int_0^t F(s, U)\, ds + \int_0^t G(s, U)\, dB_s.$$

The mapping Γ is well defined since for all $\varphi \in C(\mathbb{R}_+, \mathbb{R}^d)$

$$|F(t,\varphi)| \le |F(t,0)| + L(t)\|\varphi\|_t \quad \text{and}\quad |G(t,\varphi)| \le |G(t,0)| + \ell(t)\|\varphi\|_t\,,$$

and consequently, by the Lipschitz continuity, the stochastic processes $F(\cdot, U)$ and $G(\cdot, U)$ are progressively measurable for all $U \in S_d^p[0, T]$ and

$$F(\cdot, U) \in L^p\left(\Omega; L^1(0,T)\right) \quad \text{and} \quad G(\cdot, U) \in \Lambda^p_{d\times k}(0,T)\,.$$

Therefore

$$\int_0^{\cdot} F(r, U)\, dr \in S_d^p[0, T] \quad \text{and} \quad \int_0^{\cdot} G(r, U)\, dB_r \in S_d^p[0, T]\,.$$

With M large enough we shall prove that Γ is a strict contraction on the complete metric space $S_d^p[0, T_1]$ with the usual distance

$$d_{p,M}(U, V) = \left(\mathbb{E}\|U - V\|_{T_1}^p\right)^{1/(p\vee 1)}.$$

Let $U, V \in S_d^p[0, T_1]$. By the Burkholder–Davis–Gundy inequality we have

$$\begin{aligned}
&\mathbb{E}\|\Gamma(U) - \Gamma(V)\|_{T_1}^p \\
&\quad\le \left(1 \vee 2^{p-1}\right)\mathbb{E}\sup_{s\in[T_0,T_1]}\left|\int_{T_0}^s (F(r,U) - F(r,V))\, dr\right|^p \\
&\qquad + \left(1 \vee 2^{p-1}\right)\mathbb{E}\sup_{s\in[T_0,T_1]}\left|\int_{T_0}^s (G(r,U) - G(r,V))\, dB_r\right|^p \\
&\quad\le \left(1 \vee 2^{p-1}\right)\left[\mathbb{E}\left(\int_{T_0}^{T_1} L(r)\|U - V\|_r\, dr\right)^p\right. \\
&\qquad \left. + \mathbb{E}\left(\int_{T_0}^{T_1} \ell^2(r)\|U - V\|_r^2\, dr\right)^{p/2}\right] \\
&\quad\le \left(1 \vee 2^{p-1}\right)\boldsymbol{\alpha}\left(\tfrac{T}{M}\right)\mathbb{E}\left(\|U - V\|_{T_1}^p\right).
\end{aligned}$$

Let $M_0 \in \mathbb{N}^*$ be such that $\left(1 \vee 2^{p-1}\right) \alpha\left(\frac{T}{M_0}\right) \leq \left(\frac{1}{2}\right)^{1 \vee p}$. Then Γ is a strict contraction in $S_d^p[0, T_1]$ and consequently Eq. (3.51) has a unique solution $X \in S_d^p[0, T_1]$.

Now we consider the interval $[0, T_2]$. The mapping $\Gamma : S_d^p[0, T_2] \to S_d^p[0, T_2]$ is defined by

$$\Gamma(U)_t = \begin{cases} X_t, \text{ if } t \in [0, T_1], \\ (X_{T_1} + S_t - S_{T_1}) + \int_{T_1}^t F(s, U)\, ds + \int_{T_1}^t G(s, U)\, dB_s, \text{ if } t \in]T_1, T_2]. \end{cases}$$

By recurrence, in M_0 steps, we cover the whole interval $[0, T]$.

(II) Case $p = 0$.

Let $n \in \mathbb{N}^*$ and define the stopping time

$$\theta_n = \inf\left\{t \geq 0 : |S_t| + \int_0^t |F(s, 0)|\, ds + \int_0^t |G(s, 0)|^2 ds \geq n\right\}.$$

Clearly $\theta_n \nearrow \infty$ $a.s$. By the first part there exists a unique $X^n \in S_d^p$ solution of the approximating equation

$$X_t^n = S_{t \wedge \theta_n} \mathbf{1}_{\theta_n > 0} + \int_0^t \mathbf{1}_{[0, \theta_n]}(r)\, F(r, X^n)\, dr + \int_0^t \mathbf{1}_{[0, \theta_n]}(r)\, G(r, X^n)\, dB_r. \tag{3.53}$$

The uniqueness of the solution shows that

$$\left[X_t^{n+1}(\omega) - X_t^n(\omega)\right] \mathbf{1}_{[0, \theta_n(\omega)]}(t)\, \mathbf{1}_{(0, \infty)}(\theta_n(\omega)) = 0.$$

Hence the stochastic process $X \in S_d^0$ defined by

$$X_t(\omega) = X_t^n(\omega), \quad \text{if } 0 \leq t < \theta_n(\omega) \text{ and } \theta_n(\omega) > 0$$

is solution of the SDE (3.51). The proof is complete. ■

3.4.2 SDEs with Stieltjes Integrals

To conclude the results involving Lipschitz coefficients, we shall prove an existence and uniqueness result for the more general equation (3.20)

$$X_t = S_t + \int_0^t \Phi(s, X_s)\, dQ_s + \int_0^t G(s, X_s)\, dB_s, \tag{3.54}$$

where

$$Q : \Omega \times [0, \infty[\to \mathbb{R}_+ \text{ is } \mathcal{P}\text{–}m.i.c.s.p.,\ Q_0 = 0.$$

We shall assume
$(\mathbf{LB}_{\Phi G})$:

($\Diamond$) *the functions* $\Phi(\cdot,\cdot,x) : \Omega \times [0,\infty[\to \mathbb{R}^d$ *and* $G(\cdot,\cdot,x) : \Omega \times [0,+\infty[\to \mathbb{R}^{d\times k}$ are *$\mathcal{P}$-m.s.p. for every $x \in \mathbb{R}^d$, such that for all $T \geq 0$,*

$$\int_0^T |G(t, S_t)|^2 dt + \int_0^T |\Phi(t, S_t)| dQ_t < \infty, \ \ a.s., \tag{3.55}$$

($\Diamond\Diamond$) *there exist $\mathcal{P}$-m.s.p. $\ell, L : \Omega \times [0,\infty[\to \mathbb{R}_+$ such that for every $T \geq 0$*

$$\int_0^T (\ell_t)^2 dt + \int_0^T L_t dQ_t < \infty, \ \ \mathbb{P}\text{-a.s.}, \tag{3.56}$$

and for all $x, y \in \mathbb{R}^d$, $d\mathbb{P} \otimes dt$-a.e.:

$$\begin{aligned} (\mathbf{L}_G):&\quad |G(t,x) - G(t,y)| \leq \ell_t |x-y|, \\ (\mathbf{L}_{\Phi,G}):&\quad |\Phi(t,x) - \Phi(t,y)| \leq L_t |x-y|. \end{aligned}$$

Theorem 3.18. *If $S \in S_d^0$ and the assumptions $(\mathbf{LB}_{\Phi G})$ are satisfied then the SDE (3.54) has a unique solution $X \in S_d^0$. Moreover for every $p \geq 2$, $\lambda > 1$, there exists a constant $C_{p,\lambda}$ such that for all $T \geq 0$*

$$\begin{aligned} \mathbb{E}\left\|e^{-V}X\right\|_T^p \leq C_{p,\lambda}\Bigg[\mathbb{E}\left\|e^{-V}S\right\|_T^p + \mathbb{E}\left(\int_0^T e^{-V_r}|\Phi(r,S_r)|\,dQ_r\right)^p \\ +\mathbb{E}\left(\int_0^T e^{-2V_r}|G(r,S_r)|^2 dr\right)^{p/2}\Bigg], \end{aligned} \tag{3.57}$$

where

$$V_t \overset{def}{=} \int_0^t L_r dQ_r + \lambda\left(\frac{p-1}{2} + 9p\lambda\right)\int_0^t (\ell_r)^2 dr.$$

Proof. (I) *Uniqueness.* X is solution of the SDE (3.54) iff $Y = X - S$ is a solution of the equation

$$Y_t = \int_0^t \tilde{\Phi}(s, Y_s)\, dQ_s + \int_0^t \tilde{G}(s, Y_s)\, dB_s, \tag{3.58}$$

where

$$\tilde{\Phi}(s,x) = \Phi(s, S_s + x) \quad \text{and} \quad \tilde{G}(s,x) = G(s, S_s + x).$$

Note that $\tilde{\Phi}$ and $\tilde{G}$ satisfy the assumptions of Proposition 3.9 and Theorem 3.10. Hence Eq. (3.58) (equivalently (3.54)) has at most one solution in S_d^0 and the inequality (3.57) holds.

Moreover the uniqueness result shows us that it is sufficient to prove the existence of the solution on an arbitrary fixed interval $[0, T]$.

(*II*) *Existence for (3.58) under the condition*

$$\mathbb{E}\int_0^T e^{-2aE_s} |G(s, S_s)|^2 \, ds + \mathbb{E}\left(\int_0^T e^{-aE_s} |\Phi(s, S_s)| \, dQ_s\right)^2 < \infty,$$

where

$$E_t = \int_0^t L_r dQ_r + \int_0^t \ell_s^2 ds$$

and $a \geq 2$ will be chosen below.

Let $\mathbb{U}$ denote the Banach space of the stochastic processes Y from $S_d^0[0, T]$ such that

$$\|Y\|_{\mathbb{U}} \overset{def}{=} \left(\frac{1}{2a}\mathbb{E}\left\|e^{-aE}Y\right\|_T^2\right)^{1/2} + \left(\mathbb{E}\int_0^T e^{-2aE_s} |Y_s|^2 \, dE_s\right)^{1/2} < \infty.$$

The solution of the SDE (3.58) on $[0, T]$ is a fixed point of the mapping $\Gamma : \mathbb{U} \to \mathbb{U}$ defined by

$$\Gamma_t(Y) \overset{def}{=} \int_0^t \tilde{\Phi}(s, Y_s) \, dQ_s + \int_0^t \tilde{G}(s, Y_s) \, dB_s.$$

First we show that the mapping Γ is well defined.

Since for all $Y \in S_d^0$

$$\left|\tilde{\Phi}(s, Y_s)\right| \leq |\Phi(s, S_s)| + L_s |Y_s| \leq |\Phi(s, S_s)| + L_s \|Y\|_T$$

and

$$\left|\tilde{G}(s, Y_s)\right| \leq |G(s, S_s)| + \ell_s |Y_s| \leq |G(s, S_s)| + \ell_s \|Y\|_T,$$

it follows from (3.55) and (3.56) that $\Gamma(Y)$ is a well defined element of S_d^0.

We shall now show $\Gamma(Y) \in \mathbb{U}$, whenever $Y \in \mathbb{U}$.

Since

$$dD_s = |\Gamma_s(Y)|^2 \, dE_s = |\Gamma_s(Y)|^2 \left(\ell_s^2 ds + L_s dQ_s\right),$$

and for $p = 2$ and $\lambda > 1$,

$$\begin{aligned}&\left\langle \Gamma_s(Y), \tilde{\Phi}(s, Y_s)\, dQ_s \right\rangle + \left(\frac{1}{2}m_p + 9p\lambda\right) \left|\tilde{G}(s, Y_s)\right|^2 ds \\ &\quad\leq |\Gamma_s(Y)|\, |Y_s|\, L_s dQ_s + |\Gamma_s(Y)|\, |\Phi(s, S_s)|\, dQ_s \\ &\qquad\qquad + (1 + 36\lambda)\left(\ell_s^2 |Y_s|^2 + |G(s, S_s)|^2\right) ds \\ &\quad\leq (1 + 36\lambda)\left[|G(s, S_s)|^2 ds + |Y_s|^2 \left(\ell_s^2 ds + L_s dQ_s\right)\right] \\ &\qquad\qquad + |\Gamma_s(Y)|\, |\Phi(s, S_s)|\, dQ_s + |\Gamma_s(Y)|^2 L_s dQ_s,\end{aligned}$$

we deduce that

$$\begin{aligned}&dD_s + \left\langle \Gamma_s(Y), \tilde{\Phi}(s, Y_s)\, dQ_s \right\rangle + \left(\frac{1}{2}m_p + 9p\lambda\right) \left|\tilde{G}(s, Y_s)\right|^2 ds \\ &\leq (1 + 36\lambda)\left[|G(s, S_s)|^2 ds + |Y_s|^2 dE_s\right] + |\Gamma_s(Y)|\, |\Phi(s, S_s)|\, dQ_s \\ &+ |\Gamma_s(Y)|^2 a\, dE_s.\end{aligned}$$

Hence by (3.7) from Proposition 3.3

$$\begin{aligned}&\mathbb{E}\left\| e^{-aE} \Gamma(Y) \right\|_T^2 + \mathbb{E}\int_0^T e^{-2aE_r} |\Gamma_r(Y)|^2 dE_r \\ &\leq C_\lambda \mathbb{E}\int_0^T e^{-2aE_s}\left[|G(s, S_s)|^2 ds + |Y_s|^2 dE_s\right] \\ &+ C_\lambda \mathbb{E}\left(\int_0^T e^{-aE_r} |\Phi(s, S_s)|\, dQ_s\right)^2 \\ &< \infty,\end{aligned}$$

that is $\Gamma(Y) \in \mathbb{U}$ whenever $Y \in \mathbb{U}$. The existence of the solution in $\mathbb{U}$ will follow from the fact that Γ is a strict contraction on $\mathbb{U}$.

Let $Y, Z \in \mathbb{U}$. By Itô's formula we have

$$\begin{aligned}&e^{-2aE_t} |\Gamma_t(Y) - \Gamma_t(Z)|^2 + 2a\int_0^t e^{-2aE_r} |\Gamma_r(Y) - \Gamma_r(Z)|^2 dE_r \\ &= 2\int_0^t e^{-2aE_r} \left\langle \Gamma_r(Y) - \Gamma_r(Z), \left[\tilde{\Phi}(r, Y_r) - \tilde{\Phi}(r, Z_r)\right] dQ_r \right\rangle \\ &+ \int_0^t e^{-2aE_r} |G(r, Y_r) - G(r, Z_r)|^2 dr \\ &+ 2\int_0^t e^{-2aE_r} \langle \Gamma_r(Y) - \Gamma_r(Z), [G(r, Y_r) - G(r, Z_r)]\, dB_r \rangle.\end{aligned}$$

Then by the Burkholder–Davis–Gundy inequality

$$\mathbb{E}\left\|e^{-aE}\left[\Gamma(Y)-\Gamma(Z)\right]\right\|_t^2+2a\mathbb{E}\int_0^t e^{-2aE_r}\left|\Gamma_r(Y)-\Gamma_r(Z)\right|^2 dE_r$$

$$\leq 4\mathbb{E}\int_0^t e^{-2aE_r}\left|\Gamma_r(Y)-\Gamma_r(Z)\right|\left|Y_r-Z_r\right|\langle L_r,dQ_r\rangle$$

$$+2\mathbb{E}\int_0^t e^{-2aE_r}\left|Y_r-Z_r\right|^2(\ell_r)^2 dr$$

$$+12\mathbb{E}\left(\int_0^t e^{-4aE_r}\left|\Gamma_r(Y)-\Gamma_r(Z)\right|^2\left|Y_r-Z_r\right|^2(\ell_r)^2 dr\right)^{1/2}$$

$$\leq 2\mathbb{E}\int_0^t e^{-2aE_r}\left|\Gamma_r(Y)-\Gamma_r(Z)\right|^2 L_r dQ_r$$

$$+2\mathbb{E}\int_0^t e^{-2aE_r}\left|Y_r-Z_r\right|^2\left(L_r dQ_r+(\ell_r)^2 dr\right)$$

$$+12\mathbb{E}\left\|e^{-aE}\left[\Gamma(Y)-\Gamma(Z)\right]\right\|_t\left(\int_0^t e^{-2aE_r}\left|Y_r-Z_r\right|^2(\ell_r)^2 dr\right)^{1/2}$$

$$\leq a\mathbb{E}\int_0^t e^{-2aE_r}\left|\Gamma_r(Y)-\Gamma_r(Z)\right|^2 dE_r+2\mathbb{E}\int_0^t e^{-2aE_r}\left|Y_r-Z_r\right|^2 dE_r$$

$$+\frac{1}{2}\mathbb{E}\left\|e^{-aE}\left[\Gamma(Y)-\Gamma(Z)\right]\right\|_t^2$$

$$+72\mathbb{E}\int_0^t e^{-2aE_r}\left|Y_r-Z_r\right|^2(\ell_r)^2 dr.$$

Hence

$$\frac{1}{2a}\mathbb{E}\left\|e^{-aE}\left[\Gamma(Y)-\Gamma(Z)\right]\right\|_t^2+\mathbb{E}\int_0^t e^{-2aE_r}\left|\Gamma_r(Y)-\Gamma_r(Z)\right|^2 dE_r$$
$$\leq\frac{74}{a}\mathbb{E}\int_0^t e^{-2aE_r}\left|Y_r-Z_r\right|^2 dE_r,$$

from which it follows that, for an appropriate choice of a,

$$\left\|\Gamma(Y)-\Gamma(Z)\right\|_{\mathbb{U}}\leq\frac{1}{2}\left\|Y-Z\right\|_{\mathbb{U}}.$$

(III) Existence in S_d^0.

The argument is identical to that of step (*II*) of the proof of Theorem 3.17 so we shall not repeat it.

The proof is complete. ∎

3.5 Global Monotonicity

Consider the SDE

$$X_t = \xi + \int_0^t F(s, X_s)\, ds + \int_0^t G(s, X_s)\, dB_s, \ \ t \geq 0, \tag{3.59}$$

under the assumptions

$$\textbf{(SDE-HD}_F\textbf{)}: \tag{3.60}$$

◊ *the functions* $F(\cdot,\cdot,x) : \Omega \times [0, +\infty[\to \mathbb{R}^d$ *and* $G(\cdot,\cdot,x) : \Omega \times [0, +\infty[\to \mathbb{R}^{d\times k}$ *are* $\mathcal{P}$*-m.s.p. for every* $x \in \mathbb{R}^d$,

◊ *there exist* $\mu \in L^1_{loc}(0,\infty)$ *and* $\ell \in L^2_{loc}(0,\infty;\mathbb{R}_+)$, *such that* $d\mathbb{P} \otimes dt$*-a.e.:*

$$\begin{cases} & \textit{Continuity:} \\ (\mathbf{C}_F) & x \longrightarrow F(t,x) : \mathbb{R}^d \to \mathbb{R}^d \textit{ is continuous;} \\ & \textit{Monotonicity condition:} \\ (\mathbf{M}_F) & \langle x-y, F(t,x)-F(t,y)\rangle \leq \mu(t)\,|x-y|^2, \quad \forall\, x,y \in \mathbb{R}^d; \\ & \textit{Boundedness condition:} \\ (\mathbf{B}_F) & \displaystyle\int_0^T F^\#_\rho(t)\, dt < \infty, \quad \mathbb{P}\textit{-a.s.}, \ \forall\, \rho, T \geq 0; \end{cases}$$

□

and

$$\textbf{(SDE-HD}_G\textbf{)}: \tag{3.61}$$

$$\begin{cases} & \textit{Lipschitz condition:} \\ (\mathbf{L}_G) & |G(t,x)-G(t,y)| \leq \ell(t)\,|x-y|, \quad \forall\, x,y \in \mathbb{R}^d; \\ & \textit{Boundedness condition:} \\ (\mathbf{B}_G) & \displaystyle\int_0^T |G(t,0)|^2 dt < \infty, \quad \mathbb{P}\text{-a.s.} \ \forall\, T \geq 0. \end{cases}$$

□

3.5.1 A Deterministic Problem

In order to prove the existence of the solution, we need some preliminary results on some ordinary differential equations.

Let $f : [0, +\infty[\times \mathbb{R}^d \to \mathbb{R}^d$ be a measurable function such that for all $x, y \in \mathbb{R}^d$:

$$\textbf{(SDE-H}_f\textbf{)} \quad \begin{cases} (i) & f(t, \cdot) : \mathbb{R}^d \to \mathbb{R}^d \text{ is continuous, a.e. } t \geq 0, \\ (ii) & \langle x - y, f(t,x) - f(t,y) \rangle \leq \mu(t)\,|x-y|^2, \text{ a.e. } t \geq 0, \\ (iii) & \displaystyle\int_0^T f_\rho^{\#}(t)\,dt < \infty, \quad \forall\, \rho > 0,\ \forall\, T > 0, \end{cases} \tag{3.62}$$

where $\mu \in L^1_{loc}(0, \infty)$ and

$$f_\rho^{\#}(t) := \sup\{|f(t,x)| : |x| \leq \rho\}.$$

We define

$$\kappa(\rho) \stackrel{def}{=} \rho + \left(\int_0^T f_\rho^{\#}(s)\,ds\right) \exp\left(\int_0^T \mu^+(r)\,dr\right)$$

and

$$C(\rho) = \left[4\rho + 8\int_0^T f_{\kappa(\rho)}^{\#}(s)\,ds\right] \exp\left[4\int_0^T \mu^+(s)\,ds\right].$$

Lemma 3.19. *Let $T > 0$ be fixed, $h \in C\left([0,T];\mathbb{R}^d\right)$ and the assumption $\left(\textbf{SDE-H}_f\right)$ be satisfied. Then the ordinary differential equation*

$$x(t) = h(t) + \int_0^t f(s, x(s))\,ds, \quad t \in [0,T], \tag{3.63}$$

has a unique solution $x \in C\left([0,T];\mathbb{R}^d\right)$. Moreover

$$\|x\|_T = \sup_{t \in [0,T]} |x(t)| \leq \kappa(\|h\|_T). \tag{3.64}$$

If x, $\tilde{x}$, are two solutions corresponding to $h, \tilde{h}$ respectively, then

$$\|x - \tilde{x}\|_T^2 \leq C\left(\|h\|_T \vee \|\tilde{h}\|_T\right) \|h - \tilde{h}\|_T. \tag{3.65}$$

In particular the mapping $h \longmapsto x$ is continuous from $C\left([0,T];\mathbb{R}^d\right)$ into itself.

Proof. *Step 1. Boundedness of solutions.*

Let x be a solution of the Eq. (3.63). Then

$$\begin{aligned} \frac{1}{2}\frac{d}{dt}|x(t) - h(t)|^2 &= \langle f(t, x(t)), x(t) - h(t) \rangle \\ &\leq \langle f(t, h(t)), x(t) - h(t) \rangle + \mu(t)\,|x(t) - h(t)|^2. \end{aligned}$$

From the first part of Lemma 6.63 (Annex C) we deduce that

$$|x(t)-h(t)| \le \int_0^t |f(s,h(s))| e^{\int_s^t \mu(r)dr} ds,$$

which establishes (3.64).

Step 2. Uniqueness of the solution.

Let $x,\ \tilde{x} \in C\left([0,T];\mathbb{R}^d\right)$ be two solutions corresponding to $h, \tilde{h}$ respectively. We have

$$\begin{aligned}
&|x(t)-h(t)-\tilde{x}(t)+\tilde{h}(t)|^2 \\
&= 2\int_0^t \left\langle f(s,x(s))-f(s,\tilde{x}(s)), x(s)-h(s)-\tilde{x}(s)+\tilde{h}(s)\right\rangle ds \\
&\le 2\int_0^t \mu(s)|x(s)-\tilde{x}(s)|^2 ds + 2\|h-\tilde{h}\|_T \int_0^t [|f(s,x(s))|+|f(s,\tilde{x}(s))|]\, ds.
\end{aligned}$$

Hence, using the inequality $\frac{1}{2}|u|^2-|v|^2 \le |u-v|^2$ with $u = x(t)-\tilde{x}(t)$ and $v = h(t)-\tilde{h}(t)$ we obtain:

$$\begin{aligned}
|x(t)-\tilde{x}(t)|^2 &\le 4\int_0^t \mu^+(s)|x(s)-\tilde{x}(s)|^2 ds \\
&\quad + 2\|h-\tilde{h}\|_T^2 + 8\|h-\tilde{h}\|_T \int_0^T f^{\#}_{\kappa\left(\|h\|_T \vee \|\tilde{h}\|_T\right)}(s)\, ds.
\end{aligned}$$

The estimate (3.65) as well as uniqueness now follow from Gronwall's inequality.

Step 3. Existence of the solution.

Let $\rho \in C_0^\infty\left(\mathbb{R}^d;\mathbb{R}_+\right)$ be such that $\rho(x) = 0$ if $|x| \ge 1$ and $\int_{\mathbb{R}^d} \rho(y)\, dy = 1$. Let $C = \sup\left\{|\nabla_x \rho(x)| : x \in \mathbb{R}^d\right\}$.

Step 3.1. Approximating equation.

Define for $0 < \varepsilon \le 1$:

$$\begin{aligned}
f_\varepsilon(t,x) &= \int_{\bar{B}(0,1)} f(t,x-\varepsilon u)\, \mathbf{1}_{[0,1]}(\varepsilon|f(t,x-\varepsilon u)|)\, \rho(u)\, du \\
&= \frac{1}{\varepsilon^{d+1}} \int_{\mathbb{R}^d} \varepsilon f(t,u)\, \mathbf{1}_{[0,1]}(\varepsilon|f(t,u)|)\, \rho\left(\frac{x-u}{\varepsilon}\right) du.
\end{aligned} \tag{3.66}$$

This mollifier approximation of f satisfies the following properties:

(L_ε): for all $\varepsilon \in]0,1]$, $x, y \in \mathbb{R}^d$, a.e. $t \ge 0$:

$$\begin{aligned}
&(i)\ |f_\varepsilon(t,x)| \le \frac{1}{\varepsilon}, \\
&(ii)\ |f_\varepsilon(t,x)-f_\varepsilon(t,y)| \le \frac{C}{\varepsilon^{d+2}}|x-y|,
\end{aligned} \tag{3.67}$$

(M_R): for all $\varepsilon \in]0,1]$, $x, y \in \mathbb{R}^d$, $|y| \le R$, a.e. $t \ge 0$:

$$\begin{aligned} &(j)\ |f_\varepsilon(t,y)| \le f^{\#}_{R+1}(t), \\ &(jj)\ \langle x-y, f_\varepsilon(t,x)\rangle \le \mu(t)|x-y|^2 + f^{\#}_{R+1}(t)|x-y|, \end{aligned} \tag{3.68}$$

and

(C_R): for all $\varepsilon, \delta \in]0,1]$, $x, y \in \mathbb{R}^d$, $|x|, |y| \le R$, a.e. $t \ge 0$:

$$\begin{aligned} \langle x-y, f_\varepsilon(t,x) - f_\delta(t,y)\rangle \le\ & 2(\varepsilon+\delta)\left[f^{\#}_{R+1}(t) + 2\mu^+(t)\right] \\ & + |x-y|\, f^{\#}_{R+1}(t)\,\mathbf{1}_{[\frac{1}{\varepsilon}\wedge\frac{1}{\delta},\infty[}\left(f^{\#}_{R+1}(t)\right) \\ & + 2\mu^+(t)|x-y|^2. \end{aligned} \tag{3.69}$$

Let us prove the last inequality. We have

$$\begin{aligned} &\langle x-y, f_\varepsilon(t,x) - f_\delta(t,y)\rangle \\ &= \int_{\bar{B}(0,1)} \langle (x-\varepsilon u) - (y-\delta u) + (\varepsilon-\delta)u, f(t,x-\varepsilon u) - f(t,y-\delta u)\rangle \\ &\qquad\qquad \times \mathbf{1}_{[0,1]}(\varepsilon|f(t,x-\varepsilon u)|)\,\rho(u)\,du \\ &\quad + \int_{\bar{B}(0,1)} \langle x-y, f(t,y-\delta u)\rangle \\ &\qquad \times\left[\mathbf{1}_{[0,1]}(\varepsilon|f(t,x-\varepsilon u)|) - \mathbf{1}_{[0,1]}(\delta|f(t,y-\delta u)|)\right]\rho(u)\,du \\ &\le \int_{\bar{B}(0,1)} \mu(t)|(x-\varepsilon u) - (y-\delta u)|^2\rho(u)\,du \\ &\qquad + 2|\varepsilon-\delta|\int_{\bar{B}(0,1)} |u|\, f^{\#}_{R+1}(t)\,\rho(u)\,du \\ &\qquad + \int_{\bar{B}(0,1)} |x-y|\, f^{\#}_{R+1}(t)\,\mathbf{1}_{[\frac{1}{\varepsilon}\wedge\frac{1}{\delta},\infty[}\left(f^{\#}_{R+1}(t)\right)\rho(u)\,du \\ &\le 2\mu^+(t)|x-y|^2 + 4(\varepsilon+\delta)\mu^+(t) + 2(\varepsilon+\delta) f^{\#}_{R+1}(t) \\ &\qquad + |x-y|\, f^{\#}_{R+1}(t)\,\mathbf{1}_{[\frac{1}{\varepsilon}\wedge\frac{1}{\delta},\infty[}\left(f^{\#}_{R+1}(t)\right). \end{aligned}$$

By a classical result on ordinary differential equations with Lipschitz coefficients (which is a particular case of Theorem 3.17), for each $\varepsilon \in]0,1]$, there exists a unique $x_\varepsilon \in C\left([0,T];\mathbb{R}^d\right)$ such that

$$x_\varepsilon(t) = h(t) + \int_0^t f_\varepsilon(s, x_\varepsilon(s))\,ds, \quad t \in [0,T]. \tag{3.70}$$

Step 3.2. Boundedness of $\|x_\varepsilon\|_T$.

Since by (3.68-jj)

$$\frac{1}{2}\frac{d}{dt}|x_\varepsilon(t)-h(t)|^2 = \langle x_\varepsilon(t)-h(t), f_\varepsilon(t,x_\varepsilon(t))\rangle$$
$$\le \mu(t)|x_\varepsilon(t)-h(t)|^2 + f^{\#}_{1+\|h\|_T}(t)|x_\varepsilon(t)-h(t)|,$$

we deduce, using the first part of Lemma 6.63 (Annex C), that

$$|x_\varepsilon(t)-h(t)| \le \int_0^t f^{\#}_{1+\|h\|_T}(s)\, e^{\int_s^t \mu(r)dr} ds$$

and the boundedness

$$\|x_\varepsilon\|_T \le \kappa(1+\|h\|_T) \stackrel{def}{=} R$$

follows.
Step 3.3. Equicontinuity of $\{x_\varepsilon : 0 < \varepsilon \le 1\}$.
By (3.68-j)

$$|x_\varepsilon(t)-x_\varepsilon(s)| \le |h(t)-h(s)| + \left|\int_s^t f_\varepsilon(r,x_\varepsilon(r))dr\right|$$
$$\le |h(t)-h(s)| + \int_s^t f^{\#}_{1+R}(r)\, dr.$$

Step 3.4. Passage to the limit.
By the Arzéla–Ascoli theorem, $\{x_\varepsilon\}_{0<\varepsilon\le1}$ is relatively compact in $C([0,T];\mathbb{R}^d)$, and consequently there exists a $x \in C([0,T];\mathbb{R}^d)$ and a sequence $\varepsilon_n \to 0$ such that $x_{\varepsilon_n} \to x$ in $C([0,T];\mathbb{R}^d)$.
Clearly, for all $t \in [0,T]$:

$$f_{\varepsilon_n}(t,x_{\varepsilon_n}(t)) \to f(t,x(t))$$

and by the Lebesgue dominated convergence theorem:

$$\int_0^t f_{\varepsilon_n}(s,x_{\varepsilon_n}(s))\, ds \to \int_0^t f(s,x(s))\, ds.$$

Then passing to the limit in (3.70) we conclude that x is solution of the Eq. (3.63). ■

Remark 3.20. The uniqueness of the solution x implies that the whole sequence x_ε satisfies

$$x_\varepsilon \to x \quad \text{in } C([0,T];\mathbb{R}^d) \quad \text{as } \varepsilon \to 0.$$

In fact $\{x_\varepsilon : 0 < \varepsilon \leq 1\}$ is a Cauchy sequence in $C\left([0,T];\mathbb{R}^d\right)$, since $|x_\varepsilon(t)| \leq R$, $|x_\delta(t)| \leq R$ and therefore by (3.69)

$$\langle x_\varepsilon(t) - x_\delta(t), d(x_\varepsilon(t) - x_\delta(t))\rangle \leq dR_{\varepsilon,\delta}(t) + |x_\varepsilon(t) - x_\delta(t)|\, dN_{\varepsilon,\delta}(t) \\ + |x_\varepsilon(t) - x_\delta(t)|^2 dV(t),$$

with

$$R_{\varepsilon,\delta}(t) = 2(\varepsilon+\delta)\int_0^t \left[f_{R+1}^{\#}(s) + 2\mu^+(s)\right] ds,$$

$$N_{\varepsilon,\delta}(t) = \int_0^t f_{R+1}^{\#}(s)\, \mathbf{1}_{[\frac{1}{\varepsilon}\wedge\frac{1}{\delta},\infty[}\left(f_{R+1}^{\#}(s)\right) ds, \quad \text{and}$$

$$V(t) = 2\int_0^t \mu^+(s)\, ds.$$

Then by Proposition 6.67 from Annex C we have

$$\|x_\varepsilon - x_\delta\|_T \leq 2e^{V(T)}\left[\sqrt{R_{\varepsilon,\delta}(T)} + N_{\varepsilon,\delta}(T)\right].$$

3.5.2 Main Result

We can now establish the main result of this section.

Theorem 3.21. *If $\xi \in L^0\left(\Omega,\mathcal{F},\mathbb{P};\mathbb{R}^d\right)$ and the assumptions (3.60-**SDE-HD**$_F$) and (3.61-**SDE-HD**$_G$) are satisfied, then the SDE (3.51) has a unique solution $X \in S_d^0$. Moreover if there exists a $p \geq 2$ such that for all $T \geq 0$:*

$$\mathbb{E}|\xi|^p + \mathbb{E}\left(\int_0^T |F(t,0)|\, dt\right)^p + \mathbb{E}\left(\int_0^T |G(t,0)|^2 dt\right)^{p/2} < +\infty, \qquad (3.71)$$

then $X \in S_d^p$.

Proof. *Uniqueness* is proved exactly as in Theorem 3.8.

Existence. It suffices to prove existence on an arbitrary interval $[0,T]$. The proof will be done in two steps.

Step 1. Existence under the additional condition (3.71).
Let $U \in S_d^p[0,T]$. Clearly $G(\cdot,U_\cdot) \in \Lambda_d^p(0,T)$.

Consider the stochastic equation

$$X_t = \xi + \int_0^t F(s,X_s)\, ds + \int_0^t G(s,U_s)\, dB_s, \quad t \in [0,T]. \qquad (3.72)$$

Define the progressively measurable continuous stochastic process (p.m.c.s.p.):

$$h(t) = \xi + \int_0^t G(s, U_s)\, dB_s.$$

For fixed $\omega \in \Omega$, Eq. (3.72) is of the form (3.63), where the data depend on the parameter ω:

$$X_t(\omega) = h(\omega, t) + \int_0^t F(\omega, s, X_s(\omega))\, ds.$$

By Lemma 3.19 Eq. (3.72) has for each $\omega \in \Omega$ a unique solution $X_\cdot(\omega) \in C\left([0,T]; \mathbb{R}^d\right)$. Moreover, by Remark 3.20 $X_\cdot(\omega)$ is the limit in $C\left([0,T]; \mathbb{R}^d\right)$ of the approximating sequence

$$X_t^\varepsilon(\omega) = h(\omega, t) + \int_0^t F_\varepsilon\left(\omega, s, X_s^\varepsilon(\omega)\right) ds, \tag{3.73}$$

where F_ε is the approximation of F defined in (3.66). By Theorem 3.17 and the fact that $h \in S_d^p[0,T]$, the approximating equation (3.73) has a unique solution $X^\varepsilon \in S_d^p[0,T]$. Since as $\varepsilon \to 0$, for each $t \in [0,T]$

$$\sup_{s\in[0,t]} |X_s^\varepsilon - X_s| \to 0, \quad a.s.,$$

the progressive measurability of X follows; hence $X \in S_d^0[0,T]$.

Writing for the SDE (3.72) the inequality (3.15), with $\ell_t = 0$, $p \geq 2$, $\lambda > 1$ and

$$V_t = \int_0^t \mu(r)\, dr,$$

we have

$$\begin{aligned} e^{-\|V\|_T}\mathbb{E}\|X\|_T^p &\leq \mathbb{E}\left\|e^{-V}X\right\|_T^p \\ &\leq C_{p,\lambda}\left[\mathbb{E}|\xi|^p + \mathbb{E}\left(\int_0^T e^{-V_r}|F(r,0)|\, dr\right)^p \right. \\ &\quad \left. + \mathbb{E}\left(\int_0^T e^{-2V_r}|G(r,U_r)|^2 dr\right)^{p/2}\right] \\ &< \infty. \end{aligned}$$

Hence $X \in S_d^p[0,T]$.

We note that the Eq. (3.59) on $[0,T]$ may be written in the form

$$\Gamma(X) = X,$$

where $\Gamma : S_d^p[0,T] \to S_d^p[0,T]$ is defined by $X = \Gamma(U)$, X is the solution of the equation

$$X_t = \xi + \int_0^t F(r, X_r)\,dr + \int_0^t G(r, U_r)\,dB_r, \quad 0 \le t \le T.$$

The existence and uniqueness of a solution of (3.51) in $S_d^p[0,T]$ will follow from Banach's fixed point theorem and the fact that Γ is a strict contraction on $S_d^p[0,T]$ equipped with the equivalent norm $\|\!|\cdot|\!\|_a$ given by

$$\|\!|X|\!\|_a = \sup_{t\in[0,T]} \left[e^{-at} \left(\mathbb{E}\left\| e^{-V} X \right\|_t^p \right)^{1/p} \right],$$

with

$$V(t) = \int_0^t \mu(r)\,dr$$

and a large enough, which we now prove.

Let $U, \tilde{U} \in S_d^p[0,T]$, $X = \Gamma(U)$ and $\tilde{X} = \Gamma(\tilde{U})$. Then $X - \tilde{X}$ satisfies

$$\begin{aligned} X_t - \tilde{X}_t &= \int_0^t \left[F(r, \tilde{X}_r + (X_r - \tilde{X}_r)) - F(r, \tilde{X}_r) \right] dr \\ &\quad + \int_0^t \left[G(r, U_r) - G(r, \tilde{U}_r) \right] dB_r. \end{aligned}$$

Since the functions

$$\tilde{F}(t,x) = F(r, \tilde{X}_r + x) - F(r, \tilde{X}_r) \quad \text{and} \quad \tilde{G}(t,x) = G(r, U_r) - G(r, \tilde{U}_r)$$

satisfy the assumptions of Proposition 3.6 (with the corresponding monotonicity and Lipschitz "constants" $\tilde{\mu} = \mu$ and $\tilde{\ell} = 0$) we have, by (3.15), for $p \ge 2$ and $\lambda = 2$:

$$\begin{aligned} & e^{-pat} \times \mathbb{E}\left\| e^{-V}(X - \tilde{X}) \right\|_t^p \\ & \le e^{-pat} \times C_p \mathbb{E}\left(\int_0^t e^{-2V(r)} \left| \tilde{G}(r,0) \right|^2 dr \right)^{p/2} \\ & \le e^{-pat} \times C_p \mathbb{E}\left(\int_0^t \ell^2(r)\, e^{2ar} e^{-2ar} \left\| e^{-V}\left(U - \tilde{U}\right) \right\|_r^2 dr \right)^{p/2} \\ & \le e^{-pat} \times C_p \left(\int_0^t \ell^2(r)\, e^{2ar} e^{-2ar} \left(\mathbb{E}\left\| e^{-V}\left(U - \tilde{U}\right) \right\|_r^p \right)^{2/p} dr \right)^{p/2} \\ & \le \varphi(a) \sup_{r\in[0,T]} \left(e^{-par} \mathbb{E}\left\| e^{-V}\left(U - \tilde{U}\right) \right\|_r^p \right), \end{aligned}$$

where

$$\varphi(a) = C_p \sup_{t\in[0,T]} \left(e^{-2at} \int_0^t \ell^2(r)\, e^{2ar} dr \right)^{p/2}.$$

Taking the sup over $t \in [0, T]$, we deduce that

$$\left\|\left|\Gamma(U) - \Gamma(\tilde{U})\right|\right\|_a \leq [\varphi(a)]^{1/p} \; \left\|\left|U - \tilde{U}\right|\right\|_a.$$

Since by Proposition 6.57, Annex B, $\lim_{a\to\infty} \varphi(a) = 0$, it follows that Γ is a strict contraction for a large enough.

Step 2. Existence in S_d^0.

The argument is identical to that of step (*II*) of the proof of Theorem 3.17 so we shall not repeat it.

The proof is complete. ■

3.5.3 SDEs with Deterministic Initial Condition

We let $\{X_s^{t,x} : s \geq 0\}$ be the solution starting from x at time t, that is

$$X_s^{t,x} = x + \int_t^{s\vee t} F(r, X_r^{t,x})\, dr + \int_t^{s\vee t} G(r, X_r^{t,r})\, dB_r. \tag{3.74}$$

We can also write

$$X_s^{t,x} = x + \int_0^s \mathbf{1}_{[t,\infty[}(s)\, F(r, X_r^{t,x})\, dr + \int_0^s \mathbf{1}_{[t,\infty[}(s)\, G(r, X_r^{t,r})\, dB_r.$$

Let $p \geq 2$. Assume that the assumptions (3.60-**SDE-HD**$_F$) and (3.61-**SDE-HD**$_G$) are satisfied and for all $T > 0$,

$$\mathbb{E}\left(\int_0^T |F(t,0)|\, dt\right)^p + \mathbb{E}\left(\int_0^T |G(t,0)|^2 dt\right)^{p/2} < +\infty. \tag{3.75}$$

Then by Theorem 3.21 the stochastic differential equation (3.74) has a unique solution $X^{t,x} \in S_d^p$.

Define

$$V_t = \int_0^t \left[\mu^+(r) + \lambda\left(\frac{p-1}{2} + 9p\lambda\right) \ell^2(r)\right] dr$$

and for $1 < q \leq \infty$,

$$M_{q,p,T}(x) \overset{def}{=} T^{\frac{p}{2}\left(1-\frac{1}{q}\right)}\mathbb{E}\|F(\cdot,x)\|^p_{L^q(0,T)} + \mathbb{E}\|G(\cdot,x)\|^p_{L^{2q}(0,T)}. \tag{3.76}$$

Proposition 3.22. *Assume that (3.60-**SDE-HD**$_F$), (3.61-**SDE-HD**$_G$) and (3.75) hold. Then for every $p \geq 2$ and $\lambda > 1$, there exists a constant $C_{p,\lambda}$ such that for all $T, t, t' \geq 0$ and $x, x' \in \mathbb{R}^d$:*

$$\begin{aligned}\mathbb{E}\sup_{s\in[0,T]}|X_s^{t,x} - X_s^{t',x'}|^p &\leq C_{p,\lambda}\, e^{p\left(V_T - V_{T\wedge t\wedge t'}\right)}\Big[\,|x-x'|^p \\ &+\mathbb{E}\left|\int_{T\wedge t'}^{T\wedge t}|F(r,x)|\,dr\right|^p + \mathbb{E}\left|\int_{T\wedge t'}^{T\wedge t}|G(r,x)|^2 dr\right|^{p/2}\Big].\end{aligned} \tag{3.77}$$

Then for all $1 < q \leq \infty$, $x, x' \in \mathbb{R}$ and $T, t, t' \geq 0$:

$$\mathbb{E}\left(\sup_{0\leq s\leq T}|X_s^{t,x} - X_s^{t',x'}|^p\right) \leq C_{p,\lambda}\, e^{pV_T}\left[|x-x'|^p + M_{q,p,T}(x)\,|t-t'|^{\frac{p}{2}\left(1-\frac{1}{q}\right)}\right]. \tag{3.78}$$

Proof. We first remark that if the hypothesis (3.60-**SDE-HD**$_F$) is satisfied, then it is also satisfied replacing μ by μ^+.

Let $t' \leq t$. We have three cases:

Case 1: $T \leq t'$.
We have

$$\sup_{0\leq s\leq T}|X_s^{t,x} - X_s^{t',x'}|^p = |x' - x|^p.$$

Case 2: $t' < T \leq t$.
We have

$$\begin{aligned}\sup_{0\leq s\leq T}|X_s^{t,x} - X_s^{t',x'}|^p &\leq \sup_{0\leq s\leq t'}|X_s^{t,x} - X_s^{t',x'}|^p + \sup_{t'\leq s\leq T}|X_s^{t,x} - X_s^{t',x'}|^p \\ &= |x'-x|^p + \sup_{t'\leq s\leq T}|x - X_s^{t',x'}|^p.\end{aligned}$$

Since

$$\begin{aligned}X_s^{t',x'} - x = x' - x &+ \int_0^s \mathbf{1}_{[t',\infty[}(r)\, F\left(r, x + (X_r^{t',x'} - x)\right) dr \\ &+ \int_0^s \mathbf{1}_{[t',\infty[}(r)\, G\left(r, x + (X_r^{t',x'} - x)\right) dB_r,\end{aligned}$$

we have, by Proposition 3.6,

$$\begin{aligned}
e^{-V_T}\mathbb{E}\sup_{s\in[t',T]}\left|X_s^{t',x'}-x\right|^p &\le \mathbb{E}\sup_{s\in[t',T]}e^{-V_s}\left|X_s^{t',x'}-x\right|^p\\
&\le C_{p,\lambda}\Big[e^{-V_{t'}}\left|x'-x\right|^p+\mathbb{E}\Big(\int_{t'}^T e^{-V_r}\left|F(r,x)\right|dr\Big)^p\\
&\quad+\mathbb{E}\left(\int_{t'}^T e^{-2V_r}\left|G(r,x)\right|^2dr\right)^{p/2}\Big].
\end{aligned}$$

Hence

$$\begin{aligned}
\mathbb{E}\sup_{s\in[0,T]}\left|X_s^{t',x'}-x\right|^p &\le C_{p,\lambda}e^{p(V_T-V_{t'})}\Big[\left|x'-x\right|^p\\
&+\mathbb{E}\Big(\int_{t'}^T\left|F(r,x)\right|dr\Big)^p+\mathbb{E}\left(\int_{t'}^T\left|G(r,x)\right|^2dr\right)^{p/2}\Big].
\end{aligned}$$

Case 3: $t < T$

Note that for $s \ge t$

$$X_s^{t,x}=x+\int_0^s \mathbf{1}_{[t,\infty[}(r)\,F(r,\,X_r^{t,x})\,dr+\int_0^s \mathbf{1}_{[t,\infty[}(r)\,G(r,\,X_r^{t,x})\,dB_r$$

and

$$X_s^{t',x'}=X_t^{t',x'}+\int_0^s \mathbf{1}_{[t,\infty[}(r)\,F(r,\,X_r^{t',x'})\,dr+\int_0^s \mathbf{1}_{[t,\infty[}(r)\,G(r,\,X_r^{t',x'})\,dB_r.$$

By Theorem 3.8 we obtain

$$\begin{aligned}
e^{-V_T}\mathbb{E}\sup_{s\in[t,T]}|X_s^{t',x'}-X_s^{t,x}|^p &\le \mathbb{E}\sup_{s\in[t,T]}e^{-V_s}|X_s^{t',x'}-X_s^{t,x}|^p\\
&\le C_{p,\lambda}\,\mathbb{E}e^{-V_t}\left|X_t^{t',x'}-x\right|^p.
\end{aligned}$$

Hence

$$\begin{aligned}
\mathbb{E}\sup_{0\le s\le T}|X_s^{t,x}-X_s^{t',x'}|^p &\le \mathbb{E}\sup_{0\le s\le t'}|X_s^{t,x}-X_s^{t',x'}|^p+\mathbb{E}\sup_{t'\le s\le t}|X_s^{t,x}-X_s^{t',x'}|^p\\
&\quad+\mathbb{E}\sup_{t\le s\le T}|X_s^{t,x}-X_s^{t',x'}|^p\\
&\le\Big[|x'-x|^p+\left(1+C_{p,\lambda}e^{V_T-V_t}\right)\mathbb{E}\sup_{t'\le s\le t}|X_s^{t,x}-X_s^{t',x'}|^p\Big]\\
&\le C_{p,\lambda}e^{V_T-V_{t'}}\Big[|x'-x|^p+\mathbb{E}\Big(\int_{t'}^T|F(r,x)|\,dr\Big)^p+\mathbb{E}\left(\int_{t'}^T|G(r,x)|^2dr\right)^{p/2}\Big].
\end{aligned}$$

The inequality (3.78) clearly follows from (3.77). The proof is complete.

■

This result permits us to apply Kolmogorov's criterion (Theorem 1.40) to the $C\left([0,T];\mathbb{R}^d\right)$-valued stochastic process $\left\{X_\cdot^{t,x} : (t,x) \in [0,T] \times [-R,R]^d\right\}$, setting on $[0,T] \times [-R,R]^d$ the metric

$$\rho\left((t,x),(t',x')\right) = \left|t-t'\right|^{\frac{1}{2}-\frac{1}{2q}} + \left|x-x'\right|.$$

Assuming that for all $R > 0$

$$M_{q,p,R,T} \overset{def}{=} \sup_{|x|\le R} M_{q,p,T}(x) < \infty,$$

then by (3.78) for $|x|, |x'| \le R$ and $t,t' \in [0,T]$:

$$\mathbb{E}\left(\sup_{0\le s\le T} |X_s^{t,x} - X_s^{t',x'}|^p\right) \le C \times \left\{\rho\left((t,x),(t',x')\right)\right\}^{1+d+(p-1-d)}$$

where C is a positive constant independent of x,x',t,t'. By Kolmogorov's criterion for $p > 1+d$ and $0 < \delta < 1 - p^{-1}(1+d)$, there exists an $\eta \in L^p(\Omega,\mathcal{F},\mathbb{P})$ (a random variable independent of x,x',t,t') such that

$$\sup_{0\le s\le T} |X_s^{t,x} - X_s^{t',x'}| \le \eta\,\left\{\rho\left((t,x),(t',x')\right)\right\}^{\frac{p-1-d}{p}-\delta}.$$

Hence we have:

Corollary 3.23. *Let the assumptions (3.60-**SDE-HD**$_F$) and (3.61-**SDE-HD**$_G$) be satisfied and assume that for some $1 < q \le \infty$, $1+d < p < \infty$ and for all $R > 0$*

$$M_{q,p,T,R} \overset{def}{=} \sup_{|x|\le R} \mathbb{E}\left\| F(\cdot,x)\right\|_{L^q(0,T)}^p + \sup_{|x|\le R} \mathbb{E}\left\| G(\cdot,x)\right\|_{L^{2q}(0,T)}^p < \infty. \quad (3.79)$$

Then there exists a process $\{\tilde{X}_s^{t,x}\,;\, s,t \in \mathbb{R}_+,\ x \in \mathbb{R}^d\}$ such that:

(i) $\tilde{X}_s^{t,x} = X_s^{t,x}$, $s \ge 0$, *a.s., for each* $t \in \mathbb{R}_+$, $x \in \mathbb{R}^d$;

(ii) the mapping

$$(t,x) \to \tilde{X}_\cdot^{t,x} : [0,T] \times [0,R]^d \to C\left([0,T];\mathbb{R}^d\right)$$

is a.s. continuous and moreover for all $\frac{1+d}{p} < \varepsilon \le 1$,

$$\sup_{s\in[0,T]} \left|\tilde{X}_s^{t,x} - \tilde{X}_s^{t',x'}\right| \le \eta_\varepsilon \times \left[\left|x-x'\right|^{1-\varepsilon} + \left|t-t'\right|^{\left(\frac{1}{2}-\frac{1}{2q}\right)(1-\varepsilon)}\right]$$

with η_ε a random variable independent of x,x',t,t' such that $\mathbb{E}\,|\eta_\varepsilon|^p < \infty$;

(iii) the mapping

$$(s,t,x) \to \tilde{X}_s^{t,x} : \mathbb{R}_+ \times \mathbb{R}_+ \times \mathbb{R}^d \to \mathbb{R}^d$$

is a.s. continuous.

Remark 3.24. From now on, whenever this modification $\tilde{X}$ exists we shall write X in place of $\tilde{X}$ (that is, we shall assume that $(t,s,x) \to X_s^{t,x}$ is a.s. continuous).

3.5.4 SDEs with Stieltjes Integrals

We shall conclude this section with an existence result for the more general equation

$$X_t = \xi + \int_0^t \Phi(s, X_s)\, dQ_s + \int_0^t G(s, X_s)\, dB_s, \tag{3.80}$$

where $\xi : \Omega \to \mathbb{R}^d$ is an $\mathcal{F}_0$-measurable random vector, $\Phi : \Omega \times [0,\infty[\times \mathbb{R}^d \to \mathbb{R}^d$ and $Q : \Omega \times [0,\infty[\to \mathbb{R}$ is $\mathcal{P}$-m.i.c.s.p.

We first remark that for the differential equation

$$x(t) = h(t) + \int_0^t f(s, x(s))\, dq(s), \quad t \in [0,T], \tag{3.81}$$

we have results similar to those from Lemma 3.19.

Assume that

$$\textbf{(SDE-H}_q\textbf{)} : \quad q : \mathbb{R}_+ \to \mathbb{R}_+, \textit{is a continuous increasing function}, \; q(0) = 0$$

and $f : \mathbb{R}_+ \times \mathbb{R}^d \to \mathbb{R}^d$ *is a measurable function such that there exists a measurable function* $\mu : \mathbb{R}_+ \to \mathbb{R}$ *satisfying for all* $T \geq 0$:

$$\int_0^T |\mu(t)|\, dq(t) < \infty$$

and

$$\textbf{(SDE-H}_{fq}\textbf{)} : \begin{cases} (i) & f(t,\cdot) : \mathbb{R}^d \to \mathbb{R}^d \; \textit{are continuous a.e.} \; t \in \mathbb{R}_+; \\ (ii) & \langle x - y, f(t,x) - f(t,y) \rangle \leq \mu(t)\, |x-y|^2, \\ & \qquad\qquad\qquad a.e. \; t \in \mathbb{R}_+, \; \forall\, x, y \in \mathbb{R}^d; \\ (iii) & \displaystyle\int_0^T f_\rho^{\#}(t)\, dq(t) < \infty, \quad \textit{for all} \; T, \rho \geq 0. \end{cases}$$

We also introduce the following notation for $\rho \geq 0$:

$$R_\rho(t) \stackrel{def}{=} \int_0^t f_\rho^\#(s)\,dq(s),$$

$$\kappa(\rho) \stackrel{def}{=} \rho + 2R_\rho(T)\exp\left(\int_0^T \mu^+(s)\,dq(s)\right).$$

Note that if $x \in C([0,T];\mathbb{R}^d)$ is a solution of the Eq. (3.81) with $h \in C([0,T];\mathbb{R}^d)$ then

$$\begin{aligned}
&\langle x(t) - h(t), d(x(t) - h(t))\rangle \\
&= \langle x(t) - h(t), f(t, x(t))\,dq(t)\rangle \\
&\leq |x(t) - h(t)|\,|f(t, h(t))|\,dq(t) + |x(t) - h(t)|^2 \mu^+(t)\,dq(t)
\end{aligned}$$

and by Proposition 6.67 from Annex C we have

$$\|x - h\|_T \leq 2\left[\int_0^T |f(r, h(r))|\,dq(r)\right]\exp\int_0^T \mu^+(r)\,dq(r).$$

Hence

$$\|x\|_T \leq \kappa(\|h\|_T). \tag{3.82}$$

Also if $x, \tilde{x}$, are two solutions corresponding to $h, \tilde{h}$ respectively, then

$$\begin{aligned}
&\left\langle x(t) - h(t) - \tilde{x}(t) + \tilde{h}(t), d\left(x(t) - h(t) - \tilde{x}(t) + \tilde{h}(t)\right)\right\rangle \\
&= \left\langle x(t) - h(t) - \tilde{x}(t) + \tilde{h}(t), f(t, x(t)) - f(t, \tilde{x}(t))\,dq(t)\right\rangle \\
&\leq 2\left|h(t) - \tilde{h}(t)\right| dR_{\kappa(\|h\|_T \vee \|\tilde{h}\|_T)}(t) + |x(t) - \tilde{x}(t)|^2 \mu^+(t)\,dq(t) \\
&\leq \left[2\left|h(t) - \tilde{h}(t)\right| dR_{\kappa(\|h\|_T \vee \|\tilde{h}\|_T)}(t) + 2\left|h(t) - \tilde{h}(t)\right|^2 \mu^+(t)\,dq(t)\right] \\
&\quad + 2\left|x(t) - h(t) - \tilde{x}(t) + \tilde{h}(t)\right|^2 \mu^+(t)\,dq(t).
\end{aligned}$$

By Proposition 6.67 from Annex C we obtain

$$\begin{aligned}
\left\|x - h - \tilde{x} + \tilde{h}\right\|_T &\leq 2\left\|h - \tilde{h}\right\|_T^{1/2} \\
&\quad \times \left[2R_{\kappa(\|h\|_T \vee \|\tilde{h}\|_T)}(T) + 2\left\|h - \tilde{h}\right\|_T \int_0^T \mu^+(t)\,dq(t)\right]^{1/2} \\
&\quad \times \exp\left\{2\int_0^T \mu^+(t)\,dq(t)\right\}.
\end{aligned}$$

Hence

$$\|x - \tilde{x}\|_T \le \left\|h - \tilde{h}\right\|_T^{1/2} C\left(\|h\|_T \vee \|\tilde{h}\|_T\right). \tag{3.83}$$

With a very similar proof as for Lemma 3.19 we have:

Lemma 3.25. *Let $T > 0$ be fixed, $h \in C\left([0,T];\mathbb{R}^d\right)$ and the assumptions (**SDE-H**$_q$) and (**SDE-H**$_{fq}$) be satisfied. Then the differential equation (3.81) has a unique solution $x \in C\left([0,T];\mathbb{R}^d\right)$. Moreover the mapping $h \longmapsto x$ is continuous from $C\left([0,T];\mathbb{R}^d\right)$ into itself.*

Now we can give an existence and uniqueness result for the SDE (3.80).
Define

$$V_t \overset{def}{=} \int_0^t \mu_r dQ_r + \lambda\left(\frac{p-1}{2} + 9p\lambda\right)\int_0^t (\ell_r)^2\, dr.$$

Theorem 3.26. *If $\xi \in L^0\left(\Omega, \mathcal{F}_0, \mathbb{P}; \mathbb{R}^d\right)$ and the assumptions (3.11-**SDE-H**$_G$), (3.21-**SDE-H**$_Q$) and (3.22-**SDE-H**$_\Phi$) are satisfied, then the SDE (3.80) has a unique solution $X \in S_d^0$. Moreover for every $p \ge 2$ and $\lambda > 1$ there exists a constant $C_{p,\lambda}$ such that for all $T \ge 0$:*

$$\begin{aligned}\mathbb{E}\left\|e^{-V}X\right\|_T^p \le C_{p,\lambda}\Big[\mathbb{E}\,|\xi|^p &+ \mathbb{E}\left(\int_0^T e^{-V_r}\,|\Phi(r,0)|\,dQ_r\right)^p \\ &+ \mathbb{E}\left(\int_0^T e^{-2V_r}\,|G(r,0)|^2\,dr\right)^{p/2}\Big].\end{aligned} \tag{3.84}$$

Proof. The inequality (3.84) follows from (3.24).

Uniqueness was proved in Theorem 3.10. The uniqueness result tells us that it is sufficient to prove the existence of the solution on an arbitrary fixed interval $[0,T]$.

Existence under the condition

$$\mathbb{E}\,|\xi|^2 + \mathbb{E}\int_0^T e^{-2aE_s}\,|G(s,0)|^2\,ds + \mathbb{E}\left(\int_0^T e^{-aE_s}\,|\Phi(s,0)|\,dQ_s\right)^2 < \infty,$$

where

$$E_t = \int_0^t (\ell_r)^2\,dr + \int_0^t \mu_r^+\,dQ_r$$

and $a \ge 4$ will be chosen below.

Note that by Lemma 3.25 and with similar arguments as in the proof of Theorem 3.21, for each $U \in S_d^0[0,T]$ the SDE

$$X_t = \xi + \int_0^t \Phi(s, X_s)\, dQ_s + \int_0^t G(s, U_s)\, dB_s \tag{3.85}$$

has a unique solution $\Gamma(U) \stackrel{def}{=} X \in S_d^0[0, T]$.

We continue the proof following the steps of the proof of Theorem 3.18.

Consider the Banach space $\mathbb{U}$ of the stochastic processes $X \in S_d^0[0, T]$ such that

$$\|X\|_{\mathbb{U}} \stackrel{def}{=} \left(\frac{1}{2a}\mathbb{E}\left\|e^{-aE}X\right\|_T^2\right)^{1/2} + \left(\mathbb{E}\int_0^T e^{-2aE_s}|X_s|^2\, dE_s\right)^{1/2} < \infty.$$

The solution of the SDE (3.58) on $[0, T]$ is a fixed point of the mapping $\Gamma : S_d^0[0, T] \to S_d^0[0, T]$ defined by $\Gamma(U) = X$, where X is the solution of the SDE (3.85).

We now show that $\Gamma(U) \in \mathbb{U}$, whenever $U \in \mathbb{U}$.

Since

$$dD_s = |X_s|^2\, dE_s = |X_s|^2 \left(\ell_s^2 ds + \langle \mu_s^+, dQ_s \rangle\right),$$

and for $p = 2$ and $\lambda > 1$,

$$\begin{aligned}&\langle X_s, \Phi(s, X_s)\, dQ_s\rangle + \left(\frac{1}{2}m_p + 9p\lambda\right)|G(s, U_s)|^2\, ds\\ &\le |X_s|^2 \mu_s dQ_s + |X_s|\left(|\Phi(s, 0)|\, dQ_s\right) + (1 + 36\lambda)\left(\ell_s^2|U_s|^2 + |G(s, 0)|^2\right) ds,\end{aligned}$$

it follows that

$$\begin{aligned}&dD_s + \langle X_s, \Phi(s, X_s)\, dQ_s\rangle + \left(\frac{1}{2}m_p + 9p\lambda\right)|G(s, U_s)|^2\, ds\\ &\le (1 + 36\lambda)\left[|G(s, 0)|^2\, ds + |U_s|^2\, dE_s\right] + |X_s|\,|\Phi(s, 0)|\, dQ_s + |X_s|^2\, a\, dE_s.\end{aligned}$$

Hence by (3.7) from Proposition 3.3

$$\begin{aligned}&\mathbb{E}\left\|e^{-aE}X\right\|_T^2 + \mathbb{E}\int_0^T e^{-2aE_r}|X|^2\, dE_r\\ &\le C_\lambda \mathbb{E}|\xi|^2 + C_\lambda \mathbb{E}\int_0^T e^{-2aE_s}\left[|G(s, 0)|^2\, ds + |U_s|^2\, dE_s\right]\\ &\qquad + C_\lambda \mathbb{E}\left(\int_0^T e^{-aE_r}|\Phi(s, 0)|\, dQ_s\right)^2\\ &< \infty,\end{aligned}$$

that is $\Gamma(U) \in \mathbb{U}$. The existence of the solution in $\mathbb{U}$ will follows from the fact that Γ is a strict contraction on $\mathbb{U}$.

Let $U, \tilde{U} \in \mathbb{U}$, $X = \Gamma(U) \in \mathbb{U}$ and $\tilde{X} = \Gamma(\tilde{U}) \in \mathbb{U}$. By Itô's formula we have

$$\begin{aligned}
& e^{-2aE_t} \left|X_t - \tilde{X}_t\right|^2 + 2a \int_0^t e^{-2aE_r} \left|X_r - \tilde{X}_r\right|^2 dE_r \\
& = 2\int_0^t e^{-2aE_r} \left\langle X_r - \tilde{X}_r, \left[\Phi(r, X_r) - \Phi(r, \tilde{X}_r)\right] dQ_r\right\rangle \\
& + \int_0^t e^{-2aE_r} \left|G(r, U_r) - G\left(r, \tilde{U}_r\right)\right|^2 dr \\
& + 2\int_0^t e^{-2aE_r} \left\langle X_r - \tilde{X}_r, \left[G(r, U_r) - G\left(r, \tilde{U}_r\right)\right] dB_r\right\rangle .
\end{aligned}$$

Since, by the Burkholder–Davis–Gundy inequality,

$$\begin{aligned}
& \sup_{s\in[0,t]} \left|\int_0^s e^{-2aE_r} \left\langle X_r - \tilde{X}_r, \left[G(r, U_r) - G\left(r, \tilde{U}_r\right)\right] dB_r\right\rangle\right| \\
& \leq 3\mathbb{E} \left(\int_0^t e^{-4aE_r} \left|X_r - \tilde{X}_r\right|^2 \left|U_r - \tilde{U}_r\right|^2 (\ell_r)^2 dr\right)^{1/2} \\
& \leq 3\mathbb{E} \left\|e^{-aE}\left(X - \tilde{X}\right)\right\|_t \left(\int_0^t e^{-2aE_r} \left|U_r - \tilde{U}_r\right|^2 (\ell_r)^2 dr\right)^{1/2} \\
& \leq \frac{1}{8}\mathbb{E} \left\|e^{-aE}\left(X - \tilde{X}\right)\right\|_t^2 + 18\mathbb{E}\int_0^t e^{-2aE_r} \left|U_r - \tilde{U}_r\right|^2 (\ell_r)^2 dr,
\end{aligned}$$

it follows that for $a \geq 4$

$$\begin{aligned}
& \mathbb{E} \left\|e^{-aE}\left(X - \tilde{X}\right)\right\|_t^2 + 2a\mathbb{E}\int_0^t e^{-2aE_r} \left|X_r - \tilde{X}_r\right|^2 dE_r \\
& \leq a\mathbb{E}\int_0^t e^{-2aE_r} \left|X_r - \tilde{X}_r\right|^2 dE_r + 2\mathbb{E}\int_0^t e^{-2aE_r} \left|U_r - \tilde{U}_r\right|^2 (\ell_r)^2 dr \\
& \quad + \frac{1}{2}\mathbb{E} \left\|e^{-aE}\left(X - \tilde{X}\right)\right\|_t^2 + 36\mathbb{E}\int_0^t e^{-2aE_r} \left|U_r - \tilde{U}_r\right|^2 (\ell_r)^2 dr.
\end{aligned}$$

Hence

$$\begin{aligned}
& \frac{1}{2a}\mathbb{E} \left\|e^{-aE}\left(X - \tilde{X}\right)\right\|_t^2 + \mathbb{E}\int_0^t e^{-2aE_r} \left|X_r - \tilde{X}_r\right|^2 dE_r \\
& \qquad \leq \frac{38}{a}\mathbb{E}\int_0^t e^{-2aE_r} \left|U_r - \tilde{U}_r\right|^2 dE_r ,
\end{aligned}$$

from which it follows that, for an appropriate choice of a,

$$\left\|\Gamma(U)-\Gamma(\tilde{U})\right\|_{\mathbb{U}} \leq \frac{1}{2}\left\|U-\tilde{U}\right\|_{\mathbb{U}}.$$

(III) Existence in S_d^0.

The argument is identical to that of step *(II)* of the proof of Theorem 3.17 so we shall not repeat it.

The proof is complete. ■

3.6 Local Monotonicity

3.6.1 Locally Monotone Drift

Let us now extend the results of Theorem 3.21 to the case of "locally monotone drift and locally Lipschitz diffusion coefficient". We shall study the equation

$$X_t = \xi + \int_0^t F(s, X_s)ds + \int_0^t G(s, X_s)dB_s, \quad \forall t \geq 0, \ a.s., \tag{3.86}$$

where for all $x \in \mathbb{R}^d$ the functions

$$F(\cdot,\cdot,x) : \Omega \times [0,\infty[\to \mathbb{R}^d, \qquad G(\cdot,\cdot,x) : \Omega \times [0,\infty[\to \mathbb{R}^{d\times k}$$

are $\mathcal{P}$-measurable.

Recall the notations

$$F_\rho^{\#}(t) \overset{def}{=} \sup_{x\leq\rho} |F(t,x)| \quad \text{and} \quad m_p = 1 \vee (p-1).$$

The assumptions (3.10-**SDE-H**$_F$) and (3.11-**SDE-H**$_G$) will be modified as follows: for all $\rho \geq 0$ there exist two $\mathcal{P}$-m.s.p. $\mu^{(\rho)}$ and $\ell^{(\rho)}$ such that

$$\int_0^T \left[\left|\mu_t^{(\rho)}\right| + \left|\ell_t^{(\rho)}\right|^2\right] dt < \infty, \quad a.s.\ \forall\, T \geq 0,$$

and

(**SDE-H**$_{F,loc}$):

(C_F) *Continuity:*
$x \longrightarrow F(t,x) : \mathbb{R}^d \to \mathbb{R}^d$ *is continuous,* $d\mathbb{P} \otimes dt$*-a.e.;*

$(M_{F,loc})$ *Local monotonicity condition:*

$$\langle x - y, F(t,x) - F(t,y)\rangle \leq \mu_t^{(\rho)} |x-y|^2, \quad d\mathbb{P} \otimes dt\text{-a.e.},$$
$$\forall\, \rho \geq 0 \text{ and } \forall\, |x|, |y| \leq \rho;$$

(B_F) *Boundedness condition:*

$$\int_0^T \left| F_\rho^\#(s)\right| ds < \infty, \quad a.s., \ \forall\, T, \rho \geq 0.$$

(SDE-H$_{G,loc}$**)**:

$(L_{G,loc})$ *Local Lipschitz condition:*

$$|G(t,x) - G(t,y)| \leq \ell_t^{(\rho)} |x-y|, \quad d\mathbb{P} \otimes dt\text{-a.e.},$$
$$\forall\, \rho \geq 0 \text{ and } \forall\, |x|, |y| \leq \rho;$$

(B_G) *Boundedness condition:*

$$\int_0^T |G(t,0)|^2 dt < \infty, \quad a.s. \ \forall\, T \geq 0;$$

and the mixed boundedness condition
(B$_{FG}$**)**:
there exist $a > 1, b > 1$, *a continuous increasing function* $V : [0,\infty[\to [0,\infty[$, $V(0) = 0$ *and two* $\mathcal{P}$-m.i.c.s.p. R, N, $R_0 = N_0 = 0$ *such that, as a signed measure on* $[0,\infty[$:

$$\langle F(t,x(t)), x(t)\rangle\, dt + \left(\frac{1}{2} m_a + 9ab\right) |G(t,x(t))|^2 dt$$
$$\leq \mathbf{1}_{a\geq 2} dR_t + |x(t)|\, dN_t + |x(t)|^2 d\, V(t),$$

for all continuous functions $x : [0,\infty[\to \mathbb{R}^d$, where

$$m_a = 1 \vee (a-1).$$

Theorem 3.27. *Let the assumptions* **(SDE-H**$_{F,loc}$**)**, **(SDE-H**$_{G,loc}$**)** *and* **(B**$_{FG}$**)** *be satisfied. If* $\xi \in L^0(\Omega, \mathcal{F}_0, P; \mathbb{R}^d)$, *then the SDE (3.86) has a unique solution* $X \in S_d^0$. *Moreover for every* $p \in [1,a]$ *and* $\lambda \in]1,b]$ *there exists a constant* $C_{p,\lambda}$ *such that:*

$$\mathbb{E} \sup_{t\in[0,T]} |X_t|^p \leq C_{p,\lambda} e^{V(T)} \left[\mathbb{E}|\xi|^p + \mathbf{1}_{p\geq 2} \mathbb{E} R_T^{p/2} + \mathbb{E} N_T^p\right]. \tag{3.87}$$

Proof. By the inequality (3.7) from Proposition 3.3, the estimate (3.87) clearly follows.

Uniqueness. Let $X, \tilde{X} \in S_d^0$ be two solutions of SDE (3.86) corresponding to initial data $\xi, \tilde{\xi} \in L^0(\Omega, \mathcal{F}_0, \mathbb{P}; \mathbb{R}^d)$, respectively. Let the stopping time be

$$\tau_n(\omega) = \inf\left\{t \geq 0 : |X_t(\omega)| + |\tilde{X}_t(\omega)| \geq n\right\}.$$

We have

$$X_{t\wedge\tau_n} = \xi + \int_0^t \mathbf{1}_{[0,\tau_n]}(s) F(s \wedge \tau_n, X_{s\wedge\tau_n}) ds + \int_0^t \mathbf{1}_{[0,\tau_n]}(s) G(s \wedge \tau_n, X_{s\wedge\tau_n}) dB_s$$

and similarly for $\tilde{X}_{t\wedge\tau_n}$.

Let

$$V_t^n = \int_0^t \left[\mu_r^{(n)} + \left(\frac{1}{2}m_p + 9p\lambda\right)\left(\ell_r^{(n)}\right)^2\right] dr.$$

By Theorem 3.8 we have with $\delta = 1$ and arbitrary $p \geq 2$, $\lambda > 1$ and $T \geq 0$:

$$\mathbb{E}\frac{\|e^{-V^n}(X_{\cdot\wedge\tau_n} - \hat{X}_{\cdot\wedge\tau_n})\|_{[0,T]}^p}{\left(1+\|e^{-V^n}(X_{\cdot\wedge\tau_n} - \hat{X}_{\cdot\wedge\tau_n})\|_{[0,T]}^2\right)^{p/2}} \leq C_{p,\lambda}\mathbb{E}\frac{|(X_0 - \hat{X}_0)|^p}{\left(1+|(X_0 - \hat{X}_0)|^2\right)^{p/2}}.$$

Uniqueness follows.

Existence. Let $\sigma \in C\left(\mathbb{R}^d; [0,1]\right)$ be given by

$$\sigma(x) = \begin{cases} 1, & \text{if } |x| \leq 1, \\ 2 - |x|, & \text{if } 1 < |x| \leq 2, \\ 0, & \text{if } |x| > 2. \end{cases}$$

Let $n \in \mathbb{N}^*$ and the stopping time be

$$\theta_n(\omega) = \inf\left\{t \geq 0 : \mathbf{1}_{a\geq 2} R_t^{a/2} + N_t^a + \left(\int_0^t |F(t,0)|\, dt\right)^a + \left(\int_0^t |G(t,0)|^2 dt\right)^{a/2} \geq n^{(a-1)/2}\right\}.$$

Clearly $\theta_n \nearrow \infty$ $a.s.$, as $n \to \infty$.

Consider the approximating SDE

$$X_t^n = \xi \mathbf{1}_{[0,n^{(a-1)/(2a)}]}(|\xi|) + \int_0^t F_n\left(s, X_s^n\right) ds + \int_0^t G_n\left(s, X_s^n\right) dB_s, \qquad (3.88)$$

where

$$F_n(t,x) = \mathbf{1}_{[0,\theta_n]}(t)\,\sigma\left(\frac{x}{n}\right) F(t,x),$$
$$G_n(t,x) = \mathbf{1}_{[0,\theta_n]}(t)\,\sigma\left(\frac{x}{n}\right) G(t,x).$$

Note that for all $\rho \geq 0$

$$F^{\#}_{n,\rho}(t) \overset{def}{=} \sup_{|x|\leq\rho} |F_n(t,x)| \leq F^{\#}_{2n}(t)$$

and

$$\int_0^T F^{\#}_{n,\rho}(t)\,dt < \infty, \ a.s.$$

Moreover

$$\mathbb{E}\left(\int_0^T |F_n(t,0)|\,dt\right)^a + \mathbb{E}\left(\int_0^T |G_n(t,0)|^2\,dt\right)^{a/2} \leq n^{(a-1)/2}.$$

The function F_n is globally monotone. Indeed:

◇ for $|x| \geq 2n$, $|y| \geq 2n$:

$$\langle F_n(t,x) - F_n(t,y), x-y\rangle = 0;$$

◇ for $|x| \leq 2n$, $|y| \leq 2n$ we have:

$$\begin{aligned}
&\langle F_n(t,x) - F_n(t,y), x-y\rangle \\
&\quad = \mathbf{1}_{[0,\theta_n]}(t)\,\sigma\left(\tfrac{x}{n}\right)\langle F(t,x) - F(t,y), x-y\rangle \\
&\qquad + \left[\sigma\left(\tfrac{y}{n}\right) - \sigma\left(\tfrac{x}{n}\right)\right]\mathbf{1}_{[0,\theta_n]}(t)\,\langle F(t,y), y-x\rangle \\
&\quad \leq \left[\left(\mu_t^{(2n)}\right)^+ + \frac{1}{n}F^{\#}_{2n}(t)\right]|x-y|^2;
\end{aligned}$$

◇ for $|x| \leq 2n$ and $|y| \geq 2n$ (similarly for $|x| \geq 2n$ and $|y| \leq 2n$) we have

$$\sigma\left(\frac{x}{n}\right) = \left|\sigma\left(\frac{x}{n}\right) - \sigma\left(\frac{y}{n}\right)\right| \leq \frac{1}{n}|x-y|$$

and

$$\begin{aligned}\langle F_n(t,x)-F_n(t,y),x-y\rangle &= \mathbf{1}_{[0,\theta_n]}(t)\,\sigma\left(\frac{x}{n}\right)\langle F(t,x),x-y\rangle\\ &\leq \frac{1}{n}\,|x-y|^2\,\mathbf{1}_{[0,\theta_n]}(t)\,|F(t,x))|\\ &\leq \frac{1}{n}\,|x-y|^2\,F^{\#}_{2n}(t)\,.\end{aligned}$$

Hence for all $x, y \in \mathbb{R}^d$, $d\mathbb{P}\otimes dt$-$a.e.$:

$$\langle F_n(t,x)-F_n(t,y),x-y\rangle \leq \left[\left(\mu_t^{(2n)}\right)^+ + \frac{1}{n}F^{\#}_{2n}(t)\right]|x-y|^2.$$

Similarly we obtain that G_n is globally Lipschitz:

$$\begin{aligned}|G_n(t,x)-G_n(t,y)| &\leq \left[\ell_t^{(2n)} + \frac{1}{n}\mathbf{1}_{[0,\theta_n]}(t)\,\ell_t^{(2n)}2n)\right]|x-y|\\ &\leq \left(1+\frac{2}{n}\right)\ell_t^{(2n)}|x-y|.\end{aligned}$$

By Theorem 3.26 with $dQ_t = dt$ and $\mathbb{R}^{d\times m}$ replaced by $\mathbb{R}^d$, there exists a unique solution $X^n \in S_d^0$ of the SDE (3.88). Since

$$\begin{aligned}&\left\langle X_s^n, F_n\left(s,X_s^n\right)ds\right\rangle + \left(\frac{1}{2}m_a+9ab\right)\left|G_n\left(s,X_s^n\right)\right|^2 ds\\ &\leq \mathbf{1}_{[0,\theta_n]}(s)\,\sigma\left(\frac{|X_s^n|}{n}\right)\left[\left\langle X_s^n, F\left(s,X_s^n\right)ds\right\rangle + \left(\frac{1}{2}m_a+9ab\right)\left|G\left(s,X_s^n\right)\right|^2\right]ds\\ &\leq \mathbf{1}_{a\geq 2}\mathbf{1}_{[0,\theta_n]}(s)\,dR_s + |X_s^n|\,\mathbf{1}_{[0,\theta_n]}(s)\,dN_s + |X_s^n|^2\mathbf{1}_{[0,\theta_n]}(s)\,dV(s)\,,\end{aligned}$$

we have, by Proposition 3.3,

$$\begin{aligned}\mathbb{E}\sup_{t\in[0,T]}\left[\left|e^{-V(t\wedge\theta_n)}X_t^n\right|^a\right] \leq C_{a,b}\Bigg[&\mathbb{E}\left[|\xi|^a\,\mathbf{1}_{[0,n^{(a-1)/(2a)}]}(|\xi|)\right]\\ &+\mathbb{E}\left(\int_0^T e^{-2V(s\wedge\theta_n)}\mathbf{1}_{[0,\theta_n]}(s)\,\mathbf{1}_{a\geq 2}dR_s\right)^{a/2}\\ &+\mathbb{E}\left(\int_0^T e^{-V(s\wedge\theta_n)}\mathbf{1}_{[0,\theta_n]}(s)\,dN_s\right)^a\Bigg]\\ \leq C_{a,b}\Bigg[\mathbb{E}\left(|\xi|^a\,\mathbf{1}_{[0,n^{(a-1)/(2a)}]}(|\xi|)\right) &+ \mathbf{1}_{a\geq 2}\mathbb{E}R_{T\wedge\theta_n}^{a/2} + \mathbb{E}N_{T\wedge\theta_n}^a\Bigg].\end{aligned}$$

Hence

$$\mathbb{E}\sup_{t\in[0,T]}|X_t^n|^a \leq 2C_{a,b}e^{aV(T)}\,n^{(a-1)/2}$$

and

$$\mathbb{E}\sup_{t\in[0,T]}|X_t^n|^a \leq C_{a,b}e^{aV(T)}\Big[\mathbb{E}\left[|\xi|^a\right] + \mathbf{1}_{a\geq 2}\mathbb{E}R_T^{a/2}+\mathbb{E}N_T^a\Big].$$

Define the stopping time

$$\tau_n(\omega) = \inf\{t \geq 0 : |X_t^n(\omega)| \geq n\}, \quad n \geq 2.$$

By Corollary 1.8, we have $\tau_n \overset{a.s.}{\rightarrow} \infty$, since for all $T > 0$

$$\begin{aligned}\sum_{n=1}^{\infty}\mathbb{P}(\tau_n < T) &\leq \sum_{n=1}^{\infty}\mathbb{P}\left(\sup_{t\in[0,T]}|X_t^n| \geq n\right)\\ &\leq \sum_{n=1}^{\infty}\frac{1}{n^a}\mathbb{E}\sup_{t\in[0,T]}|X_t^n|^a\\ &\leq \sum_{n=1}^{\infty}\frac{C\,n^{(a-1)/2}}{n^a} < \infty.\end{aligned}$$

Since for all $t \geq 0$, $\mathbb{P}$-a.s. $\omega \in \Omega$:

$$\left[X_t^{n+1}(\omega) - X_t^n(\omega)\right]\mathbf{1}_{[0,\theta_n(\omega)\wedge\tau_n(\omega)]}(t)\,\mathbf{1}_{\left[0,n^{(a-1)/(2a)}\right]}(|\xi(\omega)|) = 0,$$

we deduce that the stochastic process

$$X_t(\omega) \overset{def}{=} X_t^n(\omega), \quad \text{if } 0 \leq t < \theta_n(\omega) \wedge \tau_n(\omega) \text{ and } |\xi(\omega)| \leq n^{(a-1)/(2a)}$$

belongs to S_d^0 and it is a solution of the Eq. (3.86). ■

3.6.2 Locally Lipschitz Coefficients

A particular case is where the drift is also locally Lipschitz continuous. Then the functions F and G are $(\mathcal{P},\mathbb{R}^d)$-Carathéodory functions and satisfy:

(LL): $\forall\, R > 0$ there exist $L_R \in L^1_{loc}(\mathbb{R}_+;\mathbb{R}_+)$ and $\ell_R \in L^2_{loc}(\mathbb{R}_+;\mathbb{R}_+)$ such that $d\mathbb{P}\otimes dt$-$a.e.$:

$$(i): \quad |F(t,x) - F(t,y)| \leq L_R(t)\,|x-y|, \quad \forall\ |x|,|y| \leq R,$$
$$(ii): \quad |G(t,x) - G(t,y)| \leq \ell_R(t)\,|x-y| \quad \forall\ |x|,|y| \leq R;$$

(GL): there exist $a \in L^1_{loc}(\mathbb{R}_+;\mathbb{R}_+)$, $b \in L^2_{loc}(\mathbb{R}_+;\mathbb{R}_+)$ and $\mathcal{P}$-m.s.p. α, β, $\alpha \in L^1_{loc}(\mathbb{R}_+;\mathbb{R}_+)$ and $\beta \in L^2_{loc}(\mathbb{R}_+;\mathbb{R}_+)$, $\mathbb{P}$-a.s., such that $\forall\, x \in \mathbb{R}^d$, $d\mathbb{P} \otimes dt$-$a.e.$:

$$(j) \quad |F(t,x)| \leq \alpha_t + a(t)\,|x|,$$
$$(jj) \quad |G(t,x)| \leq \beta_t + b(t)\,|x|.$$

We consider the SDE

$$X_t = S_t + \int_0^t F(s,X_s)\,ds + \int_0^t G(s,X_s)\,dB_s, \quad t \geq 0 \tag{3.89}$$

(a slightly generalized version of Eq. (3.86)). With a similar proof as for Theorem 3.27 based this time on the global existence result from Theorem 3.17, we infer that:

Proposition 3.28. *If the assumptions* (LL) *and* (GL) *are satisfied and* $S \in S^0_d$, *then the SDE (3.89) has a unique solution* $X \in S^0_d$. *Moreover for every* $p \geq 2$ *there exists a constant* C_p *such that*

$$\begin{aligned}\mathbb{E}\sup_{t\in[0,T]}|X_t|^p \leq C_p \left[\mathbb{E}\,\|S\|_T^p + \mathbb{E}\left(\int_0^T \alpha_t dt\right)^p + \mathbb{E}\left(\int_0^T \beta_t^2 dt\right)^{p/2}\right] \\ \times \exp\left[C_p \int_0^T \left(a(t) + b^2(t)\right) dt\right].\end{aligned} \tag{3.90}$$

In particular $X \in S^p_d$, *if the right-hand side of (3.90) is finite.*

However we sketch a proof with a different approximation approach.

Proof. To prove the inequality (3.90) it is easy to show that

$$\begin{aligned}&\langle X_s - S_s, F(s,X_s)\rangle ds + \left(\frac{p-1}{2} + 9p\lambda\right)|G(s,X_s)|^2 ds \\ &\leq dR_s + |X_s - S_s|dN_s + |X_s - S_s|^2 dV(s),\end{aligned}$$

where $R_t = C_p \int_0^t \left[\beta_s^2 + b^2(s)\,|S_s|^2\right] ds,$

$$N_t = \int_0^t \left[\alpha_s + a(s)\,|S_s|\right] ds \quad \text{and} \quad V(t) = C_p \int_0^t \left[a(s) + b^2(s)\right] ds.$$

Using the inequality (3.7) (Proposition 3.3) we infer

$$\mathbb{E}\left\|e^{-V}(X-S)\right\|_T^p \le C_p\left[\mathbb{E}\|S\|_T^p + \left(\int_0^T \beta_s^2 ds\right)^{p/2} + \left(\int_0^T \alpha_s ds\right)^p\right]$$

and the inequality (3.90) follows.
The uniqueness and the existence are proved in the same manner as in Theorem 3.27; now the classical approach is to consider the approximating equation (satisfying global Lipschitz conditions, Theorem 3.17):

$$\begin{aligned} X_t^n = \mathbf{1}_{\theta_n>0} S_{t\wedge\theta_n} &+ \int_0^t \mathbf{1}_{[0,\theta_n]}(s)\, F\left(s, \pi_n\left(X_s^n\right)\right) ds \\ &+ \int_0^t \mathbf{1}_{[0,\theta_n]}(s)\, G\left(s, \pi_n\left(X_s^n\right)\right) dB_s \end{aligned}$$

where

$$\theta_n = \inf\left\{t \ge 0 : |S_t|^2 + \left(\int_0^t \alpha_s ds\right)^2 + \int_0^t \beta_s^2 ds \ge \sqrt{n}\right\}$$

and $\pi_n : \mathbb{R}^d \to \overline{B}(0, n)$ is the projection operator, that is

$$\pi_n(x) = \begin{cases} x, & \text{if } |x| \le n, \\ \dfrac{n}{|x|}x, & \text{if } |x| > n. \end{cases}$$

■

Remark 3.29. The inequality (3.90) also holds for $p = 1$ and, moreover, by Proposition 6.68 from Annex C we have for all $p \ge 1$:

$$\begin{aligned} \mathbb{E}\sup_{t\in[0,T]} |X_t|^p \le C_p &\left[\mathbb{E}\|S\|_T^p + \mathbb{E}\left(\int_0^T \alpha_s ds\right)^p + \mathbb{E}\left(\int_0^T \beta_s^2 ds\right)^{p/2}\right] \\ &\times \exp\left\{C_p T^{p-1}\int_0^T \left(a^p(s) + b^{2p}(s)\right) ds\right\}. \end{aligned} \tag{3.91}$$

3.7 Markov Solutions of SDEs

3.7.1 Markov Processes

Definition 3.30. A d-dimensional stochastic process $\{X_t : t \ge 0\}$ adapted to the filtration $\{\mathcal{F}_t : t \ge 0\}$ is said to be a

a. *Markov* process if for any $t, r \geq 0$:

$$\mathbb{P}(X_{t+r} \in G|\mathcal{F}_t) = \mathbb{P}(X_{t+r} \in G|X_t), \text{ a.s., } \forall\, G \in \mathcal{B}_d; \tag{3.92}$$

b. time-*homogeneous Markov process if, moreover,* for any $t, r \geq 0$:

$$\mathbb{P}(X_{t+r} \in G|X_t) = \mathbb{P}(X_r \in G|X_0), \text{ a.s., } \forall\, G \in \mathcal{B}_d; \tag{3.93}$$

c. strong Markov process if for any stopping time $\tau < \infty$ a.s. and any real number $r \geq 0$:

$$\mathbb{P}(X_{\tau+r} \in G|\mathcal{F}_\tau) = \mathbb{P}(X_{\tau+r} \in G|X_\tau), \text{ a.s., } \forall\, G \in \mathcal{B}_d. \tag{3.94}$$

Hence a Markov process is a stochastic process for which the future and the past are conditionally independent, given the present.

Note that condition (3.92) (and similarly for (3.93) and (3.94)) is equivalent to

$$\mathbb{E}\left[g\left(X_{t+r}\right)|\mathcal{F}_t\right] = \mathbb{E}\left[g\left(X_{t+r}\right)|X_t\right], \quad \forall\, g \in B_b(\mathbb{R}^d)$$

and also with the same condition for all $g \in C_b(\mathbb{R}^d)$.

We introduce the following:

Notation 3.31. *Given a probability distribution μ on $\mathbb{R}^d$, we denote by $\mathbb{P}_\mu$ a probability on Ω such that*

$$\mathbb{P}_\mu\left(X_0 \in B\right) = \mu\left(B\right), \quad \textit{for all } B \in \mathcal{B}_d.$$

The expectation with respect to $\mathbb{P}_\mu$ will be denoted $\mathbb{E}_\mu$. If $\mu = \delta_x$ (the Dirac measure at $x \in \mathbb{R}^d$), then $\mathbb{P}_x \stackrel{def}{=} \mathbb{P}_{\delta_x}$ and $\mathbb{E}_x \stackrel{def}{=} \mathbb{E}_{\delta_x}$.
Denote by $\{X_t^x : t \geq 0\}$ the Markov process $\{X_t : t \geq 0\}$ "starting" from x, that is for all $n \in \mathbb{N}^*$, $0 < t_0 < t_1 < \ldots < t_n$, $B_1, B_2, \ldots B_n \in \mathcal{B}_d$:

$$\mathbb{P}\left(X_{t_1}^x \in B_1, \ldots, X_{t_n}^x \in B_n\right) = \mathbb{P}_x\left(X_{t_1} \in B_1, \ldots, X_{t_n} \in B_n\right).$$

By Lemma 1.1, for each $0 \leq t \leq s$ and $G \in \mathcal{B}_d$ there exists a Borel measurable function $P\left(t, \cdot; s, G\right) : \mathbb{R}^d \to [0, 1]$ such that

$$\mathbb{P}(X_s \in G|X_t) = P\left(t, X_t; s, G\right).$$

This function is called the transition probability associated to the *Markov process* $\{X_t : t \geq 0\}$; it satisfies the following:

Definition 3.32. $\left\{P\left(t, x; s, G\right) : 0 \leq t \leq s,\ x \in \mathbb{R}^d,\ G \in \mathcal{B}_d\right\}$ is a transition probability if:

(i) $P\left(t, x; t, G\right) = \mathbf{1}_G\left(x\right)$ for all $x \in \mathbb{R}^d$ and $t \geq 0$;

(ii) $P(t,x;s,\cdot):\mathcal{B}_d\to[0,1]$ is a probability measure for all $x\in\mathbb{R}^d$ and $0\le t<s$;

(iii) $P(t,\cdot;s,G):\mathbb{R}^d\to[0,1]$ is a Borel measurable function for all $G\in\mathcal{B}_d$ and $0\le t<s$;

(iv) (Chapman–Kolmogorov equation) for all $x\in\mathbb{R}^d$ and $0\le t<r<s$

$$P(t,x;s,G)=\int_{\mathbb{R}^d}P(t,x;r,dz)\,P(r,z;s,G)\,,\ \forall\,G\in\mathcal{B}_d.$$

The transition probability can be viewed as the conditional distribution of the process at time s given that at time t the process was at the position x:

$$P(t,x;s,G)=\mathbb{P}(X_s\in G|X_t=x).$$

Let $x\in\mathbb{R}^d$ and $\{X_t^x : t\ge 0\}$ be a time-homogeneous Markov process. Then for all $t,s\ge 0$ and $G\in\mathcal{B}_d$ we have

$$\begin{aligned}P(x,t,G)&\overset{def}{=}P(s,x;s+t,G)\\&=P(0,x;t,G)\end{aligned}$$

and we associate to $\{X_t^x : t\ge 0\}$ a *semigroup* $\{P_t : t\ge 0\}$ of linear bounded operators $P_t : B_b(\mathbb{R}^d)\to B_b(\mathbb{R}^d)$ defined as:

$$(P_t\varphi)(x)=\int_{\mathbb{R}^d}\varphi(y)\,P(x,t,dy)=\mathbb{E}\varphi\left(X_t^x\right).$$

It is easy to verify that for all $\varphi\in B_b(\mathbb{R}^d)$ and $t,s\ge 0$:

$$\begin{aligned}&(i) && P_0\varphi=\varphi,\\&(ii) && P_{t+s}=P_t\circ P_s,\\&(iii) && \sup_{x\in\mathbb{R}^d}|(P_t\varphi)(x)|\le\sup_{x\in\mathbb{R}^d}|\varphi(x)|.\end{aligned}$$

A subset $\mathcal{S}\subset B_b(\mathbb{R}^d)$ is a P-continuous invariant subset of $B_b(\mathbb{R}^d)$ if

$$\begin{aligned}&(i) && P_t\varphi\in\mathcal{S},\ \ \forall\,\varphi\in\mathcal{S},\forall\,t\ge 0,\\&(ii) && \lim_{t\searrow 0}P_t\varphi(x)=\varphi(x),\ \ \forall\,\varphi\in\mathcal{S},\ \forall x\in\mathbb{R}^d.\end{aligned}$$

The *infinitesimal generator* of the semigroup $\{P_t : t\ge 0\}$ on $B_b(\mathbb{R}^d)$, or equivalently of the Markov process $\{X_t : t\ge 0\}$, is the (usually unbounded) linear operator $\mathcal{A}:\mathrm{Dom}(\mathcal{A})\subset B_b(\mathbb{R}^d)\to B_b(\mathbb{R}^d)$ defined as follows

$$\operatorname{Dom}(\mathcal{A}) \stackrel{def}{=} \Big\{\varphi \in B_b(\mathbb{R}^d) : \lim_{t\to 0} \frac{1}{t} \left[(P_t\varphi)(x) - \varphi(x)\right] \text{ exists for all } x \in \mathbb{R}^d$$
$$\text{and } x \longmapsto \lim_{t\to 0} \frac{1}{t} \left[(P_t\varphi)(x) - \varphi(x)\right] \text{ belongs to } B_b(\mathbb{R}^d)\Big\}$$

and

$$(\mathcal{A}\varphi)(x) \stackrel{def}{=} \lim_{t\to 0} \frac{1}{t} \left[(P_t\varphi)(x) - \varphi(x)\right], \quad \forall\, \varphi \in \operatorname{Dom}(\mathcal{A}),\ \forall\, x \in \mathbb{R}^d.$$

One can verify that $\operatorname{Dom}(\mathcal{A})$ is a P-continuous invariant subset of $B_b(\mathbb{R}^d)$ and for all $\varphi \in \operatorname{Dom}(\mathcal{A})$

$$P_{t+h}\varphi(x) - P_t\varphi(x) = \int_t^{t+h} P_r\mathcal{A}\varphi(x)\,dr = \int_t^{t+h} \mathcal{A}P_r\varphi(x)\,dr, \quad \forall\, x \in \mathbb{R}^d.$$

If $\{X_t^x : t \geq 0\}$ is a time-homogeneous Markov process with $\{P_t : t \geq 0\}$ the associated semigroup and $\mathcal{A}$ its *infinitesimal generator*, then by the Markov property, if $\varphi \in \operatorname{Dom}(\mathcal{A})$,

$$\begin{aligned}
&\mathbb{E}\left[\varphi\left(X_{t+h}^x\right) - \varphi\left(X_t^x\right) - \int_t^{t+h} \mathcal{A}\varphi\left(X_r^x\right) dr|\mathcal{F}_t\right] \\
&= \left(P_{t+h}\varphi\right)\left(X_t^x\right) - \left(P_t\varphi\right)\left(X_t^x\right) - \int_t^{t+h} \left(P_r\mathcal{A}\varphi\right)\left(X_t^x\right) dr \\
&= 0.
\end{aligned}$$

Hence for all $\varphi \in \operatorname{Dom}(\mathcal{A})$

$$\varphi\left(X_t^x\right) - \varphi(x) - \int_0^t \mathcal{A}\varphi\left(X_r^x\right) dr$$

is an $\mathcal{F}_t$-martingale for all $x \in \mathbb{R}^d$.

Conversely:

Proposition 3.33. *Let $x \in \mathbb{R}^d$ and $\{X_t^x : t \geq 0\}$ be a time-homogeneous Markov process (starting from x) with $\{P_t : t \geq 0\}$ the associated semigroup and $\mathcal{A}$ its infinitesimal generator. Let $\varphi, \psi \in B_b(\mathbb{R}^d)$ be such that*

$$\begin{aligned}
&(i) \quad \lim\nolimits_{t\searrow 0} P_t\psi(x) = \psi(x), \ \forall x \in \mathbb{R}^d, \\
&(ii) \quad \mathbb{E}\varphi\left(X_t^x\right) = \varphi(x) + \mathbb{E}\int_0^t \psi\left(X_r^x\right) dr.
\end{aligned}$$

Then

$$\varphi \in \operatorname{Dom}(\mathcal{A}) \ \text{ and } \ \mathcal{A}\varphi = \psi.$$

Proof. Let $h > 0$. Then

$$\begin{aligned}\frac{1}{h}\left[P_h\varphi(x)-\varphi(x)\right]-\psi(x) &= \frac{1}{h}\left[\mathbb{E}\varphi\left(X_h^x\right)-\varphi(x)\right]-\psi(x)\\ &= \int_0^1 \mathbb{E}\left[\psi\left(X_{rh}^x\right)-\psi(x)\right]dr\\ &= \int_0^1 \left(P_{rh}\psi(x)-\psi(x)\right)dr\\ &\to 0,\end{aligned}$$

as $h \to 0$. Hence $\varphi \in \mathrm{Dom}(\mathcal{A})$ and $\psi = \mathcal{A}\varphi$. ∎

The semigroup $\{P_t : t \geq 0\}$ is said to be Feller (and the time-homogeneous Markov process is called a *Feller process*) if $C_0\left(\mathbb{R}^d\right)$ is a P-continuous invariant subset of $B_b(\mathbb{R}^d)$.

Proposition 3.34. *A d-dimensional $\mathcal{F}_t$-Brownian motion $\{B_t : t \geq 0\}$ is a homogeneous strong Markov process (and a Feller process) with:*

(i) transition probability

$$P(x,t,G) = \frac{1}{(2\pi t)^{d/2}}\int_G \exp\Big(-\frac{|y-x|^2}{2t}\Big)dy,$$

for all $0 \leq t < s$, $x \in \mathbb{R}^d$, $G \in \mathcal{B}_d$;

(ii) semigroup $P_t\varphi(x) = \int_{\mathbb{R}^d}\varphi(y)\,P(x,t,dy) = \mathbb{E}\varphi(x+B_t)$;

(iii) infinitesimal generator $\mathcal{A}$ satisfying:

$$C_b^2\left(\mathbb{R}^d\right) \subset \mathrm{Dom}(\mathcal{A})$$

and

$$\mathcal{A}\varphi = \frac{1}{2}\sum_{i=1}^d \frac{\partial^2\varphi}{\partial x_i^2}, \quad \textit{for all } \varphi \in C_b^2\left(\mathbb{R}^d;\mathbb{R}\right).$$

Proof. By Propositions 1.34 and 1.94 from Chap. 1 for any stopping time $\tau < \infty$ $\mathbb{P}$-a.s., and any $G \in \mathcal{B}_d$, $h > 0$,

$$\begin{aligned}&\mathbb{P}(B_{\tau+h} \in G|\mathcal{F}_\tau)\\ &\quad= \mathbb{E}(\mathbf{1}_G(B_{\tau+h}-B_\tau+B_\tau)|\mathcal{F}_\tau)\\ &\quad= \mathbb{E}(\mathbf{1}_G(B_{\tau+h}-B_\tau+B_\tau)|B_\tau)\end{aligned}$$

$$= \frac{1}{(2\pi h)^{d/2}} \int_{\mathbb{R}^d} \mathbf{1}_G (y + B_\tau) \exp\Big(- \frac{|y|^2}{2h} \Big) dy$$
$$= P(0, B_\tau; h, G).$$

Clearly $P_t \varphi(x) = \mathbb{E}\varphi(x + B_t)$. If $\varphi \in C_b(\mathbb{R}^d)$, then $\mathbb{E}\varphi(x + B_t) \to \varphi(x)$, as $t \to 0$, and by Itô's formula for $\varphi \in C_b^2(\mathbb{R}^d)$ we obtain that

$$\varphi(x + B_t) - \varphi(x) - \frac{1}{2} \int_0^t \Delta\varphi(x + B_r)\, dr$$

is an $\mathcal{F}_t$-martingale. Hence by Proposition 3.33 $\mathrm{Dom}(\mathcal{A}) \supset C_b^2(\mathbb{R}^d)$ and

$$\mathcal{A}\varphi = \frac{1}{2}\Delta\varphi, \quad \forall\, \varphi \in C_b^2(\mathbb{R}^d).$$

■

In the non-homogeneous case the transition probability $P(t, x; s, G)$, $0 \le t \le s$, $x \in \mathbb{R}^d$, $G \in \mathcal{B}_d$ defines an evolution operator $P_{t,s} : B_b(\mathbb{R}^d) \to B_b(\mathbb{R}^d)$

$$(P_{t,s}\varphi)(x) = \int_{\mathbb{R}^d} \varphi(y)\, P(t, x; s, dy).$$

Obviously, for $0 \le t \le r \le s$:

$$P_{t,t}\varphi = \varphi, \quad \text{and} \quad P_{t,s} = P_{t,r} \circ P_{r,s}.$$

The *infinitesimal generator* is defined for all $t \ge 0$ as follows: $\mathcal{A}_t : \mathrm{Dom}(\mathcal{A}_t) \subset B_b(\mathbb{R}^d) \to B_b(\mathbb{R}^d)$

$$\mathrm{Dom}(\mathcal{A}_t) \overset{def}{=} \Big\{\varphi \in B_b(\mathbb{R}^d) : \exists \lim_{h \to 0} \frac{1}{h} \left[(P_{t,t+h}\varphi)(x) - \varphi(x) \right], \ \forall\, x \in \mathbb{R}^d,$$
$$\text{and the function } x \longmapsto \lim_{h \to 0} \frac{1}{h} \left[(P_{t,t+h}\varphi)(x) - \varphi(x) \right] \text{ belongs to } B_b(\mathbb{R}^d)\Big\}$$

and

$$(\mathcal{A}_t\varphi)(x) = \lim_{h \searrow 0} \frac{1}{h} \left(P_{t,t+h}\varphi(x) - \varphi(x) \right), \quad \forall\, \varphi \in \mathrm{Dom}(\mathcal{A}_t),\ \forall\, x \in \mathbb{R}^d.$$

From the definition of a Markov process, we have that for any $\varphi \in B_b(\mathbb{R}^d)$ and $t, h \ge 0$:

$$\begin{aligned}\mathbb{E}\left(\varphi\left(X_{t+h}\right)-\varphi\left(X_t\right)|\mathcal{F}_t\right) &= \int_{\mathbb{R}^d}\varphi\left(y\right)P\left(t,X_t;t+h,dy\right)-\varphi\left(X_t\right)\\ &= P_{t,t+h}\varphi\left(X_t\right)-\varphi\left(X_t\right)\\ &= h\mathcal{A}_t\varphi\left(X_t\right)+o\left(h\right).\end{aligned}$$

3.7.2 The Markov Property of Solutions of SDEs

We now consider the stochastic differential equation with deterministic coefficients

$$X_t = x + \int_0^t f(r, X_r)dr + \int_0^t g(r, X_r)dB_r, \quad \forall\, t \geq 0,\ \mathbb{P}\text{-a.s.}, \tag{3.95}$$

where $f : [0,+\infty[\,\times\mathbb{R}^d \to \mathbb{R}^d$ *and* $g : [0,+\infty[\,\times\mathbb{R}^d \to \mathbb{R}^{d\times k}$ satisfy

$$\textbf{(SDE-H}_f\textbf{)}: \tag{3.96}$$

◊ *the functions* $f(\cdot,x) : [0,+\infty[\, \to \mathbb{R}^d$ *and* $g(\cdot,x) : [0,+\infty[\, \to \mathbb{R}^{d\times k}$ *are (Borel) measurable for every* $x \in \mathbb{R}^d$;

◊ *there exist* $\mu \in L^1_{loc}(0,\infty)$ *and* $\ell \in L^2_{loc}(0,\infty;\mathbb{R}_+)$, *such that* dt-a.e.:

$$\begin{cases} & \textit{Continuity:}\\ (\mathbf{C}_f) & x \longrightarrow f(t,x) : \mathbb{R}^d \to \mathbb{R}^d \text{ is continuous;}\\ & \textit{Monotonicity condition:}\\ (\mathbf{M}_f) & \langle x-y, f(t,x)-f(t,y)\rangle \leq \mu(t)\,|x-y|^2, \quad \forall\, x,y\in\mathbb{R}^d;\\ & \textit{Boundedness condition:}\\ (\mathbf{B}_f) & \displaystyle\int_0^T f_R^{\#}(t)\,dt < \infty,\ \forall\, R,T \geq 0;\end{cases}$$

□

and

$$\textbf{(SDE-H}_g\textbf{)}: \tag{3.97}$$

$$\begin{cases} & \textit{Lipschitz condition:}\\ (\mathbf{L}_g) & |g(t,x)-g(t,y)| \leq \ell(t)\,|x-y|, \quad \forall\, x,y\in\mathbb{R}^d;\\ & \textit{Boundedness condition:}\\ (\mathbf{B}_g) & \displaystyle\int_0^T |g(t,0)|^2 dt < \infty, \quad \forall\, T\geq 0,\end{cases}$$

□

where

$$f_R^{\#}(t) = \sup_{|x| \le R} |f(t,x)| .$$

Since the condition (3.71) is satisfied for all $p \ge 2$, we deduce, by Theorem 3.21, that the SDE (3.95) has a unique solution $X \in \bigcap_{p \ge 2} S_d^p$. Moreover, denoting by $X_s^{t,x}$ the solution of

$$\begin{cases} X_s^{t,x} = x, & \text{if } 0 \le s \le t \\ X_s^{t,x} = x + \int_t^s f(r, X_r^{t,x})\, dr + \int_t^s g(r, X_r^{t,x})\, dB_r, & \text{if } s > t, \end{cases} \tag{3.98}$$

then by Proposition 3.22 for every $p \ge 2$ and $\lambda > 1$ there exists a constant $C_{p,\lambda}$ such that, if

$$V_t = \int_0^t \left[\mu^+(r) + \lambda \left(\frac{p-1}{2} + 9p\lambda \right) \ell^2(r) \right] dr$$

and

$$M_{q,T}(x) \overset{def}{=} T^{\frac{1}{2}\left(1-\frac{1}{q}\right)} \left\| f(\cdot,x) \right\|_{L^q(0,T)} + \left\| g(\cdot,x) \right\|_{L^{2q}(0,T)}, \tag{3.99}$$

then for all $x, x' \in \mathbb{R}$, $t, t' \in [0,T]$ and $q \in \,]1,\infty]$ the following inequalities hold:

$$\mathbb{E}\left(\sup_{0 \le s \le T} |X_s^{t,x} \quad X_s^{t',x'}|^p \right) \le C_{p,\lambda}\, e^{pV_T} \left[|x - x'|^p + \left| M_{q,T}(x) \right|^p |t - t'|^{\frac{p}{2}\left(1-\frac{1}{q}\right)} \right] \tag{3.100}$$

and

$$\mathbb{E}\left(\sup_{0 \le s \le T} |X_s^{t,x} - x|^p \right) \le C_{p,\lambda}\, e^{pV_T} \left| M_{q,T}(x) \right|^p |t - T|^{\frac{p}{2}\left(1-\frac{1}{q}\right)}. \tag{3.101}$$

If moreover for all $R > 0$

$$\sup_{|x| \le R} \left| M_{q,T}(x) \right| < \infty,$$

then by the Kolmogorov criterion for all $0 < \varepsilon \le 1$, $q \in \,]1,\infty]$, there exists a random variable $\eta = \eta_{\varepsilon,q}$ and a version of $\{X_s^{t,x} : s \ge t\}$ also denoted by $X_s^{t,x}$ such that

$$\sup_{s\in[0,T]}\left|X_s^{t,x}-X_s^{t',x'}\right|\leq\eta\times\left[\left|x-x'\right|^{1-\varepsilon}+\left|t-t'\right|^{\left(\frac{1}{2}-\frac{1}{2q}\right)(1-\varepsilon)}\right]$$

with $\mathbb{E}\,|\eta|^r<\infty$ for all $r>0$.

As a consequence we have:

Theorem 3.35. *Let the assumptions (3.96-**SDE-H**$_f$) and (3.97-**SDE-H**$_g$) be satisfied. Assume there exist $q\in\,]1,\infty]$ and $M,m>0$ such that for all $x\in\mathbb{R}^d$*

$$\left\|f\left(\cdot,x\right)\right\|_{L^q(0,T)}+\left\|g\left(\cdot,x\right)\right\|_{L^{2q}(0,T)}\leq M\left(1+|x|^m\right). \tag{3.102}$$

Then (after choosing a proper version) $(s,t,x)\longrightarrow X_s^{t,x}$ is $\mathbb{P}$-a.s. continuous and the inequalities (3.100) and (3.101) hold with $\left|M_{q,T}\left(x\right)\right|^p$ replaced by $M^p\left(1+|x|^{pm}\right)$.

Fix $(t,x)\in\mathbb{R}_+\times\mathbb{R}^d$. Let

$$\mathcal{F}_{t,s}^B\overset{def}{=}\begin{cases}\sigma\left\{B_s-B_t:t\leq r\leq s\right\}\vee\mathcal{N},\ \text{if } s\geq t,\\ \left\{\Omega,\emptyset\right\}\vee\mathcal{N},\ \text{if } 0\leq s\leq t.\end{cases}$$

From Theorem 3.21, with $\xi=x$, $F\left(s,x\right)=\mathbf{1}_{[t,\infty[}\left(s\right)f\left(s,x\right)$, $G\left(s,x\right)=\mathbf{1}_{[t,\infty[}\left(s\right)g\left(s,x\right)$, and the usual filtration $\{\mathcal{F}_s\}$ replaced by $\left\{\mathcal{F}_{t,s}^B\right\}$, we obtain:

Lemma 3.36. *$X_s^{t,x}$ is $\mathcal{F}_{t,s}^B$-measurable.*

Moreover one can generalize the result to the case of a random initial time. Let $\tau<\infty$ a.s. be a stopping time and $\xi:\Omega\to\mathbb{R}^d$ be an $\mathcal{F}_\tau$-measurable random variable. We introduce the Brownian motion $\hat{B}_s=B_{\tau+s}-B_\tau$, and we define the filtration

$$\mathcal{G}_s=\mathcal{F}_{\tau,\tau+s}^{B,\xi}\overset{def}{=}\mathcal{F}^\xi\vee\mathcal{F}_{\tau,\tau+s}^B,$$

where

$$\mathcal{F}_{\tau,\tau+s}^B\overset{def}{=}\begin{cases}\sigma\left\{B_{\tau+r}-B_\tau:0\leq r\leq s\right\}\vee\mathcal{N},\\ \left\{\Omega,\emptyset\right\}\vee\mathcal{N},\ \text{if } s\leq 0.\end{cases}$$

Consider the SDE

$$X_s^{\tau,\xi}=\xi+\int_\tau^{\tau\vee s}f(r,X_r^{\tau,\xi})dr+\int_\tau^{\tau\vee s}g(r,X_r^{\tau,\xi})dB_r,\quad s\geq 0, \tag{3.103}$$

or equivalently

$$X_s^{\tau,\xi}=\xi+\int_0^s\mathbf{1}_{[\tau,\infty[}\left(r\right)f(r,X_r^{\tau,\xi})dr+\int_0^s\mathbf{1}_{[\tau,\infty[}\left(r\right)g(r,X_r^{\tau,\xi})dB_r,\quad s\geq 0.$$

In other words

$$X_s^{\tau,\xi} = Y_{(s-\tau)\vee 0},$$

where

$$Y_s = \xi + \int_0^s f(\tau + r, Y_r)dr + \int_0^s g(\tau + r, Y_r)d\hat{B}_r.$$

Again $\left\{X_s^{\tau,\xi} : s \geq 0\right\}$ is $\mathcal{G}_s$-progressively measurable. Hence:

Lemma 3.37. *$X_{\tau+s}^{\tau,\xi}$ is $\mathcal{F}^{\xi} \vee \mathcal{F}_{\tau,\tau+s}^{B}$-measurable for all $s \geq 0$.*

Proposition 3.38. *The solution $\{X_t : t \geq 0\}$ of the stochastic differential equation (3.95) is a strong Markov process with:*

(i) transition probability

$$P(t, x; s, G) = \mathbb{P}\left(X_s^{t,x} \in G\right)$$

for $t, s \geq 0$ and $G \in \mathcal{B}_d$;
(ii) evolution operator $P_{t,s} : B_b(\mathbb{R}^d) \to B_b(\mathbb{R}^d)$, $0 \leq t \leq s$,

$$(P_{t,s}\varphi)(x) = \mathbb{E}\varphi\left(X_s^{t,x}\right);$$

(iii) infinitesimal generator $\mathcal{A}_t$ satisfying $C_b^2\left(\mathbb{R}^d\right) \subset \mathrm{Dom}(\mathcal{A}_t)$ for all $t \geq 0$ and for $\varphi \in C_b^2\left(\mathbb{R}^d\right)$

$$\begin{aligned}\mathcal{A}_t(\varphi)(x) &= \frac{1}{2}\mathbf{Tr}\left[g(t,x)\,g^*(t,x)\,\varphi''_{xx}(x)\right] + \left\langle f(t,x), \varphi'_x(x)\right\rangle \\ &= \frac{1}{2}\sum_{i,j=1}^{d}(gg^*)_{ij}(t,x)\frac{\partial^2\varphi(x)}{\partial x_i \partial x_j} + \sum_{i=1}^{d} f_i(t,x)\frac{\partial\varphi(x)}{\partial x_i}.\end{aligned}$$

Proof. From the uniqueness, for all $t \leq r \leq s$ we have

$$X_s^{t,x} = X_s^{r,X_r^{t,x}}$$

and then $P(t, x; s, G) = \mathbb{P}\left(X_s^{t,x} \in G\right)$ is clearly a transition probability.

Let $\tau < \infty$ a.s. be a stopping time. It follows again from uniqueness that for any $s \geq 0$

$$X_{\tau+s} = X_{\tau+s}^{\tau,X_\tau}.$$

Let $\{X_\tau^n\}_{n\in\mathbb{N}}$ be a sequence of random variables of the form

$$X_\tau^n = \sum_{i=1}^n x^{i,n} \mathbf{1}_{\{X_\tau \in G_i^n\}},$$

where the G_i^n are Borel subsets of $\mathbb{R}^d$, which converges a.s. to X_τ, as $n \to \infty$. Since $X_{\tau+s}^{\tau,x_i^n}$ is $\mathcal{F}_{\tau,\tau+s}^B$-measurable and hence independent of $\mathcal{F}_\tau$, we deduce that for any $\varphi \in C_b(\mathbb{R}^d)$

$$\begin{aligned}
\mathbb{E}[\varphi(X_{\tau+s}^{\tau,X_\tau^n})|\mathcal{F}_\tau] &= \sum_{i=1}^n \mathbb{E}[\varphi(X_{\tau+s}^{\tau,x_i^n})]\mathbf{1}_{\{X_\tau \in G_i^n\}} \\
&= \sum_{i=1}^n \mathbb{E}[\varphi(X_{\tau+s}^{\tau,x_i^n})|X_\tau]\mathbf{1}_{\{X_\tau \in G_i^n\}} \\
&= \mathbb{E}[\varphi(X_{\tau+s}^{\tau,X_\tau^n})|X_\tau],
\end{aligned}$$

and it remains to take the limit as $n \to \infty$ in the resulting identity, using the continuity in probability of $x \longrightarrow X_{\tau+s}^{\tau,x}$ (in fact the map is continuous a.s.) and Lemma 1.37. Hence

$$\mathbb{E}[\varphi(X_{\tau+s})|\mathcal{F}_\tau] = \mathbb{E}[\varphi(X_{\tau+s})|X_\tau]$$

and $\{X_t\}_{t\geq 0}$ is a strong Makov process with the transition probability

$$P(t,x;s,G) = \mathbb{P}\left(X_s^{t,x} \in G\right).$$

The evolution operator is given by

$$\begin{aligned}
P_{t,s}(\varphi)(x) &= \int_{\mathbb{R}^d} \varphi(y)\, P(t,x;s,dy) \\
&= \mathbb{E}\varphi\left(X_s^{t,x}\right)
\end{aligned}$$

and Itô's formula with $\varphi \in C_b^2(\mathbb{R}^d)$ and $t,h \geq 0$ yields

$$\mathbb{E}\varphi(X_{t+h}^{t,x}) = \varphi(x) + \mathbb{E}\int_t^{t+h} \mathcal{A}_r\varphi(X_r^{t,x})\, dr;$$

hence

$$\lim_{h \searrow 0} \frac{1}{h} \left[P_{t,t+h} (\varphi) (x) - \varphi(x) \right] = \mathcal{A}_t \varphi(x).$$

■

Proposition 3.38 and Proposition 3.33 yield:

Corollary 3.39. *In the case where the coefficients f and g depend only on x, and not on t, $\{X_t : t \geq 0\}$ is a homogeneous Feller process with:*

(i) associated semigroup

$$(P_t \varphi)(x) = \mathbb{E}\varphi(X_t^{0,x});$$

(ii) infinitesimal generator $\mathcal{A}$ satisfying: $\mathrm{Dom}(\mathcal{A}) \supset C_c^2(\mathbb{R}^d)$ *and for* $\varphi \in C_b^2(\mathbb{R}^d)$

$$\begin{aligned}(\mathcal{A}\varphi)(x) &= \frac{1}{2}\mathbf{Tr}\left[g(x)\, g^*(x)\, \varphi''_{xx}(x)\right] + \langle f(x), \varphi'_x(x)\rangle \\ &= \frac{1}{2}\sum_{i,j=1}^{d} (gg^*)_{ij}(x) \frac{\partial^2 \varphi(x)}{\partial x_i \partial x_j} + \sum_{i=1}^{d} f_i(x) \frac{\partial \varphi(x)}{\partial x_i}.\end{aligned}$$

Proposition 3.40 (Fokker–Planck Equation). *Let $\{X_t : t \geq 0\}$ denote the Markov process which is the solution of the SDE (3.95). For each $t \geq 0$, let μ_t denote the probability law of X_t. Then $\{\mu_t : t \geq 0\}$ solves in the distributional sense the parabolic PDE:*

$$\frac{d}{dt}\mu_t(\varphi) = \mu_t(\mathcal{A}_t \varphi),\ t \geq 0,\ \varphi \in C_c^\infty(\mathbb{R}^d),$$

where $\mu_t(\varphi) \overset{def}{=} \int_{\mathbb{R}^d} \varphi(x)\mu_t(dx) = \mathbb{E}\varphi(X_t)$.

Proof. Again from Itô's formula for $\varphi \in C_b^2(\mathbb{R}^d)$

$$\mathbb{E}\varphi(X_t) = \mathbb{E}\varphi(X_0) + \mathbb{E}\int_0^t \mathcal{A}_s \varphi(X_s)\, ds.$$

■

In the case where for each $t > 0$, μ_t has a density, i.e. $\mu_t(dx) = p(t,x)\,dx$, then the family of densities $\{p(t,\cdot),\ t > 0\}$ solves (at least in a weak sense) the PDE

$$\frac{\partial p}{\partial t}(t,x) = \frac{1}{2}\sum_{i,j} \frac{\partial^2}{\partial x_i \partial x_j}((gg^*)_{i,j}\, p)(t,x) - \sum_i \frac{\partial}{\partial x_i}(fp)(t,x), \quad t \geq 0,\ x \in \mathbb{R}^d.$$

3.8 The Feynman–Kac Formula

We now wish to establish a famous identity, due originally to Richard Feynman and Mark Kac, which provides a probabilistic formula for solutions of certain linear PDEs.

3.8.1 Backward Parabolic PDEs

Let us fix a terminal time $T > 0$. Let $c, h : [0, T] \times \mathbb{R}^d \to \mathbb{R}$ and $\kappa : \mathbb{R}^d \to \mathbb{R}$ be continuous mappings which are such that for some constants $C, p > 0$,

$$|c(t,x)| \le C, \ |h(t,x)| + |\kappa(x)| \le C(1 + |x|^p), \ (t,x) \in [0,T] \times \mathbb{R}^d. \tag{3.104}$$

Again let $\{X_s^{t,x}\}$ be the process solution of the stochastic differential equation (3.98), whose coefficients f and g are assumed here to be jointly continuous with respect to both variables t and x and to have at most linear growth at infinity. For each $(t,x) \in [0,T] \times \mathbb{R}^d$, we define

$$u(t,x) \stackrel{def}{=} \mathbb{E}\left[\kappa(X_T^{t,x}) e^{\int_t^T c(s,X_s^{t,x})\,ds} + \int_t^T h(s, X_s^{t,x}) e^{\int_t^s c(r,X_r^{t,x})\,dr}\, ds\right] \tag{3.105}$$

(*Feynman–Kac formula*).

We consider the (backward) parabolic PDE in $\mathbb{R}^d$

$$\begin{cases} \dfrac{\partial u}{\partial t}(t,x) + \mathcal{A}u(t,x) + cu(t,x) + h(t,x) = 0, & (t,x) \in [0,T[\times\mathbb{R}^d, \\ u(T,x) = \kappa(x), \quad x \in \mathbb{R}^d, \end{cases} \tag{3.106}$$

where

$$\begin{aligned} (\mathcal{A}\varphi)(t,x) &= \frac{1}{2}\mathbf{Tr}\left[g(t,x)\,g^*(t,x)\,\varphi''_{xx}(x)\right] + \langle f(t,x), \varphi'_x(x)\rangle \\ &= \frac{1}{2}\sum_{i,j=1}^d (gg^*)_{ij}(t,x)\frac{\partial^2\varphi}{\partial x_i \partial x_j}(x) + \sum_{i=1}^d f_i(t,x)\frac{\partial \varphi}{\partial x_i}(x). \end{aligned}$$

The aim of this subsection is to relate equation (3.106) and the quantity defined by (3.105). The first result says that any classical solution of Eq. (3.106) is given by the formula (3.105).

Proposition 3.41. *Let $u \in C^{1,2}([0,T] \times \mathbb{R}^d)$ be a solution of (3.106) such that for some $M, q > 0$,*

$$|u(t,x)| \leq M\left(1+|x|^q\right), \quad \forall\ (t,x) \in [0,T] \times \mathbb{R}^d.$$

If moreover the above assumptions, including (3.104), are satisfied, then $u(t,x)$ satisfies the Feynman–Kac formula (3.105).

Proof. By Itô's formula (2.17) on $[t, T \wedge \theta_n]$ with

$$V_s = (s, e^{\int_t^s c(r,X_r^{t,x})\,dr}),$$

$$X_s^{t,x} = x + \int_t^s f(r,\ X_r^{t,x})\,dr + \int_t^s g(r,\ X_r^{t,x})\,dB_r$$

and

$$\theta_n = \inf\left\{s \geq t : \left|X_s^{t,x} - x\right| \geq n\right\},$$

we have

$$\begin{aligned}
&u(T \wedge \theta_n,\ X_{T\wedge\theta_n}^{t,x})e^{\int_t^{T\wedge\theta_n} c(r,X_r^{t,x})\,dr} \\
&= u(t,x) + \int_t^{T\wedge\theta_n} [\frac{\partial u}{\partial t} + \mathcal{A}u + cu](r,\ X_r^{t,x})e^{\int_t^r c(s,X_s^{t,x})\,ds}dr \\
&+ \int_t^{T\wedge\theta_n} e^{\int_t^r c(s,X_s^{t,x})\,ds} \left\langle \nabla_x u(r,\ X_r^{t,x}), g(r,\ X_r^{t,x})dB_r \right\rangle.
\end{aligned}$$

First taking the expectation, then using the fact that u is a solution of (3.106), the Feynman–Kac formula (3.105) follows by letting $n \to \infty$, using uniform integrability, which follows from the boundedness of $c(s, X_s^{t,x})$ and the polynomial growth of u and h. ■

Let us rewrite the PDE (3.106) as

$$\begin{cases} -\dfrac{\partial u}{\partial t}(t,x) + \Phi(t,x,u(t,x), Du(t,x), D^2u(t,x)) = 0, \\ u(T,x) = \kappa(x), \end{cases} \tag{3.107}$$

where

$$\Phi : [0,T] \times \mathbb{R}^d \times \mathbb{R} \times \mathbb{R}^d \times \mathbb{S}^d \to \mathbb{R},$$

is given by

$$\Phi(t,x,r,p,X) = -\frac{1}{2}\mathbf{Tr}[g(t,x)g^*(t,x)X] - < f(t,x), p > -c(t,x)r - h(t,x),$$

and $\mathbb{S}^d$ denotes the set of non-negative symmetric $d \times d$ matrices. The notion of a viscosity solution of such a parabolic PDE is made precise by Definition 6.96 in Annex D.

Recall that the results in Annex D require Φ to be proper, which implies that $c(t,x) \leq 0$. This is not a restriction since $u(t,x)$ solves the original equation iff $v(t,x) = u(t,x)e^{\lambda t}$ solves the PDE

$$\begin{cases} \dfrac{\partial v}{\partial t}(t,x) + (\mathcal{A}v)(t,x) + (c(t,x) - \lambda)v(t,x) + h(t,x)e^{\lambda t} = 0, \\ \qquad\qquad\qquad\qquad\qquad\qquad\qquad\qquad (t,x) \in [0,T[\times\mathbb{R}^d, \\ v(T,x) = \kappa(x)e^{\lambda T}, \quad x \in \mathbb{R}^d. \end{cases}$$

Hence, since c is bounded, we can always choose λ large enough such that $c(t,x) - \lambda \leq 0$, and we can study an equivalent equation with a new Φ which is proper.

We now give the main result of this section.

Theorem 3.42 (Feynman–Kac's Formula). *Let the assumptions of Theorem 3.35 and (3.104) be satisfied and all coefficients of the PDE (3.106) be jointly continuous in (t,x). Then the quantity $u(t,x)$ defined by the formula (3.105) is a continuous function of $(t,x) \in [0,T] \times \mathbb{R}^d$ which grows at most polynomially at infinity, and it is the unique viscosity solution of the PDE (3.106), among those functions u which satisfy*

$$\lim_{|x|\to+\infty} |u(t,x)|e^{-\delta[\log(|x|)]^2} = 0, \tag{3.108}$$

uniformly for $t \in [0,T]$, for some $\delta > 0$.

Proof. Uniqueness of the viscosity solution of (3.106) follows from Theorem 6.106 in Annex D. The continuity of u follows easily from (3.100) and the assumptions on the growth of h and κ. Moreover from the conclusion of Theorem 3.35 and (3.104) we deduce that for some $M > 0$, $q > 0$ and for all $(t,x) \in [0,T] \times \mathbb{R}^d$,

$$|u(t,x)| \leq M(1 + |x|^q),$$

from which (3.108) follows with $\delta = q$. Since $X_T^{T,x} = x$, clearly $u(T,x) = \kappa(x)$. We now establish the sub-solution property of u (the super-solution property is proved analogously).

Let $(t,x) \in [0,T[\times\mathbb{R}^d$ and $\varphi \in C^{1,2}([0,T[\times\mathbb{R}^d)$ be a test function such that

$$u(s,y) - \varphi(s,y) \leq 0 = u(t,x) - \varphi(t,x),$$

for all (s,y) in a neighborhood of (t,x). We argue by contradiction. Suppose that

$$-\frac{\partial\varphi}{\partial t}(t,x) - \mathcal{A}\varphi(t,x) - c\varphi(t,x) - h(t,x) > 0.$$

Then there exist $\delta > 0$ and $\varepsilon \in\,]0, T-t[$ such that for all $s \in [t, t+\varepsilon]$ and $|y-x| \le \varepsilon$

$$\begin{array}{ll} (i) & u(s,y) \le \varphi(s,y), \\ (ii) & -\varphi_t'(s,y) - \mathcal{A}_s\varphi(s,y) - c(s,y)\varphi(s,y) - h(s,y) \ge \delta. \end{array}$$

We introduce the stopping time

$$\tau = (t+\varepsilon) \wedge \inf\left\{r : r \ge t,\ \left|X_r^{t,x} - x\right| \ge \varepsilon\right\}.$$

Clearly $\mathbb{E}\,\tau > t$.

It follows from the strong Markov property of the diffusion process $X_\cdot^{t,x}$ that

$$\begin{aligned} u(t,x) &= \mathbb{E}\left[\mathbb{E}\left(\kappa(X_T^{t,x})e^{\int_t^T c(s,X_s^{t,x})\,ds} + \int_t^T h(s,X_s^{t,x})e^{\int_t^s c(r,X_r^{t,x})\,dr}\,ds\,\middle|\,\mathcal{F}_\tau\right)\right] \\ &= \mathbb{E}\left[u(\tau, X_\tau^{t,x})e^{\int_t^\tau c(s,X_s^{t,x})\,ds} + \int_t^\tau h(s,X_s^{t,x})e^{\int_t^s c(r,X_r^{t,x})\,dr}\,ds\right]. \end{aligned}$$

On the other hand, from Itô's formula (2.17) on the time interval $[t,\tau]$, as in the proof of Proposition 3.41,

$$\begin{aligned} \varphi(t,x) = \mathbb{E}\Big[&\varphi(\tau, X_\tau^{t,x})e^{\int_t^\tau c(s,X_s^{t,x})\,ds} \\ &- \int_t^\tau [\varphi_s' + \mathcal{A}_s\varphi + c\varphi](s,X_s^{t,x})e^{\int_t^s c(r,X_r^{t,x})\,dr}\,ds\Big]. \end{aligned}$$

Hence by the definition of the stopping time τ, we have

$$\begin{aligned} 0 &= \varphi(t,x) - u(t,x) \\ &= \mathbb{E}\left[\varphi(\tau, X_\tau^{t,x}) - u(\tau, X_\tau^{t,x})\right]e^{\int_t^\tau c(s,X_s^{t,x})\,ds} \\ &\quad - \mathbb{E}\int_t^\tau [\varphi_s' + \mathcal{A}_s\varphi + c\varphi + h](s,X_s^{t,x})e^{\int_t^s c(r,X_r^{t,x})\,dr}\,ds \\ &\ge \mathbb{E}\int_t^\tau \delta e^{\int_t^s c(r,X_r^{t,x})\,dr}\,ds \\ &\ge (\mathbb{E}\tau - t)\,\delta e^{-C(T-t)} \\ &> 0 \end{aligned}$$

which is impossible.

Consequently

$$-\frac{\partial \varphi}{\partial t}(t,x) - \mathcal{A}\varphi(t,x) - cu(t,x) - h(t,x) \le 0,$$

and u is a viscosity sub-solution. ∎

3.8.2 Forward Parabolic PDEs

Define the function

$$v(t,x) = \mathbb{E}\left[\kappa(X_t^x)e^{\int_0^t c(X_s^x)ds} + \int_0^t h(X_s^x)e^{\int_0^s c(X_r^x)dr}ds\right]. \tag{3.109}$$

and consider the PDE

$$\begin{cases} \frac{\partial v}{\partial t}(t,x) = \mathcal{A}v(t,x) + cu(t,x) + h(x),\ t \ge 0,\ x \in \mathbb{R}^d; \\ v(0,x) \quad = \kappa(x),\ x \in \mathbb{R}^d. \end{cases} \tag{3.110}$$

Theorem 3.43. *Let the assumptions of Theorem 3.36 be satisfied, where the coefficients of the SDE—as well as h and c—are no longer assumed to depend upon the time variable. Then $v(t,x)$ defined by the formula (3.109) is a continuous function of $(t,x) \in \mathbb{R}_+ \times \mathbb{R}^d$ which grows at most polynomially at infinity, and it is the unique viscosity solution of the PDE (3.110).*

Proof. For any fixed $T > 0$, consider $v(t,x)$ for $(t,x) \in [0,T] \times \mathbb{R}^d$ given by (3.109). Because of the time homogeneity of the SDE for X, we can rewrite $v(t,x)$ as

$$v(t,x) = \mathbb{E}\left[\kappa(X_T^{T-t,x})e^{\int_{T-t}^T c(X_s^{T-t,x})ds} + \int_{T-t}^T h(X_s^{T-t,x})e^{\int_{T-t}^s c(X_r^{T-t,x})dr}ds\right].$$

Now define, again for $(t,x) \in [0,T] \times \mathbb{R}^d$, $u(t,x) = v(T-t,x)$. We have

$$u(t,x) = \mathbb{E}\left[\kappa(X_T^{t,x})e^{\int_t^T c(X_s^{t,x})ds} + \int_t^T h(X_s^{t,x})e^{\int_t^s c(X_r^{t,x})dr}ds\right].$$

From Theorem 3.42, u is a viscosity solution of (3.106), hence v is a viscosity solution of (3.110). Uniqueness follows by the same argument as in Theorem 3.42. ∎

3.8.3 Parabolic PDEs with Dirichlet Boundary Conditions

We now give a similar result for a parabolic PDE in a connected open bounded subset $D \subset \mathbb{R}^d$, with Dirichlet boundary condition. The process $\{X_s^{t,x};\ s \ge t\}$ is defined as above. Let D be a connected open bounded subset of $\mathbb{R}^d$, whose boundary is of class C^1. For each $(t,x) \in [0,T] \times \overline{D}$, we define the stopping time

$$\tau_{t,x} = \inf\{s \ge t : X_s^{t,x} \notin \overline{D}\}.$$

We assume that the set

$$\Lambda = \{(t,x) \in [0,T] \times \partial D : \mathbb{P}(\tau_{t,x} > t) = 0\} \quad \text{is closed.} \tag{3.111}$$

Note that we have the following *zero-one law* which will be used below.

Lemma 3.44. *For all* $(t,x) \in [0,T] \times \partial D$,

$$\mathbb{P}(\tau_{t,x} > t) \in \{0,1\}.$$

Proof. To simplify the notation, we let $t = 0$ and write τ_x for $\tau_{0,x}$. Again let $\{\mathcal{F}_t : t \ge 0\}$ denote the natural filtration of the Brownian motion driving the SDE (3.98). For all $n \ge 1$,

$$\{\tau_x = 0\} = \bigcap_{k \ge n} \left\{\tau_x \le \frac{1}{k}\right\} \in \mathcal{F}_{1/n}.$$

This, together with the right continuity of the filtration $\{\mathcal{F}_t\}$ (see Proposition 1.89), implies that $\{\tau_x = 0\} \in \mathcal{F}_0$, hence the result. ■

We consider the parabolic PDE (with the same function Φ as above)

$$\begin{cases} -\dfrac{\partial u}{\partial t}(t,x) + \Phi(t,x,u(t,x),Du(t,x),D^2u(t,x)) = 0, \ (t,x) \in [0,T] \times D, \\ u(T,x) = \kappa(x), \ x \in D, \\ u(t,x) = \chi(t,x), \ (t,x) \in [0,T] \times \partial D. \end{cases} \tag{3.112}$$

Here all data are as in the preceding section, except for the new $\chi \in C([0,T] \times \partial D)$, which is assumed to be such that $\chi(T,x) = \kappa(x)$, $x \in \partial D$.

We want to show that the Feynman–Kac formula

$$\begin{aligned} u(t,x) = \mathbb{E}\Big[\Big(\kappa(X_T^{t,x})\mathbf{1}_{\{T \le \tau_{t,x}\}} + \chi(\tau_{t,x}, X_{\tau_{t,x}}^{t,x})\mathbf{1}_{\{\tau_{t,x} < T\}}\Big) e^{\int_t^{T \wedge \tau_{t,x}} c(s,X_s^{t,x})ds} \\ + \int_t^{T \wedge \tau_{t,x}} h(s,X_s^{t,x}) e^{\int_t^s c(r,X_r^{t,x})dr} ds\Big] \end{aligned} \tag{3.113}$$

defines a viscosity solution of Eq. (3.112). We again refer to Annex D for the definition of a viscosity solution of (3.112).

The fact that u, defined by (3.113), is continuous follows from some arguments in the preceding subsection together with:

Proposition 3.45. *Under the condition (3.111), the mapping* $(t,x) \to \tau_{t,x}$ *is a.s. continuous on* $\overline{D}$.

Proof. Let $\{(t_n, x_n),\ n \in \mathbb{N}\}$ be a sequence in $[0,T]\times\overline{D}$ such that $(t_n, x_n) \to (t,x)$, as $n \to \infty$.

We first show that

$$\limsup_{n\to\infty} \tau_{t_n,x_n} \le \tau_{t,x} \quad \text{a.s.} \tag{3.114}$$

Suppose that (3.114) is false. Then

$$\mathbb{P}(\tau_{t,x} < \limsup_{n\to\infty} \tau_{t_n,x_n}) > 0. \tag{3.115}$$

For each $\varepsilon > 0$, let

$$\tau^{\varepsilon}_{t,x} = \inf\{s \ge t;\ d(X^{t,x}_s, D) \ge \varepsilon\}.$$

From (3.115), there exists ε and T such that

$$\mathbb{P}(\tau^{\varepsilon}_{t,x} < \limsup_{n\to\infty} \tau_{t_n,x_n} \le T) > 0.$$

But since $X^{t_n,x_n}_{\cdot} \to X^{t,x}_{\cdot}$ uniformly on $[t,T]$ a.s., this implies that

$$\mathbb{P}(\limsup_{n\to\infty} \tau^{\varepsilon/2}_{t_n,x_n} \le \tau^{\varepsilon}_{t,x} < \limsup_{n\to\infty} \tau_{t_n,x_n} \le T) > 0,$$

which would mean that for some n, X^{t_n,x_n} exits the $\varepsilon/2$-neighborhood of D before exiting D with positive probability, which is impossible.

We next prove that

$$\liminf_{n\to\infty} \tau_{t_n,x_n} \ge \tau_{t,x} \quad \text{a.s.} \tag{3.116}$$

For this part of the proof, we will need the assumption (3.111) that Λ is closed.

It suffices to prove that (3.116) holds a.s. on $\Omega_M = \{\tau_{t,x} \le M\}$, with M arbitrary. From the result of the first step, for almost all $\omega \in \Omega_M$, there exists an $n(\omega)$ such that $n \ge n(\omega)$ implies $\tau_{t_n,x_n} \le M+1$. From the a.s. (on Ω_M) uniform convergence $X^{t_n,x_n}_{\cdot} \to X^{t,x}_{\cdot}$ on the interval $[0, M+1]$, $X^{t,x}_{\cdot}$ hits the set

$$\overline{\{(\tau_{t_n,x_n}, X^{t_n,x_n}_{\tau_{t_n,x_n}});\ n \in \mathbb{N}\}} \subset \overline{\Lambda} = \Lambda$$

on the random interval $[t, \liminf_n \tau_{t_n,x_n}]$ a.s. on Ω_M. The result follows, since $X_\cdot^{t,x}$ exits $\overline{D}$ when it hits Λ. ■

We now prove the following:

Theorem 3.46. *Assume again that the coefficients of $\mathcal{A}$, c and h are continuous on $[0,T] \times \overline{D}$, $\kappa \in C(\overline{D})$, $\chi \in C([0,T] \times \overline{D})$ with $\kappa(x) = \chi(T,x)$, $x \in \partial D$. Then $u(t,x)$, given by (3.113), is a continuous function of $(t,x) \in [0,T] \times \overline{D}$ and it is the unique viscosity solution of (3.112).*

Proof. Uniqueness of the viscosity solution follows from Theorem 6.103. Continuity of u follows from arguments similar to those in the proof of Theorem 3.42, together with the conclusion of Proposition 3.45. Let us prove that u is a viscosity sub-solution. The only new case to consider is that where $(t,x) \in [0,T[\times\partial D$ is a local maximum of $u-\varphi$, where $\varphi \in C^{1,2}([0,T[\times\overline{D})$. If $(t,x) \in \Lambda$, then $\tau_{t,x} = t$ and $u(t,x) = \chi(t,x)$. On the other hand, if $(t,x) \in [0,T[\times\partial D\backslash\Lambda$, then by Lemma 3.44, $\mathbb{P}(\tau_{t,x} > t) = 1$, and the same argument as in the proof of Theorem 3.42 (see also the proof of Theorem 3.49 below) shows that

$$-\frac{\partial\varphi}{\partial t}(t,x) + \Phi(t,x,u(t,x),D\varphi(t,x),D^2\varphi(t,x)) \le 0.$$

■

3.8.4 Elliptic Equations with Dirichlet Boundary Condition

Consider the differential operator

$$\mathcal{A} = \frac{1}{2}\sum_{i,j=1}^{d}(gg*)_{ij}(x)\frac{\partial^2}{\partial x_l \partial x_j} + \sum_{i=1}^{d} f_i(x)\frac{\partial}{\partial x_i}$$

and the linear elliptic PDE

$$\begin{cases} \mathcal{A}u(x) + c(x)u(x) + h(x) = 0, & x \in D, \\ u(x) = \chi(x), & x \in \partial D, \end{cases} \tag{3.117}$$

where D is a bounded connected domain with a boundary ∂D of class C^1, $f : \mathbb{R}^d \to \mathbb{R}^d$, $g : \mathbb{R}^d \to \mathbb{R}^{d\times k}$, $c : D \to \mathbb{R}_-$, $h : D \to \mathbb{R}$ and $\kappa : \partial D \to \mathbb{R}$ are continuous. Define the stopping time

$$\tau_x = \inf\{t \ge 0,\ X_t^x \notin \overline{D}\}.$$

We have the following:

Theorem 3.47. *Let $u \in C^2(D) \cap C(\overline{D})$ be a classical solution of Eq. (3.117). Provided that the above assumptions hold and*

$$\sup_{x\in D} \mathbb{E}\tau_x < \infty, \tag{3.118}$$

we have the Feynman–Kac formula

$$u(x) = \mathbb{E}\left[\chi(X^x_{\tau_x})e^{\int_0^{\tau_x} c(X^x_s)ds} + \int_0^{\tau_x} h(X^x_s)e^{\int_0^s c(X^x_r)dr}ds\right]. \tag{3.119}$$

Proof. It follows from the regularity of u, Eq. (3.117) and Itô's formula that

$$u(x) = \mathbb{E}\left[u(X^x_{t\wedge\tau_x})e^{\int_0^{t\wedge\tau_x} c(X^x_s)ds} + \int_0^{t\wedge\tau_x} h(X^x_s)e^{\int_0^s c(X^x_r)dr}ds\right].$$

The result follows by letting $t \to \infty$, exploiting the assumption made on τ_x. ■

Remark 3.48. A sufficient condition for (3.118) to hold is that there exists a $v \in \mathbb{R}^d$, with $|v| = 1$, such that $\inf_{x\in D} |g^*(x)v| > 0$.

Suppose now that the coefficients are as above. We need to formulate the elliptic version of condition (3.111), namely

$$\Lambda = \{x \in \partial D;\ \mathbb{P}(\tau_x > 0) = 0\} \quad \text{is closed.} \tag{3.120}$$

We rewrite our elliptic PDE as

$$\begin{cases} \Phi(x, u(x), Du(x), D^2u(x)) = 0,\ x \in D, \\ u(x) = \chi(x),\ x \in \partial D, \end{cases}$$

where

$$\Phi : D \times \mathbb{R} \times \mathbb{R}^d \times \mathbb{S}^{d\times d} \to \mathbb{R}$$

is defined as

$$\Phi(x, r, p, X) = -\frac{1}{2}\mathrm{Tr}[(gg^*)(x)X] - \langle f(x), p\rangle - c(x)r - h(x). \tag{3.121}$$

Note that the fact that c is non positive makes Φ proper.

We can now state the following:

Theorem 3.49. *Suppose that the above assumptions, including condition (3.120), are in force. Then the function $u : \overline{D} \to \mathbb{R}$ defined by (3.119) is continuous on $\overline{D}$, and it is the unique continuous viscosity solution of (3.117).*

Proof. Uniqueness follows from Corollary 6.102 in Annex D.

The continuity of u follows from arguments similar to those in the proof of Theorem 3.46. Let us check that u is a viscosity sub-solution. The proof of the super-solution property is very similar. Let $\varphi \in C^2(D)$ and $x \in \overline{D}$ be a local maximum of $u - \varphi$. If $x \in \Lambda$, then $\tau_x = 0$ a.s., hence from the Feynman–Kac formula $u(x) = \chi(x)$, and we are done. Suppose now that $x \in \overline{D}\backslash\Lambda$. We want to show that $\Phi(x, u(x), D\varphi(x), D^2\varphi(x)) \le 0$. We argue by contradiction. Suppose that $\Phi(x, u(x), D\varphi(x), D^2\varphi(x)) > 0$. To each $\varepsilon > 0$, we associate the set

$$B_{D,\varepsilon}(x) = \overline{D} \cap \{y;\ |y - x| \le \varepsilon\}.$$

From now on we fix $\varepsilon > 0$ small enough such that (recall that Λ is closed):

1. $B_{D,\varepsilon}(x) \cap \Lambda = \emptyset$;
2. $u(y) - \varphi(y) \le 0 = u(x) - \varphi(x)$, for all $y \in B_{D,\varepsilon}(x)$;
3. $\Phi(x, u(x), D\varphi(x), D^2\varphi(x)) \ge \varepsilon$, for all $y \in B_{D,\varepsilon}(x)$.

Let us introduce the stopping time

$$\theta_x := \inf\{t > 0;\ X_t^x \notin B_{D,\varepsilon}(x)\} \wedge \varepsilon.$$

From the strong Markov property of the process $\{X_t^x,\ t \ge 0\}$, we would have that

$$\begin{aligned} u(x) &= \mathbb{E}\left[\mathbb{E}\left(\kappa(X_{\tau_x}^x)e^{\int_0^{\tau_x} c(X_s^x)ds}\int_0^{\tau_x} h(X_s^x)e^{\int_0^s c(X_r^x)dr}ds \middle| \mathcal{F}_{\theta_x}\right)\right] \\ &= \mathbb{E}\left[u(X_{\theta_x}^x)e^{\int_0^{\theta_x} c(X_s^x)ds} + \int_0^{\theta_x} h(X_s^x)e^{\int_0^s c(X_r^x)dr}ds\right]. \end{aligned}$$

On the other hand, from Itô's formula,

$$\varphi(x) = \mathbb{E}\left[\varphi(X_{\theta_x}^x)e^{\int_0^{\theta_x} c(X_s^x)ds} - \int_0^{\theta_x} [\mathcal{A}\varphi + c\varphi](X_s^x)e^{\int_0^s c(X_r^x)dr}ds\right].$$

Hence by the definition of the stopping time θ_x, we have

$$\begin{aligned} 0 &= \varphi(x) - u(x) \\ &= \mathbb{E}\left([\varphi(X_{\theta_x}^x) - u(X_{\theta_x}^x)]e^{\int_0^{\theta_x} c(X_s^x)ds}\right) \\ &\quad - \mathbb{E}\int_0^{\theta_x} [\mathcal{A}\varphi + c\varphi + h](X_s^x)e^{\int_0^s c(X_r^x)dr}ds \\ &\ge \varepsilon\mathbb{E}\int_0^{\theta_x} e^{\int_0^s c(X_r^x)dr}ds \end{aligned}$$

$$\geq \varepsilon \mathbb{E}\left[\theta_x e^{-C\theta_x}\right]$$
$$> 0,$$

which is a contradiction.

3.8.5 *Elliptic PDEs in* $\mathbb{R}^d$

Consider the linear elliptic PDE in $\mathbb{R}^d$:

$$\Phi(x, u(x), Du(x), D^2u(x))) = 0, \quad x \in \mathbb{R}^d, \tag{3.122}$$

where Φ is defined by (3.121), $f : \mathbb{R}^d \to \mathbb{R}^d$, $g : \mathbb{R}^d \to \mathbb{R}^{d\times k}$, $c : \mathbb{R}^d \to \mathbb{R}$ and $h : \mathbb{R}^d \to \mathbb{R}$ are continuous functions satisfying

$$\sup_{x\in\mathbb{R}^d} (|h(x)| + |c(x)|) < \infty, \tag{3.123}$$

$$\sup_{x\in\mathbb{R}^d} c(x) \leq -\bar{c} < 0. \tag{3.124}$$

Under these conditions, the following function of $x \in \mathbb{R}^d$ is well defined

$$u(x) = \int_0^\infty \mathbb{E}\left[h(X_t^x)e^{\int_0^t c(X_s^x)ds}\right] dt. \tag{3.125}$$

We now want to prove that (3.125) is a viscosity solution of (3.122).

The notion of a viscosity solution of (3.122) is defined in Definition 6.85 of Annex D.

We have the following result, whose proof, which is very similar to and simpler than that of Theorem 3.49, is left as an exercise for the reader.

Theorem 3.50. *Under the above conditions, in particular (3.123) and (3.124), u given by (3.125) belongs to $C(\mathbb{R}^d)$, and it is the unique bounded viscosity solution of Eq. (3.122).*

3.9 Remarks on Weak and Strong Solutions

We consider the stochastic differential equation with deterministic coefficients

$$X_t = x_0 + \int_0^t f(r, X_r)dr + \int_0^t g(r, X_r)dB_r, \quad \forall\, t \geq 0,\ \mathbb{P}\text{-a.s.}, \tag{3.126}$$

where $f : [0,\infty[\times \mathbb{R}^d \to \mathbb{R}^d$ *and* $g : [0,\infty[\times \mathbb{R}^d \to \mathbb{R}^{d\times k}$ are $(\mathcal{B}_{\mathbb{R}_+}, \mathbb{R}^d)$-Carathéodory functions, i.e.

◊ *the functions* $f(\cdot, x) : [0,\infty[\to \mathbb{R}^d$ *and* $g(\cdot, x) : [0,\infty[\to \mathbb{R}^{d\times k}$ *are (Borel) measurable for every* $x \in \mathbb{R}^d$,

◊ *the functions* $f(t,\cdot) : \mathbb{R}^d \to \mathbb{R}^d$ *and* $g(t,\cdot) : \mathbb{R}^d \to \mathbb{R}^{d\times k}$ *are continuous for dt-almost all* $t \geq 0$.

We also assume to be satisfied the *boundedness condition:*

$$\int_0^T \left[f_R^{\#}(t) + |g_R^{\#}(t)|^2 \right] dt < \infty, \ \forall\, R, T \geq 0, \tag{3.127}$$

where

$$f_R^{\#}(t) = \sup_{|x|\leq R} |f(t,x)|$$

and $g_R^{\#}$ is defined in a similar manner.

In most cases, the only relevant quantity attached to the solution of a SDE is its probability law. Then the probability space $(\Omega, \mathcal{F}, \mathbb{P})$ and the Brownian motion $\{B_t : t \geq 0\}$ only have an auxiliary role. For this reason it is natural to define the notion of a weak solution.

Recall that given a stochastic basis $(\Omega, \mathcal{F}, \mathbb{P}, \mathcal{F}_t)_{t\geq 0}$ and an $\mathbb{R}^k$-valued $\mathcal{F}_t$-Brownian motion $\{B_t : t \geq 0\}$, then a $\mathcal{P}$-m.c.s.p. $X : \Omega \times [0,\infty[\to \mathbb{R}^d \times \mathbb{R}^d$ is a strong solution of the SDE (3.126) if $\mathbb{P}$-a.s. $\omega \in \Omega$:

$$X_t = x_0 + \int_0^t f(s, X_s)\, ds + \int_0^t g(s, X_s)\, dB_s, \ \forall\, t \geq 0.$$

Definition 3.51. If there exist a stochastic basis $(\Omega, \mathcal{F}, \mathbb{P}, \mathcal{F}_t)_{t\geq 0}$, an $\mathbb{R}^k$-valued $\mathcal{F}_t$-Brownian motion $\{B_t : t \geq 0\}$ and an $\mathcal{F}_t$-progressively measurable continuous stochastic process $X : \Omega \times [0,\infty[\to \mathbb{R}^d \times \mathbb{R}^d$ such that

$$X_t = x_0 + \int_0^t f(s, X_s)\, ds + \int_0^t g(s, X_s)\, dB_s, \ \forall\, t \geq 0,$$

then the collection $(\Omega, \mathcal{F}, \mathbb{P}, \mathcal{F}_t, B_t, X_t)_{t\geq 0}$ is called a weak solution of the SDE (3.126).

We now derive a result on the existence of weak solutions. To this aim we add an assumption on f and g:
there exist $p > 1$, $\lambda > 1$ and *continuous increasing functions* $V, R, N : [0,\infty[\to [0,\infty[$, $V(0) = N(0) = R(0) = 0$, *such that as a signed measure on* $[0,\infty[$:

$$\langle f(t,x(t)), x(t)\rangle\, dt + \left(\frac{1}{2}m_p + 9p\lambda\right)|g(t,x(t))|^2 dt \\ \leq \mathbf{1}_{p\geq 2} dR(t) + |x(t)|\, dN(t) + |x(t)|^2 d\,V(t) \tag{3.128}$$

for all continuous function $x : [0,\infty[\to \mathbb{R}^d$, where

$$m_p = 1 \vee (p-1).$$

From Corollary 6.75 (Annex C) we clearly have:

Lemma 3.52. *Let* $f : \mathbb{R}_+ \times \mathbb{R}^d \to \mathbb{R}^d$ *and* $g : \mathbb{R}_+ \times \mathbb{R}^d \to \mathbb{R}^{d\times k}$ *be two* $(\mathcal{B}_{\mathbb{R}_+}, \mathbb{R}^d)$*-Carathéodory functions satisfying (3.127) and (3.128). Then every weak solution* $(\Omega, \mathcal{F}, \mathbb{P}, B_t, X_t)_{t\geq 0}$ *of the SDE (3.126) satisfies for all* $p \geq 0$

$$\mathbb{E} \sup_{t\in[0,T]} |X_t^n|^p \leq L_{p,\lambda},$$

where

$$L_{a,b} = C_{p,\lambda} e^{pV(T)} \left[|x_0|^p + \mathbf{1}_{p\geq 2} R^{p/2}(T) + N^p(T) \right]$$

and $C_{p,\lambda}$ *is a positive constant depending only on* (p,λ).

Also, by elementary calculus we deduce:

Lemma 3.53. *Let* $f : \mathbb{R}_+ \times \mathbb{R}^d \to \mathbb{R}^d$ *and* $g : \mathbb{R}_+ \times \mathbb{R}^d \to \mathbb{R}^{d\times k}$ *be* $(\mathcal{B}_{\mathbb{R}_+}, \mathbb{R}^d)$*– Carathéodory functions such that there exists an* $A > 0$ *satisfying:*

$$|f(t,x)| + |g(t,x)| \leq A, \quad \text{for all } (t,x) \in \mathbb{R}_+ \times \mathbb{R}^d.$$

If $(\Omega, \mathcal{F}, \mathbb{P}, \mathcal{F}_t, B_t, X_t)_{t\geq 0}$ *is a weak solution of the SDE (3.126), then for every* $q > 0$ *there exists a constant* C_q*, such that for all* $T, \varepsilon > 0$ *and any stopping time* $\theta \geq 0$*:*

$$\mathbb{E} \sup_{t\in[0,T]} |X_{t\wedge\theta}|^q \leq C_q \left[|x_0|^q + A^q T^{q/2} + A^q T^q \right]$$

and

$$\mathbb{E}\left[\sup_{0\leq s\leq \varepsilon} \left| X_{(t+s)\wedge\theta} - X_{t\wedge\theta} \right|^q \right] \leq C_q\, A^q \left(\varepsilon^q + \varepsilon^{q/2} \right).$$

We now give the main existence result for weak solutions.

Theorem 3.54. *If $f : \mathbb{R}_+ \times \mathbb{R}^d \to \mathbb{R}^d$ and $g : \mathbb{R}_+ \times \mathbb{R}^d \to \mathbb{R}^{d\times k}$ are $\left(\mathcal{B}_{\mathbb{R}_+}, \mathbb{R}^d\right)$-Carathéodory functions satisfying (3.127) and (3.128), then the SDE (3.126) has a weak solution.*

Proof. Step 1: The assumption (3.128) is replaced by a stronger condition: there exists an $A > 0$ such that

$$|f(t,x)| + |g(t,x)| \le A, \qquad \text{for all } (t,x) \in \mathbb{R}_+ \times \mathbb{R}^d.$$

Let $\left(\hat{\Omega}, \hat{\mathcal{F}}, \hat{\mathbb{P}}, \mathcal{F}_t^{\hat{B}}, \hat{B}_t\right)_{t\ge 0}$ be a Brownian motion and $\hat{X}^n : \hat{\Omega} \times [0,\infty[\to \mathbb{R}^d$ the $\mathcal{F}_t^{\hat{B}}$-progressively measurable continuous stochastic processes defined recursively by

$$\hat{X}_t^n = x_0 + \int_0^t f\left(s, \hat{X}^n_{s-\frac{1}{n}}\right) ds + \int_0^t g\left(s, \hat{X}^n_{s-\frac{1}{n}}\right) d\hat{B}_s, \quad t \ge 0.$$

Since $\hat{X}_0^n = x_0$ and

$$\mathbb{E}_{\hat{\mathbb{P}}}\left[\sup_{0\le\theta\le\varepsilon}\left|\hat{X}^n_{t+\theta} - \hat{X}^n_t\right|^4\right] \le C\, A^4\left(\varepsilon^4 + \varepsilon^2\right),$$

we deduce, by Proposition 1.47, that $\left\{\hat{X}^n : n \ge 1\right\}$ is tight on $C\left(\mathbb{R}_+;\mathbb{R}^d\right)$. Consequently $\left\{\left(\hat{X}^n, \hat{B}\right) : n \in \mathbb{N}^*\right\}$ is tight on $\mathbb{X} = C\left(\mathbb{R}_+;\mathbb{R}^{d+k}\right)$.

Then by the Prokhorov theorem along a subsequence still denoted by $(\hat{X}^n, \hat{B})$ we have

$$(\hat{X}^n, \hat{B}) \to (\hat{X}, \hat{B}) \quad \text{in law, as } n \to \infty,$$

on $\mathbb{X}$.

By the Skorohod theorem, we can choose a probability space $(\Omega, \mathcal{F}, \mathbb{P})$, and some random pairs (X^n, B^n), (X, B) defined on that probability space, respectively having the same laws as $\left(\hat{X}^n, \hat{B}\right)$ and $(\hat{X}, \hat{B})$, such that moreover as $n \to \infty$,

$$(X^n, B^n) \xrightarrow{\mathbb{P}\text{-a.s.}} (X, B) \text{ on } \mathbb{X}.$$

Then, by Proposition 2.15, $\left(B^n, \{\mathcal{F}_t^{B^n,X^n}\}\right), n \ge 1$, and $\left(B, \{\mathcal{F}_t^{B,X}\}\right)$ are $\mathbb{R}^k$-Brownian motions.

Using the Lebesgue theorem and, once again Proposition 2.15, we infer that for $n \to \infty$

$$M_\cdot^n = x_0 + \int_0^\cdot f\left(s, X^n_{s-\frac{1}{n}}\right) ds + \int_0^\cdot g\left(s, X^n_{s-\frac{1}{n}}\right) dB^n_s$$
$$\longrightarrow M_\cdot = x_0 + \int_0^\cdot f(s, X_s)\, ds + \int_0^\cdot g(s, X_s)\, dB_s, \text{ in } S_d^p[0,T], \quad \forall\, p \geq 1.$$

If

$$\hat{M}^n_t \overset{def}{=} x_0 + \int_0^t f\left(s, \hat{X}^n_{s-\frac{1}{n}}\right) ds + \int_0^t g\left(s, \hat{X}^n_{s-\frac{1}{n}}\right) d\hat{B}_s$$

then from Corollary 2.14 it follows that

$$\mathcal{L}\left(\hat{X}^n, \hat{B}, \hat{M}^n\right) = \mathcal{L}(X^n, B^n, M^n) \quad \text{on } C\left(\mathbb{R}_+; \mathbb{R}^{d+k+d}\right).$$

Since for every $t \geq 0$

$$\hat{X}^n_t - \hat{M}^n_t = 0, \ a.s.,$$

we have by Corollary 1.18 that

$$X^n_t - M^n_t = 0, \ a.s.$$

and consequently, letting $n \to \infty$,

$$X_t - M_t = 0, \ a.s.$$

Step 2. The assumption (3.128) is satisfied.

As in the proof of Theorem 3.27 we consider $\rho \in C\left(\mathbb{R}^d; [0,1]\right)$ to be given by

$$\rho(r) = \begin{cases} 1, & \text{if } |x| \leq 1, \\ 2 - |x|, & \text{if } 1 < |x| \leq 2, \\ 0, & \text{if } |x| > 2, \end{cases}$$

and

$$\alpha_n(t,x) = \mathbf{1}_{[0,n]}\left(f^\#_{2n}(t) + g^\#_{2n}(t)\right)\rho\left(\frac{x}{n}\right).$$

Consider the approximating SDE

$$X_t = x_0 + \int_0^t f_n(s, X_s)\, ds + \int_0^t g_n(s, X_s)\, dB_s \tag{3.129}$$

where

$$f_n(t,x) = \alpha_n(t,x)\, f(t,x) \quad \text{and} \quad g_n(t,x) = \alpha_n(t,x)\, g(t,x).$$

By the first step the SDE (3.129) has at least one weak solution $(\Omega^n, \mathcal{F}^n, \mathbb{P}^n, \mathcal{F}_t^n, B_t^n, X_t^n)_{t\geq 0}$. Let $\mathbb{E}^n = \mathbb{E}_{\mathbb{P}^n}$.

Since

$$\begin{aligned}&\left\langle X_s^n, f_n\left(s, X_s^n\right) ds\right\rangle + \left(\frac{1}{2}m_p + 9pb\right)\left|g_n\left(s, X_s^n\right)\right|^2 ds \\ &\leq \alpha_n\left(s, X_s^n\right)\left[\left\langle X_s^n, f\left(s, X_s^n\right) ds\right\rangle + \left(\frac{1}{2}m_p + 9p\lambda\right)\left|g\left(s, X_s^n\right)\right|^2\right] ds \\ &\leq \mathbf{1}_{p\geq 2} dR_s + |X_s^n|\, dN_s + |X_s^n|^2 dV(s),\end{aligned}$$

we have, by Lemma 3.52, that

$$\mathbb{E}^n\left[\sup_{t\in[0,T]} |X_t^n|^p\right] \leq L_{p,\lambda}.$$

Hence

$$\mathbb{P}^n\left(\sup_{t\in[0,T]} |X_t^n| \geq N\right) \leq \frac{1}{N^p}\mathbb{E}^{(n)}\left[\sup_{t\in[0,T]} |X_t^n|^p\right] \leq \frac{L_{p,\lambda}}{N^p}.$$

Define the stopping time

$$\theta_N(\omega) = \inf\{t \geq 0 : |X_t^n(\omega)| \geq N\}.$$

It is easy to show that for every $q > 0$ there exists a constant C_q such that

$$\mathbb{E}^n\left[\sup_{0\leq s\leq\varepsilon}\left|X^n_{(t+s)\wedge\theta_N^{(n)}} - X^n_{t\wedge\theta_N^{(n)}}\right|^q\right] \leq C_q\, N^q\left(\varepsilon^q + \varepsilon^{q/2}\right),$$

for all $n, N \in \mathbb{N}^*$ and $\varepsilon > 0$.

Hence, by Proposition 1.47, $\left\{X^n_{\cdot\wedge\theta_N^{(n)}} : n \geq 1\right\}$ is tight on $C\left(\mathbb{R}_+;\mathbb{R}^d\right)$.

Since for all $\varepsilon, T, \delta > 0$

$$\left\{\mathbf{m}_{X_\cdot^n}(\varepsilon;[0,T]) > \delta\right\} \subset \left\{\sup_{t\in[0,T]} |X_t^n| \geq N\right\} \bigcup \left\{\mathbf{m}_{X^n_{\cdot\wedge\theta_N^{(n)}}}(\varepsilon;[0,T]) > \delta\right\},$$

it follows that

$$\limsup_{\varepsilon\searrow 0}\left[\sup_{n\in\mathbb{N}^*}\mathbb{P}^n\left(\mathbf{m}_{X_\cdot^n}(\varepsilon;[0,T]) > \delta\right)\right] \leq \frac{L_{p,\lambda}}{N^p},$$

for all $N > 0$, which yields that for all $T, \delta > 0$

$$\lim_{\varepsilon \searrow 0} \left[\sup_{n \in \mathbb{N}^*} \mathbb{P}^n \left(\mathbf{m}_{X^n_\cdot} (\varepsilon; [0, T]) > \delta \right) \right] = 0.$$

Consequently $\{X^n, n \geq 1\}$ is tight on $C\left(\mathbb{R}_+; \mathbb{R}^d\right)$. Moreover the tightness on $C\left(\mathbb{R}_+; \mathbb{R}^{d+k}\right)$ of $\{(X^n, B^n) : n \in \mathbb{N}^*\}$ clearly follows. Then as in *Step 1* by the Prohorov theorem and the Skorohod theorem, there exist a probability space $\left(\bar{\Omega}, \bar{\mathcal{F}}, \bar{\mathbb{P}}\right)$, and random variables

$$(\bar{X}^n, \bar{B}^n), (\bar{X}, \bar{B}) : \bar{\Omega} \to C\left(\mathbb{R}_+; \mathbb{R}^{d+k}\right), \quad n \in \mathbb{N}^*$$

such that

$$\mathcal{L}(\bar{X}^n, \bar{B}^n) = \mathcal{L}(X^n, B^n),$$

and as $n \to \infty$

$$(\bar{X}^n, \bar{B}^n) \xrightarrow{\bar{\mathbb{P}}\text{-a.s.}} (\bar{X}, \bar{B}) \text{ in } C\left(\mathbb{R}_+; \mathbb{R}^{d+k}\right).$$

Moreover by Proposition 2.15, $\left(\bar{B}^n, \{\mathcal{F}_t^{\bar{B}^n, \bar{X}^n}\}\right)$, $n \geq 1$, and $\left(\bar{B}, \{\mathcal{F}_t^{\bar{B}, \bar{X}}\}\right)$ are $\mathbb{R}^k$-Brownian motions.

Defining

$$\begin{aligned} S_t^n(Y, B) &\overset{def}{=} x_0 + \int_0^t f_n(s, Y_s)\, ds + \int_0^t g_n(s, Y_s)\, dB_s, \quad t \geq 0, \quad \text{and} \\ S_t(Y, B) &\overset{def}{=} x_0 + \int_0^t f(s, Y_s)\, ds + \int_0^t g(s, Y_s)\, dB_s, \quad t \geq 0, \end{aligned}$$

by Corollary 2.14 it follows that

$$\mathcal{L}\left(X^n, B^n, S_t^n(X^n, B^n)\right) = \mathcal{L}\left(\bar{X}^n, \bar{B}^n, S_t^n(\bar{X}^n, \bar{B}^n)\right) \quad \text{on } C\left(\mathbb{R}_+; \mathbb{R}^{d+k+d}\right).$$

Since for every $t \geq 0$,

$$X_t^n - S_t^n(X^n, B^n) = 0, \ \ a.s.,$$

we have

$$\bar{X}_t^n - S_t^n(\bar{X}^n, \bar{B}^n) = 0, \ \ a.s.,$$

and consequently, letting $n \to \infty$,

$$\bar{X}_t - S_t(\bar{X}, \bar{B}) = 0, \quad a.s.$$

Hence $\left(\bar{\Omega}, \bar{\mathcal{F}}, \bar{\mathbb{P}}, \mathcal{F}_t^{\bar{B},\bar{X}}, \bar{B}_t, \bar{X}_t\right)_{t\geq 0}$ is a weak solution of the SDE (3.126). ∎

We say that the SDE (3.126) has the *pathwise uniqueness property* if whenever $(\Omega, \mathcal{F}, \mathbb{P}, X, B)$, $(\Omega, \mathcal{F}, \mathbb{P}, Y, B)$ are solutions of (3.126), then

$$\mathbb{P}(X_t = Y_t, \ \forall\, t \geq 0) = 1$$

follows.

Theorem 3.55. *Let* $f : \mathbb{R}_+ \times \mathbb{R}^d \to \mathbb{R}^d$ *and* $g : \mathbb{R}_+ \times \mathbb{R}^d \to \mathbb{R}^{d\times k}$ *be* $\left(\mathcal{B}_{\mathbb{R}_+}, \mathbb{R}^d\right)$*-Carathéodory functions satisfying (3.127). Then*

(*i*) *the existence and uniqueness of a strong solution for the SDE (3.126) is equivalent to*
(*ii*) *the existence of a weak solution together with pathwise uniqueness.*

Proof. The following proof is based on the proof of Theorem 1.1 page 149 in Ikeda and Watanabe [38]. Let us introduce the notations $W^d = C(\mathbb{R}_+; \mathbb{R}^d)$ and

$$W_0^k = \left\{x \in W^k : x(0) = 0\right\}.$$

The only point which needs to be proved is that weak existence + pathwise uniqueness implies strong existence.

The central idea of the proof is rather simple. Let $(\Omega, \mathcal{F}, \mathbb{P}, \mathcal{F}_t, X_t, B_t)_{t\geq 0}$ and $(\Omega', \mathcal{F}', \mathbb{P}', \mathcal{F}_t', X_t', B_t')_{t\geq 0}$ denote two (weak) solutions starting from x (i.e. $X_0 = X_0' = x$). Let $\mathbb{P}_x$ (resp. $\mathbb{P}_x'$) denote the probability distribution of (X, B) (resp. (X', B')) on $W^d \times W_0^k$. If

$$\pi : W^d \times W_0^k \to W_0^k$$

denotes the projection on the second factor, then clearly

$$\pi(\mathbb{P}_x) = \pi(\mathbb{P}_x') = \mathbb{P}^W,$$

the Wiener measure. Now let $Q^{w_2}(dw_1)$ (resp. $Q'^{\,w_2}(dw_1)$) denote the regular conditional distribution of w_1 given w_2 under $\mathbb{P}_x$ (resp. $\mathbb{P}_x'$), i.e.

$$\mathbb{P}_x(A_1 \times A_2) = \int_{A_2} Q^{w_2}(A_1)\mathbb{P}^W(dw_2),$$

$$\mathbb{P}_x'(A_1 \times A_2) = \int_{A_2} Q'^{\,w_2}(A_1)\mathbb{P}^W(dw_2).$$

Define on the space $\bar{\Omega} = W^d \times W^d \times W_0^k$ the measure Q by

$$Q(dw_1, dw_2, dw_3) = Q^{w_3}(dw_1)Q'^{\,w_3}(dw_2)\mathbb{P}^W(dw_3).$$

Let $\bar{\mathcal{F}}$ denote the completion of the σ-algebra of Borel sets of $\bar{\Omega}$, with respect to Q, and $\bar{\mathcal{F}}_t = \cap_{\varepsilon>0}(\mathcal{B}_{t+\varepsilon} \vee \mathcal{N})$, where

$$\mathcal{B}_t = \mathcal{B}_t(W^d) \otimes \mathcal{B}_t(W^d) \otimes \mathcal{B}_t(W_0^k),$$

and $\mathcal{N}$ denotes the class of all Q-null sets. Denote by $\overline{\mathcal{B}_t(W_0^k)}^{\mathbb{P}^W}$ the completion of the σ-algebra $\mathcal{B}_t(W_0^k)$ with respect to $\mathbb{P}^W$.

The two technical lemmas from Ikeda and Watanabe [38], p. 151, say:

Lemma 1. *For $A \in \mathcal{B}_t(W^d)$, $w \in W_0^k \mapsto Q^w(A)$ and $Q'^{\,w}(A)$ are $\overline{\mathcal{B}_t(W_0^k)}^{\mathbb{P}^W}$-measurable.*

And

Lemma 2. *w_3 is an k-dimensional $(\mathcal{F}_t)$-Brownian motion on $(\bar{\Omega}, \bar{\mathcal{F}}, Q)$.*

It now follows from Lemma 2 that (w_1, w_3) and (w_2, w_3) are two solutions of the SDE (4.99) on the same space $(\bar{\Omega}, \bar{\mathcal{F}}, Q)$, with the same filtration $(\bar{\mathcal{F}}_t)$ and the same driving Brownian motion w_3. Hence pathwise uniqueness implies that $w_1 = w_2$, Q-a.s., that is

$$Q^w \times Q'^{\,w}(w_1 = w_2) = 1, \quad \mathbb{P}^W(dw)\text{-a.s.}$$

Now this is possible only if there exists a mapping

$$w \in W_0^k \mapsto F_x(w) \in W^d,$$

such that

$$Q^w(\cdot) = Q'^{\,w}(\cdot) = \delta_{F_x(w)}(\cdot), \quad \mathbb{P}^W(dw)\text{-a.s.}$$

In other words, the only way that two independent random variables X and Y can coincide a.s. is that their common law is a Dirac measure. By Lemma 1, $w \mapsto F_x(w)$ is $\left(\overline{\mathcal{B}_t(W_0^k)}^{\mathbb{P}^W}, \mathcal{B}_t(W^d)\right)$-measurable and $F_x(w)$ is uniquely determined up to a set of $\mathbb{P}^W$-measure 0. Consequently $\bar{X}_t(w) = F_x(w)_t$ is a strong solution for the stochastic basis $(\bar{\Omega}, \bar{\mathcal{F}}, Q, \bar{\mathcal{F}}_t)_{t\geq 0}$ and the Brownian motion $w = w_3$.

If now $(\Omega, \mathcal{F}, \mathbb{P}, \mathcal{F}_t)_{t\geq 0}$ is an arbitrary stochastic basis and $\{B_t : t \geq 0\}$ is an $(\mathcal{F}_t)$-Brownian motion, then $X_t = F_x(B)_t$ is a strong solution. ■

To complete this section, we give the proofs of Lemmas 1 and 2. The proofs are taken from Ikeda and Watanabe [38], p. 151.

Proof of Lemma 1. For $t > 0$ and $A \in \mathcal{B}_t\left(W^d\right)$, there exists a regular conditional probability $Q_t^w(A)$ such that $w \in W_0^k \longmapsto Q_t^w(A)$ is $\overline{\mathcal{B}_t(W_0^k)}^{\mathbb{P}^W}$ -measurable and

$$\mathbb{P}_x(A \times C) = \int_C Q_t^w(A)\mathbb{P}^W(dw), \quad \forall\, C \in \mathcal{B}_t(W_0^k).$$

If we can show that this equality holds for all $C \in \mathcal{B}(W_0^k)$, then

$$Q_t^w(A) = Q^w(A), \qquad \mathbb{P}^W(dw)\text{-}a.e.$$

and the assertion of the lemma holds.

Let $(\theta_t w)(s) = w(t+s) - w(t)$ and $(\rho_t w)(s) = w(t \wedge s)$. One may assume that C is of the form

$$C = \left\{w \in W_0^k : \rho_t w \in A_1,\ \theta_t w \in A_2\right\}, \quad A_1, A_2 \in \mathcal{B}(W_0^k).$$

Since $\theta_t w$ and $\mathcal{B}_t(W_0^k)$ are independent with respect to $\mathbb{P}^W$, we have

$$\begin{aligned}
\int_C Q_t^w(A)\mathbb{P}^W(dw) &= \int_{\{\rho_t w \in A_1\}} Q_t^w(A)\mathbb{P}^W(dw)\,\mathbb{P}^W(\theta_t w \in A_2) \\
&= \mathbb{P}_x(A \times \{\rho_t w \in A_1\})\,\mathbb{P}_x\left(W^d \times \{\theta_t w \in A_2\}\right) \\
&= \mathbb{P}(X \in A, \rho_t(B) \in A_1)\mathbb{P}(\theta_t(B) \in A_1) \\
&= \mathbb{P}(X \in A, \rho_t(B) \in A_1, \theta_t(B) \in A_1) \\
&= \mathbb{P}(X \in A, B \in C) \\
&= \mathbb{P}_x(A \times C).
\end{aligned}$$

This completes the proof of Lemma 1. ∎

Proof of Lemma 2. Since for every $t > s$, $u \in \mathbb{R}^k$, $A_1, A_2 \in \mathcal{B}_s\left(W^d\right)$ and $A_3 \in \mathcal{B}_s\left(W_0^k\right)$:

$$\begin{aligned}
&\mathbb{E}_Q\left[e^{i\langle u, w_3(t) - w_3(s)\rangle}\mathbf{1}_{A_1 \times A_2 \times A_3}\right] \\
&\quad= \int_{A_3} e^{i\langle u, w_3(t) - w_3(s)\rangle} Q^{w_3}(A_1)Q'^{\,w_3}(A_2)\mathbb{P}^W(dw_3) \\
&\quad= e^{-\frac{1}{2}|u|^2(t-s)}\int_{A_3} Q^{w_3}(A_1)Q'^{\,w_3}(A_2)\mathbb{P}^W(dw_3) \\
&\quad= e^{-\frac{1}{2}|u|^2(t-s)} Q(A_1 \times A_2 \times A_3),
\end{aligned}$$

we obtain that $w_3(t) - w_3(s)$ and $\bar{\mathcal{F}}_s$ are independent with respect to Q and the result follows. ∎

3.10 Exercises

Exercise 3.1 (Stabilization). Consider the controlled stochastic differential equation

$$X_t = \xi + \int_0^t F(s, X_s)\, ds + \int_0^t U_s ds + \int_0^t G(s, X_s)\, dB_s, \quad t \geq 0, \tag{3.130}$$

where $\xi \in L^p(\Omega, \mathcal{F}_0, P; \mathbb{R}^d)$, $p \geq 1$, and the assumptions (**SDE-H**$_{F,loc}$), (**SDE-H**$_{G,loc}$) and (**B**$_{FG}$) from Sect. 3.6.1 (Locally monotone drift) are satisfied. Also assume

$$F(s, 0) = 0 \quad \text{and} \quad G(s, 0) = 0.$$

1. Show that for every $a > 0$ there exists a progressively measurable control

$$U \in L^p\left(\Omega; L^1\left(0, T; \mathbb{R}^d\right)\right)$$

which stabilizes the solution with the given exponential rate a in the following sense: the corresponding solution $X = X^U \in S_d^p$ satisfies for all $\delta \geq 0$:

$$\begin{aligned} &(a)\ \mathbb{E}\frac{|X_s|^p}{\left(1+\delta|X_s|^2\right)^{p/2}} \leq e^{-as}\, \mathbb{E}\frac{|\xi|^p}{\left(1+\delta|\xi|^2\right)^{p/2}} \text{ and} \\ &(b)\ \ \mathbb{E}\int_0^\infty \frac{|X_s|^p}{\left(1+\delta|X_s|^2\right)^{p/2}} ds \leq \frac{1}{a}\mathbb{E}\frac{|\xi|^p}{\left(1+\delta|\xi|^2\right)^{p/2}}; \end{aligned} \tag{3.131}$$

and moreover

$$\mathbb{E}^{\mathcal{F}_t} \frac{|X_s|^p}{\left(1+\delta|X_s|^2\right)^{p/2}} \leq e^{-a(s-t)} \frac{|X_t|^p}{\left(1+\delta|X_t|^2\right)^{p/2}}, \quad \mathbb{P}\text{-a.s.}; \tag{3.132}$$

whenever $0 \leq t \leq s$ and $\delta \geq 0$.
2. Show that if $0 \leq \lambda < a$

$$e^{\lambda t}|X_t| \to 0 \quad \text{in probability as } t \to \infty. \tag{3.133}$$

3. Show that if moreover $\xi = x \in \mathbb{R}^d$ and $0 \leq \delta < a$, then

$$|X_t(\omega; x)|^p \leq e^{-\delta t}|x|^p, \quad \forall\, t \geq \theta(\omega),\ \mathbb{P}\text{-a.s.}, \tag{3.134}$$

where $\theta < \infty$, $\mathbb{P}$-a.s.

Exercise 3.2. Consider the ordinary differential equation

$$x(t) = m(t) + \int_0^t f(s, x(s))\, ds, \quad t \in [0, T], \tag{3.135}$$

in a Hilbert space $\mathbb{H}$. Assume $m \in C([0, T]; \mathbb{H})$ and $f : [0, T] \times \mathbb{H} \to \mathbb{H}$ is a measurable function such that for all $x, y \in H$:

$$(H_f) \begin{cases} (i) & f(t, \cdot) : \mathbb{H} \to \mathbb{H} \text{ is continuous, a.e. } t \geq 0, \\ (ii) & \langle x - y, f(t, x) - f(t, y) \rangle \leq \mu(t)\, |x - y|^2, \text{ a.e. } t \geq 0, \\ (iii) & \int_0^T \left| f_R^{\#}(t) \right|^2 dt < \infty, \quad \forall\, R > 0,\ \forall\, T > 0, \end{cases} \tag{3.136}$$

where $\mu \in L^2_{loc}(0, \infty)$, and

$$f_R^{\#}(t) \overset{def}{=} \sup \{ |f(t, x)| : |x| \leq R \}.$$

Consider the approximating equation

$$x_\varepsilon(t) = m(t) + \int_0^t f_\varepsilon(s, x_\varepsilon(s)), \quad t \in [0, T], \tag{3.137}$$

where $f_\varepsilon(t, x) = f(t, \Gamma_\varepsilon(t, x))$ and Γ_ε is the unique solution of the equation

$$\Gamma_\varepsilon + \varepsilon \left[\mu(t)\, \Gamma_\varepsilon - f(t, \Gamma_\varepsilon) \right] = x.$$

Show that the differential equation (3.135) has a unique solution $x \in C([0, T]; \mathbb{H})$ and

$$\sup_{t \in [0,T]} |x_\varepsilon(t) - x(t)| \leq C\varepsilon.$$

Exercise 3.3. Let F_ε be the Yosida approximation of F defined in (6.4) Annex B:

$$F_\varepsilon(t, x) = F(t, \Gamma_\varepsilon(t, x)) \quad \text{where} \quad \Gamma_\varepsilon + \varepsilon \left(\mu_t \Gamma_\varepsilon - F(t, \Gamma_\varepsilon) \right) = x.$$

Let $p \geq 2$, the assumptions of Theorem 3.21 be satisfied and

$$\int_0^T |\mu(t)|\, |F(t, 0)|\, dt < \infty, \quad \forall T > 0.$$

Let $X \in S_d^p$ the solution of the SDE (3.59) and X^ε, $0 < \varepsilon \leq 1$, be a solution of the approximating equation

$$X_t^\varepsilon = \xi + \int_0^t F_\varepsilon\left(s, X_s^\varepsilon\right) ds + \int_0^t G\left(s, X_s^\varepsilon\right) dB_s, \quad t \geq 0. \tag{3.138}$$

Show that for every $T > 0$

$$X^\varepsilon \longrightarrow X \quad \text{in } S_d^p\,[0, T]\,, \ \text{ as } \varepsilon \to 0.$$

Exercise 3.4. Let the assumptions $(l l)$, $(l l l)$ from Sect. 3.4 be satisfied and $(t, x) \in \mathbb{R}_+ \times \mathbb{R}^d$ be fixed. Define recursively the sequence

$$X_s^n = x, \quad \text{for } s \leq t, \quad \text{and}$$

$$X_s^n = x + \int_t^s f\left(r, X_{r-1/n}^n\right) dr + \int_t^s g\left(r, X_{r-1/n}^n\right) dB_r, \quad \text{for } s > t.$$

Show that X_s^n is $F_{t,s}^B \overset{def}{=} \sigma\left\{B_r - B_t : t \leq r \leq s\right\}$-measurable and

$$X^n \to X^{t,x} \ \text{ in } \ S_d^p\,[0, T] \ \text{ as } n \to \infty,$$

where $X^{t,x}$ is the solution of

$$X_s^{t,x} = x + \int_t^{s \vee t} f\left(r, X_r^{t,x}\right) dr + \int_t^{s \vee t} g\left(r, X_r^{t,x}\right) dB_r$$

(as a consequence $X_s^{t,x}$ is an $F_{t,s}^B$-measurable random variable).

Remark. This exercise provides an alternative proof of the existence of a solution to an SDE.

Exercise 3.5. Let $b : \mathbb{R} \to \mathbb{R}$ be uniformly Lipschitz continuous. Let x be a real number and $\{B_t,\ t \geq 0\}$ be a one-dimensional Brownian motion. $\{X_t,\ t \geq 0\}$ stands for the unique solution of the SDE

$$X_t = x + \int_0^t b(X_s) ds + B_t.$$

1. Suppose that

$$b(y) \leq 0, \ \text{ whenever } \ y > 0; \quad \text{ and } b(y) \geq 0, \ \text{ whenever } \ y < 0.$$

Show that if $x > 0$ (resp. $x < 0$), then $X_t \leq x + B_t$ (resp. $X_t \geq x + B_t$), for $t \in [0, \tau_0]$, where $\tau_0 := \inf\{t,\ X_t = 0\}$.

From now on, we assume that $x > 1$ and that there exists a $c > 0$ such that

$$b(y) = \begin{cases} -cy, & \text{if } |y| \le 1; \\ -\dfrac{c}{y}, & \text{if } |y| \ge 1. \end{cases}$$

We first assume that $c \neq 1/2$.

2. Define

$$Y_t = t + \frac{X_t^2}{2c - 1},$$

and for $b > x$,

$$\tau_{(1,b)} = \inf\{t, X_t \notin (1, b)\}.$$

Show that $\{Y_{t \wedge \tau_{(1,b)}},\ t \ge 0\}$ is a martingale, and that

$$\frac{x^2}{2c-1} = \frac{\mathbb{E}[X^2_{t \wedge \tau_{(1,b)}}]}{2c-1} - \mathbb{E}[t \wedge \tau_{(1,b)}].$$

3. Deduce that

$$\mathbb{E}[\tau_{(1,b)}] = \frac{x^2 - \mathbb{E}[X^2_{\tau_{(1,b)}}]}{2c-1}.$$

4. Show that $\{X^{2c+1}_{t \wedge \tau_{(1,b)}},\ t \ge 0\}$ is a martingale. Conclude that

$$\mathbb{E}[X^{2c+1}_{\tau_{(1,b)}}] = \mathbb{E}[X^{2c+1}_{t \wedge \tau_{(1,b)}}] = x^{2c+1}.$$

5. Deduce from the above the value of $\mathbb{E}[\tau_{(1,b)}]$.
6. Let now $\tau_1 = \inf\{t,\ X_t = 1\}$. Show that $\mathbb{E}[\tau_1] < \infty$, if $c > 1/2$, while $\mathbb{E}[\tau_1] = \infty$, if $c < 1/2$.

   From now on, we assume that $c = 1/2$.
7. Show that $\{X^2_{t \wedge \tau_{(1,b)}},\ t \ge 0\}$ is a martingale.
8. Let $Z_t = t - \varphi(X_t)$, with

$$\varphi(y) = 2\frac{y^2-1}{b^2-1}\int_1^y \frac{b^2 - z^2}{2z} dz + 2\frac{b^2 - y^2}{b^2 - 1}\int_y^b \frac{z^2 - 1}{2z} dz.$$

   Show that $\{Z_{t \wedge \tau_{(1,b)}},\ t \ge 0\}$ is a martingale.
9. Compute $\mathbb{E}[\tau_{(1,b)}]$ and $\mathbb{E}[\tau_1]$.

Exercise 3.6. Let f and g be locally Lipschitz mappings from $\mathbb{R}$ into $\mathbb{R}$. We consider the one-dimensional SDE

$$dX_t = f(X_t)dt + g(X_t)dB_t,\ t \ge 0.$$

1. Show that this SDE has unique solution, $\{X_t;\ 0 \le t \le \tau\}$, where τ is a stopping time such that $|X_t| \to \infty$, as $t \to \tau$.
2. We assume now that $xf(x) \le K(1 + |x|^2)$ and $|g(x)| \le K(1 + |x|)$. Deduce that $\tau = +\infty$ $\mathbb{P}$-a.s.
3. Suppose that $f(0) \ge 0$, $f(1) \le 0$, $g(0) = g(1) = 0$, and $0 \le X_0 \le 1$, $\mathbb{P}$-a.s. Show that $0 \le X_t \le 1$, $\mathbb{P}$-a.s. for all $t \ge 0$ and $\tau = +\infty$, $\mathbb{P}$-a.s.

Exercise 3.7. We consider the $\mathbb{R}$-valued SDE

$$dX_t = (F_t X_t + f_t)dt + \langle G_t X_t + g_t, dB_t\rangle\,,\ X_0 = x,$$

where B is a k-dimensional Brownian motion. The coefficients F_t, f_t, G_t and g_t are assumed to be adapted and bounded processes, with values resp. in $\mathbb{R}$, $\mathbb{R}$, $\mathbb{R}^k$ and $\mathbb{R}^k$. Define

$$\Phi_t = \exp\int_0^t \left(F_s ds + \langle G_s, dB_s\rangle - \frac{1}{2}|G_s|^2 ds\right).$$

1. Show that the solution of the above SDE is given by the formula

$$X_t = \Phi_t\left(x + \int_0^t \Phi_s^{-1}(f_s - \langle G_s, g_s\rangle)ds + \int_0^t \Phi_s^{-1}\langle g_s, dB_s\rangle\right).$$

2. Generalize this result to the case where X_t takes its values in $\mathbb{R}^d$, and the equation is of the form

$$dX_t = (F_t X_t + f_t)dt + \sum_{i=1}^k (G_t^i X_t + g_t^i)dB_t^i,\ X_0 = x,$$

where F_t, f_t, G_t^i, g_t^i take values resp. in $\mathbb{R}^{d\times d}$, $\mathbb{R}^d$, $\mathbb{R}^{d\times d}$ and $\mathbb{R}^d$.

Exercise 3.8. Let $\{B_t,\ t \ge 0\}$ be a standard Brownian motion, f and g Lipschitz mappings from $\mathbb{R}$ into $\mathbb{R}$, which satisfy $|g(x)| \le K$, and suppose there exist $R, L > 0$ such that

$$\begin{cases} |f(x)| \le L, & \text{if } |x| \le R; \\ |f(x)| \le L|x|, & \text{if } |x| > R. \end{cases}$$

Let $\{X_t,\ t \ge 0\}$ denote the solution of the one-dimensional SDE

$$X_t = x + \int_0^t f(X_s)ds + \int_0^t g(X_s)dB_s,\ t \ge 0,$$

where $x \in \mathbb{R}$ is arbitrary.

1. Show that if $t \leq T$,

$$\mathbb{E}[X_t^{2p}] \leq M^{2p} + 2pL\mathbb{E}\int_0^t |X_s|^{2p} ds + p(2p-1)K^2\mathbb{E}\int_0^t X_s^{2p-2} ds,$$

where $M = \sup(|x_0|, TL) + R$.
2. Show that the solution of the linear SDE

$$Y_t = M + L\int_0^t Y_s ds + KB_t$$

satisfies the same relation, but with equality.
3. Show that if x, y, u, v are four non-negative integrable functions defined on $[0, T]$ and satisfying (C is a positive constant)

$$\begin{aligned} x(t) &\leq u(t) + C\int_0^t x(s)ds \\ y(t) &= v(t) + C\int_0^t y(s)ds \\ u(t) &\leq v(t), \end{aligned}$$

then $x(t) \leq y(t)$, whenever $0 \leq t \leq T$.
4. Deduce by recurrence on the integer p that $\mathbb{E}[X_t^{2p}] \leq \mathbb{E}[Y_t^{2p}]$, whenever $0 \leq t \leq T$, and that there exists an $\alpha > 0$ such that $\mathbb{E}[\exp(\alpha X_t^2)] < \infty, 0 \leq t \leq T$.
5. Show that for all $T > 0$, $c \in \mathbb{R}$,

$$\mathbb{E}\left[\sup_{0\leq t\leq T} \exp(cX_t)\right] < \infty.$$

6. Consider the SDE

$$X_t = 1 + \int_0^t X_s dB_s,$$

and show that for all $c > 0, t > 0$,

$$\mathbb{E}\left[\exp(cX_t)\right] = \infty.$$

Exercise 3.9. Let $(\Omega, \mathcal{F}, \mathbb{P}, \mathcal{F}_t, B_t)_{t\geq 0}$ be a k-dimensional Brownian motion: $\mathcal{F}_t = \sigma\left\{B_s : 0 \leq s \leq t\right\}$ and $\mathcal{F} = \sigma\left\{\bigcup_{t\geq 0}\mathcal{F}_t\right\}$. Let $x \in \mathbb{R}^k$. Show that:

1. For all $C, t > 0$

$$
\begin{aligned}
&(i) \quad \mathbb{E} \exp\left(C\,|x+B_t|^b\right) < \infty, && \text{if } 0 \le b < 2,\\
&(ii) \quad \mathbb{E} \exp\left(C\int_0^t |x+B_s|^a\,ds\right) < \infty, && \text{if } -1 < a < 2,\\
&(iii) \quad \mathbb{E}\left\{\exp\left[C\log^2(|x+B_t|)\right]\right\} = \infty,\\
&(iv) \quad \mathbb{E}\left[\exp\left(C\int_0^t \log^2(|x+B_s|)ds\right)\right] < \infty.
\end{aligned}
$$

2. If $g : \mathbb{R}^k \to \mathbb{R}^k$ is a bounded Borel measurable function and $-1 < a < 2$, then

   (j)

$$
\begin{aligned}
M_t = \exp\Bigg(&\int_0^t |x+B_s|^{a/2}\langle g(x+B_s), dB_s\rangle\\
&-\frac{1}{2}\int_0^t |g(x+B_s)|^2\,|x+B_s|^a\,ds\Bigg),\ t \ge 0, \textit{and}\\
N_t = \exp\Bigg(&\int_0^t \log|x+B_s|\,\langle g(x+B_s), dB_s\rangle\\
&-\frac{1}{2}\int_0^t |g(x+B_s)|^2\log^2(|x+B_s|)ds\Bigg),\ t \ge 0,
\end{aligned}
$$

   are martingales.

   (jj)

$$
\hat{B}_t = B_t - \int_0^t |x+B_s|^{a/2}\,g(x+B_s)\,ds
$$

   (resp. $\tilde{B}_t = B_t - \int_0^t g(x+B_s)\log|x+B_s|\,ds$)
   is a k-dimensional $\mathcal{F}_t$-Brownian motion under a new probability measure $\hat{\mathbb{P}}$ (resp. $\tilde{\mathbb{P}}$) defined on $(\Omega, \mathcal{F})$.

   (jjj)

   $\left(\Omega, \mathcal{F}, \tilde{\mathbb{P}}, \mathcal{F}_t, \tilde{B}_t, X_t = x + B_t\right)_{t\ge 0}$ is the weak solution of the SDE

$$
\begin{cases}
dX_t = g(X_t)\log\dfrac{1}{|X_t|}dt + dB_t,\\
X_0 = x,
\end{cases}
$$

   and if $-1 < a < 2$ then $\left(\Omega, \mathcal{F}, \hat{\mathbb{P}}, \mathcal{F}_t, \hat{B}_t, X_t = x + B_t\right)_{t\ge 0}$ is the weak solution of the SDE

$$\begin{cases} dX_t = |X_t|^a \, g(X_t)\, dt + dB_t, \\ X_0 = x. \end{cases}$$

Exercise 3.10. Let $f : \mathbb{R} \to (-\infty, 0]$ and $g : \mathbb{R} \to \mathbb{R}$ be locally Lipschitz continuous functions such that $f(0) < 0$ and $|g(x)| \leq a + b|x|$ for all $x \in \mathbb{R}$, where a, b are some positive constants. Using Theorem 3.27 show that the following SDE

$$\begin{cases} dX_t = f(X_t)\sqrt{|X_t|}\operatorname{sign}(X_t)\, dt + g(X_t)\, dB_t, & t > 0, \\ X_0 = 1, \end{cases}$$

has a unique solution $X \in S_1^p$, for all $p \geq 0$. If, moreover, $|f(x)| + |g(x)| \leq a + b|x|$ for all $x \in \mathbb{R}$, can we apply Proposition 3.28 to obtain the same results? (Here $\operatorname{sign}(r) \overset{def}{=} \frac{r}{|r|}$ for $r \neq 0$ and $\operatorname{sign}(0) \overset{def}{=} 0$.)

Chapter 4
SDEs with Multivalued Drift

4.1 Introduction

In the previous chapter we studied the stochastic differential equation

$$\begin{aligned} dX_t &= F(t, X_t)dt + G(t, X_t)dB_t, \\ X_0 &= \xi \end{aligned}$$

as a mathematical model of the evolution of a state $X_t \in \mathbb{R}^d$ of a dynamic system with a drift $F(t, X_t)$ and a diffusion coefficient $G(t, X_t)$.

If G is non-degenerate (i.e. $GG^* \geq \alpha I, \ \alpha > 0$), X_t can take any values in $\mathbb{R}^d$ (X_t can be found in any open domain $\mathcal{D} \subset \mathbb{R}^d$ with a positive probability), as a consequence of Stroock–Varadhan's support theorem [69]. This may be inadequate in certain applications where the state X_t should be maintained in a (possibly convex) domain $\overline{\mathcal{O}} \subset \mathbb{R}^d$. Practically this is realized with a supplementary drift $-\partial I_{\overline{\mathcal{O}}}(X_t)$ in the equation. In this case instead of the above model, we shall consider the model:

$$\begin{cases} dX_t + \partial I_{\overline{\mathcal{O}}}(X_t)\,(dt) \ni F(t, X_t))dt + G(t, X_t)dB_t, \\ X_0 = \xi \in \overline{\mathcal{O}}. \end{cases}$$

or more general models:

- stochastic equation with subdifferential drift

$$\begin{cases} dX_t + \partial\varphi(X_t)(dt) \ni F(t, X_t))dt + G(t, X_t)dB_t, \\ X_0 = \xi, \end{cases} \tag{4.1}$$

and

E. Pardoux and A. Răşcanu, *Stochastic Differential Equations, Backward SDEs, Partial Differential Equations*, Stochastic Modelling and Applied Probability 69,
DOI 10.1007/978-3-319-05714-9_4,

- stochastic equation with maximal monotone operator in the drift

$$\begin{cases} dX_t + A(X_t)(dt) \ni F(t, X_t))dt + G(t, X_t)dB_t, \\ X_0 = \xi, \end{cases} \tag{4.2}$$

- reflected SDE in a non-convex domain

$$\begin{cases} X_t + K_t = \xi + \int_0^t F(s, X_s))ds + \int_0^t G(s, X_s)dB_s, \\ (X, K) \in C\left(\mathbb{R}_+; \overline{\mathcal{O}}\right) \times BV_{loc}\left(\mathbb{R}_+; \mathbb{R}^d\right), \ a.s., \\ K_t = \int_0^t \mathbf{n}\left(X_s\right) d \updownarrow K \updownarrow_s = \int_0^t \mathbf{n}\left(X_s\right) \mathbf{1}_{X_s \in \mathrm{Bd}(\overline{\mathcal{O}})} d \updownarrow K \updownarrow_s, \ a.s., \end{cases}$$

where $\mathbf{n}\left(X_s\right)$ is a unit outward normal to $\overline{\mathcal{O}}$ at X_s.

Here $\partial\varphi$ is the subdifferential of a convex lower semicontinuous function $\varphi : \mathbb{R}^d \to]-\infty, +\infty]$ and $A \subset \mathbb{R}^d \times \mathbb{R}^d$ is a maximal monotone operator. In the case

$$\varphi(x) = I_{\overline{\mathcal{O}}}(x) = \begin{cases} 0, & \text{if } x \in \overline{\mathcal{O}}, \\ +\infty, & \text{if } x \in \mathbb{R}^d \setminus \overline{\mathcal{O}}, \end{cases}$$

the subdifferential $\partial\varphi$ is given by

$$\begin{aligned} \partial I_{\overline{\mathcal{O}}}(x) &= \{\hat{x} \in \mathbb{R}^d : \langle \hat{x}, z - x \rangle + I_{\overline{\mathcal{O}}}(x) \le I_{\overline{\mathcal{O}}}(z), \ \forall z \in \mathbb{R}^d\} \\ &= \begin{cases} 0, & \text{if } x \in \mathcal{O}, \\ \{\hat{x} \in \mathbb{R}^d : \langle \hat{x}, z - x \rangle \le 0, \ \forall z \in \overline{\mathcal{O}}\}, & \text{if } x \in \partial\left(\overline{\mathcal{O}}\right), \\ \emptyset, & \text{if } x \in \mathbb{R}^d \setminus \overline{\mathcal{O}}, \end{cases} \\ &= \begin{cases} 0, & \text{if } x \in \mathcal{O}, \\ \mathcal{N}_{\overline{\mathcal{O}}}(x), & \text{if } x \in \partial\left(\overline{\mathcal{O}}\right), \\ \emptyset, & \text{if } x \in \mathbb{R}^d \setminus \overline{\mathcal{O}}, \end{cases} \end{aligned}$$

where $\mathrm{Bd}\left(\overline{\mathcal{O}}\right)$ is the boundary of $\overline{\mathcal{O}}$ and $\mathcal{N}_{\overline{\mathcal{O}}}(x)$ is the outward normal cone to $\overline{\mathcal{O}}$ at $x \in \mathrm{Bd}\left(\overline{\mathcal{O}}\right)$.

If $a, b \in \mathbb{R}$, $a < b$, and

$$\varphi(x) = I_{[a,b]}(x) = \begin{cases} 0, & \text{if } x \in [a, b], \\ +\infty, & \text{if } x \in \mathbb{R} \setminus [a, b], \end{cases}$$

then

$$\partial I_{[a,b]}(x) = \begin{cases} 0, & \text{if } x \in]a, b[, \\ ]-\infty, 0], & \text{if } x = a, \\ [0, +\infty[, & \text{if } x = b, \\ \emptyset, & \text{if } x \in \mathbb{R} \setminus [a, b]. \end{cases}$$

We see that this supplementary drift $-\partial I_{\overline{\mathcal{O}}}(X_t)$ is an "inward push" that prevents the process X_t from exiting the domain $\overline{\mathcal{O}}$ and this is done in a minimal way (i.e. this drift acts only when X_t is on the boundary of $\overline{\mathcal{O}}$).

These models can be viewed as generalizations of Skorohod's problem in convex domains. Indeed:

♦ Given

$$\begin{array}{ll} i) & m \in C\left([0,\infty[\,;\mathbb{R}\right), \; m\left(0\right)=0, \\ ii) & x_0 \geq 0, \end{array}$$

the classical version of Skorohod's problem is to find two functions $x, k : \mathbb{R}_+ \to \mathbb{R}$ such that

$$\begin{array}{ll} j) & x \in C\left([0,\infty[\,;\mathbb{R}\right), \; x\left(0\right)=x_0, \; x\left(t\right) \geq 0, \text{ for all } t \geq 0, \\ jj) & k \in C\left([0,\infty[\,;\mathbb{R}\right) \text{ is a decreasing function, } k\left(0\right)=0, \\ jjj) & x\left(t\right)+k\left(t\right)=x_0+m\left(t\right), \text{ for all } t \geq 0, \\ jv) & \displaystyle\int_0^T \mathbf{1}_{\{s:\,x(s)>0\}}\left(t\right) dk\left(t\right)=0, \text{ for all } T>0. \end{array} \tag{4.3}$$

In other words, one wants to apply an "upward push" to the path $x_0 + m(t)$ that keeps the resulting process $x(t)$ nonnegative, and to do this in a minimal way (i.e. the push acts only when $x(t) = 0$). It is easy to verify that the solution of this problem is given by

$$\begin{cases} k\left(t\right)=-\sup\left\{\left(x_0+m\left(r\right)\right)^- : 0 \leq r \leq t\right\}, \\ x\left(t\right)=x_0+m\left(t\right)-k\left(t\right). \end{cases}$$

In fact the problem (4.3) is equivalent to

$$\begin{cases} x\left(t\right)+k\left(t\right)=x_0+m\left(t\right), \text{ for all } t \geq 0, \\ x\left(t\right) \geq 0, \text{ for all } t \in [0,T], \\ \displaystyle\int_s^t \left(y-x(r)\right) dk\left(r\right) \leq 0, \text{ for all } y \geq 0, \text{ for all } 0<s<t, \end{cases}$$

and from Annex B this means $dk(t) \in \partial I_{[0,\infty[}(x(t))\,(dt)$, where $I_{[0,\infty[} : \mathbb{R} \to]-\infty, +\infty]$ is the convex indicator function of $[0,\infty[$, that is

$$I_{[0,\infty[}(x)=\begin{cases} 0, & \text{if } x \geq 0, \\ +\infty, & \text{if } x<0, \end{cases}$$

and $\partial I_{[0,\infty[}$ is the subdifferential mapping:

$$\partial I_{[0,\infty[}(x)=\begin{cases} 0, & \text{if } x>0, \\ ]-\infty,0], & \text{if } x=0, \\ \emptyset, & \text{if } x<0. \end{cases}$$

Hence the problem (4.3) is equivalent to the multivalued equation

$$\begin{cases} dx\,(t) + \partial I_{[0,\infty[}(x(t))\,(dt) \ni dm\,(t)\,, \\ x\,(0) = x_0. \end{cases}$$

♦ Given

$$\begin{array}{rl} i) & \mathcal{O} \text{ an open convex subset of } \mathbb{R}^d, \\ ii) & m \in C\left([0,\infty[\,;\mathbb{R}^d\right),\ m\,(0) = 0, \\ iii) & x_0 \in \overline{\mathcal{O}}, \end{array}$$

Skorohod's problem is to find two functions $x, k : \mathbb{R}_+ \to \mathbb{R}^d$ such that:

$$\begin{array}{rl} j) & x \in C\left([0,\infty[\,;\mathbb{R}^d\right),\ x\,(0) = x_0,\ x\,(t) \in \overline{\mathcal{O}},\ \text{for all } t \geq 0, \\ jj) & k \in C\left([0,\infty[\,;\mathbb{R}^d\right) \cap BV_{loc}\left([0,\infty[\,;\mathbb{R}^d\right),\ k\,(0) = 0, \\ jjj) & x\,(t) + k\,(t) = x_0 + m\,(t)\,,\ \text{for all } t \geq 0, \\ jv) & \displaystyle\int_s^t \mathbf{1}_{\{r:\, x(r) \in \mathcal{O}\}}\,(r)\, dk\,(r) = 0 \quad \text{and } k\,(t) = \int_0^t \mathbf{n}_r d\,|k|_r\,, \\ & \qquad\qquad\qquad\qquad \text{for all } 0 \leq s \leq t, \end{array} \tag{4.4}$$

where $\mathbf{n}_r$ is a unit outward normal to $\overline{\mathcal{O}}$ at $x\,(r)$ and $\updownarrow k \updownarrow_r$ stands for the total variation of k on $[0,r]$. Note that the problem (4.4) is equivalent to

$$\begin{cases} x\,(t) + k\,(t) = x_0 + m\,(t)\,,\ \text{for all } t \geq 0 \\ x\,(t) \in \overline{\mathcal{O}},\ \text{for all } t \geq 0, \\ \displaystyle\int_s^t \langle y - x(r), dk\,(r)\rangle \leq 0,\ \forall\, y \in \overline{\mathcal{O}},\ \forall\, 0 \leq s \leq t, \end{cases}$$

and by Proposition 6.35 from Annex B this means $dk\,(t) \in \partial I_{\overline{\mathcal{O}}}(x(t))\,(dt)$, where $I_{\overline{\mathcal{O}}} : \mathbb{R} \to]-\infty, +\infty]$ is the convex indicator function of $\overline{\mathcal{O}}$.

Hence we can write the problem (4.4) as the multivalued equation

$$\begin{cases} dx\,(t) + \partial I_{\overline{\mathcal{O}}}(x(t))\,(dt) \ni dm\,(t)\,, \\ x\,(0) = x_0, \end{cases}$$

with the solution defined in the sense of Skorohod's problem.

Since by Proposition 6.36 from Annex B

$$\begin{cases} x\,(t) \in \overline{\mathcal{O}},\ \text{for all } t \geq 0 \text{ and} \\ \displaystyle\int_s^t \langle y - x(r), dk\,(r)\rangle \leq 0,\ \text{for all } y \in \overline{\mathcal{O}},\ \text{for all } 0 \leq s \leq t, \end{cases}$$

is equivalent to

$$\int_0^T \langle v(t) - x(t), dk(t)\rangle + \int_0^T I_{\overline{\mathcal{O}}}(x(t))\,dt \le \int_0^T I_{\overline{\mathcal{O}}}(v(t))\,dt,$$
$$\text{for all } v \in C\left([0,\infty[\,;\mathbb{R}^d\right), \text{ for all } T > 0,$$

it is natural to consider the multivalued differential equation

$$\begin{cases} dx(t) + \partial\varphi(x(t))\,(dt) \ni dm(t), \\ x(0) = x_0 \in \mathrm{Dom}(\varphi), \end{cases} \tag{4.5}$$

where $\varphi : \mathbb{R}^d \to]-\infty, +\infty]$ is a proper convex lower-semicontinuous function and $\partial\varphi$ is the subdifferential of φ defining the solution as a pair (x, k) such that

$$\begin{array}{ll} j) & x \in C\left([0,\infty[\,;\mathbb{R}^d\right),\ x(0) = x_0, \\ jj) & k \in C\left([0,\infty[\,;\mathbb{R}^d\right) \cap BV_{loc}\left([0,\infty[\,;\mathbb{R}^d\right),\ k(0) = 0, \\ jjj) & x(t) + k(t) = x_0 + m(t),\ \text{for all } t \ge 0, \\ jv) & \displaystyle\int_0^T \langle v(t) - x(t), dk(t)\rangle + \int_0^T \varphi(x(t))\,dt \le \int_0^T \varphi(v(t))\,dt, \\ & \qquad\qquad \text{for all } v \in C\left([0,\infty[\,;\mathbb{R}^d\right). \end{array}$$

Now once again by Proposition 6.36 from Annex B, note that the last condition *jv*) is equivalent to:

$$\int_s^t \langle x(r) - z, dk(r) - \hat{z}dr\rangle \ge 0, \text{for all } (z, \hat{z}) \in \partial\varphi, \text{for all } 0 \le s \le t.$$

Then by the same generalization it is also natural to consider the multivalued differential equation

$$\begin{cases} dx(t) + A(x(t))\,(dt) \ni dm(t), \\ x(0) = x_0 \in \mathrm{Dom}(A), \end{cases}$$

where $A \subset \mathbb{R}^d \times \mathbb{R}^d$ is a maximal monotone operator and the solution is a pair (x, k) such that

$$\begin{array}{ll} j) & x \in C\left([0,\infty[\,;\mathbb{R}^d\right),\ x(0) = x_0, \\ jj) & k \in C\left([0,\infty[\,;\mathbb{R}^d\right) \cap BV_{loc}\left([0,\infty[\,;\mathbb{R}^d\right),\ k(0) = 0,\ \forall\, T > 0, \\ jjj) & x(t) + k(t) = x_0 + m(t),\ \forall\, t \ge 0, \\ jv) & \displaystyle\int_s^t \langle x(r) - z, dk(r) - \hat{z}dr\rangle \ge 0,\ \forall\ (z, \hat{z}) \in A,\ \forall\, 0 \le s \le t. \end{array}$$

These generalizations of Skorohod's problem permit us to give precise natural concepts of solutions of the stochastic equations (4.1) and (4.2).

We begin with the following natural definition:

The stochastic process $\{X_t, t > 0\}$ is a (strong) solution of the SDE (4.1) if X and K are $\mathbb{R}^d$-valued continuous progressively measurable processes with $K_0 = 0$ a.s. and $\mathbb{P}$-a.s.:

$$\begin{cases} i) & K_\cdot \in BV_{loc}\left([0,\infty[\,;\mathbb{R}^d\right), \\ ii) & X_t + K_t = \xi + \int_0^t F(s,X_s)ds + \int_0^t G(s,X_s)dB_s, \; \forall\, t \geq 0, \\ iii) & \int_s^t \langle Y_r - X_r, dK_r\rangle + \int_s^t \varphi(X_r)dr \leq \int_s^t \varphi(Y_r)\,dr, \; a.s., \\ & \qquad \text{for all } Y \text{ a } \mathcal{P}\text{-p.m.c.s.p., for all } 0 \leq s \leq t. \end{cases} \tag{4.6}$$

The condition (4.6-*iii*) can be written symbolically as

$$dK_t \in \partial\varphi(X_t)\,(dt)\,, \quad (\omega,t)\text{-}a.e.$$

Remark 4.1. The conditions (4.6-*ii*) and *iii*)) may be combined into a variational formulation, $\forall\, Y$ a $\mathcal{P}$-m.c.s.p., $\forall\, 0 \leq s \leq t$,

$$\int_s^t \left\langle Y_r - X_r, F(r,X_r)dr + d\left(\textstyle\int_0^r G(u,X_u)dB_u - X_r\right)\right\rangle + \int_s^t \varphi(X_r)\,dr \\ \leq \int_s^t \varphi(Y_r)\,dr.$$

This is why Eq. (4.1) will also be called a "stochastic variational inequality".

By Itô's formula for $|Y_t - X_t|^2$, with $Y \in W_{ad}$, where

$$W_{ad} \overset{def}{=} \left\{ Y \in S_d^2 : Y_t = Y_0 + \int_0^t \hat{Y}_s ds + \int_0^t \tilde{Y}_s dB_s, \quad \hat{Y} \in \Lambda_d^2, \tilde{Y} \in \Lambda_{d\times k}^2 \right\},$$

we arrive at the following variational-weak formulation of the solution: the stochastic process X_t is a variational-weak solution for (4.1) if $\forall\, T \geq 0$:

$$\begin{aligned} i) \quad & X \in \Lambda_d^2, \quad \mathbb{E}\textstyle\int_0^T |\varphi(X_t)|\,dt < \infty, \\ ii) \quad & \mathbb{E}\int_0^T \langle \hat{Y}_r - F(r,X_r), Y_r - X_r\rangle dr + \frac{1}{2}\mathbb{E}\int_0^T |\tilde{Y}_r - G(r,X_r)|^2 dr \\ & + \frac{1}{2}\mathbb{E}|Y_0 - \xi|^2 + \mathbb{E}\int_0^T \varphi(Y_r))dr \geq \mathbb{E}\int_0^T \varphi(X_r)dr, \;\; \forall Y \in W_{ad}\,. \end{aligned}$$

Concerning the SDE (4.2) with a maximal monotone operator in the drift, the solution is now naturally defined as an $\mathbb{R}^d$-valued continuous progressively measurable process $\{X_t, t > 0\}$ for which there exists an $\mathbb{R}^d$-valued continuous progressively measurable process K such that $\mathbb{P}$-a.s.:

$$\begin{cases} (i) \quad K_{\cdot} \in BV_{loc}\left(\mathbb{R}_+;\mathbb{R}^d\right), \quad K_0 = 0, \\ (ii) \quad X_t + K_t = \xi + \displaystyle\int_0^t F(s,X_s)ds + \int_0^t G(s,X_s)dB_s, \\ \qquad\qquad\qquad\qquad\qquad\qquad\qquad\qquad \forall\, t \geq 0, \\ (iii) \quad \displaystyle\int_s^t \langle X_r - u, dK_r - vdr\rangle \geq 0, \ \forall\ (u,v) \in A, \\ \qquad\qquad\qquad\qquad\qquad\qquad\qquad\qquad \forall\, 0 \leq s \leq t. \end{cases} \tag{4.7}$$

4.2 SDEs with a Maximal Monotone Operator in the Drift

4.2.1 Assumptions: Definitions

In the following two sections we shall study stochastic differential equations with a multivalued maximal monotone operator drift.

Let $\{B_t, t \geq 0\}$ be a k-dimensional Brownian motion with respect to a given complete stochastic basis $(\Omega, \mathcal{F}, P, \{\mathcal{F}_t\}_{t\geq 0})$. Let

$$A : \mathbb{R}^d \rightrightarrows \mathbb{R}^d, \ \ \mathrm{Dom}\,(A) = \left\{x \in \mathbb{R}^d : A\,(x) \neq \emptyset\right\}$$

be a (multivalued) operator and

$$F : \Omega \times [0,+\infty[\, \times \mathbb{R}^d \to \mathbb{R}^d, \ \ G : \Omega \times [0,+\infty[\, \times \mathbb{R}^d \to \mathbb{R}^{d\times k}.$$

Consider the stochastic differential equation

$$\begin{cases} dX_t + A(X_t)(dt) \ni F(t,X_t)dt + G(t,X_t)dB_t, \quad t \geq 0, \\ X_0 = \xi, \end{cases} \tag{4.8}$$

where we shall assume[1]

$$(\mathbf{MM\text{-}H}_A): \tag{4.9}$$

$$\begin{cases} i) \quad A : \mathbb{R}^d \rightrightarrows \mathbb{R}^d \textit{ is a maximal monotone operator}, \\ ii) \quad \mathrm{int}\,(\mathrm{Dom}\,(A)) \neq \emptyset. \end{cases}$$

□

From this assumption, ($\mathbf{MM\text{-}H}_A$) it follows that for every $u_0 \in \mathrm{int}\,(\mathrm{Dom}\,(A))$ there exists an $r_0 > 0$ such that $\bar{B}\,(u_0, r_0) \subset \mathrm{Dom}\,(A)$ and

$$A^{\#}_{u_0,r_0} \stackrel{def}{=} \sup\left\{|\hat{u}| : \hat{u} \in A\,(u_0 + u)\,,\ |u| \leq r_0\right\} < \infty.$$

[1] Note that in infinite dimension assumption $ii)$ is typically not satisfied, see e.g. [5, 8, 9] and [63].

In what follows we fix $u_0 \in \operatorname{int}(\operatorname{Dom}(A))$ and $r_0 \in]0,1]$ with these properties.

Definition 4.2. A pair (X, K) of $\mathbb{R}^d$-valued stochastic processes is a solution of the SDE (4.8) if the following conditions are satisfied:

$$\begin{aligned}
&(d_1): \quad X, K \in S_d^0, \quad K_0 = 0,\\
&(d_2): \quad X_t \in \overline{\operatorname{Dom}(A)}, \quad \forall\, t > 0, \ \mathbb{P}\text{-}a.s.,\\
&(d_3): \quad K \in BV_{loc}\left(\mathbb{R}_+; \mathbb{R}^d\right), \quad \mathbb{P}\text{-a.s.},\\
&(d_4): \quad F(\cdot, X_\cdot) \in L^1_{loc}\left(\mathbb{R}_+; \mathbb{R}^d\right) \ a.s., \quad G(\cdot, X_\cdot) \in \Lambda_d^0,\\
&(d_5): \quad X_t + K_t = \xi + \int_0^t F(s, X_s)ds + \int_0^t G(s, X_s)dB_s, \quad \forall t \geq 0, \ \mathbb{P}\text{-a.s.},\\
&(d_6): \quad \int_s^t \langle X_r - u, dK_r - \hat{u}dr\rangle \geq 0, \ \mathbb{P}\text{-}a.s., \quad \forall\, (u, \hat{u}) \in A, \quad \forall\, 0 \leq s \leq t.
\end{aligned}$$

Notation 4.3. *The notation $dK_t \in A(X_t)(dt)$ will be used to say that, $\mathbb{P}$-a.s.*

$$\begin{aligned}
&(a_1) \quad X \in C\left(\mathbb{R}_+; \overline{\operatorname{Dom}(A)}\right),\\
&(a_2) \quad K \in C\left(\mathbb{R}_+; \mathbb{R}^d\right) \bigcap BV_{loc}\left(\mathbb{R}_+; \mathbb{R}^d\right), \quad K_0 = 0,\\
&(a_3) \quad \langle X_t - u, dK_t - \hat{u}dt\rangle \geq 0, \quad \forall\, (u, \hat{u}) \in A.
\end{aligned}$$

The SDE (4.8) will also be written in the form

$$\begin{cases} X_t + K_t = \xi + \displaystyle\int_0^t F(s, X_s)ds + \int_0^t G(s, X_s)dB_s, \quad \forall\, t \geq 0,\\ dK_t \in A(X_t)(dt). \end{cases}$$

We highlight that by Proposition 6.17 the condition (d_6) from Definition 4.2 is equivalent to

$$\int_s^t \langle X(r) - u(r), dK(r) - \hat{u}(r)dr\rangle \geq 0, \quad \forall\, u, \hat{u} \in C([0,T]; \mathbb{R}^d),$$
$$(u(r), \hat{u}(r)) \in A, \ \forall\, r \in [0,T], \ \forall\, 0 \leq s \leq t \leq T.$$

Remark 4.4. If (X, K) and $(\tilde{X}, \tilde{K})$ satisfy (d_6) then from Proposition 6.17 from Annex B

$$\int_s^t \langle X_r - \tilde{X}_r, dK_r - d\tilde{K}_r\rangle \geq 0, \quad \forall\, 0 \leq s \leq t, \quad \mathbb{P}\text{-a.s.} \tag{4.10}$$

Also by Proposition 6.19 from Annex B we have:

Remark 4.5. If $y \in C\left([0,T]; \mathbb{R}^d\right)$, $0 < \varepsilon \leq 1$ and $dK_t \in A(X_t)(dt)$, then for $\bar{B}(u_0, r_0) \subset \operatorname{Dom}(A)$, $0 < r_0 \leq 1$, we have the following comparison between signed measures on $[0, \infty[$

$$r_0 d \updownarrow K \updownarrow_t \leq \langle X_t - u_0, dK_t\rangle + \left(A^{\#}_{u_0,r_0} |X_t - u_0| + A^{\#}_{u_0,r_0}\right) dt, \ \mathbb{P}\text{-a.s.} \tag{4.11}$$

and there exists a $b_0 > 0$ such that, with A_ε the Yosida approximation of A, which is defined just after Proposition 6.3 in Annex B,

$$r_0 |A_\varepsilon y(t)| dt \le \langle y(t) - u_0, A_\varepsilon y(t)\rangle dt + (A^{\#}_{u_0,r_0} |y(t) - u_0| + b_0) dt. \quad (4.12)$$

Notation 4.6. *We introduce the following notation. For $u \in \mathbb{R}^d$ and $R \ge 0$*

$$F^{\#}_{u,R}(t) \overset{def}{=} \sup\{|F(t,u+x)| : |x| \le R\}, \quad F^{\#}_R(t) \overset{def}{=} F^{\#}_{0,R}(t).$$

The basic assumptions on F and G under which we shall study the multivalued stochastic equation (4.8) are the same as in Chap. 3, Sect. 3.5. For convenience we recall them.

$$(\mathbf{MM\text{-}H}_F): \quad (4.13)$$

- *The functions $F(\cdot,\cdot,x) : \Omega \times [0,+\infty[\to \mathbb{R}^d$ and $G(\cdot,\cdot,x) : \Omega \times [0,+\infty[\to \mathbb{R}^{d\times k}$ are progressively measurable stochastic processes for every $x \in \mathbb{R}^d$.*
- *There exist $\mu \in L^1_{loc}(0,\infty)$ and $\ell \in L^2_{loc}(0,\infty;\mathbb{R}_+)$*

such that $d\mathbb{P} \otimes dt$-a.e.:

$$\begin{cases} & \textit{Continuity:} \\ (\mathbf{C}_F): & x \longrightarrow F(t,x) : \mathbb{R}^d \to \mathbb{R}^d \textit{ is continuous}, \\ & \textit{Monotonicity condition:} \\ (\mathbf{M}_F): & \langle x-y, F(t,x) - F(t,y)\rangle \le \mu(t)|x-y|^2, \quad \forall x,y \in \mathbb{R}^d, \\ & \textit{Boundedness condition:} \\ (\mathbf{B}_F): & \displaystyle\int_0^T F^{\#}_R(s)\,ds < \infty, \ \mathbb{P}\text{-a.s.} \quad \text{for all } R,T \ge 0. \end{cases}$$

□

and

$$(\mathbf{MM\text{-}H}_G). \quad (4.14)$$

$$\begin{cases} & \textit{Lipschitz condition:} \\ (\mathbf{L}_G): & |G(t,x) - G(t,y)| \le \ell(t)|x-y|, \quad \forall x,y \in \mathbb{R}^d, \\ & \textit{Boundedness condition:} \\ (\mathbf{B}_g): & \displaystyle\int_0^T |G(t,0)|^2 dt < \infty, \quad \mathbb{P}\text{-a.s.} \end{cases}$$

□

Clearly $(\mathbf{MM\text{-}H}_F)$ and $(\mathbf{MM\text{-}H}_G)$ yield $F(\cdot,\cdot,X.) \in L^1_{loc}(\mathbb{R}_+;\mathbb{R}^d)$ *a.s.*, $G(\cdot,\cdot,X.) \in \Lambda^0_d$ for all $X \in S^0_d$.

The monotonicity property of F implies that

$$\langle F(t,x), x-u_0\rangle \le \langle F(t,u_0), x-u_0\rangle + \mu(t)|x-u_0|^2.$$

Using (4.11) we deduce that for all (X,K) such that $dK_t \in A\,(X_t)\,(dt)$,

$$\begin{aligned} r_0 d\,\updownarrow K\updownarrow_t &\leq \langle X_t - u_0, dK_t - F\,(t,X_t)\,dt\rangle \\ &+ \left[A^{\#}_{u_0,r_0} + \left(A^{\#}_{u_0,r_0} + |F\,(t,u_0)|\right)|X_t - u_0| + \mu\,(t)\,|X_t - u_0|^2\right]dt, \end{aligned} \tag{4.15}$$

and for all $y \in C\left([0,T]\,;\mathbb{R}^d\right)$, $0 < \varepsilon \leq 1$,

$$\begin{aligned} r_0\,|A_\varepsilon y\,(t)| &\leq \langle y\,(t) - u_0, A_\varepsilon y\,(t) - F\,(t, y\,(t))\rangle \\ &+ \left[b_0 + \left(A^{\#}_{u_0,r_0} + |F\,(t,u_0)|\right)|y\,(t) - u_0| + \mu\,(t)\,|y\,(t) - u_0|^2\right]. \end{aligned} \tag{4.16}$$

In the remaining of this subsection, we discuss the particular case $A = \partial\varphi$ where $\varphi : \mathbb{R}^d \to]-\infty, +\infty]$ is a convex lower semicontinuous function then we can write the assumption ((4.9)-$\mathbf{MM\text{-}H}_A$) in the form

$$(\mathbf{MM\text{-}H}_\varphi): \tag{4.17}$$

$$\begin{cases} i) & \varphi : \mathbb{R}^d \to]-\infty, +\infty] \text{ is a convex l.s.c. function,} \\ ii) & \text{int}(\text{Dom}(\varphi)) \neq \emptyset, \end{cases}$$

□

since

$$\text{int}(\text{Dom}(\varphi)) = \text{int}\,(\text{Dom}\,(\partial\varphi))\,.$$

Recall that $\text{Dom}(\varphi) \overset{def}{=} \{x \in \mathbb{R}^d : \varphi(x) < +\infty\}$ and the subdifferential of the function φ at x is defined by:

$$\partial\varphi(x) = \{y \in \mathbb{R}^d : \langle y, z - x\rangle + \varphi(x) \leq \varphi(z),\ \forall\, z \in \mathbb{R}^d\}.$$

By Proposition 6.36, Annex B, the condition (d_6) from Definition 4.2 is equivalent to each of the following equivalent conditions

$$\begin{aligned} &(d_6') && \int_s^t \langle z - X_r, dK_r\rangle + \int_s^t \varphi(X_r)dr \leq (t-s)\varphi(z), \quad \mathbb{P}\text{-a.s.}, \\ &&& \qquad \forall z \in \mathbb{R}^d, \quad \forall\, 0 \leq s \leq t, \\ &(d_6'') && \int_s^t \langle y(r) - X_r, dK_r\rangle + \int_s^t \varphi(X_r)dr \leq \int_s^t \varphi(y(r))dr, \quad \mathbb{P}\text{-a.s.}, \\ &&& \qquad \forall\, y \in C([0,T];\mathbb{R}^d), \ \forall\, 0 \leq s \leq t \leq T, \\ &(d_6''') && \int_s^t \langle X_r - u,\, dK_r - \hat{u}dr\rangle \geq 0, \quad \mathbb{P}\text{-a.s.}, \\ &&& \qquad \forall\ (u,\hat{u}) \in \partial\varphi, \quad \forall\, 0 \leq s \leq t. \end{aligned} \tag{4.18}$$

Since $\overline{\text{Dom}(\partial\varphi)} = \overline{\text{Dom}(\varphi)}$, we can replace $\xi \in L^0(\Omega, \mathcal{F}_0, P; \overline{\text{Dom}\,(A)})$ by $\xi \in L^0(\Omega, \mathcal{F}_0, P; \overline{\text{Dom}(\varphi)})$ and using (d_6') we can replace (d_2) by $\mathbb{P}$-a.s.:

$$X_t \in \text{Dom}(\varphi),\ a.e.\ t > 0 \text{ and } \varphi(X) \in L^1_{loc}(0,\infty).$$

Hence we make the following definition:

Definition 4.7. A pair (X, K) of $\mathbb{R}^d$-valued stochastic processes is a solution of

$$\begin{cases} dX_t + \partial\varphi(X_t)(dt) \ni F(t, X_t)dt + G(t, X_t)dB_t, & t \geq 0, \\ X_0 = \xi, \end{cases} \tag{4.19}$$

if the following conditions are satisfied, $\mathbb{P}$-a.s.:

$$\begin{cases} d_1) & X, K \in S_d^0 \quad K_0 = 0, \\ d_2) & X_t \in \text{Dom}(\varphi),\ a.e.\ t > 0 \text{ and } \varphi(X) \in L^1_{loc}(0,\infty), \\ d_3) & \updownarrow K \updownarrow_T < \infty,\ \forall T > 0, \\ d_4) & X_t + K_t = \xi + \displaystyle\int_0^t F(s, X_s)ds + \int_0^t G(s, X_s)dB_s,\ \forall t \geq 0, \\ d_5) & \displaystyle\int_s^t \langle z - X_r, dK_r\rangle + \int_s^t \varphi(X_r)dr \leq (t-s)\varphi(z), \\ & \qquad\qquad\qquad \forall z \in \mathbb{R}^d, \quad \forall\, 0 \leq s \leq t. \end{cases} \tag{4.20}$$

The stochastic differential equation (4.19) is also called a *stochastic variational inequality*.

4.2.2 A Priori Estimates: Uniqueness

If (X, K) is a solution of the Eq. (4.8) (or (4.19)) then

$$(X_t - u_0) = (\xi - u_0) + \mathcal{K}_t + \int_0^t G_r dB_r,$$

where

$$\mathcal{K}_t = \int_0^t F(r, X_r)\, dr - K_t \quad \text{and} \quad G_r = G(r, X_r).$$

For all $\lambda > 1$

$$|G(t,x)|^2 \leq \frac{\lambda}{\lambda - 1}|G(t,u_0)|^2 + \lambda\ell^2(t)\,|x - u_0|^2.$$

Combining this with (4.15) we deduce that for all $p \geq 2$ and $\lambda > 1$:

$$\begin{aligned} dD_r + \langle X_r - u_0, d\mathcal{K}_r\rangle + \left(\frac{m_p}{2} + 9p\lambda\right)|G_r|^2\,dr \\ \leq dR_r + |X_r - u_0|dN_r + |X_r - u_0|^2 dV_r, \end{aligned} \tag{4.21}$$

where $m_p = 1 \vee (p-1) = p-1$,

$$\begin{aligned} D_t &= \ r_0 \updownarrow K \updownarrow_t, \\ R_t &= \int_0^t \left[A^\#_{u_0,r_0} + \frac{\lambda}{\lambda-1}\left(\frac{p-1}{2} + 9p\lambda\right)|G(r,u_0)|^2\right]dr, \\ N_t &= \int_0^t \left[A^\#_{u_0,r_0} + |F(r,u_0)|\right]dr, \\ V(t) &= \int_0^t \left[\mu(r) + \lambda\left(\frac{p-1}{2} + 9p\lambda\right)\ell^2(r)\right]dr. \end{aligned} \tag{4.22}$$

By Proposition 3.3 we deduce the following:

Proposition 4.8. *Let* $\xi \in L^0(\Omega, \mathcal{F}_0, P; \overline{\mathrm{Dom}(A)})$, *the assumptions ((4.9)-**MM-H**$_A$), ((4.13)-**MM-H**$_F$), ((4.14)-**MM-H**$_G$) be satisfied and* (X, K) *be a solution of the SDE (4.8). Let* $p \geq 2$, $\lambda > 1$ *be arbitrary and* V *defined as in (4.22). Then there exists a constant* $C_{p,\lambda}$ *such that for all* $0 \leq t \leq s$:

$$\begin{aligned} &\mathbb{E}^{\mathcal{F}_t}\left\|e^{-V}(X-u_0)\right\|^p_{[t,s]} + r_0\mathbb{E}^{\mathcal{F}_t}\int_t^s e^{-pV(r)}|X_r - u_0|^{p-2}\,d\updownarrow K\updownarrow_r \\ &\qquad\qquad + r_0^{p/2}\mathbb{E}^{\mathcal{F}_t}\left(\int_t^s e^{-2V(r)}d\updownarrow K\updownarrow_r\right)^{p/2} \\ &\leq C_{p,\lambda}\Bigg[e^{-pV(t)}|X_t - u_0|^p + \mathbb{E}^{\mathcal{F}_t}\left(\int_t^s e^{-2V(r)}\left[A^\#_{u_0,r_0} + |G(r,u_0)|^2\right]dr\right)^{p/2} \\ &\qquad\qquad + \mathbb{E}^{\mathcal{F}_t}\left(\int_t^s e^{-V(r)}\left[A^\#_{u_0,r_0} + |F(r,u_0)|\right]dr\right)^p\Bigg] \quad a.s. \end{aligned} \tag{4.23}$$

for all $u_0 \in \mathrm{int}(\mathrm{Dom}(A))$ *and* $r_0 > 0$ *such that* $A^\#_{u_0,r_0} < \infty$.

Since the monotonicity condition of F remains valid with μ replaced by μ^+, we can replace in (4.23) V by

$$\hat{V}(t) = \int_0^t \left[\mu^+(r) + \lambda\left(\frac{p-1}{2} + 9p\lambda\right)\ell^2(r)\right]dr.$$

We then deduce from (4.23), choosing $\lambda = 2$:

Corollary 4.9. *For every* $p \geq 2$ *there exists a constant* C_p *(depending only on* p*) such that*

$$\begin{aligned}
&\mathbb{E}\sup_{t\in[0,T]}|X_t-u_0|^p+r_0^{p/2}\mathbb{E}\updownarrow K\updownarrow_T^{p/2}\\
&\leq\Big[\mathbb{E}|\xi-u_0|^p+\mathbb{E}\Big(\int_0^T\left(A^\#_{u_0,r_0}+|F(t,u_0)|\right)dt\Big)^p\\
&\qquad+\mathbb{E}\Big(\int_0^T\left(A^\#_{u_0,r_0}+|G(t,u_0)|^2\right)dt\Big)^{p/2}\Big]\\
&\qquad\times\exp\Big\{C_p+C_p\textstyle\int_0^T\left(\mu^+(t)+\ell^2(t)\right)dt\Big\}.
\end{aligned}\tag{4.24}$$

In particular if

$$\mathbb{E}|\xi|^p+\mathbb{E}\left(\int_0^T|F(t,u_0)|\,dt\right)^p+\mathbb{E}\left(\int_0^T|G(t,u_0)|^2dt\right)^{p/2}<+\infty,$$

then $X\in S_d^p$, $K\in S_d^{p/2}$ *and* $\mathbb{E}\updownarrow K\updownarrow_T^{p/2}<\infty$.

Consider the particular case $A=\partial\varphi$. Let $(u_0,\hat{u}_0)\in\partial\varphi$, $0<r_0\leq 1$ and $\bar{B}(u_0,r_0)\subset\mathrm{Dom}(\varphi)$. Define

$$\varphi^\#_{u_0,r_0}\overset{def}{=}\sup\{\varphi(u_0+r_0v):|v|\leq 1\}$$

and observe that $\left|\varphi^\#_{u_0,r_0}\right|<\infty$. We have:

Corollary 4.10. *Let the assumptions of Proposition 4.8 be satisfied and suppose that* $A=\partial\varphi$, *where* φ *satisfies ((4.17)-**MM-H**$_\varphi$). Then*

$$\begin{aligned}
&\mathbb{E}\left(\int_0^T|\varphi(X_s)|\,ds\right)^{p/2}\leq\Big[1+\mathbb{E}|\xi-u_0|^p+\mathbb{E}\Big(\int_0^T|F(t,u_0)|\,dt\Big)^p\\
&+\mathbb{E}\Big(\int_0^T|G(t,u_0)|^2dt\Big)^{p/2}\Big]\times\exp\Big\{C+C\textstyle\int_0^T\left(\mu^+(t)+\ell^2(t)\right)dt\Big\},
\end{aligned}\tag{4.25}$$

where $C=C(p,T,u_0,r_0,\hat{u}_0)\geq 0$.

Proof. By (6.40) from Annex B we have

$$\begin{aligned}
r_0d\updownarrow K\updownarrow_t+|\varphi(X_t)-\varphi(u_0)|\,dt&\leq\langle X_t-u_0,dK_t\rangle+2|\hat{u}_0|\,|x(t)-u_0|\,dt\\
&\quad+\left(\varphi^\#_{u_0,r_0}-\varphi(u_0)\right)dt.
\end{aligned}$$

In this case the inequality (4.21) is satisfied with D,R,N,V as defined in (4.22), where $r_0\updownarrow K\updownarrow_t$ is replaced by $r_0\updownarrow K\updownarrow_t+\displaystyle\int_0^t|\varphi(x(r))-\varphi(u_0)|\,dr$ and $A^\#_{u_0,r_0}$ is replaced by $2|\hat{u}_0|\vee\left(\varphi^\#_{u_0,r_0}-\varphi(u_0)\right)$. Hence the inequality (4.25) follows from (4.24). ■

We now prove the uniqueness of the solution to SDE (4.8). The result can be applied to Eq. (4.19) as a particular case. We again write

$$m_p = 1 \vee (p-1).$$

Theorem 4.11 (Uniqueness). *Let the assumptions ((4.9)-**MM-H**$_A$), ((4.13)-**MM-H**$_F$), ((4.14)-**MM-H**$_G$) be satisfied and (X,K) and $(\hat{X},\hat{K})$ be two solutions of the SDE (4.8) corresponding to the initial conditions*

$$\xi, \hat{\xi} \in L^0(\Omega, \mathcal{F}_0, P; \overline{\mathrm{Dom}\,(A)}),$$

respectively. Let $p \geq 1$ and $\lambda > 1$ be arbitrary.

(I) With

$$V_t = \int_0^t \left(\mu(r) + \frac{1}{2} m_p \ell^2(r)\right) dr,$$

for all $\delta \geq 0$, $0 \leq t \leq s$, $\mathbb{P}$-a.s.:

$$\mathbb{E}^{\mathcal{F}_t} \frac{e^{-pV_s}|X_s - \hat{X}_s|^p}{\left(1+\delta e^{-2V_s}|X_s-\hat{X}_s|^2\right)^{p/2}} \leq \frac{e^{-pV_t}|X_t-\hat{X}_t|^p}{\left(1+\delta e^{-2V_t}|X_t-\hat{X}_t|^2\right)^{p/2}}. \tag{4.26}$$

(II) Moreover with

$$V_t = \int_0^t \left[\mu(r) + \left(\frac{1}{2} m_p + 9p\lambda\right) \ell^2(r)\right] dr,$$

there exists a constant $C_{p,\lambda}$ depending only on (p,λ) such that for all $\delta \geq 0$, $0 \leq t \leq s$:

$$\mathbb{E}^{\mathcal{F}_t} \frac{\|e^{-V}(X-\hat{X})\|_{[t,s]}^p}{\left(1+\delta\|e^{-V}(X-\hat{X})\|_{[t,s]}^2\right)^{p/2}} \leq C_{p,\lambda} \frac{|e^{-V_t}(X_t-\hat{X}_t)|^p}{\left(1+\delta|e^{-V_t}(X_t-\hat{X}_t)|^2\right)^{p/2}}, \quad a.s. \tag{4.27}$$

The uniqueness of X in S_d^0 follows from (4.27) by choosing $t = 0$, $\delta > 0$; the uniqueness of K also clearly follows.

Proof. We have

$$X_t - \hat{X}_t = (\xi - \hat{\xi}) + \int_0^t d\mathcal{K}_r + \int_0^t G_r dB_r,$$

where

$$\mathcal{K}_t = -\left(K_t - \hat{K}_t\right) + \int_0^t \left[F(r,X_r) - F(r,\hat{X}_r)\right] dr,$$

$$G_r = G(r,X_r) - G(r,\hat{X}_r).$$

In view of the assumptions (4.9-**MM-H**$_A$), (**MM-H**$_F$) and (**MM-H**$_G$), for all $p \geq 1$ and $\gamma \geq 0$:

$$\begin{aligned}
&\langle X_r - \hat{X}_r, d\mathcal{K}_r \rangle + \left(\frac{1}{2}m_p + 9p\gamma\right) |G_r|^2 \, dr \\
&\quad \leq \left|X_r - \hat{X}_r\right|^2 \left[\mu_r dr + \left(\frac{1}{2}m_p + 9p\gamma\right) (\ell_r)^2 \, dr\right].
\end{aligned}$$

Hence, by Proposition 3.3 (or by Corollary 6.77 from Annex C) the inequalities (4.26) and (4.27) follow. ∎

Proposition 4.12. *Let the assumptions ((4.9)-**MM-H**$_A$), ((4.13)-**MM-H**$_F$), ((4.14)-**MM-H**$_G$) be satisfied. Let*

$$\xi^n \in L^0(\Omega, \mathcal{F}_0, \mathbb{P}; \overline{\mathrm{Dom}\,(A)}), \quad n \in \mathbb{N}^*,$$

and (X^n, K^n) be a solution of the Eq. (4.8) corresponding to an initial condition ξ^n. If

$$\xi^n \to \xi \quad in\ L^0(\Omega, \mathcal{F}_0, \mathbb{P}; \mathbb{R}^d), \quad as\ n \to \infty,$$

and there exists $p > 0$ such that for all $T > 0$

$$\sup_{n \geq 1} \mathbb{E}\ \updownarrow K^n \updownarrow_T^p < \infty,$$

then there exists a solution $(X, K) \in S_d^0 \times S_d^0$ of the Eq. (4.8) and for every $T > 0$: $\mathbb{E}\ \updownarrow K \updownarrow_T^p < \infty$,

$$X^n \to X \text{ and } K^n \to K \quad in\ S_d^0\,[0,T]\,, \quad as\ n \to \infty.$$

Proof. From (4.27) with $s = T$ and $t = 0$ we obtain that there exists $X \in S_d^0\,[0,T]$ such that $X^n \to X$ in $S_d^0\,[0,T]$. Using the continuity and the boundedness properties of F and G it is easy to prove that all terms on the right-hand side of

$$X_t^n + K_t^n = \xi^n + \int_0^t F\left(s, X_s^n\right) ds + \int_0^t G\left(s, X_s^n\right) dW\left(s\right)$$

converge (see e.g. Proposition 6.9 and (2.5)). Hence

$$K^n \to K \quad \text{in } S_d^0\,[0,T]\,,$$

where

$$X_t + K_t = \xi + \int_0^t F\left(s, X_s\right) ds + \int_0^t G\left(s, X_s\right) dW\left(s\right).$$

By Proposition 1.20 we infer $\mathbb{E}\ \updownarrow K \updownarrow_T^p < \infty$. Since $dK_t^n \in A\left(X_t^n\right) dt$, we obtain, by Corollary 1.22, $dK_t \in A\left(X_t\right) dt$. The proof is complete. ■

4.2.3 The Generalized Convex Skorohod Problem

The aim of this subsection is to generalize the Skorohod problem on convex sets to a singular deterministic differential equation with a maximal monotone operator of the form

$$(GSP): \begin{cases} dx(t) + Ax(t)\,(dt) \ni f(t, x(t))dt + dm(t), & \\ x(0) = x_0, & t \in [0, T]. \end{cases} \tag{4.28}$$

We recall the assumptions ((4.9)-**MM-H**$_A$) and $\left(\textbf{SDE-H}_f\right)$ for the convenience of the reader:

$$(\mathbf{MM\text{-}H}_A): \tag{4.29}$$

$$\begin{cases} i) & A : \mathbb{R}^d \rightrightarrows \mathbb{R}^d \textit{ is a maximal monotone operator,} \\ ii) & \operatorname{int} D(A) \neq \emptyset, \end{cases}$$

□

and

$$(\mathbf{MM\text{-}H}_f): \tag{4.30}$$

- ♦ $f(\cdot, x) : [0, +\infty[\to \mathbb{R}^d$ *is measurable for all* $x \in \mathbb{R}^d$,
- ♦ *there exists a* $\mu \in L^1_{loc}(0, \infty)$ *such that a.e.* $t \geq 0$:

$$\begin{cases} & \textit{Continuity:} \\ (\mathbf{C}_f): & u \longrightarrow f(t, u) : \mathbb{R}^d \to \mathbb{R}^d \textit{ is continuous,} \\ & \textit{Monotonicity (right-side Lipschitz) condition:} \\ (\mathbf{M}_f): & \langle x - y, f(t, x) - f(t, y)\rangle \leq \mu(t)\,|x - y|^2,\ \forall\, x, y \in \mathbb{R}^d, \\ & \textit{Boundedness and continuity conditions:} \\ & \textit{for all } R, T \geq 0: \\ (\mathbf{B}_f): & \quad a)\ \ \displaystyle\int_0^T f_R^{\#}(s)\, ds < \infty \quad \textit{and} \\ & \quad b)\ \ \lim_{\delta \to 0} \boldsymbol{\mu}_T(\delta; f, R) = 0, \end{cases}$$

□

where

$$f_{u,R}^{\#}(t) \overset{def}{=} \sup\{|f(t,u+x)| : |x| \leq R\} \text{ and } f_R^{\#}(t) \overset{def}{=} f_{0,R}^{\#}(t).$$

and

$$\mu_T(\delta; f, R) \overset{def}{=} \sup\left\{\int_0^T |f(s+\delta, y(s)) - f(s, y(s))|\, ds : \right.$$
$$\left. y \in C\left([0,T];\mathbb{R}^d\right),\ \|y\|_T \leq R\right\}.$$

Concerning x_0 and m we assume

$$(\mathbf{MM\text{-}H}_{0,m}): \tag{4.31}$$

$$\begin{cases} (i) & x_0 \in \overline{D(A)}, \\ (ii) & m : [0,\infty[\to \mathbb{R}^d \text{ is continuous and } m(0) = 0. \end{cases}$$

□

In the sequel we fix arbitrary $u_0 \in \mathbb{R}^d$ and $0 < r_0 \leq 1$ such that

$$\bar{B}(u_0, r_0) \subset \text{Dom}(A),$$

and we note that

$$A_{u_0,r_0}^{\#} \overset{def}{=} \sup\{|\hat{u}| : \hat{u} \in A(u_0+v),\ |v| \leq r_0\} < \infty$$

and if $u_0 = 0$ then $A_{r_0}^{\#} \overset{def}{=} A_{0,r_0}^{\#}$.

Definition 4.13 (Generalized Convex Skorohod Problem). *A pair of functions (x,k) is a solution of the generalized Skorohod problem (GSP) (we shall write $(x,k) = \mathcal{GS}(A; x_0, f, m)$ or $x = \mathcal{GS}(A; x_0, f, m)$) if the following conditions hold:*

$$\begin{array}{ll} i) & x, k : [0,\infty[\to \mathbb{R}^d \text{ are continuous, } x(0) = x_0,\ k(0) = 0, \\ ii) & x(t) \in \overline{D(A)},\ \ \forall\, t \geq 0,\ \ k \in BV_{loc}\left([0,\infty[;\mathbb{R}^d\right), \\ iii) & x(t) + k(t) = x_0 + \displaystyle\int_0^t f(s, x(s))\, ds + m(t),\ \ \forall\, t \geq 0, \\ iv) & \displaystyle\int_s^t \langle x(r) - z, dk(r) - \hat{z}dr\rangle \geq 0,\ \ \forall\, (z,\hat{z}) \in A,\ \forall\, 0 \leq s \leq t. \end{array} \tag{4.32}$$

We remark that the classical Skorohod problem corresponds to the case $f = 0$ and

$$A = \partial I_{\overline{\mathcal{O}}}(x) = \begin{cases} 0, & \text{if } x \in \mathcal{O}, \\ \mathcal{N}_{\overline{\mathcal{O}}}(x), & \text{if } x \in \text{Bd}\left(\overline{\mathcal{O}}\right), \\ \emptyset, & \text{if } x \in \mathbb{R}^d \setminus \overline{\mathcal{O}}, \end{cases}$$

where $\mathcal{O}$ is an open convex subset of $\mathbb{R}^d$ and $\mathcal{N}_{\overline{\mathcal{O}}}(x)$ is the outward normal cone to $\overline{\mathcal{O}}$ at $x \in \text{Bd}\left(\overline{\mathcal{O}}\right)$. As in the classical case, in the generalized convex Skorohod problem the solution is constrained with the help of a bounded variation feedback law $k(\cdot)$ to stay in a convex set, here $\overline{D(A)}$, but moreover we add a dynamic evolution governed by the maximal operator A in $\overline{D(A)}$.

As we noted in Remark 6.18, we write $dk(t) \in A(x(t))\,dt$ when (x,k) satisfies the conditions (4.32-*i*, *ii*, *iv*). Then by Proposition 6.17 we have:

Remark 4.14. If $dk(t) \in A(x(t))\,dt$ and $d\tilde{k}(t) \in A(\tilde{x}(t))\,dt$, then

$$\int_s^t \left\langle x(r) - \tilde{x}(r), d\,k(r) - d\,\tilde{k}(r)\right\rangle \geq 0, \quad \forall\, 0 \leq s \leq t. \tag{4.33}$$

Recall that $u_0 \in \text{Dom}(A)$ and $r_0 \in]0,1]$ are fixed, such that $\bar{B}(u_0, r_0) \subset \text{Dom}(A)$ and

$$A^{\#}_{u_0,r_0} \overset{def}{=} \sup\{|\hat{u}| : \hat{u} \in A(u_0 + u),\ |u| \leq r_0\} < \infty.$$

Also recall the notation $f^{\#}_{u_0,r_0}(t) \overset{def}{=} \sup\{|f(t,u_0+x)| : |x| \leq r_0\}$.

The monotonicity property of f implies that for all $|v| \leq 1$:

$$\langle f(t, u_0 + r_0 v) - f(t,x), u_0 + r_0 v - x\rangle \leq \mu(t)\,|u_0 + r_0 v - x|^2,$$

and, consequently, $\forall\ |v| \leq 1$:

$$\begin{aligned}
&r_0 \langle f(t,x), -v\rangle + \langle f(t,x), x - u_0\rangle \\
&\quad \leq \mu(t)\,|u_0 + r_0 v - x|^2 + |f(t, u_0 + r_0 v)|\,|x - r_0 v - u_0| \\
&\quad \leq \mu(t)\left[|x-u_0|^2 - 2r_0\langle v, x - u_0\rangle + r_0^2\,|v|^2\right] + f^{\#}_{u_0,r_0}(t)\,(|x-u_0| + r_0), \\
&\quad \leq r_0\left[f^{\#}_{u_0,r_0}(t) + r_0\mu^+(t)\right] + \left[f^{\#}_{u_0,r_0}(t) + 2r_0\,|\mu(t)|\right]|x-u_0| + \mu(t)\,|x-u_0|^2.
\end{aligned}$$

Taking $\sup_{|v|\leq 1}$, then

$$\begin{aligned}
r_0\,|f(t,x)| + \langle f(t,x), x-u_0\rangle &\leq r_0\left[f^{\#}_{u_0,r_0}(t) + r_0\mu^+(t)\right] \\
&\quad + \left[f^{\#}_{u_0,r_0}(t) + 2r_0\,|\mu(t)|\right]|x-u_0| + \mu(t)\,|x-u_0|^2.
\end{aligned}$$

Combining this inequality with those from Remark 4.5 we deduce:

Remark 4.15. Under the assumptions ((4.29)-**MM-H**$_A$) and ((4.30)-**MM-H**$_f$) if (x,k) satisfies (4.32-*iv*), then as signed measures on $[0,\infty[$

$$\begin{aligned}r_0 d \updownarrow k \updownarrow_t + r_0 |f(t, x(t))| dt \le \langle x(t) - u_0, dk(t) - f(t, x(t)) dt \rangle \\ + \left[\gamma(t) + \alpha(t) |x(t) - u_0| + \mu(t) |x(t) - u_0|^2 \right] dt\end{aligned} \tag{4.34}$$

and for all $y \in C([0, T]$ (again A_ε is the Yosida approximation of A); $\mathbb{R}^d)$, $0 < \varepsilon \le 1$:

$$\begin{aligned}r_0 |A_\varepsilon y(t)| dt + r_0 |f(t, y(t))| dt \le \langle y(t) - u_0, A_\varepsilon y(t) - f(t, y(t)) \rangle dt \\ + \left[\hat{\gamma}(t) + \alpha(t) |y(t) - u_0| + \mu(t) |y(t) - u_0|^2 \right] dt,\end{aligned}$$

where

$$\begin{aligned}\alpha(t) &= A^\#_{u_0,r_0} + f^\#_{u_0,r_0}(t) + 2|\mu(t)|, \\ \gamma(t) &= A^\#_{u_0,r_0} + f^\#_{u_0,r_0}(t) + \mu^+(t), \\ \hat{\gamma}(t) &= b_0 + f^\#_{u_0,r_0}(t) + \mu^+(t), \\ \alpha, \gamma, \hat{\gamma} &\in L^1_{loc}(0, \infty),\end{aligned}$$

and $b_0 = A^\#_{u_0,r_0}$ if $0 \in Au_0$.

Recall from Annex B that if $y : [0, \infty[\to \mathbb{R}^d$ is a continuous function we define

$$\begin{aligned}\mathbf{m}_{y,T}(\delta) &= \mathbf{m}_y(\delta; [0, T]) = \sup \{|y(t) - y(s)| : |t - s| \le \delta,\ t, s \in [0, T]\}, \\ \boldsymbol{\mu}_y(\delta) &= \delta + \mathbf{m}_{y;T}(\delta).\end{aligned}$$

If $\mathcal{M}$ is a bounded and equicontinuous subset (relatively compact subset) of $C([0, T]; \mathbb{R}^d)$, we define

$$\begin{aligned}\|\mathcal{M}\|_T &= \sup \{\|y\|_T : y \in \mathcal{M}\}, \\ \mathbf{m}_{\mathcal{M},T}(\delta) &= \sup \{\mathbf{m}_{y,T}(\delta) : y \in \mathcal{M}\}, \\ \boldsymbol{\mu}_{\mathcal{M}}(\delta) &= \delta + \mathbf{m}_{\mathcal{M},T}(\delta).\end{aligned}$$

Let $\delta_0 = \delta_{0,\mathcal{M}} > 0$ be defined by

$$\delta_0 + \mathbf{m}_{\mathcal{M},T}(\delta_0) = \frac{r_0}{4}.$$

Proposition 4.16 (Uniqueness). *Let the assumptions ((4.29)-**MM-H**$_A$) and ((4.30)-**MM-H**$_f$) and ((4.31)-**MM-H**$_{0,m}$) be satisfied. Let $T > 0$ and $\mathcal{M}$ be a bounded and equicontinuous subset of $C([0, T]; \mathbb{R}^d)$. Then there exists a positive constant*

$$C_{0,\mathcal{M}} = C_0 \left(T, u_0, r_0, A^\#_{u_0,r_0}, \|\mu\|_{L^1(0,T)}, \left\| f^\#_{u_0,r_0} \right\|_{L^1(0,T)}, \delta_{0,\mathcal{M}}, \|\mathcal{M}\|_T \right)$$

such that:

I. *If $m \in \mathcal{M}$ and $(x,k) = \mathcal{GS}(A; x_0, f, m)$ then*

$$\|x\|_T^2 + \updownarrow k \updownarrow_T + \int_0^T |f(r, x(r))| \, dr \le C_{0,\mathcal{M}} \left(1 + |x_0|^2\right). \tag{4.35}$$

II. *If $m, \hat{m} \in \mathcal{M}$, $(x,k) = \mathcal{GS}(A; x_0, f, m)$ and $(\hat{x}, \hat{k}) = \mathcal{GS}(A; \hat{x}_0, f, \hat{m})$ then*

$$\|x - \hat{x}\|_T \le C_{0,\mathcal{M}} (1 + |x_0| + |\hat{x}_0|) \left(|x_0 - \hat{x}_0| + \|m - \hat{m}\|_T^{1/2}\right). \tag{4.36}$$

In particular if $x_0 = \hat{x}_0$ and $m = \hat{m}$ then $(x,k) = (\hat{x}, \hat{k})$.

III. *If for every $(x_0, m) \in \overline{D(A)} \times C\left([0,T]; \mathbb{R}^d\right)$ the Eq. (4.28) has a solution $(x,k) = \mathcal{GS}(A; x_0, f, m)$, then the mapping*

$$(x_0, m) \longmapsto x = \mathcal{GS}(A; x_0, f, m) : \overline{D(A)} \times C\left([0,T]; \mathbb{R}^d\right) \to C\left([0,T]; \overline{D(A)}\right)$$

is continuous.

Proof. I. By (6.19) in Lemma 6.20 with $k(t)$ replaced by $k(t) - \int_0^t f(r, x(r)) \, dr$, we have

$$\begin{aligned} &|x(t) - m(t) - u_0|^2 + 2\int_0^t \langle x(r) - u_0, dk(r) - f(r, x(r)) \, dr \rangle \\ &\quad = |x_0 - u_0|^2 + 2\int_0^t \langle m(r), dk(r) - f(r, x(r)) \, dr \rangle \end{aligned}$$

and using (4.34) we obtain

$$\begin{aligned} &|x(t) - m(t) - u_0|^2 + r_0 \updownarrow k \updownarrow_t + r_0 \int_0^t |f(r, x(r))| \, dr \\ &\le |x_0 - u_0|^2 + 2\int_0^t \langle m(r), dk(r) - f(r, x(r)) \, dr \rangle \\ &\quad + \int_0^t \left[\gamma(r) + \alpha(r) |x(r) - u_0| + \mu(r) |x(r) - u_0|^2\right] dr. \end{aligned}$$

Let $n_0 \in \mathbb{N}^*$ be fixed such that $\dfrac{T}{n_0} \le \delta_0$.

Let $0 = t_0 < t_1 < \ldots < t_{n_0} = T$, $t_{i+1} - t_i = \dfrac{T}{n_0}$, $i = \overline{0, n_0 - 1}$. Recall that

$$\delta_0 + \mathbf{m}_{\mathcal{M},T}(\delta_0) = \frac{r_0}{4}.$$

Then

$$\begin{aligned}
&\int_0^t \langle m(r), dk(r) - f(r, x(r))\, dr\rangle \\
&\quad = \sum_{i=0}^{k-1} \int_{t\wedge t_i}^{t\wedge t_{i+1}} \langle m(r) - m(t\wedge t_i), dk(r) - f(r, x(r))\, dr\rangle \\
&\qquad\quad + \sum_{i=0}^{k-1} \langle m(t\wedge t_i), k(t\wedge t_{i+1}) - k(t\wedge t_i) - \int_{t\wedge t_i}^{t\wedge t_{i+1}} f(r, x(r))\, dr\rangle \\
&\quad \leq \mathbf{m}_{\mathcal{M},T}(\delta_0)\left(\updownarrow k \updownarrow_t + \int_0^t |f(r, x(r))|\, dr\right) \\
&\qquad\quad + \sum_{i=0}^{k-1} \langle m(t\wedge t_i), m(t\wedge t_{i+1}) - x(t\wedge t_{i+1}) + u_0 - m(t\wedge t_i) \\
&\qquad\qquad + x(t\wedge t_i) - u_0\rangle \\
&\quad \leq \frac{r_0}{4} \updownarrow k \updownarrow_t + \frac{r_0}{4}\int_0^t |f(r, x(r))|\, dr + 2n_0 \|m\|_t \|x - u_0 - m\|_t .
\end{aligned}$$

Hence

$$\begin{aligned}
&|x(t) - m(t) - u_0|^2 + \frac{r_0}{2} \updownarrow k \updownarrow_t + \frac{r_0}{2}\int_0^t |f(r, x(r))|\, dr \\
&\leq |x_0 - u_0|^2 + 4n_0 \|m\|_t \|x - u_0 - m\|_t \\
&\quad + \int_0^t \left[\gamma(r) + a(r)|x(r) - u_0| + \mu^+(r)|x(r) - u_0|^2\right] dr.
\end{aligned}$$

Since

◇ $\frac{1}{2}|x(t) - u_0|^2 - \|m\|_t^2 \leq |x(t) - m(t) - u_0|^2$,

◇ $4n_0 \|m\|_t \|x - u_0 - m\|_t \leq \frac{1}{8}\|x - u_0\|_t^2 + \left(4n_0 + 32n_0^2\right)\|m\|_t^2$, and

◇ $\int_0^t \alpha(r)|x(r) - u_0|\, dr \leq \frac{1}{8}\|x - u_0\|_t^2 + 2\left(\int_0^T \alpha(r)\, dr\right)^2$,

we deduce that

$$\begin{aligned}
\frac{1}{4}\|x - u_0\|_t^2 &\leq |x_0 - u_0|^2 + C(n_0)\|\mathcal{M}\|_T^2 + C(\alpha, \gamma) \\
&\quad + 2\int_0^t \mu^+(r)\|x - u_0\|_r^2\, dr
\end{aligned}$$

and

$$\frac{r_0}{2} \updownarrow k \updownarrow_t + \frac{r_0}{2} \int_0^t |f(r, x(r))| \, dr \leq \frac{1}{4} \|x - u_0\|_t^2 + |x_0 - u_0|^2 + C(n_0) \|\mathcal{M}\|_T^2$$
$$+ C(\alpha, \gamma) + 2 \int_0^t \mu^+(r) \|x - u_0\|_r^2 \, dr.$$

Combining Gronwall's inequality with the first estimate, the resulting inequality with the second estimate clearly yields (4.35).

II. From ordinary differential calculus (see Lemma 6.20) we have

$$|x(t) - m(t) - \hat{x}(t) + \hat{m}(t)|^2 + 2 \int_0^t \langle x(r) - \hat{x}(r), dk(r) - d\hat{k}(r) \rangle$$
$$= |x_0 - \hat{x}_0|^2 + 2 \int_0^t \langle x(r) - \hat{x}(r), f(r, x(r)) - f(r, \hat{x}(r)) \rangle dr$$
$$+ 2 \int_0^t \langle m(r) - \hat{m}(r), dk(r) - f(r, x(r)) \, dr - d\hat{k}(r) + f(r, \hat{x}(r)) \, dr \rangle$$
$$\leq |x_0 - \hat{x}_0|^2 + 2 \int_0^t \mu^+(r) |x(r) - \hat{x}(r)|^2 dr$$
$$+ 2 \|m - \hat{m}\|_T \left[\updownarrow k \updownarrow + \updownarrow \hat{k} \updownarrow_T + \int_0^T [|f(r, x(r))| + |f(r, \hat{x}(r))|] \, dr \right]$$
$$\leq |x_0 - \hat{x}_0|^2 + 2 \int_0^t \mu^+(r) |x(r) - \hat{x}(r)|^2 dr$$
$$+ 4C_{0,\mathcal{M}} \|m - \hat{m}\|_T \left(1 + |x_0|^2 + |\hat{x}_0|^2\right).$$

On the other hand

$$|x(t) - m(t) - \hat{x}(t) + \hat{m}(t)|^2 \geq \frac{1}{2} |x(t) - \hat{x}(t)|^2 - \|m - \hat{m}\|_T^2$$
$$\geq \frac{1}{2} |x(t) - \hat{x}(t)|^2 - 2 \|\mathcal{M}\|_T \|m - \hat{m}\|_T.$$

Combining these last two inequalities with (4.33) and Gronwall's inequality, we deduce

$$|x(t) - \hat{x}(t)|^2$$
$$\leq \Big[2 |x_0 - \hat{x}_0|^2 + 4 \|\mathcal{M}\|_T \|m - \hat{m}\|_T$$
$$+ 4C_{0,\mathcal{M}} \|m - \hat{m}\|_T \left(1 + |x_0|^2 + |\hat{x}_0|^2\right) \Big] \exp\left\{ 4 \int_0^T \mu^+(r) \, dr \right\}$$

from which (4.36) follows.

II. To prove the continuity of the mapping

$$(x_0, m) \longmapsto x = \mathcal{GS}\,(A; x_0, f, m)$$

from $\overline{D(A)} \times C\left([0,T];\mathbb{R}^d\right)$ to $C\left([0,T];\overline{D(A)}\right)$, let $x_{0,n} \in \overline{D(A)}$, $m_n \in C\left([0,T];\mathbb{R}^d\right)$ and $(x_n, k_n) = \mathcal{GS}\,(A; x_{0,n}, f, m_n)$, $n \in \mathbb{N}^*$, be such that

$$x_{0,n} \to x_0 \quad \text{in } \mathbb{R}^d \quad \text{and} \quad m_n \to m \quad \text{in } C\left([0,T];\mathbb{R}^d\right).$$

Note that $\mathcal{M} = \{m, m_1, m_2, \ldots\}$ is a relatively compact subset of $C([0,T];\mathbb{R}^d)$. Then by (4.35) we have

$$\|x_n\|_T^2 + \updownarrow k_n \updownarrow_T \leq R^2 = C_{0,\mathcal{M}}\left(1 + |x_0|^2\right)$$

and by (4.36)

$$\left\|x_n - x_j\right\|_T \leq C_{0,\mathcal{M}}\left(\left|x_{0,n} - x_{0,j}\right| + \|m_n - m_j\|_T^{1/2}\right).$$

Hence there exists a $x \in C\left([0,T];\mathbb{R}^d\right)$ such that as $n \to \infty$

$$x_n \to x \quad \text{in } C\left([0,T];\overline{D(A)}\right).$$

Let

$$k(t) = x_0 + \int_0^t f(s, x(s))\,ds + m(t) - x(t).$$

Since $|f(t, x_n(t)) - f(t, x(t))| \leq 2f_R^{\#}(t)$ and $f(t, x_n(t)) \to f(t, x(t))$, *a.e.* $t \in \,]0,T[$, as $n \to \infty$ we deduce, by the Lebesgue dominated convergence theorem, that

$$k_n = x_0 + \int_0^{\cdot} f(s, x_n(s))\,ds + m_n - x_n \longrightarrow k \quad \text{in } C\left([0,T];\mathbb{R}^d\right).$$

By Proposition 6.16 from Annex B we obtain

$$k \in BV\left([0,T];\mathbb{R}^d\right), \quad \updownarrow k \updownarrow_T \leq R^2$$

and

$$0 \leq \int_s^t \langle x_n(r) - z, dk_n(r) - \hat{z}dr\rangle \longrightarrow \int_s^t \langle x(r) - z, dk(r) - \hat{z}dr\rangle,$$

for all $(z,\hat{z}) \in A$ and for all $0 \le s \le t \le T$. Hence $(x,k) = \mathcal{GS}(A;x_0,f,m)$.

The proof is complete. ■

Theorem 4.17. *Let the assumptions ((4.29)-**MM-H**$_A$) and ((4.30)-**MM-H**$_f$) and ((4.31)-**MM-H**$_{0,m}$) be satisfied. Then the generalized convex Skorohod problem (4.28) has a unique solution (x,k) and the estimates (4.35), (4.36) hold.*

Proof. The uniqueness and the estimates (4.35), (4.36) have been obtained in the above Proposition 4.16.

It suffices to prove existence on an arbitrary fixed interval $[0,T]$.

By the continuous property from Proposition 4.16 we may assume that $x_0 \in D(A)$.

Moreover for the proof of existence we can assume that $0 \in D(A)$, $0 \in A(0)$ and $f(t,0) = 0$. Indeed if $x_0 \in D(A)$ and $\hat{x}_0 \in A(x_0)$ then we can change the Eq. (4.28) into an equivalent form (in the sense of the definition of the solution)

$$\begin{aligned} &d\tilde{x}(t) + \tilde{A}(\tilde{x}(t))(dt) \ni \tilde{f}(t,\tilde{x}(t))dt + d\tilde{m}(t), \\ &\tilde{x}(0) = 0, \qquad\qquad t \in [0,T], \end{aligned}$$

with

$$\begin{aligned} \tilde{A}(x) &= A(x + x_0) - \hat{x}_0, \\ \tilde{f}(t,x) &= f(t,x+x_0) - f(t,x_0), \\ \tilde{m}(t) &= m(t) + \int_0^t f(s,x_0)ds - \hat{x}_0 t \end{aligned}$$

and then the solution $(x,k) = \mathcal{GS}(A;x_0,f,m)$ is given by

$$x = \tilde{x} + x_0, \quad \text{and} \quad k = \tilde{k} + \hat{x}_0 t,$$

where $(\tilde{x},\tilde{k}) = \mathcal{GS}(\tilde{A};0,\tilde{f},\tilde{m})$.

Hence by the above transformations we reduce the problem of existence of the solution to the case

$$x_0 = 0 \in D(A), \quad 0 \in A(0), \quad f(t,0) = 0.$$

Let $0 < \varepsilon \le 1$. We consider the penalized problem

$$(P_\varepsilon): \quad x_\varepsilon(t) + \int_0^t A_\varepsilon(x_\varepsilon(s))\,ds = \int_0^t f(s,x_\varepsilon(s))dt + m(t), \ t \ge 0,$$

where A_ε is Yosida's approximation of A defined in Annex B. Hence

$$a) \quad A_\varepsilon x = \frac{1}{\varepsilon}(x - J_\varepsilon x), \quad \text{where} \quad J_\varepsilon(x) = (I + \varepsilon A)^{-1}(x),$$

$$b) \quad |A_\varepsilon(x) - A_\varepsilon(y)| \le \frac{1}{\varepsilon}|x-y|,$$

$$c) \quad \langle A_\varepsilon(x) - A_\varepsilon(y), x-y\rangle \ge 0$$

and since $0 \in D(A)$ and $0 \in A(0)$, then $J_\varepsilon(0) = A_\varepsilon(0) = 0$.

From Proposition 6.5 the assumption $\text{int}(\text{Dom}(A)) \neq \emptyset$ yields the existence of $u_0 \in \text{int}(\text{Dom}(A))$, $r_0 \in]0,1]$ and $a_0, b_0 \ge 0$ such that

$$r_0|A_\varepsilon x| \le \langle A_\varepsilon x, x-u_0\rangle + a_0|x-u_0| + b_0, \quad \forall x \in \mathbb{R}^d.$$

By Lemma 3.19 from Chap. 3, with $f := f - A_\varepsilon$, there exists a unique continuous solution $x_\varepsilon : [0,\infty[\to \mathbb{R}^d$ of the equation (P_ε). Let

$$k_\varepsilon(t) = \int_0^t A_\varepsilon(x_\varepsilon(s))\,ds, \quad \text{and} \quad \updownarrow k_\varepsilon \updownarrow_T = \int_0^T |A_\varepsilon(x_\varepsilon(s))|\,ds.$$

We shall prove that as $\varepsilon \to 0$

$$x_\varepsilon \to x \quad \text{and} \quad k_\varepsilon \to k \qquad \text{in } C\left([0,T];\mathbb{R}^d\right)$$

and $(x,k) = \mathcal{GS}(A; x_0, f, m)$.

Repeating the proof from Proposition 4.16 we obtain for $\mathcal{M} = \{m\}$ an estimate of the form (4.35): there exists a positive constant R_0 independent of ε such that

$$\|x_\varepsilon\|_{T+1}^2 + \updownarrow k_\varepsilon \updownarrow_{T+1} + \int_0^{T+1} |f(r, x_\varepsilon(r))|\,dr \le R_0. \tag{4.37}$$

We have

$$|x_\varepsilon(t) - m(t)|^2 + 2\int_0^t \langle x_\varepsilon(s) - m(s), A_\varepsilon(x_\varepsilon(s))\rangle\,ds$$
$$= 2\int_0^t \langle x_\varepsilon(s) - m(s), f(s, x_\varepsilon(s))\rangle\,ds.$$

Since $\langle y, A_\varepsilon(y)\rangle \ge 0$ for all $y \in \mathbb{R}^d$ and

$$\langle x_\varepsilon - m, f(s, x_\varepsilon)\rangle \le |x_\varepsilon - m|\,|f(s,m)| + \mu^+(s)|x_\varepsilon - m|^2,$$

it follows that

$$|x_\varepsilon(t) - m(t)|^2$$
$$\le 2\|m\|_t \updownarrow k_\varepsilon \updownarrow_T + 2\int_0^t |f(s, m(s))|\,|x_\varepsilon(s) - m(s)|\,ds$$
$$+ 2\int_0^t \mu^+(s)|x_\varepsilon(s) - m(s)|^2\,ds$$

and by the Gronwall type inequality from Lemma 6.63-II, we deduce that

$$|x_\varepsilon(t) - m(t)| \le \left[\sqrt{2C\,\|m\|_t} + \int_0^t |f(s, m(s))|\, ds\right] \exp\left\{\int_0^T \mu^+(s)\, ds\right\}.$$

(C denotes a generic constant independent of ε.) Hence

$$\|x_\varepsilon - m\|_t \le C\left(\sqrt{\|m\|_t} + \int_0^t |f(s, m(s))|\, ds\right). \tag{4.38}$$

Let $0 \le \theta \le 1$. From (P_ε) we have

$$x_\varepsilon(t+\theta) + \int_\theta^{t+\theta} A_\varepsilon(x_\varepsilon(s))\, ds = x_\varepsilon(\theta) + \int_\theta^{t+\theta} f(s, x_\varepsilon(s))ds + m(t+\theta) - m(\theta)$$

and moreover

$$\begin{aligned} x_\varepsilon(t+\theta) - m(t+\theta) &+ \int_0^t A_\varepsilon(x_\varepsilon(s+\theta))\, ds \\ &= x_\varepsilon(\theta) - m(\theta) + \int_0^t f(s+\theta, x_\varepsilon(s+\theta))ds. \end{aligned}$$

Hence

$$\begin{aligned} &|[x_\varepsilon(t+\theta) - m(t+\theta)] - [x_\varepsilon(t) - m(t)]|^2 \\ &+ 2\int_0^t \langle x_\varepsilon(s+\theta) - m(s+\theta) - x_\varepsilon(s) + m(s), A_\varepsilon(x_\varepsilon(s+\theta)) - A_\varepsilon(x_\varepsilon(s))\rangle\, ds \\ &= |x_\varepsilon(\theta) - m(\theta)|^2 \\ &+ 2\int_0^t \langle x_\varepsilon(s+\theta) - m(s+\theta) - x_\varepsilon(s) + m(s), f(s+\theta, x_\varepsilon(s+\theta)) - f(s, x_\varepsilon(s))\rangle\, ds. \end{aligned}$$

The monotonicity of A_ε and f yield then

$$\begin{aligned} |x_\varepsilon(t+\theta) &- m(t+\theta) - x_\varepsilon(t) + m(t)|^2 \\ &\le |x_\varepsilon(\theta) - m(\theta)|^2 + 2\mathbf{m}_m(\theta)\int_0^T [|A_\varepsilon(x_\varepsilon(s+\theta))| + |A_\varepsilon(x_\varepsilon(s))|]\, ds \\ &\quad + 2\mathbf{m}_m(\theta)\int_0^T [|f(s+\theta, x_\varepsilon(s+\theta))| + |f(s, x_\varepsilon(s)|]\, ds \end{aligned}$$

$$+2\int_0^t |x_\varepsilon(s+\theta)-x_\varepsilon(s)|\,|f(s+\theta,x_\varepsilon(s))-f(s,x_\varepsilon(s))|\,ds$$
$$+2\int_0^t \mu^+(s+\theta)\,|x_\varepsilon(s+\theta)-x_\varepsilon(s)|^2\,ds.$$

Since

$$|x_\varepsilon(t+\theta)-m(t+\theta)-x_\varepsilon(t)+m(t)|^2\geq \frac{1}{2}|x_\varepsilon(t+\theta)-x_\varepsilon(t)|^2-\mathbf{m}_m^2(\theta),$$

we have, by the Gronwall type inequality from Lemma 6.63 and the boundedness (4.37), that

$$\begin{aligned}&|x_\varepsilon(t+\theta)-x_\varepsilon(t)|\\&\leq C\left[|x_\varepsilon(\theta)-m(\theta)|+\sqrt{\mathbf{m}_m(\theta)}+\int_0^T|f(s+\theta,x_\varepsilon(s))-f(s,x_\varepsilon(s))|\,ds\right].\end{aligned}$$

Since J_ε is 1-Lipschitz, it follows from (4.38) and the assumption (4.30-B$_f$-b) that

$$\begin{aligned}|J_\varepsilon(x_\varepsilon(t+\theta))-J_\varepsilon(x_\varepsilon(t))|&\leq|x_\varepsilon(t+\theta)-x_\varepsilon(t)|\\&\leq C\left[\sqrt{\mathbf{m}_m(\theta)}+\textstyle\int_0^\theta|f(s,m(s))|\,ds+\boldsymbol{\mu}_T(\theta;f,R_0)\right].\end{aligned}$$

Hence $\{x_\varepsilon:\varepsilon\in\,]0,1]\}$ and $\{J_\varepsilon(x_\varepsilon):\varepsilon\in\,]0,1]\}$ are bounded and equicontinuous subsets of $C([0,T];\mathbb{R}^d)$. By the Arzelà–Ascoli theorem we infer that there exist $x,y\in C([0,T];\mathbb{R}^d)$ and $\varepsilon_n\to0$ such that

$$\begin{aligned}x_{\varepsilon_n}&\to x&&\text{in } C([0,T];\mathbb{R}^d),\\J_{\varepsilon_n}(x_{\varepsilon_n})&\to y&&\text{in } C([0,T];\mathbb{R}^d).\end{aligned}$$

Since $J_\varepsilon(u)=u-\varepsilon A_\varepsilon(u)\in\overline{D(A)}$, for all $u\in\mathbb{R}^d$ and

$$\int_0^T|x_{\varepsilon_n}(t)-J_{\varepsilon_n}(x_{\varepsilon_n}(t))|\,dt=\varepsilon_n\int_0^T|A_{\varepsilon_n}(x_{\varepsilon_n}(t))|\,dt\leq\varepsilon_nC$$

we deduce that

$$x_{\varepsilon_n}-J_{\varepsilon_n}(x_{\varepsilon_n})\to0\quad\text{in } L^1(0,T;\mathbb{R}^d)$$

and therefore

$$\begin{aligned}&J_{\varepsilon_n}(x_{\varepsilon_n})\to x\qquad\text{in } C([0,T];\mathbb{R}^d)\quad\text{and}\\&x(t)\in\overline{D(A)}\ \forall t\in[0,T].\end{aligned}$$

Let

$$k(t) = x_0 + \int_0^t f(s, x(s))\, ds + m(t) - x(t).$$

It follows easily that

$$k_{\varepsilon_n} = x_0 + \int_0^{\cdot} f(s, x_{\varepsilon_n}(s))\, ds + m - x_{\varepsilon_n} \longrightarrow k \quad \text{in } C\left([0, T]; \mathbb{R}^d\right).$$

By Proposition 6.16 we obtain that

$$k \in BV\left([0, T]; \mathbb{R}^d\right), \quad \updownarrow k \updownarrow_T \leq C_0$$

and

$$\int_s^t \langle x_{\varepsilon_n}(r) - z, dk_{\varepsilon_n}(r) - \hat{z}dr\rangle \longrightarrow \int_s^t \langle x(r) - z, dk(r) - \hat{z}dr\rangle,$$

for all $(z, \hat{z}) \in A$ and for all $0 \leq s \leq t \leq T$.

Since $A_\varepsilon(x_\varepsilon(r)) \in A(J_\varepsilon(x_\varepsilon(r)))$ we have

$$\begin{aligned}
&\int_s^t \langle x_{\varepsilon_n}(r) - z, dk_{\varepsilon_n}(r) - \hat{z}dr\rangle \\
&\quad = \int_s^t \langle x_{\varepsilon_n}(r) - z, A_{\varepsilon_n}(x_{\varepsilon_n}(r)) - \hat{z}\rangle\, dr \\
&\quad \geq \int_s^t \langle x_{\varepsilon_n}(r) - J_{\varepsilon_n}(x_{\varepsilon_n}(r)), A_{\varepsilon_n}(x_{\varepsilon_n}(r)) - \hat{z}\rangle\, dr \\
&\quad \geq -\|x_{\varepsilon_n} - J_{\varepsilon_n}(x_{\varepsilon_n})\|_T \int_0^T (|A_{\varepsilon_n}(x_{\varepsilon_n}(r))| + |\hat{z}|)\, dr \\
&\quad \geq -\|x_{\varepsilon_n} - J_{\varepsilon_n}(x_{\varepsilon_n})\|_T (C + T|\hat{z}|) \\
&\quad \to 0.
\end{aligned}$$

Therefore for all $(z, \hat{z}) \in A$ and for all $0 \leq s \leq t \leq T$,

$$\int_s^t \langle x(r) - z, dk(r) - \hat{z}dr\rangle \geq 0.$$

Hence we can conclude that $(x, k) = \mathcal{GS}(A; x_0, f, m)$. Moreover, by uniqueness of the solution $(x, k) = \mathcal{GS}(A; x_0, f, m)$, the whole sequence $(x_\varepsilon, k_\varepsilon) \to (x, k)$ in $\left[C\left([0, T]; \mathbb{R}^d\right)\right]^2$ as $\varepsilon \to 0$. ■

4.2.4 Main Result: Existence

We recall the Eq. (4.8):

$$\begin{cases} dX_t + A(X_t)(dt) \ni F(t, X_t)dt + G(t, X_t)dB_t, & t \geq 0, \\ X_0 = \xi, \end{cases} \tag{4.39}$$

The assumptions from Sect. 4.2.1 are assumed to be in force.
We introduce the notation: let $R, T, \delta \geq 0$ and

$$\mu_T(\delta; F, R) \stackrel{def}{=} \sup \left\{ \int_0^T |F(s+\delta, y(s)) - F(s, y(s))| \, ds : \right. \\ \left. y \in C\left([0,T]; \mathbb{R}^d\right), \ \|y\|_T \leq R \right\}$$

and we add the following continuity condition on F:

$$(\mathbf{MM\text{-}H}_{cF}): \qquad \lim_{\delta \to 0} \mu_T(\delta; F, R) = 0, \ \ a.s. \tag{4.40}$$

Proposition 4.18. *Let $\xi \in L^0(\Omega, \mathcal{F}_0, P; \overline{\mathrm{Dom}(A)})$ and the assumptions ((4.9)-**MM-H**$_A$), ((4.13)-**MM-H**$_F$) and ((4.40)-**MM-H**$_{cF}$) be satisfied. Assume*

(I) If $M \in S_d^0$, $M_0 = 0$, then the stochastic differential equation

$$\begin{cases} dX_t + A(X_t)(dt) \ni F(t, X_t)dt + dM_t, & t \geq 0, \\ X_0 = \xi, \end{cases} \tag{4.41}$$

has a unique solution $(X, K) \in S_d^0 \times S_d^0$ (in the sense of Definition 4.2).

(II) If, moreover, M is an Itô integral

$$M_t = \int_0^t G_s dB_s, \ \ G \in \Lambda^0_{d\times k}$$

and there exist $p \geq 2$ and $u_0 \in \mathrm{int}(\mathrm{Dom}(A))$ such that for every $T \geq 0$:

$$\mathbb{E}|\xi|^p + \mathbb{E}\left(\int_0^T |F(t, u_0)| \, dt\right)^p + \mathbb{E}\left(\int_0^T |G_t|^2 dt\right)^{p/2} < +\infty,$$

then $X \in S_d^p$, $K \in S_d^{p/2}$, and $\mathbb{E} \updownarrow K \updownarrow_T^{p/2} < \infty$.

Proof. For every fixed ω, by Theorem 4.17, the differential equation (4.41) has a unique solution

$$(X_\cdot(\omega), K_\cdot(\omega)) \in C([0,T]; \mathbb{R}^d) \times C([0,T]; \mathbb{R}^d).$$

Since $(\omega, t) \longrightarrow M_t(\omega)$ is progressively measurable and the mapping $(\xi, M) \longrightarrow X : \mathbb{R}^d \times C([0,t];\mathbb{R}^d) \to C([0,t];\mathbb{R}^d)$ is continuous for each $t \leq T$ we deduce that X is progressively measurable; hence $X \in S_d^0$ and consequently $K \in S_d^0$. The second part clearly follows from Corollary 4.9. ■

Theorem 4.19. *If $\xi \in L^0(\Omega, \mathcal{F}_0, P; \overline{\mathrm{Dom}(A)})$ and the assumptions ((4.9)-**MM-H**$_A$), ((4.13)-**MM-H**$_F$), ((4.14)-**MM-H**$_G$) and ((4.40)-**MM-H**$_{cF}$) are satisfied, then the SDE (4.39) has a unique solution $(X, K) \in S_d^0 \times S_d^0$ (in the sense of Definition 4.2). Moreover if there exist $p \geq 2$ and $u_0 \in \mathrm{int}(\mathrm{Dom}(A))$ such that for all $T \geq 0$:*

$$\mathbb{E}|\xi|^p + \mathbb{E}\left(\int_0^T |F(t, u_0)|\, dt\right)^p + \mathbb{E}\left(\int_0^T |G(t, u_0)|^2 dt\right)^{p/2} < +\infty, \tag{4.42}$$

then $X \in S_d^p$, $K \in S_d^{p/2}$ and $\mathbb{E}\ \updownarrow K \updownarrow_T^{p/2} < \infty$.

Proof. Uniqueness was proved in Theorem 4.11. Moreover by Corollary 4.9, if $(X, K) \in S_d^0 \times S_d^0$ is a solution and (4.42) is satisfied then $X \in S_d^p$, $K \in S_d^{p/2}$ and $\mathbb{E}\ \updownarrow K \updownarrow_T^{p/2} < \infty$ for every $T \geq 0$.

It suffices to prove existence on an arbitrary interval $[0, T]$.

The proof will be split into two steps.

Step 1. Existence under the condition (4.42).

Let $U \in S_d^p[0, T]$. Clearly $G(\cdot, U_\cdot) \in \Lambda_d^p(0, T)$ and

$$M_\cdot \stackrel{def}{=} \int_0^\cdot G(s, U_s)\, dB_s \in S_d^p[0, T].$$

By Proposition 4.18 the SDE

$$\begin{cases} X_t + K_t = \xi + \displaystyle\int_0^t F(s, X_s)ds + \int_0^t G(s, U_s)dB_s, \quad \forall t \geq 0, \\ dK_t \in A(X_t)(dt), \end{cases} \tag{4.43}$$

has a unique solution (X, K) in the sense of Definition 4.2 and $X \in S_d^p$ and $K \in S_d^{p/2}$ and $\mathbb{E}\ \updownarrow K \updownarrow_T^{p/2} < \infty$ for every $T \geq 0$.

We note that the Eq. (4.39) on $[0, T]$ may be written in the form

$$\Gamma(X) = X, \tag{4.44}$$

where $\Gamma : S_d^p[0, T] \to S_d^p[0, T]$ is defined by $X = \Gamma(U)$ and $X \in S_d^p$ is the solution of Eq. (4.43).

The existence and uniqueness of a solution of (4.44) in $S_d^p[0, T]$ will follow from Banach's fixed point theorem and the fact that Γ is a strict contraction on $S_d^p[0, T]$ equipped with the equivalent norm $|||\cdot|||_a$ given by

$$\|X\|_a = \sup_{t\in[0,T]} \left[e^{-at} \left(\mathbb{E} \left\| e^{-V} X \right\|_t^p \right)^{1/p} \right],$$

with

$$V(t) = \int_0^t \mu(r)\, dr$$

and a large enough, which we now prove.

Let $U, \tilde{U} \in S_d^p[0,T]$, $X = \Gamma(U)$ and $\tilde{X} = \Gamma(\tilde{U})$. Then

$$X_t - \tilde{X}_t = \mathcal{K}_t + \int_0^t G_r dB_r,$$

where

$$\begin{aligned} \mathcal{K}_t &= -\left(K_t - \tilde{K}_t\right) + \int_0^t \left[F(r, X_r) - F(r, \tilde{X}_r)\right] dr, \\ G_t &= G(t, U_t) - G(t, \tilde{U}_t). \end{aligned}$$

Since for all $\lambda > 1$:

$$\left\langle X_t - \tilde{X}_t, d\mathcal{K}_t \right\rangle + \left(\frac{1}{2} m_p + 9p\lambda \right) |G_t|^2 dt \le dR_t + |X_t - \tilde{X}_t|^2 dV(t)$$

with

$$\begin{aligned} m_p &= 1 \vee (p-1), \\ R_t &= \left(\frac{1}{2} m_p + 9p\lambda \right) \int_0^t \ell^2(r) \left| U_r - \tilde{U}_r \right|^2 dr \quad \text{and} \\ V(t) &= \int_0^t \mu(r)\, dr, \end{aligned}$$

we have, by Corollary 6.75 (or Proposition 3.3) and Minkowski's inequality (see Exercise 1.2), for $p \ge 2$ and $\lambda = 2$:

$$\begin{aligned} & e^{-pat} \times \mathbb{E} \left\| e^{-V} (X - \tilde{X}) \right\|_t^p \\ & \le e^{-pat} \times C_p \mathbb{E} \left(\int_0^t e^{-2V(r)} \ell^2(r) \left| U_r - \tilde{U}_r \right|^2 dr \right)^{p/2} \\ & \le e^{-pat} \times C_p \mathbb{E} \left(\int_0^t \ell^2(r) e^{2ar} e^{-2ar} \left\| e^{-V} \left(U - \tilde{U} \right) \right\|_r^2 dr \right)^{p/2} \end{aligned}$$

$$\le e^{-pat} \times C_p\Big(\int_0^t \ell^2(r)\, e^{2ar} e^{-2ar} \left(\mathbb{E}\left\|e^{-V}\left(U-\tilde{U}\right)\right\|_r^p\right)^{2/p} dr\Big)^{p/2}$$

$$\le \varphi(a) \sup_{r\in[0,T]} \left(e^{-par}\mathbb{E}\left\|e^{-V}\left(U-\tilde{U}\right)\right\|_r^p\right),$$

where

$$\varphi(a) = C_p \sup_{t\in[0,T]} \left(e^{-2at}\int_0^t \ell^2(r)\, e^{2ar} dr\right)^{p/2}.$$

Taking the sup over $t \in [0,T]$, we deduce that

$$\left\|\Gamma(U) - \Gamma\left(\tilde{U}\right)\right\|_a \le [\varphi(a)]^{1/p} \ \left\|U - \tilde{U}\right\|_a.$$

Since by Proposition 6.57, Annex B, $\lim_{a\to\infty} \varphi(a) = 0$, we obtain that Γ is a strict contraction for a large enough.

Step 2. Existence in S_d^0.

Let $n \in \mathbb{N}^*$ and the stopping time

$$\theta_n = \inf\left\{t \ge 0 : |\xi| + \int_0^t \left[|F(s,u_0)| + |G(s,u_0)|^2\right] ds \ge n\right\}.$$

Clearly $\theta_n \nearrow \infty$ $a.s.$ By the first part there exists a unique solution (X^n, K^n) of the approximating equation

$$\begin{cases} X_t^n + K_t^n = \xi \mathbf{1}_{\theta_n>0} + \int_0^t \mathbf{1}_{[0,\theta_n]}(r)\, F\left(r, X_r^n\right) dr + \int_0^t \mathbf{1}_{[0,\theta_n]}(r)\, G\left(r, X_r^n\right) dB_r, \\ dK_t^n \in A\left(X_t^n\right)(dt), \end{cases}$$

and for all $p \ge 2 : X^n \in S_d^p$, $K^n \in S_d^{p/2}$ and $\mathbb{E}\ \updownarrow K \updownarrow_T^{p/2} < \infty$ for every $T \ge 0$.

By the same argument as in step (II) of the proof of Theorem 4.20 the pair of stochastic processes $X, K \in S_d^0$ defined by

$$(X_t(\omega), K_t(\omega)) = \left(X_t^n(\omega), K_t^n(\omega)\right), \quad \text{if } 0 \le t < \theta_n(\omega) \text{ and } \theta_n(\omega) > 0$$

is a solution of the SDE (4.39). The proof is complete. ■

4.2.5 SDEs with a Subdifferential Operator in the Drift

In this section we consider a particular case of a maximal monotone operator in the drift: a subdifferential operator, that is an SDE of the form (4.19):

$$\begin{cases} dX_t + \partial\varphi(X_t)(dt) \ni F(t,X_t)dt + G(t,X_t)dB_t, & t \geq 0, \\ X_0 = \xi \in L^0(\Omega,\mathcal{F}_0,P;\overline{\mathrm{Dom}(\varphi)}). \end{cases} \tag{4.45}$$

The problem becomes more complicated if we replace $\partial\varphi(X_t)$ by $H(X_t)\partial\varphi(X_t)$, where $H(x)$ is a rotation matrix, see [33].

A direct proof by a penalized approximating procedure also permits us to give a precise approximating procedure with an estimate of the speed of convergence. Our approach here is based on [1].

In the sense of Definition 4.7 we can change the SDE (4.19) into the following equation for $\tilde{X}_t := X_t - v_0$,

$$\begin{cases} d\tilde{X}_t + \partial\tilde{\varphi}(\tilde{X}_t)(dt) \ni \tilde{F}(t,\tilde{X}_t)dt + \tilde{G}(t,\tilde{X}_t)dB_t, & t \geq 0, \\ \tilde{X}_0 = \tilde{\xi}, \end{cases}$$

where v_0, $\hat{v}_0$, r_0 are such that $\overline{B}(v_0,r_0) \subset \mathrm{int}(\mathrm{Dom}(\varphi))$ and $\hat{v}_0 \in \partial\varphi(v_0)$, and for $x \in \mathbb{R}^d$, $t \geq 0$,

$$\begin{aligned} &\tilde{\varphi}(x) := \varphi(x+v_0) - \varphi(v_0) - \langle \hat{v}_0, x\rangle, \\ &\tilde{F}(t,x) := F(t,x+v_0) - \hat{v}_0 \quad \text{and} \quad \tilde{G}(t,x) := G(t,x+v_0). \end{aligned}$$

Clearly $\tilde{\varphi}$, $\tilde{F}$, $\tilde{G}$ and $\tilde{\xi} = \xi - v_0$ satisfy the assumptions $(\mathbf{MM\text{-}H}_\varphi)$, $(\mathbf{MM\text{-}H}_F)$, $(\mathbf{MM\text{-}H}_G)$ and

$$0 \in \mathrm{int}(\mathrm{Dom}(\tilde{\varphi})), \qquad 0 = \tilde{\varphi}(0) \leq \tilde{\varphi}(x), \ \forall x \in \mathbb{R}^d,$$

that is $0 \in \partial\varphi(0)$. The solution (X,K) of Eq. (4.45) is given by: $X_t = \tilde{X}_t + v_0$, $K_t = \tilde{K}_t - \hat{v}_0 t$.

To prove the existence of the solution we shall need some additional assumptions (required by the method of proof). We shall assume that there exists a stochastic process $\alpha : \Omega \times [0,\infty[\to [0,\infty[$ such that for some $\rho > 0$

$$(\mathbf{MM\text{-}H}_{ad}): \tag{4.46}$$

$$\begin{cases} (i) & 0 \in \mathrm{int}(\mathrm{Dom}(\varphi)), \qquad 0 = \varphi(0) \leq \varphi(x), \ \forall x \in \mathbb{R}^d, \\ (ii) & \alpha, \mu^+, \ell^2, |G(\cdot,0)|^2 \in L^{1+\rho}_{loc}(\mathbb{R}_+), \ a.s., and \\ (iii) & \langle \hat{x}, F(t,x)\rangle \leq \alpha_t |\hat{x}| \left(1+|x|^3\right), \quad d\mathbb{P}\otimes dt\text{-}a.e., \forall\ (x,\hat{x}) \in \partial\varphi. \end{cases}$$

□

We mention that in the case of the convex indicator $\varphi(x) = I_{\overline{\mathcal{O}}}(x)$ the last assumption (iii) is equivalent to

$$\langle \mathbf{n}_x, F(t,x)\rangle \leq \alpha_t \left(1+|x|^3\right), \quad \forall x \in \mathrm{Bd}\left(\overline{\mathcal{O}}\right)$$

for all $x \in \mathrm{Bd}\left(\overline{\mathcal{O}}\right)$ and $\mathbf{n}_x$ any unit outward normal vector to $\overline{\mathcal{O}}$ at x. The main existence and uniqueness result, inspired by [13], is given in the next theorem.

Theorem 4.20. *If $\xi \in L^0(\Omega, \mathcal{F}_0, P; \overline{\mathrm{Dom}(\varphi)})$ and the assumptions ((4.17)-**MM-H**$_\varphi$), ((4.13)-**MM-H**$_F$), ((4.14)-**MM-H**$_G$), ((4.46)-**MM-H**$_{ad}$) are satisfied, then the SDE (4.45) has a unique solution $(X, K) \in S_d^0 \times S_d^0$. Moreover if there exist $p \geq 2$ and $u_0 \in \mathrm{int}(\mathrm{Dom}(\varphi))$ such that for all $T \geq 0$*

$$\mathbb{E}|\xi|^p + \mathbb{E}|\varphi(\xi)| + \mathbb{E}\left(\int_0^T |F(t,u_0)|\,dt\right)^p + \mathbb{E}\left(\int_0^T |G(t,u_0)|^2 dt\right)^{p/2} < +\infty, \tag{4.47}$$

then $X \in S_d^p$, $K \in S_d^{p/2}$ and $\mathbb{E}\, \updownarrow K \updownarrow_T^{p/2} < \infty$.

Proof. Uniqueness was proved in Theorem 4.11. Moreover by Corollary 4.9, if $(X, K) \in S_d^0 \times S_d^0$ is a solution and (4.47) is satisfied then $X \in S_d^p$, $K \in S_d^{p/2}$ and $\mathbb{E}\, \updownarrow K \updownarrow_T^{p/2} < \infty$ for every $T \geq 0$.

It suffices to prove existence on an arbitrary interval $[0, T]$. The proof of existence will be done in two steps.

We write

$$\Theta_t = \mu^+(t) + \ell^2(t) + \alpha_t + |G(t,0)|^2.$$

Step 1. Existence under the additional assumption: there exists an $M > 0$ such that $\mathbb{P}$-a.s.:

$$|\xi| + |\varphi(\xi)| + \int_0^T \left[|F(t,0)| + \Theta_t^{1+\rho}\right] dt \leq M < \infty. \tag{4.48}$$

Let $\varepsilon \in]0,1]$. We shall consider the penalized problem

$$\begin{cases} dX_t^\varepsilon + \nabla\varphi_\varepsilon(X_t^\varepsilon)dt = F(t, X_t^\varepsilon)dt + G(t, X_t^\varepsilon)dB_t, \\ X_0^\varepsilon = \xi, \end{cases} \tag{4.49}$$

where $\nabla\varphi_\varepsilon$ is the gradient of the Yosida's regularization φ_ε of the function φ, that is

$$\varphi_\varepsilon(x) = \inf\{\frac{1}{2\varepsilon}|z - x|^2 + \varphi(z) : z \in \mathbb{R}^d\}.$$

We write $J_\varepsilon x = x - \varepsilon\nabla\varphi_\varepsilon(x)$. For the convenience of the reader we recall from Annex B that $\varphi_\varepsilon : \mathbb{R}^d \to \mathbb{R}$ is a convex differentiable function and for all $x, y \in \mathbb{R}^d$, $\varepsilon > 0$:

$$\begin{array}{ll} a) & \nabla\varphi_\varepsilon(x) = \partial\varphi_\varepsilon(x) \in \partial\varphi(J_\varepsilon x), \text{ and } \varphi(J_\varepsilon x) \leq \varphi_\varepsilon(x) \leq \varphi(x), \\ b) & |\nabla\varphi_\varepsilon(x) - \nabla\varphi_\varepsilon(y)| \leq \dfrac{1}{\varepsilon}|x - y|, \\ c) & \langle \nabla\varphi_\varepsilon(x) - \nabla\varphi_\varepsilon(y), x - y\rangle \geq 0, \\ d) & \langle \nabla\varphi_\varepsilon(x) - \nabla\varphi_\delta(y), x - y\rangle \geq -(\varepsilon + \delta)\langle \nabla\varphi_\varepsilon(x), \nabla\varphi_\delta(y)\rangle. \end{array} \tag{4.50}$$

Since $0 = \varphi(0) \leq \varphi(x)$ for all $x \in \mathbb{R}^d$, we have, by (6.26), that

$$\begin{array}{ll}(a) & 0 = \varphi_\varepsilon(0) \leq \varphi_\varepsilon(x) \quad \text{and} \quad J_\varepsilon(0) = \nabla\varphi_\varepsilon(0) = 0, \\ (b) & \dfrac{\varepsilon}{2}|\nabla\varphi_\varepsilon(x)|^2 \leq \varphi_\varepsilon(x) \leq \langle\nabla\varphi_\varepsilon(x), x\rangle, \quad \forall x \in \mathbb{R}^d.\end{array} \tag{4.51}$$

The stochastic differential equation (4.49) satisfies the assumptions of Theorem 3.21 with $F(t,x)$ replaced by $F(t,x) - \nabla\varphi_\varepsilon(x)$. Hence Eq. (4.49) has a unique solution $X^\varepsilon \in S_d^p[0,T]$ for all $p \geq 2$. Moreover Eq. (4.49) is of the form (4.19) and as a consequence the estimates from Proposition 4.8, Corollaries 4.9 and 4.10 hold for all $p \geq 2$ and with $dK_t^\varepsilon = \nabla\varphi_\varepsilon(X_t^\varepsilon)dt$:

$$\begin{aligned}&\mathbb{E}\sup_{t\in[0,T]}|X_t^\varepsilon|^p + \mathbb{E}\left(\int_0^T|\nabla\varphi_\varepsilon(X_s^\varepsilon)|\,ds\right)^{p/2} + \mathbb{E}\left(\int_0^T\varphi_\varepsilon(X_s^\varepsilon)ds\right)^{p/2} \\ &\quad + \mathbb{E}\left(\int_0^T|F(s,X_s^\varepsilon)|\,ds\right)^{p/2} + \mathbb{E}\left(\int_0^T|G(s,X_s^\varepsilon)|^2\,ds\right)^{p/2} \leq C\end{aligned} \tag{4.52}$$

(in the proof we shall denote by C, C' generic constants independent of $\varepsilon, \delta \in]0,1]$, which can change from one line to another).

To pass to the limit as $\varepsilon \searrow 0$ some supplementary estimates are necessary.

A. *A priori estimate.*

Let ρ be the exponent appearing in the assumption ((4.46)-**MM-H**$_{ad}$). Since φ_ε is of class C^1 and $\nabla\varphi_\varepsilon$ is $\dfrac{1}{\varepsilon}$-Lipschitz continuous, from (2.38) we have the following inequality:

$$\begin{aligned}&\varphi_\varepsilon^{1+2\rho}(X_t^\varepsilon) + (1+2\rho)\int_0^t\varphi_\varepsilon^{2\rho}(X_s^\varepsilon)|\nabla\varphi_\varepsilon(X_s^\varepsilon)|^2ds \\ &\quad\leq \varphi_\varepsilon^{1+2\rho}(\xi) + (1+2\rho)\int_0^t\varphi_\varepsilon^{2\rho}(X_s^\varepsilon)\langle\nabla\varphi_\varepsilon(X_s^\varepsilon), F(s,X_s^\varepsilon)\rangle\,ds \\ &\qquad + (1+2\rho)\,\rho\int_0^t\varphi_\varepsilon^{2\rho-1}(X_s^\varepsilon)\,|\nabla\varphi_\varepsilon(X_s^\varepsilon)|^2|G(s,X_s^\varepsilon)|^2ds \\ &\qquad + \frac{(1+2\rho)}{2\varepsilon}\int_0^t\varphi_\varepsilon^{2\rho}(X_s^\varepsilon)|G(s,X_s^\varepsilon)|^2ds + \\ &\qquad + (1+2\rho)\int_0^t\varphi_\varepsilon^{2\rho}(X_s^\varepsilon)\langle\nabla\varphi_\varepsilon(X_s^\varepsilon), G(s,X_s^\varepsilon)dB_s\rangle\,.\end{aligned}$$

Then by Burkholder–Davis–Gundy's inequality and (4.51-b) we have

$$\begin{aligned}&\mathbb{E}\sup_{t\in[0,T]}(1+2\rho)\left|\int_0^t\varphi_\varepsilon^{2\rho}(X_s^\varepsilon)\langle\nabla\varphi_\varepsilon(X_s^\varepsilon), G(s,X_s^\varepsilon)dB_s\rangle\right| \\ &\quad\leq 3(1+2\rho)\,\mathbb{E}\left(\int_0^t\varphi_\varepsilon^{4\rho}(X_s^\varepsilon)\,|\nabla\varphi_\varepsilon(X_s^\varepsilon)|^2|G(s,X_s^\varepsilon)|^2ds\right)^{1/2}\end{aligned}$$

$$\leq C\ \mathbb{E}\left(\int_0^t \frac{1}{\varepsilon}\varphi_\varepsilon^{4\rho+1}\left(X_s^\varepsilon\right)\left|G\left(s,X_s^\varepsilon\right)\right|^2 ds\right)^{1/2}$$

$$\leq C\ \mathbb{E}\sup_{t\in[0,T]}\varphi_\varepsilon^{\frac{1}{2}+\rho}\left(X_t^\varepsilon\right)\left(\int_0^t \frac{1}{\varepsilon}\varphi_\varepsilon^{2\rho}\left(X_s^\varepsilon\right)\left|G\left(s,X_s^\varepsilon\right)\right|^2\right)^{1/2}$$

$$\leq \frac{1}{4}\mathbb{E}\sup_{t\in[0,T]}\varphi_\varepsilon^{1+2\rho}\left(X_t^\varepsilon\right)+\frac{C}{\varepsilon}\mathbb{E}\int_0^t \varphi_\varepsilon^{2\rho}\left(X_s^\varepsilon\right)\left|G\left(s,X_s^\varepsilon\right)\right|^2 ds$$

and, since $0<\varepsilon\leq 1$, from assumptions (**MM-H**$_F$) and (**MM-H**$_{ad}$) we deduce that

$$\begin{aligned}
&\langle\nabla\varphi_\varepsilon(X_s^\varepsilon),F(s,X_s^\varepsilon)\rangle\\
&\quad=\frac{1}{\varepsilon}\left\langle X_s^\varepsilon-J_\varepsilon(X_s^\varepsilon),F(s,X_s^\varepsilon)\right\rangle\\
&\quad=\frac{1}{\varepsilon}\left\langle X_s^\varepsilon-J_\varepsilon(X_s^\varepsilon),F(s,X_s^\varepsilon)-F(s,J_\varepsilon\left(X_s^\varepsilon\right))\right\rangle+\left\langle\nabla\varphi_\varepsilon(X_s^\varepsilon),F(s,J_\varepsilon\left(X_s^\varepsilon\right))\right\rangle\\
&\quad\leq\frac{1}{\varepsilon}\mu\left(s\right)\left|X_s^\varepsilon-J_\varepsilon(X_s^\varepsilon)\right|^2+\alpha_s\left|\nabla\varphi_\varepsilon(X_s^\varepsilon)\right|\left[1+\left|J_\varepsilon(X_s^\varepsilon)\right|^3\right]\\
&\quad\leq\frac{1}{\varepsilon}\mu^+\left(s\right)\left|X_s^\varepsilon\right|^2+\frac{1}{\varepsilon}\alpha_s\left|X_s^\varepsilon\right|+\frac{1}{\varepsilon}\alpha_s\left|X_s^\varepsilon\right|^4.
\end{aligned}$$

We also have

$$\left|G\left(s,X_s^\varepsilon\right)\right|^2\leq 2\left|G\left(s,0\right)\right|^2+2\ell^2\left(s\right)\left|X_s^\varepsilon\right|^2.$$

Then

$$\begin{aligned}
&\mathbb{E}\sup_{t\in[0,T]}\varphi_\varepsilon^{1+2\rho}\left(X_t^\varepsilon\right)+\mathbb{E}\int_0^T\varphi_\varepsilon^{2\rho}\left(X_s^\varepsilon\right)\left|\nabla\varphi_\varepsilon\left(X_s^\varepsilon\right)\right|^2 ds\\
&\qquad\leq C\ \mathbb{E}\varphi^{1+2\rho}(\xi)+\frac{C}{\varepsilon}\mathbb{E}\int_0^T\varphi_\varepsilon^{2\rho}\left(X_s^\varepsilon\right)Q_s^\varepsilon ds,
\end{aligned}\tag{4.53}$$

where

$$\begin{aligned}
Q_s^\varepsilon&=\left|G\left(s,0\right)\right|^2+\alpha_s\left|X_s^\varepsilon\right|+\left[\mu^+\left(s\right)+\ell^2\left(s\right)\right]\left|X_s^\varepsilon\right|^2+\alpha_s\left|X_s^\varepsilon\right|^4\\
&\leq C\left(1+\left\|X^\varepsilon\right\|_T^4\right)\Theta_s.
\end{aligned}$$

Let $r=\frac{1+\rho}{\rho}$. By (4.51-b) we infer

$$\begin{aligned}
&\frac{C}{\varepsilon}\varphi_\varepsilon^{2\rho}\left(X^\varepsilon\right)Q^\varepsilon\\
&\quad=\left(\varphi_\varepsilon\left(X^\varepsilon\right)\right)^{2\rho-\frac{2}{r}}\left(\varphi_\varepsilon\left(X^\varepsilon\right)\right)^{\frac{2}{r}}\left|Q^\varepsilon\right|\frac{C}{\varepsilon}
\end{aligned}$$

$$
\begin{aligned}
&\leq (\varphi_\varepsilon (X^\varepsilon))^{2\rho-\frac{2}{r}} |\nabla\varphi_\varepsilon (X^\varepsilon)|^{\frac{2}{r}} |X^\varepsilon|^{\frac{2}{r}} |Q^\varepsilon| \frac{C}{\varepsilon} \\
&\leq \frac{1}{r}\left[(\varphi_\varepsilon (X^\varepsilon))^{2\rho-\frac{2}{r}} |\nabla\varphi_\varepsilon (X^\varepsilon)|^{\frac{2}{r}}\right]^r + \frac{r-1}{r}\left[|X^\varepsilon|^{\frac{2}{r}} |Q^\varepsilon| \frac{C}{\varepsilon}\right]^{r/(r-1)} \\
&= \frac{\rho}{1+\rho}\varphi_\varepsilon^{2\rho} (X^\varepsilon) |\nabla\varphi_\varepsilon (X^\varepsilon)|^2 + \frac{C'}{\varepsilon^{1+\rho}} |X^\varepsilon|^{2\rho} |Q^\varepsilon|^{1+\rho},
\end{aligned}
$$

and it follows that

$$
\begin{aligned}
&\mathbb{E} \sup_{t\in[0,T]} \varphi_\varepsilon^{1+2\rho} (X_t^\varepsilon) + \frac{1}{1+\rho}\mathbb{E}\int_0^T \varphi_\varepsilon^{2\rho} (X_s^\varepsilon) \left|\nabla\varphi_\varepsilon (X_s^\varepsilon)\right|^2 ds \\
&\leq C\, \mathbb{E}\varphi^{1+2\rho}(\xi) + \frac{C}{\varepsilon^{1+\rho}}\mathbb{E}\int_0^T |X_s^\varepsilon|^{2\rho} |Q_s^\varepsilon|^{1+\rho} ds \\
&\leq C' + \frac{C'}{\varepsilon^{1+\rho}}\mathbb{E}\left[\left(1+\|X^\varepsilon\|_T^{4+6\rho}\right)\int_0^T \Theta_s^{1+\rho} ds\right].
\end{aligned}
$$

Hence

$$
\mathbb{E} \sup_{t\in[0,T]} \left(\varphi_\varepsilon^{1+2\rho} (X_t^\varepsilon)\right) + \frac{1}{1+\rho}\mathbb{E}\int_0^T \varphi_\varepsilon^{2\rho} (X_s^\varepsilon) \left|\nabla\varphi_\varepsilon (X_s^\varepsilon)\right|^2 ds \leq \frac{C}{\varepsilon^{1+\rho}}. \tag{4.54}
$$

By (4.51-b)

$$
\left(\frac{\varepsilon}{2}\right)^{1+2\rho} \left|\nabla\varphi_\varepsilon (X_t^\varepsilon)\right|^{2(1+2\rho)} \leq \varphi_\varepsilon^{1+2\rho} (X_t^\varepsilon).
$$

Hence

$$
\begin{cases}
a) \quad \mathbb{E} \sup\limits_{t\in[0,T]} \left|\nabla\varphi_\varepsilon (X_t^\varepsilon)\right|^{2+4\rho} \leq \dfrac{C}{\varepsilon^{2+3\rho}}, \\
b) \quad \mathbb{E} \sup\limits_{t\in[0,T]} \left|X_t^\varepsilon - J_\varepsilon (X_t^\varepsilon)\right|^{2+4\rho} \leq C\, \varepsilon^\rho.
\end{cases} \tag{4.55}
$$

B. X^ε is a Cauchy sequence in $S_d^2\,[0,T]$.

Let $\varepsilon, \delta \in]0,1]$. Itô's formula for $\left|X_t^\varepsilon - X_t^\delta\right|^2$ gives:

$$
\begin{aligned}
&\left|X_t^\varepsilon - X_t^\delta\right|^2 + 2\int_0^t \left\langle \nabla\varphi_\varepsilon (X_s^\varepsilon) - \nabla\varphi_\delta (X_s^\delta), X_s^\varepsilon - X_s^\delta \right\rangle ds \\
&= \int_0^t \left[2\left\langle X_s^\varepsilon - X_s^\delta, F(s, X_s^\varepsilon) - F(s, X_s^\delta)\right\rangle + \left|G(s, X_s^\varepsilon) - G(s, X_s^\delta)\right|^2\right] ds \\
&\quad +2\int_0^t \left\langle X_s^\varepsilon - X_s^\delta, (G(s, X_s^\varepsilon) - G(s, X_s^\delta))\, dB_s\right\rangle.
\end{aligned}
$$

By (**MM-H**$_F$), (**MM-H**$_G$) and (4.50) we have

$$\left|X_t^\varepsilon - X_t^\delta\right|^2 \leq 2\left(\varepsilon+\delta\right)\int_0^T \left|\nabla\varphi_\varepsilon\left(X^\varepsilon\right)\right|\left|\nabla\varphi_\delta\left(X^\delta\right)\right|ds \\ + \int_0^t\left[2\mu^+\left(s\right)+\ell^2\left(s\right)\right]\left|X_s^\varepsilon - X_s^\delta\right|^2 + \int_0^t G_s^{\varepsilon,\delta}dB_s,$$

where

$$G_s^{\varepsilon,\delta} = \left(X_s^\varepsilon - X_s^\delta\right)^*\left(G\left(s,X_s^\varepsilon\right) - G\left(s,X_s^\delta\right)\right).$$

Since $\left|G_s^{\varepsilon,\delta}\right| \leq \ell\left(s\right)\left|X_s^\varepsilon - X_s^\delta\right|^2$ it follows, by the stochastic Gronwall inequality, Proposition 6.68, with $q=1$ and $X_t = \left|X_t^\varepsilon - X_t^\delta\right|^2$, that

$$\mathbb{E}\sup_{s\in[0,T]}\left|X_s^\varepsilon - X_s^\delta\right|^2 \\ \leq 2\left(\varepsilon+\delta\right)\left[\int_0^T\left|\nabla\varphi_\varepsilon\left(X^\varepsilon\right)\right|\left|\nabla\varphi_\delta\left(X^\delta\right)\right|ds\right]\times e^{C+C\int_0^T\left[\mu^+(s)+\ell^2(s)\right]ds}.$$

But by (4.55) and (4.52)

$$\left(\varepsilon+\delta\right)\mathbb{E}\int_0^T\left|\nabla\varphi_\varepsilon\left(X_s^\varepsilon\right)\right|\left|\nabla\varphi_\delta\left(X_s^\delta\right)\right|ds$$

$$\leq \varepsilon\mathbb{E}\sup_{t\in[0,T]}\left|\nabla\varphi_\varepsilon\left(X_t^\varepsilon\right)\right|\int_0^T\left|\nabla\varphi_\delta\left(X_s^\delta\right)\right|ds \\ + \delta\mathbb{E}\sup_{t\in[0,T]}\left|\nabla\varphi_\delta\left(X_t^\delta\right)\right|\int_0^T\left|\nabla\varphi_\varepsilon\left(X_s^\varepsilon\right)\right|ds$$

$$\leq \varepsilon\left(\mathbb{E}\sup_{t\in[0,T]}\left|\nabla\varphi_\varepsilon\left(X_t^\varepsilon\right)\right|^{2+4\rho}\right)^{\frac{1}{2+4\rho}}\left[\mathbb{E}\left(\int_0^T\left|\nabla\varphi_\delta\left(X_s^\delta\right)\right|ds\right)^{\frac{2+4\rho}{1+4\rho}}\right]^{\frac{1+4\rho}{2+4\rho}} \\ + \delta\left(\mathbb{E}\sup_{t\in[0,T]}\left|\nabla\varphi_\delta\left(X_t^\delta\right)\right|^{2+4\rho}\right)^{\frac{1}{2+4\rho}}\left[\mathbb{E}\left(\int_0^T\left|\nabla\varphi_\varepsilon\left(X_s^\varepsilon\right)\right|ds\right)^{\frac{2+4\rho}{1+4\rho}}\right]^{\frac{1+4\rho}{2+4\rho}}$$

$$\leq \varepsilon\left(\frac{C}{\varepsilon^{2+3\rho}}\right)^{\frac{1}{2+4\rho}}\left[\mathbb{E}\left(\int_0^T\left|\nabla\varphi_\delta\left(X_s^\delta\right)\right|ds\right)^{2+4\rho}\right]^{\frac{1}{2+4\rho}} \\ + \delta\left(\frac{C}{\delta^{2+3\rho}}\right)^{\frac{1}{2+4\rho}}\left[\mathbb{E}\left(\int_0^T\left|\nabla\varphi_\varepsilon\left(X_s^\varepsilon\right)\right|ds\right)^{2+4\rho}\right]^{\frac{1}{2+4\rho}}$$

$$\leq C_1\left(\varepsilon^{\frac{\rho}{2+4\rho}} + \delta^{\frac{\rho}{2+4\rho}}\right),$$

since $\frac{2+4\rho}{1+4\rho} \le 2 + 4\rho$ and $\left(\mathbb{E}\,|\eta|^a\right)^{1/a} \le \left(\mathbb{E}\,|\eta|^b\right)^{1/b}$ for all $0 < a \le b$.

Hence

$$\mathbb{E} \sup_{s\in[0,t]} \left|X_s^\varepsilon - X_s^\delta\right|^2 \le C\left(\varepsilon^{\frac{\rho}{2+4\rho}} + \delta^{\frac{\rho}{2+4\rho}}\right). \tag{4.56}$$

D. Passing to the limit.

Write $K_t^\varepsilon = \int_0^t \nabla\varphi_\varepsilon(X_s^\varepsilon)ds$ and $F_t^\varepsilon = \int_0^t F\left(s, X_s^\varepsilon\right) ds$.

From (4.56) it follows that there exists an $X \in S_d^2\,[0,T]$ such that

$$\lim_{\varepsilon\searrow 0} X^\varepsilon = X \quad \text{in } S_d^2\,[0,T]$$

and by (4.55-b)

$$\lim_{\varepsilon\searrow 0} J_\varepsilon\left(X^\varepsilon\right) = X \quad \text{in } S_d^2\,[0,T]\,.$$

By the assumptions (**MM-H**$_F$) and (**MM-H**$_G$) we can pass to the limit in the approximating equation

$$X_t^\varepsilon + K_t^\varepsilon = \xi + \int_0^t F\left(s, X_s^\varepsilon\right) ds + \int_0^t G\left(s, X_s^\varepsilon\right) dW\left(s\right).$$

Hence there exists a $K \in S_d^0\,[0,T]$ such that

$$\lim_{\varepsilon\searrow 0} K^\varepsilon = K \quad \text{in } S_d^0\,[0,T]$$

and

$$X_t + K_t = \xi + \int_0^t F\left(s, X_s\right) ds + \int_0^t G\left(s, X_s\right) dW\left(s\right).$$

Finally

$$\|X^\varepsilon - X\|_T + \|J_\varepsilon\left(X^\varepsilon\right) - X\|_T + \|K^\varepsilon - K\|_T \xrightarrow{prob.} 0, \quad \text{as } \varepsilon \to 0.$$

Using Propositions 1.17 and 1.20 and $0 \le \varphi(J_\varepsilon\left(X^\varepsilon\right)) \le \varphi_\varepsilon(X^\varepsilon)$, we infer from (4.52) that for all $p \ge 2$

$$\mathbb{E} \sup_{t\in[0,T]} |X_s|^p + \mathbb{E}\left(\int_0^T \varphi(X_s)ds\right)^{p/2} + \mathbb{E}\, \updownarrow K \updownarrow_T^{p/2} \le C < \infty.$$

Since $dK_t^\varepsilon = \nabla\varphi_\varepsilon(X_t^\varepsilon)dt \in \partial\varphi\left(J_\varepsilon\left(X_t^\varepsilon\right)\right)dt$, it follows, by Corollary 1.22, that $dK_t \in \partial\varphi(X_t)dt$. Hence (X,K) is a solution of the Eq. (4.19) and for all $p \geq 2$: $X \in S_d^p$, $K \in S_d^{p/2}$ and $\mathbb{E}\ \updownarrow K \updownarrow_T^{p/2} < \infty$ for every $T \geq 0$.

Step 2. Existence without the boundedness condition (4.48).

Let $n \in \mathbb{N}^*$ and the stopping time

$$\theta_n = \inf\left\{t \geq 0 : |\xi| + |\varphi(\xi)| + \int_0^t \left[|F(s,0)| + \Theta_s^{1+\rho}\right] ds \geq n\right\}$$

where

$$\Theta_t = \mu^+(t) + \ell^2(t) + \alpha_t + |G(t,0)|^2.$$

Clearly $\theta_n \nearrow \infty$ *a.s.* By the first part there exists a unique solution (X^n, K^n) of the approximating equation

$$\begin{cases} X_t^n + K_t^n = \xi \mathbf{1}_{\theta_n>0} + \int_0^t \mathbf{1}_{[0,\theta_n]}(r) F\left(r, X_r^n\right) dr + \int_0^t \mathbf{1}_{[0,\theta_n]}(r) G\left(r, X_r^n\right) dB_r, \\ dK_t^n \in \partial\varphi\left(X_t^n\right)(dt), \end{cases}$$

and $X^n \in S_d^p$, $K^n \in S_d^{p/2}$ and $\mathbb{E}\ \updownarrow K^n \updownarrow_T^{p/2} < \infty$ for every $T \geq 0$.

It is easy to deduce from uniqueness that for all $l \in \mathbb{N}^*$,

$$\left(X^{n+i}, K^{n+i}\right) \mathbf{1}_{\theta_n>0}\mathbf{1}_{[0,\theta_n]} = \left(X^n, K^n\right) \mathbf{1}_{\theta_n>0}\mathbf{1}_{[0,\theta_n]}.$$

Consequently the pair of stochastic processes $X, K \in S_d^0$ defined by

$$\left(X_t(\omega), K_t(\omega)\right) = \mathbf{1}_{\theta_n>0}\left(X_t^n(\omega), K_t^n(\omega)\right), \quad \text{if } 0 \leq t < \theta_n(\omega)$$

is a solution of the SDE (3.51). The proof is complete. ■

From the proof of Theorem 4.20 we infer:

Remark 4.21. It follows from the above proof that whenever $|\xi| + |\varphi(\xi)|$ is bounded and condition (4.48) holds there exists a constant C such that

$$\mathbb{E} \sup_{t\in[0,T]} |X_t^\varepsilon - X_t|^2 \leq C\, \varepsilon^{\frac{1}{4}\left(1-\frac{1}{1+2\rho}\right)}. \tag{4.57}$$

A particular case of SDE (4.148) is the reflected SDE

$$\begin{cases} dX_t + \partial I_{\overline{\mathcal{O}}}(X_t)(dt) \ni F(t,X_t)dt + G(t,X_t)dB_t, \\ X_0 = x_0 \in \overline{\mathcal{O}}, \quad t \in [0,T], \end{cases} \tag{4.58}$$

where $\mathcal{O}$ is an open convex subset of $\mathbb{R}^d$ and $I_{\overline{\mathcal{O}}}$ is the convex indicator of $\overline{\mathcal{O}}$:

$$I_{\overline{D}}(x) = \begin{cases} 0, & \text{if } x \in \overline{\mathcal{O}}, \\ +\infty, & \text{if } x \in \mathbb{R}^d \setminus \overline{\mathcal{O}}. \end{cases}$$

Then

$$\partial I_{\overline{\mathcal{O}}}(x) = \begin{cases} 0, & \text{if } \ x \in \mathcal{O}, \\ \left\{\mathbf{n} \in \mathbb{R}^d : \langle \mathbf{n}, y - x \rangle \le 0, \ \forall \ y \in \overline{\mathcal{O}}\right\}, & \text{if } \ x \in \mathrm{Bd}\,(\mathcal{O}), \\ \emptyset, & \text{if } \ x \in \mathbb{R}^d \setminus \overline{\mathcal{O}}, \end{cases}$$

that is $\partial I_{\overline{\mathcal{O}}}(x)$ is the exterior normal cone to $\overline{\mathcal{O}}$ if $x \in \mathrm{Bd}\,(\mathcal{O})$.

Assuming that $\mathbb{P}$-a.s.

$$|F(t,x)| \le a + a\,|x|^3, \quad \forall x \in \mathrm{Bd}\left(\overline{\mathcal{O}}\right)$$

and for some $\rho > 0$

$$\int_0^T \left[|F(t,0)| + \left[\mu^+(t) + \ell^2(t) + |G(t,0)|^2\right]^{1+\rho} \right] dt \le M < \infty,$$

then by Theorem 4.20, there exists a unique pair $X \in S_d^p$, $K \in S_d^{p/2}$ and $\mathbb{E}\ \updownarrow K \updownarrow_T^{p/2} < \infty$ for every $T \ge 0$ for all $p \ge 2$, such that

$$X_t(\omega) \in \overline{\mathcal{O}}, \ \forall \ t \in [0,T], \ a.s. \ \omega \in \Omega$$

and

$$\begin{cases} j) \quad X_t + K_t = x_0 + \displaystyle\int_0^t F(s, X_s)ds + \int_0^t G(s, X_s)dB_s \\ \qquad\qquad\qquad\qquad\qquad\qquad \forall t \in [0,T]; \ a.s. \ \omega \in \Omega, \\ jj) \quad \displaystyle\int_s^t (z - X_r, dK_r) \le 0, \quad \forall z \in \overline{\mathcal{O}}, \\ \qquad\qquad\qquad\qquad\qquad\qquad \forall\, 0 \le s \le t \le T; \ a.s. \ \omega \in \Omega. \end{cases} \tag{4.59}$$

Note that the condition *jj*) is equivalent to:

$$jj') \quad K_t - K_s = \int_s^t \mathbf{n}(X_r)\, d \ \updownarrow K \updownarrow_r \ , \ \text{and} \qquad \int_s^t 1_{\mathcal{O}}(X_r) d \ \updownarrow K \updownarrow_r = 0$$

$\forall\, 0 \le s \le t \le T;\ a.s.\omega \in \Omega$, where $\mathbf{n}(x)$ is the unitary exterior normal vector to $\overline{\mathcal{O}}$ at $x \in \mathrm{Bd}\left(\overline{\mathcal{O}}\right)$ and $\mathbf{n}(x) = 0$ if $x \in \mathcal{O}$.

In this case the estimate of the speed of convergence (4.57) holds with X^ε defined by the approximating equation (4.49) of the following particular form

$$\begin{cases} dX_t^\varepsilon + \dfrac{1}{\varepsilon}\left(X_t^\varepsilon - \Pr_{\overline{\mathcal{O}}}(X_t^\varepsilon)\right) dt = F(t, X_t^\varepsilon)dt + G(t, X_t^\varepsilon)dB_t, \\ X_0^\varepsilon = \xi, \end{cases} \tag{4.60}$$

where $\Pr_{\overline{\mathcal{O}}}(x)$ is the projection of $x \in \mathbb{R}^d$ onto $\overline{\mathcal{O}}$.

4.3 Reflected SDEs

4.3.1 *The Generalized Skorohod Problem*

4.3.1.1 Preliminaries

For convenience we recall from Annex *B* some definitions and remarks.

Let E be a non-empty closed subset of $\mathbb{R}^d$ and $N_E(x)$ be the closed external normal cone of E at $x \in \mathrm{Bd}(E)$, i.e.

$$N_E(x) \overset{def}{=} \left\{ u \in \mathbb{R}^d : \lim_{\delta \searrow 0} \frac{d_E(x + \delta u)}{\delta} = |u| \right\},$$

where

$$d_E(z) \overset{def}{=} \inf\{|z - x| : x \in E\}$$

is the distance of a point $z \in \mathbb{R}^d$ to E.

Let $\varepsilon > 0$. We denote by

$$U_\varepsilon(E) \overset{def}{=} \left\{ y \in \mathbb{R}^d : d_E(y) < \varepsilon \right\}$$

the open ε-neighbourhood of E and

$$\overline{U}_\varepsilon(E) \overset{def}{=} \left\{ z \in \mathbb{R}^d : d_E(z) \leq \varepsilon \right\}$$

the closed ε-neighbourhood of E.

Given $z \in \mathbb{R}^d$ and E a non-empty closed subset of $\mathbb{R}^d$, we denote by $\Pi_E(z)$ the set of elements $x \in E$ with $|z - x| = d_E(z)$. We note that $\Pi_E(z)$ is non-empty since E is non-empty and closed.

Recall the notation $B(x, r) = \left\{ y \in \mathbb{R}^d : |y - x| < r \right\}$.

Definition 4.22. Let $r_0 > 0$. We say that E satisfies the r_0-uniform exterior ball condition, abbreviated r_0-$UEBC$, if

- $N_E(x) \neq \{0\}$ for all $x \in \mathrm{Bd}(E)$,
- $\forall\, x \in \mathrm{Bd}(E)$ and $\forall\, u \in N_E(x)$, $|u| = r_0$, we have:

$$d_E(x+u) = r_0$$

(or equivalently $B(x+u, r_0) \cap E = \emptyset$).

Note that for all $v \in N_E(x)$, $|v| \leq r_0$, we also have

$$d_E(x+v) = |v|\,. \tag{4.61}$$

Definition 4.23. Let $\gamma \geq 0$. A set $E = \overline{E} \subset \mathbb{R}^d$ is γ–semiconvex if for all $x \in \mathrm{Bd}(E)$ there exists a $\hat{x} \in \mathbb{R}^d \setminus \{0\}$ such that

$$\langle \hat{x}, y - x \rangle \leq \gamma\, |\hat{x}|\, |y-x|^2\,; \quad \forall y \in E.$$

We have the following equivalence (for other equivalences see Lemma 6.47):

Lemma 4.24. *Let $r_0 > 0$ and $E = \overline{E} \subset \mathbb{R}^d$. Then E satisfies the r_0-$UEBC$ if and only if E is $\frac{1}{2r_0}$–semiconvex.*

It is clear that, under the uniform exterior ball condition with radius r_0, for all $z \in \mathbb{R}^d$ with $d_E(z) < r_0$, the set $\Pi_E(z)$ is a singleton. The unique element of $\Pi_E(z)$ is called the projection of z on E, and it is denoted by $\pi_E(z)$. We have, see Corollary 6.49 and Lemma 6.47 in the Annex B.

Lemma 4.25. *Let the uniform exterior ball condition with radius r_0 be satisfied and $\varepsilon \in]0, r_0[$. Then*

$$N_E(x) = \left\{ \hat{x} : \langle \hat{x}, y - x \rangle \leq \frac{1}{2r_0}\, |\hat{x}|\, |y-x|^2\,; \quad \forall y \in E \right\},$$

the projection π_E restricted to the closed ε-neighbourhood of E, $\overline{U}_\varepsilon(E)$, is Lipschitz with Lipschitz constant $L_\varepsilon = r_0/(r_0 - \varepsilon)$, and the function d_E^2 is of class C^1 on $\overline{U}_\varepsilon(E)$ with

$$\frac{1}{2}\nabla d_E^2(z) = z - \pi_E(z)\,, \quad \textit{and} \quad z - \pi_E(z) \in N_E(\pi_E(z))$$

for all $z \in \overline{U}_\varepsilon(E)$.

A function $\varphi : \mathbb{R}^d \to]-\infty, +\infty]$ is proper if

$$\mathrm{Dom}(\varphi) \overset{def}{=} \left\{ v \in \mathbb{R}^d : \varphi(v) < +\infty \right\} \neq \emptyset$$

and $\mathrm{Dom}(\varphi)$ has no isolated point.

Definition 4.26. The (Fréchet) subdifferential of φ at $x \in \mathbb{R}^d$ is defined by $\partial^- \varphi(x) = \emptyset$, if $x \notin \mathrm{Dom}(\varphi)$ and for $x \in \mathrm{Dom}(\varphi)$,

$$\partial^- \varphi(x) = \left\{ \hat{x} \in \mathbb{R}^d : \quad \liminf_{y \to x} \frac{\varphi(y) - \varphi(x) - \langle \hat{x}, y - x \rangle}{|y - x|} \geq 0 \right\}.$$

We moreover write

$$\begin{aligned} &a) \quad \mathrm{Dom}(\partial^- \varphi) = \left\{ x \in \mathbb{R}^d : \partial^- \varphi(x) \neq \emptyset \right\}, \\ &b) \quad \partial^- \varphi = \left\{ (x, \hat{x}) : \ x \in \mathrm{Dom}(\partial^- \varphi), \ \hat{x} \in \partial^- \varphi(x) \right\}. \end{aligned}$$

If E is a non-empty closed subset of $\mathbb{R}^d$ and

$$\varphi(x) = I_E(x) = \begin{cases} 0, & \text{if } x \in E, \\ +\infty, & \text{if } x \notin E, \end{cases}$$

then φ is l.s.c. and

$$\partial^- I_E(x) = \left\{ \hat{x} \in \mathbb{R}^d : \quad \limsup_{y \to x,\ y \in E} \frac{\langle \hat{x}, y - x \rangle}{|y - x|} \leq 0 \right\}$$

is the *Fréchet normal cone* at E in x. By a result of Colombo and Goncharov [17] we have for any closed subset E of a Hilbert space

$$\partial^- I_E(x) = N_E(x).$$

Definition 4.27. $\varphi : \mathbb{R}^d \to]-\infty, +\infty]$ is a semiconvex function if there exist $\rho, \gamma \geq 0$ such that

(*i*) $\overline{\mathrm{Dom}(\varphi)}$ is γ–semiconvex.
(*ii*) $\mathrm{Dom}(\partial^- \varphi) \neq \emptyset$.
(*iii*) For all $(x, \hat{x}) \in \partial^- \varphi$, $y \in \mathbb{R}^d$:

$$\langle \hat{x}, y - x \rangle + \varphi(x) \leq \varphi(y) + (\rho + \gamma |\hat{x}|) |y - x|^2.$$

A function φ satisfying the properties of this definition will sometimes be called a (ρ, γ)–semiconvex function, or a γ–semiconvex function (since the second parameter is the most important one).

Note that $\varphi = I_E$ is $(0, \gamma)$–semiconvex iff E is γ–semiconvex.

A convex function is a (ρ, γ)–semiconvex function for all $\rho \geq 0$ and $\gamma \geq 0$.

The set $E = \overline{E} \subset \mathbb{R}^d$ is 0–semiconvex if and only if E is convex.

If $\varphi : \mathbb{R}^d \to]-\infty, +\infty]$ is a semiconvex function, then there exists an $a \in \mathbb{R}$ such that

$$\varphi(y) + a |y|^2 + a \geq 0 \quad \text{for all } y \in \mathbb{R}^d.$$

Example 4.28. If E is a closed bounded subset of $\mathbb{R}^d$ satisfying the uniform exterior ball condition and $g \in C^2(\mathbb{R}^d)$ (or $g \in C(\mathbb{R}^d)$ is a convex function), then $f : \mathbb{R}^d \to]-\infty, +\infty]$, $f(x) = I_E(x) + g(x)$ is a l.s.c. semiconvex function. Moreover

$$|f(x) - f(y)| \leq L|x - y|, \ \forall\, x, y \in \mathrm{Dom}(f) = E.$$

4.3.1.2 The Generalized Skorohod Problem

Our aim is to solve a Cauchy type ordinary differential equation, written formally as

$$\begin{cases} dx(t) + \partial^- \varphi(x(t))(dt) \ni dm(t), & t > 0, \\ x(0) = x_0, \end{cases} \tag{4.62}$$

where

$$\begin{cases} (i) & x_0 \in \overline{\mathrm{Dom}(\varphi)}, \\ (ii) & m \in C(\mathbb{R}_+; \mathbb{R}^d), \quad m(0) = 0, \end{cases} \tag{4.63}$$

and

$$\varphi : \mathbb{R}^d \to]-\infty, +\infty] \text{ is a proper l.s.c. } (\rho, \gamma)\text{–semiconvex function.} \tag{4.64}$$

Definition 4.29 (Generalized Skorohod Problem). A pair (x, k) of continuous functions $x, k : [0, \infty[\to \mathbb{R}^d$ is a solution of Eq. (4.62) if

$$\begin{cases} (j) & x(t) \in \overline{\mathrm{Dom}(\varphi)}, \ \forall\, t \geq 0, \ \varphi(x(\cdot)) \in L^1_{loc}(\mathbb{R}_+), \\ (jj) & k \in BV_{loc}(\mathbb{R}_+; \mathbb{R}^d), \quad k(0) = 0, \\ (jjj) & x(t) + k(t) = x_0 + m(t), \quad \forall\, t \geq 0, \\ (jv) & \forall\, 0 \leq s \leq t, \ \forall y : \mathbb{R}_+ \to \mathbb{R}^d \text{ continuous:} \\ & \displaystyle\int_s^t \langle y(r) - x(r), dk(r)\rangle + \int_s^t \varphi(x(r))\, dr \\ & \displaystyle\qquad \leq \int_s^t \varphi(y(r))\, dr + \int_s^t |y(r) - x(r)|^2 (\rho dr + \gamma d\, \updownarrow k \updownarrow_r). \end{cases} \tag{4.65}$$

A solution of (4.65) is called a solution of the generalized Skorohod problem $(\partial^- \varphi; x_0, m)$ (and we write $(x, k) = \mathcal{SP}(\partial^- \varphi; x_0, m,))$. If $\partial^- \varphi = \partial^- I_E = N_E$ then we say that (x, k) is a solution of the Skorohod problem $(E; x_0, m)$ and we write $(x, k) = \mathcal{SP}(E; x_0, m)$.

Remark 4.30. The relation

$$dk(t) \in \partial^- \varphi(x(t))(dt)$$

will mean that $x, k : [0, \infty[\to \mathbb{R}^d$ are continuous functions satisfying the conditions (4.65-j, jj, jv).

Lemma 4.31. *Let the assumption (4.64) be satisfied. Let $x, k : [0, \infty[\to \mathbb{R}^d$ be continuous functions. Then the following assertions are equivalent:*

(a_1) *(x, k) satisfies (4.65-j, jj, jv)*

(a_2) *(x, k) satisfies (4.65-j, jj) and there exists a continuous increasing function $A : [0, \infty[\to [0, \infty[$ such that*

$$\begin{aligned}(\text{jv}') \quad & \forall\, 0 \leq s \leq t,\ \forall y : [0, \infty[\to \mathbb{R}^d \text{ continuous:} \\ & \int_s^t \langle y(r) - x(r), dk(r)\rangle + \int_s^t \varphi(x(r))\, dr \\ & \quad \leq \int_s^t \varphi(y(r))\, dr + \int_s^t |y(r) - x(r)|^2\, dA(r).\end{aligned} \tag{4.66}$$

Proof. Clearly $(a_1) \Rightarrow (a_2)$ setting $A(t) = \rho t + \gamma \updownarrow k \updownarrow_t$.
Let us prove that $(a_2) \Rightarrow (a_1)$.

Define

$$a(r) := r + \updownarrow k \updownarrow_r + A(r)$$

and let the measurable functions θ, λ, η (given by the Radon–Nikodym theorem) such that

$$dk(r) = \theta(r)\, da(r), \quad dr = \lambda(r)\, da(r) \quad \text{and} \quad dA_r = \eta(r)\, da(r).$$

Clearly $d \updownarrow k \updownarrow_r = |\theta(r)|\, da(r)$ and $0 \leq \lambda(r) \leq 1$, $da(r)$-a.e., and, from (4.66) we deduce that, for all $0 \leq s \leq t$ and $z \in \text{Dom}(\varphi)$

$$\begin{aligned}&\left\langle z, \int_s^t \theta(r)\, dQ_r\right\rangle - \int_s^t \langle x(r), \theta(r)\rangle\, dQ_r + \int_s^t \varphi(x(r))\, \lambda(r)\, dQ_r \\ &\quad \leq \varphi(z) \int_s^t \lambda(r)\, dQ_r + |z|^2 \int_s^t \eta(r)\, dQ_r - 2\left\langle z, \int_s^t x(r)\, \eta(r)\, dQ_r\right\rangle \\ &\qquad + \int_s^t |x(r)|^2\, \eta(r)\, dQ_r.\end{aligned}$$

Since for any locally bounded measurable function $f : \mathbb{R}_+ \to \mathbb{R}^d$,

$$\frac{1}{\varepsilon}\int_t^{a^{-1}(a(t)+\varepsilon)} f(r)\, dQ(r) = \frac{1}{\varepsilon}\int_{a(t)}^{a(t)+\varepsilon} f\left(a^{-1}(s)\right) ds \xrightarrow[\varepsilon \to 0]{} f(t), \quad \text{a.e. } t \geq 0$$

it follows that for all $z \in \text{Dom}(\varphi)$,

$$\langle z - x(t), \theta(t)\rangle + \varphi(x(t))\lambda(t) \le \varphi(z)\lambda(t) + |z - x(t)|^2 \eta(t), \quad \text{a.e. } t \ge 0,$$

and from the definition of the Fréchet subdifferential we obtain

$$\frac{\theta(r)}{\lambda(r)} \in \partial^- \varphi(x(r)), \ \forall\, r \in \mathbb{R}_+ \setminus \Gamma,$$

where $\Gamma = \{r \ge 0 : \lambda(r) = 0\}$ with $\int_\Gamma dr = 0$.

Since φ is a (ρ,γ)–semiconvex function, we have for any continuous function $y : \mathbb{R}_+ \to \mathbb{R}^d$,

$$\left\langle y(r) - x(r), \frac{\theta(r)}{\lambda(r)}\right\rangle + \varphi(x(r)) \le \varphi(y(r)) + |y(r) - x(r)|^2 \left(\rho + \gamma\left|\frac{\theta(r)}{\lambda(r)}\right|\right),$$
$$\forall\, r \in \mathbb{R}_+ \setminus \Gamma.$$

Therefore (with the convention $0 \cdot (+\infty) = 0$) for all $r \in \mathbb{R}_+$:

$$\begin{aligned}\langle y(r) - x(r), \theta(r)\rangle + \varphi(x(r))\lambda(r) &\le \varphi(y(r))\lambda(r) \\ &\quad + |y(r) - x(r)|^2 (\rho\lambda(r) + \gamma|\theta(r)|)\end{aligned}$$

(we also use that $\overline{\text{Dom}(\varphi)}$ is γ–semiconvex).

Integrating on $[s,t]$ with respect to the measure $da(r)$ we infer that (4.65-jv) holds. ■

Lemma 4.32. *If $dk(t) \in \partial^- \varphi(x(t))(dt)$ and $d\hat{k}(t) \in \partial^- \varphi(\hat{x}(t))(dt)$, then for all $0 \le s \le t$:*

$$\begin{aligned}&\int_s^t \left\langle x(r) - \hat{x}(r), dk(r) - d\hat{k}(r)\right\rangle \\ &\qquad + \int_s^t |x(r) - \hat{x}(r)|^2 \left(2\rho dr + \gamma d\,\updownarrow k \updownarrow_r + \gamma d \updownarrow \hat{k} \updownarrow_r\right) \ge 0.\end{aligned} \tag{4.67}$$

Proof. We write (4.65–*iv*) for (x,k) with $y = \hat{x}$ and for $(\hat{x},\hat{k})$ with $y = x$; and add the two resulting inequalities. (4.67) follows. ■

Proposition 4.33 (Uniqueness). *Let the assumptions (4.63) and (4.64) be satisfied. If $(x,k) = \mathcal{SP}(\partial^-\varphi; x_0, m)$ and $(\hat{x},\hat{k}) = \mathcal{SP}(\partial^-\varphi; \hat{x}_0, \hat{m})$ then for all $t \ge 0$:*

$$\begin{aligned}\|x - \hat{x}\|_t^2 \le 2\Big[\, &|x_0 - \hat{x}_0|^2 + \|m - \hat{m}\|_t^2 \\ &+ 2\|m - \hat{m}\|_t \updownarrow k - \hat{k} \updownarrow_t \Big] e^{4(2\rho t + \gamma\updownarrow k\updownarrow_t + \gamma\updownarrow\hat{k}\updownarrow_t)}.\end{aligned} \tag{4.68}$$

In particular $\mathcal{SP}(\partial^-\varphi; x_0, m)$ has at most one solution.

Proof. We clearly have

$$\begin{aligned} &|x(t) - m(t) - \hat{x}(t) + \hat{m}(t)|^2 \\ &\quad = |x_0 - \hat{x}_0|^2 + 2\int_0^t \left\langle m(r) - \hat{m}(r), dk(r) - d\hat{k}(r)\right\rangle \\ &\qquad\qquad -2\int_0^t \left\langle x(r) - \hat{x}(r), dk(r) - d\hat{k}(r)\right\rangle. \end{aligned} \tag{4.69}$$

Then by (4.67) it follows that

$$\begin{aligned} &\frac{1}{2}|x(t) - \hat{x}(t)|^2 - |m(t) - \hat{m}(t)|^2 \\ &\le |x(t) - m(t) - \hat{x}(t) + \hat{m}(t)|^2 \\ &\le |x_0 - \hat{x}_0|^2 + 2\|m - \hat{m}\|_t \updownarrow k - \hat{k} \updownarrow_t \\ &+ 2\int_s^t |x(r) - \hat{x}(r)|^2 \left(2\rho dr + \gamma d\updownarrow k \updownarrow_r + \gamma d\updownarrow \hat{k} \updownarrow_r\right) \end{aligned}$$

which implies (4.68) via Gronwall's inequality from Corollary 6.60. The proof is complete. ■

To derive some uniform boundedness and continuity properties of the solution of the generalized Skorohod problem we introduce additional assumptions:

$$|\varphi(x) - \varphi(y)| \le L + L|x - y|, \quad \forall\, x, y \in \mathrm{Dom}(\varphi) \tag{4.70}$$

and

$$\mathrm{Dom}(\varphi) \text{ satisfies the } \gamma\text{-SUIBC (shifted uniform interior ball condition).} \tag{4.71}$$

Definition 4.34. $E \subset \mathbb{R}^d$ satisfies the shifted uniform interior ball condition (abbreviated SUIBC) if there exist $\gamma \ge 0$ and $\delta, \sigma > 0$, and for every $y \in E$ there exist $\lambda_y \in]0,1]$ and $v_y \in \mathbb{R}^d$, $|v_y| \le 1$, such that

$$\begin{cases} (i) & \lambda_y - \left(|v_y| + \lambda_y\right)^2 \gamma \ge \sigma, \\ (ii) & \overline{B}\left(x + v_y, \lambda_y\right) \subset E, \quad \forall\, x \in E \cap \overline{B}(y, \delta) \end{cases} \tag{4.72}$$

(we also say that E satisfies γ-SUIBC or (γ, δ, σ)-SUIBC).

Below we shall give easily verifiable conditions which imply (4.71).

Note that (4.71) implies that $\mathrm{int}(\mathrm{Dom}(\varphi)) \ne \emptyset$.

Also observe that the lower semicontinuity of φ and the assumption (4.70) clearly yield

$$\mathrm{Dom}(\varphi) = \overline{\mathrm{Dom}(\varphi)}. \tag{4.73}$$

Let $E = \overline{E} \subset \mathbb{R}^d$. Let $E^c = \mathbb{R}^d \setminus E$ and

$$E_\varepsilon = \{x \in E : \operatorname{dist}(x, E^c) \geq \varepsilon\}$$

the ε-interior of E.

Let $x, v \in \mathbb{R}^d$, $r > 0$. The set

$$\begin{aligned} D_x(v, r) &= \operatorname{conv}\left\{x, \overline{B}(x + v, r)\right\} \\ &= \left\{x + t(u - x) : u \in \overline{B}(x + v, r),\ t \in [0, 1]\right\} \end{aligned}$$

is called the $(|v|, r)$*-drop with vertex* x *and running direction* v. Note that if $|v| \leq r$, then $D_x(v, r) = \overline{B}(x + v, r)$.

Proposition 4.35. *Let* $E = \overline{E} \subset \mathbb{R}^d$ *and* $\operatorname{int}(E) \neq \emptyset$. *Each of the following conditions implies that* E *satisfies the* γ*-shifted uniform interior ball condition (SUIBC) for all* $\gamma \geq 0$. *(*E *satisfies (4.72).)*

(A_1)

$$\lim_{\varepsilon \searrow 0} \frac{1}{\sqrt{\varepsilon}} \sup_{x \in E} \operatorname{dist}(x, E_\varepsilon) = 0.$$

(A_2) *There exist* $N, \varepsilon_0 > 0$ *such that for all* $0 < \varepsilon \leq \varepsilon_0$*:*

$$E \subset \bigcup_{y \in E_\varepsilon} B(y, \varepsilon N).$$

(A_3) *(uniform interior* (h_0, r_0)*-drop condition) There exist* $r_0, h_0 > 0$ *and for all* $x \in E$ *there exists a* $v_x \in \mathbb{R}^d$, $|v_x| \leq h_0$, *such that*

$$D_x(v_x, r_0) \subset E.$$

(A_4) *There exists an* $r_0 > 0$ *such that* E *satisfies the* r_0*-uniform interior ball condition (i.e., by definition,* E^c *satisfies the* r_0*-uniform exterior ball condition).*

Proof. We shall show that $(A_4) \Rightarrow (A_3) \Rightarrow (A_2) \Rightarrow (A_1) \Rightarrow$(4.72).

$(A_1) \Rightarrow$(4.72): Let $\gamma \geq 0$. Let $\varepsilon > 0$ be fixed (sufficiently small) such that

$$d_\varepsilon = \sup_{x \in E} \operatorname{dist}(x, E_\varepsilon) \leq 1 \quad \text{and} \quad \left(\frac{d_\varepsilon}{\sqrt{\varepsilon}} + \frac{\sqrt{\varepsilon}}{4}\right)\sqrt{\gamma} < \frac{1}{2}.$$

For $y \in E$, let $y_\varepsilon \in E_\varepsilon$ be such that

$$|y - y_\varepsilon| \leq d_\varepsilon.$$

Let $\delta = \lambda = \varepsilon/4$, $v_y = y_\varepsilon - y$. Then

$$\lambda - \left(|v_y| + \lambda\right)^2 \gamma \geq \varepsilon \left[\frac{1}{4} - \left(\frac{d_\varepsilon}{\sqrt{\varepsilon}} + \frac{\sqrt{\varepsilon}}{4}\right)^2 \gamma\right] \overset{def}{=} \sigma > 0.$$

If $x \in E \cap \overline{B}(y, \delta)$ then

$$\overline{B}\left(x + v_y, \lambda\right) \subset \overline{B}\left(y_\varepsilon, \frac{\varepsilon}{2}\right) \subset E.$$

Indeed for all $|u| \leq 1$:

$$\left|\left(x + v_y + \lambda u\right) - y_\varepsilon\right| = |x - y + \lambda u| \leq \delta + \lambda = \frac{\varepsilon}{2}.$$

$(A_2) \Rightarrow (A_1)$: If $x \in E$ then there exists an $x_\varepsilon \in E_\varepsilon$ such that $x \in B(x_\varepsilon, N\varepsilon)$. Hence

$$\begin{aligned}\frac{1}{\sqrt{\varepsilon}} \sup_{x \in E} \operatorname{dist}(x, E_\varepsilon) &\leq \frac{1}{\sqrt{\varepsilon}} \sup_{x \in E} |x - x_\varepsilon| \\ &\leq \frac{1}{\sqrt{\varepsilon}} N\,\varepsilon \to 0, \quad \text{as } \varepsilon \to 0.\end{aligned}$$

$(A_3) \Rightarrow (A_2)$: Let $0 < \varepsilon \leq r_0$. We show that

$$E \subset \bigcup_{y \in E_\varepsilon} \overline{B}\left(y, \varepsilon \frac{h_0}{r_0}\right).$$

Let $x \in E$ and $D_x(v_x, r_0) \subset E$ with $|v_x| \leq h_0$. Then clearly $x \in \overline{B}(x + \frac{\varepsilon}{r_0} v_x, \varepsilon \frac{h_0}{r_0})$. Moreover $x + \frac{\varepsilon}{r_0} v_x \in E_\varepsilon$ since $\overline{B}\left(x + \frac{\varepsilon}{r_0} v_x, \varepsilon\right) \subset D_x(v_x, r_0) \subset E$. Indeed for all $z \in \overline{B}\left(x + \frac{\varepsilon}{r_0} v_x, \varepsilon\right)$ there exists a u with $|u| \leq 1$ such that

$$\begin{aligned}z &= x + \frac{\varepsilon}{r_0} v_x + \varepsilon u \\ &= \left(1 - \frac{\varepsilon}{r_0}\right) x + \frac{\varepsilon}{r_0}\left(x + v_x + r_0 u\right) \\ &\in D_x(v_x, r_0).\end{aligned}$$

$(A_4) \Rightarrow (A_3)$: Let $x \in \operatorname{Bd}(E^c) = \operatorname{Bd}(E)$ and $u_x \in N_{E^c}(x)$, $|u_x| = r_0$. Then $D_x\left(u_x, \frac{r_0}{2}\right) \subset D_x(u_x, r_0) = \overline{B(x + u_x, r_0)} \subset E$.

The proof is complete. ■

Remark 4.36. The *shifted uniform interior ball condition does not involve any of the conditions* (A_1), (A_2), (A_3), (A_4). Indeed the set $E = \{(x_1, x_2) \in \mathbb{R}^2 : x_1 \geq 0,\ x_1^2 + (x_2 \pm 2)^2 \geq 4\}$ satisfies SUIBC with $\lambda_y = \frac{1}{5}$, $v_y = (1, 0)$, $\delta > 0$, $\sigma = \frac{1}{25}$ and $\gamma = \frac{1}{36}$ because $\lambda_y - \left(|v_y| + \lambda_y\right)^2 \gamma \geq \sigma$ and the distance between $(0, \pm 2)$ and $(x_1 + 1, x_2)$ is greater than $2 + \frac{1}{5}$, for all $(x_1, x_2) \in E$, and therefore $\overline{B}\left((x_1, x_2) + (1, 0); \frac{1}{5}\right) \subset E$ for all $(x_1, x_2) \in E$. But the set E does not satisfy (A_1): we have

$$\begin{aligned}\frac{1}{\sqrt{\varepsilon}} \sup_{x \in E} \operatorname{dist}(x, E_\varepsilon) &= \frac{1}{\sqrt{\varepsilon}} \sqrt{(2+\varepsilon)^2 - 2^2} \\ &= \sqrt{2+\varepsilon} \\ &\to \sqrt{2}\end{aligned}$$

as $\varepsilon \to 0$.

Example 4.37. Let $E = \overline{E} \subset \mathbb{R}^d$ be defined by

$$\begin{cases} (i) & E = \left\{x \in \mathbb{R}^d : \phi(x) \leq 0\right\}, \text{ where } \phi \in C_b^2\left(\mathbb{R}^d\right), \\ (ii) & \operatorname{int}(E) = \left\{x \in \mathbb{R}^d : \phi(x) < 0\right\}, \\ (iii) & \operatorname{Bd}(E) = \left\{x \in \mathbb{R}^d : \phi(x) = 0\right\} \text{ and } |\nabla\phi(x)| = 1\ \forall\ x \in \operatorname{Bd}(E). \end{cases}$$

The set E satisfies both the uniform exterior ball condition (UEBC) and the shifted uniform interior ball condition (SUIBC).

Proof. Note that at any boundary point $x \in \operatorname{Bd}(E)$, $\nabla\phi(x)$ is a unit normal vector to the boundary, pointing towards the exterior of E. Hence for all $x \in \operatorname{Bd}(E)$ we have $N_E(x) = \{c\ \nabla\phi(x) : c \geq 0\}$ and $N_{E^c}(x) = \{-c\ \nabla\phi(x) : c \geq 0\}$.
Since for all $y \in E$ and $x \in \operatorname{Bd}(E)$, we see that $\phi(y) < 0$, $\phi(x) = 0$,

$$\begin{aligned}\langle \nabla\phi(x), y - x\rangle &= \phi(y) - \phi(x) - \int_0^1 \langle \nabla\phi(x + \lambda(y - x)) - \nabla\phi(x), y - x\rangle\, d\lambda \\ &\leq M\, |y - x|^2,\end{aligned}$$

and by Lemma 4.24 E satisfies (UEBC).

If $y \in E^c$ and $x \in \operatorname{Bd}(E)$, then $\phi(y) > 0$, $\phi(x) = 0$ and

$$\begin{aligned}\langle -\nabla\phi(x), y - x\rangle &= -\phi(y) + \int_0^1 \langle \nabla\phi(x + \lambda(y - x)) - \nabla\phi(x), y - x\rangle\, d\lambda \\ &\leq M\, |y - x|^2,\end{aligned}$$

that is E^c satisfies (UEBC) and consequently, by Proposition 4.35, E satisfies (SUIBC). ∎

Example 4.38. If $E \subset \mathbb{R}^d$ is a closed convex set such that

$$\exists\, r_0 > 0,\ E_{r_0} \neq \emptyset \quad \text{and} \quad h_0 = \sup_{z \in E} d\left(z, E_{r_0}\right) < \infty$$

(in particular if E is a bounded closed convex set), then E satisfies the uniform interior (h_0, r_0)-drop condition and, consequently the *γ-shifted uniform interior ball condition (SUIBC) for all* $\gamma \geq 0$. Moreover for every $0 < \delta \leq \frac{r_0}{2(1+h_0)} \wedge 1$, E satisfies 0-*SUIBC* with $\lambda_y = \sigma = \delta$.

Proof. Recall that E_ε denotes the ε-interior of E that is

$$E_\varepsilon = \{y \in E : \text{dist}\,(y, E^c) \geq \varepsilon\}\,.$$

Let $y \in E$, $\hat{y} = \text{Pr}_{E_{r_0}}(y)$ and $v_y = \frac{1}{1+h_0}(\hat{y} - y)$. Then $|\hat{y} - y| \leq h_0$, $\left|v_y\right| \leq 1$ and for all $x \in E \cap \overline{B}\,(y, \delta)$

$$\overline{B}\left(x + v_y, \delta\right) \subset \overline{B}\left(y + v_y, \frac{r_0}{1 + h_0}\right) \subset \text{conv}\left\{y, \overline{B}\left(\hat{y}, r_0\right)\right\} = D_y\left(\hat{y} - y, r_0\right) \subset E.$$

■

Let $(x,k) = \mathcal{SP}(\partial^- \varphi; x_0, m)$ and $y \in C([0,\infty[\,; E)$, where $E = \text{Dom}\,(\varphi)$. From (4.70), for all $0 \leq s \leq t$,

$$\begin{aligned}\int_s^t \langle y(r) - x(r), dk(r)\rangle \leq L(t-s) + L\int_s^t |y(r) - x(r)|\, dr \\ + \int_s^t |y(r) - x(r)|^2 \left(\rho dr + \gamma d \updownarrow k \updownarrow_r\right).\end{aligned} \tag{4.74}$$

Suppose that $x(r) \in \text{int}\,(\text{Dom}\,(\varphi))$ for all $r \in [s,t]$, and let

$$0 < b \leq \inf_{r \in [s,t]} \text{dist}\,(x(r), \text{Bd}\,(E))\,.$$

Let $y(r) = x(r) + \lambda b \alpha(r)$ with $\alpha \in C\left(\mathbb{R}_+; \mathbb{R}^d\right)$, $\|\alpha\|_{[s,t]} \leq 1$ and $0 < \lambda < 1$. From (4.74) we deduce that, for $\lambda = 1/\left[(1+\gamma)(1+b)^2\right]$

$$\begin{aligned}\lambda b \int_s^t \langle \alpha(r), dk(r)\rangle &\leq (L + Lb)(t-s) + \lambda^2 b^2 \left[\rho(t-s) + \gamma\left(\updownarrow k \updownarrow_t - \updownarrow k \updownarrow_s\right)\right] \\ &\leq \left(L + Lb + \lambda^2 b^2 \rho\right)(t-s) + \frac{\lambda b}{1+b}\left(\updownarrow k \updownarrow_t - \updownarrow k \updownarrow_s\right).\end{aligned}$$

Taking the supremum over all α such that $\|\alpha\|_{[s,t]} \leq 1$ we have

$$\frac{\lambda b^2}{1+b}\left(\updownarrow k \updownarrow_t - \updownarrow k \updownarrow_s\right) \leq \left(L + Lb + \lambda^2 b^2 \rho\right)(t-s)\,. \tag{4.75}$$

Hence:

Lemma 4.39. *Let* $(x, k) = \mathcal{SP}(\partial^- \varphi; x_0, m)$ *and the assumption (4.70) be satisfied. If* $x(r) \in \operatorname{int}(\operatorname{Dom}(\varphi))$ *for all* $r \in [s, t]$*, then there exists a positive constant* $C = C(L, \rho, \gamma, b)$ *such that*

$$\updownarrow k \updownarrow_t - \updownarrow k \updownarrow_s \leq C \ (t - s)$$

with

$$0 < b \leq \inf_{r \in [s,t]} \operatorname{dist}(x(r), \operatorname{Bd}(E)).$$

In general we have:

Lemma 4.40. *Let* $(x, k) = \mathcal{SP}(\partial^- \varphi; x_0, m)$ *with* φ *a* (ρ, γ)*–semiconvex function* $(\gamma \geq 0)$*. Assume that (4.70) holds and* $\operatorname{Dom}(\varphi)$ *satisfies the* (γ, σ, δ)*-SUIBC (4.72). If* $0 \leq s \leq t$ *and*

$$\sup_{r \in [s,t]} |x(r) - x(s)| \leq \delta,$$

then

$$\updownarrow k \updownarrow_t - \updownarrow k \updownarrow_s \leq \frac{1}{\sigma} |k(t) - k(s)| + \frac{3L + 4\rho}{\sigma} (t - s). \tag{4.76}$$

Proof. Let $\alpha \in C\left([0, \infty[; \mathbb{R}^d\right)$, $\|\alpha\|_{[s,t]} \leq 1$, be arbitrary. From (4.72), if

$$y(r) = x(r) + v_{x(s)} + \lambda_{x(s)} \alpha(r), \quad r \in [s, t],$$

then $y(r) \in E$. Moreover

$$|y(r) - x(r)| \leq \left|v_{x(s)}\right| + \lambda_{x(s)} \leq 2$$

and

$$|\varphi(y(r)) - \varphi(x(r))| \leq 3L.$$

From (4.74) we deduce that

$$\begin{aligned} \lambda_{x(s)} \int_s^t \langle \alpha(r), dk(r) \rangle \leq & - \int_s^t \left\langle v_{x(s)}, dk(r) \right\rangle + (3L + 4\rho)(t - s) \\ & + \gamma \int_s^t \left(\left|v_{x(s)}\right| + \lambda_{x(s)}\right)^2 d \updownarrow k \updownarrow_r . \end{aligned}$$

Taking the $\sup_{\|\alpha\|_{[s,t]}\leq 1}$ we have, using (4.71),

$$\sigma\left(\updownarrow k\updownarrow_t - \updownarrow k\updownarrow_s\right) \leq |k(t) - k(s)| + (3L + 4\rho)(t - s)$$

that is (4.76). ■

For the convenience of the reader we recall the notations for $y \in C\left([0,T];\mathbb{R}^d\right)$ and $\varepsilon > 0$,

$$\boldsymbol{\mu}_y(\varepsilon) = \varepsilon + \mathbf{m}_y(\varepsilon) = \varepsilon + \sup\{|y(t) - y(s)| : |t - s| \leq \varepsilon,\ t, s \in [0,T]\}.$$

Note that $\boldsymbol{\mu}_y : [0,T] \to \left[0, \boldsymbol{\mu}_y(T)\right]$ is a strictly increasing continuous function and, then, the inverse function $\boldsymbol{\mu}_y^{-1} : \left[0, \boldsymbol{\mu}_y(T)\right] \to [0,T]$ is well defined and is also a strictly increasing continuous function.

Lemma 4.41. *If $\psi : \mathbb{R}_+ \to \mathbb{R}_+$ is a continuous function, $\psi(r) > 0$ for all $r \geq 0$, and $\alpha_\psi : C\left([0,T];\mathbb{R}^d\right) \to \mathbb{R}_+$*

$$\alpha_\psi(x) = \|x\|_T + \frac{1}{\boldsymbol{\mu}_x^{-1}\left(\psi\left(\|x\|_T\right)\right)},$$

then for any compact subset $\mathcal{K} \subset C\left([0,T];\mathbb{R}^d\right)$,

$$\sup_{x\in\mathcal{K}} \alpha_\psi(x) < \infty.$$

Proof. If $R = \sup_{x\in K} \|x\|_T$ then $0 < R < \infty$ and

$$a \overset{def}{=} \inf\{\psi(r) : 0 \leq r \leq R\} > 0.$$

Since

$$\lim_{\varepsilon\searrow 0}\left[\sup_{y\in\mathcal{K}} \boldsymbol{\mu}_y(\varepsilon)\right] = 0,$$

there exists a $b > 0$ such that

$$\boldsymbol{\mu}_x\left(\frac{1}{b}\right) \leq \sup_{y\in K} \boldsymbol{\mu}_y\left(\frac{1}{b}\right) < a \leq \psi\left(\|x\|_T\right),\quad \forall\, x \in \mathcal{K}.$$

Hence

$$\frac{1}{b} < \boldsymbol{\mu}_x^{-1}\left(\psi\left(\|x\|_T\right)\right),\quad \forall\, x \in \mathcal{K},$$

and consequently

$$\sup_{x\in\mathcal{K}} \alpha_{\psi}(x) \le R + b < \infty.$$

■

Theorem 4.42. *Assume that the assumptions (4.63), (4.64), (4.70) and (4.71) are satisfied. Then there exists a constant C depending only upon the constants* $L, \delta, \sigma, \rho, \gamma$ *from our assumptions such that if* $m \in C([0,T];\mathbb{R}^d)$, $x_0 \in \mathrm{Dom}(\varphi)$, $(x,k) = \mathcal{SP}(\partial^-\varphi; x_0, m)$ *and*

$$\Delta_m \stackrel{def}{=} 1/\boldsymbol{\mu}_m^{-1}\left(\delta^2 e^{-C(1+T+\|m\|_T)}\right),$$

$$C_{T,m} \stackrel{def}{=} \exp\left[C\left(1+T+\|m\|_T+\Delta_m\right)\right],$$

then

$$\begin{array}{ll} (a) & \|k\|_{BV([0,T];\mathbb{R}^d)} = \updownarrow k \updownarrow_T \le C_{T,m}, \\ (b) & \|x\|_{C([0,T];\mathbb{R}^d)} = \|x\|_T \le |x_0| + C_{T,m}, \\ (c) & \begin{array}{l} |x(t)-x(s)| + \updownarrow k \updownarrow_t - \updownarrow k \updownarrow_s \le C_{T,m} \times \sqrt{\boldsymbol{\mu}_m(t-s)}, \\ \qquad \forall\, 0 \le s \le t \le T. \end{array} \end{array} \tag{4.77}$$

If moreover $\hat{m} \in C([0,T];\mathbb{R}^d)$, $\hat{x}_0 \in \mathrm{Dom}(\varphi)$ *and* $(\hat{x},\hat{k}) = \mathcal{SP}(\partial^-\varphi; \hat{x}_0, \hat{m})$, *then*

$$\|x-\hat{x}\|_T + \|k-\hat{k}\|_T \le C_T \times \left[|x_0-\hat{x}_0| + \sqrt{\|m-\hat{m}\|_T}\right], \tag{4.78}$$

with $C_T = \kappa\left(C_{T,m}, C_{T,\hat{m}}\right)$, *where* κ *is a continuous function.*

Proof. We denote by C, C', C'' generic constants independent of x_0, $\hat{x}_0$, m, $\hat{m}$ and T (C, C', C'' depend on $L, \delta, \sigma, \rho, \gamma$).
Step 1: Some estimates of $\mathbf{m}_x$.
Let $0 \le s \le t \le T$. Since

$$|x(t)-x(s)-m(t)+m(s)| = |k(t)-k(s)| \le \updownarrow k \updownarrow_t - \updownarrow k \updownarrow_s,$$

it follows that

$$|x(t)-x(s)| \le |m(t)-m(s)| + \updownarrow k \updownarrow_t - \updownarrow k \updownarrow_s. \tag{4.79}$$

We clearly have

$$\begin{aligned} &|x(t)-x(s)-m(t)+m(s)|^2 \\ &\quad = 2\int_s^t \langle m(r)-m(s), dk(r)\rangle + 2\int_s^t \langle x(s)-x(r), dk(r)\rangle. \end{aligned}$$

From (4.65-*iv*) with $y(r) \equiv x(s)$ and (4.70) we have

$$\int_s^t \langle x(s) - x(r), dk(r)\rangle \le L(t-s) + L\int_s^t |x(s) - x(r)|\, dr + \int_s^t |x(s) - x(r)|^2 \left(\rho dr + \gamma d \updownarrow k \updownarrow_r\right).$$

Combining this inequality with the previous identity and $\frac{1}{2}|x|^2 \le |x-y|^2 + |y|^2$, we obtain

$$\frac{1}{2}|x(t) - x(s)|^2 \le |m(t) - m(s)|^2 + 2\mathbf{m}_m(t-s)\left(\updownarrow k \updownarrow_t - \updownarrow k \updownarrow_s\right) + C(t-s) + C\int_s^t |x(r) - x(s)|^2 \left(dr + d\updownarrow k \updownarrow_r\right),$$

which implies, by Gronwall's inequality (Proposition 6.59):

$$\begin{aligned} |x(t) - x(s)|^2 \le & \left[\mathbf{m}_m^2(t-s) + \mathbf{m}_m(t-s)\left(\updownarrow k \updownarrow_t - \updownarrow k \updownarrow_s\right) + (t-s)\right] \\ & \times \exp\left[C\left(1 + t - s + \updownarrow k \updownarrow_t - \updownarrow k \updownarrow_s\right)\right], \end{aligned} \tag{4.80}$$

for all $0 \le s \le t \le T$.

Step 2: Estimates of $|x(t) - x(s)|$ and $\updownarrow k \updownarrow_t - \updownarrow k \updownarrow_s$ *under the assumption* $|x(t) - x(s)| \le \delta$*, where* δ *is one of the parameters entering the SUIBC condition.*

Let $0 \le s \le r \le t \le T$ such that $|x(t) - x(s)| \le \delta$. From (4.76) we have

$$\begin{aligned} \updownarrow k \updownarrow_t - \updownarrow k \updownarrow_s &\le C\ |k(t) - k(s)| + C\ (t-s) \\ &\le C\ |x(t) - x(s) - m(t) + m(s)| + C\ (t-s) \\ &\le C\ |x(t) - x(s)| + C\mathbf{m}_m(t-s) + C\ (t-s) \\ &\le C\delta + C\boldsymbol{\mu}_m(t-s). \end{aligned}$$

Plugging this estimate into (4.80), it clearly follows that

$$|x(t) - x(s)|^2 \le \boldsymbol{\mu}_m(t-s)\ e^{C(1+T+\|m\|_T)}, \quad \text{for all } 0 \le s \le t \le T,$$

since $(t-s) + \mathbf{m}_m(t-s) \le \boldsymbol{\mu}_m(t-s) \le T + 2\|m\|_T$.

Hence if $0 \le s \le t \le T$ and $|x(t) - x(s)| \le \delta$ then

$$|x(t) - x(s)| + \updownarrow k \updownarrow_t - \updownarrow k \updownarrow_s \le \sqrt{\boldsymbol{\mu}_m(t-s)} \times e^{C(1+T+\|m\|_T)}. \tag{4.81}$$

Step 3: Adapted time partition and local estimates.

Define the sequence

$$\begin{aligned} t_0 &= T_0 = 0, \\ T_1 &= \inf\{t \in [t_0, T] : \operatorname{dist}(x(t), \partial E) \le \delta/4\}, \end{aligned}$$

$$\begin{aligned}
t_1 &= \inf\{t \in [T_1, T] : |x(t) - x(T_1)| > \delta/2\}, \\
T_2 &= \inf\{t \in [t_1, T] : \operatorname{dist}(x(t), \partial E) \le \delta/4\}, \\
&\cdots\ \cdots\ \cdots \\
t_i &= \inf\{t \in [T_i, T] : |x(t) - x(T_i)| > \delta/2\} \\
T_{i+1} &= \inf\{t \in [t_i, T] : \operatorname{dist}(x(t), \partial E) \le \delta/4\} \\
&\cdots\ \cdots\ \cdots
\end{aligned}$$

Clearly

$$0 = T_0 = t_0 \le T_1 < t_1 \le T_2 < \cdots < t_i \le T_{i+1} < t_{i+1} \le \cdots \le T.$$

Let $t_i \le r \le T_{i+1}$. Then $x(r) \in int(E)$ and $dist(x(r), \partial E) \ge \delta/4$ or $t_i = T_{i+1}$. By Lemma 4.39 we have

$$|k(t) - k(s)| \le \updownarrow k \updownarrow_t - \updownarrow k \updownarrow_s \le C\ (t-s) \ \text{ for } t_i \le s \le t \le T_{i+1}.$$

Also for $t_i \le s \le t \le T_{i+1}$:

$$\begin{aligned}
|x(t) - x(s)| &\le |k(t) - k(s)| + |m(t) - m(s)| \\
&\le C(t-s) + |m(t) - m(s)| \\
&\le C\ \boldsymbol{\mu}_m(t-s)
\end{aligned}$$

and then

$$|x(t) - x(s)| + \updownarrow k \updownarrow_t - \updownarrow k \updownarrow_s \le C\ \boldsymbol{\mu}_m(t-s).$$

On each of the intervals $[T_i, t_i]$, we have

$$|x(t) - x(s)| \le \delta, \quad \text{for all } T_i \le s \le t \le t_i,$$

and consequently, by (4.81), for all $T_i \le s \le t \le t_i$:

$$|x(t) - x(s)| + \updownarrow k \updownarrow_t - \updownarrow k \updownarrow_s \le \sqrt{\boldsymbol{\mu}_m(t-s)} \times e^{C(1+T+\|m\|_T)}.$$

If $T_i \le s \le t_i \le t \le T_{i+1}$ then

$$\begin{aligned}
|x(t) - x(s)| + \updownarrow k \updownarrow_t - \updownarrow k \updownarrow_s &\le |x(t) - x(t_i)| + \updownarrow k \updownarrow_t - \updownarrow k \updownarrow_{t_i} \\
&\quad + |x(t_i) - x(s)| + \updownarrow k \updownarrow_{t_i} - \updownarrow k \updownarrow_s \\
&\le C\ \boldsymbol{\mu}_m(t - t_i) + \sqrt{\boldsymbol{\mu}_m(t_i - s)} \times e^{C(1+T+\|m\|_T)} \\
&\le \sqrt{\boldsymbol{\mu}_m(t-s)} \times e^{C'(1+T+\|m\|_T)}.
\end{aligned}$$

Consequently for all $i \in \mathbb{N}$ and $T_i \le s \le t \le T_{i+1}$:

$$|x(t) - x(s)| + \updownarrow k \updownarrow_t - \updownarrow k \updownarrow_s \le \sqrt{\mu_m(t-s)} \times e^{C(1+T+\|m\|_T)}.$$

Step 4: Conclusion (4.77).
Since $\mu_m : [0,T] \to [0, \mu_m(T)]$ is a strictly increasing continuous function, the inverse function $\mu_m^{-1} : [0, \mu_m(T)] \to [0,T]$ is well defined and is also a strictly increasing continuous function. We have

$$\begin{aligned}\frac{\delta}{2} &\le |x(t_i) - x(T_i)| \\ &\le \sqrt{\mu_m(t_i - T_i)} \times e^{C(1+T+\|m\|_T)} \\ &\le \sqrt{\mu_m(T_{i+1} - T_i)} \times e^{C(1+T+\|m\|_T)}\end{aligned}$$

and consequently

$$\begin{aligned}T_{i+1} - T_i &\ge \mu_m^{-1}\left(\frac{\delta^2}{4} e^{-2C(1+T+\|m\|_T)}\right) \\ &\ge \mu_m^{-1}\left(\delta^2 e^{-C'(1+T+\|m\|_T)}\right) > 0.\end{aligned}$$

Hence the bounded increasing sequence $(T_i)_{i\ge 0}$ is finite. Let j be such that $T = T_j$. Then

$$T = T_j = \sum_{i=1}^{j} (T_i - T_{i-1}) \ge \frac{j}{\Delta_m},$$

where

$$\Delta_m \stackrel{def}{=} 1/\mu_m^{-1}\left(\delta^2 e^{-C'(1+T+\|m\|_T)}\right).$$

Let $0 \le s \le t \le T$. We have

$$\begin{aligned}\updownarrow k \updownarrow_t - \updownarrow k \updownarrow_s &= \sum_{i=1}^{j} \left(\updownarrow k \updownarrow_{(t\wedge T_i)\vee s} - \updownarrow k \updownarrow_{(t\wedge T_{i-1})\vee s}\right) \\ &\le \sum_{i=1}^{j} \sqrt{\mu_m\Big((t\wedge T_i)\vee s - (t \wedge T_{i-1}) \vee s\Big)} \times e^{C(1+T+\|m\|_T)} \\ &\le j \times \sqrt{\mu_m(t-s)} \times e^{C(1+T+\|m\|_T)} \\ &\le T\, \Delta_m \sqrt{\mu_m(t-s)} \times e^{C(1+T+\|m\|_T)}\end{aligned}$$

and consequently

$$\updownarrow k \updownarrow_T \leq T\ \Delta_m \sqrt{\boldsymbol{\mu}_m (T)} \times e^{C(1+T+\|m\|_T)}$$
$$\leq \exp\left[C' (1 + T + \|m\|_T + \Delta_m)\right]$$

and

$$|x(t)| = |x_0 + m(t) - k(t)|$$
$$\leq |x_0| + \|m\|_t + \updownarrow k \updownarrow_t$$
$$\leq |x_0| + \|m\|_T + \updownarrow k \updownarrow_T .$$

Hence there exists a positive constant $C = C(L, \delta, \sigma, \rho, \gamma)$ such that if

$$\Delta_m = 1/\boldsymbol{\mu}_m^{-1}\left(\delta^2 e^{-C(1+T+\|m\|_T)}\right), \quad \text{and}$$
$$C_{T,m} = \exp\left[C(1 + T + \|m\|_T + \Delta_m)\right],$$

then

$$\updownarrow k \updownarrow_T \leq C_{T,m} \ \text{ and } \ \|x\|_T \leq |x_0| + C_{T,m}$$

that is (4.77-a,b).

By (4.80) $\forall\, 0 \leq s \leq t \leq T$:

$$|x(t) - x(s)|^2 \leq \left[\mathbf{m}_m^2 (t-s) + \mathbf{m}_m (t-s)\, C_{T,m} + (t-s)\right]$$
$$\times \exp\left[C(1 + C_{T,m})\right]$$
$$\leq C_{T,m} \times \boldsymbol{\mu}_m (t-s)$$

that is (4.77-c) holds.

Step 5: Conclusion (4.78).

Now, since $\updownarrow k \updownarrow_T + \updownarrow \hat{k} \updownarrow_T \leq C_{T,m} + C_{T,\hat{m}}$, from (4.68) we have

$$\|x - \hat{x}\|_T^2 \leq 2\Big[|x_0 - \hat{x}_0|^2 + \|m - \hat{m}\|_T^2$$
$$+2\, \|m - \hat{m}\|_T\, \|k - \hat{k}\|_T\Big]\, e^{4\gamma\left(2t + \updownarrow k \updownarrow_T + \updownarrow \hat{k} \updownarrow_T\right)}$$
$$\leq C_T \times \left[|x_0 - \hat{x}_0|^2 + \|m - \hat{m}\|_T\right],$$

which, combined with $k - \hat{k} = x_0 - \hat{x}_0 + m - \hat{m} - (x - \hat{x})$, yields (4.78). The proof is complete. ■

We can now derive the following continuity result for the mapping $(x_0, m) \mapsto (x, k) = \mathcal{SP}(\partial^-\varphi; x_0, m)$.

Corollary 4.43. *Assume that the assumptions (4.63), (4.64), (4.70) and (4.71) are satisfied. If* $x_{0n}, x_0 \in \overline{\mathrm{Dom}(\varphi)}$, $m_n, m \in C([0,\infty[; \mathbb{R}^d)$, $m_n(0) = 0$ *and*

$$\begin{array}{ll} i) & (x_n, k_n) = \mathcal{SP}(\partial^-\varphi; x_{0n}, m_n), \\ ii) & x_{0n} \to x_0, \\ iii) & m_n \to m \quad \text{in } C([0,T]; \mathbb{R}^d), \ \forall T \geq 0, \end{array}$$

then

$$\sup_{n \in \mathbb{N}^*} \updownarrow k_n \updownarrow_T < \infty, \ \forall T \geq 0,$$

and there exist $x, k \in C([0,\infty[; \mathbb{R}^d)$ *such that for all* $T \geq 0$:

$$\begin{array}{ll} (a) & \|x_n - x\|_T + \|k_n - k\|_T \to 0, \\ (b) & (x, k) = \mathcal{SP}(\partial^-\varphi; x_0, m). \end{array}$$

Proof. Let $T > 0$ be arbitrary. The set $\mathcal{M} = \{m, m_n : n \in \mathbb{N}^*\}$ is a compact subset of $C([0,T]; \mathbb{R}^d)$. Let $C_{T,m}$ be the constant from Theorem 4.42. By Lemma 4.41,

$$C_{T,\mathcal{M}} \overset{def}{=} \sup_{x \in \mathcal{M}} C_{T,x} < \infty.$$

Also

$$\boldsymbol{\mu}_{\mathcal{M}}(\varepsilon) \overset{def}{=} \sup_{x \in \mathcal{M}} \boldsymbol{\mu}_x(\varepsilon) \searrow 0, \ \text{ as } \varepsilon \searrow 0.$$

Let $a > 0$ be such that $|x_{0n}| \leq a$. By Theorem 4.42, for all $n, i \in \mathbb{N}^*$ and for all $s, t \in [0,T]$, $s \leq t$, we have

$$\|x_n\|_T + \updownarrow k_n \updownarrow_T \leq a + C_{T,\mathcal{M}},$$

$$|x_n(t) - x_n(s)| + \updownarrow k_n \updownarrow_t - \updownarrow k_n \updownarrow_s \leq C_{T,\mathcal{M}} \times \sqrt{\boldsymbol{\mu}_{\mathcal{M}}(t-s)}$$

and

$$\|x_n - x_i\|_T + \|k_n - k_i\|_T \leq C_{T,\mathcal{M}} \times \left[|x_{0n} - x_{0i}| + \sqrt{\|m_n - m_i\|_T} \right].$$

Hence there exist $x, k, A \in C([0,\infty[; \mathbb{R}^d)$ such that

$$x_n \to x, \quad k_n \to k \quad \text{in } C([0,T]; \mathbb{R}^d), \quad \text{as } n \to \infty$$

and by Arzelà–Ascoli's Theorem, on a subsequence also denoted by $\updownarrow k_n \updownarrow$,

$$\updownarrow k_n \updownarrow \to A \quad \text{in } C\left([0,T];\mathbb{R}^d\right), \quad \text{as } n \to \infty,$$

A is an increasing function and $A_0 = 0$. Clearly (x,k) satisfies (4.65-i,ii,iii) and (4.66-iv′). Hence, by Lemma 4.31, $(x,k) = \mathcal{SP}(\partial^-\varphi; x_0, m)$. ∎

Theorem 4.44. *Assume that*

$$\begin{array}{ll} (i) & \varphi : \mathbb{R}^d \to]-\infty,+\infty] \textit{ is a proper l.s.c. } (\rho,\gamma)\textit{–semiconvex function,} \\ (ii) & |\varphi(x) - \varphi(y)| \le L + L|x-y|, \quad \forall\, x,y \in \mathrm{Dom}(\varphi), \\ (iii) & \mathrm{Dom}(\varphi) \textit{ satisfies the } \gamma\textit{-SUIBC,} \\ (iv) & x_0 \in \mathrm{Dom}(\varphi), \\ (v) & m \in C\left(\mathbb{R}_+;\mathbb{R}^d\right), \quad m(0) = 0. \end{array} \tag{4.82}$$

Then the generalized Skorohod problem

$$\begin{cases} x(t) + k(t) = x_0 + m(t), & t \ge 0, \\ dk(t) \in \partial^-\varphi(x(t))(dt) \end{cases}$$

has a unique solution (x,k) (see Definition 4.29) denoted by $(x,k) = \mathcal{SP}(\partial^-\varphi; x_0, m)$.

Proof. Uniqueness was proved in Proposition 4.33. To prove existence, let $m_n \in C^1\left([0,\infty[;\mathbb{R}^d\right)$, $m_n(0) = 0$ be such that $\|m_n - m\|_T \to 0$ for all $T \ge 0$. From Proposition 6.55, there exists a unique solution (x_n,k_n) of $\mathcal{SP}(\partial^-\varphi; x_0, m_n)$, and by Corollary 4.43 there exist $x,k \in C\left([0,\infty[;\mathbb{R}^d\right)$ such that for all $T \ge 0$

$$\begin{aligned} &\|x_n - x\|_T + \|k_n - k\|_T \to 0, \quad \text{as } n \to \infty, \text{ and} \\ &(x,k) = \mathcal{SP}(\partial^-\varphi; x_0, m). \end{aligned}$$

The proof is complete. ∎

Corollary 4.45. *Let $(\Omega,\mathcal{F},\mathbb{P},\{\mathcal{F}_t\}_{t\ge 0})$ be a stochastic basis. If*

$$\xi \in L^0\left(\Omega,\mathcal{F}_0,\mathbb{P};\overline{\mathrm{Dom}(\varphi)}\right)$$

and $M \in S_d^0$, $M_0 = 0$, then there exists a unique solution $(X,K) \in S_d^0 \times S_d^0$ of the problem

$$(X_\cdot(\omega), K_\cdot(\omega)) = \mathcal{SP}(\partial^-\varphi; \xi(\omega), M_\cdot(\omega))$$

$\mathbb{P}$-a.s. $\omega \in \Omega$.

Proof. For every fixed ω, by Theorem 4.44, the Skorohod problem

$$(X_\cdot(\omega), K_\cdot(\omega)) = \mathcal{SP}(\partial^- \varphi; \xi(\omega), M_\cdot(\omega))$$

has a unique solution

$$(X_\cdot(\omega), K_\cdot(\omega)) \in C([0,\infty[;\mathbb{R}^d) \times C([0,\infty[;\mathbb{R}^d).$$

Since $(\omega, t) \longrightarrow M_t(\omega)$ is progressively measurable and the mapping

$$(\xi, M) \longmapsto X : \overline{\mathrm{Dom}(\varphi)} \times C([0,t];\mathbb{R}^d) \to C([0,t];\mathbb{R}^d)$$

is continuous for each $0 \le t \le T$, we see that X is progressively measurable. Hence $X \in S_d^0$ and consequently $K \in S_d^0$. ■

4.3.2 The Classical Skorohod Problem

Let $E \subset \mathbb{R}^d$ be a non-empty closed subset of $\mathbb{R}^d$.

If E satisfies the r_0-uniform exterior ball condition (r_0-$UEBC$), then by Lemmas 4.24 and 4.25 the set E is $\frac{1}{2r_0}$–semiconvex and the indicator function

$$\varphi(x) = I_E(x) = \begin{cases} 0, & \text{if } x \in E, \\ +\infty, & \text{if } x \notin E, \end{cases}$$

is a $(0, \frac{1}{2r_0})$–semiconvex function; the assumptions (4.64), (4.70) are satisfied.

We state the following:

Definition 4.46. Let $E = \overline{E} \subset \mathbb{R}^d$ satisfy the r_0-UEB condition. A pair (x,k) is a solution of the Skorohod problem (and we write $(x,k) = \mathcal{SP}(E; x_0, m)$) if

- $x, k : [0,\infty[\to \mathbb{R}^d$ are continuous functions and
- for all $0 \le s \le t \le T$:

$$\begin{cases} j) & x(t) \in E, \\ jj) & k \in BV_{loc}([0,\infty[;\mathbb{R}^d), \quad k(0) = 0, \\ jjj) & x(t) + k(t) = x_0 + m(t), \\ jv) & \forall\, y \in C(\mathbb{R}_+; E) : \\ & \displaystyle\int_s^t \langle y(r) - x(r), dk(r)\rangle \le \frac{1}{2r_0} \int_s^t |y(r) - x(r)|^2 \, d\updownarrow k \updownarrow_r . \end{cases} \tag{4.83}$$

We highlight that if $(x,k) = \mathcal{SP}(E; x_0, m)$ and $(\hat{x},\hat{k}) = \mathcal{SP}(E; \hat{x}_0, \hat{m})$, then from (4.83) we get

$$\left\langle x(t) - \hat{x}(t), dk(t) - d\hat{k}(t)\right\rangle + \frac{1}{2r_0}\left|x(t) - \hat{x}(t)\right|^2 \left(d \updownarrow k \updownarrow_t + d \updownarrow \hat{k} \updownarrow_t\right) \geq 0. \tag{4.84}$$

Theorem 4.47. *Let $x_0 \in E = \overline{E} \subset \mathbb{R}^d$ and $m : \mathbb{R}_+ \to \mathbb{R}^d$ be a continuous function such that $m(0) = 0$. If there exists an $r_0 > 0$ such that E satisfies the r_0-UEBC and $\frac{1}{2r_0}$-SUIBC (see Definitions 4.22 and 4.34) then the Skorohod problem (4.83) has a unique solution. Moreover the two following sets of conditions are equivalent.*

$$\begin{cases} (a) \quad \updownarrow k \updownarrow_t = \displaystyle\int_0^t \mathbf{1}_{x(s)\in \mathrm{Bd}(E)} d \updownarrow k \updownarrow_s, \\ (b) \quad k(t) = \displaystyle\int_0^t n_{x(s)} d \updownarrow k \updownarrow_s, \quad \text{where } n_{x(s)} \in N_E(x(s)) \\ \qquad\qquad \text{and } \left|n_{x(s)}\right| = 1, \quad d \updownarrow k \updownarrow_s \text{-a.e.,} \end{cases} \tag{4.85}$$

and

$$\begin{cases} \exists\, \beta > 0 \text{ such that } \forall\, y : [0,\infty[\to E \text{ continuous:} \\ \displaystyle\int_s^t \langle y(r) - x(r), dk(r)\rangle \leq \beta \int_s^t |y(r) - x(r)|^2 d \updownarrow k \updownarrow_r. \end{cases} \tag{4.86}$$

Proof. Uniqueness was proved in Proposition 4.33 using the corresponding inequality (4.84).

(4.85) $\Longrightarrow$ (4.83-jv):

By Lemma 4.24 we have

$$\begin{aligned} \int_s^t \langle y(r) - x(r), dk(r)\rangle &= \int_s^t \left\langle y(r) - x(r), n_{x(r)} d \updownarrow k \updownarrow_r\right\rangle \\ &= \int_s^t \left\langle y(r) - x(r), n_{x(r)} \mathbf{1}_{x(r)\in \mathrm{Bd}(E)} d \updownarrow k \updownarrow_r\right\rangle \\ &\leq \frac{1}{2r_0}\int_s^t \left|n_{x(r)}\right| |y(r) - x(r)|^2 \mathbf{1}_{x(s)\in \mathrm{Bd}(E)} d \updownarrow k \updownarrow_r \\ &\leq \frac{1}{2r_0}\int_s^t |y(r) - x(r)|^2 d \updownarrow k \updownarrow_r. \end{aligned}$$

Clearly (4.83-jv) $\Longrightarrow$ (4.86).

(4.86) $\Longrightarrow$ (4.85):

Let $[s,t]$ be an interval such that $x(r) \in Int(E)$ for all $r \in [s,t]$. Then there exists a $\delta = \delta_{s,t} > 0$ such that

$$\inf_{r\in[s,t]} d_{\mathrm{Bd}(E)}(x(r)) \geq \delta.$$

Let $\lambda \in [0,\delta]$ and $\alpha \in C\left([0,T];\mathbb{R}^d\right)$, $\|\alpha\|_T \leq 1$. Setting $y(r) = x(r) + \lambda\alpha(r)$ in (4.83-jv) we obtain

$$\int_s^t \langle \alpha(r), dk(r)\rangle \leq \beta\lambda \int_s^t d \updownarrow k \updownarrow_r .$$

Clearly passing to the limit as $\lambda \to 0$ and taking $\sup_{\|\alpha\|_T \leq 1}$, we have

$$x(r) \in Int(E), \ \forall\, r \in [s,t] \implies \updownarrow k \updownarrow_t - \updownarrow k \updownarrow_s = 0.$$

Hence (4.85-a) holds.

Let $\ell(r)$ be a measurable function such that $|\ell(r)| = 1$, $d \updownarrow k \updownarrow_r$ -*a.e.*, and

$$k(t) = \int_0^t \ell(r)\, d \updownarrow k \updownarrow_r .$$

Since (4.86) holds for all $0 \leq s \leq t$ we deduce that

$$\langle \ell(r), y(r) - x(r)\rangle \leq \beta\, |y(r) - x(r)|^2, \quad d \updownarrow k \updownarrow_r \text{-}a.e.$$

for all $y \in C([0,T];E)$. By Lemma 4.24 we infer

$$\ell(r) \in N_E(x(r)), \quad d \updownarrow k \updownarrow_r \text{-}a.e.$$

Hence (4.85-b) holds.

Existence. The existence was proved in Theorem 4.44, but for the convenience of the reader we reproduce the proof from Lions and Sznitman [43], see also [67].

Step 1. Case $m \in C^1\left([0,\infty[;\mathbb{R}^d\right)$. We know that the uniform exterior ball condition with ball radius r_0 yields the Lipschitz continuity of the projection π_E while restricted to the closed ε_0-neighbourhood ($0 < \varepsilon_0 < r_0$) of E, $\overline{U}_{\varepsilon_0}(E)$ and moreover

$$\frac{1}{2}\nabla d_E^2(z) = z - \pi_E(z), \quad \forall\, z \in \overline{U}_{\varepsilon_0}(E).$$

Let $0 < \varepsilon_0 < (1 \wedge r_0)/2$ and $\alpha \in C^\infty\left(\mathbb{R}^d\right)$ be such that

$$\alpha(z) = \begin{cases} 1, & \text{if } z \in \overline{U}_{\varepsilon_0}(E), \\ 0, & \text{if } z \notin \overline{U}_{2\varepsilon_0}(E), \\ \in [0,1], & \text{otherwise.} \end{cases}$$

Let

$$\psi(z) = \frac{1}{2} d_E^2(z)\,\alpha(z) + (1 - \alpha(z)).$$

Consider the penalized problem

$$\begin{cases} \dfrac{dx_\varepsilon}{dt} + \dfrac{1}{\varepsilon} \nabla \psi \left(x_\varepsilon \right) = \dfrac{dm}{dt}, \\ x_\varepsilon \left(0 \right) = x_0, \end{cases}$$

or equivalently

$$x_\varepsilon \left(t \right) + k_\varepsilon \left(t \right) = x_0 + m \left(t \right)$$

where

$$k_\varepsilon \left(t \right) = \frac{1}{\varepsilon} \int_0^t \nabla \psi \left(x_\varepsilon \left(s \right) \right) ds.$$

Let $m' = \dfrac{dm}{dt}$. Then

$$\begin{aligned} \psi \left(x_\varepsilon \left(t \right) \right) + \frac{1}{\varepsilon} \int_0^t \left| \nabla \psi \left(x_\varepsilon \left(s \right) \right) \right|^2 ds &= \int_0^t \left\langle m' \left(s \right), \nabla \psi \left(x_\varepsilon \left(s \right) \right) \right\rangle ds \\ &\leq \frac{\varepsilon}{2} \int_0^t \left| m' \left(s \right) \right|^2 ds + \frac{1}{2\varepsilon} \int_0^t \left| \nabla \psi \left(x_\varepsilon \left(s \right) \right) \right|^2 ds. \end{aligned}$$

Consequently for an arbitrary fixed $T > 0$:

$$\begin{cases} (a) & \displaystyle\int_0^T \left| \frac{1}{\varepsilon} \nabla \psi \left(x_\varepsilon \left(s \right) \right) \right|^2 ds \leq T \left\| m' \right\|_T^2, \\ (b) & 0 \leq 1 - \alpha \left(x_\varepsilon \left(t \right) \right) \leq \psi \left(x_\varepsilon \left(t \right) \right) \leq \varepsilon T \left\| m' \right\|_T^2, \\ (c) & \left\| k_\varepsilon \right\|_T \leq \ \updownarrow k_\varepsilon \updownarrow_T = \displaystyle\int_0^T \left| \frac{1}{\varepsilon} \nabla \psi \left(x_\varepsilon \left(s \right) \right) \right| ds \leq T \left\| m' \right\|_T, \end{cases} \tag{4.87}$$

and for $0 \leq s \leq t \leq T$:

$$\begin{aligned} \left| k_\varepsilon \left(t \right) - k_\varepsilon \left(s \right) \right| &\leq \updownarrow k_\varepsilon \updownarrow_t - \updownarrow k_\varepsilon \updownarrow_s \\ &= \int_s^t \left| \frac{1}{\varepsilon} \nabla \psi \left(x_\varepsilon \left(r \right) \right) \right| dr \\ &\leq \left\| m' \right\|_T \left(t - s \right). \end{aligned}$$

On the other hand

$$\begin{aligned} \left| x_\varepsilon \left(t \right) \right| &\leq \left| x_0 \right| + \left| m \left(t \right) \right| + \left| -k_\varepsilon \left(t \right) \right| \\ &\leq \left| x_0 \right| + \left\| m \right\|_T + T \left\| m' \right\|_T \end{aligned}$$

and for $0 \le s \le t \le T$:

$$\begin{aligned} |x_\varepsilon(t) - x_\varepsilon(s)| &\le |m(t) - m(s)| + |k_\varepsilon(t) - k_\varepsilon(s)| \\ &\le |m(t) - m(s)| + \|m'\|_T (t-s). \end{aligned}$$

Hence $\left\{\frac{1}{\varepsilon}\nabla\psi(x_\varepsilon)\right\}_{\varepsilon>0}$ is bounded in $L^2\left(0,T;\mathbb{R}^d\right)$ and $\{x_\varepsilon\}_{\varepsilon>0}$, $\{k_\varepsilon\}_{\varepsilon>0}$ are uniformly bounded and uniformly equicontinuous on $[0,T]$. Consequently, there exist $h \in L^2\left(0,T;\mathbb{R}^d\right)$, $a \in L^2(0,T)$ and $x,k \in C\left([0,T];\mathbb{R}^d\right)$ such that (eventually on a subsequence $\varepsilon = \varepsilon_n \to 0$):

$$\begin{aligned} \frac{1}{\varepsilon}\nabla\psi(x_\varepsilon) &\rightharpoonup h, \quad \text{weakly in } L^2\left(0,T;\mathbb{R}^d\right), \\ \frac{1}{\varepsilon}|\nabla\psi(x_\varepsilon)| &\rightharpoonup a, \quad \text{weakly in } L^2(0,T), \\ x_\varepsilon &\longrightarrow x \quad \text{in } C\left([0,T];\mathbb{R}^d\right), \\ \psi(x_\varepsilon) &\longrightarrow 0 \quad \text{in } C\left([0,T];\mathbb{R}^d\right), \\ \alpha(x_\varepsilon) &\longrightarrow 1 \quad \text{in } C\left([0,T];\mathbb{R}^d\right), \\ k_\varepsilon &\longrightarrow \int_0^\cdot h(s)\,ds = k \quad \text{in } C\left([0,T];\mathbb{R}^d\right). \end{aligned}$$

Then

$$\begin{aligned} \frac{1}{2}d_E^2(x(t)) &= \lim_{\varepsilon\to 0}\psi(x_\varepsilon(t)) = 0, \\ x(t) + k(t) &= x_0 + m(t) \end{aligned}$$

and $x(t) \in E$, for all $t \in [0,T]$.

Let $0 < \varepsilon < \varepsilon_0$ be such that

$$\|x_\varepsilon - x\|_T \le \varepsilon_0.$$

Then $x_\varepsilon(t) \in \overline{U}_{\varepsilon_0}(E)$ for all $t \in [0,T]$ and

$$\psi(x_\varepsilon) = \frac{1}{2}d_E^2(x_\varepsilon).$$

Let $y \in C([0,\infty[;E)$ be arbitrary. Since $x_\varepsilon - \pi_E(x_\varepsilon) \in N_E(\pi_E(x_\varepsilon))$ we have

$$\begin{aligned} &\left\langle y(s) - x_\varepsilon(s), \frac{1}{\varepsilon}\nabla\psi(x_\varepsilon(s))\right\rangle \\ &= \left\langle y(s) - x_\varepsilon(s), \frac{1}{\varepsilon}\left[x_\varepsilon(s) - \pi_E(x_\varepsilon(s))\right]\right\rangle \end{aligned}$$

$$\leq \left\langle y(s) - \pi_E(x_\varepsilon(s)), \frac{1}{\varepsilon}[x_\varepsilon(s) - \pi_E(x_\varepsilon(s))]\right\rangle$$

$$\leq \frac{1}{2r_0}\frac{1}{\varepsilon}|x_\varepsilon(s) - \pi_E(x_\varepsilon(s))|\,|y(s) - \pi_E(x_\varepsilon(s))|^2.$$

Passing to the limit as $\varepsilon = \varepsilon_n \to 0$ we obtain

$$\langle y(s) - x(s), h(s)\rangle \leq \frac{1}{2r_0}a(s)|y(s) - x(s)|^2, \quad a.e.$$

which implies $h(s) \in N_E(x(s))$, *a.e.* and so

$$\langle y(s) - x(s), h(s)\rangle \leq \frac{1}{2r_0}|h(s)|\,|y(s) - x(s)|^2, \quad a.e.$$

i.e.

$$\langle y(s) - x(s), dk(s)\rangle \leq \frac{1}{2r_0}|y(s) - x(s)|^2 d\updownarrow k\updownarrow_s.$$

Hence $(x,k) = \mathcal{SP}(x_0, m, E)$.
Step 2. Case $m \in C(\mathbb{R}_+;\mathbb{R}^d)$.

There exists an $m_n \in C^1(\mathbb{R}_+;\mathbb{R}^d)$ such that $\|m_n - m\|_T \to 0$ for all $T \geq 0$. By the first step there exists a unique solution $(x_n, k_n) = \mathcal{SP}(E; x_0, m_n)$.

Let $\mathcal{M} = \{m, m_n : n \in \mathbb{N}^*\}$. The set $\mathcal{M}$ is a compact subset of $C([0,T];\mathbb{R}^d)$. If $C_{T,m}$ is the constant and κ is the function from Theorem 4.42, then by Lemma 4.41,

$$C_{T,\mathcal{M}} \stackrel{def}{=} \sup_{x\in\mathcal{M}} C_{T,x} + \sup_{x,y\in\mathcal{M}} \kappa\left(C_{T,x}, C_{T,y}\right) < \infty.$$

Hence for all $n, i \in \mathbb{N}^*$ we have $\updownarrow k_n\updownarrow_T \leq C_{T,\mathcal{M}}$ and

$$\|x_n - x_i\|_T \leq C_{T,\mathcal{M}}\sqrt{\|m_n - m_i\|_T}.$$

Now using Corollary 4.43 we infer that there exist $x, k \in C(\mathbb{R}_+;\mathbb{R}^d)$ such that $(x,k) = \mathcal{SP}(E; x_0, m)$. The proof is complete. ■

4.3.3 Skorohod Equations

Consider the generalized Skorohod differential equation

$$\begin{cases} x(t) + k(t) = x_0 + \int_0^t f(s, x(s))\,ds + m(t), & t \geq 0, \\ dk(t) \in \partial^-\varphi(x(t))(dt), \end{cases} \tag{4.88}$$

where we assume that (4.82) is satisfied.

We recall that $dk(t) \in \partial^{-}\varphi(x(t))(dt)$ means

$$\begin{array}{ll} a) & x \in C\left([0,\infty[\,;\overline{\mathrm{Dom}(\varphi)}\right),\ \varphi(x) \in L^1_{loc}(0,\infty),\\ b) & k \in C\left([0,\infty[\,;\mathbb{R}^d\right) \bigcap BV_{loc}\left([0,\infty[\,;\mathbb{R}^d\right),\ \ k(0)=0,\\ c) & \forall\, 0 \le s \le t,\ \forall y : [0,\infty[\, \to \mathbb{R}^d \text{ continuous:}\\ & \displaystyle\int_s^t \langle y(r) - x(r), dk(r)\rangle + \int_s^t \varphi(x(r))\,dr\\ & \displaystyle\qquad \le \int_s^t \varphi(y(r))\,dr + \int_s^t |y(r)-x(r)|^2 \left(\rho dr + \gamma d \updownarrow k \updownarrow_r\right). \end{array} \tag{4.89}$$

We state the assumptions:

- the function $f(\cdot,x) : [0,+\infty[\, \to \mathbb{R}^d$ is measurable for all $x \in \mathbb{R}^d$,
- there exists a $\mu \in L^1_{loc}(0,\infty)$ such that $a.e.\ t \ge 0$:

$$\left\{\begin{array}{l} (C_f) - \text{Continuity:}\\ \quad u \longrightarrow f(t,u) : \mathbb{R}^d \to \mathbb{R}^d \text{ is continuous,}\\ (M_f) - \text{Monotonicity condition:}\\ \quad \langle x-y, f(t,x)-f(t,y)\rangle \le \mu(t)\,|x-y|^2,\ \forall\, x,y \in \mathbb{R}^d,\\ (B_f) - \text{Boundedness:}\\ \quad \displaystyle\int_0^T f^{\#}(s)\,ds < \infty, \quad \forall\, T \ge 0, \end{array}\right. \tag{4.90}$$

where

$$f^{\#}(t) \stackrel{def}{=} \sup\left\{|f(t,u)| : u \in \overline{\mathrm{Dom}(\varphi)}\right\}.$$

Proposition 4.48 (Generalized Skorohod Equation). *Assume that the assumptions (4.82) and (4.90) are satisfied. If $x_0 \in \overline{\mathrm{Dom}(\varphi)}$ and $m \in C\left([0,T];\mathbb{R}^d\right)$, $m(0)=0$, then the problem (4.88) has a unique solution.*

Proof. Uniqueness. Let (x,k) and $(\hat{x},\hat{k})$ be two solutions. Then

$$\begin{aligned} &|x(t)-\hat{x}(t)|^2 + 2\int_0^t \langle x(r)-\hat{x}(r), dk(r) - d\hat{k}(r)\rangle\\ &\quad = 2\int_0^t \langle x(r)-\hat{x}(r), f(r,x(r)) - f(r,\hat{x}(r))\rangle dr\\ &\quad \le 2\int_0^t \mu^{+}(r)\,|x(r)-\hat{x}(r)|^2 dr, \end{aligned}$$

and using Lemma 4.32 it follows that

$$|x(t) - \hat{x}(t)|^2 \leq 2\int_0^t |x(r) - \hat{x}(r)|^2 \, dA_r$$

with

$$A_t = 2\rho t + \gamma \updownarrow k \updownarrow_t + \gamma \updownarrow \hat{k} \updownarrow_t + \int_0^t \mu^+(r)\, dr,$$

which yields $x = \hat{x}$ via Gronwall's inequality from Corollary 6.60, Annex C.

Existence. We shall obtain the solution (x, k) as the limit in $C\left([0, T]; \mathbb{R}^d\right) \times C\left([0, T]; \mathbb{R}^d\right)$ of the sequence $(x_n, k_n)_{n \in \mathbb{N}^*}$ defined by an approximate Skorohod problem

$$\begin{cases} x_n(t) = x_0, \quad \text{for } t < 0, \\ x_n(t) + k_n(t) = x_0 + \displaystyle\int_0^t f\left(s, x_n(s - \frac{1}{n})\right) ds + m(t), \quad \text{for } t \geq 0, \\ dk_n(t) \in \partial^- \varphi(x_n(t))(dt). \end{cases} \tag{4.91}$$

Since for $t \in \left[\frac{i}{n}, \frac{i+1}{n}\right]$, $i \in \mathbb{N}$, we can write

$$x_n(t) + \left[k_n(t) - k_n\left(\frac{i}{n}\right)\right] = x_n\left(\frac{i}{n}\right) + \int_{\frac{i}{n}}^t f\left(s, x_n(s - \frac{1}{n})\right) ds$$
$$+ m(t) \quad m\left(\frac{i}{n}\right),$$

then by recurrence on the intervals $\left[\frac{i}{n}, \frac{i+1}{n}\right]$ there exists (via Theorem 4.44) a unique pair $(x_n, k_n) = \mathcal{SP}(\partial^- \varphi; x_0, m_n)$, with

$$m_n(t) = \int_0^t f\left(s, x_n(s - \frac{1}{n})\right) ds + m(t).$$

Let $T > 0$ and

$$\mathcal{M} = \{m_n : n \in \mathbb{N}^*\}.$$

$\mathcal{M}$ is a relatively compact subset of $C\left([0, T]; \mathbb{R}^d\right)$ since it is a bounded and equicontinuous subset of $C\left([0, T]; \mathbb{R}^d\right)$. Indeed

$$\|m_n\|_T \leq \int_0^T f^\#(s)\, ds + \|m\|_T$$

and for $s < t$

$$|m_n(t) - m_n(s)| \leq \int_s^t f^\#(r)\,dr + |m(t) - m(s)|.$$

Then by Theorem 4.42

$$\|x_n\|_T + \updownarrow k_n \updownarrow_T \leq |x_0| + C_{T,\mathcal{M}}$$

and for all $0 \leq s \leq t$:

$$|x(t) - x(s)| + \updownarrow k \updownarrow_t - \updownarrow k \updownarrow_s \leq C_{T,\mathcal{M}} \sqrt{\mu_{\mathcal{M}}(t-s)}.$$

Hence, again by the Arzelà–Ascoli theorem, $\mathcal{X} = \{x_n : n \in \mathbb{N}^*\}$ is a relatively compact subset of $C([0,T];\mathbb{R}^d)$. Let $x \in C([0,T];\mathbb{R}^d)$ be such that along a sequence still denoted by $\{x_n : n \in \mathbb{N}^*\}$ as an abuse of notation

$$\|x_n - x\|_T \to 0, \quad \text{as } n \to \infty.$$

Then, uniformly with respect to $t \in [0,T]$,

$$m_n(t) \to \int_0^t f(s, x(s))\,ds + m(t), \quad \text{as } n \to \infty,$$

and

$$k_n(t) \to k(t) = x_0 + \int_0^t f(s, x(s))\,ds + m(t) - x(t), \quad \text{as } n \to \infty.$$

Using Corollary 4.43 we infer that

$$(x,k) = \mathcal{SP}\left(\partial^-\varphi; x_0, \int_0^\cdot f(s, x(s))\,ds + m\right) \tag{4.92}$$

that is (x,k) is a solution of the problem (4.88). The uniqueness of the solution of (4.92) implies that the whole sequence (x_n, k_n) converges to the solution (x,k). The proof is complete. ■

If in the above Theorem 4.48 we put $\varphi = I_E$, where $E = \overline{E} \subset \mathbb{R}^d$, we get (via Theorem 4.47):

Corollary 4.49 (Skorohod Equation). *Let $x_0 \in E$ and $m : [0,\infty[\to \mathbb{R}^d$ be a continuous function such that $m(0) = 0$. If f satisfies the assumption (4.90) and E satisfies the r_0-UEBC and $\frac{1}{2r_0}$-SUIBC for some $r_0 > 0$, then the following problem has a unique solution (x,k):*

$$
\begin{aligned}
&j) \quad x,k \in C\left([0,\infty[\,;E\right), \quad k(0)=0,\\
&jj) \quad k \in BV_{loc}\left([0,\infty[\,;\mathbb{R}^d\right),\\
&jjj) \quad x(t)+k(t)=x_0+\int_0^t f(s,x(s))\,ds+m(t),\\
&jv) \quad \updownarrow k\updownarrow_t=\int_0^t \mathbf{1}_{x(s)\in \mathrm{Bd}(E)}d\updownarrow k\updownarrow_s,\\
&\qquad k(t)=\int_0^t n_{x(s)}d\updownarrow k\updownarrow_s, \quad \textit{where } n_{x(s)}\in N_E(x(s))\\
&\qquad\qquad \textit{and } \left|n_{x(s)}\right|=1, \quad d\updownarrow k\updownarrow_s\textit{-a.e.}
\end{aligned}
\tag{4.93}
$$

In the second part of this section we shall study the multivalued SDE (called the *stochastic variational inequality in a non-convex domain* or *generalized stochastic Skorohod equation*)

$$
\begin{cases}
X_t+K_t=\xi+\int_0^t F(s,X_s)\,ds+\int_0^t G(s,X_s)\,dB_s, \quad t\geq 0,\\
dK_t(\omega)\in\partial^-\varphi(X_t(\omega))(dt),
\end{cases}
\tag{4.94}
$$

where φ is a (ρ,γ)–semiconvex function and as usual for most of the SDEs in this book we have

- ▲ $\{B_t : t\geq 0\}$ is an $\mathbb{R}^k$-valued Brownian motion with respect to the given stochastic basis $(\Omega,\mathcal{F},\mathbb{P},\{\mathcal{F}_t\}_{t\geq 0})$,
- ▲ *(Carathéodory conditions)* $F(\cdot,\cdot,\cdot):\Omega\times[0,+\infty[\,\times\mathbb{R}^d\to\mathbb{R}^d$ and $G(\cdot,\cdot,\cdot):\Omega\times[0,+\infty[\,\times\mathbb{R}^d\to\mathbb{R}^{d\times k}$ are $\left(\mathcal{P},\mathbb{R}^d\right)$-Carathéodory functions, that is

$$
\begin{cases}
(a) \quad F(\cdot,\cdot,x) \text{ and } G(\cdot,\cdot,x) \text{ are p.m.s.p.}, \forall\, x\in\mathbb{R}^d,\\
(b) \quad F(\omega,t,\cdot) \text{ and } G(\omega,t,\cdot) \text{ are continuous function } d\mathbb{P}\otimes dt\text{-}a.e.
\end{cases}
\tag{4.95}
$$

Defining

$$
F^{\#}(s)=\sup\left\{|F(t,u)| : u\in\overline{\mathrm{Dom}(\varphi)}\right\},
$$

$$
G^{\#}(s)=\sup\left\{|G(t,u)| : u\in\overline{\mathrm{Dom}(\varphi)}\right\}
$$

we assume that the following conditions are satisfied:

- ▲ *(Boundedness conditions)* For all $T\geq 0$:

$$
\begin{cases}
(a) \quad \int_0^T F^{\#}(s)\,ds<\infty, \ \mathbb{P}\text{-a.s.}\\
(b) \quad \int_0^T \left|G^{\#}(s)\right|^2 ds<\infty, \ \mathbb{P}\text{-a.s.}
\end{cases}
\tag{4.96}
$$

▲ *(Monotonicity and Lipschitz conditions)* There exist $\mu \in L^1_{loc}(0,\infty)$ and $\ell \in L^2_{loc}(0,\infty;\mathbb{R}_+)$ such that $d\mathbb{P}\otimes dt$-*a.e.*:

$$\begin{cases} (\mathbf{M}_F) \quad \text{Monotonicity condition:} \\ \quad \langle x-y, F(t,x)-F(t,y)\rangle \leq \mu(t)|x-y|^2, \quad \forall\, x,y\in\mathbb{R}^d, \\ (\mathbf{L}_G) \quad \text{Lipschitz condition:} \\ \quad |G(t,x)-G(t,y)| \leq \ell(t)|x-y|, \quad \forall\, x,y\in\mathbb{R}^d. \end{cases} \tag{4.97}$$

To study the SDE (4.94) we begin by giving the definitions of strong and weak solutions.

Definition 4.50. (*I*) Given a stochastic basis $(\Omega,\mathcal{F},\mathbb{P},\mathcal{F}_t)_{t\geq 0}$ and an $\mathbb{R}^k$-valued $\mathcal{F}_t$-Brownian motion $\{B_t : t\geq 0\}$, a pair $(X,K) : \Omega\times[0,\infty[\,\to\mathbb{R}^d\times\mathbb{R}^d$ of continuous $\mathcal{F}_t$-progressively measurable stochastic processes is a strong solution of the SDE (4.94) if $\mathbb{P}$-a.s. $\omega\in\Omega$:

$$\begin{cases} j) \quad X_t\in\overline{\mathrm{Dom}(\varphi)}, \ \forall\, t\geq 0, \ \varphi(X_\cdot)\in L^1_{loc}(0,\infty), \\ jj) \quad K_\cdot\in BV_{loc}\left([0,\infty[;\mathbb{R}^d\right), \quad K_0=0, \\ jjj) \quad X_t+K_t=\xi+\displaystyle\int_0^t F(s,X_s)\,ds+\int_0^t G(s,X_s)\,dB_s, \ \forall\, t\geq 0, \\ jv) \quad \forall\, 0\leq s\leq t, \ \forall y:[0,\infty[\,\to\mathbb{R}^d \text{ continuous:} \\ \quad \displaystyle\int_s^t\langle y(r)-X_r, dK_r\rangle+\int_s^t\varphi(X_r)\,dr \\ \qquad \leq \displaystyle\int_s^t\varphi(y(r))\,dr+\int_s^t|y(r)-X_r|^2\left(\rho dr+\gamma d\updownarrow K\updownarrow_r\right), \end{cases} \tag{4.98}$$

that is

$$(X_\cdot(\omega),K_\cdot(\omega)) = \mathcal{SP}(\partial^-\varphi;\xi(\omega),M_\cdot(\omega)), \quad \mathbb{P}\text{-a.s. } \omega\in\Omega,$$

with

$$M_t=\int_0^t F(s,X_s)\,ds+\int_0^t G(s,X_s)\,dB_s.$$

(*II*) Let $F(\omega,t,x) = f(t,x)$, $G(\omega,t,x) = g(t,x)$ and $\xi(\omega) = x_0$ be independent of ω. If there exist a stochastic basis $(\Omega,\mathcal{F},\mathbb{P},\mathcal{F}_t)_{t\geq 0}$, an $\mathbb{R}^k$-valued $\mathcal{F}_t$-Brownian motion $\{B_t : t\geq 0\}$ and a pair $(X_\cdot,K_\cdot):\Omega\times[0,\infty[\,\to\mathbb{R}^d\times\mathbb{R}^d$ of $\mathcal{F}_t$-p.m.c.s.p. such that

$$(X_\cdot(\omega),K_\cdot(\omega)) = \mathcal{SP}(\partial^-\varphi;x_0,M_\cdot(\omega)), \quad \mathbb{P}\text{-a.s. } \omega\in\Omega,$$

with

$$M_t=\int_0^t f(s,X_s)\,ds+\int_0^t g(s,X_s)\,dB_s,$$

then the collection $(\Omega, \mathcal{F}, \mathbb{P}, \mathcal{F}_t, B_t, X_t, K_t)_{t\geq 0}$ is called a weak solution of the SDE (4.94).

Since the stochastic process K is uniquely determined from (X, B) by the Eq. (4.98-jjj), we also say that X is a strong solution (and respectively $(\Omega, \mathcal{F}, \mathbb{P}, \mathcal{F}_t, B_t, X_t)_{t\geq 0}$ is a weak solution).

We first give a uniqueness result for strong solutions.

Proposition 4.51 (Pathwise Uniqueness). *Let $(\Omega, \mathcal{F}, \mathbb{P}, \mathcal{F}_t, B_t)_{t\geq 0}$ be given and the assumption (4.82) be satisfied. Assume that*

$$F(\cdot,\cdot,\cdot) : \Omega \times [0, +\infty[\times \mathbb{R}^d \to \mathbb{R}^d \quad \textit{and} \quad G(\cdot,\cdot,\cdot) : \Omega \times [0, +\infty[\times \mathbb{R}^d \to \mathbb{R}^{d\times k}$$

satisfy (4.95), (4.96) and (4.97). Then the SDE (4.94) has at most one strong solution.

Proof. Let (X, K) and $(\hat{X}, \hat{K})$ be two solutions corresponding to ξ and respectively $\hat{\xi}$. Since

$$dK_t \in \partial^{-}\varphi(X_t)(dt) \quad \text{and} \quad d\hat{K}_t \in \partial^{-}\varphi(\hat{X}_t)(dt)$$

we deduce by Lemma 4.32, for $p \geq 1$ and $\lambda > 0$, that

$$\begin{aligned}&\left\langle X_t - \hat{X}_t, (F(t, X_t)\,dt - dK_t) - \left(F(t, \hat{X}_t)dt - d\hat{K}_t\right)\right\rangle \\ &+ \left(\frac{1}{2}m_p + 9p\lambda\right)\left|G(t, X_t) - G(t, \hat{X}_t)\right|^2 dt \leq |X_t - \hat{X}_t|^2 dV_t\end{aligned}$$

with

$$V_t = \int_0^t \left[\mu(s)\,ds + \left(\frac{1}{2}m_p + 9p\lambda\right)\ell^2(s)\,ds + 2\rho ds + \gamma d \updownarrow K \updownarrow_s + \gamma d \updownarrow \hat{K} \updownarrow_s\right].$$

Therefore, by Corollary 6.78 (Annex C), we get

$$\mathbb{E}\left[1 \wedge \left\|e^{-V}\left(X - \hat{X}\right)\right\|_T^p\right] \leq C_{p,\lambda}\mathbb{E}\left[1 \wedge \left|\xi - \hat{\xi}\right|^p\right].$$

Hence the uniqueness follows. ■

Note also that in the case of additive noise (i.e. G does not depend upon X) we have the existence of a strong solution.

Lemma 4.52. *Let $(\Omega, \mathcal{F}, \mathbb{P}, \mathcal{F}_t, B_t)_{t\geq 0}$ be given and the assumption (4.82) be satisfied. Assume*

$$F(\cdot,\cdot,\cdot) : \Omega \times [0, +\infty[\times \mathbb{R}^d \to \mathbb{R}^d$$

is a $(\mathcal{P}, \mathbb{R}^d)$*-Carathéodory function satisfying the boundedness condition (4.96-a) and the monotonicity condition (4.97-*$\boldsymbol{M}_F$*). If*

$$\xi \in L^0\left(\Omega, \mathcal{F}_0, \mathbb{P}; \overline{\mathrm{Dom}\,(\varphi)}\right)$$

and $M \in S_d^0$, $M_0 = 0$, *then there exists a unique solution* $(X, K) \in S_d^0 \times S_d^0$ *of the problem*

$$\begin{cases} X_t(\omega) + K_t(\omega) = \xi(\omega) + \int_0^t F(\omega, s, X_s(\omega))\, ds + M_t(\omega), & t \geq 0, \\ dK_t(\omega) \in \partial^- \varphi(X_t(\omega))(dt), \end{cases}$$

$\mathbb{P}$*-a.s.* $\omega \in \Omega$, *that is*

$$(X_\cdot(\omega), K_\cdot(\omega)) = \mathcal{SP}(\partial^- \varphi; \xi(\omega), M_\cdot(\omega)), \quad \mathbb{P}\text{-a.s. } \omega \in \Omega.$$

Proof. By Corollary 4.45 the approximating problem

$$\begin{cases} X_t^n(\omega) + K_t^n(\omega) = \xi(\omega) + \int_0^t F\left(\omega, s, X_{s-\frac{1}{n}}^n(\omega)\right) ds + M_t(\omega), & t \geq 0, \\ dK_t^n(\omega) \in \partial^- \varphi\left(X_t^n(\omega)\right)(dt), \end{cases}$$

has a unique solution $(X^n, K^n) \in S_d^0 \times S_d^0$. Now the solution (X, K) is the limit of the sequence (X^n, K^n), exactly as in Proposition 4.48. ■

To study the general SDE (4.94) we consider only the case when F, G and ξ are independent of ω. Hence the SDE (4.94) becomes

$$\begin{cases} X_t + K_t = x_0 + \int_0^t f(s, X_s)\, ds + \int_0^t g(s, X_s)\, dB_s, & t \geq 0, \\ dK_t(\omega) \in \partial^- \varphi(X_t(\omega))(dt), \end{cases} \tag{4.99}$$

where $f(\cdot,\cdot) : [0, +\infty[\times \mathbb{R}^d \to \mathbb{R}^d$ and $g(\cdot,\cdot) : [0, +\infty[\times \mathbb{R}^d \to \mathbb{R}^{d \times k}$.

As above, define

$$f^\#(s) \stackrel{def}{=} \sup\left\{|f(t,u)| : u \in \overline{\mathrm{Dom}\,(\varphi)}\right\},$$

$$g^\#(s) \stackrel{def}{=} \sup\left\{|g(t,u)| : u \in \overline{\mathrm{Dom}\,(\varphi)}\right\}.$$

Theorem 4.53. *Assume that the assumption (4.82) holds. Let* $(t, x) \mapsto f(t, x)$ *and* $(t, x) \mapsto g(t, x)$ *be* $(\mathcal{B}_1, \mathbb{R}^d)$*-Carathéodory functions satisfying the boundedness conditions*

$$\int_0^T \left[f^\#(s)^2 + g^\#(s)^4\right] ds < \infty, \quad \forall\, T \geq 0.$$

If $x_0 \in \overline{\mathrm{Dom}\,(\varphi)}$, then the problem (4.99) has a weak solution $(\Omega, \mathcal{F}, \mathbb{P}, \mathcal{F}_t, X_t, K_t, B_t)_{t\geq 0}$.

Proof. Step 1. Approximating sequence. Let $\left(\Omega, \mathcal{F}, \mathbb{P}, \mathcal{F}_t^B, B_t\right)_{t\geq 0}$ be a Brownian motion. By Lemma 4.52 there exists a unique pair $(X^n, K^n) : \Omega \times [0,\infty[\to \mathbb{R}^d \times \mathbb{R}^d$ of $\mathcal{F}_t^B$-progressively measurable continuous stochastic processes, forming a solution of the approximating problem

$$\begin{cases} X_t^n + K_t^n = x_0 + \displaystyle\int_0^t f\left(s, X_{s-\frac{1}{n}}^n\right) ds + \int_0^t g\left(s, X_{s-\frac{1}{n}}^n\right) dB_s, \quad t \geq 0, \\ dK_t^n(\omega) \in \partial^- \varphi\left(X_t^n(\omega)\right)(dt). \end{cases} \tag{4.100}$$

Let

$$M_t^n = \int_0^t f\left(s, X_{s-\frac{1}{n}}^n\right) ds + \int_0^t g\left(s, X_{s-\frac{1}{n}}^n\right) dB_s.$$

Since

$$\begin{aligned} &\mathbb{E}\left[\sup_{0\leq\theta\leq\varepsilon}\left|M_{t+\theta}^n - M_t^n\right|^4\right] \\ &\leq C\left[\left(\int_t^{t+\varepsilon} f^\#(s)\,ds\right)^4 + \left(\int_t^{t+\varepsilon}\left|g^\#(s)\right|^2 ds\right)^2\right] \\ &\leq \varepsilon C\left[\sup_{t\in[0,T]}\left(\int_t^{t+\varepsilon}\left|f^\#(s)\right|^2 ds\right)^2 + \sup_{t\in[0,T]}\int_t^{t+\varepsilon}\left|g^\#(s)\right|^4 ds\right], \end{aligned}$$

it follows, by Proposition 1.47, that the family of laws of $\{M^n : n \geq 1\}$ is tight on $C\left([0,\infty[\,;\mathbb{R}^d\right)$.

Therefore by Theorem 1.46 for all $T \geq 0$

$$\lim_{N\nearrow\infty}\left[\sup_{n\geq 1}\mathbb{P}\left(\|M^n\|_T \geq N\right)\right] = 0,$$

and for all $a > 0$ and $T > 0$:

$$\lim_{\varepsilon\searrow 0}\left[\sup_{n\geq 1}\mathbb{P}\big(\left(\mathbf{m}_{M^n}(\varepsilon;[0,T]) \geq a\right)\big)\right] = 0. \tag{4.101}$$

Defining

$$\begin{aligned} \boldsymbol{\mu}_{M^n} &= \varepsilon + \mathbf{m}_{M^n}(\varepsilon;[0,T]) \\ &= \varepsilon + \sup\left\{\left|M_t^n - M_s^n\right| : 0 \leq s \leq t \leq T,\ t - s \leq \varepsilon\right\}, \end{aligned}$$

we can replace in (4.101) $\mathbf{m}_{M^n}$ by $\boldsymbol{\mu}_{M^n}$.

Step 2. Tightness. Let $T \geq 0$ be arbitrary. We now show that the family of laws of the random variables $U^n = (X^n, K^n, \updownarrow K^n \updownarrow)$ is tight on $C\left([0,T];\mathbb{R}^{2d+1}\right)$.

From (4.77-c) we deduce

$$\mathbf{m}_{U^n}(\varepsilon;[0,T]) \leq G(M^n)\sqrt{\boldsymbol{\mu}_{M^n}(\varepsilon)}, \quad \text{a.s.}$$

where $G : C\left([0,T];\mathbb{R}^d\right) \to \mathbb{R}_+$,

$$\begin{aligned} G(x) &= C_{T,x} = \exp\left[C\left(1+T+\|x\|_T+B_x\right)\right], \qquad \text{with} \\ B_x &= 1/\boldsymbol{\mu}_x^{-1}\left(\delta^2 e^{-C(1+T+\|x\|_T)}\right). \end{aligned}$$

By Lemma 4.41 G is bounded on compact subset of $C\left([0,T];\mathbb{R}^d\right)$ and therefore by Proposition 1.48, $\{U^n; n \in \mathbb{N}^*\}$ is tight on $\mathbb{X}$ (recall that $U_0 = (x_0, 0, 0, 0)$).

Then by the Prohorov theorem there exists a subsequence (also denoted by n) such that as $n \to \infty$

$$(X^n, K^n, \updownarrow K^n \updownarrow, B) \to (X, K, V, B) \quad \text{in law}$$

on $C\left([0,T];\mathbb{R}^{2d+1+k}\right)$ and by the Skorohod theorem, we can choose a probability space $(\Omega, \mathcal{F}, \mathbb{P})$, and some random quadruples $(\bar{X}^n, \bar{K}^n, \bar{V}^n, \bar{B}^n)$, $(\bar{X}, \bar{K}, \bar{V}, \bar{B})$ defined on $(\Omega, \mathcal{F}, \mathbb{P})$, having the same laws as resp. $(X^n, K^n, \updownarrow K^n \updownarrow, B)$ and (X, K, V, B), such that, in $C\left([0,T];\mathbb{R}^{2d+1+k}\right)$, as $n \to \infty$,

$$(\bar{X}^n, \bar{K}^n, \bar{V}^n, \bar{B}^n) \xrightarrow{\mathbb{P}\text{-a.s.}} (\bar{X}, \bar{K}, \bar{V}, \bar{B}).$$

Note that by Proposition 2.15, $\left(\bar{B}^n, \{\mathcal{F}_t^{\bar{X}^n, \bar{K}^n, \bar{V}^n, \bar{B}^n}\}\right), n \geq 1$, and $\left(\bar{B}, \{\mathcal{F}_t^{\bar{X}, \bar{K}, \bar{V}, \bar{B}}\}\right)$ are $\mathbb{R}^k$-Brownian motions.

Step 3. Passing to the limit.

Since we also have $(X^n, K^n, \updownarrow K^n \updownarrow, B) \to (\bar{X}, \bar{K}, \bar{V}, \bar{B})$, in law, we deduce, by Corollary 1.18, that for all $0 \leq s \leq t$, $\mathbb{P}$-a.s.

$$\begin{aligned} &\bar{X}_0 = x_0, \qquad \bar{K}_0 = 0, \qquad \bar{X}_t \in E, \\ &\updownarrow \bar{K} \updownarrow_t - \updownarrow \bar{K} \updownarrow_s \leq \bar{V}_t - \bar{V}_s \quad \text{and} \quad 0 = \bar{V}_0 \leq \bar{V}_s \leq \bar{V}_s. \end{aligned} \tag{4.102}$$

Moreover, since for all $0 \leq s < t$, $n \in \mathbb{N}^*$

$$\begin{aligned} \int_s^t \varphi(X_r^n)\, dr \leq \int_s^t \varphi(y(r))\, dr - \int_s^t \langle y(r) - X_r^n, dK_r^n \rangle \\ + \int_s^t |y(r) - X_r^n|^2 \left(\rho dr + \gamma d \updownarrow K^n \updownarrow_r\right), \quad a.s., \end{aligned}$$

then by Proposition 1.19 we infer

$$\begin{aligned}\int_s^t \varphi\left(\bar{X}_r\right) dr \leq \int_s^t \varphi\left(y\left(r\right)\right) dr - \int_s^t \left\langle y\left(r\right) - \bar{X}_r, d\bar{K}_r\right\rangle \\ + \int_s^t \left|y\left(r\right) - \bar{X}_r\right|^2 \left(\rho dr + \gamma d\bar{V}_r\right).\end{aligned} \tag{4.103}$$

Hence, based on (4.102), (4.103) and Lemma 4.31 we have

$$d\bar{K}_r \in \partial^- \varphi\left(\bar{X}_r\right)(dr).$$

Now as in the proof of Theorem 3.54 we obtain that $\mathbb{P}$-a.s.

$$\bar{X}_t + \bar{K}_t = x_0 + \int_0^t f\left(s, \bar{X}_s\right) ds + \int_0^t g\left(s, \bar{X}_s\right) d\bar{B}_s, \ \forall\, t \in [0, T],$$

and consequently $\left(\bar{\Omega}, \bar{\mathcal{F}}, \bar{\mathbb{P}}, \mathcal{F}_t^{\bar{B},\bar{X}}, \bar{X}_t, \bar{K}_t, \bar{B}_t\right)_{t\geq 0}$ is a weak solution. The proof is complete. ■

Since the stochastic process K is uniquely determined by (X, B) via the Eq. (4.99), a weak solution for the SDE is a sextuple $\left(\Omega, \mathcal{F}, \mathbb{P}, \{\mathcal{F}_t\}_{t\geq 0}, X, B\right)$. Exactly the same arguments as used in Theorem 3.55 allow us to show that again *weak existence + pathwise uniqueness* implies strong existence. Hence we deduce from Theorem 4.53 and Proposition 4.51 the following:

Theorem 4.54. *Assume that the assumption (4.82) holds. Let* $(t, x) \mapsto f(t, x)$ *and* $(t, x) \mapsto g(t, x)$ *be* $\left(\mathcal{B}_1, \mathbb{R}^d\right)$*-Carathéodory functions satisfying (4.97) and*

$$\int_0^T \left[f^{\#}(s)^2 + g^{\#}(s)^4\right] ds < \infty, \quad \forall\, T \geq 0.$$

If $x_0 \in \overline{\mathrm{Dom}(\varphi)}$*, then the problem (4.99) has a unique strong solution* $(X_t, K_t)_{t\geq 0}$.

4.3.4 Markov Solutions of Reflected SDEs

In this subsection we study the Markov property of the solutions of reflected SDEs, when the set E has a particular form. Assume that

$$\begin{cases} (i) & E = \left\{x \in \mathbb{R}^d : \phi(x) \leq 0\right\}, \text{ where } \phi \in C_b^2\left(\mathbb{R}^d\right), \\ (ii) & \mathrm{int}(E) = \left\{x \in \mathbb{R}^d : \phi(x) < 0\right\}, \\ (iii) & \mathrm{Bd}(E) = \left\{x \in \mathbb{R}^d : \phi(x) = 0\right\} \text{ and } |\nabla\phi(x)| = 1 \ \forall\, x \in \mathrm{Bd}(E).\end{cases} \tag{4.104}$$

In Example 4.37 it is shown that this set E satisfies the (r_0-$UEBC$) (see Definition 4.22) and the $\left(\frac{1}{2r_0}\text{-}SUIBC\right)$ (see Definition 4.34).

Note that at any boundary point $x \in \text{Bd}$, $\nabla\phi(x)$ is a unit normal vector to the boundary pointing towards the exterior of E.

Let $t \in [0,\infty[$ and $x \in E$. Consider the equation

$$\begin{cases} X_s^{t,x} + K_s^{t,x} = x + \int_t^s f\left(r, X_r^{t,x}\right) dr + \int_t^s g\left(r, X_r^{t,x}\right) dB_r, \quad s \geq t, \\ dK_r(\omega) \in \partial^- I_E\left(X_r^{t,x}(\omega)\right)(dr), \end{cases} \tag{4.105}$$

where $f(\cdot,\cdot) : [0,+\infty[\times \mathbb{R}^d \to \mathbb{R}^d$ and $g(\cdot,\cdot) : [0,+\infty[\times \mathbb{R}^d \to \mathbb{R}^{d\times k}$ are continuous functions and satisfy: there exist $\mu \in \mathbb{R}$ and $\ell > 0$ such that for all $u, v \in \mathbb{R}^d$

$$\begin{aligned} &(i) \quad \langle u - v, f(t,u) - f(t,v)\rangle \leq \mu|u-v|^2, \\ &(ii) \quad |g(t,u) - g(t,v)| \leq \ell|u-v|. \end{aligned} \tag{4.106}$$

If follows from Theorem 4.54 that there exists a unique pair $(X^{t,x}, K^{t,x}) : \Omega \times [0,\infty[\to \mathbb{R}^d \times \mathbb{R}^d$ of continuous progressively measurable stochastic processes which is a strong solution of the SDE (4.105) that is $\mathbb{P}$-a.s. $\omega \in \Omega$:

$$\begin{cases} (j) \quad X_s^{t,x} \in E \text{ and } X_{s\wedge t}^{t,x} = x \text{ for all } s \geq 0, \\ (jj) \quad K_\cdot^{t,x} \in BV_{loc}\left([0,\infty[;\mathbb{R}^d\right), \quad K_s^{t,x} = 0 \text{ for all } 0 \leq s \leq t, \\ (jjj) \quad X_s^{t,x} + K_s^{t,x} = x + \int_t^s f\left(r, X_r^{t,x}\right) dr + \int_t^s g\left(r, X_r^{t,x}\right) dB_r, \ \forall\, s \geq t, \\ (jv) \quad \updownarrow K^{t,x} \updownarrow_s = \int_t^s \mathbf{1}_{\text{Bd}(E)}\left(X_r^{t,x}\right) d\updownarrow K^{t,x}\updownarrow_r, \ \forall\, s \geq t, \\ (v) \quad K_s^{t,x} = \int_t^s \nabla\phi\left(X_r^{t,x}\right) d\updownarrow K^{t,x}\updownarrow_r, \ \forall\, s \geq t. \end{cases} \tag{4.107}$$

Note that by Theorem 4.47 the conditions (jv) & (v) are equivalent to

$$\begin{cases} \exists\, \gamma > 0 \text{ such that } \forall\, y : [0,\infty[\to E \text{ continuous:} \\ \left\langle y(r) - X_r^{t,x}, dK_r^{t,x}\right\rangle \leq \gamma\, |y(r) - X_r^{t,x}|^2\, d\updownarrow K^{t,x}\updownarrow_r. \end{cases}$$

If $(X^{t,x}, K^{t,x})$ and $(X^{t,x'}, K^{t,x'})$ are two solutions, then from this last inequality it follows that

$$\begin{aligned} &-\left\langle X_r^{t,x} - X_r^{t,x'}, dK_r^{t,x} - dK_r^{t,x'}\right\rangle \\ &\qquad \leq \gamma \left|X_r^{t,x} - X_r^{t,x'}\right|^2 \left(d\updownarrow K^{t,x}\updownarrow_r + d\updownarrow K^{t,x'}\updownarrow_r\right), \ a.s. \end{aligned} \tag{4.108}$$

Proposition 4.55. *Let the assumptions (4.104) and (4.106) be satisfied and*

$$\sup_{(t,x)\in[0,T]\times E} \left[|f(t,x)| + |g(t,x)| \right] < \infty.$$

Then for all $p \geq 1$, $\lambda > 0$ *and* $s \geq t$, $x, x' \in E$,

$$\begin{aligned}
&(j) \quad \mathbb{E}^{\mathcal{F}_t} \sup_{r\in[t,s]} \left| X_r^{t,x} - X_r^{t,x'} \right|^p \leq C\,|x-x'|^p \; \exp\left[C\,(s-t)\right], \; a.s., \\
&(jj) \quad \mathbb{E}^{\mathcal{F}_t} \sup_{r\in[t,s]} |K_r^{t,x}|^p \leq \mathbb{E}^{\mathcal{F}_t} \updownarrow K^{t,x} \updownarrow_s^p \leq C\,(1 + (s-t)^p), \; a.s., \\
&(jjj) \quad \mathbb{E}^{\mathcal{F}_t} \sup_{r\in[t,s]} |X_r^{t,x}|^p \leq C\left(1 + (s-t)^p + |x|^p\right), \\
&(jv) \quad \mathbb{E}^{\mathcal{F}_t} e^{\lambda \left|K_s^{t,x}\right|} \leq \mathbb{E}^{\mathcal{F}_t} e^{\lambda \updownarrow K^{t,x} \updownarrow_s} \leq \exp\left(C\lambda + C\lambda t + \frac{C^2\lambda^2}{2} t\right), \; a.s.,
\end{aligned} \tag{4.109}$$

where C *is a constant independent of* x, x', s, t *and* λ.
If the monotonicity condition (4.106-i) is replaced by

$$|f(t,u) - f(t,v)| \leq \mu |u-v|,$$

and $\phi \in C_b^3\left(\mathbb{R}^d\right)$, *then moreover we have*

$$\begin{aligned}
\mathbb{E}^{\mathcal{F}_t} \sup_{r\in[t,s]} \left| K_r^{t,x} - K_r^{t,x'} \right|^p + \mathbb{E}^{\mathcal{F}_t} \sup_{r\in[t,s]} \left| \updownarrow K^{t,x} \updownarrow_r - \Big\updownarrow K^{t,x'} \Big\updownarrow_r \right|^p& \\
\leq C e^{C(s-t)} \left| x - x' \right|^p,&
\end{aligned} \tag{4.110}$$

where C *is a constant independent of* x, x', s, t.

Proof. Let $(X, K) = (X^{t,x}, K^{t,x})$ and $(\hat{X}, \hat{K}) = \left(X^{t,x'}, K^{t,x'}\right)$. By Itô's formula we have

$$\begin{aligned}
d\phi\,(X_r) = &\left\langle \nabla\phi\,(X_r)\,, f\,(r, X_r)\,dr - dK_r + g\,(r, X_r)\,dB_r \right\rangle \\
&+ \mathbf{Tr}\left[\nabla^2\phi\,(X_r)\, g^*\,(r, X_r)\, g\,(r, X_r)\right] dr
\end{aligned}$$

and if we define

$$\mathcal{L}\phi\,(\cdot) = \mathbf{Tr}\left[\nabla^2\phi\,(\cdot)\, g^*\,(r,\cdot)\, g\,(r,\cdot)\right] + \left\langle \nabla\phi\,(\cdot)\,, f\,(r,\cdot) \right\rangle$$

then

$$d\phi\,(X_r) = \left[\mathcal{L}\phi\,(X_r)\,dr - d\updownarrow K \updownarrow_r\right] + \left\langle \nabla\phi\,(X_r)\,, g\,(r, X_r)\,dB_r \right\rangle. \tag{4.111}$$

Let $\Phi_r = \exp(\gamma\phi(X_r))$ and $\hat{\Phi}_r = \exp\left(\gamma\phi\left(\hat{X}_r\right)\right)$. Once again by Itô's formula we have

$$d\Phi_r = \gamma\Phi_r\Big[\,\mathcal{L}\phi(X_r)\,dr - \,d\updownarrow K\updownarrow_r + \frac{\gamma}{2}\left|g^*(r,X_r)\nabla\phi(X_r)\right|^2 dr\Big] \\ +\gamma\Phi_r\left\langle\nabla\phi(X_r), g(r,X_r)\,dB_r\right\rangle$$

and therefore

$$\begin{aligned} d\left(\Phi_r\hat{\Phi}_r\right) &= d\left(\Phi_r\right)\hat{\Phi}_r + \Phi_r d(\hat{\Phi}_r) + d\Phi_r\, d\hat{\Phi}_r \\ &= \Phi_r\hat{\Phi}_r\left\{b_r dr + \sigma_r^* dB_r - \gamma\left[d\updownarrow K\updownarrow_r + d\updownarrow\hat{K}\updownarrow_r\right]\right\}, \end{aligned}$$

where $b:\Omega\times[0,\infty[\,\to\mathbb{R}$ and $\sigma:\Omega\times[0,\infty[\,\to\mathbb{R}^k$ are bounded $\mathcal{P}$-m.s.p.

Let

$$\begin{aligned} Y_r &= \Phi_r\hat{\Phi}_r\left(X_r-\hat{X}_r\right), \\ G_r &= \Phi_r\hat{\Phi}_r\left[\left(X_r-\hat{X}_r\right)\sigma_r^* + g(r,X_r) - g\left(r,\hat{X}_r\right)\right] \end{aligned}$$

and

$$\mathcal{K}_s = \int_t^s \Phi_r\hat{\Phi}_r\Big[(X_r-\hat{X}_r)b_r + f(r,X_r) - f(r,\hat{X}_r) + \left(g(r,X_r)-g(r,\hat{X}_r)\right)\sigma_r\Big]dr \\ + \int_t^s \Phi_r\hat{\Phi}_r\Big[-\gamma(X_r-\hat{X}_r)\left(d\updownarrow K\updownarrow_r + d\updownarrow\hat{K}\updownarrow_r\right) - (dK_r - d\hat{K}_r)\Big].$$

Then

$$Y_s = \left(x-x'\right) + \int_t^s d\mathcal{K}_r + \int_t^s G_r dB_r.$$

Let $p\geq 1$, $\lambda>1$ and $m_p = 1\vee(p-1)$.

Using (4.108) we deduce that

$$\langle Y_r, d\mathcal{K}_r\rangle + \left(\frac{1}{2}m_p + 9p\lambda\right)|G_r|^2\,dr \leq C_p'\,|Y_r|^2.$$

Hence by (3.7) we infer that, for $\lambda=2$,

$$\mathbb{E}^{\mathcal{F}_t}\sup_{r\in[t,s]}\left|X_r^{t,x}-X_r^{t,x'}\right|^p \leq C_p\left|x-x'\right|^p\ \exp\left[C_p(s-t)\right],$$

that is (4.109-j).

Now from (4.111) and the Eq. (4.105) we have for $s\geq t$

$$\updownarrow K^{t,x}\updownarrow_s = \int_t^s \mathcal{L}\phi(X_r^{t,x})dr + \int_t^s \langle \nabla\phi(X_r^{t,x}), g(r, X_r^{t,x})dB_r\rangle \\ - \left[\phi(X_s^{t,x}) - \phi(x)\right]. \tag{4.112}$$

Clearly, by the Burkholder–Davis–Gundy inequality and Hölder's inequality the estimate (4.109-jj) follows. Using Lemma 2.24 we infer (4.109-jv). The inequality (4.109-jjj) is a consequence of (4.109-jj) and

$$X_s^{t,x} = x + \int_t^s f(r, X_r^{t,x})dr + \int_t^s g(r, X_r^{t,x})dB_r - K_s^{t,x}.$$

The inequality (4.110) is obtained from (4.112), the Burkholder–Davis–Gundy inequality and Hölder's inequality. ■

Corollary 4.56. *Let the assumptions of Proposition 4.55 be satisfied and E be a bounded set. Then for every $T > 0$ and $p \geq 1$ the mapping*

$$(t,x) \mapsto \left(X_\cdot^{t,x}, K_\cdot^{t,x}, \updownarrow K^{t,x}\updownarrow\right) : [0,T] \times E \to S_d^p[0,T] \times S_d^p[0,T] \times S_1^p[0,T]$$

is continuous and if $h_1, h_2 : [0,T] \times E \to \mathbb{R}$ are continuous functions, then

$$(t,x) \mapsto \mathbb{E}\int_t^T h_1(s, X_s^{t,x})ds + \mathbb{E}\int_t^T h_2(s, X_s^{t,x})\, d\updownarrow K^{t,x}\updownarrow_s : [0,T] \times E \to \mathbb{R} \tag{4.113}$$

is continuous.

Proof. Let $(t', x') \in \mathbb{R}_+ \times E$. We can assume that $t' \leq t$. By the uniqueness of the solution we have $\mathbb{P}$-a.s.

$$X_s^{t',x'} + \left(K_s^{t',x'} - K_t^{t',x'}\right) = X_t^{t',x'} + \int_t^s f(r, X_r^{t',x'})dr + \int_t^s g(r, X_r^{t',x'})dB_r,$$

for all $s \geq t$ and by (4.109-j)

$$\begin{aligned} \mathbb{E}\sup_{r\in[t,T]}\left|X_r^{t',x'} - X_r^{t,x}\right|^p &\leq C\, e^{C(T-t)}\mathbb{E}\left|X_t^{t',x'} - x\right|^p \\ &\leq C' e^{CT}\left[\mathbb{E}\left|X_t^{t',x'} - x'\right|^p + \left|x' - x\right|^p\right]. \end{aligned}$$

Therefore

$$\begin{aligned} &\mathbb{E}\sup_{r\in[0,T]}\left|X_r^{t',x'} - X_r^{t,x}\right|^p \\ &\leq \left|x' - x\right|^p + \mathbb{E}\sup_{r\in[t',t]}\left|X_r^{t',x'} - x\right|^p + \mathbb{E}\sup_{r\in[t,T]}\left|X_r^{t',x'} - X_r^{t,x}\right|^p \end{aligned}$$

$$\leq C_p \left|x' - x\right|^p + C_p \mathbb{E} \sup_{r \in [t',t]} \left|X_r^{t',x'} - x'\right|^p .$$

Since $X_\cdot^{t',x'} \in S_d^0 [0,T]$, $X_{t'}^{t',x'} = x'$ and (4.109-jjj) holds for all $p \geq 1$, the continuity of $(t,x) \mapsto X_\cdot^{t,x} : [0,T] \times E \to S_d^p [0,T]$ follows. The continuity of $(t,x) \mapsto \left(K_\cdot^{t,x}, \updownarrow K^{t,x} \updownarrow\right)$ is now a consequence of the relations

$$K_s^{t,x} = \int_t^s f(r, X_r^{t,x}) dr + \int_t^s g(r, X_r^{t,x}) dB_r - \left(X_s^{t,x} - x\right)$$

and (4.112).

The continuity (4.113) is now a consequence of Proposition 1.20 and the uniformly integrability of the random variables

$$\int_t^T h_1(s, X_s^{t,x}) ds \quad \text{and} \quad \int_t^T h_2(s, X_s^{t,x}) \, d \updownarrow K^{t,x} \updownarrow_s .$$

■

We close for this section with the following:

Proposition 4.57. *Let the assumptions (4.104) and (4.106) be satisfied and E be a bounded set. Then the solution* $\left\{X_s^{0,x} : s \geq 0\right\}$ *of the SDE (4.105) is a strong Markov process with:*

(i) transition probability

$$P(t,x;s,G) = \mathbb{P}\left(X_s^{t,x} \in G\right)$$

for $t, s \geq 0$ *and* $G \in \mathcal{B}_d$;

(ii) evolution operator $P_{t,s} : B_b(\mathbb{R}^d) \to B_b(\mathbb{R}^d)$, $0 \leq t \leq s$,

$$\left(P_{t,s}\psi\right)(x) = \mathbb{E}\psi\left(X_s^{t,x}\right);$$

(iii) infinitesimal generator $\mathcal{A}_t$ *satisfying*

$$\mathcal{D} \stackrel{def}{=} \left\{\psi \in C_b^2\left(\mathbb{R}^d\right) : \langle \nabla\psi(x), \nabla\phi(x)\rangle = 0 \text{ if } x \in \operatorname{Bd}(E)\right\} \subset \operatorname{Dom}(\mathcal{A}_t), \ \forall\, t \geq 0$$

and for $\psi \in \mathcal{D}$

$$\begin{aligned}\mathcal{A}_t(\psi)(x) &= \frac{1}{2}\mathbf{Tr}\left[g(t,x)\, g^*(t,x)\, \psi''_{xx}(x)\right] + \left\langle f(t,x), \psi'_x(x)\right\rangle \\ &= \frac{1}{2}\sum_{i,j=1}^d (gg^*)_{ij}(t,x)\frac{\partial^2\psi(x)}{\partial x_i \partial x_j} + \sum_{i=1}^d f_i(t,x)\frac{\partial\psi(x)}{\partial x_i}.\end{aligned}$$

Proof. The proof is the same as for Proposition 3.38 except for one difference regarding the infinitesimal generator: the evolution operator is given by

$$P_{t,s}(\psi)(x) = \int_{\mathbb{R}^d} \psi(y) P(t,x;s,dy) = \mathbb{E}\psi\left(X_s^{t,x}\right)$$

and Itô's formula with $\psi \in C_c^2(\mathbb{R}^d)$ and $t, h \geq 0$ yields

$$\begin{aligned}\mathbb{E}\psi(X_{t+h}^{t,x}) =& \psi(x) + \mathbb{E}\int_t^{t+h} \mathcal{A}_r\psi(X_r^{t,x})dr \\ &- \mathbb{E}\int_t^{t+h} \left\langle \nabla\psi\left(X_r^{t,x}\right), \nabla\phi\left(X_r^{t,x}\right)\right\rangle \mathbf{1}_{\mathrm{Bd}(E)}\left(X_r^{t,x}\right) d\updownarrow K^{t,x}\updownarrow_r .\end{aligned}$$

Assuming $\langle \nabla\psi(x), \nabla\phi(x)\rangle = 0$ if $x \in \mathrm{Bd}\,(E)$ then

$$\lim_{h\searrow 0} \frac{1}{h}\left[P_{t,t+h}(\psi)(x) - \psi(x)\right] = \mathcal{A}_t\psi(x).$$

■

4.3.5 SDEs with Oblique Reflection

In this section we shall assume that E is a non-empty closed subset of $\mathbb{R}^d$ and there exists a $\gamma \geq 0$ such that:

(γ_1) E is a γ–semiconvex set (see Definition 4.23); and
(γ_2) E satisfies γ-SUIBC (see Definition 4.34).

Recall from Annex B, Lemma 6.47, that E is 0–semiconvex if and only if E is a convex set and, for $\gamma > 0$, E is γ–semiconvex if and only if E satisfies the $\frac{1}{2\gamma}$*-uniform exterior ball condition*. From Proposition 4.35 if E satisfies the *uniform interior drop condition* (that is, there exist $r_0, h_0 > 0$ and for all $x \in E$ there exists $v_x \in \mathbb{R}^d$, $|v_x| \leq h_0$, such that

$$D_x(v_x, r_0) \overset{def}{=} conv\left\{x, \overline{B(x+v_x, r_0)}\right\} \subset E\ )$$

then E satisfies the γ*-shifted uniform interior ball condition* for all $\gamma \geq 0$.

To each $x \in \mathrm{Bd}(E)$, we associate a unit vector ν_x and a unit vector $n_x \in N_E(x)$ such that for some fixed $\mu > 0$, $\langle n_x, \nu_x\rangle \geq \mu$. Recall that $N_E(x)$ is the closed external normal cone of E at $x \in \mathrm{Bd}\,(E)$. Note that if we consider the symmetric matrix

$$H(x) = \langle \nu_x, n_x\rangle I_{d\times d} - \nu_x \otimes n_x - n_x \otimes \nu_x + \frac{2}{\langle \nu_x, n_x\rangle}\nu_x \otimes \nu_x, \tag{4.114}$$

then

$$\nu_x = H(x)\, n_x, \quad x \in \mathrm{Bd}(E).$$

Moreover if the maps $x \longmapsto \nu_x, n_x$ are smooth (i.e. can be extended to functions from $C_b^2(\mathbb{R}^d;\mathbb{R}^d)$), then $H \in C_b^2(\mathbb{R}^d;\mathbb{R}^{2d})$.

Let $H = (h_{i,j})_{d\times d} \in C_b^2(\mathbb{R}^d;\mathbb{R}^{2d})$ be such that for some constant $c \geq 1$ and for all $x \in \mathbb{R}^d$,

$$\begin{array}{ll} (i) & h_{i,j}(x) = h_{j,i}(x), \quad \text{ and } i,j \in \overline{1,d}, \\ (ii) & \dfrac{1}{c}|u|^2 \leq \langle H(x)\, u, u\rangle \leq c\,|u|^2, \quad \forall\, u \in \mathbb{R}^d. \end{array} \tag{4.115}$$

We denote by $[H(x)]^{-1}$ the inverse matrix of $H(x)$. Then $[H(x)]^{-1}$ has the same properties (4.115) as $H(x)$. Define

$$b = \sup_{x,y\in\mathbb{R}^d,\ x\neq y} \frac{|H(x) - H(y)|}{|x-y|} + \sup_{x,y\in\mathbb{R}^d,\ x\neq y} \frac{\left|[H(x)]^{-1} - [H(y)]^{-1}\right|}{|x-y|}.$$

We introduce the *oblique directions*

$$\nu_x = H(x)\, n_x, \quad x \in \mathrm{Bd}(E),$$

where $n_x \in N_E(x)$.

Consider the differential equation

$$\begin{cases} dx(t) + H(x(t))\, \partial^- \varphi(x(t))\,(dt) \ni dm(t), & t > 0, \\ x(0) = x_0, \end{cases} \tag{4.116}$$

where

$$\begin{cases} (i) & x_0 \in \overline{\mathrm{Dom}(\varphi)} \\ (ii) & m \in C(\mathbb{R}_+;\mathbb{R}^d), \quad m(0) = 0 \end{cases} \tag{4.117}$$

and

$$\varphi : \mathbb{R}^d \to]-\infty, +\infty] \text{ is a proper l.s.c. } (\rho,\gamma)\text{–semiconvex function.} \tag{4.118}$$

Our aim in this section is to solve the differential equation (4.116) in the case

$$\varphi(x) = I_E(x) = \begin{cases} 0, & \text{if } x \in E, \\ +\infty, & \text{if } x \notin E, \end{cases}$$

where $E \subset \mathbb{R}^d$ is a *bounded closed subset satisfying* (r_0-*UEBC*) *and* $\left(\frac{1}{2r_0}\text{-}SUIBC\right)$ for some $r_0 > 0$. Recall that $\partial^- I_E(x) = \emptyset$ if $x \notin E$ and

$$\partial^- I_E(x) = \left\{ \hat{x} \in \mathbb{R}^d : \quad \limsup_{y \to x,\ y \in E} \frac{\langle \hat{x}, y - x \rangle}{|y - x|} \leq 0 \right\} = N_E(x), \text{ if } x \in E.$$

We introduce the following (the oblique reflection problem was initiated in Lions and Sznitman [43], see also Dupuis and Ishii [24]):

Definition 4.58. Let $H = (h_{i,j})_{d \times d} \in C_b^2(\mathbb{R}^d; \mathbb{R}^{2d})$. A pair (x, k) is a solution of the H-oblique reflection Skorohod problem $(E; x_0, m)$ (and we write $(x, k) = \mathcal{SP}(E; x_0, m; H)$) if:

- $x, k : \mathbb{R}_+ \to \mathbb{R}^d$ are continuous functions; and
- for all $0 \leq s \leq t$:

$$\begin{cases} (j) & x(t) \in E, \\ (jj) & k \in BV_{loc}(\mathbb{R}_+; \mathbb{R}^d), \quad k(0) = 0, \\ (jjj) & x(t) + \displaystyle\int_0^t H(x(r))\, dk(r) = x_0 + m(t), \\ (jv) & \updownarrow k \updownarrow_t = \displaystyle\int_0^t \mathbf{1}_{x(r) \in \mathrm{Bd}(E)} d \updownarrow k \updownarrow_r, \\ (v) & \begin{array}{l} k(t) = \displaystyle\int_0^t n_{x(r)} \mathbf{1}_{x(r) \in \mathrm{Bd}(E)} d \updownarrow k \updownarrow_r, \quad \text{where } n_{x(r)} \in N_E(x(r)) \\ \qquad \text{and } \left| n_{x(r)} \right| = 1, \ d \updownarrow k \updownarrow_r \text{-}a.e. \end{array} \end{cases} \tag{4.119}$$

Note that by Theorem 4.47 the conditions (jv) & (v) are equivalent to

$$\begin{cases} \forall\, y : [0, \infty[\to E \text{ continuous and for all } 0 \leq s \leq t, \\ \displaystyle\int_s^t \langle y(r) - x(r), dk(r) \rangle \leq \gamma \int_s^t |y(r) - x(r)|^2 d \updownarrow k \updownarrow_r, \end{cases} \tag{4.120}$$

and also to

$$\begin{cases} \exists\, \beta > 0 \text{ such that } \forall\, y : [0, \infty[\to E \text{ continuous and for all } 0 \leq s \leq t, \\ \displaystyle\int_s^t \langle y(r) - x(r), dk(r) \rangle \leq \beta \int_s^t |y(r) - x(r)|^2 d \updownarrow k \updownarrow_r. \end{cases}$$

Definition 4.59. We say that $dk(t) \in \partial^- I_E(x(t))(dt)$ if $x, k : [0, \infty[\to \mathbb{R}^d$ are continuous functions and for all $0 \leq s \leq t$,

$$\begin{cases} (a) & x(t) \in E, \\ (b) & k \in BV_{loc}(\mathbb{R}_+; \mathbb{R}^d), \quad k(0) = 0, \\ (c) & \displaystyle\int_s^t \langle y(r) - x(r), dk(r) \rangle \leq \gamma \int_s^t |y(r) - x(r)|^2 d \updownarrow k \updownarrow_r, \\ & \qquad \forall\, y : [0, \infty[\to E \text{ continuous.} \end{cases}$$

Hence $(x,k) = \mathcal{SP}(E;x_0,m;H)$ means

$$\begin{cases} x(t) + \displaystyle\int_0^t H(x(s))\,dk(s) \ni x_0 + m(t), & t \geq 0, \\ dk(t) \in \partial^- I_E(x(t))(dt). \end{cases}$$

Note that if $dk(t) \in \partial^- I_E(x(t))(dt)$ and $d\hat{k}(t) \in \partial^- I_E(\hat{x}(t))(dt)$ then for all $0 \leq s \leq t$:

$$\int_s^t \left\langle x(r) - \hat{x}(r), dk(r) - d\hat{k}(r)\right\rangle \\ +\gamma \int_s^t |x(r) - \hat{x}(r)|^2 \left(d\updownarrow k\updownarrow_r + d\updownarrow \hat{k}\updownarrow_r\right) \geq 0. \tag{4.121}$$

From (4.119-iv) we infer that

$$\text{if } x(r) \in \text{int}(E) \text{ for all } r \in [s,t], \text{ then } \updownarrow k\updownarrow_t - \updownarrow k\updownarrow_s = 0.$$

Proposition 4.60. *If $m \in BV_{loc}(\mathbb{R}_+;\mathbb{R}^d)$ then the "oblique reflection" Skorohod problem $\mathcal{SP}(E;x_0,m;H)$ has at most one solution.*

Proof. Let (x,k) and $(\hat{x},\hat{k})$ be two solutions. Consider the symmetric and strict positive matrix $U(r) = [H(x(r))]^{-1} + [H(\hat{x}(r))]^{-1}$. Note that

$$U(r)\left[H(\hat{x}(r))\,d\hat{k}(r) - H(x(r))\,dk(r)\right] \\ = \left([H(x(r))]^{-1} - [H(\hat{x}(r))]^{-1}\right)\left[H(\hat{x}(r))\,d\hat{k}(r) + H(x(r))\,dk(r)\right] \\ + 2\left[d\hat{k}(r) - dk(r)\right].$$

Let $u(r) = U^{1/2}(r)(x(r) - \hat{x}(r))$. Then

$$\begin{aligned} du(r) &= \left[dU^{1/2}(r)\right](x(r) - \hat{x}(r)) + U^{1/2}(r)\,d[x(r) - \hat{x}(r)] \\ &= [\alpha(r)\,dx(r) + \beta(r)\,d\hat{x}(r)](x(r) - \hat{x}(r)) \\ &\quad + U^{1/2}(r)\left[H(\hat{x}(r))\,d\hat{k}(r) - H(x(r))\,dk(r)\right], \end{aligned}$$

where $\alpha, \beta : \mathbb{R}_+ \to \mathbb{R}^{d\times d}$ are some bounded continuous functions.

Using (4.121) and the assumptions on the matrix-valued functions $x \longmapsto H(x)$ and $x \longmapsto [H(x)]^{-1}$ we have for some positive constants C_1, C_2, C

$$\frac{2}{c}|x(t) - \hat{x}(t)|^2 \\ \leq |u(t)|^2$$

$$
\begin{aligned}
&= 2\int_0^t \langle u(r), du(r)\rangle \\
&\leq C_1 \int_0^t |x(r) - \hat{x}(r)|^2 \left(d \updownarrow x \updownarrow_r + d \updownarrow \hat{x} \updownarrow_r\right) \\
&+ 2\int_0^t \left\langle x(r) - \hat{x}(r), U(r)\left[H(\hat{x}(r))\, d\hat{k}(r) - H(x(r))\, dk(r)\right]\right\rangle \\
&\leq \left(C_1 + C_2 + \frac{2}{r_0}\right) \int_0^t |x(r) - \hat{x}(r)|^2 \left(d \updownarrow x \updownarrow_r + d \updownarrow \hat{x} \updownarrow_r + d \updownarrow k \updownarrow_r + d \updownarrow \hat{k} \updownarrow_r\right).
\end{aligned}
$$

Now by the Stieltjes–Gronwall inequality 6.59 (from Annex C) we conclude $x = \hat{x}$ and consequently $k = \hat{k}$. ■

We recall the notation for the modulus of continuity of a function $g : [0, T] \to \mathbb{R}^d$:

$$
\mathbf{m}_g(\varepsilon) = \sup\{|g(u) - g(v)| : u, v \in [0, T],\ |u - v| \leq \varepsilon\}.
$$

Lemma 4.61. *If* $(x, k) = \mathcal{SP}(E; x_0, m; H)$*, then for all* $0 \leq s \leq t \leq T$*,*

$$
\begin{aligned}
\mathbf{m}_x(t - s) \leq C &\left[\mathbf{m}_m(t - s) + \sqrt{\mathbf{m}_m(t - s)\left(\updownarrow k \updownarrow_t - \updownarrow k \updownarrow_s\right)}\right] \\
&\times \exp\left[C\left(\updownarrow k \updownarrow_t - \updownarrow k \updownarrow_s + 1\right)\left(\updownarrow k \updownarrow_t - \updownarrow k \updownarrow_s\right)\right],
\end{aligned}
\tag{4.122}
$$

where C *is a constant depending on* (b, c, γ).

Proof. Let $0 \leq s \leq t$ and

$$
\begin{aligned}
h(t) = \big\langle H^{-1}(x(t))\,[x(t) - m(t) - x(s) + m(s)]&, \\
x(t) - m(t) - x(s) + m(s)&\big\rangle.
\end{aligned}
$$

We have

$$
\begin{aligned}
h(t) &= 2\int_s^t \big\langle H^{-1}(x(t))\,[x(r) - m(r) - x(s) + m(s)], \\
&\qquad\qquad d_r[x(r) - m(r) - x(s) + m(s)]\big\rangle \\
&= -2\int_s^t \big\langle H^{-1}(x(t))\,[x(r) - m(r) - x(s) + m(s)], H(x(r))\, dk(r)\big\rangle \\
&= 2\int_s^t \big\langle H^{-1}(x(t))\,[m(r) - m(s)], H(x(r))\, dk(r)\big\rangle
\end{aligned}
$$

$$+2\int_s^t \langle x(s) - x(r), dk(r)\rangle$$

$$+2\int_s^t \left\langle \left[H^{-1}(x(r)) - H^{-1}(x(t))\right][x(r) - x(s)], H(x(r))\, dk(r)\right\rangle.$$

Since

$$\int_s^t \langle x(s) - x(r), dk(r)\rangle \leq \gamma \int_s^t |x(s) - x(r)|^2 d \updownarrow k \updownarrow_r$$

and

$$\frac{1}{c}|x(t) - x(s)|^2 - \frac{2}{c}|m(t) - m(s)|^2 \leq h(t),$$

it follows that

$$\begin{aligned}
&|x(t) - x(s)|^2 \\
&\leq 2\,\mathbf{m}_m^2(t-s) + 2c^3\,\mathbf{m}_m(t-s)(\updownarrow k \updownarrow_t - \updownarrow k \updownarrow_s) \\
&+ \int_s^t \left[c\gamma|x(r) - x(s)|^2 + 2bc^2|x(r) - x(t)|\,|x(r) - x(s)|\right] d \updownarrow k \updownarrow_r.
\end{aligned}$$

Here we continue the estimates by

$$\begin{aligned}
&2bc^2 \int_s^t |x(r) - x(t)|\,|x(r) - x(s)|\, d \updownarrow k \updownarrow_r \\
&\leq 2bc^2|x(s) - x(t)| \int_s^t |x(r) - x(s)|\, d \updownarrow k \updownarrow_r + 2bc^2 \int_s^t |x(r) - x(s)|^2 d \updownarrow k \updownarrow_r \\
&\leq \frac{1}{2}|x(s) - x(t)|^2 + \frac{1}{2}\left(2bc^2\right)^2 \left(\int_s^t |x(r) - x(s)|\, d \updownarrow k \updownarrow_r\right)^2 \\
&+ 2bc^2 \int_s^t |x(r) - x(s)|^2 d \updownarrow k \updownarrow_r
\end{aligned}$$

and we obtain

$$\begin{aligned}
|x(t) - x(s)|^2 &\leq 4\,\mathbf{m}_m^2(t-s) + 4c^3\,\mathbf{m}_m(t-s)(\updownarrow k \updownarrow_t - \updownarrow k \updownarrow_s) \\
&+ \left[4b^2c^4(\updownarrow k \updownarrow_t - \updownarrow k \updownarrow_s) + 4bc^2 + 2c\gamma\right] \int_s^t |x(r) - x(s)|^2 d \updownarrow k \updownarrow_r.
\end{aligned}$$

By the Stieltjes–Gronwall inequality (6.59-Annex C), from this last inequality, the estimate (4.122) follows. ■

In the next statement, σ is associated to E by Definition 3.34, b and c are related to H (see (4.115) and the formula three lines below).

Lemma 4.62. *Let* $(x,k) \in \mathcal{SP}(E;x_0,m;H)$, $0 \le s \le t \le T$ *and*

$$\sup_{r\in[s,t]} |x(r) - x(s)| \le \delta < \frac{\sigma}{2bc}.$$

Then

$$\updownarrow k \updownarrow_t - \updownarrow k \updownarrow_s \le \frac{1}{\sigma} |k(t) - k(s)| \tag{4.123}$$

and

$$|x(t) - x(s)| + \updownarrow k \updownarrow_t - \updownarrow k \updownarrow_s \le \sqrt{\mathbf{m}_m(t-s)} \times e^{C\left(1+\|m\|_T^2\right)} \tag{4.124}$$

with $C = C(b,c,\sigma,\gamma)$.

Proof. Let $\alpha \in C\left([0,\infty[\,;\mathbb{R}^d\right)$, $\|\alpha\|_{[s,t]} \le 1$, be arbitrary. Since E satisfies the (γ-*SUIBC*), we have that for all $r \in [s,t]$, $y(r) = x(r) - v_{x(s)} + \lambda_{x(s)}\alpha(r) \in E$. Note that

$$\begin{aligned}
&\lambda_{x_s}\int_s^t \left\langle \alpha(r), n_{x(r)}\right\rangle \mathbf{1}_{x(r)\in \mathrm{Bd}(E)} d\updownarrow k \updownarrow_r \\
&\quad = \int_s^t \left\langle y(r) - x(r), n_{x(r)}\right\rangle \mathbf{1}_{x(r)\in \mathrm{Bd}(E)} d\updownarrow k \updownarrow_r \\
&\qquad + \left\langle v_{x(s)}, \int_s^t n_{x(r)} \mathbf{1}_{x(r)\in \mathrm{Bd}(E)} d\updownarrow k \updownarrow_r\right\rangle \\
&\quad \le \gamma \int_s^t |y(r) - x(r)|^2 \mathbf{1}_{x(r)\in \mathrm{Bd}(E)} d\updownarrow k \updownarrow_r + \left\langle v_{x(s)}, k(t) - k(s)\right\rangle \\
&\quad \le \gamma\left(\left|v_{x(s)}\right| + \lambda_{x(s)}\right)^2 \left(\updownarrow k \updownarrow_t - \updownarrow k \updownarrow_s\right) + |k(t) - k(s)|.
\end{aligned}$$

Taking the $\sup_{\|\alpha\|_{[s,t]}\le 1}$ and using the (γ,σ,δ)-*SUIBC* of E we infer

$$\sigma\left(\updownarrow k \updownarrow_t - \updownarrow k \updownarrow_s\right) \le |k(t) - k(s)|,$$

that is (4.123).

From (4.123) we have

$$\begin{aligned}
&\updownarrow k \updownarrow_t - \updownarrow k \updownarrow_s \\
&\quad \le \frac{1}{\sigma} |k(t) - k(s)| \\
&\quad = \frac{1}{\sigma}\int_s^t \left[H^{-1}(x(r)) - H^{-1}(x(s))\right] H(x(r))\, dk(r)
\end{aligned}$$

$$
\begin{aligned}
&+\frac{1}{\sigma}H^{-1}(x(s))\int_s^t H(x(r))\,dk(r)\\
&\leq \frac{bc}{\sigma}\int_s^t |x(r)-x(s)|\,d\updownarrow k\updownarrow_r + \frac{c}{\sigma}|x(t)-x(s)-m(t)+m(s)|\\
&\leq \frac{bc}{\sigma}\delta\left(\updownarrow k\updownarrow_t - \updownarrow k\updownarrow_s\right) + \frac{c}{\sigma}|x(t)-x(s)| + \frac{c}{\sigma}\mathbf{m}_m(t-s)
\end{aligned}
$$

and for δ sufficiently small ($\delta < \frac{\sigma}{2bc}$), we obtain

$$
\begin{aligned}
\updownarrow k\updownarrow_t - \updownarrow k\updownarrow_s &\leq \frac{2c}{\sigma}|x(t)-x(s)| + \frac{2c}{\sigma}\mathbf{m}_m(t-s)\\
&\leq \frac{1}{b} + \frac{2c}{\sigma}\mathbf{m}_m(t-s)\\
&\leq C_1\left(1+\|m\|_T\right)
\end{aligned}
\tag{4.125}
$$

with $C_1 = C_1(b,c,\sigma)$.

Now plugging this estimate into (4.122), it follows that

$$
\begin{aligned}
&\mathbf{m}_x(t-s)\\
&\quad\leq C\left[\mathbf{m}_m(t-s) + \sqrt{\mathbf{m}_m(t-s)\left(\updownarrow k\updownarrow_t - \updownarrow k\updownarrow_s\right)}\right]\\
&\qquad\qquad\times\exp\left[C\left(\updownarrow k\updownarrow_t - \updownarrow k\updownarrow_s + 1\right)\left(\updownarrow k\updownarrow_t - \updownarrow k\updownarrow_s\right)\right]\\
&\quad\leq \left[\mathbf{m}_m(t-s) + \sqrt{\mathbf{m}_m(t-s)}\right]\exp\left[C_2\left(1+\|m\|_T^2\right)\right]\\
&\quad\leq \sqrt{\mathbf{m}_m(t-s)}\exp\left[C_3\left(1+\|m\|_T^2\right)\right]
\end{aligned}
$$

with $C_3 = C_3(b,c,\sigma,\gamma)$. Therefore from (4.125) we have

$$
\updownarrow k\updownarrow_t - \updownarrow k\updownarrow_s \leq \frac{2c}{\sigma}\sqrt{\mathbf{m}_m(t-s)}\exp\left[C_3\left(1+\|m\|_T^2\right)\right] + \frac{2c}{\sigma}\mathbf{m}_m(t-s)
$$

and (4.124) follows. ■

Let $\boldsymbol{\mu}_m(\varepsilon) = \varepsilon + \mathbf{m}_m(\varepsilon)$, $\varepsilon \geq 0$.

Proposition 4.63. *If* $(x,k) \in \mathcal{SP}(E;x_0,m;H)$, *then there exists a positive constant* $C = C(b,c,\sigma,\gamma)$ *such that if*

$$
\Delta_m = 1/\boldsymbol{\mu}_m^{-1}\left(\delta^2 e^{-C(1+T+\|m\|_T)}\right) \text{ and } C_{T,m} = \exp\left[C\left(1+T+\|m\|_T+\Delta_m\right)\right]
$$

then for all $0 \leq s \leq t \leq T$*:*

$$\begin{aligned} &(a) \quad \|x\|_T \le |x_0| + C_{T,m}, \\ &(b) \quad \updownarrow k \updownarrow_T \le C_{T,m}, \\ &(c) \quad |x(t) - x(s)| + \updownarrow k \updownarrow_t - \updownarrow k \updownarrow_s \le C_{T,m} \times \sqrt{\mathbf{m}_m(t-s)}. \end{aligned} \tag{4.126}$$

Proof. We mimic the proof of Theorem 4.42. As there we define the sequence

$$\begin{aligned} t_0 &= T_0 = 0, \\ T_1 &= \inf\{t \in [t_0, T] : \mathrm{dist}(x(t), \partial E) \le \delta/4\}, \\ t_1 &= \inf\{t \in [T_1, T] : |x(t) - x(T_1)| > \delta/2\}, \\ T_2 &= \inf\{t \in [t_1, T] : \mathrm{dist}(x(t), \partial E) \le \delta/4\}, \\ &\cdots\cdots\cdots \\ t_i &= \inf\{t \in [T_i, T] : |x(t) - x(T_i)| > \delta/2\} \\ T_{i+1} &= \inf\{t \in [t_i, T] : \mathrm{dist}(x(t), \partial E) \le \delta/4\} \\ &\cdots\cdots\cdots \end{aligned}$$

Clearly

$$0 = T_0 = t_0 \le T_1 < t_1 \le T_2 < \cdots < t_i \le T_{i+1} < t_{i+1} \le \cdots \le T.$$

As in Step 3 from the proof of Theorem 4.42 we have:

- for $t_i \le s \le t \le T_{i+1}$:

$$|x(t) - x(s)| + \updownarrow k \updownarrow_t - \updownarrow k \updownarrow_s = |m(t) - m(s)|,$$

since for $t_i \le r \le T_{i+1}$, $x(r) \in \mathrm{int}(E)$ and $\mathrm{dist}(x(r), \partial E) \ge \delta/4$; hence

$$|k(t) - k(s)| \le \updownarrow k \updownarrow_t - \updownarrow k \updownarrow_s = 0 \text{ for } t_i \le s \le t \le T_{i+1};$$

- $k(r) = 0$ for all $r \in [0, T_1]$;
- for $T_i \le s \le t \le t_i$, by (4.124)

$$\begin{aligned} |x(t) - x(s)| + \updownarrow k \updownarrow_t - \updownarrow k \updownarrow_s &\le \sqrt{\mathbf{m}_m(t-s)} \times e^{C\left(1+\|m\|_T^2\right)} \\ &\le \sqrt{\mu_m(t-s)} \times e^{C\left(1+\|m\|_T^2\right)}; \end{aligned}$$

- for $T_i \le s \le t_i \le t \le T_{i+1}$

$$\begin{aligned} |x(t) - x(s)| + \updownarrow k \updownarrow_t - \updownarrow k \updownarrow_s &\le |x(t) - x(t_i)| + \updownarrow k \updownarrow_t - \updownarrow k \updownarrow_{t_i} \\ &\quad + |x(t_i) - x(s)| + \updownarrow k \updownarrow_{t_i} - \updownarrow k \updownarrow_s. \end{aligned}$$

Hence for all $T_i \le s \le t \le T_{i+1}$:

$$\begin{aligned}|x(t) - x(s)| + \updownarrow k\updownarrow_t - \updownarrow k\updownarrow_s &\le \sqrt{\mathbf{m}_m(t-s)} \times e^{C\left(1+\|m\|_T^2\right)}\\ &\le \sqrt{\boldsymbol{\mu}_m(t-s)} \times e^{C\left(1+\|m\|_T^2\right)},\end{aligned}$$

with $C = C(b, c, \sigma, r_0)$.

Now the proof continues exactly as in *Step 4* of Theorem 4.42. ■

Theorem 4.64. *Let $\gamma \ge 0$ and E be a bounded closed γ–semiconvex subset of $\mathbb{R}^d$ satisfying γ-SUIBC. If $x_0 \in E$ and $m \in C\left([0,T];\mathbb{R}^d\right)$, $m(0) = 0$, then the Skorohod problem $\mathcal{SP}(E; x_0, m; H)$ with oblique reflection defined in (4.119) has at least one solution. If moreover $m \in BV_{loc}\left(\mathbb{R}_+;\mathbb{R}^d\right)$ then the solution is unique.*

Proof. We only sketch the proof since it repeats, in a new context, the arguments from the proof of Theorem 4.47.

By Lemma 6.47 we know that E satisfies r_0-$UEBC$ with $r_0 = \frac{1}{2\gamma}$ if $\gamma > 0$ and an arbitrary $r_0 > 0$ if $\gamma = 0$.

Step 1. Case $m \in C^1\left([0,\infty[;\mathbb{R}^d\right)$. As in the proof of Theorem 4.47 we consider for $0 < \varepsilon < \varepsilon_0 < (1 \wedge r_0)/2$ the penalized problem

$$x_\varepsilon(t) + \int_0^t H(x_\varepsilon(s))\, dk_\varepsilon(s) = x_0 + m(t), \tag{4.127}$$

where

$$\begin{aligned}k_\varepsilon(t) &= \frac{1}{\varepsilon}\int_0^t \nabla\psi(x_\varepsilon(s))\, ds,\\ \psi(z) &= \frac{1}{2}d_E^2(z)\,\alpha(z) + (1 - \alpha(z))\end{aligned}$$

and $\alpha \in C^\infty\left(\mathbb{R}^d\right)$,

$$\alpha(z) = \begin{cases} 1, & \text{if } z \in \overline{U}_{\varepsilon_0}(E),\\ 0, & \text{if } z \notin \overline{U}_{2\varepsilon_0}(E),\\ \in [0,1], & \text{otherwise.}\end{cases}$$

By Lemma 6.49 the function d_E^2 is of class C^1 on $\overline{U}_{r_0/2}(E)$ (the closed $r_0/2$-neighbourhood of E) and

$$\frac{1}{2}\nabla d_E^2(z) = z - \pi_E(z) \in N_E(\pi_E(z)), \quad \text{for all } z \in \overline{U}_{r_0/2}(E).$$

Moreover the projection π_E restricted to $\overline{U}_{r_0/2}(E)$ is Lipschitz with Lipschitz constant $L = \frac{r_0}{r_0 - r_0/2} = 2$. Hence ψ is of class C^1 with $\nabla\psi$ a Lipschitz function.

Consequently there exists a unique solution $x_\varepsilon \in C\left([0,T];\mathbb{R}^d\right)$ of Eq. (4.127). We have

$$\psi\left(x_\varepsilon(t)\right)+\frac{1}{\varepsilon}\int_0^t \left\langle H\left(x_\varepsilon(s)\right)\nabla\psi\left(x_\varepsilon(s)\right),\nabla\psi\left(x_\varepsilon(s)\right)\right\rangle ds = \int_0^t \left\langle \nabla\psi\left(x_\varepsilon(s)\right),dm(s)\right\rangle$$

and since

$$c\left|\nabla\psi\left(x_\varepsilon\right)\right|^2 \le \left\langle H\left(x_\varepsilon\right)\nabla\psi\left(x_\varepsilon\right),\nabla\psi\left(x_\varepsilon\right)\right\rangle,$$

$$\int_0^t \left\langle \nabla\psi\left(x_\varepsilon(s)\right),dm(s)\right\rangle \le \frac{\varepsilon}{2c}\int_0^t \left|m'(s)\right|^2 ds+\frac{c}{2\varepsilon}\int_0^t \left|\nabla\psi\left(x_\varepsilon(s)\right)\right|^2 ds$$

then

$$\begin{cases} (a) & \displaystyle\int_0^T \left|\frac{1}{\varepsilon}\nabla\psi\left(x_\varepsilon(s)\right)\right|^2 ds \le T\left\|m'\right\|_T^2, \\ (b) & 0\le 1-\alpha\left(x_\varepsilon(t)\right)\le\psi\left(x_\varepsilon(t)\right)\le\varepsilon T\left\|m'\right\|_T^2, \\ (c) & \left\|k_\varepsilon\right\|_T \le\ \updownarrow k_\varepsilon \updownarrow_T = \displaystyle\int_0^T \left|\frac{1}{\varepsilon}\nabla\psi\left(x_\varepsilon(s)\right)\right| ds \le T\left\|m'\right\|_T. \end{cases}$$

Moreover for $0\le s\le t\le T$:

$$\begin{aligned} \left|k_\varepsilon(t)-k_\varepsilon(s)\right| &\le \updownarrow k_\varepsilon\updownarrow_t-\updownarrow k_\varepsilon\updownarrow_s \\ &= \int_s^t \left|\frac{1}{\varepsilon}\nabla\psi\left(x_\varepsilon(r)\right)\right| dr \\ &\le \left\|m'\right\|_T(t-s) \end{aligned}$$

and

$$\begin{aligned} \left|x_\varepsilon(t)-x_\varepsilon(s)\right| &\le \left|m(t)-m(s)\right|+\left|\int_s^t H\left(x_\varepsilon(r)\right)dk_\varepsilon(r)\right| \\ &\le \left|m(t)-m(s)\right|+c\times\left(\updownarrow k_\varepsilon\updownarrow_t-\updownarrow k_\varepsilon\updownarrow_s\right) \\ &\le \left|m(t)-m(s)\right|+c\times\left\|m'\right\|_T(t-s). \end{aligned}$$

Hence

$$\begin{aligned} \left|x_\varepsilon(t)\right|+\left|k_\varepsilon(t)\right| &\le \left|x_\varepsilon(t)\right|+\updownarrow k_\varepsilon\updownarrow_t \\ &\le \left|x_0\right|+\left|m(t)\right|+(c+1)T\left\|m'\right\|_T \end{aligned}$$

and we can continue exactly as in the proof of Theorem 4.47, obtaining $(x,k) = \mathcal{SP}(E;x_0,m;H)$ as the limit of $(x_{\varepsilon_n},k_{\varepsilon_n})$ in $C\left([0,T];\mathbb{R}^{2d}\right)$. Note that passing to the limit in $\int_0^t H(x_\varepsilon(s))\,dk_\varepsilon(s)$ is based on Helly–Bray's Proposition 6.16. In this case, by Proposition 4.60 the solution $(x,k) = \mathcal{SP}(E;x_0,m;H)$ is unique and therefore the whole sequence x_ε satisfies

$$x_\varepsilon \to x \qquad \text{in } C\left([0,T];\mathbb{R}^d\right) \quad \text{as } \varepsilon \to 0.$$

Step 2. Case $m \in C\left(\mathbb{R}_+;\mathbb{R}^d\right)$.

There exists an $m_n \in C^1\left(\mathbb{R}_+;\mathbb{R}^d\right)$ such that $\|m_n - m\|_T \to 0$ for all $T \geq 0$. By the first step there exists a unique solution $(x_n,k_n) = \mathcal{SP}(E;x_0,m_n;H)$.

Let $\mathcal{M} = \{m, m_n : n \in \mathbb{N}^*\}$. The set $\mathcal{M}$ is a compact subset of $C\left([0,T];\mathbb{R}^d\right)$. By Proposition 4.63 and Arzelà–Ascoli's Theorem (6.10) there exists a subsequence also denoted by (x_n,k_n) and $(x,k) \in C\left([0,T];\mathbb{R}^{2d}\right)$ such that

$$\|x_n - x\|_T + \|k_n - k\|_T \to 0, \quad \text{as } n \to \infty$$

and (see Example 6.30)

$$\updownarrow k \updownarrow_T \leq \liminf_{n\to+\infty} \updownarrow k_n \updownarrow_T \leq \sup_{n\in\mathbb{N}^*} \updownarrow k_n \updownarrow_T < \infty.$$

Now passing to the limit as in *Step 1* we infer that $(x,k) = \mathcal{SP}(E;x_0,m;H)$. Taking into account Proposition 4.60 the proof is complete. ■

Corollary 4.65. *If* $\xi \in L^0\left(\Omega,\mathcal{F}_0,\mathbb{P};\mathbb{R}^d\right)$ *and* $M \in S_d^0[0,T]$, $\updownarrow M \updownarrow_T < \infty$, $\mathbb{P}$*-a.s., then the SDE*

$$\begin{cases} X_t + \displaystyle\int_0^t H(X_s)\,dK_s \ni \xi + M_t, & t \in [0,T], \\ dK_t \in \partial^- I_E(X_t)(dt) \end{cases}$$

has a unique solution (X,K) *with* $X,K \in S_d^0[0,T]$ *and* $\updownarrow K \updownarrow_T < \infty$, $\mathbb{P}$*-a.s.*

Proof. By Proposition 3.28 with $G = 0$ the approximating equation (4.127) has a unique solution $X^\varepsilon \in S_d^0[0,T]$. Since as $\varepsilon \to 0$, for each $t \in [0,T]$

$$\sup_{s\in[0,t]} |X_s^\varepsilon - X_s| \to 0, \quad a.s.,$$

the progressive measurability of X follows; hence $X \in S_d^0[0,T]$. K is defined by

$$K_t = \int_0^t [H(X_s)]^{-1}\,d(X_s - M_s)$$

and consequently $K \in S_d^0[0,T]$. ■

Let $(\Omega, \mathcal{F}, \mathbb{P}, \{\mathcal{F}_t\}_{t\geq 0})$ be a stochastic basis and $\{B_t : t \geq 0\}$ an $\mathbb{R}^k$-valued Brownian motion. We consider the SDE

$$\begin{cases} dX_t + H(X_t)\,\partial^- I_E(X_t)(dt) \ni f(t, X_t)\,dt + g(t, X_t)\,dB_t, & a.e.\ t \in [0,T] \\ X_0 = x_0, \end{cases} \tag{4.128}$$

where $f : \mathbb{R}_+ \times \mathbb{R}^d \to \mathbb{R}^d$ and $g : \mathbb{R}_+ \times \mathbb{R}^d \to \mathbb{R}^{d\times k}$. We define

$$f^{\#}(s) = \sup\{|f(t,u)| : u \in E\} \quad \text{and} \quad g^{\#}(s) = \sup\{|g(t,u)| : u \in E\}.$$

We make the assumptions:

OR1 $(t,x) \mapsto f(t,x)$ and $(t,x) \mapsto g(t,x)$ are $(\mathcal{B}_1, \mathbb{R}^d)$-Carathéodory functions (that is, measurable in t and continuous in x);

OR2 there exist $L \in L^1_{loc}(0,\infty)$ and $\ell \in L^2_{loc}(0,\infty;\mathbb{R}_+)$ such that for all $x, y \in \mathbb{R}^d$

$$\begin{cases} |f(t,x) - f(t,y)| \leq L(t)\,|x-y| \quad \text{and,} \\ |g(t,x) - g(t,y)| \leq \ell(t)\,|x-y|; \end{cases}$$

OR3

$$\int_0^T \left[\left(f^{\#}(t)\right)^2 + \left(g^{\#}(t)\right)^4\right] dt < \infty, \quad \forall\, T \geq 0.$$

Definition 4.66. A pair of stochastic processes $X, K \in S^0_d$ is a solution of the stochastic oblique reflection problem (4.128) if for all $0 \leq s \leq t$, $\mathbb{P}$-a.s. $\omega \in \Omega$

$$\begin{cases} j) & X_t(\omega) \in E, \\ jj) & K_\cdot(\omega) \in BV_{loc}\left(\mathbb{R}_+;\mathbb{R}^d\right), \quad K_0 = 0, \\ jjj) & X_t + \displaystyle\int_0^t H(X_r)\,dK_r = x_0 + \int_0^t f(r,X_r)\,dr + \int_0^t g(r,X_r)\,dB_r, \\ jv) & \updownarrow K \updownarrow_t = \displaystyle\int_0^t \mathbf{1}_{X_r \in \mathrm{Bd}(E)} d \updownarrow K \updownarrow_r, \\ v) & K_t = \displaystyle\int_0^t n(X_r)\,\mathbf{1}_{X_r \in \mathrm{Bd}(E)} d \updownarrow K \updownarrow_r, \quad \text{where } n(X_r) \in N_E(X_r) \\ & \qquad \text{and } |n(X_r)| = 1, \quad d \updownarrow K \updownarrow_r\text{-}a.e., \end{cases}$$

or equivalently $\mathbb{P}$-a.s. $\omega \in \Omega$

$$\begin{cases} \ell) & X_t + \displaystyle\int_0^t H(X_r)\,dK_r = x_0 + \int_0^t f(r,X_r)\,dr + \int_0^t g(r,X_r)\,dB_r, \\ & \qquad\qquad \forall\, t \geq 0, \\ \ell\ell) & dK_t \in \partial^- I_E(X_t)(dt). \end{cases}$$

Theorem 4.67. *Let $E = \overline{E} \subset \mathbb{R}^d$ and $x_0 \in E$. Assume that there exists a $\gamma \geq 0$ such that E is a closed γ–semiconvex subset of $\mathbb{R}^d$ and E satisfies γ-SUIBC. Let the assumptions (OR1), (OR2) and (OR3) be satisfied. Then there exists a unique solution $(X, K) \in S_d^0 \times S_d^0$ of the stochastic oblique reflection problem (4.128).*

Proof. Uniqueness.

Let (X, K), $(\hat{X}, \hat{K}) \in S_d^0 \times S_d^0$ be two solutions of the Skorohod problem (4.128) with oblique reflection. Denote by $g^{(j)}(t, x)$ the column j of the matrix $g(t, x)$. Consider the symmetric and strict positive matrix $U_t = H^{-1}(X_t) + H^{-1}(\hat{X}_t)$. Let $J_t = U_t^{1/2}\left(X_t - \hat{X}_t\right)$. Since

$$d\,U_t^{1/2} = dQ_t + \sum_{j=1}^{k} \beta_t^{(j)} dB_t^{(j)}$$

where Q is an $\mathbb{R}^{d\times d}$-valued $\mathcal{P}$-m.b-v.c.s.p., $Q_0 = 0$ and $\beta_t^j \in \Lambda_{d\times d}^0$, for all $j \in \overline{1,k}$, it follows that

$$dJ_t = d\mathcal{K}_t + G_t dB_t$$

with

$$\begin{aligned} &d\mathcal{K}_t \\ &= (dQ_t)\,U_t^{-1/2}J_t + U_t^{1/2}\left[H\left(\hat{X}_t\right)d\hat{K}_t - H(X_t)\,dK_t + f(t, X_t) - f\left(t, \hat{X}_t\right)\right]dt \\ &+ \sum_{j=1}^{k} \beta_t^{(j)}\left(g^{(j)}(t, X_t) - g^{(j)}(t, \hat{X}_t)\right) \end{aligned}$$

and G_t is an $\mathbb{R}^{d\times k}$ matrix with the columns $\beta_t^{(1)}(X_t - \hat{X}_t), \ldots, \beta_t^{(k)}(X_t - \hat{X}_t)$.

Using (4.121) and the properties of H and H^{-1} we have

$$\begin{aligned} &\left\langle J_t, U_t^{1/2}\left[H(\hat{X}_t)d\hat{K}_t - H(X_t)\,dK_t\right]\right\rangle \\ &= \left\langle X_t - \hat{X}_t, \left([H(X_t)]^{-1} - \left[H(\hat{X}_t)\right]^{-1}\right)\left[H(\hat{X}_t)d\hat{K}_t + H(X_t)\,dK_t\right]\right\rangle \\ &- 2\left\langle X_t - \hat{X}_t, dK_t - d\hat{K}_t\right\rangle \\ &\leq (bc + 2\gamma)\left|X_t - \hat{X}_t\right|^2\left(d\updownarrow K\updownarrow_t + d\updownarrow \hat{K}\updownarrow_t\right). \end{aligned}$$

Hence

$$\langle J_t, d\mathcal{K}_t\rangle + \frac{1}{2}|G_t|^2\,dt \leq |J_t|^2 dV_t$$

where

$$dV_t = C \times \left(d \updownarrow Q \updownarrow_t + d \updownarrow K \updownarrow_t + d \updownarrow \hat{K} \updownarrow_t + L(t)\, dt \right) + C \sum_{j=1}^{k} \left| \beta_t^{(j)} \right|^2 dt$$

and $C = C(b, c, \gamma) > 0$. By the estimate (6.75) from Proposition 6.71 we infer

$$\mathbb{E} \frac{e^{-2V_t} |J_t|^2}{1 + e^{-2V_t} |J_t|^2} \leq \mathbb{E} \frac{e^{-2V_0} |J_0|^2}{1 + e^{-2V_0} |J_0|^2} = 0.$$

Consequently $U_t^{1/2} \left(X_t - \hat{X}_t \right) = J_t = 0$, $\mathbb{P}$-a.s., for all $t \geq 0$ and by the continuity of X and $\hat{X}$ we conclude that $\mathbb{P}$-a.s.,

$$X_t = \hat{X}_t \quad \text{for all } t \geq 0.$$

Existence.

It is sufficient to prove the existence on an arbitrary interval $[0, T]$. Let $n \in \mathbb{N}^*$. We define $X_t^n = x_0$ and $M_t^n = 0$ for $t \leq 0$. For $t \geq 0$ we define the regularization

$$\begin{aligned} M_t^n &= n \int_0^t \Big[\int_0^s f(r, X_{r-1/n}^n) dr + \int_0^s g(r, X_{r-1/n}^n) dB_r \Big] e^{-n(t-s)} ds \\ &= \int_0^\infty \Big[\int_0^{(t-u/n)\vee 0} f(r, X_{r-1/n}^n) dr + \int_0^{(t-u/n)\vee 0} g(r, X_{r-1/n}^n) dB_r \Big] e^{-u} du. \end{aligned}$$

By Corollary 4.65 there exists a unique solution $(X^n, K^n) \in \left(S_d^0 [0, T] \right)^2$, $\updownarrow K^n \updownarrow_T < \infty$, $\mathbb{P}$-a.s., of the SDE

$$\begin{cases} X_t^n + \displaystyle\int_0^t H\left(X_s^n\right) dK_s^n \ni x_0 + M_t^n, & t \in [0, T], \\ dK_t^n \in \partial^- I_E\left(X_t^n\right)(dt) \end{cases}$$

(the solution is defined recursively on the intervals $\left[0, \frac{1}{n}\right]$, $\left[\frac{1}{n}, \frac{2}{n}\right]$, $\left[\frac{2}{n}, \frac{3}{n}\right]$, ...).

Since by the convexity of the function $x \longmapsto |x|^4$,

$$\begin{aligned} &\mathbb{E} \left[\sup_{0 \leq \theta \leq \varepsilon} \left| M_{t+\theta}^n - M_t^n \right|^4 \right] \\ &\leq \mathbb{E} \int_0^\infty \left[\sup_{0 \leq \theta \leq \varepsilon} \left| \int_{(t-\frac{s}{n})\vee 0}^{(t+\theta-\frac{s}{n})\vee 0} \left(f(r, X_{r-\frac{1}{n}}^n) dr + g(r, X_{r-\frac{1}{n}}^n) dB_r \right) \right|^4 e^{-s} ds \right] \end{aligned}$$

$$\leq C\left[\sup_{t\in[0,T]}\left(\int_t^{t+\varepsilon}\left|f^{\#}(s)\right|ds\right)^4+\sup_{t\in[0,T]}\left(\int_t^{t+\varepsilon}\left|g^{\#}(s)\right|^2ds\right)^2\right]$$

$$\leq \varepsilon C\left[\sup_{t\in[0,T]}\left(\int_t^{t+\varepsilon}\left|f^{\#}(s)\right|^2ds\right)^2+\sup_{t\in[0,T]}\int_t^{t+\varepsilon}\left|g^{\#}(s)\right|^4ds\right],$$

we deduce, by Proposition 1.47, that the family of laws of $\{M^n : n \geq 1\}$ is tight on $C\left([0,\infty[;\mathbb{R}^d\right)$.

From now, based on the estimates (4.126), the proof follows exactly the same steps as those of Theorem 4.53 with $\varphi = I_E$.

(a) By Lemma 4.41, C_{T,M^n}, $n \in \mathbb{N}^*$, is bounded on compact subset of $C\left([0,T];\mathbb{R}^d\right)$.

(b) By Proposition 1.48 $U^n = (X^n, K^n, \updownarrow K^n \updownarrow)$, $n \in \mathbb{N}^*$, is tight on $C([0,T];\mathbb{R}^{2d+1})$ since by (4.126-c) we deduce

$$\mathbf{m}_{U^n}(\varepsilon;[0,T]) \leq \tilde{C}_T\sqrt{\varepsilon+\mathbf{m}_{M^n}(\varepsilon)}, \quad \text{a.s.}$$

(c) By Prohorov's Theorem there exists a subsequence such that as $n \to \infty$

$$(X^n, K^n, \updownarrow K^n \updownarrow, B) \to (X, K, V, B) \quad \text{in law}$$

on $C\left([0,T];\mathbb{R}^{2d+1+k}\right)$ and by the Skorohod theorem, we can choose a probability space $(\Omega,\mathcal{F},\mathbb{P})$, and some random quadruples $(\bar{X}^n,\bar{K}^n,\bar{V}^n,\bar{B}^n)$, $(\bar{X},\bar{K},\bar{V},\bar{B})$ defined on $(\Omega,\mathcal{F},\mathbb{P})$, having the same laws as respectively $(X^n, K^n, \updownarrow K^n \updownarrow, B)$ and (X, K, V, B), such that moreover as $n \to \infty$,

$$(\bar{X}^n,\bar{K}^n,\bar{V}^n,\bar{B}^n) \xrightarrow{\mathbb{P}\text{-a.s.}} (\bar{X},\bar{K},\bar{V},\bar{B})$$

in $C\left([0,T];\mathbb{R}^{2d+1+k}\right)$.

Note that by Proposition 2.15, $\left(\bar{B}^n,\{\mathcal{F}_t^{\bar{X}^n,\bar{K}^n,\bar{V}^n,\bar{B}^n}\}\right), n \geq 1$, and $\left(\bar{B},\{\mathcal{F}_t^{\bar{X},\bar{K},\bar{V},\bar{B}}\}\right)$ are $\mathbb{R}^k$-Brownian motions.

(d) Since we also have $(X^n, K^n, \updownarrow K^n \updownarrow, B) \to (\bar{X},\bar{K},\bar{V},\bar{B})$, in law, we deduce, by Corollary 1.18, that for all $0 \leq s \leq t$, $\mathbb{P}$-a.s.

$$\bar{X}_0 = x_0, \qquad \bar{K}_0 = 0, \qquad \bar{X}_t \in E,$$
$$\updownarrow\bar{K}\updownarrow_t - \updownarrow\bar{K}\updownarrow_s \leq \bar{V}_t - \bar{V}_s \quad \text{and} \quad 0 = \bar{V}_0 \leq \bar{V}_s \leq \bar{V}_s.$$

Since for all $0 \leq s < t$, $n \in \mathbb{N}^*$

$$\int_s^t \langle y(r) - X_r^n, dK_r^n\rangle \leq \gamma\int_s^t |y(r) - X_r^n|^2 d\updownarrow K^n\updownarrow_r, \quad a.s.,$$

it follows, by Proposition 1.19, that

$$\int_s^t \langle y(r) - \bar{X}_r, d\bar{K}_r \rangle \le \gamma \int_s^t |y(r) - \bar{X}_r|^2 d\bar{V}_r.$$

Hence $d\bar{K}_r \in \partial^- I_E(\bar{X}_r)(dr)$.

(*e*) As in the proof of Theorem 3.54 we obtain $\mathbb{P}$-a.s.

$$\bar{X}_t + \int_0^t H(X_s)\, d\bar{K}_s = x_0 + \int_0^t f(s, \bar{X}_s)\, ds + \int_0^t g(s, \bar{X}_s)\, d\bar{B}_s, \ \forall t \in [0, T],$$

and consequently $\left(\bar{\Omega}, \bar{\mathcal{F}}, \bar{\mathbb{P}}, \mathcal{F}_t^{\bar{B},\bar{X}}, \bar{X}_t, \bar{K}_t, \bar{B}_t\right)_{t \ge 0}$ is a weak solution of Eq. (4.128).

(*f*) The same arguments as used in Theorem 3.55 show again that *weak existence + pathwise uniqueness* imply strong existence.

The proof is complete. ∎

4.4 The Feynman–Kac Formula

4.4.1 Parabolic PDEs with Neumann Boundary Conditions

We consider the following backward parabolic PDE in $[0, T] \times D$

$$\begin{cases} -\dfrac{\partial u}{\partial t}(t,x) + \Phi(t, x, u(t,x), Du(t,x), D^2u(t,x)) = 0, \ (t,x) \in [0,T] \times D, \\ u(T,x) = \kappa(x), \ x \in D, \\ \dfrac{\partial u}{\partial n}(t,x) = \chi(t,x), \ (t,x) \in [0,T] \times \partial D, \end{cases} \tag{4.129}$$

where Φ is as in Eq. (3.107) and D is a bounded open connected subset of $\mathbb{R}^d$, whose boundary ∂D is assumed to be of class C^2. Note that $E = \overline{D}$ satisfies (4.104). We assume that all the coefficients are continuous.

Let $X^{t,x}$ be the solution of the reflected stochastic differential equation for all $s \in [t, T]$, $\mathbb{P}$-a.s.,

$$\begin{cases} X_s^{t,x} + K_s^{t,x} = x + \displaystyle\int_t^s f(r, X_r^{t,x})dr + \int_t^s g(r, X_r^{t,x})dB_r, \\ X_s^{t,x} \in \overline{D}, \\ K_s^{t,x} = \displaystyle\int_t^s n(X_r^{t,x}) \mathbf{1}_{\partial D}\left(X_r^{t,x}\right) d\updownarrow K^{t,x} \updownarrow_r. \end{cases} \tag{4.130}$$

For each $(t,x) \in [0,T] \times \overline{D}$, we define

$$u(t,x) \overset{def}{=} \mathbb{E}\Big[\kappa(X_T^{t,x})e^{\int_t^T c(s,X_s^{t,x})\,ds} + \int_t^T h(s,X_s^{t,x})e^{\int_t^s c(r,X_r^{t,x})\,dr}\,ds$$
$$+\int_t^T \chi(s,X_s^{t,x})e^{\int_t^s c(r,X_r^{t,x})\,dr}\,d\updownarrow K^{t,x}\updownarrow_s\Big] \tag{4.131}$$

(Feynman–Kac formula).

The aim of this subsection is to relate equation (4.129) and the quantity defined by (4.131). The first result says that any classical solution of Eq. (4.129) is given by the formula (4.131).

Proposition 4.68. *Under the above assumptions, let $u \in C^{1,2}([0,T] \times D) \cap C^{0,1}([0,T] \times \overline{D})$ be a bounded solution of (4.129). Then $u(t,x)$ satisfies the Feynman–Kac formula (4.131).*

Proof. By Itô's formula (2.17) on $[t, T \wedge \theta_n]$ with

$$V_s = (s, e^{\int_t^s c(r,X_r^{t,x})\,dr}),$$

$$X_s^{t,x} + K_s^{t,x} = x + \int_t^s f(r, X_r^{t,x})\,dr + \int_t^s g(r, X_r^{t,x})\,dB_r$$

and

$$\theta_n = \inf\left\{s \geq t : \left|X_s^{t,x} - x\right| \geq n\right\},$$

we have

$$u(T \wedge \theta_n, X_{T\wedge\theta_n}^{t,x})e^{\int_t^{T\wedge\theta_n} c(r,X_r^{t,x})\,dr}$$
$$= u(t,x) + \int_t^{T\wedge\theta_n} [\frac{\partial u}{\partial t} + \mathcal{A}u + cu](r, X_r^{t,x})e^{\int_t^r c(s,X_s^{t,x})\,ds}dr$$
$$+ \int_t^{T\wedge\theta_n} e^{\int_t^r c(s,X_s^{t,x})\,ds}\left\langle \nabla_x u(r, X_r^{t,x}), g(r, X_r^{t,x})dB_r\right\rangle$$
$$- \int_t^{T\wedge\theta_n} \frac{\partial u}{\partial n}(r, X_r^{t,x})e^{\int_t^r c(s,X_s^{t,x})\,ds}\mathbf{1}_{X_r \in \partial D}\,d\updownarrow K\updownarrow_r.$$

Taking first the expectation, then using the fact that u is a solution of (4.129), the Feynman–Kac formula (4.131) follows by letting $n \to \infty$, using uniform integrability, which follows from the boundedness of c, u and h on $[0,T] \times \overline{D}$. ■

We next show that the function $u(t,x)$, defined by (4.131), is the viscosity solution of (4.129). The corresponding definition of a viscosity solution is an obvious combination of Definitions 6.95 and 6.96 in Annex D.

Theorem 4.69. *Assume again that f, g, c and h are continuous on* $[0,T]\times\overline{D}$, $\kappa \in C(\overline{D})$, $\chi \in C([0,T]\times\overline{D})$. *Then $u(t,x)$, given by* (4.131), *is a continuous function of* $(t,x)\in[0,T]\times\overline{D}$ *and it is the unique viscosity solution of* (4.129).

Proof. Uniqueness of the viscosity solution can be proved by arguments similar to, but simpler than those of Theorem 6.112 in Annex D.

Continuity follows from Corollary 4.56.

Most of the proof of the sub-solution property of u is analogous to that the proof of Theorem 3.42, using the strong Markov property of the process $X^{t,x}$, see Proposition 4.57.

Note that if $(t,x)\in[0,T]\times D$ is a local maximum of $u-\varphi$, we should choose the radius ε such that the ball centered at x with radius ε is contained in D. Consider now the case where $(t,x)\in[0,T]\times\partial D$ is a local maximum of $u-\varphi$ and $u(t,x)=\varphi(t,x)$. We argue by contradiction. Suppose that

$$-[\frac{\partial\varphi}{\partial t}+\mathcal{A}\varphi+c\varphi+h](t,x)>0,\quad [\langle\varphi'_x,n\rangle-\chi](t,x)>0.$$

Then there exist $\delta>0$ and $\varepsilon\in\,]0,T-t[$ such that for all $s\in[t,t+\varepsilon]$ and $|y-x|\le\varepsilon$,

$$\begin{array}{ll}(i) & u(s,y)\le\varphi(s,y),\\ (ii) & -[\varphi'_t+\mathcal{A}\varphi+c\varphi+h](s,y)\ge\delta,\\ (iii) & [\langle\varphi'_x,n\rangle-\chi](s,y)\ge\delta.\end{array}$$

We introduce the stopping time

$$\tau=(t+\varepsilon)\wedge\inf\left\{r: r\ge t,\ \left|X_r^{t,x}-x\right|\ge\varepsilon\right\}.$$

From the strong Markov property of the process $X_\cdot^{t,x}$,

$$\begin{aligned}u(t,x)=\mathbb{E}\Big[&u(\tau,X_\tau^{t,x})e^{\int_t^\tau c(s,X_s^{t,x})\,ds}+\int_t^\tau h(s,X_s^{t,x})e^{\int_t^s c(r,X_r^{t,x})\,dr}\,ds\\&+\int_t^\tau \chi(s,X_s^{t,x})e^{\int_t^s c(r,X_r^{t,x})\,dr}\,d\updownarrow K^{t,x}\updownarrow_s\Big].\end{aligned}$$

Now from Itô's formula applied to the function φ,

$$\begin{aligned}\varphi(t,x)=\mathbb{E}\Big[&\varphi(\tau,X_\tau^{t,x})e^{\int_t^\tau c(s,X_s^{t,x})\,ds}\\&-\int_t^\tau[\varphi'_s+\mathcal{A}_s\varphi+c\varphi](s,X_s^{t,x})e^{\int_t^s c(r,X_r^{t,x})\,dr}\,ds\\&+\int_t^\tau\langle\varphi'_x,n\rangle(s,X_s^{t,x})e^{\int_t^s c(r,X_r^{t,x})\,dr}\,d\updownarrow K^{t,x}\updownarrow_s\Big].\end{aligned}$$

Taking the difference, and exploiting the above assumptions on the stopping time τ, we deduce that

$$\begin{aligned}
0 &= \varphi(t,x) - u(t,x) \\
&= \mathbb{E}\left[\varphi(\tau, X_\tau^{t,x}) - u(\tau, X_\tau^{t,x})\right] e^{\int_t^\tau c(s,X_s^{t,x})\,ds} \\
&\quad - \mathbb{E}\int_t^\tau [\varphi_s' + \mathcal{A}_s\varphi + c\varphi + h](s, X_s^{t,x}) e^{\int_t^s c(r,X_r^{t,x})\,dr}\,ds \\
&\quad + \int_t^\tau [\langle \varphi_x', n\rangle - \chi](s, X_s^{t,x}) e^{\int_t^s c(r,X_r^{t,x})dr}\, d\updownarrow K^{t,x}\updownarrow_s \\
&\geq \mathbb{E}\int_t^\tau \delta e^{\int_t^s c(r,X_r^{t,x})\,dr}\left[ds + d\updownarrow K^{t,x}\updownarrow_s\right] \\
&\geq (\mathbb{E}\tau - t)\,\delta e^{-C(T-t)} \\
&> 0
\end{aligned}$$

which is impossible. ■

4.4.2 Elliptic Equations with Neumann Boundary Conditions

Consider the differential operator

$$\mathcal{A} = \frac{1}{2}\sum_{i,j=1}^{d} (gg*)_{ij}(x)\frac{\partial^2}{\partial x_i \partial x_j} + \sum_{i=1}^{d} f_i(x)\frac{\partial}{\partial x_i}$$

and the linear elliptic PDE

$$\begin{cases} \mathcal{A}u(x) + c(x)u(x) + h(x) = 0, \ x \in D, \\ \dfrac{\partial u}{\partial n}(x) = \chi(x), \ x \in \partial D, \end{cases} \tag{4.132}$$

where D is a bounded connected open subset of $\mathbb{R}^d$ with a boundary ∂D of class C^2, $f : \mathbb{R}^d \to \mathbb{R}^d$, $g : \mathbb{R}^d \to \mathbb{R}^{d\times k}$, $c : D \to \mathbb{R}_-$, $h : D \to \mathbb{R}$ and $\kappa : \partial D \to \mathbb{R}$ are continuous.

Let $\{(X_t^x, K_t^x),\ t \geq 0\}$ denote the solution of the reflected SDE, for all $t \geq 0$, $\mathbb{P}$-a.s.,

$$\begin{cases} X_t^x + K_t^x = x + \displaystyle\int_0^t f(X_s^x)ds + \int_0^t g(X_s^x)dB_s, \\ X_t^x \in \overline{D}, \\ K_t^x = \displaystyle\int_0^t n(X_s^x)\mathbf{1}_{X_s^x \in \partial D} d\updownarrow K^x \updownarrow_s . \end{cases}$$

For each $x \in \overline{D}$, we define

$$u(x) \stackrel{def}{=} \mathbb{E}\Big[\int_0^{\infty} h(X_t^x)e^{\int_0^t c(X_r^x)\,dr}\,dt + \int_0^{\infty} \chi(X_t^x)e^{\int_0^t c(X_r^x)\,dr}\,d\updownarrow K^x\updownarrow_t\Big]. \quad (4.133)$$

We have the following:

Theorem 4.70. *Under the above assumptions, if moreover* $c(x) \leq \bar{c} < 0$*, for all* $x \in \overline{D}$*, then* $u \in C(\overline{D})$ *and it is the unique viscosity solution of the elliptic PDE* (4.132).

Proof. Let us just prove that the right-hand side of (4.133) is well defined. The rest of the proof is very similar to that of Theorem 4.69.

The assumption c makes the dt-integral on the right of (4.133) clearly convergent. Consider the second integral. It is sufficient to show that

$$\mathbb{E}\int_0^{\infty} \exp(\bar{c}t)d\updownarrow K^x\updownarrow_t < \infty. \quad (4.134)$$

By integration by parts,

$$\mathbb{E}\int_0^{T} \exp(\bar{c}t)d\updownarrow K^x\updownarrow_t \leq \exp(\bar{c}T)\mathbb{E}\updownarrow K^x\updownarrow_T - \bar{c}\int_0^{T} \exp(\bar{c}t)\mathbb{E}\updownarrow K^x\updownarrow_t\,dt.$$

Finally from (4.109-jj), $\exp(\bar{c}t)\mathbb{E}\updownarrow K^x\updownarrow_t$ goes to zero at exponential speed, as $t \to \infty$, hence (4.134). ■

4.5 Invariant Sets of SDEs

Given a non-empty closed set $E \subset \mathbb{R}^d$, a starting moment $t \geq 0$ and a starting point $x \in E$, we saw in previous sections that with a supplementary source on the stochastic equation

$$X_s^{tx} = x + \int_t^{s\vee t} f(r, X_r^{tx})dr + \int_t^{s\vee t} g(r, X_r^{tx})dB_r, \quad t, s \geq 0, \quad (4.135)$$

the solution X_s^{tx} can be maintained in E for all $s \geq t$. It is natural to ask the question: given the Eq. (4.135), what are the conditions on the drift and diffusion coefficients such that the evolution of the state satisfies the constraint $X_s^{tx} \in E$, for all $s \geq t$? The main ideas of this section are based upon [14].

Let $\mathcal{E} = \{E(t) : t \geq 0\}$ be a family of non-empty closed subsets $E(t) \subset \mathbb{R}^d$.

Definition 4.71. d_1) The family $\mathcal{E}$ is strong invariant for the SDE (4.135) if for all $t \geq 0$, $x \in E(t)$ and for all solutions $\{X_s^{tx} : s \geq t\}$ it follows that

$$X_s^{tx} \in E(s), \quad \mathbb{P}\text{-a.s.}\ \forall\, s \geq t.$$

d_2) The family $\mathcal{E}$ is weak invariant (viable) for SDE (4.135) if for every $t \geq 0$ and $x \in E(t)$ there exists a solution $\{X_s^{tx} : s \geq t\}$ such that

$$X_s^{tx} \in E(s), \quad \mathbb{P}\text{-a.s. } \forall\, s \geq t.$$

Remark 4.72. If for every $(t,x) \in \mathbb{R}_+ \times E(t)$ the Eq. (4.135) has a unique solution, then the two notions coincide and in this case we shall say that $\mathcal{E}$ is invariant for SDE (4.135).

Remark 4.73. If the interval $\mathbb{R}_+$ is replaced by $[0,T]$ we shall say 'invariant on $[0,T]$'.

Our goal, here, is to give a characterization of the invariance of the moving sets $E(t),\ t \geq 0$.

Assume that $f : [0,\infty[\, \times \mathbb{R}^d \to \mathbb{R}^d$ and $g : [0,\infty[\, \times \mathbb{R}^d \to \mathbb{R}^{d\times k}$ and for every $T > 0$ there exist $L, M, m \geq 0$ and $\mu \in \mathbb{R}$ such that $\forall t \in [0,T],\ \forall\, x,y \in \mathbb{R}^d$:

$$\begin{cases} i) & \langle x-y, f(t,x) - f(t,y)\rangle \leq \mu|x-y|^2, \\ ii) & |g(t,x) - g(t,y)| \leq L|x-y|, \\ iii) & f \text{ and } g \text{ are continuous on } [0,\infty[\, \times \mathbb{R}^d, \\ iv) & \sup\limits_{t\in[0,T]} |f(t,x)| \leq M\left(1+|x|^m\right). \end{cases} \tag{4.136}$$

By Theorem 3.21 it follows that for every $(t,x) \in \mathbb{R}_+ \times \mathbb{R}^d$ the Eq. (4.135) has a unique solution $X_\cdot^{tx} \in S_d^p$ for all $p \geq 1$. Note that by Proposition 3.6 and Proposition 3.22:

$$\begin{aligned} a)\quad & \mathbb{E} \sup_{s\in[0,T]} |X_s^{tx}|^p \leq C_{p,T}(1+|x|^p), \\ b)\quad & \mathbb{E} \sup_{s\in[0,T]} |X_s^{tx} - X_s^{t'x'}|^p \leq C_{p,T}(1+|x|^p)(|t-t'|^{p/2} + |x-x'|^p). \end{aligned} \tag{4.137}$$

Recall the notations

- the distance from x to $E(t)$:

$$d(t,x) = d_{E(t)}(x) = \inf\{|x-y| : y \in E(t)\},$$

- $\mathbb{S}^d \subset \mathbb{R}^{d\times d}$ is the set of $d\times d$ symmetric non-negative matrices,
- $C_{pol}^{k,n}([0,T]\times\mathbb{R}^d)$ is the set of functions $h : [0,T]\times\mathbb{R}^d \to \mathbb{R}$ of class $C^{k,n}$ such that the function h and its derivatives $D_t^j h(t,x)$, $j \in \overline{0,k}$, and $D_x^\alpha h(t,x)$, $\alpha = (\alpha_1,\ldots,\alpha_d)$, $0 \leq \alpha_1 + \cdots + \alpha_d \leq n$, are polynomially increasing to infinity in the space variable, that is there exist $C = C_T \geq 0$ and $p = p_T \in \mathbb{N}^*$ such that

$$\sum_{i,\,\alpha}\left[\left|D_t^i h\,(t,x)\right| + \left|D_x^\alpha h\,(t,x)\right|\right] \le C\left(1+|x|^p\right),$$

for all $(t,x) \in [0,T] \times \mathbb{R}^d$,

♦ the infinitesimal generator associated to $\{X_s^{tx} : s \ge t\}$:

$$\begin{aligned}\mathcal{A}(t)\varphi(x) &= \frac{1}{2}\mathbf{Tr}[D_x^2\varphi(x)g(t,x)g^*(t,x)] + \langle f(t,x), \nabla_x\varphi(x)\rangle \\ &= \frac{1}{2}\sum_{j,\ell=1}^{d}(gg^*)_{j\ell}(t,x)\frac{\partial^2\varphi(x)}{\partial x_j\partial x_\ell} + \sum_{j=1}^{d} f_j(t,x)\frac{\partial\varphi(x)}{\partial x_j}.\end{aligned}$$

For the convenience of the reader we recall the definition of a viscosity solution for the particular PDE

$$\begin{cases}\dfrac{\partial u(t,x)}{\partial t} + \mathcal{A}(t)u(t,x) + G(t,x) = 0 \\ u(T,x) = H(x), \quad (t,x) \in [0,T] \times \mathbb{R}^d,\end{cases} \tag{4.138}$$

where $G \in C_{pol}([0,T] \times \mathbb{R}^d \times \mathbb{R} \times \mathbb{R}^k)$ and $H \in C_{pol}\left(\mathbb{R}^d;\mathbb{R}\right)$.

Definition 4.74. Let $u : [0,T] \times \mathbb{R}^d \to \mathbb{R}$ be an upper semicontinuous function and $(t,x) \in \left]0,T\right[\times \mathbb{R}^d$. We denote by $\mathcal{P}^{2+}v(t,x)$ (the parabolic superjet of u at (t,x)) the set of triples $(p,q,S) \in \mathbb{R} \times \mathbb{R}^d \times \mathbb{S}^d$ such that

$$\begin{aligned}u(s,y) \le u(t,x) + p(s-t) + \langle q, y-x\rangle + \\ +\frac{1}{2}\langle S(y-x), y-x\rangle + o(|s-t| + |y-x|^2).\end{aligned}$$

Let $u : [0,T] \times \mathbb{R}^d \to \mathbb{R}$ be a lower semicontinuous function and $(t,x) \in \left]0,T\right[\times \mathbb{R}^d$. We denote by $\mathcal{P}^{2-}u(t,x)$ (the parabolic subjet of u at (t,x)) the set of triples $(p,q,S) \in \mathbb{R} \times \mathbb{R}^d \times \mathbb{S}^d$ such that

$$\begin{aligned}u(s,y) \ge u(t,x) + p(s-t) + \langle q, y-x\rangle + \\ +\frac{1}{2}\langle S(y-x), y-x\rangle + o(|s-t| + |y-x|^2).\end{aligned}$$

We can now give the definition of a viscosity solution of the parabolic equation (4.138).

Definition 4.75. a) A lower semicontinuous $u : [0,T] \times \mathbb{R}^d \to \mathbb{R}$ is a viscosity super-solution of (4.138) if:

$$u\,(T,x) \ge H\,(x)\,, \forall x \in \mathbb{R}^d, \forall\, i = \overline{1,n}$$

and for any point $(t,x) \in \left]0,T\right[\times \mathbb{R}^d$ and for any $(p,q,S) \in \mathcal{P}^{2-}u(t,x)$

$$p + \frac{1}{2}Tr\,(gg^*(t,x)S) + (f(t,x),q) + G(t,x) \le 0.$$

b) An upper semicontinuous function $u : [0,T] \times \mathbb{R}^d \to \mathbb{R}$ is a viscosity sub-solution of (4.138) if:

$$u\,(T,x) \le H\,(x)\,,\forall x \in \mathbb{R}^d$$

and for any point $(t,x) \in \left]0,T\right[\times \mathbb{R}^d$ and for any $(p,q,S) \in \mathcal{P}^{2+}u(t,x)$

$$p + \frac{1}{2}Tr\,((gg^*)(t,x)S) + (f(t,x),q) + G(t,x) \ge 0.$$

c) $u \in C([0,T] \times \mathbb{R}^d)$ is a viscosity solution of (4.138) if it is both a viscosity sub- and super-solution.

We also have the equivalent definition:

Definition 4.76. a) A lower semicontinuous $u : [0,T] \times \mathbb{R}^d \to \mathbb{R}$ is a viscosity super-solution of (4.138) if $u\,(T,x) \ge H\,(x)$, $\forall x \in \mathbb{R}^d$, and for any $\varphi \in C^2(\mathbb{R}^d)$ and any local minimum $(t,x) \in \left]0,T\right[\times \mathbb{R}^d$ of $u - \varphi$,

$$\frac{\partial}{\partial t}\varphi(t,x) + A(t)\varphi(t,x) + G(t,x) \le 0.$$

b) An upper semicontinuous function $u : [0,T] \times \mathbb{R}^d \to \mathbb{R}$ is a viscosity sub-solution of the system (4.138) if $u(T,x) \le H\,(x)$, $\forall x \in \mathbb{R}^d$, and for any $\varphi \in C^2(\mathbb{R}^d)$ and any local maximum $(t,x) \in \left]0,T\right[\times \mathbb{R}^d$ of $u - \varphi$,

$$\frac{\partial}{\partial t}\varphi(t,x) + A(t)\varphi(t,x) + G(t,x) \ge 0.$$

c) A continuous function $u : [0,T] \times \mathbb{R}^d \to \mathbb{R}^n$ is a viscosity solution of (4.138) if it is a viscosity super-solution and a viscosity sub-solution.

By Theorem 3.42 the Eq. (4.138) has a unique viscosity solution u and $\underline{u} \le u \le \overline{u}$ for all viscosity sub-solutions $\underline{u}$ and viscosity super-solutions $\overline{u}$.

Now we can give the main result of this section.

Theorem 4.77. *Assume that for every $T > 0$, $x \in \mathbb{R}^d$ the function $t \mapsto d^2\,(t,x) : [0,T] \to \mathbb{R}$ is lower semicontinuous (l.s.c.) and there exists a $b_T \ge 0$ such that $d^2(t,0) \le b_T\ \ \forall\, t \in [0,T]$. Then the following assertions are equivalent:*

(I) Equation (4.135) is $\mathcal{E}$-invariant on $\mathbb{R}_+$.
(II) For every $T > 0$ there exists a constant $C = C_T \in \mathbb{R}$ such that the square distance function $h\,(t,x) = d^2(t,x)$ is a viscosity super-solution of

the equation

$$\begin{cases} \dfrac{\partial u(t,x)}{\partial t} + A(t)u(t,x) - Cd^2(t,x) = 0, & (t,x) \in]0,T[\times \mathbb{R}^d \\ u(T,x) = d^2(T,x), \quad x \in \mathbb{R}^d, \end{cases} \tag{4.139}$$

that is

$$p + \frac{1}{2}\mathbf{Tr}\,(S\,gg^*(t,x)) + \langle f(t,x), q\rangle \le Cd^2(t,x), \tag{4.140}$$

for all $(p,q,S) \in \mathcal{P}^{2-}d^2(t,x)$, $(t,x) \in]0,T[\times \mathbb{R}^d$.

Example 4.78 (Control Security Tube). Let $\rho \in C^1([0,T];\mathbb{R}_+)$, $\rho > 0$, $a \in C^1([0,T];\mathbb{R}^d)$ and

$$E(t) = \{x \in \mathbb{R}^d : |x - a(t)| \le \rho(t)\}.$$

Then for $(t,x) \in]0,T[\times \mathbb{R}^d$

$$h(t,x) = d^2(t,x) = ((|x - a(t)| - \rho(t))^+)^2$$

and

$$p = \frac{\partial h}{\partial t}(t,x) = -2(|x - a(t)| - \rho(t))^+ \left[\frac{1}{|x - a(t)|}\langle x - a(t), a'(t)\rangle + \rho'(t)\right],$$

$$q = \nabla_x h(t,x) = \begin{cases} 0, & \text{if } |x \quad a(t)| \le \rho(t), \\ 2\dfrac{|x - a(t)| - \rho(t)}{|x - a(t)|}(x - a(t)), & \text{if } |x - a(t)| > \rho(t), \end{cases}$$

$$S = \begin{cases} 0, \quad \text{if } |x - a(t)| \le \rho(t), \\ 2\dfrac{|x - a(t)| - \rho(t)}{|x - a(t)|}\mathbf{I}_{d\times d} + \dfrac{2\rho(t)}{|x - a(t)|^3}[x - a(t)] \otimes [x - a(t)], \\ \qquad\qquad\qquad\qquad\qquad \text{if } |x - a(t)| > \rho(t). \end{cases}$$

The SDE (4.135) is $\mathcal{E}$-invariant on $[0,T]$ iff the distance function $d^2(t,x)$ is a viscosity super-solution of the Eq. (4.139), or equivalent, for $(t,x) \in]0,T[\times \mathbb{R}^d$ with $|x| > \rho(t)$:

$$\begin{aligned} &\frac{|x - a(t)| - \rho(t)}{|x - a(t)|}\Big[2\langle x - a(t), f(t,x)\rangle + |g(t,x)|^2 - 2\langle x - a(t), a'(t)\rangle \\ &\qquad - 2|x - a(t)|\rho'(t)\Big] + \frac{\rho(t)}{|x - a(t)|^3}|g^*(t,x)(x - a(t))|^2 \\ &\le C(|x - a(t)| - \rho(t))^2. \end{aligned}$$

By taking the limit as $|x - a(t)| \searrow \rho(t)$, we obtain for all $t \in [0, T]$ and for all $x \in \mathbb{R}^d$ with $|x - a(t)| = \rho(t)$:

$$\begin{cases} g^*(t,x)(x - a(t)) = 0, \text{ and} \\ 2\langle x - a(t), f(t,x)\rangle + |g(t,x)|^2 \le 2\langle x - a(t), a'(t)\rangle + 2\rho(t)\rho'(t). \end{cases}$$

This condition is also sufficient for the $\mathcal{E}$-invariance. Indeed since from (4.136) for all $x \in \mathbb{R}^d$ and $0 < \lambda < 1$

$$\begin{aligned} |g(t,x)|^2 &\le [|g(t,\lambda(x-a)+a)| + L|x - \lambda(x-a) - a|]^2 \\ &\le \frac{1}{\lambda}|g(t,\lambda(x-a)+a)|^2 + (1-\lambda)L^2|x-a|^2, \end{aligned}$$

(in the proof we write a and ρ in place of $a(t)$ and respectively $\rho(t)$) and

$$\begin{aligned} &\langle x - a(t), f(t,x)\rangle \\ &= \frac{1}{1-\lambda}\langle x - \lambda(x-a) - a, f(t,x)\rangle \\ &\le \frac{1}{1-\lambda}\langle x - \lambda(x-a) - a, f(t,\lambda(x-a)+a)\rangle \\ &+ \frac{1}{1-\lambda}\mu^+|x - \lambda(x-a) - a|^2 \\ &\le \frac{1}{\lambda}\langle \lambda(x-a), f(t,\lambda(x-a)+a)\rangle + (1-\lambda)\mu^+|x-a|^2, \end{aligned}$$

then for all $x \in \mathbb{R}^d$ with $|x - a(t)| > \rho(t)$ and $\lambda = \dfrac{\rho(t)}{|x - a(t)|}$:

$$\begin{aligned} &2\langle x - a(t), f(t,x)\rangle + |g(t,x)|^2 \\ &\le \frac{1}{\lambda}\left(2\langle \lambda(x-a), f(t,\lambda(x-a)+a)\rangle + |g(t,\lambda(x-a)+a)|^2\right) \\ &+ (1-\lambda)\left(2\mu^+ + L^2\right)|x-a|^2 \\ &\le \frac{2}{\lambda}\left[\langle \lambda(x-a), a'(t)\rangle + \rho(t)\rho'(t)\right] + (1-\lambda)\left(2\mu^+ + L^2\right)|x-a|^2 \\ &= \left[2\langle x-a, a'(t)\rangle + 2|x-a|\rho'(t) + \left(2\mu^+ + L^2\right)|x-a|(|x-a| - \rho)\right]. \end{aligned}$$

We also have

$$\begin{aligned} &|g^*(t,x)(x - a(t))|^2 \\ &= \frac{|x-a|^2}{\rho^2}\left|g^*(t,x)\frac{\rho(x-a)}{|x-a|} - g^*(t, \frac{\rho}{|x-a|}(x-a)+a)\frac{\rho(x-a)}{|x-a|}\right|^2 \end{aligned}$$

$$\leq \frac{|x-a|^2}{\rho^2}L^2\left|x-\frac{\rho}{|x-a|}(x-a)-a\right|^2\left|\frac{\rho(x-a)}{|x-a|}\right|^2$$
$$= L^2|x-a|^2(|x-a|-\rho)^2$$

which yields

$$\frac{|x-a(t)|-\rho(t)}{|x-a(t)|}\Big[2\langle x-a(t),f(t,x)\rangle+|g(t,x)|^2-2\big\langle x-a(t),a'(t)\big\rangle$$
$$-2|x-a(t)|\rho'(t)\Big]+\frac{\rho(t)}{|x-a(t)|^3}|g^*(t,x)(x-a(t))|^2$$
$$\leq 2\left(\mu^++L^2\right)(|x-a(t)|-\rho(t))^2.$$

Hence Eq. (4.135) is $\bar{B}(a(t),\rho(t))$-invariant if and only if

$$\begin{aligned}&\forall\,(t,x)\in[0,T]\times\mathbb{R}^d \text{ with } |x|=\rho(t):\\&\begin{cases}g^*(t,x)(x-a(t))=0, \text{ and}\\2\langle x-a(t),f(t,x)\rangle+|g(t,x)|^2\leq 2\langle x-a(t),a'(t)\rangle+2\rho(t)\rho'(t).\end{cases}\end{aligned} \tag{4.141}$$

Consequently, a nondegenerate SDE cannot have the $\{\bar{B}(a(t),\rho(t)):\ t\in[0,T]\}$-invariance property.

Example 4.79 (Comparison of the Solutions). Consider the two dimensional system

$$\begin{pmatrix}X_s^{t,x,y}\\Y_s^{t,x,y}\end{pmatrix}=\begin{pmatrix}x\\y\end{pmatrix}+\int_t^s\begin{pmatrix}f_1\left(r,X_r^{t,x,y},Y_r^{t,x,y}\right)\\f_2\left(r,X_r^{t,x,y},Y_r^{t,x,y}\right)\end{pmatrix}dr$$
$$+\sum_{j=1}^k\int_t^s\begin{pmatrix}g_{1,j}\left(r,X_r^{t,x,y},Y_r^{t,x,y}\right)\\g_{2,j}\left(r,X_r^{t,x,y},Y_r^{t,x,y}\right)\end{pmatrix}dB_r^{(j)}$$

and

$$\mathcal{E}=\{E(t):t\in[0,T]\},\quad \text{where } E(t)=E=\left\{(x,y)^*\in\mathbb{R}^2:x\geq y\right\}.$$

The viability of $\mathcal{E}$ means that for all $t\in[0,T]$,

$$\text{if } x\geq y \quad \text{then } X_s^{t,x,y}\geq Y_s^{t,x,y}, \quad \text{for all } s\geq t.$$

In this case

$$h(t,x,y)=d^2(x,y)=\frac{1}{2}\left[(y-x)^+\right]^2$$

and $(p, q, S) \in \mathcal{P}^{2-}h(t, x, y)$ are as follows

$$p = \frac{\partial h}{\partial t}(t, x, y) = 0,$$
$$q = \nabla_{(x,y)} h(t, x, y) = \left(-(y-x)^{+},\ (y-x)^{+}\right)^{*},$$
$$S = \begin{cases} 0, & \text{if } y \le x, \\ \begin{pmatrix} 1 & -1 \\ -1 & 1 \end{pmatrix}, & \text{if } y > x. \end{cases}$$

The SDE (4.135) is $\mathcal{E}$-invariant on $[0, T]$ iff the distance function $d^2(x, y)$ is a viscosity super-solution of the Eq. (4.139), that is

$$\frac{1}{2}\sum_{j=1}^{k}\left(g_{1j}(t, x, y) - g_{2j}(t, x, y)\right)^2 - f_1(t, x, y)(y - x) + f_2(t, x, y)(y - x)$$
$$\le C(y - x)^2$$

for all $0 < t < T$ and $x < y$, or equivalently (using the Lipschitz properties of f and g) that

$$\begin{cases} (c_1) & g_{1j}(t, a, a) = g_{2j}(t, a, a), \quad \forall\, a \in \mathbb{R},\ \forall\ j \in \overline{1, k}, \\ (c_2) & f_1(t, a, a) \ge f_2(t, a, a), \quad \forall\, a \in \mathbb{R}. \end{cases}$$

Theorem 4.77 follows clearly from the next two lemmas.

Lemma 4.80. *The following statements are equivalent:*

(i) Equation (4.135) is $\mathcal{E}$-invariant on $\mathbb{R}_+$.
(ii) For all $T > 0$ there exists a $C = C_T \in \mathbb{R}$ such that

$$\mathbb{E}d^2(s, X_s^{tx}) \le e^{C(s-t)} d^2(t, x), \quad \forall\, 0 \le t \le s \le T,\ \forall\, x \in \mathbb{R}^d.$$

Proof. $(ii) \Longrightarrow (i)$: This is obvious.
$(i) \Longrightarrow (ii)$: Let $(t, x) \in\ ]0, +\infty[\ \times \mathbb{R}^d$ be fixed, and $\hat{x} \in \mathrm{Pr}_{E(t)}(x)$, the projection of x on $E(t)$. Let Y^{tx} be the solution of the equation

$$Y_s^{tx} = \hat{x} + \int_t^{s \vee t} f(r, Y_r^{tx})dr + \int_t^{s \vee t} g(r, Y_r^{tx})dW_r\,.$$

Then

$$Y_s^{tx} \in E(s) \quad \forall s \ge t,$$

and by (3.18) we have

$$\mathbb{E}d^2\left(s, X_s^{tx}\right) \leq \mathbb{E}|Y_s^{tx} - X_s^{tx}|^2 \leq e^{(2\mu+L^2)(s-t)}|\hat{x} - x|^2 = e^{C(s-t)}d^2(t,x).$$

■

Lemma 4.81. *Let $T > 0$, $C = C_T \in \mathbb{R}$, and $\psi : [0,T] \times \mathbb{R}^d \to \mathbb{R}$ be a lower semicontinuous function which is polynomially increasing at infinity in the space variable. Then the following assertions are equivalent:*

(i) $\mathbb{E}\psi(s, X_s^{tx}) \leq e^{C(s-t)}\psi(t,x), \quad \forall\, 0 \leq t \leq s \leq T.$

(ii) The function ψ is a viscosity super-solution for

$$\begin{cases} \dfrac{\partial u(t,x)}{\partial t} + A(t)u(t,x) - C\psi\,(t,x) = 0, & (t,x) \in \,]0,T[\, \times \mathbb{R}^d, \\ u\,(T,x) = \psi(T,x), \quad x \in \mathbb{R}^d. \end{cases} \tag{4.142}$$

Proof. $(i) \Longrightarrow (ii)$:
Let $\varphi \in C_{pol}^{1,2}([0,T] \times \mathbb{R}^d)$ and $(t,x) \in \,]0,T[\, \times \mathbb{R}^d$ be a minimum point of

$$(t',x') \mapsto \psi(t',x') - \varphi(t',x').$$

Hence

$$\begin{aligned} \mathbb{E}[\varphi(t+\varepsilon, X_{t+\varepsilon}^{tx}) - \varphi(t,x)] &\leq \mathbb{E}[\psi(t+\varepsilon, X_{t+\varepsilon}^{tx}) - \psi(t,x)] \\ &< (e^{C\varepsilon} - 1)\psi(t,x). \end{aligned}$$

Then

$$\begin{aligned} &\frac{\partial\varphi(t,x)}{\partial t} + \mathbb{E}\langle\nabla_x\varphi(t,x), \frac{1}{\varepsilon}(X_{t+\varepsilon}^{tx} - x)\rangle + \\ &\quad + \frac{1}{2\varepsilon}\mathbb{E}\left\langle D_x^2\varphi(t,x)(X_{t+\varepsilon}^{tx} - x), X_{t+\varepsilon}^{tx} - x\right\rangle + \\ &\quad + \frac{1}{\varepsilon}\mathbb{E}\gamma^{tx}(t+\varepsilon, X_{t+\varepsilon}^{tx}) \leq \frac{e^{C\varepsilon} - 1}{\varepsilon}\psi(t,x) \end{aligned} \tag{4.143}$$

where $\gamma^{tx} \in C_{pol}\left([0,T] \times \mathbb{R}^d\right)$ and

$$\lim_{\substack{t' \to t \\ x' \to x}} \frac{\gamma^{tx}(t',x')}{|t'-t| + |x'-x|^2} = 0.$$

By Lemma 4.82 below we have

$$\lim_{\varepsilon \searrow 0} \frac{1}{\varepsilon}\mathbb{E}\left|\gamma^{tx}(t+\varepsilon, X_{t+\varepsilon}^{tx})\right| = 0.$$

Hence by passing to the limit in (4.143) as $\varepsilon \to 0$, we obtain

$$\frac{\partial\varphi(t,x)}{\partial t} + \mathbb{E}\left\langle \nabla_x\varphi(t,x), f(t,x)\right\rangle + \frac{1}{2}Tr[D_x^2\varphi(t,x)g(t,x)g^*(t,x)] \le C\psi(t,x)$$

that is ψ is a viscosity super-solution of (4.142).
$(ii) \Longrightarrow (i)$:
Since $\psi(t,x)$ is a viscosity super-solution of the equation

$$\begin{cases} \dfrac{\partial u(t,x)}{\partial t} + A(t)u(t,x) - C\psi(t,x) = 0, \ (t,x) \in]0,R[\times \mathbb{R}^d, \\ u(R,x) = \psi(R,x), \quad x \in \mathbb{R}^d \end{cases}$$

for all $R \in]0,T]$, the upper semicontinuous function

$$h(t,x) = -e^{-2tC}\psi(t,x)$$

is a viscosity sub-solution of the equation

$$\begin{cases} \dfrac{\partial v(t,x)}{\partial t} + A(t)v(t,x) + Ch_k(t,x) = 0, \ (t,x) \in]0,R[\times \mathbb{R}^d, \\ v(R,x) = h_k(R,x), \quad x \in \mathbb{R}^d \end{cases} \tag{4.144}$$

for all $R \in]0,T]$, where $h_k \in C_{pol}([0,T] \times \mathbb{R}^d)$, $h_k(t,x) \searrow h(t,x)$, as $k \to \infty$.

Then by the Feyman–Kac formula (Theorem 3.42 from Chap. 3) the viscosity solution of the Eq. (4.144) is

$$v(t,x) = \mathbb{E}h_k(R, X_R^{tx}) + \mathbb{E}\int_t^R C\, h_k(r, X_r^{tx})dr.$$

Hence

$$\mathbb{E}h(s, X_s^{tx}) \le \mathbb{E}v(s, X_s^{tx}) = \mathbb{E}h_k(R, X_R^{tx}) + C\int_s^R \mathbb{E}h_k(r, X_r^{tx})dr,$$

which implies for $k \to \infty$

$$\mathbb{E}h(s, X_s^{tx}) \le e^{C(R-s)}\mathbb{E}h(R, X_R^{tx}).$$

Setting here $s = t$ and $h(r,u) = -e^{-2rC}\psi(r,u)$ we have

$$\psi(t,x) \ge e^{-C(R-t)}\mathbb{E}\psi(R, X_R^{tx}),$$

for all $R \ge t$. The proof of Lemma 4.81 is complete. ■

Lemma 4.82.

$$\lim_{\varepsilon \searrow 0} \frac{1}{\varepsilon} \mathbb{E} \left| \gamma^{tx}(t+\varepsilon, X^{tx}_{t+\varepsilon}) \right| = 0.$$

Proof. First we remark that

$$\begin{aligned} \gamma^{tx}(t+\varepsilon, X^{tx}_{t+\varepsilon}) &= \varepsilon[\varphi(t+\varepsilon, X^{tx}_{t+\varepsilon}) - \varphi(t,x)] - \varepsilon \frac{\partial \varphi(t,x)}{\partial t} \\ &- \left\langle \nabla_x \varphi(t,x), (X^{tx}_{t+\varepsilon} - x) \right\rangle - \frac{1}{2} \mathbb{E} \left\langle D^2_x \varphi(t,x)(X^{tx}_{t+\varepsilon} - x), X^{tx}_{t+\varepsilon} - x \right\rangle \end{aligned}$$

and $\varphi \in C^{1,2}_{pol}([0,T] \times \mathbb{R}^d)$. Hence by (4.137) there exists a positive constant C independent of ε, t, x, $0 < \varepsilon \le 1$, such that

$$\mathbb{E} \left| \gamma^{tx}(t+\varepsilon, X^{tx}_{t+\varepsilon}) \right|^2 \le C.$$

Since for all $\delta > 0$ there exists an $a = a(\delta, t, x) > 0$ such that

$$\left| \gamma^{tx}(t', x') \right| \le \left(|t'-t| + |x'-x|^2 \right) \delta$$

if $|t'-t| + |x'-x| \le a$, we deduce that for $\varepsilon \in]0,1]$ and $A = \{\omega \in \Omega : \varepsilon + \left| X^{tx}_{t+\varepsilon} - x \right| > a\}$

$$\begin{aligned} &\frac{1}{\varepsilon} \mathbb{E} \left| \gamma^{tx}(t+\varepsilon, X^{tx}_{t+\varepsilon}) \right| \\ &\le \frac{1}{\varepsilon} \mathbb{E} \left[\left| \gamma^{tx}(t+\varepsilon, X^{tx}_{t+\varepsilon}) \right| \mathbf{1}_{A^c} \right] + \frac{1}{\varepsilon} \mathbb{E} \left[\left| \gamma^{tx}(t+\varepsilon, X^{tx}_{t+\varepsilon}) \right| \mathbf{1}_{A} \right] \\ &\le \frac{\delta}{\varepsilon} \left[\varepsilon + \mathbb{E} \left| X^{tx}_{t+\varepsilon} - x \right|^2 \right] + \frac{1}{\varepsilon} \sqrt{\mathbb{E} \left| \gamma^{tx}(t+\varepsilon, X^{tx}_{t+\varepsilon}) \right|^2} \sqrt{\mathbb{P}(A)} \\ &\le C\delta + \frac{C}{\varepsilon} \sqrt{\frac{\mathbb{E} \left[\left(\varepsilon + \left| X^{tx}_{t+\varepsilon} - x \right| \right)^8 \right]}{a^8}} \\ &\le C\delta + \frac{C'}{a^4} \varepsilon, \end{aligned}$$

where C, C' are positive constants independent of ε and δ. We used here Chebyshev inequality

$$\mathbb{P}(|\eta| > a) \le \frac{\mathbb{E}|\eta|^r}{a^r}, \quad \forall\ a, r > 0$$

for $\eta = \varepsilon + \left| X^{tx}_{t+\varepsilon} - x \right|$, and $r = 8$. Then

$$\overline{\lim_{\varepsilon \searrow 0}} \frac{1}{\varepsilon} \mathbb{E} \left| \gamma^{tx}(t+\varepsilon, X^{tx}_{t+\varepsilon}) \right| \le C\delta, \quad \text{for all } \delta > 0$$

and hence

$$\lim_{\varepsilon \searrow 0} \frac{1}{\varepsilon} \mathbb{E} \left| \gamma^{tx}(t+\varepsilon, X_{t+\varepsilon}^{tx}) \right| = 0.$$

■

We discuss now the particular case of a fixed closed convex set $E \subset \mathbb{R}^d$. We recall that, in this case $x \to d_E^2(x) = \min\{|x-y|^2 : y \in E\} = |x - \pi_E(x)|^2$ is a convex continuously differentiable function with Lipschitz gradient and

$$\frac{1}{2} \nabla d_E^2(x) = x - \pi_E(x)$$

(see Exercise 4.6).

Moreover, due to Alexandroff's Theorem, d_E^2 admits a.e. on $\mathbb{R}^d$ a second order Taylor development. More precisely, a function $\varphi : \mathbb{R}^d \to \mathbb{R}$ admits a second Taylor expansion in $x \in \mathbb{R}^d$ if there exist a vector in $\mathbb{R}^d$ denoted $\nabla_x \varphi(x)$, a matrix in $\mathbb{R}^{d \times d}$ denoted $D_x^2 \varphi(x)$, and a function $\gamma(\cdot; x) \in C(\mathbb{R}^d; \mathbb{R})$ such that, for all $y \in \mathbb{R}^d$,

$$\begin{cases} \varphi(y) = \varphi(x) + \langle \nabla \varphi(x), y - x \rangle + \dfrac{1}{2} \left\langle D^2 \varphi(x)(y-x), y - x \right\rangle + \gamma(y; x), \\ \lim\limits_{y \to x} \dfrac{\gamma(y; x)}{|y-x|^2} = 0. \end{cases}$$

We underline that this may hold true even if the first derivative is not continuous.

We denote by Θ_E the set of points $x \in \mathbb{R}^d$ where d_E^2 admits a second Taylor expansion. Then the Lebesgue measure $\lambda_d\left(\mathbb{R}^d \setminus \Theta_E\right) = 0$.

We have the following viability criterion:

Theorem 4.83. *Let E be a non-empty closed convex set in $\mathbb{R}^d$. Then, under the assumption (4.136) on f and g, the following assertions are equivalent*

(j) *Equation (4.135) is E-viable on $\mathbb{R}_+$.*
(jj) *There exists a constant $C \geq 0$ such that for all $(t, x) \in \mathbb{R}_+ \times \Theta_E$:*

$$\frac{1}{2} \mathrm{Tr}[gg^*(t,x) D^2 d_E^2(x)] + 2 \langle f(t,x), x - \pi_E(x) \rangle \leq C d_E^2(x). \tag{4.145}$$

Proof. By Theorem 4.77 it is easy to see that (j) implies (jj). Actually for a direct proof we proceed as follows:

- by Lemma 4.80 the viability of E yields

$$\mathbb{E} d_E^2(X_s^{tx}) \leq e^{C(s-t)} d_E^2(x), \quad \forall\, 0 \leq t \leq s \leq T,\ \forall\, x \in \mathbb{R}^d,$$

- starting with this inequality we continue the proof exactly as in Lemma 4.81, but now with ψ and φ replaced by d_E^2.

It only remains to show that (jj) implies (j). Let $\rho \in C^\infty(\mathbb{R}^d)$ be a nonnegative function such that $\rho(y) = 0$ for $|y| \geq 1$, and $\int_{\mathbb{R}^d} \rho(y)dy = 1$. For $\varepsilon > 0$, we put $\rho_\varepsilon(x) = \varepsilon^{-d}\rho(\varepsilon^{-1}x)$ and introduce the mollification of d_E^2:

$$\psi_\varepsilon(x) = \int_{\mathbb{R}^d} d_E^2(x - \varepsilon y)\rho(y)dy = \int_{\mathbb{R}^d} d_E^2(y)\rho_\varepsilon(x - y)dy.$$

Then, by Itô's formula and Exercise 4.6 we have, for all $\tau \geq t$,

$$\mathbb{E}\psi_\varepsilon(X_\tau^{t,x}) = \psi_\varepsilon(x) + \mathbb{E}[\int_t^\tau \int_{\mathbb{R}^d} \langle \nabla d_E^2(y)\rho_\varepsilon(X_s^{t,x} - y), f(s, X_s^{t,x})\rangle dyds]$$
$$+ \frac{1}{2}\mathbb{E}[\int_t^\tau \int_{\mathbb{R}^d} \mathrm{Tr}[gg^*(s, X_s^{t,x})D^2 d_E^2(y)\rho_\varepsilon(X_s^{t,x} - y)dyds]$$

(we extend $D^2 d_E^2(y) = 0$ for $y \notin \Theta_E$). Then, in virtue of (4.145), this yields

$$\begin{aligned}
&\mathbb{E}\psi_\varepsilon(X_\tau^{t,x})\\
&\leq \psi_\varepsilon(x) + C\mathbb{E}[\int_t^\tau \int_{\mathbb{R}^d} d_E^2(y)\rho_\varepsilon(X_s^{t,x} - y)dyds\\
&+ \mathbb{E}[\int_t^\tau \int_{\mathbb{R}^d} \langle \nabla d_E^2(y), f(s, X_s^{t,x}) - f(s, y)\rangle \rho_\varepsilon(X_s^{t,x} - y)dyds]\\
&+ \frac{1}{2}\mathbb{E}[\int_t^\tau \int_{\mathbb{R}^d} \mathrm{Tr}\left\{\left[\left|gg^*(s, X_s^{t,x}) - gg^*(s, y)\right|\right] D^2 d_E^2(y)\right\} \rho_\varepsilon(X_s^{t,x} - y)dyds].\\
&\leq \psi_\varepsilon(x) + C\mathbb{E}[\int_t^\tau \int_{\mathbb{R}^d} d_E^2(X_s^{t,x} - \varepsilon y)\rho(y)dyds\\
&+ \mathbb{E}[\int_t^\tau \int_{\mathbb{R}^d} \left|\nabla d_E^2(X_s^{t,x} - \varepsilon y)\right| \left|f(s, X_s^{t,x}) - f\left(s, X_s^{t,x} - \varepsilon y\right)\right| \rho(y)dyds]\\
&+ \mathbb{E}[\int_t^\tau \int_{\mathbb{R}^d} \left|gg^*(s, X_s^{t,x}) - gg^*(s, X_s^{t,x} - \varepsilon y)\right| \rho(y)dyds].
\end{aligned}$$

We pass here to the limit as $\varepsilon \to 0_+$ and using the Lebesgue dominated convergence theorem and $x \in E$, we infer

$$0 \leq \mathbb{E}d_E^2(X_\tau^{t,x}) \leq 0 + C\mathbb{E}[\int_t^\tau d_E^2(X_s^{t,x})ds$$

for all $\tau \geq t$. Consequently

$$\mathbb{E}d_E^2(X_\tau^{t,x}) = 0, \quad \forall\, \tau \geq t.$$

The proof is complete. ■

Consider the control problem:

- find a Lipschitz feedback law $U(t,x)$ which yields an invariant $E(t)$ for the SDE:

$$X_s^{tx}=x+\int_t^{s\vee t} f(r,X_r^{tx})dr+\int_t^{s\vee t} U(r,X_r)\,dr+\int_t^{s\vee t} g(r,X_r^{tx})dB_r,\quad t,s\geq 0. \tag{4.146}$$

We shall consider only the very simple case of

$$E(t)=\overline{B}(0,\rho)=\left\{x\in\mathbb{R}^d : |x|\leq\rho\right\}.$$

Of course by the results on SDEs with maximal monotone operators (Theorem 4.19) or a subdifferential operator in the drift (Theorem 4.20) there exists a feedback law $K\in S_d^0$, $K_\cdot(\omega)\in BV_{loc}\left([0,\infty[;\mathbb{R}^d\right)$, $\mathbb{P}$-a.s. $\omega\in\Omega$, such that for all $x\in\overline{B}(0,\rho)$

$$\begin{aligned}
&X_s^{tx}=x,\quad \forall\, 0\leq s\leq t,\\
&X_s^{tx}=x+\textstyle\int_t^s f(r,X_r^{tx})dr-(K_s-K_t)+\int_t^s g(r,X_r^{tx})dB_r,\quad t\leq s,\\
&X_s^{tx}\in\overline{B}(0,\rho),\quad \forall\, s\geq 0,\\
&dK_t\in\partial I_{\overline{B}(0,\rho)}(X_t)(dt).
\end{aligned}$$

The problem posed here is to find an absolutely continuous control

$$K_t=\int_0^t U(r,X_r)\,dr. \tag{4.147}$$

Assuming that a such control exists then by (4.141) $\overline{B}(0,\rho)$ is invariant if and only if for all $t\geq 0$ and $|x|=\rho$:

$$\begin{cases} g^*(t,x)x=0, \text{ and}\\ 2\langle x, f(t,x)+U(t,x)\rangle+|g(t,x)|^2\leq 0.\end{cases}$$

Hence in general a $\overline{B}(0,\rho)$-invariant control of the form (4.147) does not exist. But if $g^*(t,x)x=0$ for all $t\geq 0$ and $|x|=\rho$, then the feedback

$$U(t,x)=-f(t,x)-\frac{1}{2\rho^2}|g(t,x)|^2x$$

yields a $\overline{B}(0,\rho)$-invariant for the SDE (4.146).

If we consider the invariance only on $[0,T]$ and we request to realize it with a linear feedback $U(x)=\lambda x$, then we can take

$$\lambda=-\sup\left\{\frac{1}{\rho}|f(t,x)|+\frac{1}{2\rho^2}|g(t,x)|^2 : t\in[0,T],\ |x|=\rho\right\}.$$

4.6 Exercises

Exercise 4.1 (One Dimensional Diffusion Reflected at 0). Let $x \geq 0$. For each $n \in \mathbb{N}$, denote by $\{X_t^n,\ 0 \leq t \leq T\}$ the solution of the SDE

$$X_t^n = x + \int_0^t f(X_s^n)ds + n\int_0^t (X_s^n)^- ds + \int_0^t g(X_s^n)dB_s,$$

where $f, g : \mathbb{R} \to \mathbb{R}$ are globally Lipschitz and $\{B_t : t \geq 0\}$ is a scalar Brownian motion. We define $K_t^n = n\int_0^t (X_s^n)^- ds$.

1. Show that $X_t^{n+1} \geq X_t^n, 0 \leq t \leq T$.
2. Show that $\sup_{n\geq 1}\left[\mathbb{E}\left(\sup_{0\leq t\leq T}|X_t^n|^2\right) + n\mathbb{E}\int_0^T |\left(X_t^n\right)^-|^2 dt\right] < \infty$.
3. Show that $\mathbb{E}\sup_{0\leq t\leq T}|\left(X_t^n\right)^-|^2 \to 0$, as $n \to \infty$.
4. Show that $\{X^n : n \geq 1\}$ is a Cauchy sequence in $S^2[0, T]$.
5. Deduce that there exists a progressively measurable process $\{X_t,\ 0 \leq t \leq T\}$ such that $X_t^n \to X_t$, $\mathbb{P}$-a.s. for all $t \in [0, T]$, $\mathbb{E}\left(\sup_{0\leq t\leq T}|X_t|^2\right) < \infty$ and $X^- = 0$.
6. Deduce that there exists a progressively measurable process $\{K_t,\ 0 \leq t \leq T\}$ which is increasing and continuous, such that $K_0 = 0$, $\mathbb{E}(|K_T|^2) < \infty$, $K_t^n \to K_t$ in probability for all $t \in [0, T]$ and $\displaystyle\int_0^T X_t dK_t = 0$.
7. Deduce that the pair $\{(X_t, K_t),\ 0 \leq t \leq T\}$ is the unique solution of the reflected SDE, formulated as:

$$X_t = x + \int_0^t f(X_s)ds + \int_0^t g(X_s)dB_s + K_t,\ 0 \leq t \leq T,$$

$$X_t \geq 0,\ K_t \text{ continuous and increasing, } \int_0^T X_t dK_t = 0.$$

8. Show that this result is a consequence of Theorem 4.20.
9. If $g(0) = 0$, prove that

$$K_t(\omega) = (f(0))^- \int_0^t \mathbf{1}_{\{X_s(\omega)=0\}} ds.$$

10. If there exists a $y \in \mathbb{R}^d$ such that $g(y) \neq 0$, show that $\mathbb{P}$-a.s. $\omega \in \Omega$, for all $t \geq 0$,

$$\int_0^t \mathbf{1}_{\{X_s(\omega)=y\}} ds = 0$$

(the Lebesgue measure of the time spent by the stochastic process X at y is zero $\mathbb{P}$-a.s.).

Exercise 4.2 (Splitting of the Drift). Consider the stochastic differential equation

$$\begin{cases} dX_t + \partial\varphi(X_t)(dt) \ni F(t, X_t)dt + G(t, X_t)dB_t, & 0 < t \le T, \\ X_0 = H_0 \in L^4(\Omega, \mathcal{F}_0, P; \overline{\mathrm{Dom}(\varphi)}), \end{cases} \tag{4.148}$$

where

$$(\mathbf{MM\text{-}H}_\varphi) : \begin{cases} i) & \varphi : \mathbb{R}^d \to]-\infty, +\infty] \text{ is a convex l.s.c. function,} \\ ii) & \mathrm{int}(\mathrm{Dom}(\varphi)) \ne \emptyset, \end{cases}$$

and the functions $F(\cdot,\cdot,x) : \Omega \times [0,T] \to \mathbb{R}^d$, $G(\cdot,\cdot,x) : \Omega \times [0,T] \to \mathbb{R}^{d\times k}$ are progressively measurable stochastic processes for every $x \in \mathbb{R}^d$, such that for all $x, y \in \mathbb{R}^d$, $d\mathbb{P} \otimes dt$-*a.e.*

$$(\mathbf{LB}_{F,G}) : \begin{cases} i) & |F(t,x) - F(t,y)| + |G(t,x) - G(t,y)| \le L|x-y|, \\ ii) & |F(t,0)|^2 + |G(t,0)|^2 \le B < +\infty, \end{cases}$$

for some positive constants L, B.
Let $n \in N^*$, $\varepsilon = T/n$, $t_i = i\varepsilon$

$$\mathbf{I}_i^n = \begin{cases} [t_i, t_{i+1}[, & i = \overline{1, n-2} \\ [t_{n-1}, T], & i = n-1 \end{cases}$$

and

$$F(t,x) = F_1(t,x) + F_2(t,x) + F_3(t,x)$$

with F_1, F_2 and F_3 satisfying $(\mathbf{LB}_{F,G})$ with the constants L_1, L_2, L_3, B_1, B_2, B_3 respectively.
Decompose the Eq. (4.148) into three simpler equations extracting the unbounded operator $\partial\varphi$ in a deterministic equation as follows:
for $i \in \{1, 2, \ldots, n-1\}$, we define

$$\begin{cases} X_{0-}^n = U_{0+}^n = H_0, \\ \dfrac{d}{dt}U_t^n + \partial\varphi(U_t^n) \ni F_1(t, U_t^n), \quad U_{i\varepsilon+}^n = X_{i\varepsilon-}^n \quad \text{for } t \in \mathbf{I}_i^n, \\ V_t^n = U_t^n + \displaystyle\int_{i\varepsilon}^t F_2(s, V_s^n)ds, \quad \text{for } t \in \mathbf{I}_i^n, \\ Y_t^n = V_t^n + \displaystyle\int_{i\varepsilon}^t F_3(s, Y_s^n)ds + \int_{i\varepsilon}^t G(s, Y_s^n)dB_s, \quad \text{for } t \in \mathbf{I}_i^n, \\ X_t^n = \mathrm{Pr}_{\overline{D}} Y_t^n, \quad \text{for } t \in \mathbf{I}_i^n. \end{cases} \tag{4.149}$$

Prove that $X^n \in S_d^0[0,T]$ and

$$\mathbb{E}\sup_{t\in[0,T]}|X_t^n - X_t|^2 \le \frac{C}{n^{1/4}}\left(1+\mathbb{E}|H_0|^4\right). \tag{4.150}$$

Remark. If $d = 1$, $F_1(t,x) = x$, $a \le 0 \le b$ and

$$\varphi(x) = I_{[a,b]}(x) = \begin{cases} 0, & \text{if } a \le x \le b \\ +\infty, & \text{otherwise,} \end{cases}$$

then for $t \in [i\varepsilon, (i+1)\varepsilon[$,

$$U_t^n = X_{i\varepsilon-}^n e^{t-i\varepsilon}\mathbf{1}_{[a,b]}(X_{i\varepsilon-}^n e^{t-i\varepsilon}) + a\,\mathbf{1}_{]-\infty,a[}(X_{i\varepsilon-}^n e^{t-i\varepsilon}) + b\,\mathbf{1}_{]b,\infty[}(X_{i\varepsilon-}^n e^{t-i\varepsilon}).$$

Exercise 4.3 (Approximating Procedures). Let $p > 2$ and the assumptions of Theorem 4.19 and the condition (4.42) be satisfied. Let $X \in S_d^p$ be the solution of the SDE

$$\begin{cases} dX_t + A(X_t)(dt) \ni F(t,X_t))dt + G(t,X_t)dB_t \\ X_0 = \xi. \end{cases}$$

Consider for $0 < \varepsilon \le 1$ the approximating equations

$$X_t^\varepsilon + \int_0^t A_\varepsilon\left(X_s^\varepsilon\right)ds = \xi + \int_0^t F\left(s, X_s^\varepsilon\right)ds + \int_0^t G\left(s, X_s^\varepsilon\right)dB_s, \quad t \ge 0, \tag{4.151}$$

and

$$\hat{X}_t^\varepsilon + \int_0^t A_\varepsilon(\hat{X}_s^\varepsilon)ds = \xi + \int_0^t F_\varepsilon(s,\hat{X}_s^\varepsilon)ds + \int_0^t G(s,\hat{X}_s^\varepsilon)dB_s, \quad t \ge 0, \tag{4.152}$$

where A_ε is the Yosida approximation of the maximal monotone operator A (i.e. A_ε is defined by $A_\varepsilon(x) \in A(J_\varepsilon(x))$, where $J_\varepsilon + \varepsilon A(J_\varepsilon) \ni x$) and F_ε is the Yosida approximation of F (i.e. $F_\varepsilon(t,x) = F(t,\Gamma_\varepsilon(t,x))$ where $\Gamma_\varepsilon + \varepsilon[\mu(t)\Gamma_\varepsilon - F(t,\Gamma_\varepsilon)] = x$). Note that $x \longrightarrow A_\varepsilon(x)$ and $x \longrightarrow \mu(t)x - F_\varepsilon(t,x)$ are single valued maximal monotone operators and $\dfrac{1}{\varepsilon}$-Lipschitz continuous on $\mathbb{R}^d$. Let

$$K_t^\varepsilon = \int_0^t A_\varepsilon\left(X_s^\varepsilon\right)ds \quad \text{and} \quad \hat{K}_t^\varepsilon = \int_0^t A_\varepsilon\left(\hat{X}_s^\varepsilon\right)ds.$$

Prove that:

1. Equation (4.151) has a unique solution $X^\varepsilon \in S_d^p$ and for all $q \in [1,p[$,

$$\lim_{\varepsilon\to 0+}\left[\mathbb{E}\sup_{t\in[0,T]}|X_t^\varepsilon - X_t|^q + \mathbb{E}\sup_{t\in[0,T]}|K_t^\varepsilon - K_t|^{q/2}\right] = 0$$

and

$$\lim_{\varepsilon \to 0+} \mathbb{E}\left(\int_0^T \left|F\left(t, X_t^\varepsilon\right) - F\left(t, X_t\right)\right| dt\right)^{q/2} = 0,$$

$$\lim_{\varepsilon \to 0+} \mathbb{E}\left(\int_0^T \left|G\left(t, X_t^\varepsilon\right) - G\left(t, X_t\right)\right|^2 dt\right)^{q/2} = 0.$$

2. If in addition

$$\mathbb{E}\left(\int_0^T |\mu(t)| \, |F(t,0)| \, dt\right)^p < +\infty$$

then the same convergence results hold with X^ε replaced by $\hat{X}^\varepsilon$ and K^ε replaced by $\hat{K}^\varepsilon$; moreover

$$\mathbb{E}\left(\int_0^T \left|F_\varepsilon\left(t, \hat{X}_t^\varepsilon\right) - F\left(t, X_t\right)\right| dt\right)^{q/2} \to 0, \quad \text{as } \varepsilon \to 0.$$

Exercise 4.4 (A Stabilization Problem). The problem is to find a $\mathbb{P}$-m.s.p. (a control process) U which stabilizes with a given exponential rate a the solution X of the multivalued SDE:

$$\begin{cases} dX_t + A(X_t)(dt) \ni (F(t, X_t) + U_t)\, dt + G(t, X_t) dB_t, & t \geq 0, \\ X_0 = \xi \in L^0(\Omega, \mathcal{F}_0, P; \overline{\mathrm{Dom}\,(A)}). \end{cases} \tag{4.153}$$

Let $p \geq 2$ and the assumptions of Theorem 4.19 be satisfied. Also assume there exists an $x_0 \in \mathrm{Dom}\,(A)$ such that $0 \in Ax_0$, $F\,(\omega, t, x_0) \equiv 0$ and $G\,(\omega, t, x_0) \equiv 0$.

1. Show that for every $a > 0$ there exists a $\mathbb{P}$-measurable control $U = U^{p,a} \in L^p\left(\Omega; L^1\left(0, T; \mathbb{R}^d\right)\right)$ which stabilizes the solution X with the given exponential rate a, that is the corresponding solution $X = X^U \in S_d^p$ satisfies for all $s \geq 0$:

$$\mathbb{E}\left|X_s - x_0\right|^p \leq e^{-as}\mathbb{E}\left|\xi - x_0\right|^p,$$

and moreover for all $0 \leq t \leq s$,

$$\mathbb{E}^{\mathcal{F}_t} e^{as}\left|X_s - x_0\right|^p \leq e^{at}\left|X_t - x_0\right|^p, \quad a.s.$$

2. Show that for all $\delta < a$,

$$\lim_{t \to \infty} e^{\delta t}\left|X_t - x_0\right|^p = 0 \quad \text{in } L^1\left(\Omega, \mathcal{F}, \mathbb{P}\right) \text{ and } \mathbb{P}\text{-a.s.}$$

3. Prove that

$$\mathbb{E}\int_0^\infty \|X_t - x_0\|^p dt \le \frac{1}{a}\mathbb{E}\,|\xi - x_0|^p .$$

4. Prove that: if $\xi = x \in \mathbb{R}^d$ and $0 \le \delta < a$, then

$$|X_t^x(\omega) - x_0|^p \le e^{-\delta t}|x - x_0|^p, \quad \forall\, t \ge \theta(\omega),\ \mathbb{P}\text{-a.s.},$$

   where $\theta < \infty$, $a.s.$

5. Show that for every $\lambda > 1$ and $a > 0$ there exist a constant $C_{p,\lambda}$ and a $\mathbb{P}$-measurable control $\tilde{U} = \tilde{U}^{p,\lambda,a} \in L^p\left(\Omega; L^1\left(0,T;\mathbb{R}^d\right)\right)$ such that the corresponding solution $\tilde{X} = X^{\tilde{U}} \in S_d^p$ of the SDE (4.153) satisfies

$$\mathbb{E}\sup_{r\in[n,\infty[}\left|\tilde{X}_r - x_0\right|^p \le C_{p,\lambda}e^{-an}\mathbb{E}\,|\xi - x_0|^p, \quad \forall\, n \ge 0.$$

Exercise 4.5 (A Comparison Result). Consider the following one dimensional SDEs

$$X_s^{t,x} = x + \int_t^s f(r, X_r^{t,x})\,dr + \int_t^s \langle g(r, X_r^{t,x}), dB_r\rangle, \quad \text{if } t \le s \le T,$$

$$\tilde{X}_s^{t,\tilde{x}} = \tilde{x} + \int_t^s \tilde{f}(r, \tilde{X}_r^{t,x})\,dr + \int_t^s \langle \tilde{g}(r, \tilde{X}_r^{t,x}), dB_r\rangle, \quad \text{if } t \le s \le T,$$

where $f, \tilde{f} : [0,T]\times\mathbb{R} \to \mathbb{R}$ and $g, \tilde{g} : [0,T]\times\mathbb{R} \to \mathbb{R}^k$ are continuous functions and there exists an $L > 0$ such that

$$|f(t,x) - f(t,y)| + \left|\tilde{f}(t,x) - \tilde{f}(t,y)\right| \le L\,|x-y|, \quad \text{and}$$

$$|g(t,x) - g(t,y)| + |\tilde{g}(t,x) - \tilde{g}(t,y)| \le L\,|x-y|,$$

for all $t \in [0,T]$ and $x, y \in \mathbb{R}$. Show that the following assertions are equivalent:

(a) $X_s^{t,x} \ge \tilde{X}_s^{t,\tilde{x}}$ a.s., for all $x \ge \tilde{x}$, $0 \le t \le s \le T$;
(b) $f(t,y) \ge \tilde{f}(t,y)$, $g(t,y) = \tilde{g}(t,y)$, for all $(t,y) \in [0,T]\times\mathbb{R}$.

Exercise 4.6. Let $E \subset \mathbb{R}^d$ be a non-empty closed set and $d_E(x) = \inf\{|y-x| : y \in E\}$ be the distance to E. The projection map is defined by

$$x \longmapsto \Pi_E(x) \overset{def}{=} \{\hat{x} \in E : |x - \hat{x}| = d_E(x)\} : \mathbb{R}^d \rightrightarrows \mathbb{R}^d.$$

Show that:

1. For all $x, y \in \mathbb{R}^d$ and $a \in E$,

$$\begin{array}{ll} (a) & |d_E(x) - d_E(y)| \leq |x - y|\,, \\ (b) & |d_E^2(x) - d_E^2(y)| \leq 4(|a| + |x| + |y|)\,|x - y|\,. \end{array}$$

2. $d_E^2(x+h) - d_E^2(x) \leq |x|^2 + 2\left\langle h - \hat{h}, x\right\rangle$, $\forall\, x, h \in \mathbb{R}^d$ and $\hat{h} \in \Pi_E(h)$.
3. The function $x \mapsto |x|^2 - d_E^2(x)$ is convex.
4. The function $x \mapsto d_E^2(x)$ is twice differentiable[2] almost everywhere in $\mathbb{R}^d$, i.e. there exists a full measure subset $\Theta_E \subseteq \mathbb{R}^d$ such that for every $x \in \Theta_E$, there exist a vector in $\mathbb{R}^d$ denoted $\nabla_x d_E^2(x)$, a matrix in $\mathbb{R}^{d\times d}$ denoted $D_x^2 d_E^2(x)$, and a function $\gamma(\cdot; x) \in C(\mathbb{R}^m)$ such that, for all $h \in \mathbb{R}^m$,

$$\begin{cases} d_E^2(x+h) = d_E^2(x) + \left\langle \nabla d_E^2(x), h\right\rangle + \dfrac{1}{2}\left\langle D^2 d_E^2(x)h, h\right\rangle + \gamma(h; x) \\ \lim\limits_{h\to 0} \dfrac{\gamma(h;x)}{|h|^2} = 0. \end{cases} \tag{4.154}$$

   Moreover for any $x \in \Theta_E$,

$$\begin{array}{ll} (a) & \Pi_E(x) \text{ is a singleton and } \nabla d_E^2(x) = x - \Pi_E(x)\,, \\ (b) & |\gamma(h;x)| \leq |h|^2 \left(1 + \left|D^2 d_E^2(x)\right|\right), \text{ for all } x, h \in \mathbb{R}^d. \end{array}$$

5. If $\{B_t : t \geq 0\}$ is an $\mathbb{R}^k$-valued Brownian motion, then for all $0 < r < t - \varepsilon < t$ and $x \in \mathbb{R}^d$, $z \in \mathbb{R}^{d\times k}$,

$$\begin{aligned} &\mathbb{E}^{\mathcal{F}_r}\left[d_E^2\left(x + z\left(B_t - B_{t-\varepsilon}\right)\right)\right] \\ &\quad = d_E^2(x) + \frac{\varepsilon}{2}\mathrm{Tr}\left(z^* D^2 d_E^2(x) z\right) + \mathbb{E}^{\mathcal{F}_r}[\gamma(\sqrt{\varepsilon} z B_1; x)]. \end{aligned}$$

6. The set E is convex iff $x \mapsto d_E^2(x) : \mathbb{R}^d \to \mathbb{R}$ is a convex function.
7. The set E is convex iff the projection map $x \longmapsto \Pi_E(x) : \mathbb{R}^d \rightrightarrows \mathbb{R}^d$ is single-valued (Motzkin's Theorem) (in this case we write $\pi_E = \Pi_E$).
8. If E is convex, then for all $x, h, y \in \mathbb{R}^d$,

$$\begin{array}{ll} (a) & \nabla d_E^2(x) = 2\left(x - \pi_E(x)\right), \\ (b) & |\pi_E(x+h) - \pi_E(x)| \leq |h|\,, \\ (c) & \left\langle \nabla d_E^2(x), h\right\rangle \leq d_E^2(x+h) - d_E^2(x), \\ (d) & 0 \leq \left\langle D^2 d_E^2(x)h, h\right\rangle \leq 2\,|h|^2\,. \end{array}$$

[2] By twice differentiable, we mean that the function admits a second order Taylor expansion.

9. Let E be a convex set and $\psi_\varepsilon \in C^\infty\left(\mathbb{R}^d;\mathbb{R}_+\right)$ be given by

$$\psi_\varepsilon(x) = \int_{\mathbb{R}^d} d_E^2(x-\varepsilon u)\rho(u)\,du = \int_{\mathbb{R}^d} d_E^2(u)\rho_\varepsilon(x-u)\,du$$

where $\rho \in C^\infty\left(\mathbb{R}^d;\mathbb{R}_+\right)$, $\rho(u) = 0$ if $|u| \geq 1$, $\int_{\mathbb{R}^d}\rho(u)\,du$ and $\rho_\varepsilon(u) = \varepsilon^{-d}\rho\left(\varepsilon^{-1}u\right)$. Show that for all $x, h \in \mathbb{R}^d$,

$$\begin{array}{ll}
(a) & 0 \leq \psi_\varepsilon(x) \leq (\varepsilon + d_E(x))^2, \\
(b) & \left|\psi_\varepsilon(x) - d_E^2(x)\right| \leq C\,(1+|x|)\,\varepsilon, \\
(c) & \nabla\psi_\varepsilon(x) = \displaystyle\int_{\mathbb{R}^d} \nabla d_E^2(u)\rho_\varepsilon(x-u)\,du, \\
(d) & |\nabla\psi_\varepsilon(x)| \leq 2(\varepsilon + d_E(x)), \\
(e) & \left|\nabla\psi_\varepsilon(x) - \nabla d_E^2(x)\right| \leq 2\varepsilon, \\
(f) & D^2\psi_\varepsilon(x) = \displaystyle\int_{\mathbb{R}^d} D^2 d_E^2(u)\rho_\varepsilon(x-u)\,du, \\
(g) & 0 \leq \left\langle D^2\psi_\varepsilon(x)\,h, h\right\rangle \leq 2\,|h|^2,
\end{array}$$

and

$$\psi_\varepsilon(x+h) = \psi_\varepsilon(x) + \langle\int_{\mathbb{R}^d} \nabla d_E^2(u)\rho_\varepsilon(x-u)\,du, h\rangle$$
$$+\frac{1}{2}\langle\left(\int_{\mathbb{R}^d} D^2[d_E^2(u)]\rho_\varepsilon(x-u)\,du\right)h, h\rangle + \delta_\varepsilon(h,x),$$

where $\delta_\varepsilon(h,x) = \int_{\mathbb{R}^m}\gamma(h,u)\rho_\varepsilon(x-u)du$ satisfies

$$\lim_{h\to 0}\frac{\delta_\varepsilon(h,x)}{|h|^2} = 0.$$

Chapter 5
Backward Stochastic Differential Equations

5.1 Introduction

In this chapter we discuss so-called "backward stochastic differential equations", BSDEs for short. Linear BSDEs first appeared a long time ago, both as the equations for the adjoint process in stochastic control, as well as the model behind the Black and Scholes formula for the pricing and hedging of options in mathematical finance. These linear BSDEs can be solved more or less explicitly (see Proposition 5.31 below). However, the first published paper on nonlinear BSDEs, appeared only in 1990, see Pardoux and Peng [51]. Since then, the interest in BSDEs has increased regularly, due to the connections of this subject with mathematical finance, stochastic control, and partial differential equations. We refer the interested reader to El Karoui et al. [29] and [30], Pham [60] and the references therein for developments on the use of BSDEs as models in mathematical finance, as well as the connection of BSDEs with stochastic control (see also [28] and [37]). BSDEs are also an efficient tool for constructing Γ-martingales on manifolds with prescribed limit, see Darling [19]. The connection of BSDEs with semi linear PDEs was initiated in Pardoux, Peng [54], see also among the now vast literature on the subject [6, 48, 52] and [53].

We shall present both the abstract theory of BSDEs, and the connection of BSDEs with semilinear PDEs (both parabolic and elliptic). Let us motivate the notion of a BSDE via an associated semilinear parabolic PDE.

To each $(t,x) \in \mathbb{R}_+ \times \mathbb{R}^d$, we associate the Markov diffusion process $\{X_s^{t,x}, s \geq t\}$ which is a solution of the SDE

$$X_s^{t,x} = x + \int_t^s f(r, X_r^{t,x})dr + \int_t^s g(r, X_r^{t,x})dB_r,$$

where the Brownian motion B has dimension k. The associated infinitesimal generator reads

E. Pardoux and A. Răşcanu, *Stochastic Differential Equations, Backward SDEs, Partial Differential Equations*, Stochastic Modelling and Applied Probability 69,
DOI 10.1007/978-3-319-05714-9_5,

$$\mathcal{A}_t\varphi(x) = \frac{1}{2}\mathbf{Tr}\left[gg^*(t,x)D^2\varphi(x)\right] + \langle f(t,x), \nabla\varphi(x)\rangle.$$

Let $T > 0$ be an arbitrary final time, $\kappa \in C(\mathbb{R}^d)$ and $F \in C([0,T]\times\mathbb{R}^d\times\mathbb{R}\times\mathbb{R}^k)$. We consider the following backward semilinear second order PDE

$$\begin{cases} \dfrac{\partial u}{\partial t}(t,x) + \mathcal{A}_t u(t,x) + F(t,x,u(t,x),(\nabla ug)(t,x)) = 0, \\ \qquad\qquad\qquad\qquad\qquad\qquad\qquad (t,x) \in [0,T]\times\mathbb{R}^d, \\ u(T,x) = \kappa(x), \quad x \in \mathbb{R}^d. \end{cases}$$

Suppose that this equation has a classical solution $u \in C^{1,2}([0,T]\times\mathbb{R}^d)$. It then follows from Itô's formula that for any $0 \le t < s \le T$,

$$u(s, X_s^{t,x}) = \kappa(X_T^{t,x}) + \int_s^T F(r, X_r^{t,x}, u(r, X_r^{t,x}), (\nabla ug)(r, X_r^{t,x})) - \int_s^T (\nabla ug)(r, X_r^{t,x})dB_r.$$

Considering the pair of adapted processes

$$\left(Y_r^{t,x}, Z_r^{t,x}\right) = \left(u(r, X_r^{t,x}), (\nabla ug)(r, X_r^{t,x})\right),$$

we have that for each $(t,x) \in [0,T]\times\mathbb{R}^d$,

$$Y_s^{t,x} = \kappa(X_T^{t,x}) + \int_s^T F(r, X_r^{t,x}, Y_r^{t,x}, Z_r^{t,x})dr - \int_s^T Z_r^{t,x}dB_r, \; s \le r \le T,$$

and $Y_t^{t,x}$ is a deterministic quantity which equals $u(t,x)$. The solution u of the above semilinear parabolic PDE is expressed in terms of the solution of this last backward stochastic differential equation (BSDE). We will see below that this is indeed an extension of the Feynman–Kac formula (in the sense that if F is affine, then the Feynman–Kac formula is a consequence of the above representation). Note that the above computation can be applied to a system of PDEs, rather than a single PDE. We shall consider only the case where the same second order PDE operator $\mathcal{A}$ is applied to each coordinate u_i of u. A probabilistic representation for more general systems of semilinear PDEs, with a different $\mathcal{A}$ for each coordinate of u, can be found in [55], see also [52] and [58].

Let us now write an abstract version of the above BSDE. Let $t = 0$, and forget about the superscript x. Suppose now that we are given a probability space with filtration $(\Omega, \mathcal{F}, \mathcal{F}_t, \mathbb{P})$ and for each $(y,z) \in \mathbb{R}\times\mathbb{R}^k$, a measurable process $\{F(t,y,z),\ 0 \le t \le T\}$ (F being jointly measurable), together with an $\mathcal{F}_T$ random variable η.

We formulate the problem of solving a BSDE as follows: find a pair of adapted processes $\{(Y_t, Z_t),\ 0 \le t \le T\}$ such that

$$Y_t = \eta + \int_t^T F(s, Y_s, Z_s)ds - \int_t^T Z_s dB_s,\ 0 \le t \le T, \text{ a.s.}$$

Note that, since the boundary condition for $\{Y_t : t \in [0, T]\}$ is given at the terminal time T, it is not really natural for the solution $\{Y_t\}$ to be adapted at each time t to the past of the Brownian motion $\{B_s\}$ before time t. The price we have to pay for such a severe constraint to be satisfied is to have the coefficient of the Brownian motion – the process $\{Z_t\}$ – to be chosen independently of $\{Y_t\}$, hence the solution of the BSDE is a *pair* of processes. Note that in the case $F \equiv 0$, $Y_t = \mathbb{E}(\eta|\mathcal{F}_t)$ and Z is given by the martingale representation theorem from Sect. 2.4.

One may also think of a "backward SDE" as an inverse problem for an SDE, namely we are looking for a point $y \in \mathbb{R}$, and an adapted process $\{Z_t\}$, such that the solution $\{Y_t\}$ of

$$Y_t = y - \int_0^t F(s, Y_s, Z_s)ds + \int_0^t Z_s dB_s$$

satisfies $Y_T = \eta$.

Finally, note that while the above presentation treats T as a deterministic quantity, an important alternative is to replace it by a stopping time (or else by $+\infty$). This is essential when giving probabilistic representations of semilinear elliptic PDEs.

In this chapter, we suppose given a stochastic basis $(\Omega, \mathcal{F}, \mathbb{P}, \{\mathcal{F}_t\}_{t\ge 0})$ with $\{B_t ; t \ge 0\}$ a k-dimensional Brownian motion and the filtration $\{\mathcal{F}_t\}_{t\ge 0}$ being the natural filtration of $\{B_t : t \ge 0\}$, i.e. for all $t \ge 0$:

$$\mathcal{F}_t = \mathcal{F}_t^B \overset{def}{=} \sigma\left(\{B_s : 0 \le s \le t\}\right) \vee \mathcal{N}.$$

5.2 Basic Inequalities

For convenience we rewrite in this context the Itô formula (2.14) and we give a basic inequality. First we introduce a notation used in this chapter.

Notation 5.1. *For $p \ge 1$ we define*

$$n_p \overset{def}{=} 1 \wedge (p-1).$$

Let $(Y, Z) \in S_m^0[0, T] \times \Lambda_{m\times k}^0(0, T)$ satisfy for all $t \in [0, T]$, $\mathbb{P}$-*a.s.*:

$$Y_t = Y_T + \int_t^T dK_s - \int_t^T Z_s dB_s,$$

where

◇ $K \in S_m^0$,
◇ $K_\cdot(\omega) \in BV_{loc}([0,\infty[;\mathbb{R}^m)$, $\mathbb{P}$-a.s. $\omega \in \Omega$.

5.2.1 Backward Itô's Formula

If $\varphi \in C^{1,2}([0,T]\times\mathbb{R}^m)$, then $\mathbb{P}$-a.s., for all $t \in [0,T]$:

$$\begin{aligned}\varphi(t,Y_t) &+ \int_t^T \left\{\frac{\partial\varphi}{\partial t}(s,Y_s) + \frac{1}{2}\mathbf{Tr}\left[Z_s Z_s^* \varphi''_{xx}(s,Y_s)\right]\right\} ds \\ &= \varphi(T,Y_T) + \int_t^T \langle \varphi'_x(s,Y_s), dK_s\rangle - \int_t^T \langle \varphi'_x(s,Y_s), Z_s dB_s\rangle .\end{aligned} \tag{5.1}$$

From Corollary 2.29 we get for all $p \in \mathbb{R}$,

$$\begin{aligned}\left(|Y_t|^2+\varepsilon\right)^{p/2} + \frac{p}{2}\int_t^T R_s^{(p,\varepsilon)} ds + \frac{p}{2}\left(L_T^{(p,\varepsilon)} - L_t^{(p,\varepsilon)}\right) &= \left(|Y_T|^2+\varepsilon\right)^{p/2} \\ +p\int_t^T \left\langle U_s^{(p,\varepsilon)}, dK_s\right\rangle &- p\int_t^T \left\langle U_s^{(p,\varepsilon)}, Z_s dB_s\right\rangle,\end{aligned} \tag{5.2}$$

where

$$\begin{aligned}&(j) \quad U_s^{(p,\varepsilon)} = \left(|Y_s|^2+\varepsilon\right)^{(p-2)/2} Y_s, \\ &(jj) \quad R_s^{(p,\varepsilon)} = \left[|Z_s|^2|Y_s|^2 + (p-2)\left|Z_s^* Y_s\right|^2\right]\left(|Y_s|^2+\varepsilon\right)^{(p-4)/2}, \\ &(jjj) \quad L_t^{(p,\varepsilon)} = \varepsilon\int_0^t |Z_s|^2\left(|Y_s|^2+\varepsilon\right)^{(p-4)/2} ds.\end{aligned}$$

We have $\left|U_s^{(p,\varepsilon)}\right| \le \left(|Y_s|^2+\varepsilon\right)^{(p-1)/2}$ and

$$n_p |Z_s|^2|Y_s|^2\left(|Y_s|^2+\varepsilon\right)^{(p-4)/2} \le R_s^{(p,\varepsilon)} \le m_p |Z_s|^2|Y_s|^2\left(|Y_s|^2+\varepsilon\right)^{(p-4)/2},$$

where $n_p \overset{def}{=} 1\wedge(p-1)$ and $m_p \overset{def}{=} 1\vee(p-1)$.

Moreover

$$\frac{1}{\sqrt{\varepsilon}}\int_0^t |Z_s|^2 \mathbf{1}_{|Y_s|\le\sqrt{\varepsilon}} ds \le 2\sqrt{2}L_t^{(1,\varepsilon)}.$$

In particular for $p \geq 1$ and $\varepsilon \searrow 0$ we obtain

$$\begin{aligned}
&|Y_t|^p + \frac{1}{2}\int_t^T R_s^{(p)} ds + \frac{1}{2}(L_T - L_t)\mathbf{1}_{p\neq 1} = |Y_T|^p \\
&+\int_t^T |Y_s|^{p-1}\langle \operatorname{sgn}(Y_s), dK_s\rangle - \int_t^T |Y_s|^{p-1}\langle \operatorname{sgn}(Y_s), Z_s dB_s\rangle,
\end{aligned} \tag{5.3}$$

where

$$\operatorname{sgn}: \mathbb{R}^m \to \mathbb{R}^m, \quad \operatorname{sgn}(x) = \begin{cases} 0, & \text{if } x = 0, \\ \dfrac{x}{|x|}, & \text{if } x \neq 0, \end{cases}$$

$$R_s^{(p)} = \begin{cases} 0, & \text{if } Y_s = 0, \\ \left(|Z_s|^2 + (p-2)|Z_s^* \operatorname{sgn}(Y_s)|^2\right)|Y_s|^{p-2}, & \text{if } Y_s \neq 0, \end{cases}$$

and $\{L_t : t \geq 0\}$ is an increasing continuous progressively measurable stochastic process such that for all $t \geq 0$ (in the sense of convergence in probability)

$$L_t = \frac{1}{2}\lim_{\varepsilon\to 0+}\int_0^t \frac{\varepsilon|Z_s|^2}{\left(|Y_s|^2 + \varepsilon\right)^{3/2}} ds.$$

The stochastic process $\{L_t : t \geq 0\}$ has the following property:

$$L_t(\omega) = L_s(\omega), \quad \mathbb{P}\text{-a.s.},$$

for every interval $[s,t] \subset \{r \geq 0 : Y_r(\omega) = 0\}$, or $[s,t] \subset \operatorname{int}\{r \geq 0 : Y_r(\omega) \neq 0\}$.

Moreover, we have

$$\limsup_{\delta\to 0+}\frac{1}{\delta}\int_0^T |Z_s|^2 \mathbf{1}_{|Y_s|\leq\delta} ds \leq 2\sqrt{2}L_T \quad \text{and} \quad \int_0^T |Z_s|^2 \mathbf{1}_{Y_s=0} ds = 0.$$

Since

$$0 \leq \frac{p}{2}n_p\int_s^t |Y_r|^{p-2}\mathbf{1}_{Y_r\neq 0}|Z_r|^2 dr \leq \int_s^t R_r^{(p)} dr < \infty, \text{ for all } 0 \leq s < t \leq T, \ a.s.,$$

it follows that for every $p \geq 1$ and $0 \leq t \leq T$:

$$\begin{aligned}
&|Y_t|^p + \frac{p}{2}n_p\int_t^T |Y_s|^{p-2}\mathbf{1}_{Y_s\neq 0}|Z_s|^2 ds \leq |Y_T|^p \\
&+p\int_t^T |Y_s|^{p-2}\mathbf{1}_{Y_s\neq 0}\langle Y_s, dK_s\rangle - p\int_t^T |Y_s|^{p-2}\mathbf{1}_{Y_s\neq 0}\langle Y_s, Z_s dB_s\rangle, \ a.s.
\end{aligned} \tag{5.4}$$

In fact we deduce from Lemma 2.37 a more general inequality:

$\diamondsuit$ *if* $\psi : [0,T]\times\mathbb{R}^m \to \mathbb{R}$ *is a function of class* C^1, *convex in its second argument, then a.s.,* $\forall\ t \in [0,T]$:

$$\psi(t,Y_t) + \int_t^T \frac{\partial\psi}{\partial t}(s,Y_s)\,ds \leq \psi(T,Y_T) + \int_t^T \langle\nabla\psi(s,Y_s),\, dK_s\rangle \\ - \int_t^T \langle\nabla\psi(s,Y_s),\, Z_s dB_s\rangle\,. \tag{5.5}$$

5.2.2 A Fundamental Inequality

Let $(Y,Z) \in S_m^0\,[0,T] \times \Lambda_{m\times k}^0\,(0,T)$ satisfy an identity of the form

$$Y_t = Y_T + \int_t^T dK_s - \int_t^T Z_s dB_s, \quad t \in [0,T]\,, \quad \mathbb{P}\text{-a.s.}, \tag{5.6}$$

where

$\diamondsuit$ $K \in S_m^0\,([0,T])$ and $K.(\omega) \in BV\,([0,T];\mathbb{R}^m)\,,$ $\mathbb{P}$-a.s. $\omega \in \Omega$.

$\diamondsuit$ *Assume that there exist*

(*a*) D, R, N $\mathcal{P}$-*m.i.c.s.p.*, $D_0 = R_0 = N_0 = 0$;
(*b*) V $\mathcal{P}$–*m.b-v.c.s.p.* $V_0 = 0$;
(*c*) $\lambda < 1 \leq p$,

such that as measures on $[0,T]$, *a.s.*

$$dD_t + \langle Y_t, dK_t\rangle \leq \left[\mathbf{1}_{p\geq 2}dR_t + |Y_t|dN_t + |Y_t|^2 dV_t\right] + \frac{n_p}{2}\lambda\,|Z_t|^2\,dt, \tag{5.7}$$

where

$$n_p \overset{def}{=} 1 \wedge (p-1)\,.$$

By Proposition 6.80, Corollary 6.81 and Corollary 6.82 from Annex C we have:

Proposition 5.2. *Let (5.6) and (5.7) be satisfied and moreover*

$$\mathbb{E}\,\left\|Ye^V\right\|_T^p < \infty.$$

(A) If $p > 1$, *then there exists a positive constant* $C_{p,\lambda}$, *depending only upon* (p,λ), *such that,* $\mathbb{P}$-*a.s., for all* $t \in [0,T]$:

$$\mathbb{E}^{\mathcal{F}_t} \sup_{r\in[t,T]} \left|e^{V_r}Y_r\right|^p + \mathbb{E}^{\mathcal{F}_t}\left(\int_t^T e^{2V_r}dD_r\right)^{p/2} + \mathbb{E}^{\mathcal{F}_t}\left(\int_t^T e^{2V_r}\,|Z_r|^2\,dr\right)^{p/2} \tag{5.8}$$

$$\leq C_{p,\lambda}\, \mathbb{E}^{\mathcal{F}_t} \left[\left|e^{V_T} Y_T\right|^p + \left(\int_t^T e^{2V_r} \mathbf{1}_{p\geq 2} dR_r \right)^{p/2} + \left(\int_t^T e^{V_r} dN_r \right)^p \right].$$

(B) If $p = 1$ (and $n_p = 0$), then $\mathbb{P}$-a.s., for all $0 \leq t \leq T$

$$e^{V_t} |Y_t| \leq \mathbb{E}^{\mathcal{F}_t} e^{V_T} |Y_T| + \mathbb{E}^{\mathcal{F}_t} \int_t^T e^{V_r} dN_r$$

and for all $0 < \alpha < 1$ there exists a positive constant C_α, depending only upon α such that

$$\sup_{r\in[t,T]} \left[\mathbb{E}\left(e^{V_r} |Y_r|\right)\right]^\alpha + \mathbb{E}\left(\sup_{r\in[t,T]} \left|e^{V_r} Y_r\right|^\alpha \right) + \mathbb{E}\left(\int_t^T e^{2V_r} |Z_r|^2 dr \right)^{\alpha/2}$$
$$+\mathbb{E}\left(\int_t^T e^{2V_r} |D_r|^2 dr \right)^{\alpha/2}$$
$$\leq C_\alpha \left[\left(\mathbb{E}\left(e^{V_T} |Y_T|\right)\right)^\alpha + \left(\mathbb{E} \int_t^T e^{V_r} dN_r \right)^\alpha \right].$$

(C) If $p \geq 1$ and $R = N = 0$, then $\mathbb{P}$-a.s., for all $t \in [0, T]$:

$$e^{pV_t} |Y_t|^p \leq \mathbb{E}^{\mathcal{F}_t} e^{pV_T} |Y_T|^p . \tag{5.9}$$

Corollary 5.3. *Under the assumptions of Proposition 5.2, if there exists a $c \geq 0$ such that $\sup_{s\in[0,T]} |V_s| \leq c$, then $\mathbb{P}$-a.s., for all $t \in [0, T]$:*

$$\mathbb{E}^{\mathcal{F}_t} \sup_{s\in[t,T]} |Y_s|^p + \mathbb{E}^{\mathcal{F}_t} \left(\int_t^T |Z_s|^2 ds \right)^{p/2}$$
$$\leq C_{p,\lambda}\, e^{2c}\, \mathbb{E}^{\mathcal{F}_t} \left[|Y_T|^p + \left(\int_t^T \mathbf{1}_{p\geq 2} dR_s \right)^{p/2} + \left(\int_t^T dN_s \right)^p \right].$$

5.3 BSDEs with Deterministic Final Time

Our main goal in this section is to study backward stochastic differential equations (abbreviated BSDEs) of the form

$$\begin{cases} -dY_t = F(t, Y_t, Z_t)\, dt + G(t, Y_t)\, dA_t - Z_t dB_t, & 0 \leq t < T, \\ Y_T = \eta, \end{cases} \tag{5.10}$$

or equivalently, a.s. for all $t \in [0, T]$:

$$Y_t = \eta + \int_t^T F(s, Y_s, Z_s)\, ds + \int_t^T G(s, Y_s)\, dA_s - \int_t^T Z_s dB_s,$$

whose solution $(Y_t, Z_t)_{t\in[0,T]}$ takes values in $\mathbb{R}^m \times \mathbb{R}^{m\times k}$, and where we assume in this section that:

- $T > 0$ is a fixed final deterministic time;
- $\eta : \Omega \to \mathbb{R}^m$, the final condition, is an $\mathcal{F}_T$-measurable random vector;
- $F : \Omega\times[0,T]\times\mathbb{R}^m\times\mathbb{R}^{m\times k} \to \mathbb{R}^m$ is a $\left(\mathcal{P}, \mathbb{R}^m \times \mathbb{R}^{m\times k}\right)$-Carathéodory function, that is

$$\begin{aligned}&F(\cdot,\cdot,y,z) \text{ is } \mathcal{P}\text{-m.s.p.}, \ \forall\ (y,z) \in \mathbb{R}^m\times\mathbb{R}^{m\times k};\\ &F(\omega,t,\cdot,\cdot) \text{ is a continuous function, } d\mathbb{P}\otimes dt\text{-}a.e.;\end{aligned}$$

- $G : \Omega \times [0,T] \times \mathbb{R}^m \to \mathbb{R}^m$ is a $(\mathcal{P}, \mathbb{R}^m)$-Carathéodory function, i.e.

$$\begin{aligned}&G(\cdot,\cdot,y) \text{ is } \mathcal{P}\text{-m.s.p.}, \ \forall\ y \in \mathbb{R}^m;\\ &G(\omega,t,\cdot,\cdot) \text{ is a continuous function, } d\mathbb{P}\otimes dt\text{-}a.e.;\end{aligned}$$

- A is a $\mathcal{P}$-m.i.c.s.p., $A_0 = 0$.

Note that, by Exercise 1.1, F is $(\mathcal{P} \otimes \mathcal{B}_m \otimes \mathcal{B}_{m\times k}, \mathcal{B}_m)$-measurable and G is $(\mathcal{P} \otimes \mathcal{B}_m, \mathcal{B}_m)$-measurable.

We state the following definition:

Definition 5.4. A pair $(Y,Z) \in S_m^0[0,T] \times \Lambda^0_{m\times k}(0,T)$ is a solution of (5.10) if

$$\int_0^T |F(t, Y_t, Z_t)|\, dt + \sum_{i=1}^N \int_0^T |G(t, Y_t)|\, dA_t < \infty, \ \ \mathbb{P}\text{-a.s.}$$

and, a.s. for all $t \in [0,T]$:

$$Y_t = \eta + \int_t^T F(s, Y_s, Z_s)\, ds + \int_t^T G(s, Y_s)\, dA_s - \int_t^T Z_s dB_s. \tag{5.11}$$

5.3.1 A Priori Estimates and Uniqueness

We now consider the BSDE

$$Y_t = \eta + \int_t^T \Phi(s, Y_s, Z_s)\, dQ_s - \int_t^T Z_s dB_s, ,\ t \in [0,T],\ a.s., \tag{5.12}$$

where

- $\eta : \Omega \to \mathbb{R}^m$, the final condition, is an $\mathcal{F}_T$-measurable random vector;
- $(\omega, t, y, z) \longmapsto \Phi(\omega, t, y, z) : \Omega \times [0, T] \times \mathbb{R}^m \times \mathbb{R}^{m \times k} \to \mathbb{R}^m$;
- $(\omega, t) \longmapsto Q_t(\omega) : \Omega \times [0, T] \to \mathbb{R}$ is a $\mathcal{P}$-m.i.c.s.p. such that $Q_0 = 0$.

Note that the BSDE (5.10) can be written in this form with

$$\begin{aligned} \Phi(\omega, t, y, z) &= \alpha_t(\omega) F(\omega, t, y, z) + \beta_t(\omega) G(\omega, y), \quad \text{and} \\ Q_t(\omega) &= t + A_t(\omega) \end{aligned}$$

where $\{\alpha_t : t \geq 0\}$ and $\{\beta_t : t \geq 0\}$ are two real positive $\mathcal{P}$-m.s.p. (given by the Radon–Nikodym representation theorem), $\alpha_t + \beta_t = 1$, such that

$$dt = \alpha_t dQ_t \quad \text{and} \quad dA_t = \beta_t dQ_t.$$

We define for any $\rho \geq 0$

$$\Phi_\rho^\#(t) \stackrel{def}{=} \sup_{|y| \leq \rho} |\Phi(t, y, 0)|; \quad \text{in particular } \Phi_0^\#(t) = |\Phi(t, 0, 0)|.$$

The basic assumptions on Φ are the following

$$\textbf{(BSDE-H}_\Phi\textbf{)}: \tag{5.13}$$

♦ *$\forall\, y \in \mathbb{R}^m$, $z \in \mathbb{R}^{m \times k}$, the function $\Phi(\cdot, \cdot, y, z) : \Omega \times [0, T] \to \mathbb{R}^m$ is $\mathcal{P}$-measurable*;

♦ *there exist three $\mathcal{P}$-m.s.p. $\mu : \Omega \times [0, T] \to \mathbb{R}$ and $\ell, \alpha : \Omega \times [0, T] \to \mathbb{R}_+$ such that, $\mathbb{P}$-a.s.*

$$\begin{aligned} &(i) \quad \alpha_t dQ_t = dt, \\ &(ii) \quad \int_0^T \Big[|\mu_t| dQ_t + (\ell_t)^2 dt \Big] < \infty, \end{aligned} \tag{5.14}$$

and for all $y, y' \in \mathbb{R}^m$ and $z, z' \in \mathbb{R}^{m \times k}$, $d\mathbb{P} \otimes dQ_t$*-a.e.*:

$$\begin{aligned} & \quad \textit{Continuity:} \\ (C_y) \quad & y \longrightarrow \Phi(t, y, z) : \mathbb{R}^m \to \mathbb{R}^m \textit{ is continuous}; \\ & \quad \textit{Monotonicity condition:} \\ (M_y) \quad & \langle y' - y, \Phi(t, y', z) - \Phi(t, y, z) \rangle \leq \mu_t |y' - y|^2; \\ & \quad \textit{Lipschitz condition:} \\ (L_z) \quad & |\Phi(t, y, z') - \Phi(t, y, z)| \leq \alpha_t\, \ell_t |z' - z|; \\ & \quad \textit{Boundedness condition:} \\ (B_y) \quad & \int_0^T \Phi_\rho^\#(s)\, dQ_s < \infty, \quad \forall\, \rho \geq 0. \end{aligned} \tag{5.15}$$

The assumptions on Φ yield a continuity behaviour result which we leave as an exercise for the reader.

Lemma 5.5. *Under the assumption (5.15)*

$$\int_0^T |\Phi(t, Y_t, Z_t)|\, dQ_t < \infty, \quad \mathbb{P}\text{-a.s.}, \quad \forall\ (Y, Z) \in S_m^0\,[0, T] \times \Lambda_{m\times k}^0\,(0, T)\,,$$

and the mapping

$$(Y, Z) \longrightarrow \int_0^{\cdot} \Phi(s, Y_s, Z_s) dQ_s$$

is continuous from $S_m^0\,[0, T] \times \Lambda_{m\times k}^0\,(0, T)$ *into* $S_m^0\,[0, T]$.

We shall show that the monotonicity of Φ yields an inequality of the form (5.7). Let (with $a > 1$ arbitrary)

$$n_p \stackrel{def}{=} 1 \wedge (p-1) \quad \text{and} \quad \gamma_s \stackrel{def}{=} \mu_s + \frac{a}{2n_p}\,(\ell_s)^2\,\alpha_s.$$

We have:

Lemma 5.6. *Let* $a, p > 1$, $r_0 \geq 0$ *and the assumptions (5.13-**BSDE-H**$_\Phi$) be satisfied. Let* (Y, Z), $(\tilde{Y}, \tilde{Z}) \in S_m^0\,[0, T] \times \Lambda_{m\times k}^0\,(0, T)$. *Then, in the sense of signed measures on* $[0, T]$:

$$dD_t^{(r_0)} + \langle Y_t, \Phi\,(t, Y_t, Z_t)\, dQ_t\rangle \leq \left[dR_t^{(r_0)} + |Y_t|dN_t^{(r_0)} + |Y_t|^2 dV_t\right] + \frac{n_p}{2a}\,|Z_t|^2\, dt, \tag{5.16}$$

and

$$\left\langle Y_t - \tilde{Y}_t, \Phi\,(t, Y_t, Z_t) - \Phi(t, \tilde{Y}_t, \tilde{Z}_t)\right\rangle dQ_t \leq |Y_t - \tilde{Y}_t|^2 dV_t + \frac{n_p}{2a}\left|Z_t - \tilde{Z}_t\right|^2 dt \tag{5.17}$$

where

$$\begin{aligned} D_t^{(r_0)} &= r_0 \int_0^t |\Phi\,(s, Y_s, Z_s)|\, dQ_s, & R_t^{(r_0)} &= r_0 \int_0^t \Phi_{r_0}^{\#}\,(s)\, dQ_s + r_0^2 \int_0^t \gamma_s^+ dQ_s, \\ V_t &= \int_0^t \gamma_s dQ_s, & N_t^{(r_0)} &= \int_0^t \Phi_{r_0}^{\#}\,(s)\, dQ_s + 2r_0 \int_0^t |\gamma_s|\, dQ_s. \end{aligned} \tag{5.18}$$

Proof. The monotonicity property of Φ implies that for any $\mathbb{R}^m$-valued stochastic process $\{U_s : s \geq 0\}$, $|U_s| \leq 1$:

$$\langle r_0 U_s - Y_s, \Phi\,(s, r_0 U_s, Z_s) - \Phi\,(s, Y_s, Z_s)\rangle\, dQ_s \leq \mu_s\, |r_0 U_s - Y_s|^2\, dQ_s.$$

Since

$$|\Phi(s, r_0 U_s, Z_s)|\, dQ_s \le \left[\Phi^{\#}_{r_0}(s) + \alpha_s \ell_s |Z_s|\right] dQ_s = \Phi^{\#}_{r_0}(s)\, dQ_s + \ell_s |Z_s|\, ds$$

it follows that

$$\begin{aligned}
&r_0 \langle U_s, -\Phi(s, Y_s, Z_s)\rangle\, dQ_s + \langle Y_s, \Phi(s, Y_s, Z_s)\rangle\, dQ_s \\
&\quad \le |r_0 U_s - Y_s|^2 \mu_s dQ_s + |r_0 U_s - Y_s| \left[\Phi^{\#}_{r_0}(s)\, dQ_s + \ell_s |Z_s|\, ds\right] \\
&\quad \le |r_0 U_s - Y_s|^2 \mu_s dQ_s + (r_0 + |Y_s|)\, \Phi^{\#}_{r_0}(s)\, dQ_s \\
&\qquad\quad + \frac{a}{2n_p} |r_0 U_s - Y_s|^2 (\ell_s)^2\, ds + \frac{n_p}{2a} |Z_s|^2\, ds.
\end{aligned}$$

Hence

$$\begin{aligned}
&r_0 \langle U_s, -\Phi(s, Y_s, Z_s)\rangle\, dQ_s + \langle Y_s, \Phi(s, Y_s, Z_s)\rangle\, dQ_s \\
&\le (r_0 + |Y_s|)\, \Phi^{\#}_{r_0}(s)\, dQ_s \\
&\qquad + \left(r_0^2 |U_s|^2 - 2r_0 \langle U_s, Y_s\rangle + |Y_s|^2\right) \gamma_s dQ_s + \frac{n_p}{2a} |Z_s|^2\, ds \\
&\le \left[r_0 \Phi^{\#}_{r_0}(s) + r_0^2 \gamma_s^+\right] dQ_s + |Y_s| \left(\Phi^{\#}_{r_0}(s) + 2r_0 |\gamma_s|\right) dQ_s + |Y_s|^2 \gamma_s dQ_s \\
&\qquad + \frac{n_p}{2a} |Z_s|^2\, ds.
\end{aligned}$$

(5.16) follows if we choose

$$U_s = \begin{cases} \frac{-\Phi(s, Y_s, Z_s)}{|\Phi(s, Y_s, Z_s)|}, & \text{if } \Phi(s, Y_s, Z_s) \neq 0, \\ 0, & \text{if } \Phi(s, Y_s, Z_s) = 0. \end{cases}$$

The inequality (5.17) is obtained as follows:

$$\begin{aligned}
&\langle Y_s - \tilde{Y}_s, \Phi(s, Y_s, Z_s) - \Phi(s, \tilde{Y}_s, \tilde{Z}_s)\rangle\, dQ_s \\
&\le \left[\mu_s \left|Y_s - \tilde{Y}_s\right|^2 + \ell_s \alpha_s \left|Y_s - \tilde{Y}_s\right| \left|Z_s - \tilde{Z}_s\right|\right] dQ_s \\
&\le \mu_s \left|Y_s - \tilde{Y}_s\right|^2 dQ_s + \left|Y_s - \tilde{Y}_s\right| \left|Z_s - \tilde{Z}_s\right| \ell_s ds \\
&\le \left(\mu_s dQ_s + \frac{a}{2n_p} (\ell_s)^2\, ds\right) \left|Y_s - \tilde{Y}_s\right|^2 + \frac{n_p}{2a} \left|Z_s - \tilde{Z}_s\right|^2 ds.
\end{aligned}$$

■

Taking into account Proposition 5.2 with $dK_s = \Phi(s, Y_s, Z_s)\, dQ_s$, we deduce from (5.16), first with $r_0 = 0$ and then with $r_0 > 0$:

Proposition 5.7. *Let the assumptions (5.13-**BSDE-H**$_\Phi$) be satisfied. Then for every $a, p > 1$ there exists a constant $C_{a,p}$ such that for all solutions $(Y,Z) \in S_m^0[0,T] \times \Lambda_{m\times k}^0(0,T)$ of the BSDE (5.12) satisfying*

$$\mathbb{E}\left\|Ye^V\right\|_T^p < \infty,$$

where again

$$V_t \overset{def}{=} V_t^{a,p} = \int_0^t \mu_s dQ_s + \frac{a}{2n_p}\int_0^t (\ell_s)^2\, ds,$$

the following inequality holds, $\mathbb{P}$-a.s., for all $t \in [0,T]$:

$$\begin{aligned}&\mathbb{E}^{\mathcal{F}_t}\left(\sup_{s\in[t,T]}\left|e^{V_s}Y_s\right|^p\right) + \mathbb{E}^{\mathcal{F}_t}\left(\int_t^T e^{2V_s}\left|Z_s\right|^2 ds\right)^{p/2}\\ &\leq C_{a,p}\left[\mathbb{E}^{\mathcal{F}_t}\left|e^{V_T}\eta\right|^p + \mathbb{E}^{\mathcal{F}_t}\left(\int_t^T e^{V_s}\left|\Phi(s,0,0)\right| dQ_s\right)^p\right].\end{aligned} \tag{5.19}$$

Moreover, if $p \geq 2$, then for all $r_0 > 0$:

$$\begin{aligned}&\mathbb{E}\left(r_0\int_0^T e^{2V_s}\left|\Phi(s,Y_s,Z_s)\right| dQ_s\right)^{p/2} \leq C_{a,p}\left[\mathbb{E}\left|e^{V_T}\eta\right|^p\right.\\ &\left.+\mathbb{E}\left(\int_0^T e^{2V_s}dR_s^{(r_0)}\right)^{p/2} + \mathbb{E}\left(\int_0^T e^{V_s}dN_s^{(r_0)}\right)^p\right].\end{aligned} \tag{5.20}$$

Corollary 5.8. *Let $p = 1$. Let the assumptions (5.13-**BSDE-H**$_\Phi$) be satisfied and Φ be independent of $z \in \mathbb{R}^{m\times k}$ ($\ell_t \equiv 0$ and $V_t = \bar{\mu}_t = \int_0^t \mu_s dQ_s$). If $(Y,Z) \in S_m^0[0,T] \times \Lambda_{m\times k}^0(0,T)$ is a solution of the BSDE (5.12) satisfying*

$$\mathbb{E}\sup_{s\in[0,T]} e^{\bar{\mu}_s}\left|Y_s\right| < \infty,$$

then the following inequality holds $\mathbb{P}$-a.s., for all $t \in [0,T]$:

$$e^{\bar{\mu}_t}\left|Y_t\right| \leq \mathbb{E}^{\mathcal{F}_t} e^{\bar{\mu}_T}\left|\eta\right| + \mathbb{E}^{\mathcal{F}_t}\int_t^T e^{\bar{\mu}_s}\left|\Phi(s,0)\right| dQ_s.$$

Moreover for all $0 < q < 1$

$$\begin{aligned}&\sup_{s\in[0,T]}\left(\mathbb{E}\left(e^{\bar{\mu}_s}\left|Y_s\right|\right)\right)^q + \mathbb{E}\sup_{s\in[0,T]}\left|e^{\bar{\mu}_s}Y_s\right|^q + \mathbb{E}\left(\int_0^T e^{2\bar{\mu}_s}\left|Z_s\right|^2 ds\right)^{q/2}\\ &\leq C_q\left[\left(\mathbb{E}\left(e^{\bar{\mu}_T}\left|\eta\right|\right)\right)^{q/2} + \left(\mathbb{E}\int_0^T e^{\bar{\mu}_s}\left|\Phi(s,0)\right| dQ_s\right)^{q/2}\right].\end{aligned}$$

Proof. Since

$$\langle Y_t, \Phi(t, Y_t, Z_t)\, dQ_t\rangle \leq |Y_t|\, |\Phi(t,0)|\, dQ_t + |Y_t|^2\, d\bar{\mu}_t$$

the conclusions follow by Corollary 6.81. ■

From (5.19) we immediately have:

Corollary 5.9. *Let $a, p > 1$. If*

$$\mathbb{E} \sup_{t\in[0,T]} \left|Y_t e^{V_t}\right|^p < \infty$$

and there exists a constant $A \geq 0$ such that for all $t \in [0, T]$:

$$\mathbb{E}^{\mathcal{F}_t}\left[\left|e^{V_T - V_t}\eta\right|^p + \left(\int_t^T e^{V_s - V_t}\, |\Phi(s,0,0)|\, dQ_s\right)^p\right] \leq A, \quad a.s.,$$

then for all $t \in [0, T]$:

$$|Y_t|^p + \mathbb{E}^{\mathcal{F}_t}\left(\int_t^T e^{2(V_s - V_t)}\, |Z_s|^2\, ds\right)^{p/2} \leq A\, C_{a,p},\ a.s.$$

Let $(Y, Z) \in S_m^0[0, T] \times \Lambda_{m\times k}^0(0, T)$ be a solution of the BSDE

$$Y_t = \eta + \int_t^T \Phi(s, Y_s, Z_s)\, dQ_s - \int_t^T Z_s dB_s, \tag{5.21}$$

where Φ satisfies (5.13–**BSDE-H$_\Phi$**) and $(\hat{Y}, \hat{Z}) \in S_m^0[0, T] \times \Lambda_{m\times k}^0(0, T)$ is a solution of the BSDE

$$\hat{Y}_t = \hat{\eta} + \int_t^T \hat{\Phi}(s, \hat{Y}_s, \hat{Z}_s) dQ_s - \int_t^T \hat{Z}_s dB_s, \tag{5.22}$$

where $\hat{\Phi}(\cdot,\cdot,\cdot,\cdot) : \Omega \times [0, T] \times \mathbb{R}^m \times \mathbb{R}^{m\times k} \to \mathbb{R}^m$ is $\mathcal{P}$-measurable with respect to $(\omega, t) \in \Omega \times [0, T]$ and continuous with respect to $(y, z) \in \mathbb{R}^m \times \mathbb{R}^{m\times k}$. We clearly need to assume that

$$\int_0^T \left|\hat{\Phi}(s, \hat{Y}_s, \hat{Z}_s)\right| dQ_s < \infty,\ \mathbb{P}\text{-a.s.}$$

Note that

$$Y_t - \hat{Y}_t = (\eta - \hat{\eta}) + \int_t^T dK_s - \int_t^T (Z_s - \hat{Z}_s) dB_s,$$

where

$$K_t = \int_0^t \left[\Phi\left(s, Y_s, Z_s\right) - \hat{\Phi}(s, \hat{Y}_s, \hat{Z}_s)\right] dQ_s,$$

and by the assumptions (5.13–**BSDE-H**$_{\Phi}$)

$$\left\langle Y_t - \hat{Y}_t, dK_t \right\rangle \leq \left|Y_t - \hat{Y}_t\right| \left|\Phi(t, \hat{Y}_t, \hat{Z}_t) - \hat{\Phi}(t, \hat{Y}_t, \hat{Z}_t)\right| dQ_t + \left|Y_t - \hat{Y}_t\right|^2 dV_t$$
$$+ \frac{n_p}{2a} \left|Z_t - \tilde{Z}_t\right|^2 dt$$

with, as above,

$$dV_t = \mu_t dQ_t + \frac{a}{2n_p} (\ell_t)^2 dt.$$

Hence by Proposition 5.2 we have:

Theorem 5.10 (Continuity and Uniqueness). *Let $a, p > 1$ and the assumptions (5.13–**BSDE-H**$_{\Phi}$) be satisfied. Let*

$$(Y, Z), \ (\hat{Y}, \hat{Z}) \in S_m^0 [0, T] \times \Lambda_{m\times k}^0 (0, T)$$

be solutions of the BSDEs (5.21) and (5.22) respectively. If

$$\mathbb{E} \sup_{t\in[0,T]} \left(e^{pV_t} \left|Y_t - \hat{Y}_t\right|^p\right) < \infty, \tag{5.23}$$

then there exists a positive $C_{a,p}$ such that:

$$\begin{aligned} &\mathbb{E}\left(\sup_{s\in[0,T]} e^{pV_s} \left|Y_s - \hat{Y}_s\right|^p\right) + \mathbb{E}\left[\left(\int_0^T e^{2V_s} \left|Z_s - \hat{Z}_s\right|^2 ds\right)^{p/2}\right] \\ &\leq C_{a,p}\, \mathbb{E}\left[e^{pV_T} \left|\eta - \hat{\eta}\right|^p + \left(\int_0^T e^{V_s} \left|\Phi(s, \hat{Y}_s, \hat{Z}_s) - \hat{\Phi}(s, \hat{Y}_s, \hat{Z}_s)\right| dQ_s\right)^p\right]. \end{aligned} \tag{5.24}$$

If $\Phi = \hat{\Phi}$, then for all $0 \leq t \leq s \leq T$,

$$e^{pV_t} \left|Y_t - \hat{Y}_t\right|^p \leq \mathbb{E}^{\mathcal{F}_t} \left(e^{pV_s} \left|Y_s - \hat{Y}_s\right|^p\right), \ \mathbb{P}\text{-a.s.} \tag{5.25}$$

In particular uniqueness follows in the space $S_m^p\left([0, T]; e^V\right) \times \Lambda_{m\times k}^0 (0, T)$, where

$$S_m^p\left([0, T]; e^V\right) \overset{def}{=} \left\{Y \in S_m^0 [0, T] : \mathbb{E} \sup_{s\in[0,T]} \left|e^{V_s} Y_s\right|^p < \infty\right\}.$$

Recall the notation

$$\bar{\mu}_t = \int_0^t \mu_s dQ_s.$$

Theorem 5.11 (Continuity and Uniqueness). *Let $p = 1$. Assume that $\Phi, \hat{\Phi} : \Omega \times [0,T] \times \mathbb{R}^m \times \mathbb{R}^{m\times k} \to \mathbb{R}^m$ satisfy assumptions (5.13) and both are independent of $z \in \mathbb{R}^{m\times k}$ ($\ell_t = \hat{\ell}_t \equiv 0$). If $(Y,Z), (\hat{Y},\hat{Z}) \in S_m^0[0,T] \times \Lambda_{m\times k}^0(0,T)$ are two solutions of the BSDE (5.107) corresponding respectively to (η,Φ) and $(\hat{\eta},\hat{\Phi})$ such that*

$$\mathbb{E} \sup_{s\in[0,T]} e^{\bar{\mu}_s} \left|Y_s - \hat{Y}_s\right| < \infty,$$

and $\Delta_s \overset{def}{=} \Phi(s,\hat{Y}_s,\hat{Z}_s) - \hat{\Phi}(s,\hat{Y}_s,\hat{Z}_s)$, then $\mathbb{P}$-a.s., for all $t \in [0,T]$:

$$e^{\bar{\mu}_t}\left|Y_t - \hat{Y}_t\right| \le \mathbb{E}^{\mathcal{F}_t}\left(e^{\bar{\mu}_T}|\eta - \hat{\eta}|\right) + \mathbb{E}^{\mathcal{F}_t}\int_t^T e^{\bar{\mu}_s}|\Delta_s|\,dQ_s$$

and for every $q \in (0,1)$ there exists a constant C_q such that

$$\begin{aligned}
&\sup_{s\in[0,T]}\left(\mathbb{E}\left(e^{\bar{\mu}_s}\left|Y_s - \hat{Y}_s\right|\right)\right)^q + \mathbb{E}\sup_{s\in[0,T]} e^{q\bar{\mu}_s}\left|Y_s - \hat{Y}_s\right|^q \\
&\qquad + \mathbb{E}\left(\int_0^T e^{2\bar{\mu}_s}\left|Z_s - \hat{Z}_s\right|^2 ds\right)^{q/2} \\
&\le C_q\left[\left(\mathbb{E}\left(e^{\bar{\mu}_T}|\eta - \hat{\eta}|\right)\right)^q + \left(\mathbb{E}\int_0^T e^{\bar{\mu}_s}|\Delta_s|\,dQ_s\right)^q\right].
\end{aligned}$$

Proof. Since

$$\begin{aligned}
&\left\langle Y_t - \hat{Y}_t, \left[\Phi(t,Y_t,Z_t) - \hat{\Phi}(t,\hat{Y}_t,\hat{Z}_t)\right] dQ_t\right\rangle \\
&\le \left|Y_t - \hat{Y}_t\right| \left|\Phi(t,\hat{Y}_t,\hat{Z}_t) - \hat{\Phi}(t,\hat{Y}_t,\hat{Z}_t)\right| dQ_t + \left|Y_t - \hat{Y}_t\right|^2 d\bar{\mu}_t,
\end{aligned}$$

the conclusions follow by Corollary 6.81. ■

5.3.2 Complementary Results

In this subsection we generalize the uniqueness result and we shall give a scheme to obtain the solution as a limit of uniformly bounded solutions of approximate BSDEs.

Let $a, p > 1$ and

$$V_t^{a,p} = \int_0^t \mu_s dQ_s + \frac{a}{2n_p}\int_0^t (\ell_s)^2 ds.$$

Define

$$S_m^p\left([0,T]; e^{V^{a,p}}\right) \stackrel{def}{=} \left\{ Y \in S_m^0[0,T] : \mathbb{E} \sup_{s\in[0,T]} \left| e^{V_s^{a,p}} Y_s \right|^p < \infty \right\}.$$

Note that if $1 < a_1 < a_2$ then $V_t^{a_1,p} \le V_t^{a_2,p}$ and consequently

$$S_m^p\left([0,T]; e^{V^{a_2,p}}\right) \subset S_m^p\left([0,T]; e^{V^{a_1,p}}\right). \tag{5.26}$$

Let

$$S_m^{1+,p}\left([0,T]; e^V\right) \stackrel{def}{=} \bigcup_{a>1} S_m^p\left([0,T]; e^{V^{a,p}}\right) \quad \text{and}$$
$$S_m^{1+,1+}\left([0,T]; e^V\right) \stackrel{def}{=} \bigcup_{a,\ p>1} S_m^p\left([0,T]; e^{V^{a,p}}\right).$$

Remark 5.12. If Q, μ and ℓ are deterministic functions, then for all $a,\ p > 1$:

$$S_m^{1+,p}\left([0,T]; e^V\right) = S_m^p\left([0,T]; e^{V^{a,p}}\right) = S_m^p[0,T]$$

and

$$S_m^{1+,1+}\left([0,T]; e^V\right) = S_m^{1+}[0,T] \stackrel{def}{=} \bigcup_{p>1} S_m^p[0,T].$$

Corollary 5.13. *Let the assumptions* (**BSDE-H$_\Phi$**) *be satisfied. Then for each $p > 1$, the BSDE (5.12) has at most one solution*

$$(Y, Z) \in S_m^{1+,p}\left([0,T]; e^V\right) \times \Lambda_{m\times k}^0(0,T).$$

If, moreover,

$$\mathbb{E}\exp\left(\lambda \int_0^T (\ell_s)^2 ds\right) < \infty, \quad \textit{for all } \lambda > 0,$$

then the BSDE (5.12) has at most one solution

$$(Y, Z) \in S_m^{1+,1+}\left([0,T]; e^V\right) \times \Lambda_{m\times k}^0(0,T).$$

Proof. Let $(Y,Z),(\hat{Y},\hat{Z}) \in S_m^0([0,T]) \times \Lambda_{m\times k}^0(0,T)$ be two solutions of the BSDE (5.12) corresponding to η.

(A) Let $p > 1$ be such that $(Y,Z),(\hat{Y},\hat{Z}) \in S_m^{1+,p}([0,T];e^V) \times \Lambda_{m\times k}^0(0,T)$. Then from (5.26) and the definition of $S_m^{1+,p}([0,T];e^V)$ there exists an $a > 1$ such that

$$\mathbb{E}\sup_{t\in[0,T]}\left|e^{V_t^{a,p}}Y_t\right|^p < \infty \text{ and } \mathbb{E}\sup_{t\in[0,T]}\left|e^{V_t^{a,p}}\hat{Y}_t\right|^p < \infty,$$

i.e. the condition (5.23) is satisfied; consequently the estimate (5.24) follows and uniqueness too.

(B) If $(Y,Z),(\hat{Y},\hat{Z}) \in S_m^{1+,1+}([0,T];e^V) \times \Lambda_{m\times k}^0(0,T)$ then there exist $a_1, a_2, p_1, p_2 > 1$ such that

$$\mathbb{E}\sup_{t\in[0,T]}\left|e^{V_t^{a_1,p_1}}Y_t\right|^{p_1} < \infty \text{ and } \mathbb{E}\sup_{t\in[0,T]}\left|e^{V_t^{a_2,p_2}}\hat{Y}_t\right|^{p_2} < \infty.$$

Let $a > 1$ and $1 < p < p_1 \wedge p_2$. Put

$$b_i = \frac{a}{2n_p} - \frac{a_i}{2n_{p_i}}.$$

Since

$$\begin{aligned} V_t^{a,p} &= \int_0^t \mu_s dQ_s + \frac{a}{2n_p}\int_0^t (\ell_s)^2 ds \\ &= V_t^{a_i,p_i} + b_i\int_0^t (\ell_s)^2 ds, \end{aligned}$$

we get

$$\begin{aligned} \mathbb{E}\sup_{t\in[0,T]}\left|e^{V_t^{a,p}}Y_t\right|^p &= \mathbb{E}\left\{\sup_{t\in[0,T]}\left|e^{V_t^{a_i,p_i}}Y_t\right|^p \exp\left[pb_i\int_0^T (\ell_s)^2 ds\right]\right\} \\ &\le \left(\mathbb{E}\sup_{t\in[0,T]}\left|e^{V_t^{a_i,p_i}}Y_t\right|^{p_i}\right)^{\frac{p}{p_i}} \left\{\mathbb{E}\exp\left[\tfrac{p_i p b_i}{p_i-p}\int_0^T (\ell_s)^2 ds\right]\right\}^{\frac{p_i-p}{p_i}} \\ &< \infty. \end{aligned}$$

Similar we have

$$\mathbb{E}\sup_{t\in[0,T]}\left|e^{V_t^{a,p}}\hat{Y}_t\right|^p < \infty.$$

Hence the estimate (5.24) holds and the uniqueness follows. ∎

The next Proposition will allow us to extend existence results from situations where the data satisfy the following strong boundedness condition: there exists a positive constant C such that for all $t \in [0,T]$:

$$|\eta| + |\Phi(t,0,0)| + \left|e^{\hat{V}_T}\eta\right| + \int_0^T e^{\hat{V}_s}|\Phi(s,0,0)|\,dQ_s \le C < \infty, \quad \mathbb{P}\text{-a.s.}$$

where

$$\hat{V}_t = \int_0^t \mu_s^+ dQ_s + \frac{a}{2n_p}\int_0^t (\ell_s)^2\,ds.$$

Let

$$V_t \overset{def}{=} V_t^{a,p} = \int_0^t \mu_s dQ_s + \frac{a}{2n_p}\int_0^t (\ell_s)^2\,ds \quad \text{and}$$

$$\beta_t \overset{def}{=} Q_t + \int_0^t |\mu_s|\,dQ_s + \int_0^t (\ell_s)^2\,ds + \int_0^t |\Phi(s,0,0)|\,dQ_s.$$

We have $V_s - V_t \le \hat{V}_s - \hat{V}_t$ for all $0 \le t \le s \le T$.

Define, for $n \in \mathbb{N}^*$,

$$\eta^n = \eta\,\mathbf{1}_{[0,n]}(\beta_T + |\eta|),$$
$$\Phi^n(t,y,z) = \Phi(t,y,z) - \Phi(t,0,0)\ \mathbf{1}_{[n,\infty[}(\beta_t + |\Phi(t,0,0)|),$$

and the stochastic processes

$$H_t^n = \left|e^{V_T - V_t}\eta^n\right| + \int_t^T e^{V_s - V_t}|\Phi^n(s,0,0)|\,dQ_s,$$

$$\hat{H}_t^n = \left|e^{\hat{V}_T - \hat{V}_t}\eta^n\right| + \int_t^T e^{\hat{V}_s - \hat{V}_t}|\Phi^n(s,0,0)|\,dQ_s.$$

It is easy to verify that there exists a positive constant $M_{n,p,a}$ such that

$$0 \le H_0^n \le \|H^n\|_T \le \left\|\hat{H}^n\right\|_T \le M_{n,T}, \quad \mathbb{P}\text{-a.s.}$$

Proposition 5.14. *Let $a, p > 1$ and the assumptions (5.13-**BSDE-H$_\Phi$**) be satisfied. Also assume that*

$$\mathbb{E}e^{pV_T}|\eta|^p + \mathbb{E}\Big(\int_0^T e^{V_s}|\Phi(s,0,0)|\,dQ_s\Big)^p < \infty. \tag{5.27}$$

If for each $n \in \mathbb{N}^*$, $(Y^n, Z^n) \in S_m^0[0,T] \times \Lambda_{m\times k}^0(0,T)$ is a solution of the BSDE

$$Y_t^n = \eta^n + \int_t^T \Phi^n\left(s, Y_s^n, Z_s^n\right) dQ_s - \int_t^T Z_s^n dB_s$$

such that $e^V Y^n \in S_m^p[0,T]$, then

$$\|Y^n\|_T + \left\|e^V Y^n\right\|_T \le M'_{n,p,a}, \quad a.s., \tag{5.28}$$

and there exists (a unique!) $(Y,Z) \in S_m^p\left([0,T]; e^V\right) \times \Lambda_{m\times k}^p\left(0,T; e^V\right)$ such that

$$\lim_{n\to\infty}\left[\mathbb{E}\,\left\|e^V(Y^n - Y)\right\|_T^p + \mathbb{E}\left(\int_0^T e^{2V_s}\left|Z_s^n - Z_s\right|^2 ds\right)^{p/2}\right] = 0 \tag{5.29}$$

and, $\mathbb{P}$-a.s. for all $t \in [0,T]$:

$$Y_t = \eta + \int_t^T \Phi(s, Y_s, Z_s)\, dQ_s - \int_t^T Z_s dB_s. \tag{5.30}$$

Proof. In view of (5.19) we have for all $t \in [0,T]$:

$$e^{pV_t}\left|Y_t^n\right|^p + \left|Y_t^n\right|^p \le M'_{n,p,a}, \quad \mathbb{P}\text{-a.s.}$$

and (5.28) follows.

For all $n, i \in \mathbb{N}^*$:

$$Y_t^n - Y_t^{n+i} = \eta^n - \eta^{n+i} + \int_t^T d\left(K_s^n - K_s^{n+i}\right) - \int_t^T \left(Z_s^n - Z_s^{n+i}\right) dB_s,$$

where

$$K_t^n = \int_0^t \Phi^n\left(s, Y_s^n, Z_s^n\right) dQ_s$$

and similarly for K_t^{n+i}.

Since

$$\begin{aligned}\left\langle Y_s^n - Y_s^{n+i}, d\left(K_s^n - K_s^{n+i}\right)\right\rangle \le{} & \left|Y_s^n - Y_s^{n+i}\right| \left|\Phi(s,0,0)\right| \mathbf{1}_{\beta_s + |\Phi(s,0,0)| \ge n} dQ_s \\ & + \left|Y_s^n - Y_s^{n+i}\right|^2 dV_s + \frac{n_p}{2a}\left|Z_s^n - Z_s^{n+i}\right|^2 ds,\end{aligned}$$

we deduce from Proposition 5.2 that

$$\mathbb{E} \sup_{s\in[0,T]} e^{pV_s} \left|Y_s^n - Y_s^{n+i}\right|^p + \mathbb{E}\left(\int_0^T e^{2V_s} \left|Z_s^n - Z_s^{n+i}\right|^2 ds\right)^{p/2}$$

$$\leq C_{a,p}\mathbb{E}\left(e^{pV_T} |\eta|^p \mathbf{1}_{\beta_T+|\eta|\geq n}\right) + C_{a,p}\mathbb{E}\left(\int_0^T e^{V_s}\mathbf{1}_{\beta_s+|\Phi(s,0,0)|\geq n} |\Phi(s,0,0)|\, dQ_s\right)^p.$$

Hence there exists $(Y,Z) \in S_m^p([0,T];e^V) \times \Lambda_{m\times k}^p(0,T;e^V)$ such that (5.29) holds. The last assertion follows from Lemma 5.16 below, whose proof is left as an exercise for the reader. ■

Clearly from the construction in Proposition 5.14 we have:

Corollary 5.15. *Suppose that the assumptions from (5.13-**BSDE-H**$_\Phi$) are satisfied. Then the existence of a solution under the conditions (5.27) with $p = 2$ and some $a > 1$ implies existence under the same conditions for any $p > 1$.*

We end this subsection with a continuity result, the easy proof of which is left as an exercise for the reader.

Lemma 5.16. *Let the assumptions (5.13-**BSDE-H**$_\Phi$) be satisfied. If $(Y,Z) \in S_m^0[0,T] \times \Lambda_{m\times k}^0(0,T)$, then*

$$\int_0^T |\Phi(t,Y_t,Z_t)|\, dQ_t < \infty, \quad \mathbb{P}\text{-a.s.}$$

and the mapping

$$(U,V) \longrightarrow \int_0^\cdot \Phi(s,U_s,V_s)dQ_s : S_m^0[0,T] \times \Lambda_{m\times k}^0(0,T) \to S_m^0[0,T]$$

is continuous.

5.3.3 BSDEs with Lipschitz Coefficients

5.3.3.1 BSDEs with Deterministic Lipschitz Conditions

Consider the backward stochastic differential equation: $\mathbb{P}$-*a.s.*, for all $t \in [0,T]$

$$Y_t = \eta + \int_t^T F(s,Y_s,Z_s)\, ds - \int_t^T Z_s dB_s, \tag{5.31}$$

under the assumptions

◊ $p > 1$,

$$\eta \in L^p\left(\Omega, \mathcal{F}_T, \mathbb{P}; \mathbb{R}^m\right), \tag{5.32}$$

◊ *the function* $F(\cdot,\cdot,y,z) : \Omega \times [0,T] \to \mathbb{R}^m$ *is* $\mathcal{P}$*-measurable for every* $(y,z) \in \mathbb{R}^m \times \mathbb{R}^{m\times k}$,

◊ *there exist* $L \in L^1(0,T)$, $\ell \in L^2(0,T)$ *such that*

$$\begin{cases} (I) \;\; \text{Lipschitz conditions:} \\ \text{for all } y,y' \in \mathbb{R}^m,\; z,z' \in \mathbb{R}^{m\times k},\; d\mathbb{P}\otimes dt\text{-a.e.:} \\ \qquad (L_y) \quad |F(t,y',z) - F(t,y,z)| \le L(t)\,|y'-y|\,, \\ \qquad (L_z) \quad |F(t,y,z') - F(t,y,z)| \le \ell(t)\,|z'-z|; \\ (II) \;\; \text{Boundedness condition:} \\ \qquad (B_F) \quad \mathbb{E}\left(\int_0^T |F(t,0,0)|\,dt\right)^p < \infty. \end{cases} \tag{5.33}$$

We recall the notation

$$S_m^{1+}[0,T] \overset{def}{=} \bigcup_{p>1} S_m^p[0,T].$$

Theorem 5.17. *Let* $p > 1$ *and the assumptions (5.32) and (5.33) be satisfied. Then the BSDE (5.31) has a unique solution* $(Y,Z) \in S_m^p[0,T] \times \Lambda^p_{m\times k}(0,T)$*. Moreover uniqueness holds in* $S_m^{1+}[0,T] \times \Lambda^0_{m\times k}(0,T)$.

Proof. We first remark that if $(Y,Z) \in S_m^p[0,T] \times \Lambda^p_{m\times k}(0,T)$ then

$$K_\cdot \overset{def}{=} \int_0^\cdot F(r,Y_r,Z_r)\,dr \;\in\; S_m^p[0,T] \text{ and } \mathbb{E}\,\updownarrow K \updownarrow_T^p < \infty.$$

Indeed, since

$$|F(r,Y_r,Z_r)| \le |F(r,0,0)| + L(r)\,|Y_r| + \ell(r)\,|Z_r|\,,$$

then

$$\begin{aligned} \mathbb{E}\sup_{t\in[0,T]} |K_t|^p &\le \mathbb{E}\,\updownarrow K \updownarrow_T^p \\ &= \mathbb{E}\left(\int_0^T |F(r,Y_r,Z_r)|\,dr\right)^p \\ &\le C_p\mathbb{E}\left(\int_0^T |F(r,0,0)|\,dr\right)^p + C_p\left(\int_0^T L(r)\,dr\right)^p \mathbb{E}\,\|Y\|_T^p \end{aligned}$$

$$+ C_p \left(\int_0^T \ell^2(r)\, dr \right)^{p/2} \mathbb{E} \left(\int_0^T |Z_r|^2 dr \right)^{p/2}$$
$$< \infty.$$

Uniqueness follows from Corollary 5.13.

We prove existence.

Note that a solution of the Eq. (5.31) is a fixed point of the mapping Γ : $S_m^p[0,T] \times \Lambda_{m\times k}^p(0,T) \to S_m^p[0,T] \times \Lambda_{m\times k}^p(0,T)$ defined by

$$(Y, Z) = \Gamma(X, U),$$

where

$$Y_t = \eta + \int_t^T F(r, X_r, U_r)\, dr - \int_t^T Z_r dB_r, \ a.s. \ t \in [0,T].$$

By Corollary 2.45 the mapping Γ is well defined.

Let $M \in \mathbb{N}^*$ and $0 = T_0 < T_1 < \cdots < T_M = T$, with $T_i = \dfrac{iT}{M}$. Then

$$\alpha\left(\tfrac{T}{M}\right) \stackrel{def}{=} \sup_{0<s-t<\frac{T}{M}} \int_t^s \left[L(r) + \ell^2(r)\right] dr \to 0, \quad \text{as } M \to \infty.$$

We show that Γ is a strict contraction on the Banach space $S_m^p[T_{M-1}, T] \times \Lambda_{m\times k}^p(T_{M-1}, T)$ with the norm

$$|||(X,U)|||_M = \left[\mathbb{E} \sup_{r\in[T_{M-1},T]} |X_r|^p + \mathbb{E}\left(\int_{T_{M-1}}^T |U_r|^2 dr \right)^{p/2} \right]^{1/p}$$

for M large enough.

Let $(X,U), (X',U') \in S_m^p[T_{M-1}, T] \times \Lambda_{m\times k}^p(T_{M-1}, T)$. Then

$$Y_t - Y_t' = \int_t^T dK_r - \int_t^T (Z_r - Z_r')\, dB_r, \quad t \in [0,T],$$

where

$$K_t = \int_0^t \left[F(r, X_r, U_r) - F\left(r, X_r', U_r'\right)\right] dr.$$

Since

$$\begin{aligned} \langle Y_r - Y_r', dK_r \rangle &\leq \left|F(r, X_r, U_r) - F\left(r, X_r', U_r'\right)\right| |Y_r - Y_r'|\, dr \\ &\leq \left[L(r)\,|X_r - X_r'| + \ell(r)\,|U_r - U_r'|\right] |Y_r - Y_r'|\, dr \end{aligned}$$

and

$$\mathbb{E}\left(\sup_{r\in[T_{M-1},T]}\left|Y_r - Y_r'\right|^p\right) < \infty,$$

we have by (5.8), with $D = 0$, $R = V = 0$, $\lambda = 0$,

$$\begin{aligned}
&\left\|\left|(Y,Z) - (Y',Z')\right\|\right|_M^p \\
&= \mathbb{E}\left(\sup_{r\in[T_{M-1},T]}\left|Y_r - Y_r'\right|^p\right) + \mathbb{E}\left(\int_{T_{M-1}}^T \left|Z_r - Z_r'\right|^2 dr\right)^{p/2} \\
&\leq C_p\mathbb{E}\left(\int_{T_{M-1}}^T \left[L(r)\left|X_r - X_r'\right| + \ell(r)\left|U_r - U_r'\right|\right] dr\right)^p \\
&\leq C_p'\left(\int_{T_{M-1}}^T L(r)\,dr\right)^p \mathbb{E}\sup_{r\in[T_{M-1},T]}\left|X_r - X_r'\right|^p \\
&\quad + C_p'\left(\int_{T_{M-1}}^T \ell^2(r)\,dr\right)^{p/2} \mathbb{E}\left(\int_{T_{M-1}}^T \left|U_r - U_r'\right|^2 dr\right)^{p/2} \\
&\leq C_p'\left[\boldsymbol{\alpha}^p\left(\tfrac{T}{M}\right) + \boldsymbol{\alpha}^{p/2}\left(\tfrac{T}{M}\right)\right]\left\|\left|(X,U) - (X',U')\right\|\right|_M^p.
\end{aligned}$$

Let $M_0 \in \mathbb{N}^*$ be such that

$$C_p'\left[\boldsymbol{\alpha}^p\left(\tfrac{T}{M_0}\right) + \boldsymbol{\alpha}^{p/2}\left(\tfrac{T}{M_0}\right)\right] \leq \frac{1}{2^p}.$$

Then Γ is a strict contraction on $S_m^p[T_{M_0-1},T]\times\Lambda_{m\times k}^p(T_{M_0-1},T)$ and consequently the Eq. (5.31) has a unique solution $(Y,Z) \in S_m^p[T_{M_0-1},T]\times\Lambda_{m\times k}^p(T_{M_0-1},T)$. The next step is to solve the equation on the interval $[T_{M_0-2},T_{M_0-1}]$ with the final value $Y(T_{M_0-1})$. Repeating the same arguments, the proof is completed in M_0 steps. ∎

Corollary 5.18. *Consider the BSDE:* $\forall\, t \in [0,T]$, $\mathbb{P}$-*a.s.*

$$Y_t = \eta + S_T - S_t + \int_t^T F(s,Y_s,Z_s)\,ds - \int_t^T Z_s dB_s. \tag{5.34}$$

If $p > 1$, $S \in S_m^p[0,T]$, $\eta \in L^p(\Omega,\mathcal{F}_T,\mathbb{P};\mathbb{R}^m)$ *and* F *satisfies the assumptions (5.33), then the Eq. (5.34) has a unique solution* $(Y,Z) \in S_m^p[0,T] \times \Lambda_{m\times k}^p(0,T)$.

Proof. By the substitutions $\hat{Y}_t = Y_t + S_t$, $\hat{\eta} = \eta + S_T$ and $\hat{F}(t,y,z) = F(t,y-S_t,z)$ the Eq. (5.34) is transformed into

$$\hat{Y}_t = \hat{\eta} + \int_t^T \hat{F}\left(s, \hat{Y}_s, Z_s\right) ds - \int_t^T Z_s dB_s,$$

which satisfies the assumptions of Theorem 5.17. ■

We now study the case $p = 1$, where we restrict ourselves to the case where F does not depend on z.

Corollary 5.19. *If $S \in S_d^1[0,T]$, $\eta \in L^1(\Omega, \mathcal{F}_T, \mathbb{P}; \mathbb{R}^m)$ and $F(t,y,z) \equiv F(t,y)$ satisfies the assumptions (5.33) with $p = 1$, then the BSDE*

$$Y_t = \eta + S_T - S_t + \int_t^T F(s, Y_s)\, ds - \int_t^T Z_s dB_s \tag{5.35}$$

has a unique solution $(Y,Z) \in S_m^0[0,T] \times \Lambda_{m\times k}^0(0,T)$ such that

$$M_t = \int_0^t Z_s dB_s \quad \text{is a martingale}$$

and

$$\sup_{t\in[0,T]} \mathbb{E}|Y_t| + \mathbb{E}\sup_{t\in[0,T]} |Y_t|^q + \mathbb{E}\left(\int_0^T |Z_t|^2 dt\right)^{q/2} < \infty,\ \forall\, 0 < q < 1.$$

Proof. As in the proof of Corollary 5.18 we can reduce the problem to the case $S = 0$.

Let $n, i \in \mathbb{N}^*$. By Theorem 5.17 there exists a unique pair (Y^n, Z^n) such that for all $p \geq 1$

$$(Y^n, Z^n) \in S_m^p[0,T] \times \Lambda_{m\times k}^p(0,T) \tag{5.36}$$

and (Y^n, Z^n) is solution of the equation

$$Y_t^n = \eta \mathbf{1}_{|\eta|\leq n} + \int_t^T \left[F\left(r, Y_r^n\right) - \mathbf{1}_{|F(r,0)|\geq n} F(r,0)\right] dr - \int_t^T Z_r^n dB_r. \tag{5.37}$$

Note that

$$\beta_n \stackrel{def}{=} \mathbb{E}|\eta|\,\mathbf{1}_{|\eta|>n} + \mathbb{E}\int_0^T |F(s,0)|\,\mathbf{1}_{|F(r,0)|\geq n} ds \to 0 \quad \text{as } n \to \infty.$$

By (5.4) for $Y_s^n - Y_s^{n+i}$ and $p = 1$ we infer

$$\begin{aligned} \left|Y_t^n - Y_t^{n+i}\right| &\leq \beta_n + \int_t^T L(s)\left|Y_s^n - Y_s^{n+i}\right| ds \\ &- \int_t^T \left|Y_s^n - Y_s^{n+i}\right|^{-1} \mathbf{1}_{Y_s^n - Y_s^{n+i} \neq 0} \left\langle Y_s^n - Y_s^{n+i}, \left(Z_s^n - Z_s^{n+i}\right) dB_s\right\rangle. \end{aligned} \tag{5.38}$$

Denote by C, C' generic constants independent of n and i.

From (5.36)

$$M_t^{n,n+i} = \int_0^t \left|Y_s^n - Y_s^{n+i}\right|^{-1} \mathbf{1}_{Y_s^n - Y_s^{n+i} \neq 0} \left\langle Y_s^n - Y_s^{n+i}, \left(Z_s^n - Z_s^{n+i}\right) dB_s \right\rangle$$

is a martingale. Then $\mathbb{E} M_t^{n,n+i} = 0$ and, taking the expectation in (5.38) we deduce from the backward Gronwall inequality (Corollary 6.62):

$$\mathbb{E}\left|Y_t^n - Y_t^{n+i}\right| \leq C\ \beta_n. \tag{5.39}$$

Using the Burkholder–Davis–Gundy inequality (1.18) and Doob's inequality (1.11-A_3), we deduce for $0 < q < 1$:

$$\begin{aligned}
&\mathbb{E}\left[\left(\int_0^T \left|Z_s^n - Z_s^{n+i}\right|^2 ds\right)^{q/2}\right] \\
&\leq \frac{1}{c_q}\mathbb{E}\left(\sup_{0\leq t\leq T}\left|\int_0^t \left(Z_s^n - Z_s^{n+i}\right) dB_s\right|^q\right) \\
&\leq \frac{1}{c_q\,(1-q)}\left[\mathbb{E}\left|\int_0^T \left(Z_s^n - Z_s^{n+i}\right) dB_s\right|\right]^q \\
&\leq \frac{1}{c_q\,(1-q)}\left[\mathbb{E}\left|Y_0^n - Y_0^{n+i}\right| + \mathbb{E}\int_0^T \left|F\left(s, Y_s^n\right) - F\left(s, Y_s^{n+i}\right)\right|\right]^q \\
&\leq \frac{1}{c_q\,(1-q)}\left[\mathbb{E}\left|Y_0^n - Y_0^{n+i}\right| + L\,(s)\int_0^T \mathbb{E}\left|Y_s^n - Y_s^{n+i}\right| ds\right]^q \\
&\leq C\ \beta_n^q.
\end{aligned}$$

Recalling that $M_t^{n,n+i}$ is a martingale then, once again by the Burkholder–Davis–Gundy inequality

$$\begin{aligned}
\mathbb{E}\sup_{t\in[0,T]}\left|M_T^{n,n+i} - M_t^{n,n+i}\right|^q &\leq 2^q\mathbb{E}\sup_{t\in[0,T]}\left|M_t^{n,n+i}\right|^q \\
&\leq C_q\mathbb{E}\left[\left(\int_0^T \left|Z_s^n - Z_s^{n+i}\right|^2 ds\right)^{q/2}\right] \\
&\leq C'\ \beta_n^q,
\end{aligned}$$

then from (5.38) and (5.39) we obtain for every $0 < q < 1$

$$\begin{aligned}\mathbb{E}\sup_{t\in[0,T]}\left|Y_t^n - Y_t^{n+i}\right|^q &\le \beta_n^q + \mathbb{E}\left(\int_0^T L(s)\left|Y_s^n - Y_s^{n+i}\right|ds\right)^q + C'\,\beta_n^q \\ &\le \beta_n^q + \left(\int_0^T L(s)\,\mathbb{E}\,\left|Y_s^n - Y_s^{n+i}\right|ds\right)^q + C'\,\beta_n^q \\ &\le C\,\beta_n^q.\end{aligned}$$

Hence, there exist Y and Z such that

$$\begin{aligned}Y^n &\to Y, \quad \text{in } S_m^q[0,T]\cap C\left([0,T];L^1(\Omega,\mathcal{F},\mathbb{P})\right), \text{ and} \\ Z^n &\to Z, \quad \text{in } \Lambda_{m\times k}^q(0,T).\end{aligned}$$

Passing to the limit in (5.37) we deduce that the pair (Y,Z) solves the problem.

Now, by Corollary 2.47, $M_t = \int_0^t Z_s dB_s$ is a martingale, because

$$S_\cdot + \int_0^\cdot F(s,Y_s)\,ds \in S_m^1[0,T].$$

Finally, the uniqueness is obtained in the same manner as the estimates for $Y^n - Y^{n+i}$ and $Z^n - Z^{n+i}$. ■

5.3.3.2 BSDEs with Random Lipschitz Conditions

We now generalize Theorem 5.17 to a class of BSDEs with random Lipschitz constants.

We consider the BSDE (5.31) in a more general form:

$$Y_t = \eta + \int_t^T \Phi(s,Y_s,Z_s)\,dQ_s - \int_t^T Z_s dB_s,\ t\in[0,T],\ a.s. \tag{5.40}$$

We assume that

(BSDE-A0) :

(i) $\eta:\Omega\to\mathbb{R}^m$ *is an* $\mathcal{F}_T$*-measurable random vector,*
(ii) Q is a $\mathcal{P}$*-m.i.c.s.p. such that* $Q_0 = 0$;

and the function $\Phi:\Omega\times[0,T]\times\mathbb{R}^m\to\mathbb{R}^m$ satisfies

(BSDE-LH$_\Phi$) ▲ *for all* $y\in\mathbb{R}^m$, $z\in\mathbb{R}^{m\times k}$, *the function* $\Phi(\cdot,\cdot,y,z):\Omega\times[0,T]\to\mathbb{R}^m$ *is* $\mathcal{P}$*-measurable*;

▲ *there exist $\mathcal{P}$-m.s.p. $L, \ell, \alpha : \Omega \times [0,T] \to \mathbb{R}_+$, such that*

$$\alpha_t dQ_t = dt \qquad \text{and} \qquad \int_0^T \left(L_t dQ_t + (\ell_t)^2 \, dt \right) < \infty, \ \ \mathbb{P}\text{-a.s.};$$

for all $t \in [0,T]$, $y, y' \in \mathbb{R}^m$ and *$z, z' \in \mathbb{R}^{m\times k}$, $\mathbb{P}$-a.s.:*

$$\begin{array}{ll} & \textit{Lipschitz conditions} \\ (i) & |\Phi(t,y',z) - \Phi(t,y,z)| \le L_t |y'-y|, \\ (ii) & |\Phi(t,y,z') - \Phi(t,y,z)| \le \alpha_t \ell_t |z'-z|, \\ & \textit{Boundedness condition:} \\ (iii) & \displaystyle\int_0^T \Phi_\rho^{\#}(t)\, dQ_t < \infty, \ \ \forall\, \rho \ge 0, \end{array} \tag{5.41}$$

where

$$\Phi_\rho^{\#}(t) \overset{def}{=} \sup_{|y|\le\rho} |\Phi(t,y,0)| \, .$$

Note that the condition $\alpha_t dQ_t = dt$ implies that $\Phi(t, Y_t, Z_t)\, dQ_t = F(t, Y_t, Z_t)\, dt + G(t, Y_t)\, dA_t$, where G does not depend upon z.

We recall the following notations. For each fixed $p > 1$ let $n_p = 1 \wedge (p-1)$ and

$$V_t = V_t^{(p)} = \int_0^t L_s dQ_s + \frac{1}{n_p}\int_0^t (\ell_s)^2 \, ds. \tag{5.42}$$

The stochastic process V is that from Lemma 5.6 with $a = 2$. Therefore for all (Y,Z), $(\tilde{Y},\tilde{Z}) \in S_m^0[0,T] \times \Lambda_{m\times k}^0(0,T)$ we have

$$\langle Y_t, \Phi(t, Y_t, Z_t)\, dQ_t \rangle \le |Y_t|\, |\Phi(t,0,0)|\, dQ_t + |Y_t|^2 dV_t + \frac{n_p}{4} |Z_t|^2 \, dt, \tag{5.43}$$

and

$$\left\langle Y_t - \tilde{Y}_t, \Phi(t, Y_t, Z_t) - \Phi(t, \tilde{Y}_t, \tilde{Z}_t) \right\rangle dQ_t \le |Y_t - \tilde{Y}_t|^2 dV_t + \frac{n_p}{4} \left| Z_t - \tilde{Z}_t \right|^2 dt. \tag{5.44}$$

Lemma 5.20. *Let $p \ge 2$ and the assumptions* **(BSDE-A0)**, **(BSDE-LH$_\Phi$)** *be satisfied. If moreover there exists a constant $b > 0$ such that*

$$\begin{array}{ll} (i) & \displaystyle\int_0^T L_s dQ_s \le b \quad \textit{and} \quad \int_0^T (\ell_s)^2 \, ds \le b, \quad \mathbb{P}\text{-a.s.}, \\ (ii) & \displaystyle\mathbb{E}\, |\eta|^p + \mathbb{E} \left(\int_0^T |\Phi(s,0,0)|\, dQ_s \right)^p < \infty, \end{array} \tag{5.45}$$

then the BSDE (5.40) has a unique solution $(Y,Z) \in S_m^p[0,T] \times \Lambda_{m\times k}^p(0,T)$.

Proof. We have

$$V_t = \int_0^t \left(L_s dQ_s + \frac{1}{n_p}(\ell_s)^2 ds\right) = \int_0^t \left(L_s dQ_s + (\ell_s)^2 ds\right).$$

Since

$$0 \le V_t \le 2b, \quad \text{for all } t \in [0,T],$$

it follows that for every $\delta > 0$ we can define on $S_m^p[0,T] \times \Lambda_{m\times k}^p(0,T)$ an equivalent norm by

$$\|(Y,Z)\|_{\delta V} \stackrel{def}{=} \left[\mathbb{E}\sup_{s\in[0,T]} e^{\delta p V_s}|Y_s|^p + \mathbb{E}\left(\int_0^T e^{2\delta V_s}|Y_s|^2 L_s dQ_s\right)^{p/2} + \mathbb{E}\left(\int_0^T e^{2\delta V_s}|Z_s|^2 ds\right)^{p/2}\right]^{1/p}.$$

Let $\Gamma : S_m^p[0,T] \times \Lambda_{m\times k}^p(0,T) \to S_m^p[0,T] \times \Lambda_{m\times k}^p(0,T)$ be defined by

$$(Y,Z) = \Gamma(X,U)$$
$$Y_t = \eta + \int_t^T \Phi(s,X_s,U_s)\,dQ_s - \int_t^T Z_s dB_s.$$

We remark that for all $X,U \in S_m^p[0,T] \times \Lambda_{m\times k}^p(0,T)$,

$$\begin{aligned}\int_0^t |\Phi(s,X_s,U_s)|\,dQ_s &\le \int_0^t |\Phi(s,0,0)|\,dQ_s + \int_0^t |X_s|\,L_s dQ_s + \int_0^t |U_s|\,\ell_s ds \\ &\le \int_0^t |\Phi(s,0,0)|\,dQ_s + b\sup_{s\in[0,t]}|X_s| + b\left(\int_0^t |U_s|^2 ds\right)^{1/2}\end{aligned}$$

and consequently $S_\cdot = \int_0^\cdot \Phi(s,X,U)\,dQ_s \in S_m^p[0,T]$. By the martingale representation result from Corollary 2.45 it follows that Γ is well defined.

The fact that BSDE (5.40) has a unique solution $(Y,Z) \in S_m^p[0,T]\times\Lambda_{m\times k}^p(0,T)$ will be a consequence of the fact that Γ is a strict contraction on the Banach space $\left(S_m^p[0,T] \times \Lambda_{m\times k}^p(0,T), \|\cdot\|_\delta\right)$, for some $\delta > 0$.

Let $(Y,Z) = \Gamma(X,U)$ and $(Y',Z') = \Gamma(X',U')$. We have

$$Y_t - Y_t' = \int_t^T dK_s - \int_t^T (Z_s - Z_s')\,dB_s,$$

where $K_t = \int_0^t \left[\Phi(s, X_s, U_s) - \Phi\left(s, X_s', U_s'\right)\right] dQ_s$ and for all $\delta > 1$

$$\begin{aligned}
&\left|Y_s - Y_s'\right|^2 L_s dQ_s + \left\langle Y_s - Y_s', dK_s\right\rangle \\
&\quad \leq \left|Y_s - Y_s'\right|^2 L_s dQ_s + \left|Y_s - Y_s'\right| \left[\left|X_s - X_s'\right| L_s dQ_s + \left|U_s - U_s'\right| \ell_s ds\right] \\
&\quad \leq \left|Y_s - Y_s'\right|^2 L_s dQ_s + \left[\frac{1}{4(\delta-1)}\left|X_s - X_s'\right|^2 + (\delta - 1)\left|Y_s - Y_s'\right|^2\right] L_s dQ_s \\
&\qquad + \left(\frac{1}{4\delta}\left|U_s - U_s'\right|^2 + \delta\left|Y_s - Y_s'\right|^2 \ell_s^2\right) ds \\
&\quad \leq \frac{1}{4\delta}\left|U_s - U_s'\right|^2 ds + \frac{1}{4(\delta-1)}\left|X_s - X_s'\right|^2 L_s dQ_s + \left|Y_s - Y_s'\right|^2 \delta dV_s.
\end{aligned}$$

Then by Proposition 5.2-A,

$$\begin{aligned}
&\mathbb{E}\left(\sup_{s\in[0,T]} e^{p\delta V_s}\left|Y_s - Y_s'\right|^p\right) + \mathbb{E}\left(\int_0^T e^{2\delta V_s}\left|Y_s - Y_s'\right|^2 L_s dQ_s\right)^{p/2} \\
&\qquad + \mathbb{E}\left(\int_0^T e^{2\delta V_s}\left|Z_s - Z_s'\right|^2 ds\right)^{p/2} \\
&\quad \leq \frac{C_p}{\delta^{p/2}}\mathbb{E}\left(\int_0^T e^{2\delta V_s}\left|U_s - U_s'\right|^2 ds\right)^{p/2} \\
&\qquad + \frac{C_p}{(\delta-1)^{p/2}}\mathbb{E}\left(\int_0^T e^{2\delta V_s}\left|X_s - X_s'\right|^2 L_s dQ_s\right)^{p/2} \\
&\quad \leq \frac{C_p}{(\delta-1)^{p/2}}\left\|(X,U) - \left(X',U'\right)\right\|_{\delta V}^p \\
&\quad \leq \frac{1}{2^p}\left\|(X,U) - \left(X',U'\right)\right\|_{\delta V}^p
\end{aligned}$$

for $\delta \geq 1 + 4C_p^{2/p}$. Hence

$$\left\|\Gamma(X,U) - \Gamma\left(X',U'\right)\right\|_{\delta V} \leq \frac{1}{2}\left\|(X,U) - \left(X',U'\right)\right\|_{\delta V}$$

and the result follows. ■

Theorem 5.21. *Let $p > 1$, $n_p = 1 \wedge (p-1)$ and the assumptions* **(BSDE-A0)**, **(BSDE-LH$_\Phi$)** *be satisfied. Let*

$$V_t = V_t^{(p)} \stackrel{def}{=} \int_0^t \left(L_s dQ_s + \frac{1}{n_p}(\ell_s)^2 ds\right).$$

Assume also that there exists $\delta > \frac{p}{p-1}$ *such that for* $q = \frac{p\delta}{p+\delta}$ *and* $n_q = 1 \wedge (q-1)$,

$$\begin{aligned}
&(i) \quad \mathbb{E}\, e^{pV_T} |\eta|^p + \mathbb{E}\left(\int_0^T e^{V_s} |\Phi(s,0,0)|\, dQ_s\right)^p < \infty, \\
&(ii) \quad \mathbb{E}\left(\int_0^T L_s dQ_s\right)^\delta + \mathbb{E}\left(\int_0^T (\ell_s)^2\, ds\right)^{\delta/2} < \infty, \\
&(iii) \quad \mathbb{E}\, \exp\left[\delta\left(\frac{1}{n_q} - \frac{1}{n_p}\right)\int_0^T (\ell_s)^2\, ds\right] < \infty.
\end{aligned} \tag{5.46}$$

Then the BSDE (5.40) has a unique solution $(Y, Z) \in S_m^0 \times \Lambda_{m\times k}^0$ *such that*

$$\mathbb{E}\left(\sup_{t\in[0,T]} e^{pV_t} |Y_t|^p\right) < \infty. \tag{5.47}$$

Moreover there exists a positive constant C_p *depending only on* p *such that for all* $t \in [0,T]$

$$\begin{aligned}
&\mathbb{E}^{\mathcal{F}_t} \sup_{s\in[t,T]} \left|e^{V_s} Y_s\right|^p + \mathbb{E}^{\mathcal{F}_t}\left(\int_t^T e^{2V_s} |Z_s|^2\, ds\right)^{p/2} \\
&\leq C_p\, \mathbb{E}^{\mathcal{F}_t}\left[\left|e^{V_T}\eta\right|^p + \left(\int_t^T e^{V_s} |\Phi(s,0,0)|\, dQ_s\right)^p\right].
\end{aligned} \tag{5.48}$$

Remark 5.22. We remark that $q = \frac{p\delta}{p+\delta}$ defined in Theorem 5.21 satisfies $1 < q < p$. If $q \geq 2$ then $n_q = n_p = 1$ and the condition (5.46-*iii*) is clearly satisfied.

Proof of Theorem 5.21. Uniqueness follows from Theorem 5.10.

Existence. Let $t \in [0,T]$ and

$$\begin{aligned}
\beta_t &= t + Q_t + \int_0^t L_s dQ_s + \int_0^t (\ell_s)^2\, ds + \int_0^t |\Phi(s,0,0)|\, dQ_s, \\
\gamma_t &= \beta_t + |\Phi(t,0,0)| + L_t + \ell_t.
\end{aligned}$$

Define, for $n \in \mathbb{N}^*$,

$$\begin{aligned}
L_t^n &= L_t \mathbf{1}_{[0,n]}(\gamma_t), \\
\ell_t^n &= \ell_t \mathbf{1}_{[0,n]}(\gamma_t), \\
\eta_n &= \eta \mathbf{1}_{[0,n]}(\beta_T + |\eta|), \\
\Phi_n(t,y,z) &= \Phi\left(t, y\mathbf{1}_{[0,n]}(\gamma_t), z\mathbf{1}_{[0,n]}(\gamma_t)\right) - \Phi(t,0,0)\ \mathbf{1}_{(n,\infty)}(\gamma_t).
\end{aligned}$$

By Lemma 5.20 we infer that the approximating BSDE

$$Y_t^n = \eta_n + \int_t^T \Phi_n\left(s, Y_s^n, Z_s^n\right) dQ_s - \int_t^T Z_s^n dB_s \tag{5.49}$$

has a unique solution $(Y^n, Z^n) \in S_m^q [0,T] \times \Lambda_{m\times k}^q (0,T)$, for all $q \geq 2$.

Let

$$V_t^n \overset{def}{=} \int_0^t \left(L_s^n dQ_s + \frac{1}{n_p} \left(\ell_s^n\right)^2 ds \right).$$

We have for all $n, i \in \mathbb{N}$

$$0 \leq V_t^n \leq V_t^{n+i} \leq (n+i)^2 + \frac{1}{n_p} (n+i)^3 .$$

Therefore

$$\mathbb{E} \sup_{t\in[0,T]} e^{pV_t^{n+i}} |Y_t^n|^p \leq C_{n,i,p} \left(\mathbb{E} \sup_{t\in[0,T]} |Y_t^n|^{2p} \right)^{1/2} < \infty,$$

and since

$$\begin{aligned} &\left\langle Y_t^n, \Phi_n\left(t, Y_t^n, Z_t^n\right) dQ_t \right\rangle \\ &\leq |Y_t^n| \left|\Phi(t,0,0)\right| \mathbf{1}_{[0,n]}(\gamma_t)\, dQ_t + |Y_t^n|^2 dV_t^n + \frac{n_p}{4} |Z_t^n|^2 dt \\ &\leq |Y_t^n| \left|\Phi(t,0,0)\right| \mathbf{1}_{[0,n]}(\gamma_t)\, dQ_t + |Y_t^n|^2 dV_t^{n+i} + \frac{n_p}{4} |Z_t^n|^2 dt, \end{aligned}$$

we obtain, by Proposition 5.2-A, that

$$\begin{aligned} &\mathbb{E}^{\mathcal{F}_t} \sup_{s\in[t,T]} e^{pV_s^{n+i}} |Y_s^n|^p + \mathbb{E}^{\mathcal{F}_t} \left(\int_t^T e^{2V_s^{n+i}} |Z_s^n|^2 ds \right)^{p/2} \\ &\leq C_q\, \mathbb{E}^{\mathcal{F}_t} \left[e^{pV_T^{n+i}} |\eta_n|^p + \left(\int_t^T e^{V_s^{n+i}} \left|\Phi_n(s,0,0)\right| dQ_s \right)^p \right] \\ &\leq C_q\, \mathbb{E}^{\mathcal{F}_t} \left[e^{pV_T^{n+i}} |\eta|^p + \left(\int_t^T e^{V_s^{n+i}} \left|\Phi(s,0,0)\right| dQ_s \right)^p \right]. \end{aligned}$$

By Beppo Levi's monotone convergence Theorem 1.9 it follows by letting $i \to \infty$ that for all $t \in [0,T]$,

$$\begin{aligned} &\mathbb{E}^{\mathcal{F}_t} \sup_{s\in[t,T]} e^{pV_s} |Y_s^n|^p + \mathbb{E}^{\mathcal{F}_t} \left(\int_t^T e^{2V_s} |Z_s^n|^2 ds \right)^{p/2} \\ &\leq C_p\, \mathbb{E}^{\mathcal{F}_t} \left[e^{pV_T} |\eta|^p + \left(\int_t^T e^{V_s} \left|\Phi(s,0,0)\right| dQ_s \right)^p \right]. \end{aligned} \tag{5.50}$$

Consequently by (5.46–i) for all $n \in \mathbb{N}^*$,

$$\mathbb{E} \sup_{s\in[0,T]} e^{pV_s} |Y_s^n|^p + \mathbb{E} \left(\int_0^T e^{2V_s} |Z_s^n|^2 ds \right)^{p/2} \le C < \infty.$$

Let $\delta > \frac{p}{p-1}$, $q = \frac{p\delta}{p+\delta} \in (1, p)$, $n_q = 1 \wedge (q-1)$ and $n_p = 1 \wedge (p-1)$ satisfy (5.46–ii, iii). If we define

$$\Delta_t = \left(\frac{1}{n_q} - \frac{1}{n_p} \right) \int_0^t (\ell_s)^2 ds \quad \text{and}$$

$$V_t^{(q)} = \int_0^t \left[L_s dQ_s + \frac{1}{n_q} (\ell_s)^2 ds \right] = V_t + \Delta_t,$$

we have for all $n \in \mathbb{N}^*$

$$\begin{aligned}
&\mathbb{E} \sup_{s\in[0,T]} e^{qV_s^{(q)}} |Y_s^n|^q \\
&\le \mathbb{E}\Big[e^{q\Delta_T} \sup_{s\in[0,T]} e^{qV_s} |Y_s^n|^q \Big] \\
&\le \Big[\mathbb{E} e^{\frac{pq}{p-q}\Delta_T} \Big]^{(p-q)/p} \Big[\mathbb{E} \sup_{s\in[0,T]} e^{pV_s} |Y_s^n|^p \Big]^{q/p} \\
&= \Big[\mathbb{E} \exp(\delta \Delta_T) \Big]^{(p-q)/p} \left(\mathbb{E} \sup_{s\in[0,T]} e^{pV_s} |Y_s^n|^p \right)^{q/p} \\
&< \infty.
\end{aligned}$$

Hence for all $n, i \in \mathbb{N}^*$

$$\mathbb{E} \sup_{s\in[0,T]} e^{qV_s^{(q)}} \left| Y_s^n - Y_s^{n+i} \right|^q < \infty.$$

Since

$$\begin{aligned}
&\left\langle Y_s^n - Y_s^{n+i}, \Phi_n\left(t, Y_s^n, Z_s^n\right) - \Phi_{n+i}\left(s, Y_s^{n+i}, Z_s^{n+i}\right)\right\rangle dQ_s \\
&\le \left\langle Y_s^n - Y_s^{n+i}, \Phi\left(s, Y_s^n \mathbf{1}_{[0,n]}(\gamma_s), Z_s^n \mathbf{1}_{[0,n]}(\gamma_s)\right)\right. \\
&\qquad \left. - \Phi\left(s, Y_s^{n+i} \mathbf{1}_{[0,n+i]}(\gamma_s), Z_s^{n+i} \mathbf{1}_{[0,n+i]}(\gamma_s)\right)\right\rangle dQ_s \\
&\qquad - \left\langle Y_s^n - Y_s^{n+i}, \Phi(t,0,0)\right\rangle \left[\mathbf{1}_{(n,\infty)}(\gamma_s) - \mathbf{1}_{(n+i,\infty)}(\gamma_s) \right] dQ_s
\end{aligned}$$

$$\le \left|Y_s^n - Y_s^{n+i}\right| \Big[\left|\Phi(t,0,0)\right| \mathbf{1}_{(n,\infty)}(\gamma_s)\, dQ_s$$

$$+ \left(L_s \left|Y_s^n\right| dQ_s + \ell_s \left|Z_s^n\right| ds\right) \left|\mathbf{1}_{[0,n]}(\gamma_s) - \mathbf{1}_{[0,n+i]}(\gamma_s)\right| \Big]$$

$$+ \left|Y_s^n - Y_s^{n+i}\right|^2 \left(L_s dQ_s + \frac{1}{n_q}\ell_s^2 ds\right) + \frac{n_q}{4}\left|Z_s^n - Z_s^{n+i}\right|^2 ds,$$

by Proposition 5.2-A, we infer that

$$\mathbb{E} \sup_{s\in[0,T]} e^{qV_s^{(q)}} \left|Y_s^n - Y_s^{n+i}\right|^q + \mathbb{E}\left(\int_0^T e^{2V_s^{(q)}} \left|Z_s^n - Z_s^{n+i}\right|^2 ds\right)^{q/2}$$

$$\le C_q\, \mathbb{E}\, e^{qV_T^{(q)}} \left|\eta_n - \eta_{n+i}\right|^q$$

$$+ C_q\, \mathbb{E}\left[\int_0^T \mathbf{1}_{(n,\infty)}(\gamma_s)\, e^{V_s^{(q)}} \left(\left|\Phi(t,0,0)\right| dQ_s + L_s \left|Y_s^n\right| dQ_s + \ell_s \left|Z_s^n\right| ds\right)\right]^q$$

$$\le C_q\, \mathbb{E}\left[e^{qV_T^{(q)}} |\eta|^q\, \mathbf{1}_{(n,\infty)}(\beta_T + |\eta|)\right]$$

$$+ C_q'\, \mathbb{E}\left(\int_0^T e^{V_s^{(q)}} \left|\Phi(t,0,0)\right| \mathbf{1}_{(n,\infty)}(\gamma_s)\, dQ_s\right)^q$$

$$+ C_q'\, \mathbb{E}\left[\left(\int_0^T L_s \mathbf{1}_{(n,\infty)}(\gamma_s)\, ds\right)^q \sup_{s\in[0,T]} e^{qV_s^{(q)}} \left|Y_s^n\right|^q\right]$$

$$+ C_q'\mathbb{E}\left[\left(\int_0^T \ell_s^2 \mathbf{1}_{(n,\infty)}(\gamma_s)\, ds\right)^{q/2} \left(\int_0^T e^{2V_s^{(q)}} \left|Z_s^n\right|^2 ds\right)^{q/2}\right].$$

Note that

(a)

$$\mathbb{E}\left[e^{qV_T^{(q)}} |\eta|^q\, \mathbf{1}_{(n,\infty)}(\beta_T + |\eta|)\right] + \mathbb{E}\left(\int_0^T e^{V_s^{(q)}} \left|\Phi(t,0,0)\right| \mathbf{1}_{(n,\infty)}(\gamma_s)\, dQ_s\right)^q$$

$$\le \left[\mathbb{E}\exp(\delta\Delta_T)\right]^{(p-q)/p} \left(\mathbb{E}\left[e^{pV_T} |\eta|^p\, \mathbf{1}_{(n,\infty)}(\beta_T + |\eta|)\right]\right)^{q/p}$$

$$+ \left[\mathbb{E}\exp(\delta\Delta_T)\right]^{(p-q)/p} \left[\mathbb{E}\left(\int_0^T e^{V_s} \left|\Phi(t,0,0)\right| \mathbf{1}_{(n,\infty)}(\gamma_s)\, dQ_s\right)^p\right]^{q/p}$$

$$\to 0, \quad \text{as } n \to \infty;$$

(b)

$$\begin{aligned}
&\mathbb{E}\left[\left(\int_0^T L_s \mathbf{1}_{(n,\infty)}(\gamma_s)\, ds\right)^q \sup_{s\in[0,T]} e^{qV_s} |Y_s^n|^q\right] \\
&\le \left[\mathbb{E}\left(\int_0^T L_s \mathbf{1}_{(n,\infty)}(\gamma_s)\, ds\right)^{\frac{qp}{p-q}}\right]^{(p-q)/p} \left(\mathbb{E} \sup_{s\in[0,T]} e^{pV_s} |Y_s^n|^p\right)^{q/p} \\
&= \left[\mathbb{E}\left(\int_0^T L_s \mathbf{1}_{(n,\infty)}(\gamma_s)\, ds\right)^{\delta}\right]^{p/(p+\delta)} \left(\mathbb{E} \sup_{s\in[0,T]} e^{pV_s} |Y_s^n|^p\right)^{\delta/(p+\delta)} \\
&\to 0, \quad \text{as } n \to \infty;
\end{aligned}$$

(c)

$$\begin{aligned}
&\mathbb{E}\left[\left(\int_0^T \ell_s^2 \mathbf{1}_{(n,\infty)}(\gamma_s)\, ds\right)^{q/2} \left(\int_0^T e^{2V_s} |Z_s^n|^2\, ds\right)^{q/2}\right] \\
&\le \left[\mathbb{E}\left(\int_0^T \ell_s^2 \mathbf{1}_{(n,\infty)}(\gamma_s)\, ds\right)^{\delta/2}\right]^{p/(p+\delta)} \left[\mathbb{E}\left(\int_0^T e^{2V_s} |Z_s^n|^2\, ds\right)^{p/2}\right]^{\delta/(p+\delta)} \\
&\to 0, \quad \text{as } n \to \infty.
\end{aligned}$$

Taking into account (a), (b) and (c) we deduce that there exists a pair $(Y, Z) \in S_m^0[0, T] \times \Lambda_{m\times k}^0(0, T)$ such that for $q = \frac{p\delta}{p+\delta}$

$$\lim_{n\to\infty} \mathbb{E}\left(\sup_{s\in[0,T]} e^{qV_s} |Y_s^n - Y_s|^q\right) + \mathbb{E}\left[\left(\int_0^T e^{2V_s} |Z_s^n - Z_s|^2\, ds\right)^{q/2}\right] = 0.$$

Now the inequality (5.48) clearly follows from (5.50) by Fatou's Lemma.

Finally passing to the limit in (5.49) we deduce using Lemma 5.16 that (Y, Z) is a solution of BSDE (5.40). ∎

5.3.3.3 BSDEs with Locally Lipschitz Coefficients

For a (forward) SDE, it is not hard to deduce from existence and uniqueness under global Lipschitz conditions an existence and uniqueness result under local Lipschitz conditions, at least until a possible explosion time. The reason is that one just needs to follow each path of the solution.

For BSDEs, the situation is dramatically different. Indeed, in a sense, solving a BSDE amounts to combining the flow of a backward ODE with the operation of taking continuously in time the conditional expectation, given the current σ-algebra

$\mathcal{F}_t$. A backward stochastic differential equation is not solved by following each individual path of the solution. Consequently one cannot a priori deduce an existence and uniqueness result under local Lipschitz conditions from the same result under global Lipschitz conditions. However, this is in fact possible, because we have an a priori bound on the solution. This is what we shall explain in this section.

We consider the BSDE

$$Y_t = \eta + \int_t^T F\left(s, Y_s, Z_s\right) ds - \int_t^T Z_s dB_s, \ t \in [0, T], \ a.s. \tag{5.51}$$

Assume that

▲ $\eta : \Omega \to \mathbb{R}^m$ *is an* $\mathcal{F}_T$*-measurable random vector;*

▲ $F : \Omega \times [0, T] \times \mathbb{R}^m \to \mathbb{R}^m$ *satisfies*

(BSDE-LL):

▲ *for all* $y \in \mathbb{R}^m, z \in \mathbb{R}^{m\times k}$, *the function* $F(\cdot, \cdot, y, z) : [0, T] \to \mathbb{R}^m$ *is* $\mathcal{P}$-m.s.p.

▲ *there exist measurable functions* $\ell, \kappa, \rho : [0, T] \to \mathbb{R}_+$ and $L : [0, T] \times \mathbb{R}_+ \to \mathbb{R}_+$ *satisfying:*

- *L is continuous and increasing in the second variable,*
- $$\int_0^T \left[L(t, q) + \ell^2(t) + \kappa(t) + \rho(t)\right] dt < \infty, \quad \text{for all } q \in \mathbb{R}_+,$$
- *for all* $y, y' \in \mathbb{R}^m, z, z' \in \mathbb{R}^{m\times k}$, dt*-a.e.*

$$\begin{aligned}
&(i) \quad |F(t, y', z) - F(t, y, z)| \le L(t, |y| \vee |y'|)\, |y' - y|, \\
&(ii) \quad |F(t, y, z') - F(t, y, z)| \le \ell(t)\, |z' - z|, \\
&(iii) \quad |F(t, y, 0)| \le \rho(t) + \kappa(t)\, |y|.
\end{aligned} \tag{5.52}$$

Let $p > 1$ and $n_p = 1 \wedge (p - 1)$. Define

$$\tilde{V}(t) = \int_0^t \left(\kappa(s) + \frac{1}{n_p}\ell^2(s)\right) ds.$$

Observe that for all $Y, Y' \in S_m^0[0, T]$ satisfying $|Y'| \le |Y|$ and all $Z \in \Lambda_{m\times k}^0(0, T)$,

$$\begin{aligned}
\left\langle Y_t, \Phi\left(t, Y_t', Z_t\right) dt\right\rangle &\le \left\langle Y_t, \Phi\left(t, Y_t', 0\right) dt\right\rangle + |Y_t|\, \ell(t)\, |Z_t|\, dQ_t \\
&\le |Y_t|\, \rho(t)\, dt + |Y_t|^2\, d\tilde{V}_t + \frac{n_p}{4} |Z_t|^2\, dt.
\end{aligned} \tag{5.53}$$

Note that if $(Y, Z) \in S_m^p[0,T] \times \Lambda_{m\times k}^0(0,T)$ is a solution of (5.51), then by Proposition 5.2-A and the inequality (5.53) there exists a $C_p > 0$ such that, $\mathbb{P}$-$a.s.$, for all $t \in [0,T]$

$$\mathbb{E}^{\mathcal{F}_t} \sup_{r\in[t,T]} \left|e^{\tilde{V}_r} Y_r\right|^p + \mathbb{E}^{\mathcal{F}_t}\left(\int_t^T e^{2\tilde{V}_r}|Z_r|^2 dr\right)^{p/2} \\ \leq C_p\, \mathbb{E}^{\mathcal{F}_t}\left[\left|e^{\tilde{V}(T)}\eta\right|^p + \left(\int_t^T e^{\tilde{V}(r)}\rho(r)\,dr\right)^p\right],$$

which yields $\mathbb{P}$-$a.s.$, for all $t \in [0,T]$:

$$|Y_t| \leq (C_p)^{1/p} e^{\tilde{V}(T)}\left[\left(\mathbb{E}^{\mathcal{F}_t}\,|\eta|^p\right)^{1/p} + \int_0^T \rho(r)\,dr\right] \overset{def}{=} R_t. \tag{5.54}$$

Define the continuous stochastic processes

$$\beta_t = e^{\tilde{V}(T)}\left[\left(\mathbb{E}^{\mathcal{F}_t}\,|\eta|^p\right)^{1/p} + \int_0^T \rho(s)\,ds\right] \tag{5.55}$$

and

$$\Gamma_t(\lambda) = \int_0^t L\left(s, \lambda + \lambda\left(\mathbb{E}^{\mathcal{F}_s}\,|\eta|^p\right)^{1/p}\right)ds, \quad \lambda \geq 1. \tag{5.56}$$

Theorem 5.23. *Let $p > 1$ and the assumption* (**BSDE-LL**) *be satisfied. If there exists a $\delta > \frac{p}{p-1}$ such that for all $\lambda \geq 1$*

$$\mathbb{E}\left[(\Gamma_T(\lambda))^\delta\right] + \mathbb{E}\left|e^{\Gamma_T(\lambda)}\eta\right|^p + \mathbb{E}\left(\int_0^T e^{\Gamma_t(\lambda)}|\Phi(t,0,0)|\,dt\right)^p < \infty, \tag{5.57}$$

then the BSDE (5.51) has a unique solution $(Y, Z) \in S_m^0[0,T] \times \Lambda_{m\times k}^0(0,T)$ such that for all $\lambda \geq 1$,

$$\mathbb{E}\sup_{s\in[0,T]} e^{p\lambda\Gamma_s(\lambda)}|Y_s|^p + \mathbb{E}\left(\int_0^T e^{2\lambda\Gamma_s(\lambda)}|Z_s|^2 ds\right)^{p/2} < \infty. \tag{5.58}$$

In particular $(Y, Z) \in S_m^p[0,T] \times \Lambda_{m\times k}^p(0,T)$ and (5.54) holds; if η is a bounded random variable then there exists a constant $C > 0$ such that $\mathbb{P}$-a.s. $\omega \in \Omega$

$$|Y_t(\omega)| \leq C, \quad \forall t \in [0,T].$$

Proof. Consider the projection operator $\pi : \Omega \times [0,T] \times \mathbb{R}^m \to \mathbb{R}^m$,

$$\pi_t(\omega, y) = \pi(\omega, t, y) = \begin{cases} y, & \text{if } |y| \le R_t(\omega), \\ \dfrac{y}{|y|} R_t(\omega) & \text{if } |y| > R_t(\omega). \end{cases}$$

Note that for all $y, y' \in \mathbb{R}^m$, $\pi(\cdot,\cdot,y)$ is a $\mathcal{P}$-m.c.s.p., $|\pi_t(y)| \le R_t$ and

$$\left|\pi_t(y) - \pi_t(y')\right| \le \left|y - y'\right|.$$

The function $\tilde{\Phi}(s,y,z) \overset{def}{=} \Phi(s, \pi_s(y), z)$ is globally Lipschitz with respect to (y,z):

$$\begin{aligned} \left|\tilde{\Phi}(s,y,z) - \tilde{\Phi}(s,y',z)\right| &= \left|\Phi(s,\pi_s(y),z) - \Phi(s,\pi_s(y'),z)\right| \\ &\le L_s\left(|\pi_s(y)| \vee |\pi_s(y')|\right) |\pi_s(y) - \pi_s(y')| \\ &\le L(s,R_s)\left|y - y'\right| \\ &\le [\kappa(s) + L(s,R_s)]\left|y - y'\right|, \end{aligned}$$

and

$$\begin{aligned} \left|\tilde{\Phi}(s,y,z) - \tilde{\Phi}(s,y,z')\right| &= \left|\Phi(s,\pi_s(y),z) - \Phi(s,\pi_s(y),z')\right| \\ &\le \ell(s)\,|z - z'|. \end{aligned}$$

Then by Theorem 5.21 the BSDE

$$Y_t = \eta + \int_t^T \tilde{\Phi}(s, Y_s, Z_s)\,dQ_s - \int_t^T Z_s dB_s,\ t \in [0,T], \tag{5.59}$$

has a unique solution $(Y,Z) \in S_m^0[0,T] \times \Lambda^0_{m\times k}(0,T)$ satisfying (5.58). Since by (5.53)

$$\begin{aligned} \left\langle Y_t, \tilde{\Phi}(t,Y_t,Z_t)\,dQ_t\right\rangle &= \langle Y_t, \Phi(t,\pi_t(Y_t),Z_t)\,dQ_t\rangle \\ &\le |Y_t|\,\rho_t dQ_t + |Y_t|^2 \kappa_t dQ_t + \frac{n_p}{4}|Z_t|^2 dt \end{aligned}$$

we infer by 5.54 that $|Y_t| \le R_t$ and consequently $\tilde{\Phi}(t,Y_t,Z_t) = \Phi(t,Y_t,Z_t)$, that is (Y,Z) is a solution of the Eq. (5.51). The solution is unique since any solution (Y,Z) of (5.51) satisfies $|Y_t| \le R_t$ and consequently it is a solution of (5.59). ■

5.3.4 BSDEs with Monotone Coefficients

5.3.4.1 The First BSDE: Monotone Coefficient $\Phi(s, Y_s)\, dQ_s$

We first consider the BSDE

$$Y_t = \eta + \int_t^T \Phi(s, Y_s)\, dQ_s - \int_t^T Z_s dB_s, \quad t \in [0, T], \ a.s. \tag{5.60}$$

We assume that

$$\textbf{(BSDE-MH0}_{\Phi}\textbf{)} : \tag{5.61}$$

▲ $\eta : \Omega \to R^m$ *is an* $\mathcal{F}_T$*-measurable random vector;*
▲ Q is a $\mathcal{P}$-m.i.c.s.p. *such that* $Q_0 = 0$;
▲ $\Phi : \Omega \times [0, \infty[\times \mathbb{R}^m \to \mathbb{R}^m$ *satisfies:*

(a) $\forall\ y \in \mathbb{R}^m,\ \Phi(\cdot, \cdot, y) : \Omega \times [0, T] \to \mathbb{R}^m$ *is* $\mathcal{P}$*-measurable;*
(b) *the mapping* $y \longrightarrow \Phi(t, y) : \mathbb{R}^m \to \mathbb{R}^m$ *is continuous;*
(c) *there exist a* $\mathcal{P}$*-m.s.p.* $\mu : \Omega \times [0, T] \to \mathbb{R}$ *such that*

$$\int_0^T |\mu_t| dQ_t < \infty, \ \mathbb{P}\text{-a.s.},$$

and for all $y, y' \in \mathbb{R}^m$, $d\mathbb{P} \otimes dQ_t$*-a.e.*

$$\left\langle y' - y, \Phi(t, y') - \Phi(t, y) \right\rangle \leq \mu_t \left| y' - y \right|^2; \tag{5.62}$$

(d) *for all* $\rho \geq 0$

$$\int_0^T \Phi_\rho^{\#}(s)\, dQ_s < \infty, \ a.s.$$

□

where

$$\Phi_\rho^{\#}(t) \stackrel{def}{=} \sup_{|y| \leq \rho} |\Phi(t, y)|.$$

We recall the notations

$$S_m^{1+}\left([0, T]; e^{\bar{\mu}}\right) = \bigcup\nolimits_{p>1} S_m^p\left([0, T]; e^{\bar{\mu}}\right)$$

and

$$\bar{\mu}_t = \int_0^t \mu_s dQ_s, \quad \hat{\mu}_t = \int_0^t \mu_s^+ dQ_s.$$

Proposition 5.24. *Let $p \geq 1$ and the assumptions (5.61-**BSDE-MH0**$_\Phi$) be satisfied. If for all $\rho > 0$*

$$\mathbb{E} \left|e^{\bar{\mu}_T}\eta\right|^p + \mathbb{E} \left(\int_0^T e^{\hat{\mu}_s} \Phi_\rho^\# (s)\, dQ_s \right)^p < \infty \tag{5.63}$$

then the BSDE (5.60) has a unique solution $(Y, Z) \in S_m^1\left([0,T]; e^{\bar{\mu}}\right) \times \Lambda_{m\times k}^0\left(0, T; e^{\bar{\mu}}\right)$. Moreover

$$(j)\quad |Y_t| \leq \mathbb{E}^{\mathcal{F}_t} \left|e^{\bar{\mu}_T - \bar{\mu}_t}\eta\right| + \mathbb{E}^{\mathcal{F}_t} \int_t^T e^{\bar{\mu}_s - \bar{\mu}_t} \left|\Phi(s,0)\right| dQ_s,$$
$$\forall t \in [0,T], \ \mathbb{P}\text{-a.s.},$$

$$\begin{aligned}(jj)\quad & \sup_{s\in[0,T]} \left(\mathbb{E}\, e^{\bar{\mu}_s} |Y_s|\right)^q + \mathbb{E} \sup_{s\in[0,T]} \left|e^{\bar{\mu}_s} Y_s\right|^q + \mathbb{E}\left(\int_0^T e^{2\bar{\mu}_s} |Z_s|^2 ds\right)^{q/2}, \\ & \leq C_q \left(\mathbb{E}\left(e^{\bar{\mu}_T}|\eta|\right)\right)^q + \left(\mathbb{E}\int_0^T e^{\bar{\mu}_s} \left|\Phi(s,0)\right| dQ_s\right)^q, \quad \forall\, q \in (0,1),\end{aligned} \tag{5.64}$$

and for $p > 1$ there exists a positive C_p (depending only on p) such that, $\mathbb{P}$-a.s., for all $t \in [0,T]$:

$$\begin{aligned}& \mathbb{E}^{\mathcal{F}_t} \sup_{s\in[t,T]} \left|e^{\bar{\mu}_s} Y_s\right|^p + \mathbb{E}^{\mathcal{F}_t} \left(\int_t^T e^{2\bar{\mu}_s} |Z_s|^2 ds\right)^{p/2} \\ & \leq C_p \left[\mathbb{E}^{\mathcal{F}_t} \left|e^{\bar{\mu}_T}\eta\right|^p + \mathbb{E}^{\mathcal{F}_t} \left(\int_t^T e^{\bar{\mu}_s} \left|\Phi(s,0)\right| dQ_s\right)^p\right].\end{aligned} \tag{5.65}$$

Remark 5.25. If $(\bar{\mu}_t)_{t\geq 0}$ is a deterministic process then the assumption (5.63) is equivalent to

$$\mathbb{E}\left(|\eta|^p\right) + \mathbb{E}\left(\int_0^T \Phi_\rho^\#(s) dQ_s\right)^p < \infty,$$

and the inequality (5.65) yields: for all $t \in [0,T]$

$$|Y_t| \leq e^{2\|\bar{\mu}\|_T} \left[\mathbb{E}^{\mathcal{F}_t} |\eta| + \mathbb{E}^{\mathcal{F}_t} \int_t^T \left|\Phi(s,0)\right| dQ_s\right].$$

Proof of Proposition 5.24. (I) Uniqueness follows from Theorem 5.10 and Theorem 5.11. If $(Y,Z) \in S_m^1\left([0,T]\,;e^{\bar{\mu}}\right) \times \Lambda_{m\times k}^0\left(0,T;e^{\bar{\mu}}\right)$ is a solution, then by Proposition 5.2 and

$$\langle Y_s, \Phi(s,Y_s)\,dQ_s\rangle \leq |\Phi(s,0)|\,|Y_s|\,dQ_s + \mu_s\,|Y_s|^2\,dQ_s$$

the inequalities (5.64-j,jj) follow.

To prove the existence of the solution we write the equation in the form, $\mathbb{P}$-a.s.

$$Y_t = \eta + \int_t^T [F(s,Y_s) + \mu_s Y_s]\,dQ_s - \int_t^T Z_s dB_s, \quad t \in [0,T], \tag{5.66}$$

where

$$F(s,y) = \Phi(s,y) - \mu_s y.$$

We remark that $\hat{\mu}_t - \hat{\mu}_s = \int_s^t \mu_r^+ dQ_r \geq \int_s^t \mu_r dQ_r = \bar{\mu}_t - \bar{\mu}_s$.

$(II\text{-}a)$ *Existence in the case: there exist $b, c > 0$ such that for all $t \in [0,T]$*

$$|\eta| + |\Phi(t,0)| + \left|e^{\hat{\mu}_T}\eta\right| + \int_0^T e^{\hat{\mu}_s}\,|\Phi(s,0)|\,dQ_s \leq b, \ a.s., \tag{5.67}$$

and

$$Q_t + |\bar{\mu}_t| + |\mu_t| + \Phi_b^{\#}(t) \leq c, \ a.s. \tag{5.68}$$

Step 1. Yosida approximation of $-F$.

Since $y \mapsto -F(t,y) = \mu_t y - \Phi(t,y) : \mathbb{R}^m \to \mathbb{R}^m$ is a monotone continuous operator (hence also a maximal monotone operator), it follows that for every $(\omega,t,y) \in \Omega \times [0,T] \times \mathbb{R}^m$ and $\varepsilon > 0$ there exists a unique $F_\varepsilon = F_\varepsilon(\omega,t,y) \in \mathbb{R}^m$ such that

$$F(\omega,t,y+\varepsilon F_\varepsilon) = F_\varepsilon.$$

From Annex B, Propositions 6.7 and 6.8, recall that $F_\varepsilon(\cdot,\cdot,y) : \Omega\times[0,T] \to \mathbb{R}^m$ is $\mathcal{P}$-m.s.p. for every $y \in \mathbb{R}^m$ and
$\forall\, \varepsilon,\delta > 0,\ \forall\, t \in [0,T], \forall\, y,y' \in \mathbb{R}^m, \quad a.s.$

$$\begin{aligned}
&(a) \quad \langle F_\varepsilon(t,y) - F_\varepsilon(t,y'), y - y'\rangle \leq 0,\\
&(b) \quad |F_\varepsilon(t,y) - F_\varepsilon(t,y')| \leq \frac{2}{\varepsilon}|y - y'|,\\
&(c) \quad |F_\varepsilon(t,y)| \leq |F(t,y)|, \qquad \lim_{\varepsilon\to 0} F_\varepsilon(t,y) = F(t,y),
\end{aligned}$$

and

$$\left\langle y-y',F_{\varepsilon}(t,y)-F_{\delta}(t,y')\right\rangle \leq (\varepsilon+\delta)\left\langle F_{\varepsilon}(t,y),F_{\delta}(t,y')\right\rangle .$$

Moreover, if $|y|\leq b$ then

$$|F_{\varepsilon}(t,y)|\leq |\mu_t|\,b+\Phi_b^{\#}(t)\,. \tag{5.69}$$

Step 2. Approximating equation.
Let $0<\varepsilon\leq 1$. Since $y\longmapsto F_{\varepsilon}(r,y)+\mu_r y$ is a Lipschitz function with the Lipschitz constants $L_t=\dfrac{2}{\varepsilon}+c$ and $\ell_t=0$ we infer by Theorem 5.21 that the approximating equation

$$Y_t^{\varepsilon}=\eta+\int_t^T\left[F_{\varepsilon}\left(r,Y_r^{\varepsilon}\right)+\mu_r Y_r^{\varepsilon}\right]dQ_r-\int_t^T Z_r^{\varepsilon}dB_r \tag{5.70}$$

has a unique solution $(Y^{\varepsilon},Z^{\varepsilon})\in S_m^q[0,T]\times\Lambda_{m\times k}^q(0,T)$ for all $q>1$.
Step 3. Boundedness of $(Y^{\varepsilon},Z^{\varepsilon})_{0<\varepsilon\leq 1}$.
We denote by C,C' generic constants independent of $\varepsilon,\delta\in]0,1]$. Since

$$\mathbb{E}\sup_{t\in[0,T]}e^{q\bar{\mu}_t}\left|Y_t^{\varepsilon}\right|^q\leq e^{qc}\mathbb{E}\sup_{t\in[0,T]}\left|Y_t^{\varepsilon}\right|^q<\infty$$

and

$$\begin{aligned}\left\langle Y_s^{\varepsilon},\left[F_{\varepsilon}\left(s,Y_s^{\varepsilon}\right)+\mu_s Y_s^{\varepsilon}\right]dQ_s\right\rangle &\leq |F_{\varepsilon}(s,0)|\,|Y_s^{\varepsilon}|\,dQ_s+\mu_s\left|Y_s^{\varepsilon}\right|^2dQ_s\\ &\leq |\Phi(s,0)|\,|Y_s^{\varepsilon}|\,dQ_s+\mu_s\left|Y_s^{\varepsilon}\right|^2dQ_s\\ &\leq |\Phi(s,0)|\,|Y_s^{\varepsilon}|\,dQ_s+\mu_s^+\left|Y_s^{\varepsilon}\right|^2dQ_s\end{aligned}$$

we deduce, by Proposition 5.2, that for $p>1$ and for all $t\in[0,T]$:

$$\begin{aligned}&\mathbb{E}^{\mathcal{F}_t}\sup_{s\in[t,T]}\left|e^{\hat{\mu}_s}Y_s^{\varepsilon}\right|^p+\mathbb{E}^{\mathcal{F}_t}\left(\int_t^T e^{\hat{\mu}_s}\left|Z_s^{\varepsilon}\right|^2ds\right)^{p/2}\\ &\quad\leq C_p\,\mathbb{E}^{\mathcal{F}_t}\left[\left|e^{\hat{\mu}_T}\eta\right|^p+\left(\int_t^T e^{\hat{\mu}_s}\left|\Phi(s,0)\right|dQ_s\right)^p\right]\leq C_p b^p,\end{aligned} \tag{5.71}$$

and (5.65) with (Y,Z) replaced by $(Y^{\varepsilon},Z^{\varepsilon})$.
By Corollary 6.81, for $p=1$, we have $\mathbb{P}$-$a.s.$, for all $t\in[0,T]$:

$$\left|Y_t^{\varepsilon}\right|\leq e^{\hat{\mu}_t}\left|Y_t^{\varepsilon}\right|\leq\mathbb{E}^{\mathcal{F}_t}\left|e^{\hat{\mu}_T}\eta\right|+\mathbb{E}^{\mathcal{F}_t}\int_t^T e^{\hat{\mu}_s}\left|\Phi(s,0)\right|dQ_s\leq b. \tag{5.72}$$

Now from (5.69) we deduce

$$\begin{aligned}\left|F_\varepsilon\left(r,Y_r^\varepsilon\right)+\mu_r Y_r^\varepsilon\right| &\le \left|F_\varepsilon\left(r,Y_r^\varepsilon\right)\right|+\left|\mu_r Y_r^\varepsilon\right|\\ &\le \Phi_b^{\#}(t)+2\left|\mu_r\right|b\\ &\le c+2cb.\end{aligned}$$

Step 4. $(Y^\varepsilon, Z^\varepsilon)$ *is a Cauchy sequence in* $S_m^q[0,T]\times\Lambda_{m\times k}^q(0,T)$, $q>0$.
Let $0<\varepsilon,\delta\le 1$. We have

$$Y_t^\varepsilon - Y_t^\delta = \int_t^T dK_s^{\varepsilon,\delta} - \int_t^T \left(Z_s^\varepsilon - Z_s^\delta\right) dB_s,$$

where

$$K_t^{\varepsilon,\delta} = \int_0^t \left(F_\varepsilon(s,Y_s^\varepsilon)+\mu_s Y_s^\varepsilon - F_\delta(s,Y_s^\delta)-\mu_s Y_s^\delta)\right) dQ_s.$$

Note that

$$\begin{aligned}&\left\langle Y_s^\varepsilon - Y_s^\delta, dK_s^{\varepsilon,\delta}\right\rangle\\ &=\left\langle Y_s^\varepsilon - Y_s^\delta, F_\varepsilon(s,Y_s^\varepsilon)-F_\delta(s,Y_s^\delta)\right\rangle dQ_s + \mu_s\left|Y_s^\varepsilon - Y_s^\delta\right|^2 dQ_s\\ &\le(\varepsilon+\delta)\left\langle F_\varepsilon(s,Y_s^\varepsilon), F_\delta(s,Y_s^\delta)\right\rangle dQ_s + \mu_s\left|Y_s^\varepsilon - Y_s^\delta\right|^2 dQ_s\\ &\le(\varepsilon+\delta)\,c^2 dQ_s + c\left|Y_s^\varepsilon - Y_s^\delta\right|^2 dQ_s.\end{aligned}$$

Since $0\le Q_t\le c$ and for every $q>1$

$$\mathbb{E}\sup_{t\in[0,T]} e^{qcQ_t}\left|Y_t^\varepsilon - Y_s^\delta\right|^q<\infty,$$

we infer from Proposition 5.2 with $D=N=0$, $\lambda=0$, that for $q\ge 2$,

$$\mathbb{E}\sup_{s\in[0,T]}\left|Y_s^\varepsilon - Y_s^\delta\right|^q+\mathbb{E}\left(\int_0^T\left|Z_s^\varepsilon - Z_s^\delta\right|^2 ds\right)^{q/2}\le C\left(\varepsilon+\delta\right)^{q/2}.$$

For $0<q<2$ we have

$$\begin{aligned}&\mathbb{E}\sup_{s\in[0,T]}\left|Y_s^\varepsilon - Y_s^\delta\right|^q+\mathbb{E}\left(\int_0^T\left|Z_s^\varepsilon - Z_s^\delta\right|^2 ds\right)^{q/2}\\ &\le\left(\mathbb{E}\sup_{s\in[0,T]}\left|Y_s^\varepsilon - Y_s^\delta\right|^{q+2}\right)^{q/(q+2)}+\left[\mathbb{E}\left(\int_0^T\left|Z_s^\varepsilon - Z_s^\delta\right|^2 ds\right)^{\frac{q+2}{2}}\right]^{q/(q+2)}\\ &\le C'\left(\varepsilon+\delta\right)^{q/2}.\end{aligned}$$

Hence there exists $(Y, Z) \in \bigcap_{q>0} S_m^q [0, T] \times \Lambda_{m\times k}^q (0, T)$ such that

$$\mathbb{E} \sup_{s\in[0,T]} |Y_s^\varepsilon - Y_s|^q + \mathbb{E} \left(\int_0^T |Z_s^\varepsilon - Z_s|^2 ds \right)^{q/2} \le C\, \varepsilon^{q/2}.$$

Note that

$$F_\varepsilon \left(r, Y_r^\varepsilon\right) + \mu_r Y_r^\varepsilon = \Phi \left(r, Y_r^\varepsilon + \varepsilon F_\varepsilon \left(r, Y_r^\varepsilon\right)\right) - \varepsilon \mu_r F_\varepsilon \left(r, Y_r^\varepsilon\right)$$

and $\left|F_\varepsilon \left(r, Y_r^\varepsilon\right)\right| + |\mu_r Y_r^\varepsilon| \le C$.

Passing to the limit as $\varepsilon \to 0_+$ in the approximating equation (5.70), we infer, by Lebesgue's dominated convergence theorem, that (Y, Z) is a solution of the BSDE (5.60). Moreover passing to the limit on a subsequence, by Fatou's Lemma we clearly infer that (Y, Z) satisfies (5.65), (5.64) and

$$\begin{aligned} &(j) \quad \mathbb{E} \sup_{s\in[0,T]} \left|e^{\hat{\mu}_s} Y_s\right|^p + \mathbb{E} \left(\int_0^T e^{\hat{\mu}_s} |Z_s|^2 ds \right)^{p/2} \le C_p b^p, \text{ if } p > 1, \\ &(jj) \quad |Y_t| \le \left|e^{\hat{\mu}_t} Y_t\right| \le b, \quad \text{for all } t \in [0, T], \mathbb{P}\text{-a.s.}, \end{aligned} \tag{5.73}$$

since the same inequalities hold for $(Y^\varepsilon, Z^\varepsilon)$.
(II-b) Existence under the assumption (5.67), but without (5.68).
Let

$$\tau_n = \inf\{t \in [0, T] : Q_t \ge n\} \quad \text{and} \quad Q_t^n = Q_{t\wedge\tau_n}.$$

Let $\zeta_t = Q_t + |\bar{\mu}_t| + |\mu_t| + \Phi_b^\#(t)$ and $\hat{\mu}_t = \int_0^t \mu_s^+ dQ_s$.
Since

$$\begin{aligned} \left\langle u - v, \Phi(r, u) \mathbf{1}_{\zeta_r<n} - \Phi(r, v) \mathbf{1}_{\zeta_r<n} \right\rangle &\le \mu_r \mathbf{1}_{\zeta_r<n} |u - v|^2 \\ &\le \mu_r^+ |u - v|^2 \end{aligned}$$

by the step *(II-a)* the BSDE

$$\begin{aligned} Y_t^n &= \eta + \int_t^T \Phi\left(r, Y_r^n\right) \mathbf{1}_{\zeta_r<n} dQ_r - \int_t^T Z_r^n dB_r \\ &= \eta + \int_t^T \Phi\left(r, Y_r^n\right) \mathbf{1}_{\zeta_r<n} dQ_r^n - \int_t^T Z_r^n dB_r, \ t \in [0, T] \end{aligned}$$

has a unique solution $(Y^n, Z^n) \in \bigcap_{q>0} S_m^q [0, T] \times \Lambda_{m\times k}^q (0, T)$. The solution (Y^n, Z^n) satisfies (5.65), (5.64) and (5.73) with (Y, Z) replaced by (Y^n, Z^n). Note that from (5.73) written for (Y^n, Z^n) we have for $p \ge 1$,

$$\mathbb{E}\sup_{s\in[0,T]}\left|e^{\hat{\mu}_s}\left(Y_s^n - Y_s^{n+i}\right)\right|^p < \infty.$$

Since

$$\begin{aligned}
&\left\langle Y_t^n - Y_t^{n+i}, \left[\Phi\left(t, Y_t^n\right)\mathbf{1}_{\zeta_t<n} - \Phi\left(t, Y_t^{n+i}\right)\mathbf{1}_{\zeta_t<n+i}\right]dQ_t\right\rangle \\
&\quad\leq \left\langle Y_t^n - Y_t^{n+i}, \Phi\left(t, Y_t^n\right)\left(\mathbf{1}_{\zeta_t<n} - \mathbf{1}_{\zeta_t<n+i}\right)\right\rangle dQ_t + \mu_t \mathbf{1}_{\zeta_t<n+i}\left|Y_t^n - Y_t^{n+i}\right|^2 dQ_t \\
&\quad\leq \left|Y_t^n - Y_t^{n+i}\right|\mathbf{1}_{\zeta_t\geq n}\Phi_b^{\#}(t)\,dQ_t + \mu_t^+\left|Y_t^n - Y_t^{n+i}\right|dQ_t,
\end{aligned}$$

we conclude by Corollary 6.81, that for $q = p$ if $p > 1$ and $q \in (0,1)$ if $p = 1$ there exists a constant C_q such that $\mathbb{P}$-$a.s.$,

$$\begin{aligned}
&\mathbb{E}\sup_{s\in[0,T]} e^{q\hat{\mu}_s}\left|Y_s^n - Y_s^{n+i}\right|^q + \mathbb{E}\left(\int_0^T e^{2\hat{\mu}_s}\left|Z_s^n - Z_s^{n+i}\right|^2 ds\right)^{q/2} \\
&\leq C_q\Big[\left(\mathbb{E}\int_0^T e^{\hat{\mu}_s}\mathbf{1}_{\zeta_s\geq n}\Phi_b^{\#}(s)\,dQ_s\right)^q \mathbf{1}_{p=1} + \mathbb{E}\left(\int_0^T .e^{\hat{\mu}_s}\mathbf{1}_{\zeta_s\geq n}\Phi_b^{\#}(s)\,dQ_s\right)^q \mathbf{1}_{p>1}\Big].
\end{aligned}$$

Taking into account (5.63) we deduce that there exists a pair $(Y,Z) \in S_m^q\left([0,T];e^{\hat{\mu}}\right) \times \Lambda_{m\times k}^q\left(0,T;e^{\hat{\mu}}\right)$ such that as $n \to \infty$

$$\lim_{n\to\infty}\mathbb{E}\sup_{s\in[0,T]} e^{q\hat{\mu}_s}\left|Y_s^n - Y_s\right|^q + \mathbb{E}\left(\int_0^T e^{2\hat{\mu}_s}\left|Z_s^n - Z_s\right|^2 ds\right)^{q/2} = 0.$$

Now using Lemma 5.16 we infer that (Y,Z) is a solution of the BSDE (5.76). We deduce by Fatou's Lemma from the inequalities (5.65), (5.64) and (5.73) written for (Y^n, Z^n) that the same inequalities hold for the limit (Y,Z).

(II-c) Existence without the two assumptions (5.67) and (5.68).
Let

$$\beta_t \overset{def}{=} Q_t + \int_0^t |\mu_s|\,dQ_s + \int_0^t |\Phi(s,0)|\,dQ_s.$$

Define, for $n \in \mathbb{N}^*$,

$$\begin{aligned}
\eta_n &= \eta\,\mathbf{1}_{[0,n]}\left(\beta_T + |\eta|\right), \\
\Phi_n(t,y) &= \Phi(t,y) - \Phi(t,0)\;\mathbf{1}_{[n,\infty[}\left(\beta_t + |\Phi(t,0)|\right).
\end{aligned}$$

The condition (5.67) is satisfied:

$$\begin{aligned}
&|\eta_n| + |\Phi_n(t,0)| + \left|e^{\hat{\mu}_T - \hat{\mu}_t}\eta_n\right| + \int_t^T e^{\hat{\mu}_s - \hat{\mu}_t}\left|\Phi_n(s,0)\right|dQ_s \\
&\leq b_n = n + n + e^n n + e^n nT
\end{aligned}$$

and consequently by part (*II-b*) of this proof there exists a unique pair $(Y^n, Z^n) \in S_m^p\left([0,T]; e^{\hat{\mu}}\right) \times \Lambda_{m\times k}^p\left(0,T; e^{\hat{\mu}}\right)$ such that

$$Y_t^n = \eta_n + \int_t^T \Phi_n\left(s, Y_s^n\right) dQ_s - \int_t^T Z_s^n dB_s,\ t \in [0,T],\ a.s. \tag{5.74}$$

and the inequalities (5.65), (5.64) for (Y^n, Z^n) in the place of (Y, Z) hold. Since

$$\begin{aligned}&\left\langle Y_s^n - Y_s^{n+i}, \Phi_n\left(s, Y_s^n\right) dQ_s - \Phi_{n+i}\left(s, Y_s^{n+i}\right) dQ_s\right\rangle \\ &\le \left|Y_s^n - Y_s^{n+i}\right| \left|\Phi(s,0)\right| \mathbf{1}_{[n,\infty[}\left(\beta_s + \left|\Phi(s,0)\right|\right) dQ_s + \mu_s \left|Y_s^n - Y_s^{n+i}\right|^2 dQ_s\end{aligned}$$

we deduce from Corollary 6.81 that in the case $p > 1$ we have

$$\begin{aligned}&\mathbb{E} \sup_{s\in[0,T]} e^{p\bar{\mu}_s} \left|Y_s^n - Y_s^{n+i}\right|^p + \mathbb{E}\left(\int_0^T e^{2\bar{\mu}_s} \left|Z_s^n - Z_s^{n+i}\right|^2 ds\right)^{p/2} \\ &\le C_p \mathbb{E}\left(e^{p\bar{\mu}_T} \left|\eta_n - \eta_{n+i}\right|^p\right) + C_p \mathbb{E}\left(\int_0^T e^{\bar{\mu}_s} \mathbf{1}_{\beta_s + |\Phi(s,0)| \ge n} \left|\Phi(s,0)\right| dQ_s\right)^p\end{aligned}$$

and in the case $p = 1$

$$\begin{aligned}&\sup_{s\in[0,T]} \left(\mathbb{E}\, e^{\bar{\mu}_s} \left|Y_s^n - Y_s^{n+i}\right|\right)^q + \mathbb{E} \sup_{s\in[0,T]} e^{q\bar{\mu}_s} \left|Y_s^n - Y_s^{n+i}\right|^q \\ &+ \mathbb{E}\left(\int_0^T e^{2\bar{\mu}_s} \left|Z_s^n - Z_s^{n+i}\right|^2 ds\right)^{q/2} \\ &\le C_q \left(\mathbb{E} e^{\bar{\mu}_T} \left|\eta_n - \eta_{n+i}\right|\right)^q + C_q \left(\mathbb{E} \int_0^T e^{\bar{\mu}_s} \mathbf{1}_{\beta_s + |\Phi(s,0)| \ge n} \left|\Phi(s,0)\right| dQ_s\right)^q\end{aligned}$$

for all $0 < q < 1$.

Hence for every $p \ge 1$ there exists $(Y, Z) \in S_m^q\left([0,T]; e^{\bar{\mu}}\right) \times \Lambda_{m\times k}^q\left(0,T; e^{\bar{\mu}}\right)$ (with $q = p$ if $p > 1$, and $0 < q < 1$ if $p = 1$) such that

$$\lim_{n\to\infty} \mathbb{E} \sup_{s\in[0,T]} e^{q\bar{\mu}_s} \left|Y_s^n - Y_s\right|^p + \mathbb{E}\left(\int_0^T e^{2\bar{\mu}_s} \left|Z_s^n - Z_s\right|^2 ds\right)^{q/2} = 0.$$

Using Fatou's Lemma, the inequalities (5.65) and (5.64) follow from the same inequalities written for (Y^n, Z^n). By Lemma 5.16 we infer that (Y, Z) is a solution of the BSDE (5.60). ■

Corollary 5.26. *Let* $p \ge 1$. *If in Proposition 5.24 we replace the assumption (5.63) by*

$$\mathbb{E}\, e^{p\bar{\mu}_T} |\eta|^p + \mathbb{E}\left(\int_0^T \sup_{|y|\le\rho} \left|e^{\bar{\mu}_s}\Phi\left(s, e^{-\bar{\mu}_s}y\right) - \mu_s y\right| dQ_s\right)^p < \infty, \ \forall \rho \ge 0, \tag{5.75}$$

then the same conclusions follow.

Proof. We remark that (Y, Z) solves the BSDE (5.60) if and only if $(\tilde{Y}_t, \tilde{Z}_t) := (e^{\bar{\mu}_t} Y_t, e^{\bar{\mu}_t} Z_t)$ is solution of the BSDE

$$\tilde{Y}_t = \tilde{\eta} + \int_t^T \tilde{\Phi}\left(s, \tilde{Y}_s\right) dQ_s - \int_t^T \tilde{Z}_s dB_s, \ \ t \in [0, T], \ a.s.$$

with

$$\begin{aligned}\tilde{\eta} &= e^{\bar{\mu}_T}\eta,\\ \tilde{\Phi}(t, y) &= -\mu_t y + e^{\bar{\mu}_t}\Phi(t, e^{-\bar{\mu}_t} y).\end{aligned}$$

Note that $\tilde{\eta}$ and $\tilde{\Phi}$ satisfy the same assumptions (5.61-**BSDE-MH0**$_\Phi$) as η and Φ, respectively, but with (5.62) replaced by

$$\left\langle y' - y, \tilde{\Phi}(t, y') - \tilde{\Phi}(t, y)\right\rangle \le 0, \ \ \mathbb{P}\text{-a.s.}$$

and consequently the corresponding $\bar{\mu}$ and $\hat{\mu}$ for $\tilde{\Phi}$ are equal to 0. Therefore the condition (5.63) for $(\tilde{\eta}, \tilde{\Phi})$ means precisely (5.75). ■

5.3.4.2 The Second BSDE: Monotone Coefficient $F\,(t, Y_t, Z_t)\,dt$

In this subsection we study the BSDE

$$Y_t = \eta + \int_t^T F\left(s, Y_s, Z_s\right) ds - \int_t^T Z_s dB_s, \ \ a.s., \ \ t \in [0, T]. \tag{5.76}$$

We shall assume:

$$(\textbf{BSDE-MH}_F): \tag{5.77}$$

- ♦ $\eta : \Omega \to \mathbb{R}^m$ *is an* $\mathcal{F}_T$*-measurable random vector;*
- ♦ *the function* $F(\cdot, \cdot, y, z) : \Omega \times [0, T] \to \mathbb{R}^m$ *is* $\mathcal{P}$*-measurable for every* $(y, z) \in \mathbb{R}^m \times \mathbb{R}^{m\times k}$*;*
- ♦ *there exist some deterministic functions* $\mu \in L^1(0, T; \mathbb{R})$ *and* $\ell \in L^2(0, T; \mathbb{R})$ *such that*

$$\left| \begin{array}{ll} (I) & \text{for all } y, y' \in \mathbb{R}^m,\ z, z' \in \mathbb{R}^{m\times k},\ d\mathbb{P} \otimes dt\text{-a.e.}: \\ & \quad \textit{Continuity:} \\ (C_y) & \quad y \longrightarrow F(t, y, z) : \mathbb{R}^m \to \mathbb{R}^m \textit{ is continuous;} \\ & \quad \textit{Monotonicity condition:} \\ (M_y) & \quad \langle y' - y, F(t, y', z) - F(t, y, z) \rangle \le \mu(t) |y' - y|^2; \\ & \quad \textit{Lipschitz condition:} \\ (L_z) & \quad |F(t, y, z') - F(t, y, z)| \le \ell(t)\ |z' - z|; \\ (II) & \textit{Boundedness condition:} \\ & (B_F)\ \int_0^T F_\rho^\#(t)\, dt < \infty,\ \ a.s.,\quad \forall \rho \ge 0, \end{array} \right. \tag{5.78}$$

□

where

$$F_\rho^\#(t) = \sup\{|F(t, y, 0)| : |y| \le \rho\}.$$

Theorem 5.27. *Let $p > 1$ and the assumptions (5.77-**BSDE-MH**$_F$) be satisfied. If for all $\rho \ge 0$:*

$$\mathbb{E}\ |\eta|^p + \mathbb{E}\left(\int_0^T F_\rho^\#(t)dt\right)^p < \infty,$$

then the BSDE (5.76):

$$Y_t = \eta + \int_t^T F(s, Y_s, Z_s)\, ds - \int_t^T Z_s dB_s,\quad a.s.$$

has a unique solution $(Y, Z) \in S_m^p[0, T] \times \Lambda_{m\times k}^p(0, T)$. Moreover, uniqueness holds in $S_m^{1+}[0, T] \times \Lambda_{m\times k}^0(0, T)$, where

$$S_m^{1+}[0, T] \stackrel{def}{=} \bigcup_{p>1} S_m^p[0, T].$$

Proof. The uniqueness is proved in Corollary 5.13. Let us prove existence.

We use again a contraction argument, which is slightly different from that in the proof of Theorem 5.17.

Note that a solution of the Eq. (5.76) is a fixed point of the mapping $\Gamma : S_m^p[0, T] \times \Lambda_{m\times k}^p(0, T) \to S_m^p[0, T] \times \Lambda_{m\times k}^p(0, T)$ defined by

$$(Y, Z) = \Gamma(X, U),$$

where

$$Y_t = \eta + \int_t^T F(r, Y_r, U_r)\, dr - \int_t^T Z_r dB_r,\ a.s.\ t \in [0, T].$$

By Proposition 5.24 and Remark 5.25, with $\Phi(\omega,t,y) = F(\omega,t,y,U_t(\omega))$, the mapping Γ is well defined since for all $\rho \geq 0$

$$\begin{aligned}
&\mathbb{E}\left(\int_0^T \Phi_\rho^\#(t)\,dt\right)^p \\
&\leq \mathbb{E}\left(\int_0^T \sup_{|y|\leq\rho} |F(t,y,0)|\,dt + \int_0^T \ell(t)\,|U_t|\,dt\right)^p \\
&\leq 2^{p-1}\mathbb{E}\left(\int_0^T F_\rho^\#(t)\,dt\right)^p + 2^{p-1}\left(\int_0^T \ell^2(t)\,dt\right)^{p/2} \mathbb{E}\left(\int_0^T |U_t|^2\,dt\right)^{p/2} \\
&< \infty.
\end{aligned}$$

Let $M \in \mathbb{N}^*$ and $0 = T_0 < T_1 < \cdots < T_M = T$, with $T_i = \dfrac{iT}{M}$. Since the function $t \longmapsto \int_0^t \ell^2(r)\,dr : [0,T] \to \mathbb{R}_+$ is uniformly continuous, we see that

$$\boldsymbol{\alpha}\left(\tfrac{T}{M}\right) \stackrel{def}{=} \sup_{0<s-t<\frac{T}{M}} \int_t^s \ell^2(r)\,dr \to 0, \quad \text{as } M \to \infty.$$

First, we show that the Eq. (5.76) has a unique solution on $[T_{M-1},T]$ in the Banach space $S_m^p[T_{M-1},T] \times \Lambda_{m\times k}^p(T_{M-1},T)$.

Let

$$\bar{\mu}(t) = \int_0^t \mu(r)\,dr.$$

To this end it is sufficient to prove that Γ is a strict contraction on the space $S_m^p[T_{M-1},T] \times \Lambda_{m\times k}^p(T_{M-1},T)$ with respect to the (equivalent) norm $|||(Y,Z)|||_M$

$$|||(Y,Z)|||_M^p \stackrel{def}{=} \mathbb{E}\left[\left(\sup_{r\in[T_{M-1},T]} e^{p\bar{\mu}(r)}|Y_r|^p\right) + \left(\int_{T_{M-1}}^T e^{2\bar{\mu}(r)}|Z_r|^2\,dr\right)^{p/2}\right],$$

for M large enough.

Let $(X,U),(X',U') \in S_m^p[0,T] \times \Lambda_{m\times k}^p(0,T)$. Then

$$Y_t - Y_t' = \int_t^T dK_r - \int_t^T (Z_r - Z_r')\,dB_r, \quad t \in [T_{M-1},T],$$

where

$$K_t = \int_0^t \left[F(r,Y_r,U_r) - F\left(r,Y_r',U_r'\right)\right]dr.$$

Since

$$\begin{aligned}\langle Y_r - Y_r', dK_r\rangle &\le \left\langle Y_r - Y_r', F\left(r, Y_r', U_r\right) - F\left(r, Y_r', U_r'\right)\right\rangle dr + \mu(r)\left|Y_r - Y_r'\right|^2 dr \\ &\le \ell(r)\left|U_r - U_r'\right|\left|Y_r - Y_r'\right| dr + \left|Y_r - Y_r'\right|^2 d\bar{\mu}(r)\end{aligned}$$

and

$$\mathbb{E}\left(\sup_{r\in[0,T]} e^{p\bar{\mu}(r)}\left|Y_r - Y_r'\right|^p\right) < \infty,$$

we have, by Proposition 5.2 with $[t,T]$ replaced by $[T_{M-1},T]$ and $D = R = 0$, $\lambda = 0$,

$$\begin{aligned}&\mathbb{E}\left(\sup_{r\in[T_{M-1},T]} e^{p\bar{\mu}(r)}\left|Y_r - Y_r'\right|^p\right) + \mathbb{E}\left(\int_{T_{M-1}}^T e^{2\bar{\mu}(r)}\left|Z_r - Z_r'\right|^2 dr\right)^{p/2} \\ &\le C_p \mathbb{E}\left(\int_{T_{M-1}}^T e^{\bar{\mu}(r)}\ell(r)\left|U_r - U_r'\right| dr\right)^p \\ &\le C_p\left(\int_{T_{M-1}}^T \ell^2(r)\, dr\right)^{p/2} \mathbb{E}\left(\int_{T_{M-1}}^T e^{2\bar{\mu}(r)}\left|U_r - U_r'\right|^2 dr\right)^{p/2} \\ &\le C_p\left[\alpha\left(\tfrac{T}{M}\right)\right]^{p/2} \left\|\left|(X,U) - (X',U')\right|\right\|_M^p.\end{aligned}$$

Let $M_0 \in \mathbb{N}^*$ be such that

$$C_p\left[\alpha\left(\tfrac{T}{M_0}\right)\right]^{p/2} \le \frac{1}{2^p}.$$

Then

$$\left\|\left|\Gamma(X,U) - \Gamma(X',U')\right|\right\|_{M_0} \le \frac{1}{2}\left\|\left|(X,U) - (X',U')\right|\right\|_{M_0}.$$

Hence the Eq. (5.76) has a unique solution in the space $S_m^p[T_{M_0-1},T] \times \Lambda_{m\times k}^p(T_{M_0-1},T)$. The next step is to solve the equation on the interval $[T_{M_0-2},T_{M_0-1}]$ with the final value $Y(T_{M_0-1})$. Repeating the same arguments, the proof is completed in M_0 steps. ■

Corollary 5.28. *Consider the BSDE:* $\forall\, t \in [0,T]$, $\mathbb{P}$-*a.s.*

$$Y_t = \eta + S_T - S_t + \int_t^T F(s, Y_s, Z_s)\, ds - \int_t^T Z_s dB_s. \tag{5.79}$$

If $p > 1$, $S \in S_m^p[0,T]$, $\eta \in L^p(\Omega, \mathcal{F}_T, \mathbb{P}; \mathbb{R}^m)$, F *satisfies the assumptions* (**MH**$_F$), *and for all* $\rho \ge 0$

$$\mathbb{E}\left(\int_0^T \sup_{|y|\leq\rho} |F(t, y - S_t, 0)|\, dt\right)^p < \infty$$

then the Eq. (5.79) has a unique solution $(Y, Z) \in S_m^p[0, T] \times \Lambda_{m\times k}^p(0, T)$.

Proof. By the substitutions $\hat{Y}_t = Y_t + S_t$, $\hat{\eta} = \eta + S_T$ and $\hat{F}(t, y, z) = F(t, y - S_t, z)$ the Eq. (5.34) is transformed into

$$\hat{Y}_t = \hat{\eta} + \int_t^T \hat{F}\left(s, \hat{Y}_s, Z_s\right) ds - \int_t^T Z_s dB_s,$$

which satisfies the assumptions of Theorem 5.27. ∎

5.3.4.3 The Third BSDE: Monotone Coefficient $\Phi(s, Y_s, Z_s)\, dQ_s$

We now generalize Theorem 5.27 to the case of the general BSDE (5.12) which we recall here:

$$Y_t = \eta + \int_t^T \Phi(s, Y_s, Z_s)\, dQ_s - \int_t^T Z_s dB_s,\ t \in [0, T],\ a.s. \tag{5.80}$$

The assumptions will be those from the beginning of Sect. 5.3.1.

Let $p, a > 1$ and $n_p = 1 \wedge (p-1)$. Define

$$\bar{\mu}_t = \int_0^t \mu_s dQ_s \quad \text{and} \quad V_t = V_t^{(a,p)} = \int_0^t \mu_s dQ_s + \frac{a}{2n_p}\int_0^t (\ell_s)^2\, ds.$$

We say that $Y \in S_m^p\left([0, T]; e^{\bar{\mu}}\right)$ if $Y \in S_m^0[0, T]$ and

$$\mathbb{E} \sup_{s\in[0,T]} e^{p\bar{\mu}_s} |Y_s|^p < \infty.$$

In the same manner $Z \in \Lambda_{m\times k}^p\left(0, T; e^{\bar{\mu}}\right)$ if $Z \in \Lambda_{m\times k}^0(0, T)$ and

$$\mathbb{E}\left(\int_0^T e^{2\bar{\mu}_s} |Z_s|^2\, ds\right)^{p/2} < \infty.$$

We first prove the following:

Lemma 5.29. *Let $p > 1$ and the assumptions (5.13-**BSDE-H**$_\Phi$) (i.e. (5.14) and (5.15)) from Sect. 5.3.1 be satisfied. Moreover assume*

$$\begin{aligned} &(i) \quad \ell \in L^2(0, T) \text{ is a positive deterministic process,} \\ &(ii) \quad \mathbb{E}\left|e^{\bar{\mu}_T}\eta\right|^p + \mathbb{E}\left(\int_0^T e^{\bar{\mu}_s} |\Phi(s, 0, 0)|\, dQ_s\right)^p < \infty. \end{aligned} \tag{5.81}$$

If in addition

(h_1) $\mathbb{E}\left(\int_0^T \sup_{|y|\leq\rho} \left|e^{\bar{\mu}_t}\Phi\left(s, e^{-\bar{\mu}_t}y, 0\right) - \mu_t y\right| dQ_t\right)^p < \infty$, *for all* $\rho \geq 0$, *or*

(h_2) $\mu \geq 0$ *and* $\mathbb{E}\left(\int_0^T e^{\bar{\mu}_t} \sup_{|y|\leq\rho} |\Phi(t, y, 0)| \, dQ_s\right)^p < \infty$, *for all* $\rho \geq 0$,

then the BSDE (5.80) has a unique solution

$$(Y, Z) \in S_m^p\left([0, T]; e^{\bar{\mu}}\right) \times \Lambda_{m\times k}^p\left(0, T; e^{\bar{\mu}}\right).$$

Proof. Uniqueness follows from Theorem 5.10. To prove the existence we shall use the Banach fixed point theorem. Let $\Gamma : S_m^p\left([0, T]; e^{\bar{\mu}}\right) \times \Lambda_{m\times k}^p\left(0, T; e^{\bar{\mu}}\right) \to S_m^p\left([0, T]; e^{\bar{\mu}}\right) \times \Lambda_{m\times k}^p\left(0, T; e^{\bar{\mu}}\right)$ be defined by $(Y, Z) = \Gamma(X, U)$, where

$$Y_t = \eta + \int_t^T \Phi(s, Y_s, U_s) \, dQ_s - \int_t^T Z_s dB_s. \tag{5.82}$$

(A) Γ *is well defined.*

Let $(X, U) \in S_m^p\left([0, T]; e^{\bar{\mu}}\right) \times \Lambda_{m\times k}^p\left(0, T; e^{\bar{\mu}}\right)$. The function $\tilde{\Phi}(\omega, t, y) = \Phi(\omega, t, y, U_t(\omega))$ is monotone

$$\left\langle y - y', \Phi(r, y, U_r) - \Phi\left(r, y', U_r\right)\right\rangle \leq \mu_r \left|y - y'\right|^2.$$

Under (h_1) the assumptions of Corollary 5.26 are satisfied, because we have

$$\begin{aligned}
&\mathbb{E}\left(\int_0^T \sup_{|y|\leq\rho} \left|e^{\bar{\mu}_t}\Phi(t, e^{\bar{\mu}_t}y, U_t) - \mu_t y\right| dQ_t\right)^p \\
&\quad \leq \mathbb{E}\left(\int_0^T \sup_{|y|\leq\rho} \left|e^{\bar{\mu}_t}\Phi(t, e^{\bar{\mu}_t}y, 0) - \mu_t y\right| dQ_t + \int_0^T e^{\bar{\mu}_t}\ell(t) |U_t| \, dt\right)^p
\end{aligned}$$

and

$$\begin{aligned}
&\mathbb{E}\left(\int_0^T e^{\bar{\mu}_t} |U_t| \, \ell(t) \, dt\right)^p \\
&\quad \leq \left(\int_0^T \ell^2(t) \, dt\right)^{p/2} \mathbb{E}\left(\int_0^T e^{2\bar{\mu}_t} |U_t|^2 \, dt\right)^{p/2} \\
&\quad < \infty.
\end{aligned}$$

Under (h_2) the assumptions of Proposition 5.24 are satisfied because for $\mu \geq 0$ we have $\hat{\mu}_t = \int_0^t \mu_s^+ ds = \int_0^t \mu_s ds = \bar{\mu}_t$ and

$$\mathbb{E}\left(\int_0^T e^{\hat{\mu}_t} \sup_{|y|\le\rho} \Phi(t,y,U_t)dQ_t\right)^p$$
$$\le \mathbb{E}\left(\int_0^T e^{\bar{\mu}_t} \sup_{|y|\le\rho} \Phi(t,y,0)dQ_t + \int_0^T e^{\bar{\mu}_t}\ell(t)\,|U_t|\,dt\right)^p < \infty.$$

(B) *Γ is a strict contraction.* Let $M \in \mathbb{N}^*$ and $0{=}T_0 < T_1 < \cdots < T_M{=}T$, with $T_i = \dfrac{iT}{M}$. To prove the existence on $[T_{M-1},T]$ of the solution it is sufficient to prove that Γ is a strict contraction on the space $S_m^p\left([T_{M-1},T]\,;e^{\bar{\mu}}\right) \times \Lambda^p_{m\times k}\left(T_{M-1},T;e^{\bar{\mu}}\right)$ with respect to the norm $\|\!|(Y,Z)|\!\|_M$

$$\|\!|(Y,Z)|\!\|_M^p \overset{def}{=} \mathbb{E}\left[\left(\sup_{r\in[T_{M-1},T]} e^{p\bar{\mu}_r}\,|Y_r|^p\right) + \left(\int_{T_{M-1}}^T e^{2\bar{\mu}_r}\,|Z_r|^2\,dr\right)^{p/2}\right],$$

for M large enough. The proof continues exactly as in Theorem 5.27. Iteratively the existence follows on every interval $[T_{i-1},T_i]$, for $i = M, M-1, \ldots, 2, 1$, and finally we get the existence on $[0,T]$. ■

Theorem 5.30. *Let the assumptions (5.13-**BSDE-H**$_\Phi$) (i.e. (5.14) and (5.15)) from Sect. 5.3.1 be satisfied. Let $p,a > 1$ be fixed, $n_p = 1 \wedge (p-1)$*

$$\bar{\mu}_t = \int_0^t \mu_s dQ_s \quad \text{and} \quad V_t \overset{def}{=} V_t^{(a,p)} = \int_0^t \mu_s dQ_s + \frac{a}{2n_p}\int_0^t (\ell_s)^2\,ds.$$

Assume there exists a $\delta > \frac{p}{p-1}$ such that for $q = \frac{p\delta}{p+\delta}$

$$\begin{aligned}
&(i) \quad \mathbb{E}\, e^{pV_T}\,|\eta|^p + \mathbb{E}\left(\int_0^T e^{V_s}\,|\Phi(s,0,0)|\,dQ_s\right)^p < \infty,\\
&(ii) \quad \mathbb{E}\left(\int_0^T (\ell_s)^2\,ds\right)^{\delta/2} < \infty,\\
&(iii) \quad \mathbb{E}\,\exp\left[\tfrac{\delta a}{2}\left(\tfrac{1}{n_q}-\tfrac{1}{n_p}\right)\int_0^T (\ell_s)^2\,ds\right] < \infty.
\end{aligned} \tag{5.83}$$

If in addition

(h_1) $\mathbb{E}\left(\int_0^T \sup_{|y|\le\rho} \left|e^{\bar{\mu}_t}\Phi\left(s,e^{-\bar{\mu}_t}y,0\right) - \mu_t y\right| dQ_t\right)^p < \infty$, *for all $\rho \ge 0$, or*

(h_2) $\mu \ge 0$ *and* $\mathbb{E}\left(\int_0^T e^{\bar{\mu}_t} \sup_{|y|\le\rho} |\Phi(t,y,0)|\,dQ_s\right)^p < \infty$, *for all $\rho \ge 0$,*

then the BSDE (5.80) has a unique solution $(Y,Z) \in S_m^0([0,T]) \times \Lambda^0_{m\times k}(0,T)$ such that

$$\mathbb{E}\sup_{t\in[0,T]} e^{pV_s}|Y_s|^p + \mathbb{E}\left(\int_0^T e^{2V_s}|Z_s|^2 ds\right)^{p/2} < \infty.$$

Moreover, for all $t \in [0,T]$*:*

$$\begin{aligned}
&\mathbb{E}^{\mathcal{F}_t}\sup_{s\geq t} e^{pV_s}|Y_s|^p + \mathbb{E}^{\mathcal{F}_t}\left(\int_t^T e^{2V_s}|Z_s|^2 ds\right)^{p/2} \\
&\leq C_p\,\mathbb{E}^{\mathcal{F}_t}\left[e^{pV_T}|\eta|^p + \left(\int_t^T e^{V_s}|\Phi(s,0,0)|\,dQ_s\right)^p\right].
\end{aligned} \tag{5.84}$$

Proof. *Uniqueness* follows from Theorem 5.10.

Existence. By Lemma 5.29 we infer that the approximating BSDE

$$Y_t^n = \eta + \int_t^T \Phi\left(s, Y_s^n, Z_s^n \mathbf{1}_{[0,n]}(\ell_s)\right) dQ_s - \int_t^T Z_s^n dB_s \tag{5.85}$$

has a unique solution $(Y^n, Z^n) \in S_m^p\left([0,T]; e^{\bar{\mu}}\right) \times \Lambda_{m\times k}^p\left(0,T; e^{\bar{\mu}}\right)$.

Let $\ell_s^n = \ell_s \mathbf{1}_{[0,n]}(\ell_s)$ and

$$V_t^n \overset{def}{=} \int_0^t \left(\mu_s dQ_s + \frac{a}{2n_p}\left(\ell_s^n\right)^2 ds\right).$$

We have for all $n, i \in \mathbb{N}$

$$\bar{\mu}_t \leq V_t^n \leq V_t^{n+i} \leq \bar{\mu}_t + \frac{a}{2n_p}(n+i)^2 T.$$

Therefore

$$\mathbb{E}\sup_{t\in[0,T]} e^{pV_t^{n+i}}|Y_t^n|^p \leq C_{n,i}\left(\mathbb{E}\sup_{t\in[0,T]} e^{2p\bar{\mu}_t}|Y_t^n|^{2p}\right)^{1/2} < \infty.$$

Since

$$\begin{aligned}
&\left\langle Y_t^n, \Phi\left(t, Y_t^n, Z_t^n \mathbf{1}_{[0,n]}(\ell_t)\right) dQ_t\right\rangle \\
&\leq |Y_t^n|\,|\Phi(t,0,0)|\,dQ_t + |Y_t^n|^2 dV_t^n + \frac{n_p}{2a}|Z_t^n|^2 dt \\
&\leq |Y_t^n|\,|\Phi(t,0,0)|\,dQ_t + |Y_t^n|^2 dV_t^{n+i} + \frac{n_p}{2a}|Z_t^n|^2 dt
\end{aligned}$$

we obtain, by Proposition 5.2-A, that

$$\begin{aligned} &\mathbb{E}^{\mathcal{F}_t} \sup_{s\in[t,T]} e^{pV_s^{n+i}} |Y_s^n|^p + \mathbb{E}^{\mathcal{F}_t} \left(\int_t^T e^{2V_s^{n+i}} |Z_s^n|^2 ds\right)^{p/2} \\ &\quad \le C_q\, \mathbb{E}^{\mathcal{F}_t}\left[e^{pV_T^{n+i}} |\eta|^p + \left(\int_t^T e^{V_s^{n+i}} |\Phi(s,0,0)|\, dQ_s\right)^p\right]. \end{aligned}$$

By Beppo Levi's monotone convergence Theorem 1.9 it follows for $i \to \infty$ that for all $t \in [0,T]$,

$$\begin{aligned} &\mathbb{E}^{\mathcal{F}_t} \sup_{s\ge t} e^{pV_s} |Y_s^n|^p + \mathbb{E}^{\mathcal{F}_t} \left(\int_t^T e^{2V_s} |Z_s^n|^2 ds\right)^{p/2} \\ &\quad \le C_p\, \mathbb{E}^{\mathcal{F}_t}\left[e^{pV_T} |\eta|^p + \left(\int_t^T e^{V_s} |\Phi(s,0,0)|\, dQ_s\right)^p\right]. \end{aligned} \tag{5.86}$$

Consequently by (5.83-i) for all $n \in \mathbb{N}^*$,

$$\mathbb{E} \sup_{s\in[0,T]} e^{pV_s} |Y_s^n|^p + \mathbb{E} \left(\int_0^T e^{2V_s} |Z_s^n|^2 ds\right)^{p/2} \le C < \infty.$$

Let $\delta > \frac{p}{p-1}$, $q = \frac{p\delta}{p+\delta}$, $n_q \stackrel{def}{=} 1 \wedge (q-1)$ and $n_p \stackrel{def}{=} 1 \wedge (p-1)$ satisfy (5.83-ii, iii). Clearly $1 < q < p$ and $0 < n_q \le n_p$. If we define

$$\begin{aligned} \Delta_t &= \frac{a}{2}\left(\frac{1}{n_q} - \frac{1}{n_p}\right) \int_0^t (\ell_s)^2 ds \quad \text{and} \\ V_t^{(a,q)} &= \int_0^t \left[\mu_s dQ_s + \frac{a}{2n_q} (\ell_s)^2 ds\right] = V_t + \Delta_t, \end{aligned}$$

we have, for all $n \in \mathbb{N}^*$,

$$\begin{aligned} &\mathbb{E} \sup_{s\in[0,T]} e^{qV_s^{(a,q)}} |Y_s^n|^q + \mathbb{E} \left(\int_0^T e^{2V_s^{(a,q)}} |Z_s^n|^2 ds\right)^{q/2} \\ &\le \mathbb{E}\left(e^{q\Delta_T} \sup_{s\in[0,T]} e^{qV_s} |Y_s^n|^q\right) + \mathbb{E}\left[e^{q\Delta_T} \left(\int_0^T e^{2V_s} |Z_s^n|^2 ds\right)^{q/2}\right] \\ &\le \left(\mathbb{E} e^{\delta\Delta_T}\right)^{\frac{p}{p+\delta}} \left[\left(\mathbb{E} \sup_{s\in[0,T]} e^{pV_s} |Y_s^n|^p\right)^{\frac{\delta}{p+\delta}} + \left(\mathbb{E}\left(\int_0^T e^{2V_s} |Z_s^n|^2 ds\right)^{p/2}\right)^{\frac{\delta}{p+\delta}}\right] \\ &\le C < \infty. \end{aligned}$$

Hence for all $n, i \in \mathbb{N}^*$

$$\mathbb{E} \sup_{s\in[0,T]} e^{qV_s^{(a,q)}} \left|Y_s^n - Y_s^{n+i}\right|^q < \infty.$$

Since

$$\begin{aligned}
&\left\langle Y_s^n - Y_s^{n+i}, \Phi\left(s, Y_s^n, Z_s^n \mathbf{1}_{[0,n]}(\ell_s)\right) - \Phi\left(s, Y_s^{n+i}, Z_s^{n+i}\mathbf{1}_{[0,n+i]}(\ell_s)\right)\right\rangle dQ_s \\
&\le \left|Y_s^n - Y_s^{n+i}\right|^2 \mu_s dQ_s + \left|Y_s^n - Y_s^{n+i}\right| \ell_s \left|Z_s^n \mathbf{1}_{[0,n]}(\ell_s) - Z_s^{n+i}\mathbf{1}_{[0,n+i]}(\ell_s)\right| ds \\
&\le \left|Y_s^n - Y_s^{n+i}\right| \ell_s \left|Z_s^n\right| \left|\mathbf{1}_{[0,n]}(\ell_s) - \mathbf{1}_{[0,n+i]}(\ell_s)\right| ds \\
&\quad + \left|Y_s^n - Y_s^{n+i}\right|^2 \left(\mu_s dQ_s + \frac{a}{2n_q}\ell_s^2 ds\right) + \frac{n_q}{2a}\left|Z_s^n - Z_s^{n+i}\right|^2 ds,
\end{aligned}$$

by Proposition 5.2-A, we infer that

$$\begin{aligned}
&\mathbb{E}\left(\sup_{s\in[0,T]} e^{qV_s^{(a,q)}} \left|Y_s^n - Y_s^{n+i}\right|^q\right) + \mathbb{E}\left(\int_0^T e^{2V_s^{(a,q)}} \left|Z_s^n - Z_s^{n+i}\right|^2 ds\right)^{q/2} \\
&\le C_q\, \mathbb{E}\left(\int_0^T \mathbf{1}_{(n,\infty)}(\ell_s)\, e^{V_s^{(a,q)}} \ell_s \left|Z_s^n\right| ds\right)^q \\
&\le C_q \mathbb{E}\left[\left(\int_0^T \ell_s^2 \mathbf{1}_{(n,\infty)}(\ell_s)\, ds\right)^{q/2} \left(\int_0^T e^{2V_s^{(a,q)}} \left|Z_s^n\right|^2 ds\right)^{q/2}\right] \\
&\le C_q \left[\mathbb{E}\left(\int_0^T \ell_s^2 \mathbf{1}_{(n,\infty)}(\ell_s)\, ds\right)^{\delta/2}\right]^{\frac{p}{p+\delta}} \left[\mathbb{E}\left(\int_0^T e^{2V_s^{(a,q)}} \left|Z_s^n\right|^2 ds\right)^{p/2}\right]^{\frac{\delta}{p+\delta}} \\
&\to 0,
\end{aligned}$$

as $n \to \infty$.

We deduce that there exists a pair $(Y, Z) \in S_m^0[0,T] \times \Lambda_{m\times k}^0(0,T)$ such that for $q = \frac{p\delta}{p+\delta}$

$$\lim_{n\to\infty} \mathbb{E}\left(\sup_{s\ge 0} e^{qV_s^{(a,q)}} \left|Y_s^n - Y_s\right|^q\right) + \mathbb{E}\left[\left(\int_0^T e^{2V_s^{(a,q)}} \left|Z_s^n - Z_s\right|^2 ds\right)^{q/2}\right] = 0.$$

Now the inequality (5.84) clearly follows from (5.86) by Fatou's Lemma.

Finally passing to the limit in (5.85) we deduce via Lemma 5.16 that (Y, Z) is a solution of BSDE (5.80). ■

5.3.5 Linear BSDEs

Let $m = 1$ and consider the BSDE

$$Y_t = \eta + \int_t^T [(a_s Y_s + b_s)\, dQ_s + \langle c_s, Z_s\rangle\, ds] - \int_t^T \langle Z_s, dB_s\rangle\,, \tag{5.87}$$

where

- η is an $\mathcal{F}_T$-measurable random variable;
- Q is a $\mathcal{P}$-m.i.c.s.p. such that $Q_0 = 0$;
- $(a_t)_{t\geq 0}$, $(b_t)_{t\geq 0}$ are $\mathbb{R}$-valued $\mathcal{P}$-m.s.p. and $(c_t)_{t\geq 0}$ is an $\mathbb{R}^k$-valued $\mathcal{P}$-m.s.p.;
- for some $p > 1$ and for all $\lambda \geq 0$,

$$\begin{aligned} &(j) \quad \mathbb{E}\left[\left(1 + |\eta|^p\right)\exp\left(\lambda V_T\right)\right] < \infty, \\ &(jj) \quad \mathbb{E}\Big(\int_0^T |b_s| \exp\left(\lambda V_s\right) dQ_s\Big)^p < \infty, \end{aligned} \tag{5.88}$$

  where

$$V_t = \int_0^t |a_s|\, dQ_s + \frac{1}{n_p}\int_0^t |c_s|^2\, ds.$$

By Theorem 5.21 the BSDE (5.87) has a unique solution satisfying

$$\begin{aligned} &\mathbb{E} \sup_{s\in[0,T]} \left|e^{V_s} Y_s\right|^p + \mathbb{E}\left(\int_0^T e^{2V_s}\,|Z_s|^2\, ds\right)^{p/2} \\ &\quad \leq C_p\, \mathbb{E}\Big[\, \left|e^{V_T}\eta\right|^p + \Big(\int_0^T e^{V_s}\,|b_s|\, dQ_s\Big)^p\Big]. \end{aligned}$$

Let

$$\Gamma_t = \exp\left[\int_0^t \left(a_r dQ_r - \frac{1}{2}|c_r|^2\, dr\right) + \int_0^t \langle c_r, dB_r\rangle\right].$$

Then

$$\begin{aligned} d\Gamma_t &= \Gamma_t a_t dQ_t + \Gamma_t \langle c_t, dB_t\rangle\,, \\ d\Gamma_t^{-1} &= \Gamma_t^{-1}\left(-a_t dQ_t + |c_t|^2\, dt\right) - \Gamma_t^{-1}\langle c_t, dB_t\rangle\,. \end{aligned}$$

Since for all $\delta > 0$, $\mathbb{E}\left[\exp\left(\delta\Lambda_T\right)\right] < \infty$ we have $\mathbb{E} \sup_{s\in[0,T]} |\Gamma_s|^\delta < \infty$ for all $\delta > 0$. Consequently there exists $1 < q < p$ such that

$$\mathbb{E}\left|\Gamma_T\eta + \int_0^T \Gamma_s b_s dQ_s\right|^q < \infty. \tag{5.89}$$

By the representation Theorem 2.42 there exists a unique stochastic process $R \in \Lambda^q_{1\times k}(0,T)$ such that

$$\Gamma_T\eta + \int_0^T \Gamma_s b_s dQ_s = \mathbb{E}\left(\Gamma_T\eta + \int_0^T \Gamma_s b_s dQ_s\right) + \int_0^T \langle R_s, dB_s\rangle .$$

Proposition 5.31. *Let the assumption (5.88) be satisfied. Then the solution of the BSDE (5.87) is given by*

$$\begin{aligned} (a) \quad Y_t &= \Gamma_t^{-1}\,\mathbb{E}^{\mathcal{F}_t}\left[\Gamma_T\eta + \int_t^T \Gamma_s b_s dQ_s\right], \\ (b) \quad Z_t &= \Gamma_t^{-1} R_t - c_t Y_t. \end{aligned} \tag{5.90}$$

Proof. It is sufficient to verify that (Y,Z) given by (5.90) is a solution of (5.87). We have

$$\begin{aligned} Y_t &= \Gamma_t^{-1}\,\mathbb{E}^{\mathcal{F}_t}\left[\Gamma_T\eta + \int_t^T \Gamma_s b_s dQ_s\right] \\ &= \Gamma_t^{-1}\left[\mathbb{E}\left(\Gamma_T\eta + \int_0^T \Gamma_s b_s dQ_s\right) + \int_0^t \langle R_s, dB_s\rangle - \int_0^t \Gamma_s b_s dQ_s\right] \\ &= \Gamma_t^{-1}\left[\mathbb{E}\left(\Gamma_T\eta + \int_0^T \Gamma_s b_s dQ_s\right) + \int_0^t \langle \Gamma_s Y_s c_s + \Gamma_s Z_s, dB_s\rangle - \int_0^t \Gamma_s b_s dQ_s\right]. \end{aligned}$$

Consequently, from Itô's formula,

$$\begin{aligned} dY_t &= \left[\Gamma_t^{-1}\left(-a_t\, dQ_t + |c_t|^2\, dt\right) - \Gamma_t^{-1}\langle c_t, dB_t\rangle\right]\Gamma_t Y_t \\ &\quad + \Gamma_t^{-1}\left[\langle \Gamma_t Y_t c_t + \Gamma_t Z_t, dB_t\rangle - \Gamma_t b_t\, dQ_t\right] - \Gamma_t^{-1}\langle c_t, c_t\Gamma_t Y_t + \Gamma_t Z_t\rangle\, dt \\ &= \left[-a_t Y_t\, dQ_t - b_t\, dQ_t - \langle c_t, Z_t\rangle\, dt\right] + \langle Z_t, dB_t\rangle . \end{aligned}$$

Since, moreover, $Y_T = \eta$, we conclude that (Y,Z) is a solution of the BSDE (5.87). ■

5.3.6 Comparison Results

In this section we again restrict ourselves to the case $m = 1$.

5.3.6.1 Lipschitz Case

Let $(Y, Z) \in S^0[0,T] \times \Lambda_k^0(0,T)$ be a solution of the BSDE

$$Y_t = \eta + \int_t^T \Phi(s, Y_s, Z_s)\, dQ_s - \int_t^T \langle Z_s, dB_s\rangle \tag{5.91}$$

and $(\tilde{Y}, \tilde{Z}) \in S^0[0,T] \times \Lambda_k^0(0,T)$ a solution of the BSDE

$$\tilde{Y}_t = \tilde{\eta} + \int_t^T \tilde{\Phi}(s, \tilde{Y}_s, \tilde{Z}_s)\, dQ_s - \int_t^T \langle \tilde{Z}_s, dB_s\rangle. \tag{5.92}$$

Assume that the functions $\Phi, \tilde{\Phi} : \Omega \times [0,\infty[\times \mathbb{R} \times \mathbb{R}^k \to \mathbb{R}$ are $(\mathcal{P}, \mathbb{R} \times \mathbb{R}^k)$-Carathéodory functions ($\mathcal{P}$-m.s.p. with respect to (ω, t) and continuous with respect to $(x, z) \in \mathbb{R} \times \mathbb{R}^k$) such that

$$\int_0^T |\Phi(s, Y_s, Z_s)|\, dQ_s + \int_0^T \left|\tilde{\Phi}(s, \tilde{Y}_s, \tilde{Z}_s)\right| dQ_s < \infty, \quad \text{a.s.} \tag{5.93}$$

We give a comparison result in the case when one of the two functions Φ and $\tilde{\Phi}$ satisfies some Lipschitz conditions.

Let $p > 1$. Without loss of generality we assume that Φ satisfies the assumptions of Theorem 5.21. Then the Eq. (5.91) has a unique solution (Y, Z) satisfying

$$\begin{aligned} &\mathbb{E} \sup_{s\in[0,T]} \left|e^{V_s} Y_s\right|^p + \mathbb{E}\left(\int_0^T e^{2V_s} |Z_s|^2\, ds\right)^{p/2} \\ &\quad \le C_p\, \mathbb{E}\left[\left|e^{V_T}\eta\right|^p + \left(\int_0^T e^{V_s} |\Phi(s,0,0)|\, dQ_s\right)^p\right]. \end{aligned}$$

Proposition 5.32. *Let $p > 1$ and the assumptions of Theorem 5.21 be satisfied. Assume that $(\tilde{Y}, \tilde{Z})$ is a solution of the BSDE (5.92) and for all $\delta \ge 0$,*

$$\mathbb{E}\left(|\eta - \tilde{\eta}| \exp(\delta V_T) + \int_0^T \left|\Phi(s, \tilde{Y}_s, \tilde{Z}_s) - \tilde{\Phi}(s, \tilde{Y}_s, \tilde{Z}_s)\right| \exp(\delta V_s)\, dQ_s\right)^p < \infty.$$

If

$$\begin{aligned} &(i) \quad \eta \ge \tilde{\eta}, \quad \mathbb{P}\text{-a.s. and} \\ &(ii) \quad \Phi(t, \tilde{Y}_t, \tilde{Z}_t) \ge \tilde{\Phi}(t, \tilde{Y}_t, \tilde{Z}_t),\ d\mathbb{P} \otimes dQ_t\text{-a.e. on } \Omega \times \mathbb{R}_+. \end{aligned}$$

(a) Then $\mathbb{P}$-a.s. $\omega \in \Omega$, $Y_t(\omega) \ge \tilde{Y}_t(\omega)$, for all $t \in [0,T]$.
(b) If moreover there exists a $t_0 \in [0, T[$ such that $\mathbb{P}$-a.s.

$$(\eta - \tilde{\eta}) + \int_{t_0}^{T} \left[\Phi(s, \tilde{Y}_s, \tilde{Z}_s) - \tilde{\Phi}(s, \tilde{Y}_s, \tilde{Z}_s)\right] dQ_s > 0$$

then $Y_{t_0} > \tilde{Y}_{t_0}$ $\mathbb{P}$*-a.s. In particular if* $\eta > \tilde{\eta}$, $\mathbb{P}$*-a.s., then* $Y_t(\omega) > \tilde{Y}_t(\omega)$*, for all* $t \in [0, T]$, $\mathbb{P}$*-a.s.* $\omega \in \Omega$.

Proof. Observe that $Y_t - \tilde{Y}_t$ can be written in the form

$$Y_t - \tilde{Y}_t = (\eta - \tilde{\eta}) + \int_t^T \left\{\left[a_s\left(Y_s - \tilde{Y}_s\right) + b_s\right] dQ_s + \left\langle c_s, Z_s - \tilde{Z}_s\right\rangle ds\right\} - \int_t^T \left(Z_s - \tilde{Z}_s\right) dB_s,$$

where

$$a_s = \begin{cases} \dfrac{1}{Y_s - \tilde{Y}_s}\left[\Phi(s, Y_s, Z_s) - \Phi(s, \tilde{Y}_s, Z_s)\right], & \text{if } Y_s - \tilde{Y}_s \neq 0, \\ 0, \ \text{if } Y_s - \tilde{Y}_s = 0, \end{cases}$$

$b_s = \Phi(s, \tilde{Y}_s, \tilde{Z}_s) - \tilde{\Phi}(s, \tilde{Y}_s, \tilde{Z}_s)$, and

$$c_s = \begin{cases} \dfrac{Z_s - \tilde{Z}_s}{\alpha_s \left|Z_s - \tilde{Z}_s\right|^2}\left[\Phi(s, \tilde{Y}_s, Z_s) - \Phi(s, \tilde{Y}_s, \tilde{Z}_s)\right], & \text{if } \alpha_s\left(Z_s - \tilde{Z}_s\right) \neq 0, \\ 0, \ \text{if } \alpha_s\left(Z_s - \tilde{Z}_s\right) = 0, \end{cases}$$

(recall that α is a $\mathcal{P}$-m.s.p. such that $\alpha_s dQ_s = ds$).

From $|a_s| \leq L_s$, $|c_s| \leq \ell_s$, the assumption of the Proposition, and the argument of the preceding section, we deduce that

$$\sup_{s\in[0,T]}\left[\left|Y_s - \tilde{Y}_s\right|^p \exp(\delta V_s)\right] + \mathbb{E}\left(\int_0^T \left|Z_s - \tilde{Z}_s\right|^2 \exp(\delta V_s)\, ds\right)^{p/2} < \infty,$$

for all $\delta \geq 0$. Hence by Proposition 5.31

$$Y_t - \tilde{Y}_t = \Gamma_t^{-1}\, \mathbb{E}^{\mathcal{F}_t}\left[\Gamma_T(\eta - \tilde{\eta}) + \int_t^T \Gamma_s b_s dQ_s\right],$$

which clearly yields the conclusions of Proposition 5.32. ∎

5.3.6.2 Monotone Case

We now give a comparison result for the solutions of the Eqs. (5.91) and (5.92) in the case when one of the two functions Φ and $\tilde{\Phi}$ satisfies a monotonicity condition. To be precise we assume without loss of generality that Φ satisfies the assumptions (5.13-**BSDE-H**$_\Phi$). Let

$$V_t = \int_0^t \left(\mu_r^+ dQ_r + \frac{a}{2n_p} (\ell_r)^2 dr\right).$$

Then for $a, p > 1$ and $n_p = (p-1) \wedge 1$,

$$\begin{aligned}
&\left(\tilde{Y}_r - Y_r\right)^+ \left[\Phi\left(r, \tilde{Y}_r, \tilde{Z}_r\right) - \Phi\left(r, Y_r, Z_r\right)\right] dQ_r \\
&\le \left[\mu_r^+ \left(\left(Y_r - \tilde{Y}_r\right)^+\right)^2 + \ell_r \alpha_r \left(\tilde{Y}_r - Y_r\right)^+ \left|Z_r - \tilde{Z}_r\right|\right] dQ_r \\
&\le \left[\left(Y_r - \tilde{Y}_r\right)^+\right]^2 dV_r + \frac{n_p}{2a} \mathbf{1}_{\tilde{Y}_r - Y_r > 0} \left|Z_r - \tilde{Z}_r\right|^2 dr.
\end{aligned}$$

Proposition 5.33. *Let the assumptions (5.13-**BSDE-H**$_\Phi$) be satisfied. Let $(Y, Z) \in S^0[0,T] \times \Lambda_k^0(0,T)$ be a solution of (5.91) and $\left(\tilde{Y}, \tilde{Z}\right) \in S^0[0,T] \times \Lambda_k^0(0,T)$ be a solution of (5.92), such that (5.93) and the condition*

$$\mathbb{E} \left\|\left(\tilde{Y} - Y\right)^+ e^V\right\|_T^p < \infty$$

are satisfied. Assume that $\mathbb{P}$-a.s.:

(i) $\eta \ge \tilde{\eta}$,
(ii) $\Phi(t, y, z) \ge \tilde{\Phi}(t, y, z)$, *for all* $(t, y, z) \in [0,T] \times \mathbb{R} \times \mathbb{R}^k$.

Then $\mathbb{P}$-a.s., $Y_t(\omega) \ge \tilde{Y}_t(\omega)$, for all $t \in [0,T]$.

Proof. Recall from Proposition 2.33 that if

$$dX_t = dK_t + \langle G_t, dB_t \rangle,$$

then

$$dX_t^+ = \theta(X_t)\, dK_t + \theta(X_t) \langle G_t, dB_t \rangle + dP_t,$$

where

$$\theta(x) = \begin{cases} 0, & \text{if } x < 0, \\ \frac{1}{2}, & \text{if } x = 0, \\ 1, & \text{if } x > 0, \end{cases}$$

and $\{P_t : t \ge 0\}$, $P_0 = 0$, is an increasing continuous stochastic process defined by (2.33).

We have

$$d\left(\tilde{Y}_t - Y_t\right) = -\left[\tilde{\Phi}\left(t, \tilde{Y}_t, \tilde{Z}_t\right) - \Phi\left(t, Y_t, Z_t\right)\right] dQ_t + \left\langle \tilde{Z}_t - Z_t, dB_t \right\rangle,$$

and therefore

$$\left(\tilde{Y}_t - Y_t\right)^+ = (\tilde{\eta} - \eta)^+ + \int_t^T dK_r - \int_t^T \theta\left(\tilde{Y}_r - Y_r\right)\left\langle \tilde{Z}_r - Z_r, dB_r \right\rangle,$$

with

$$dK_r = \theta\left(\tilde{Y}_r - Y_r\right)\left[\left(\tilde{\Phi}\left(r, \tilde{Y}_r, \tilde{Z}_r\right) - \Phi\left(r, Y_r, Z_r\right)\right) dQ_r\right] - dP_r,$$

and (see (2.33))

$$P_t = \lim_{\varepsilon \to 0_+} \frac{1}{2\varepsilon} \int_0^t \rho\left(\frac{\tilde{Y}_s - Y_s}{\varepsilon}\right) \left|\tilde{Z}_s - Z_s\right|^2 ds.$$

Since for $a, p > 1$ and $n_p = (p-1) \wedge 1$,

$$\begin{aligned}\left(\tilde{Y}_r - Y_r\right)^+ dK_r &\leq \left(\tilde{Y}_r - Y_r\right)^+ \left[\Phi\left(r, \tilde{Y}_r, \tilde{Z}_r\right) - \Phi\left(r, Y_r, Z_r\right)\right] dQ_r \\ &\leq \left[\left(Y_r - \tilde{Y}_r\right)^+\right]^2 dV_r + \frac{n_p}{2a} \theta\left(\tilde{Y}_r - Y_r\right) \left|Z_r - \tilde{Z}_r\right|^2 dr,\end{aligned}$$

we obtain, by the inequality (6.107) from Proposition 6.80, that for all $0 \leq t \leq T$:

$$e^{pV_t}\left[\left(\tilde{Y}_t - Y_t\right)^+\right]^p \leq \mathbb{E}^{\mathcal{F}_t} e^{pV_T}\left[(\tilde{\eta} - \eta)^+\right]^p = 0, \quad \mathbb{P}\text{-a.s.}$$

Consequently for all $0 \leq t \leq T$:

$$Y_t \geq \tilde{Y}_t, \quad \mathbb{P}\text{-a.s.}$$

■

We now give a strict comparison result in the case of monotone coefficients. Namely, we consider a solution $(Y, Z) \in S^0[0,T] \times \Lambda_k^0(0,T)$ of the BSDE

$$Y_t = \eta + \int_t^T \Phi\left(s, Y_s, Z_s\right) dQ_s - \int_t^T \left\langle Z_s, dB_s \right\rangle, \ \ a.s., \ \ t \in [0,T], \tag{5.94}$$

and a solution $\left(\tilde{Y}, \tilde{Z}\right) \in S^0[0,T] \times \Lambda_k^0(0,T)$ of the BSDE

$$\tilde{Y}_t = \tilde{\eta} + \int_t^T \tilde{\Phi}\left(s, \tilde{Y}_s, \tilde{Z}_s\right) dQ_s - \int_t^T \left\langle \tilde{Z}_s, dB_s \right\rangle, \tag{5.95}$$

where $\Phi, \tilde{\Phi} : \Omega \times [0,T] \times \mathbb{R} \times \mathbb{R}^k \to \mathbb{R}$ satisfy:
(**CR1**)

- $\Phi(\cdot, t, y, z), \tilde{\Phi}(\cdot, t, y, z)$ *are* $\mathcal{F}_t$*-measurable for all* $(t, y, z) \in [0,T] \times \mathbb{R} \times \mathbb{R}^k$,
- $\Phi(\omega, \cdot, \cdot, \cdot), \tilde{\Phi}(\omega, \cdot, \cdot, \cdot)$ *are continuous* $\mathbb{P}$-a.s. $\omega \in \Omega$,
- Φ *satisfies the assumption* (5.13-**BSDE-H**$_\Phi$).

We have

$$Y_t - \tilde{Y}_t = (\eta - \tilde{\eta}) + \int_t^T \left[b_s dQ_s + \langle c_s, Z_s - \tilde{Z}_s \rangle ds \right] - \int_t^T (Z_s - \tilde{Z}_s) \, dB_s,$$

with $b_s = \Phi(s, Y_s, \tilde{Z}_s) - \tilde{\Phi}(s, \tilde{Y}_s, \tilde{Z}_s)$ and

$$c_s = \begin{cases} \dfrac{Z_s - \tilde{Z}_s}{\alpha_s \left| Z_s - \tilde{Z}_s \right|^2} \left[\Phi(s, Y_s, Z_s) - \Phi\left(s, Y_s, \tilde{Z}_s\right) \right], & \text{if } \alpha_s \left(Z_s - \tilde{Z}_s \right) \neq 0, \\ 0, & \text{if } \alpha_s \left(Z_s - \tilde{Z}_s \right) = 0, \end{cases}$$

(recall that α is a $\mathcal{P}$-m.s.p. such that $\alpha_s dQ_s = ds$). Note that $|c_s| \leq \ell_s$.

Assume that
(**CR2**)

- η *and* $\tilde{\eta}$ *are* $\mathcal{F}_T$*-measurable random variables;*
- *for some* $p > 1$ *and for all* $\delta \geq 0$,

$$(j) \quad \mathbb{E}\left[\left(1 + |\eta - \tilde{\eta}|^p\right) \exp\left(\delta \int_0^T (\ell_r)^2 dr \right) \right] < \infty,$$
$$(jj) \quad \mathbb{E}\left[\int_0^T \left| \Phi(s, Y_s, \tilde{Z}_s) - \tilde{\Phi}(s, \tilde{Y}_s, \tilde{Z}_s) \right| \exp\left(\delta \int_0^s (\ell_r)^2 dr \right) dQ_s \right]^p < \infty.$$

Then by Proposition 5.31, for all $\delta \geq 0$

$$\mathbb{E} \sup_{s \in [0,T]} \left[\left| Y_s - \tilde{Y}_s \right|^p e^{\delta \int_0^s (\ell_r)^2 dr} \right] + \mathbb{E} \left(\int_0^T \left| Z_s - \tilde{Z}_s \right|^2 e^{\delta \int_0^s (\ell_r)^2 dr} ds \right)^{p/2} < \infty,$$

and for any stopping times $0 \leq \theta \leq \sigma \leq T$

$$\Gamma_\theta \left(Y_\theta - \tilde{Y}_\theta \right) = \mathbb{E}^{\mathcal{F}_\theta} \left[\Gamma_\sigma \left(Y_\sigma - \tilde{Y}_\sigma \right) + \int_\theta^\sigma \Gamma_s \left(\Phi(s, Y_s, \tilde{Z}_s) - \tilde{\Phi}(s, \tilde{Y}_s, \tilde{Z}_s) \right) dQ_s \right], \tag{5.96}$$

where

$$\Gamma_t = \exp\left[\int_0^t \langle c_r, dB_r \rangle - \int_0^t \frac{1}{2} |c_r|^2 dr \right].$$

Proposition 5.34. *Let $(Y,Z) \in S^0[0,T] \times \Lambda_k^0(0,T)$ be a solution for (5.94) and $(\tilde{Y},\tilde{Z}) \in S^0[0,T] \times \Lambda_k^0(0,T)$ be a solution for (5.95), such that*

$$\mathbb{E}\left\|(\tilde{Y}-Y)^+ \exp\left(\int_0^{\cdot} \mu_r^+ dQ_r\right)\right\|_T^p < \infty.$$

Assume that the assumptions (**CR1**), (**CR2**) *are satisfied and $\tilde{Z}$ is a continuous stochastic process.*
If $0 \le t_0 < T$, $A \in \mathcal{F}_{t_0}$ and

$$\begin{array}{ll} (i) & \eta \ge \tilde{\eta}, \quad \mathbb{P}\text{-a.s.}, \\ (ii) & \Phi(t,y,z) \ge \tilde{\Phi}(t,y,z), \ \forall \ (t,y,z) \in [0,T]\times\mathbb{R}\times\mathbb{R}^k, \quad \mathbb{P}\text{-a.s.}, \\ (iii) & \Phi\left(\omega,t_0,Y_{t_0},\tilde{Z}_{t_0}\right) > \tilde{\Phi}\left(\omega,t_0,Y_{t_0},\tilde{Z}_{t_0}\right), \ \mathbb{P}\text{-a.s. } \omega \in A, \\ (iv) & Q_{t_0} < Q_t, \quad \text{for } t_0 < t \le T, \quad \mathbb{P}\text{-a.s.}, \end{array}$$

then

$$\begin{array}{ll} (j) & Y_t(\omega) \ge \tilde{Y}_t(\omega), \ \forall\, t \in [0,T], \quad \mathbb{P}\text{-a.s. } \omega \in \Omega, \quad \text{and} \\ (jj) & Y_{t_0}(\omega) > \tilde{Y}_{t_0}(\omega), \quad \mathbb{P}\text{-a.s. } \omega \in A. \end{array}$$

Proof. By Theorem 5.33 we have

$$\mathbb{P}\text{-a.s.}, \quad Y_t(\omega) \ge \tilde{Y}_t(\omega), \quad \text{for all } t \in [0,T].$$

Assume that $\mathbb{P}\left(\{Y_{t_0} = \tilde{Y}_{t_0}\} \cap A\right) > 0$. Let the stopping time

$$\begin{aligned} \tau = \inf\Big\{s \in [t_0,T] : &\Gamma_s\left[\Phi(s,Y_s,\tilde{Z}_s) - \tilde{\Phi}(s,\tilde{Y}_s,\tilde{Z}_s)\right] \\ &\le \frac{1}{2}\Gamma_{t_0}\left[\Phi(\omega,t_0,Y_{t_0},\tilde{Z}_{t_0}) - \tilde{\Phi}(\omega,t_0,\tilde{Y}_{t_0},\tilde{Z}_{t_0})\right]\Big\}, \end{aligned}$$

if the set under inf is non-empty and $\tau = T$ if the set is empty. Clearly $\tau > t_0$ a.s. on $\{Y_{t_0} = \tilde{Y}_{t_0}\}$. Setting in (5.96) $\theta = t_0$ and $\sigma = \tau$ we obtain

$$\begin{aligned} 0 &\ge \mathbb{E}^{\mathcal{F}_{t_0}}\left(\mathbf{1}_{\{Y_{t_0}=\tilde{Y}_{t_0}\}\cap A}\int_{t_0}^{\tau} \Gamma_s\left(\Phi(s,Y_s,\tilde{Z}_s) - \tilde{\Phi}(s,\tilde{Y}_s,\tilde{Z}_s)\right) dQ_s\right) \\ &\ge \frac{1}{2}\mathbf{1}_{\{Y_{t_0}=\tilde{Y}_{t_0}\}\cap A}\Gamma_{t_0}\left[\Phi(\omega,t_0,Y_{t_0},\tilde{Z}_{t_0}) - \tilde{\Phi}(\omega,t_0,Y_{t_0},\tilde{Z}_{t_0})\right]\mathbb{E}^{\mathcal{F}_{t_0}}\left(Q_\tau - Q_{t_0}\right) \\ &> 0, \ \text{a.s. on } \{Y_{t_0} = \tilde{Y}_{t_0}\} \cap A, \end{aligned}$$

which is a contradiction. Hence $\mathbb{P}\left(\{Y_{t_0} = \tilde{Y}_{t_0}\} \cap A\right) = 0$ and the conclusion (jj) follows. ■

Unlike in the Lipschitz case $\eta > \tilde{\eta}$ does not imply that $Y_t > \tilde{Y}_t$ for all $t \in [0, T]$, as the following example will show. Let $F(x) = \tilde{F}(x) = -\sqrt{x^+}$.

Clearly

$$(Y_t, Z_t) = \left(t^2, 0\right), \quad t \geq 0,$$

is the unique solution of the BSDE

$$Y_t = 1 + \int_t^1 \left(-2\sqrt{Y_s^+}\right) ds - \int_t^1 Z_s \, dB_s, \quad t \in [0, 1],$$

and $(\tilde{Y}_t, \tilde{Z}_t) = (0, 0)$, $t \geq 0$, is the unique solution of

$$\tilde{Y}_t = 0 + \int_t^1 \left(-2\sqrt{\tilde{Y}_s^+}\right) ds - \int_t^1 Z_s \, dB_s, \quad t \in [0, 1].$$

We have $Y_1 = 1 > 0 = \tilde{Y}_1$, but $Y_0 = \tilde{Y}_0$.

5.4 Semilinear Parabolic PDEs

We need to put our BSDE into a Markovian framework: the final condition η and the coefficient F of the BSDE will be functionals of B as "explicit" functions of the solution of a forward SDE driven by $\{B_t\}$.

Let $f : [0, T] \times \mathbb{R}^d \to \mathbb{R}^d$ be continuous and globally monotone in x, uniformly with respect to t, $g : [0, T] \times \mathbb{R}^d \to \mathbb{R}^{d \times d}$ be continuous and globally Lipschitz in x uniformly with respect to t. Let $\{X_s^{t,x} \,;\, t \leq s \leq T\}$ denote the solution of the SDE

$$X_s^{t,x} = x + \int_t^s f(r, X_r^{t,x}) \, dr + \int_t^s g(r, X_r^{t,x}) dB_r, \; t \leq s \leq T, \tag{5.97}$$

and consider the backward SDE

$$Y_s^{t,x} = \kappa(X_T^{t,x}) + \int_s^T F(r, X_r^{t,x}, Y_r^{t,x}, Z_r^{t,x}) \, dr - \int_s^T Z_r^{t,x} \, dB_r, \; t \leq s \leq T, \tag{5.98}$$

where $\kappa : \mathbb{R}^d \to \mathbb{R}^m$ and $F : [0, T] \times \mathbb{R}^d \times \mathbb{R}^m \times \mathbb{R}^{m \times d} \to \mathbb{R}^m$ are continuous and such that for some K, μ, $p > 0$,

$$|\kappa(x)| \leq K(1 + |x|^p),$$

$$\sup_{|y| \leq \rho} |F(t, x, y, 0)| \leq \gamma(\rho, x),$$

$$\langle y - y', F(t,x,y,z) - F(t,x,y',z)\rangle \le \mu(t,x)|y-y'|^2,$$
$$|F(t,x,y,z) - F(t,x,y,z')| \le \ell(t,x)\|z-z'\|,$$

where for each $\rho > 0$, there exists a $K_\rho > 0$ such that $\gamma(\rho,x) \le K_\rho(1+|x|^p)$ and one of the two following conditions hold:

- $|f(t,x)| + |g(t,x)| \le K(1+|x|)$ and $|\mu(t,x)| + \ell^2(t,x) \le K$;
- $|f(t,x)| + |\mu(t,x)| + \ell^2(t,x) \le K(1+|x|)$ and $|g(t,x)| \le K$.

In the case $m > 1$ we reinforce one of the above conditions into

$$|F(t,x,y,z) - F(t,x,y',z)| \le \ell(t,x)|y-y'].$$

This is necessary for our uniqueness proof of the viscosity solution of systems of PDEs, see Theorem 6.106 in Annex D.

Finally the following additional assumption is needed again for the uniqueness of viscosity solutions

$$|F(t,x,r,p) - F(t,y,r,p)| \le \mathbf{m}_R(|x-y|(1+|p|)),$$

for all $x, y \in \mathbb{R}^d$ such that $|x| \le R$, $|y| \le R$, $r \in \mathbb{R}^m$, $p \in \mathbb{R}^d$, where for each $R > 0$, $\mathbf{m}_R \in C(\mathbb{R}_+)$ is increasing and $\mathbf{m}_R(0) = 0$.

Remark 5.35. (i) Clearly, for each $t \le s \le T$, $Y_s^{t,x}$ is $\mathcal{F}_s^t = g\{B_r - B_t,\ t \le r \le s\} \vee \mathcal{N}$ measurable, where $\mathcal{N}$ is the class of the $\mathbb{P}$-null sets of $\mathcal{F}$. Hence $Y_t^{t,x}$ is a.s. constant (i.e. deterministic).

(ii) It is not hard to see, using uniqueness for BSDEs, that $Y_{t+h}^{t,x} = Y_{t+h}^{t+h,X_{t+h}^{t,x}}$, $h > 0$.

We shall denote by

$$\mathcal{A}_t = \frac{1}{2}\sum_{i,j}(gg^*)_{ij}(t,x)\frac{\partial^2}{\partial x_i \partial x_j} + \sum_i f_i(t,x)\frac{\partial}{\partial x_i}$$

the infinitesimal generator of the Markov process $\{X_s^{t,x};\ t \le s \le T\}$.

5.4.1 Parabolic Systems in the Whole Space

We first consider the following system of backward semilinear parabolic PDEs

$$\begin{cases} \dfrac{\partial u_i}{\partial t}(t,x) + \mathcal{A}_t u_i(t,x) + F_i(t,x,u(t,x),(\nabla u g)(t,x)) = 0, \\ \qquad\qquad (t,x) \in [0,T]\times\mathbb{R}^d, \quad 0 \le i \le m; \\ u(T,x) = \kappa(x), \quad x \in \mathbb{R}^d; \end{cases} \tag{5.99}$$

where $F \in C([0,T] \times \mathbb{R}^d \times \mathbb{R}^m \times \mathbb{R}^{m\times d}; \mathbb{R}^m)$, and $\kappa \in C(\mathbb{R}^d, \mathbb{R}^m)$ grows at most polynomially at infinity.

We can first establish the following:

Theorem 5.36. *Let $u \in C^{1,2}([0,T] \times \mathbb{R}^d; \mathbb{R}^m)$ be a classical solution of (5.99). Then for each $(t,x) \in [0,T] \times \mathbb{R}^d$, $\{(u(s, X_s^{t,x}), (\nabla ug)(s, X_s^{t,x})); t \le s \le T\}$ is the solution of the BSDE (5.98). In particular, $u(t,x) = Y_t^{t,x}$.*

Proof. The result follows by applying Itô's formula to $u(s, X_s^{t,x})$. ∎

We now want to connect (5.97)–(5.98) with (5.99) in the other direction, i.e. prove that (5.97)–(5.98) provides a solution of (5.99). In order to avoid restrictive assumptions on the coefficients in (5.97)–(5.98), we will consider (5.99) in the viscosity sense. This imposes just one restriction. Indeed for the notion of viscosity solution of the system of PDEs (5.99) to make sense, we need to make the following restriction: *for $0 \le i \le k$, the i-th coordinate of F depends only on the i-th row of the matrix z.* Then the first line in (5.99) reads

$$\frac{\partial u_i}{\partial t}(t,x) + \mathcal{A}_t u_i(t,x) + F_i(t,x,u(t,x),(\nabla u_i g)(t,x)) = 0,$$

which we rewrite in the form

$$-\frac{\partial u_i}{\partial t}(t,x) + \Phi_i(t,x,u(t,x),Du_i(t,x),D^2 u_i(t,x)) = 0,$$

where

$$\Phi : \mathbb{R}_+ \times \mathbb{R}^d \times \mathbb{R}^m \times \mathbb{R}^d \times \mathbb{S}^d \to \mathbb{R}^m$$

is defined by

$$\Phi_i(t,x,r,p,X) = -\frac{1}{2}\mathbf{Tr}[(gg^*)(t,x)X] - \langle f, p\rangle - F_i(t,x,r,pg(t,x)),$$

for all $1 \le i \le m$, $(t,x) \in [0,T] \times \mathbb{R}^d$, $r \in \mathbb{R}^m$, $p \in \mathbb{R}^d$, $X \in \mathbb{S}^d$.

We add the following assumptions. For each $\rho > 0$, there exists a K_ρ such that for some $p > 1$, all $(t,x) \in [0,T] \times \mathbb{R}^d$, $\rho > 0$,

$$\sup_{\{|y| \le \rho\}} |F(t,x,y,0)| \le K_\rho(1 + |x|^p),$$

and there exists a $K > 0$ such that for all $(t,x) \in [0,T] \times \mathbb{R}^d$, $y, y' \in \mathbb{R}^m$, $z, z' \in \mathbb{R}^d$,

$$|F(t,x,y,z) - F(t,x,y',z)| + |F(t,x,y,z) - F(t,x,y,z')| \le K(|y-y'| + |z-z'|).$$

The definition of the viscosity solution of a system of elliptic PDEs is given in Definition 6.94 in Annex D. The adaptation to systems of parabolic PDEs is obvious.

We now establish the main result of this section.

Theorem 5.37. *Under the above assumptions,* $u(t,x) \stackrel{def}{=} Y_t^{t,x}$ *is a continuous function of* (t,x) *and it is the unique viscosity solution of (5.99) which grows at most polynomially at infinity.*

Proof. Uniqueness follows from Theorem 6.106 in Annex D.

The continuity follows from the mean-square continuity of $\{Y_s^{t,x},\ x \in \mathbb{R}^d,\ 0 \le t \le s \le T\}$, which in turn follows from the continuity of $X_\cdot^{t,x}$ with respect to t, x and Theorem 5.10. The polynomial growth follows from classical moment estimates for $X_\cdot^{t,x}$, the assumptions on the growth of f and g, and Proposition 5.7.

To prove that u is a viscosity sub-solution, take any $1 \le i \le k$, $\varphi \in C^{1,2}([0,T] \times \mathbb{R}^d)$ and $(t,x) \in [0,T) \times \mathbb{R}^d$ such that $u_i - \varphi$ has a local maximum at (t,x). We assume without loss of generality that

$$u_i(t,x) = \varphi(t,x).$$

We suppose that

$$-\frac{\partial\varphi}{\partial t}(t,x) + \Phi_i(t,x,u(t,x),D\varphi(t,x),D^2\varphi(t,x)) > 0,$$

and we will find a contradiction.

Let $0 < \alpha \le T - t$ be such that for all $t \le s \le t + \alpha$, $|y - x| \le \alpha$,

$$u_i(s,y) \le \varphi(s,y),$$

$$-\frac{\partial\varphi}{\partial t}(s,y) + \Phi_i(s,y,u(s,y),D\varphi(s,y),D^2\varphi(s,y)) > 0,$$

and define

$$\tau = \inf\{s \ge t;\ |X_s^{t,x} - x| \ge \alpha\} \wedge (t + \alpha).$$

Let now

$$(\overline{Y}_s, \overline{Z}_s) = ((Y_{s\wedge\tau}^{t,x})^i, \mathbf{1}_{[0,\tau]}(s)(Z_s^{t,x})^i),\ t \le s \le t + \alpha.$$

It follows from the statement in Remark 5.35(ii) that

$$Y_{t+h}^{t,x} = u(t+h, X_{t+h}^{t,x}).$$

We hence have that (first approximating τ by a sequence of stopping times taking at most finitely many values)

$$Y_\tau^{t,x} = u(\tau, X_\tau^{t,x}).$$

Consequently $(\overline{Y}, \overline{Z})$ solves the one-dimensional BSDE

$$\overline{Y}_s = u_i(\tau, X_\tau^{t,x}) + \int_s^{t+\alpha} \mathbf{1}_{[0,\tau]}(r) F_i(r, X_r^{t,x}, u(r, X_r^{t,x}), \overline{Z}_r) dr \\ - \int_s^{t+\alpha} \overline{Z}_r dB_r, \ t \le s \le t+\alpha.$$

On the other hand, from Itô's formula,

$$(\hat{Y}_s, \hat{Z}_s) = (\varphi(s, X_{s\wedge\tau}^{t,x}), \mathbf{1}_{[0,\tau]}(s)(\nabla\varphi g)(s, X_s^{t,x})), \ t \le s \le t+\alpha$$

solves the one-dimensional BSDE, for all $s \in [t, t+\alpha]$

$$\hat{Y}_s = \varphi(\tau, X_\tau^{t,x}) - \int_s^{t+\alpha} \mathbf{1}_{[0,\tau]}(r)(\frac{\partial\varphi}{\partial r} + \mathcal{A}\varphi)(r, X_r^{t,x}) dr - \int_s^{t+\alpha} \hat{Z}_r dB_r.$$

From $u_i(\tau, X_\tau^{t,x}) \le \varphi(\tau, X_\tau^{t,x})$ and the choices of α and τ, we deduce from Proposition 5.34 that $\overline{Y}_t < \hat{Y}_t$, i.e. $u_i(t,x) < \varphi(t,x)$, which contradicts our standing assumption. ■

Remark 5.38. Suppose that $k = 1$ and F has the special form:

$$F(t,x,r,z) = c(t,x)r + h(t,x).$$

In that case, the BSDE is linear:

$$Y_s^{t,x} = \kappa(X_T^{t,x}) + \int_s^T [c(r, X_r^{t,x}) Y_s^{t,x} + h(r, X_r^{t,x})] \, dr - \int_s^T Z_r^{t,x} \, dB_r,$$

hence it has an explicit solution (see Proposition 5.31):

$$Y_s^{t,x} = \kappa(X_T^{t,x}) e^{\int_s^T c(r, X_r^{t,x}) \, dr} + \int_s^T h(r, X_r^{t,x}) e^{\int_s^r c(\alpha, X_\alpha^{t,x}) \, d\alpha} \, dr \\ - \int_s^T e^{\int_s^r c(\alpha, X_\alpha^{t,x}) d\alpha} Z_r^{t,x} \, dB_r.$$

Now $Y_t^{t,x} = \mathbb{E}(Y_t^{t,x})$, so that

$$Y_t^{t,x} = \mathbb{E}\left[\kappa(X_T^{t,x}) e^{\int_t^T c(s, X_s^{t,x}) \, ds} + \int_t^T h(s, X_s^{t,x}) e^{\int_t^s c(r, X_r^{t,x}) dr} ds\right],$$

which is the well-known Feynman–Kac formula.

Clearly, Theorem 5.37 can be considered as a nonlinear extension of the Feynman–Kac formula.

Remark 5.39. We have proved that a certain function of (t, x), defined via the solution of a probabilistic problem, is the solution of a system of backward parabolic partial differential equations. Suppose that b, g and f do not depend on t, and let

$$v(t,x) = u(T-t,x),\ (t,x) \in [0,T] \times \mathbb{R}^d.$$

Then v solves the system of forward parabolic PDEs:

$$\frac{\partial v_i}{\partial t}(t,x) = \mathcal{A}v_i(t,x) + F_i(x, v(t,x), (\nabla v_i g)(t,x)),\ 1 \le i \le m, t > 0, x \in \mathbb{R}^d;$$
$$v(0,x) = \kappa(x),\ x \in \mathbb{R}^d.$$

On the other hand, we have that

$$v(t,x) = Y^{T-t,x}_{T-t} = \bar{Y}^{t,x}_0,$$

where $\{(\bar{Y}^{t,x}_s, \bar{Z}^{t,x}_s);\ 0 \le s \le t\}$, solves the BSDE

$$\bar{Y}^{t,x}_s = \kappa(X^x_t) + \int_s^t F(X^x_r, \bar{Y}^{t,x}_r, \bar{Z}^{t,x}_r)dr$$
$$- \int_s^t \bar{Z}^{t,x}_r dB_r, 0 \le s \le t.$$

So we have a probabilistic representation for a system of forward parabolic PDEs, which is valid on $\mathbb{R}_+ \times \mathbb{R}^d$.

5.4.2 Parabolic Dirichlet Problem

We now combine the situation of the preceding subsection with that of Sect. 3.8.3, and we consider the following system of parabolic semilinear PDEs with Dirichlet boundary condition

$$\begin{cases} -\dfrac{\partial u_i}{\partial t}(t,x) + \Phi_i(t,x,u(t,x), Du_i(t,x), D^2u_i(t,x)) = 0, \\ \qquad\qquad (t,x) \in [0,T] \times D, \quad 0 \le i \le m; \\ u(T,x) = \kappa(x), \quad x \in \overline{D}; \\ u(t,x) = \chi(t,x), \quad (t,x) \in [0,T] \times \partial D; \end{cases} \tag{5.100}$$

where in addition to the situation in the previous subsection, we give ourselves a function $\chi \in C([0,T] \times \partial D)$. We assume that

$$\chi(T,x) = \kappa(x),\ \forall x \in \partial D.$$

Now, together with the SDE (5.97), for each $(t,x) \in [0,T] \times \overline{D}$ we consider the BSDE, for all $s \in [t,T]$

$$Y_s^{t,x} = \chi(\tau_{t,x} \wedge T, X_{\tau_{t,x}\wedge T}^{t,x}) + \int_s^T \mathbf{1}_{\{r<\tau_{t,x}\}} F(r, X_r^{t,x}, Y_r^{t,x}, Z_r^{t,x})dr - \int_s^T Z_r^{t,x} dB_r. \tag{5.101}$$

Again Itô's formula allows us to establish the following:

Theorem 5.40. *Suppose that the above conditions on the coefficients and the domain D are satisfied. Let $u \in C^{1,2}([0,T] \times D; \mathbb{R}^m) \cap C([0,T] \times \overline{D}; \mathbb{R}^m)$ be a classical solution of (5.100). Then for each $(t,x) \in [0,T] \times D$,*

$$\{(u(s \wedge \tau_{t,x}, X_{s\wedge\tau_{t,x}}^{t,x}, \mathbf{1}_{\{s<\tau_{t,x}\}}(\nabla u g)(s, X_s^{t,x})),\ t \le s \le T\}$$

is the solution of the BSDE (5.101).

We now wish to prove that $u(t,x) := Y_t^{t,x}$ is a viscosity solution of (5.100). From the discussion in Sect. 3.8.3, we deduce that the condition (3.111) is necessary for u to be continuous.

We now prove the following:

Theorem 5.41. *Under the above conditions, including those of Theorem 5.37 and (3.111), $u(t,x) := Y_t^{t,x}$ is a continuous function from $[0,T] \times \overline{D}$ into $\mathbb{R}^m$, and it is the unique viscosity solution of (5.100).*

Proof. Uniqueness follows from the arguments developed in Annex D. The continuity of u follows from Proposition 3.45 and the argument at the beginning of the proof of Theorem 5.37.

Let us prove that u is a viscosity sub-solution. Let $1 \le i \le k$, $\varphi \in C^{1,2}([0,T] \times \mathbb{R}^d)$ and $(t,x) \in [0,T) \times \overline{D}$ be such that $u_i - \varphi$ has a local maximum at (t,x), and $u_i(t,x) = \varphi(t,x)$. If $x \in D$, then the argument in the proof of Theorem 5.37 (with $\alpha < d(x, \partial D)$) establishes the required inequality. The same is true if $(t,x) \in [0,T] \times \partial D \backslash \Lambda$ (this time choosing $\alpha < d((t,x), \Lambda)$). Finally if $(t,x) \in \Lambda$, then $\tau_{t,x} = t$, a.s., hence $u(t,x) = \chi(t,x)$. The result follows. ∎

5.4.3 Parabolic Neumann Problem

We use again the notations from Sect. 5.4.1, and we add a nonlinear Neumann condition on the boundary of the bounded open connected subset D of $\mathbb{R}^d$, whose boundary ∂D is assumed to be of class C^2.

Let

$$G \in C([0,T] \times \partial D \times \mathbb{R}^m; \mathbb{R}^m)$$

be such that for some $\rho > 0$,

$$|G(t,x,y) - G(t,x,y')| \le K|y-y'|, \tag{5.102}$$

for all $(t,x) \in [0,T] \times \partial D$, $y, y' \in \mathbb{R}^m$.

We now consider the following system of semilinear parabolic PDEs with nonlinear Neumann boundary condition:

$$\begin{cases} -\dfrac{\partial u_i}{\partial t}(t,x) + \Phi_i(t,x,u(t,x), Du_i(t,x), D^2u_i(t,x)) = 0, \\ \qquad\qquad (t,x) \in [0,T] \times D, \quad 0 \le i \le m; \\ u(T,x) = \kappa(x), \quad x \in \overline{D}; \\ \dfrac{\partial u_i}{\partial n}(t,x) - G_i(t,x,u(t,x)) = 0, \ 1 \le i \le m, \ (t,x) \in [0,T] \times \partial D. \end{cases} \tag{5.103}$$

Let $X^{t,x}$ be the process solution of the reflected stochastic differential equation, for all $s \in [t, T,]$, $\mathbb{P}$-$a.s.$

$$\begin{cases} X_s^{t,x} + K_s^{t,x} = x + \displaystyle\int_t^s f(r, X_r^{t,x})dr + \int_t^s g(r, X_r^{t,x})dB_r, \\ X_s^{t,x} \in \overline{D}, \\ K_s^{t,x} = \displaystyle\int_t^s n(X_r^{t,x})\mathbf{1}_{\partial D}\left(X_r^{t,x}\right) d \updownarrow K^{t,x} \updownarrow_r . \end{cases} \tag{5.104}$$

To each $(t,x) \in [0,T] \times \overline{D}$ we associate the BSDE

$$Y_s^{t,x} = \kappa(X_T^{t,x}) + \int_s^T F(r, X_r^{t,x}, Y_r^{t,x}, Z_r^{t,x})dr \tag{5.105}$$

$$+ \int_s^T G(r, X_r^{t,x}, Y_r^{t,x}) d \updownarrow K^{t,x} \updownarrow_r - \int_s^T Z_r^{t,x} dB_r, \ t \le s \le T.$$

Itô's formula again allows us to establish the following:

Theorem 5.42. *Under the above assumptions on the coefficients and the domain D, if $u \in C^{1,2}([0,T] \times D; \mathbb{R}^m) \cap C^{0,1}([0,T] \times \overline{D}; \mathbb{R}^m)$ is a classical solution of (5.103), then for each $(t,x) \in [0,T] \times \mathbb{R}^d$, $\{(u(s, X_s^{t,x}), (\nabla ug)(s, X_s^{t,x})); t \le s \le T\}$ is the solution of the BSDE (5.104).*

Inspired by [59] we now prove:

Theorem 5.43. *Under the above conditions, including those of Theorem 5.37, if in addition either G does not depend upon its third argument, or else the additional assumptions from Proposition 5.83 below are satisfied, then $u(t,x) := Y_t^{t,x}$ is a continuous function from $[0,T] \times \overline{D}$ into $\mathbb{R}^m$, and it is the unique viscosity solution of (5.103).*

Proof. Uniqueness follows from a combination of the arguments in the proofs of Theorems 6.106 and 6.112. If G does not depend upon its third argument, then the continuity of u follows from Corollary 4.56 combined with the argument at the beginning of the proof of Theorem 5.37. In the other case, we refer to [46] for the proof of the continuity of u. We now prove that u is a viscosity sub-solution. Let $1 \le i \le k$, $\varphi \in C^{1,2}([0,T] \times \mathbb{R}^d)$ and $(t,x) \in [0,T) \times \overline{D}$ be such that $u_i - \varphi$ has a local maximum at (t,x), and $u_i(t,x) = \varphi(t,x)$. The case where $x \in D$ is treated as in the proof of Theorem 5.37. Suppose now that $x \in \partial D$. As usual we argue by contradiction. Suppose that for some $\alpha > 0$, all $(s,y) \in B((t,x),\alpha) \cap \overline{D}$ satisfy

$$\begin{cases} -\dfrac{\partial \varphi}{\partial t}(s,y) + \Phi_i(s,y,u(s,y),D\varphi(s,y),D^2\varphi(s,y)) > 0, \\ \qquad\qquad \dfrac{\partial \varphi}{\partial n}(s,y) - G_i(s,y,u(s,y)) > 0, \text{ if } y \in \partial D. \end{cases}$$

The contradiction can now be established as in the proof of Theorem 5.37, making use of the strict comparison result from Proposition 5.34. ■

5.5 BSDEs with a Subdifferential Coefficient

5.5.1 *Uniqueness*

We extend the estimates and the uniqueness result in the case of the multivalued BSDE

$$\begin{cases} -dY_t + \partial\varphi(Y_t)\,dt + \partial\psi(Y_t)\,dA_t \\ \qquad\qquad \ni F(t,Y_t,Z_t)\,dt + G(t,Y_t)\,dA_t - Z_t dB_t, \quad 0 \le t < T, \\ Y_T = \eta, \end{cases} \tag{5.106}$$

where again $T > 0$ is a fixed deterministic time and $\partial\varphi$ and $\partial\psi$ are subdifferential operators attached to the convex lower semicontinuous functions $\varphi, \psi : \mathbb{R}^m \to]-\infty, +\infty]$.

Such multivalued backward stochastic differential equations are also called *backward stochastic variational inequalities (BSVI).*

It is natural here to assume there exists a $u_0 \in \mathbb{R}^m$ such that $\partial\varphi(u_0) \neq \emptyset$ and $\partial\psi(u_0) \neq \emptyset$.

If $Q_t(\omega) \stackrel{def}{=} t + A_t(\omega)$ and $\{\alpha_t : t \in [0,T]\}$ is a real positive $\mathcal{P}$-m.s.p. (given by the Radon–Nikodym representation theorem) such that $0 \le \alpha \le 1$ and

$$dt = \alpha_t dQ_t \quad \text{and} \quad dA_t = (1-\alpha_t)\,dQ_t,$$

then the Eq. (5.106) becomes

$$\begin{cases} -dY_t + \partial_y \Psi(t,Y_t)\,dQ_t \ni \Phi(t,Y_t,Z_t)\,dQ_t - Z_t dB_t, \quad 0 \le t < T, \\ Y_T = \eta, \end{cases} \tag{5.107}$$

where

$$\Phi(\omega,t,y,z) \overset{def}{=} \alpha_t(\omega) F(\omega,t,y,z) + (1-\alpha_t(\omega)) G(\omega,y),$$

$$\Psi(\omega,t,y) \overset{def}{=} \alpha_t(\omega) \varphi(y) + (1-\alpha_t(\omega)) \psi(y),$$

(we use the convention $0 \cdot \infty = 0$ and write $\partial\Psi$ for $\partial_y \Psi$).

We also remark that if $u_0 \in \mathrm{Dom}(\partial\varphi) \bigcap \mathrm{Dom}(\partial\psi)$, $\hat{u}_{01} \in \partial\varphi(u_0)$ and $\hat{u}_{02} \in \partial\psi(u_0)$, then

$$\hat{u}_t(\omega) = \alpha_t(\omega)\hat{u}_{01} + (1-\alpha_t(\omega))\hat{u}_{02} \in \partial_y \Psi(\omega,t,u_0).$$

We shall assume that the following assumptions hold:

$$\left(\mathbf{BSVI\text{-}H}_{\eta,\Psi,\Phi}\right): \tag{5.108}$$

(*i*) $\eta : \Omega \to \mathbb{R}^m$ *is an* $\mathcal{F}_T$*-measurable random vector;*
(*ii*) Q is a $\mathcal{P}$*-m.i.c.s.p. such that* $Q_0 = 0$;
(*iii*) $(\omega,t) \longmapsto \alpha_t(\omega) : \Omega \times [0,T] \to [0,1]$ *is* $\mathcal{P}$*-m.s.p. such that* $\alpha_t dQ_t = dt$;
(*iv*) $\Phi : \Omega \times [0,T] \times \mathbb{R}^m \times \mathbb{R}^{m\times k} \to \mathbb{R}^m$ *satisfies the assumptions* (5.13-**BSDE-H**$_\Phi$);
(*v*) $\Psi : \Omega \times [0,T] \times \mathbb{R}^m \to]-\infty,+\infty]$ *satisfies*

- ▲ $\Psi(\cdot,\cdot,y)$ is $\mathcal{P}$*-m.s.p. for all* $y \in \mathbb{R}^m$,
- ▲ $y \longmapsto \Psi(\omega,t,y) : \mathbb{R}^m \to]-\infty,+\infty]$ *is a proper convex l.s.c. function,*
- ▲ $\exists\, u_0 \in \mathbb{R}^m$ *and an* $\mathbb{R}^m$*-valued* $\mathcal{P}$*-m.s.p.* $(\hat{u}_t)_{t\in[0,T]}$ such that

$$(u_0,\hat{u}_t) \in \partial_y \Psi(\omega,t,\cdot), \quad d\mathbb{P}\otimes dt\text{-}a.e.\ (\omega,t) \in \Omega \times [0,T].$$

□

Definition 5.44. A pair $(Y,Z) \in S_m^0[0,T] \times \Lambda^0_{m\times k}(0,T)$ of stochastic processes is a solution of the backward stochastic variational inequality (5.107) if there exist $K \in S_m^0[0,T]$, $K_0 = 0$, such that

$$\begin{aligned}
&(a)\quad \updownarrow K \updownarrow_T + \int_0^T |\Psi(t,Y_t)|\, dQ_t + \int_0^T |\Phi(t,Y_t,Z_t)|\, dQ_t < \infty, \text{ a.s.,}\\
&(b)\quad dK_t \in \partial_y \Psi(t,Y_t)\, dQ_t, \text{ a.s. that is: } \mathbb{P}\text{-a.s.,}\\
&\qquad \int_t^s \langle y(r) - Y_r, dK_r\rangle + \int_t^s \Psi(r,Y_r)dQ_r \le \int_t^s \Psi(r,y(r))dQ_r,\\
&\qquad\qquad\qquad\qquad \forall y \in C([0,T];\mathbb{R}^m),\ \forall\, 0 \le t \le s \le T,
\end{aligned}$$

and $\mathbb{P}$-a.s., for all $t \in [0,T]$:

$$Y_t + K_T - K_t = \eta + \int_t^T \Phi(s,Y_s,Z_s)\, dQ_s - \int_t^T Z_s dB_s, \text{ a.s.} \tag{5.109}$$

(we also say that the triple (Y,Z,K) is a solution of the Eq. (5.107)).

Remark 5.45. If K is absolutely continuous with respect to dQ_t, i.e. there exists a progressively measurable stochastic process U such that

$$\int_0^T |U_t|\, dQ_t < \infty, \text{ a.s. and } K_t = \int_0^t U_s dQ_s, \text{ for all } t \in [0,T],$$

then $dK_t \in \partial\Psi(t, Y_t)\, dQ_t$ means, $\mathbb{P}$-a.s. $\omega \in \Omega$,

$$U_t \in \partial\Psi_y(t, Y_t), \quad dQ_t\text{-a.e.}$$

In this case we also say that the triple (Y, Z, U) is a solution of the Eq. (5.107).

If $dK_t \in \partial\Psi_y(t, Y_t)\, dQ_t$, $d\tilde{K}_t \in \partial\Psi_y(t, \tilde{Y}_t) dQ_t$ and

$$\int_0^T |\Psi(t, Y_t)|\, dQ_t + \int_0^T |\Psi(t, \tilde{Y}_t)| dQ_t < \infty, \text{ a.s.},$$

then, using the subdifferential inequalities

$$\begin{aligned} &\int_t^s \langle \tilde{Y}_r - Y_r, dK_r\rangle + \int_t^s \Psi(r, Y_r) dQ_r \le \int_t^s \Psi(r, \tilde{Y}_r) dQ_r, \\ &\int_t^s \langle Y_r - \tilde{Y}_r, d\tilde{K}_r\rangle + \int_t^s \Psi(r, \tilde{Y}_r) dQ_r \le \int_t^s \Psi(r, Y_r) dQ_r, \end{aligned}$$

we infer that, for all $0 \le t \le s \le T$

$$\int_t^s \langle Y_r - \tilde{Y}_r, dK_r - d\tilde{K}_r\rangle \ge 0, \ a.s. \tag{5.110}$$

Let $a, p > 1$ and

$$V_t = V_t^{a,p} \stackrel{def}{=} \int_0^t \left[\mu_s dQ_s + \frac{a}{2n_p} (\ell_s)^2 ds\right] \quad \text{and} \quad \bar{\mu}_t = \int_0^t \mu_s dQ_s.$$

Recall the notations

$$S_m^p\left([0,T]; e^{\bar{\mu}}\right) = \left\{Y \in S_m^0([0,T]) : e^{\bar{\mu}} Y \in S_m^p([0,T])\right\}$$

and

$$S_m^{1+}\left([0,T]; e^{\bar{\mu}}\right) = \bigcup_{p>1} S_m^p\left([0,T]; e^{\bar{\mu}}\right).$$

Note that if μ is a determinist process then $S_m^p\left([0,T]; e^{\bar{\mu}}\right) = S_m^p([0,T])$.

Proposition 5.46. *Let the assumptions* $\left(\mathbf{BSVI\text{-}H}_{\eta,\Psi,\Phi}\right)$ *be satisfied. Then for every* $a, p > 1$ *there exists a constant* $C_{a,p}$ *such that for all solutions* $(Y, Z) \in S_m^0[0,T] \times \Lambda_{m\times k}^0(0,T)$ *of the BSDE (5.107) satisfying*

$$\mathbb{E} \sup_{s\in[0,T]} e^{pV_s} |Y_s - u_0|^p < \infty,$$

the following inequality holds $\mathbb{P}$*-a.s., for all* $t \in [0,T]$*:*

$$\begin{aligned}
&\mathbb{E}^{\mathcal{F}_t} \sup_{s\in[t,T]} e^{pV_s} |Y_s - u_0|^p + \mathbb{E}^{\mathcal{F}_t} \left(\int_t^T e^{2V_s} |Z_s|^2 ds \right)^{p/2} \\
&\qquad + \mathbb{E}^{\mathcal{F}_t} \left(\int_t^T e^{2V_s} |\Psi(s,Y_s) - \Psi(s,u_0)| \, dQ_s \right)^{p/2} \\
&\le C_{a,p} \left[\mathbb{E}^{\mathcal{F}_t} e^{pV_T} |\eta - u_0|^p + \mathbb{E}^{\mathcal{F}_t} \left(\int_t^T e^{V_s} \left[|\hat{u}_s| + |\Phi(s,u_0,0)| \right] dQ_s \right)^p \right].
\end{aligned} \tag{5.111}$$

Proof. We have

$$Y_t - u_0 = \eta - u_0 + \int_t^T \left[\Phi(s,Y_s,Z_s) \, dQ_s - dK_s \right] - \int_t^T Z_s dB_s.$$

Note that

$$\begin{aligned}
&\langle Y_t - u_0, \Phi(t,Y_t,Z_t) \rangle \, dQ_t \\
&\quad = \langle Y_t - u_0, (\Phi(t,Y_t,Z_t) - \Phi(t,u_0,Z_t)) \rangle \, dQ_t \\
&\qquad + \langle Y_t - u_0, \Phi(t,u_0,Z_t) - \Phi(t,u_0,0) \rangle \, dQ_t + \langle Y_t - u_0, \Phi(t,u_0,0) \rangle \, dQ_t \\
&\quad \le |Y_t - u_0|^2 \mu_t dQ_t + |Y_t - u_0| \, |Z_t| \, \ell_t dt + |Y_t - u_0| \, |\Phi(t,u_0,0)| \, dQ_t \\
&\quad \le |Y_t - u_0| \, |\Phi(t,u_0,0)| \, dQ_t + |Y_t - u_0|^2 dV_t + \frac{n_p}{2a} |Z_t|^2 dt,
\end{aligned}$$

where $n_p = (p-1) \wedge 1$.

From the subdifferential inequalities we have

$$\begin{aligned}
&|\Psi(t,Y_t) - \Psi(t,u_0)| \le \Psi(t,Y_t) - \Psi(t,u_0) + 2 |\hat{u}_t| \, |Y_t - u_0|, \quad \text{and} \\
&[\Psi(t,Y_t) - \Psi(t,u_0)] \, dQ_t \le \langle Y_t - u_0, dK_t \rangle,
\end{aligned}$$

so

$$|\Psi(t,Y_t) - \Psi(t,u_0)| \, dQ_t \le \langle Y_t - u_0, dK_t \rangle + 2 |\hat{u}_t| \, |Y_t - u_0| \, dQ_t.$$

Hence

$$\begin{aligned}
&|\Psi(t,Y_t) - \Psi(t,u_0)| \, dQ_t + \langle Y_t - u_0, \Phi(t,Y_t,Z_t) \, dQ_t - dK_t \rangle \\
&\le |Y_t - u_0| \left[2 |\hat{u}_t| + |\Phi(t,u_0,0)| \right] dQ_t + |Y_t - u_0|^2 dV_t + \frac{n_p}{2a} |Z_t|^2 dt.
\end{aligned}$$

Now (5.111) follows from Proposition 5.2. ∎

Corollary 5.47. *Let $p = 1$. Let the assumptions $(\mathbf{BSVI\text{-}H}_{\eta,\Psi,\Phi})$ be satisfied and $\Phi(t,y,z) \equiv \Phi(t,y)$ for all $t \in [0,T]$, $y \in \mathbb{R}^m$ and $z \in \mathbb{R}^{m\times k}$ (Φ is independent of z; $\ell_t \equiv 0$ and $V_t = \bar{\mu}_t = \int_0^t \mu_s dQ_s$). Let*

$$dN_t = [|\hat{u}_t| + |\Phi(t,u_0,0)|]\, dQ_t.$$

If $(Y,Z) \in S_m^0[0,T] \times \Lambda_{m\times k}^0(0,T)$ is a solution of the BSDE (5.107) satisfying

$$\mathbb{E} \sup_{s\in[0,T]} e^{\bar{\mu}_s} |Y_s - u_0| < \infty,$$

then the following inequality holds $\mathbb{P}$-a.s., for all $t \in [0,T]$:

$$e^{\bar{\mu}_t} |Y_t - u_0| \le \mathbb{E}^{\mathcal{F}_t} e^{\bar{\mu}_T} |\eta - u_0| + \mathbb{E}^{\mathcal{F}_t} \int_t^T e^{\bar{\mu}_s} dN_s.$$

Moreover for every $q \in (0,1)$ there exists a constant C_q such that

$$\begin{aligned} &\sup_{s\in[0,T]} \left(\mathbb{E}\left(e^{\bar{\mu}_s}|Y_s|\right)\right)^q + \mathbb{E} \sup_{s\in[0,T]} e^{q\bar{\mu}_s}|Y_s|^q \\ &\quad + \mathbb{E}\left(\int_0^T e^{2\bar{\mu}_s}|Z_s|^2 ds\right)^{q/2} + \mathbb{E}\left(\int_0^T e^{2\bar{\mu}_s}|\Psi(s,Y_s) - \Psi(s,u_0)|\, dQ_s\right)^{q/2} \\ &\le C_q \left[\left(\mathbb{E}\left(e^{\bar{\mu}_T}|\eta - u_0|\right)\right)^q + \left(\mathbb{E}\int_0^T e^{\bar{\mu}_s} dN_s\right)^q\right]. \end{aligned}$$

Proof. From the proof of Proposition 5.46 we have

$$\begin{aligned} &|\Psi(t,Y_t) - \Psi(t,u_0)|\, dQ_t + \langle Y_t - u_0, \Phi(t,Y_t,Z_t)\, dQ_t - dK_t\rangle \\ &\le |Y_t - u_0|\,[2|\hat{u}_t| + |\Phi(t,u_0,0)|]\, dQ_t + |Y_t - u_0|^2\, d\bar{\mu}_t \end{aligned}$$

and the conclusions follow by Corollary 6.81. ■

Remark 5.48. A consequence of (5.111) is the following. Denoting

$$\Theta = e^{V_T}|\eta - u_0| + \int_0^T e^{V_s}[|\hat{u}_s| + |\Phi(s,u_0,0)|]\, dQ_s,$$

then for all $t \in [0,T]$:

$$|Y_t| \le |u_0| + C_{a,p}^{1/p} e^{-V_t} \left(\mathbb{E}^{\mathcal{F}_t}\Theta^p\right)^{1/p}, \quad a.s. \tag{5.112}$$

Corollary 5.49. *Let $p \ge 2$, $r_0 > 0$ and*

$$\Psi^{\#}_{u_0,r_0}(t) \overset{def}{=} \sup\{\Psi(t, u_0 + r_0 v) : |v| \le 1\}.$$

Then

$$\begin{aligned}
r_0^{p/2}\mathbb{E}\left(\int_0^T e^{2V_s} d \updownarrow K \updownarrow_s\right)^{p/2} &\le C_{a,p}^{(r_0)}\Big[\mathbb{E}\, e^{pV_T}\,|\eta - u_0|^p \\
&+\mathbb{E}\left(\int_0^T e^{2V_s}\left[\Psi^{\#}_{u_0,r_0}(s) - \Psi(s,u_0)\right]dQ_s\right)^{p/2} \\
&+\mathbb{E}\left(\int_0^T e^{V_s}\left[|\hat{u}_s| + |\Phi(s,u_0,0)|\right]dQ_s\right)^{p}\Big].
\end{aligned} \tag{5.113}$$

Proof. Let $v \in C([0,T];\mathbb{R}^m)$ be arbitrary. From the subdifferential inequality

$$\langle u_0 + r_0 v(t) - Y_t, dK_t\rangle + \Psi(t,Y_t)dQ_t \le \Psi(t, u_0 + r_0 v(t))\, dQ_t,$$

we deduce

$$r_0 d \updownarrow K \updownarrow_t + \Psi(t,Y_t)dQ_t \le \langle Y_t - u_0, dK_t\rangle + \Psi^{\#}_{u_0,r_0}(t)\, dQ_t.$$

Since

$$\langle Y_t - u_0, \hat{u}_t\rangle + \Psi(t,u_0) \le \Psi(t,Y_t),$$

we see that

$$r_0 d \updownarrow K \updownarrow_t \le \langle Y_t - u_0, dK_t\rangle + |\hat{u}_t|\,|Y_t - u_0|\, dQ_t + \left[\Psi^{\#}_{u_0,r_0}(t) - \Psi(t,u_0)\right]dQ_t.$$

Therefore

$$\begin{aligned}
&r_0 d \updownarrow K \updownarrow_t + \langle Y_t - u_0, \Phi(t,Y_t,Z_t)\, dQ_t - dK_t\rangle \\
&\quad\le \left[\Psi^{\#}_{u_0,r_0}(t) - \Psi(t,u_0)\right]dQ_t + |Y_t - u_0|\left[|\hat{u}_t| + |\Phi(t,u_0,0)|\right]dQ_t \\
&\quad+ |Y_t \quad u_0|^2\, dV_t + \frac{n_p}{2a}|Z_t|^2\, dt.
\end{aligned}$$

(5.113) now follows by Proposition 5.2. ■

Proposition 5.50 (Uniqueness). *Let $a, p > 1$. Let the assumptions (5.108-**BSVI-$H_{\eta,\Psi,\Phi}$**) be satisfied. If $(Y,Z), (\hat{Y},\hat{Z}) \in S_m^0[0,T] \times \Lambda^0_{m\times k}(0,T)$ are two solutions of the BSDE (5.107) corresponding respectively to η and $\hat{\eta}$ such that*

$$\mathbb{E}\sup_{s\in[0,T]} e^{pV_s}\left|Y_s - \hat{Y}_s\right|^p < \infty,$$

then $\mathbb{P}$-a.s., for all $t \in [0,T]$:

$$e^{pV_t}\left|Y_t - \hat{Y}_t\right|^p \le \mathbb{E}^{\mathcal{F}_t}\left(e^{pV_T}\,|\eta - \hat{\eta}|^p\right) \tag{5.114}$$

and there exists a constant $C_{a,p}$ such that, $\mathbb{P}$-a.s., for all $t \in [0,T]$:

$$\mathbb{E}^{\mathcal{F}_t} \sup_{s\in[t,T]} e^{pV_s} \left|Y_s - \hat{Y}_s\right|^p + \mathbb{E}^{\mathcal{F}_t} \left(\int_t^T e^{2V_s} \left|Z_s - \hat{Z}_s\right|^2 ds\right)^{p/2} \le C_{a,p}\ \mathbb{E}^{\mathcal{F}_t}\ e^{pV_T} \left|\eta - \hat{\eta}\right|^p . \tag{5.115}$$

Uniqueness in the space $S_m^p\left([0,T]\,; e^V\right) \times \Lambda^0_{m\times k}(0,T)$ follows. Moreover, if $(\ell_t)_{t\in[0,T]}$ is a deterministic process, uniqueness of the solution (Y,Z) of the BSDE (5.107) holds in $S_m^{1+}\left([0,T]\,; e^{\bar{\mu}}\right) \times \Lambda^0_{m\times k}(0,T)$.

Proof. Let (Y,Z), $(\hat{Y},\hat{Z}) \in S_m^0([0,T]\,0) \times \Lambda^0_{m\times k}(0,T)$ be two solutions corresponding to η and $\hat{\eta}$ respectively. Then

$$Y_t - \hat{Y}_t = \eta - \hat{\eta} + \int_t^T dL_s - \int_t^T \left(Z_s - \hat{Z}_s\right) dB_s,$$

where

$$L_t = \int_0^t \left[\left(\Phi(s, Y_s, Z_s) - \Phi\left(s, \hat{Y}_s, \hat{Z}_s\right)\right) dQ_s - \left(dK_s - d\hat{K}_s\right)\right].$$

Since by (5.110) $\left\langle Y_s - \hat{Y}_s, dK_s - d\hat{K}_s\right\rangle \ge 0$, we have for all $a > 1$:

$$\begin{aligned}\left\langle Y_t - \hat{Y}_t, dL_t\right\rangle &\le \left|Y_t - \hat{Y}_t\right|^2 \mu_t dQ_t + \left|Y_t - \hat{Y}_t\right| \left|Z_t - \hat{Z}_t\right| \ell_t dt \\ &\le \left|Y_t - \hat{Y}_t\right|^2 \left[\mu_t dQ_t + \frac{a}{2n_p}(\ell_t)^2 dt\right] + \frac{n_p}{2a}\left|Z_t - \hat{Z}_t\right|^2 dt,\end{aligned}$$

where $n_p = (p-1) \wedge 1$. (5.114) and (5.115) follow from Proposition 5.2 and, consequently, uniqueness follows, too.

Let now $(\ell_t)_{t\in[0,T]}$ be a deterministic process. If (Y,Z), $(\hat{Y},\hat{Z}) \in S_m^{1+}([0,T]\,; e^{\bar{\mu}}) \times \Lambda^0_{m\times k}(0,T)$, then there exists a $p > 1$ such that $Y, \hat{Y} \in S_m^p\left([0,T]\,; e^{\bar{\mu}}\right)$ and the uniqueness follows from the first step. ■

Proposition 5.51 (Uniqueness). *Let $p = 1$. Let the assumptions (5.108-**BSVI-H**$_{\eta,\Psi,\Phi}$) be satisfied and Φ be independent of $z \in \mathbb{R}^{m\times k}$ ($\ell_t \equiv 0$ and $V_t = \bar{\mu}_t = \int_0^t \mu_s dQ_s$). If $(Y,Z), (\hat{Y},\hat{Z}) \in S_m^0[0,T] \times \Lambda^0_{m\times k}(0,T)$ are two solutions of the BSDE (5.107) corresponding respectively to η and $\hat{\eta}$ such that*

$$\mathbb{E} \sup_{s\in[0,T]} e^{\bar{\mu}_s} \left|Y_s - \hat{Y}_s\right| < \infty,$$

then $\mathbb{P}$-a.s., for all $t \in [0,T]$:

$$e^{\bar{\mu}_t} \left|Y_t - \hat{Y}_t\right| \le \mathbb{E}^{\mathcal{F}_t}\left(e^{\bar{\mu}_T} \left|\eta - \hat{\eta}\right|\right)$$

and for every $q \in (0,1)$ there exists a constant C_q such that

$$\sup_{s\in[0,T]}\left(\mathbb{E}\left(e^{\bar{\mu}_s}\left|Y_s-\hat{Y}_s\right|\right)\right)^q+\mathbb{E}\sup_{s\in[0,T]}e^{q\bar{\mu}_s}\left|Y_s-\hat{Y}_s\right|^q$$
$$+\mathbb{E}\left(\int_0^T e^{2\bar{\mu}_s}\left|Z_s-\hat{Z}_s\right|^2ds\right)^{q/2}$$
$$\leq C_q\left(\mathbb{E}\left(e^{\bar{\mu}_T}\left|\eta-\hat{\eta}\right|\right)\right)^q.$$

Proof. Following the proof of Proposition 5.50 we now have

$$\left\langle Y_t-\hat{Y}_t,dL_t\right\rangle\leq\left|Y_t-\hat{Y}_t\right|^2\mu_t dQ_t$$

and the conclusions follow by Corollary 6.81. ■

5.5.2 Existence

We consider the following backward stochastic variational inequality (BSVI)

$$\begin{cases}-dY_t+\partial\varphi(Y_t)\,dt\ni F(t,Y_t,Z_t)\,dt-Z_tdB_t,\ 0\leq t<T,\\ Y_T=\eta,\end{cases}\tag{5.116}$$

and we suppose that the following assumptions hold:

(A_1) *$\eta:\Omega\to\mathbb{R}^m$ is an $\mathcal{F}_T$-measurable random vector.*
(A_2) *$F:\Omega\times[0,T]\times\mathbb{R}^m\times\mathbb{R}^{m\times k}\to\mathbb{R}^m$ satisfies the assumptions* (5.77-**BSDE-MH**$_F$) *(from Sect. 5.3.4).*
(A_3) *$\varphi:\mathbb{R}^m\to(-\infty,+\infty]$ is a proper, convex l.s.c. function.*

Recall that the subdifferential of φ is given by

$$\partial\varphi(y)=\{\hat{y}\in\mathbb{R}^m:\langle\hat{y},v-y\rangle+\varphi(y)\leq\varphi(v),\ \forall\, v\in\mathbb{R}^m\},$$

and by $(y,\hat{y})\in\partial\varphi$ we understand that $y\in\mathrm{Dom}(\partial\varphi)$ and $\hat{y}\in\partial\varphi(y)$.
We define

$$\begin{aligned}\mathrm{Dom}(\varphi)&=\{y\in\mathbb{R}^m:\varphi(y)<\infty\},\\ \mathrm{Dom}(\partial\varphi)&=\{y\in\mathbb{R}^m:\partial\varphi(y)\neq\emptyset\}\subset\mathrm{Dom}(\varphi).\end{aligned}$$

Let $\varepsilon>0$ and denote the Moreau regularization of φ by

$$\varphi_\varepsilon(y)\overset{def}{=}\inf\left\{\frac{1}{2\varepsilon}|y-v|^2+\varphi(v):v\in\mathbb{R}^m\right\}=\frac{1}{2\varepsilon}|y-J_\varepsilon(y)|^2+\varphi(J_\varepsilon(y)),\tag{5.117}$$

where $J_\varepsilon(y) = (I_{m\times m} + \varepsilon\partial\varphi)^{-1}(y)$. Note that φ_ε is a C^1 convex function and J_ε is a 1-Lipschitz function.

We mention some properties (see Annex B: Convex Functions): for all $x, y \in \mathbb{R}^m$

$$\begin{array}{ll} (a) & \nabla\varphi_\varepsilon(y) = \partial\varphi_\varepsilon(y) = \dfrac{y - J_\varepsilon(y)}{\varepsilon} \in \partial\varphi(J_\varepsilon y), \\ (b) & |\nabla\varphi_\varepsilon(x) - \nabla\varphi_\varepsilon(y)| \le \dfrac{1}{\varepsilon}|x-y|, \\ (c) & \langle\nabla\varphi_\varepsilon(x) - \nabla\varphi_\varepsilon(y), x-y\rangle \ge 0, \\ (d) & \langle\nabla\varphi_\varepsilon(x) - \nabla\varphi_\delta(y), x-y\rangle \ge -(\varepsilon+\delta)\langle\nabla\varphi_\varepsilon(x), \nabla\varphi_\delta(y)\rangle. \end{array} \tag{5.118}$$

Throughout this subsection we fix a pair $(u_0, \hat{u}_0) \in \partial\varphi$. Then by (6.26) from Annex B we have

$$\begin{cases} (j) & |\nabla\varphi_\varepsilon(u_0)| \le |\hat{u}_0|, \\ (jj) & \dfrac{|y - J_\varepsilon(y)|^2}{2\varepsilon} \le \varphi_\varepsilon(y) - \varphi(u_0) + |\hat{u}_0|\,|y-u_0| + \varepsilon|\hat{u}_0|^2. \end{cases} \tag{5.119}$$

We will make the following assumption:

(A_4) *There exist $p \ge 2$, a positive stochastic process $\beta \in L^1(\Omega\times(0,T))$, a positive function $b \in L^1(0,T)$ and real numbers $\kappa \ge 0$, $\lambda \in]0,1[$ such that*

$$\begin{array}{c} \textit{for all } (u,\hat{u}) \in \partial\varphi \textit{ and } z \in \mathbb{R}^{m\times k}: \\ \langle\hat{u}, F(t,u,z)\rangle \le \lambda|\hat{u}|^2 + \beta_t + b(t)|u|^p + \kappa|z|^2 \\ d\mathbb{P}\otimes dt\text{-a.e.},\ (\omega,t)\in\Omega\times[0,T]. \end{array} \tag{5.120}$$

We note that if $\langle\hat{u}, F(t,u,z)\rangle \le 0$ for all $(u,\hat{u}) \in \partial\varphi$, then the condition (5.120) is satisfied with $\beta_t = b(t) = \kappa = 0$. If for example $\varphi = I_{\overline{D}}$ (the convex indicator of the closed convex set $\overline{D}$) and $\mathbf{n}_y$ denotes any unit outward normal vector to $\overline{D}$ at $y \in \mathrm{Bd}(\overline{D})$, then the condition $\langle\mathbf{n}_y, F(t,y,z)\rangle \le 0$ for all $y \in \mathrm{Bd}(\overline{D})$ yields (5.120) with $\beta_t = b(t) = \kappa = 0$ (for example). In this last case by Itô's formula for $\psi(\tilde{Y}) = [\mathrm{dist}_{\overline{D}}(\tilde{Y})]^2$, where

$$\begin{cases} -d\tilde{Y}_t = F(t,\tilde{Y}_t,\tilde{Z}_t)\,dt - \tilde{Z}_t dB_t, & 0 \le t < T, \\ \tilde{Y}_T = \eta, \end{cases}$$

and by the uniqueness of the triple (Y,Z,U) satisfying (5.107) we infer that $(Y,Z,U) = (\tilde{Y},\tilde{Z},0)$.

Theorem 5.52 (Existence - Uniqueness). *Let $p \ge 2$ and assumptions (A_1–A_4) be satisfied with this p. Suppose moreover that, for all $\rho \ge 0$,*

$$\mathbb{E}|\eta|^p + \mathbb{E}\varphi^+(\eta) + \mathbb{E}\left(\int_0^T F_\rho^\#(s)ds\right)^p < \infty.$$

Then there exists a unique pair $(Y,Z) \in S_m^p[0,T] \times \Lambda_{m\times k}^p(0,T)$ *and a unique stochastic process* $U \in \Lambda_m^2(0,T)$ *such that*

$$\begin{aligned} &(a) \quad \int_0^T |F(t,Y_t,Z_t)|\,dt < \infty, \ \mathbb{P}\text{-a.s.},\\ &(b) \quad Y_t(\omega) \in \mathrm{Dom}(\partial\varphi), \ d\mathbb{P}\otimes dt\text{-a.e. } (\omega,t) \in \Omega\times[0,T],\\ &(c) \quad U_t(\omega) \in \partial\varphi(Y_t(\omega)), \ d\mathbb{P}\otimes dt\text{-a.e. } (\omega,t) \in \Omega\times[0,T], \end{aligned}$$

and for all $t \in [0,T]$:

$$Y_t + \int_t^T U_s ds = \eta + \int_t^T F(s,Y_s,Z_s)\,ds - \int_t^T Z_s dB_s, \ a.s. \tag{5.121}$$

Moreover, uniqueness holds in $S_m^{1+}[0,T] \times \Lambda_{m\times k}^0(0,T)$*, where*

$$S_m^{1+}[0,T] \overset{def}{=} \bigcup_{p>1} S_m^p[0,T].$$

Proof. Let (Y,Z), $(\tilde{Y},\tilde{Z}) \in S_m^{1+}[0,T] \times \Lambda_{m\times k}^0(0,T)$ be two solutions. Then $Y,\tilde{Y} \in S_m^p[0,T]$, for some p. Uniqueness follows from Proposition 5.50.

The proof of the existence will be split into several steps.

Step 1. Approximating problem.

For $\varepsilon \in (0,1]$ consider the approximating equation: $\mathbb{P}$-*a.s.*, for all $t \in [0,T]$,

$$Y_t^\varepsilon + \int_t^T \nabla\varphi_\varepsilon(Y_s^\varepsilon)\,ds = \eta + \int_t^T F(s,Y_s^\varepsilon,Z_s^\varepsilon)\,ds - \int_t^T Z_s^\varepsilon dB_s, \tag{5.122}$$

where $\nabla\varphi_\varepsilon$ is the gradient of the Moreau regularization φ_ε of φ. It follows (*without assumption* (A$_4$)) from Theorem 5.27 that Eq. (5.122) has a unique solution $(Y^\varepsilon, Z^\varepsilon) \in S_m^p[0,T] \times \Lambda_{m\times k}^p(0,T)$.

Step 2. Boundedness of Y^ε *and* Z^ε.

Let $(u_0,\hat{u}_0) \in \partial\varphi$, $a > 1$ and

$$V(t) = V_t^{a,p} \overset{def}{=} \int_0^t \left[\mu(s) + \frac{a}{2n_p}\ell^2(s)\right] ds = \int_0^t \left[\mu(s) + \frac{a}{2}\ell^2(s)\right] ds$$

($p \geq 2$ yields $n_p = 1 \wedge (p-1) = 1$).

Let $(u_0,\hat{u}_0) \in \partial\varphi$ be fixed. From Proposition 5.46 with Ψ replaced by φ_ε and dQ_s by ds, there exists a constant $C_{a,p}$ (depending only on a and p) such that the following inequality holds $\mathbb{P}$-*a.s.*, for all $t \in [0,T]$:

$$\mathbb{E}^{\mathcal{F}_t} \sup_{s\in[t,T]} e^{pV_s} |Y_s^\varepsilon - u_0|^p + \mathbb{E}^{\mathcal{F}_t} \left(\int_t^T e^{2V_s} |\varphi_\varepsilon(Y_s^\varepsilon) - \varphi_\varepsilon(u_0)| \, ds\right)^{p/2}$$
$$+\mathbb{E}^{\mathcal{F}_t} \left(\int_t^T e^{2V_s} |Z_s^\varepsilon|^2 ds\right)^{p/2}$$
$$\leq C_{a,p} \left[\mathbb{E}^{\mathcal{F}_t} e^{pV_T} |\eta - u_0|^p + \mathbb{E}^{\mathcal{F}_t} \left(\int_t^T e^{V_s} [|\nabla\varphi_\varepsilon(u_0)| + |F(s,u_0,0)|] \, ds\right)^p\right]. \tag{5.123}$$

Note that $|\nabla\varphi_\varepsilon(u_0)| \leq |\hat{u}_0|$ and $|\varphi_\varepsilon(u_0)| \leq \varphi(u_0) + |\hat{u}_0|^2$. Hence there exists a constant C independent of ε such that

$$\begin{aligned}
&(a) \quad \mathbb{E}\,\|Y^\varepsilon\|_T^2 \leq \left(\mathbb{E}\,\|Y^\varepsilon\|_T^p\right)^{2/p} \leq C,\\
&(b) \quad \mathbb{E}\int_0^T |Z_s^\varepsilon|^2 ds \leq \left[\mathbb{E}\Big(\int_0^T |Z_s^\varepsilon|^2 ds\Big)^{p/2}\right]^{2/p} \leq C,\\
&(c) \quad \mathbb{E}\int_0^T |\varphi_\varepsilon(Y_s^\varepsilon)| \, ds \leq \left[\mathbb{E}\Big(\int_0^T |\varphi_\varepsilon(Y_s^\varepsilon)| \, ds\Big)^{p/2}\right]^{2/p} \leq C.
\end{aligned} \tag{5.124}$$

Throughout the proof we shall fix $a = 2$ and therefore

$$V_t = \int_0^t \left[\mu(s) + \ell^2(s)\right] ds.$$

Step 3. Boundedness of $\nabla\varphi_\varepsilon(Y^\varepsilon)$.

Using the following stochastic subdifferential inequality given by Lemma 2.38

$$\varphi_\varepsilon(Y_t^\varepsilon) + \int_t^T \langle\nabla\varphi_\varepsilon(Y_s^\varepsilon), dY_s^\varepsilon\rangle \leq \varphi_\varepsilon(Y_T^\varepsilon) = \varphi_\varepsilon(\eta) \leq \varphi(\eta),$$

we deduce that, for all $t \in [0,T]$,

$$\varphi_\varepsilon(Y_t^\varepsilon) + \int_t^T |\nabla\varphi_\varepsilon(Y_s^\varepsilon)|^2 ds \leq \varphi(\eta) + \int_t^T \left\langle\nabla\varphi_\varepsilon(Y_s^\varepsilon), F\left(s, Y_s^\varepsilon, Z_s^\varepsilon\right)\right\rangle ds$$
$$-\int_t^T \langle\nabla\varphi_\varepsilon(Y_s^\varepsilon), Z_s^\varepsilon dB_s\rangle. \tag{5.125}$$

Since $|\nabla\varphi_\varepsilon(y)| \leq |\nabla\varphi_\varepsilon(y) - \nabla\varphi_\varepsilon(u_0)| + |\nabla\varphi_\varepsilon(u_0)| \leq \dfrac{1}{\varepsilon}|y - u_0| + |\hat{u}_0|$ and

$$\mathbb{E}\left(\int_0^T |\nabla\varphi_\varepsilon(Y_s^\varepsilon)|^2 |Z_s^\varepsilon|^2 ds\right)^{1/2}$$
$$\leq \frac{1}{\varepsilon}\mathbb{E}\left[\sup_{s\in[0,T]} |\nabla\varphi_\varepsilon(Y_s^\varepsilon)| \Big(\int_0^T |Z_s^\varepsilon|^2 ds\Big)^{1/2}\right]$$

$$
\begin{aligned}
&\le \left[\frac{2}{\varepsilon^2}\mathbb{E}\sup_{s\in[0,T]}|Y_s^\varepsilon-u_0|^2+2|\hat{u}_0|^2\right]+\mathbb{E}\left(\int_0^T|Z_s^\varepsilon|^2ds\right)\\
&<\infty,
\end{aligned}
$$

we have

$$
\mathbb{E}\int_t^T\langle\nabla\varphi_\varepsilon(Y_s^\varepsilon),Z_s^\varepsilon dB_s\rangle=0.
$$

Under assumption (A$_4$), since $\nabla\varphi_\varepsilon(Y_s^\varepsilon)\in\partial\varphi\left(J_\varepsilon\left(Y_s^\varepsilon\right)\right)$, it follows that

$$
\begin{aligned}
&\left\langle\nabla\varphi_\varepsilon(Y_s^\varepsilon),F\left(s,Y_s^\varepsilon,Z_s^\varepsilon\right)\right\rangle\\
&=\frac{1}{\varepsilon}\left\langle Y_s^\varepsilon-J_\varepsilon\left(Y_s^\varepsilon\right),F\left(s,Y_s^\varepsilon,Z_s^\varepsilon\right)-F\left(s,J_\varepsilon\left(Y_s^\varepsilon\right),Z_s^\varepsilon\right)\right\rangle\\
&\qquad\qquad+\left\langle\nabla\varphi_\varepsilon(Y_s^\varepsilon),F\left(s,J_\varepsilon\left(Y_s^\varepsilon\right),Z_s^\varepsilon\right)\right\rangle\\
&\le\frac{1}{\varepsilon}\mu^+(s)\left|Y_s^\varepsilon-J_\varepsilon\left(Y_s^\varepsilon\right)\right|^2+\lambda\left|\nabla\varphi_\varepsilon(Y_s^\varepsilon)\right|^2+\beta_s+b(s)\left|J_\varepsilon\left(Y_s^\varepsilon\right)\right|^p+\kappa\left|Z_s^\varepsilon\right|^2.
\end{aligned}
\tag{5.126}
$$

Using here the inequalities (5.119), then from (5.125) we infer that for all $t\in[0,T]$,

$$
\begin{aligned}
\mathbb{E}\varphi_\varepsilon(Y_t^\varepsilon)+(1-\lambda)\,\mathbb{E}\int_t^T|\nabla\varphi_\varepsilon(Y_s^\varepsilon)|^2ds&\le\mathbb{E}\varphi(\eta)+2\int_t^T\mu^+(s)\,\mathbb{E}\varphi_\varepsilon(Y_s^\varepsilon)ds\\
&\quad+C\,\mathbb{E}\int_t^T\left(\left[1+\beta_s+b(s)\left(1+|Y_s^\varepsilon-u_0|^p\right)+\kappa\,|Z_s^\varepsilon|^2\right]\right)ds
\end{aligned}
$$

which yields, via estimates (5.124) and the backward Gronwall inequality (Corollary 6.62), that there exists a constant $C>0$ independent of $\varepsilon\in(0,1]$ such that

$$
\begin{aligned}
&(a)\quad \mathbb{E}\varphi_\varepsilon(Y_t^\varepsilon)+\mathbb{E}\int_0^T|\nabla\varphi_\varepsilon(Y_s^\varepsilon)|^2ds\le C,\\
&(b)\quad \mathbb{E}\left|Y_t^\varepsilon-J_\varepsilon\left(Y_t^\varepsilon\right)\right|^2\le C\varepsilon.
\end{aligned}
\tag{5.127}
$$

Step 4. Cauchy sequence and convergence.

Let $\varepsilon,\delta\in(0,1]$.
We can write

$$
Y_t^\varepsilon-Y_t^\delta=\int_t^T dK_s^{\varepsilon,\delta}-\int_t^T Z_s^\varepsilon dB_s,
$$

where

$$
K_t^{\varepsilon,\delta}=\int_0^t\left[F\left(s,Y_s^\varepsilon,Z_s^\varepsilon\right)-F\left(s,Y_s^\delta,Z_s^\delta\right)-\nabla\varphi_\varepsilon\left(Y_s^\varepsilon\right)+\nabla\varphi_\delta\left(Y_s^\delta\right)\right]ds.
$$

Then

$$\langle Y_t^\varepsilon - Y_t^\delta, dK_t^{\varepsilon,\delta}\rangle \leq (\varepsilon+\delta)\langle \nabla\varphi_\varepsilon(Y_t^\varepsilon), \nabla\varphi_\delta(Y_t^\delta)\rangle dt + |Y_t^\varepsilon - Y_t^\delta|^2 dV_t + \frac{1}{4}|Z_t^\varepsilon - Z_t^\delta|^2 dt,$$

and by Proposition 5.2, with $a = p = 2$,

$$\begin{aligned}
&\mathbb{E} \sup_{s\in[0,T]} \left|Y_s^\varepsilon - Y_s^\delta\right|^2 + \mathbb{E}\int_0^T \left|Z_s^\varepsilon - Z_s^\delta\right|^2 ds \\
&\quad \leq C\,\mathbb{E}\int_0^T (\varepsilon+\delta)\left\langle \nabla\varphi_\varepsilon(Y_s^\varepsilon), \nabla\varphi_\delta(Y_s^\delta)\right\rangle ds \\
&\quad \leq \frac{1}{2}C(\varepsilon+\delta)\left[\mathbb{E}\int_0^T |\nabla\varphi_\varepsilon(Y_s^\varepsilon)|^2 ds + \mathbb{E}\int_0^T \left|\nabla\varphi_\delta(Y_s^\delta)\right|^2 ds\right] \\
&\quad \leq C'(\varepsilon+\delta).
\end{aligned}$$

Hence there exist $(Y,Z,U) \in S_m^2[0,T] \times \Lambda_{m\times k}^2(0,T) \times \Lambda_m^2(0,T)$ and a sequence $\varepsilon_n \searrow 0$ such that

$$\begin{aligned}
&Y^{\varepsilon_n} \to Y, \text{ in } S_m^2[0,T] \text{ and a.s. in } C([0,T];\mathbb{R}^m), \\
&Z^{\varepsilon_n} \to Z, \text{ in } \Lambda_{m\times k}^2(0,T) \text{ and a.s. in } L^2\left(0,T;\mathbb{R}^{m\times k}\right), \\
&\nabla\varphi_\varepsilon(Y^\varepsilon) \rightharpoonup U, \text{ weakly in } \Lambda_m^2(0,T), \\
&J_{\varepsilon_n}(Y^{\varepsilon_n}) \to Y, \text{ in } \Lambda_m^2(0,T) \text{ and a.s. in } L^2(0,T;\mathbb{R}^m).
\end{aligned}$$

Passing to the limit in (5.122) we conclude that

$$Y_t + \int_t^T U_s ds = \eta + \int_t^T F(s,Y_s,Z_s)\,ds - \int_t^T Z_s dB_s, \text{ a.s.}$$

Since $\nabla\varphi_\varepsilon(Y_s^\varepsilon) \in \partial\varphi\left(J_\varepsilon\left(Y_s^\varepsilon\right)\right)$ it follows that for all $A \in \mathcal{F}$, $0 \leq s \leq t \leq T$ and $v \in S_m^2[0,T]$,

$$\mathbb{E}\int_s^t \mathbf{1}_A \langle \nabla\varphi_\varepsilon(Y_r^\varepsilon), v_r - Y_r^\varepsilon\rangle dr + \mathbb{E}\int_s^t \mathbf{1}_A\varphi(J_\varepsilon\left(Y_r^\varepsilon\right))dr \leq \mathbb{E}\int_s^t \mathbf{1}_A\varphi(v_r)dr.$$

Passing to lim inf for $\varepsilon = \varepsilon_n \searrow 0$ in the above inequality we obtain that $U_s \in \partial\varphi(Y_s)$. Hence $(Y,Z,U) \in S_m^p[0,T] \times \Lambda_{m\times k}^p(0,T) \times \Lambda_m^2(0,T)$ and (Y,Z,K), with $K_t = \displaystyle\int_0^t U_s ds$, is the solution of BSVI (5.116). The proof is complete. ■

Remark 5.53. The existence Theorem 5.52 is well adapted to the Hilbert space setting, since we do not impose an assumption of the form

$$\text{int}\,(\text{Dom}\,(\varphi)) \neq \emptyset,$$

which is very restrictive for infinite dimensional spaces. In the context of Hilbert spaces Theorem 5.52 holds in the same form (see [57] where some examples of partial differential backward stochastic variational inequalities are given too).

Let $(u_0, \hat{u}_0) \in \partial\varphi$ be fixed. From the inequality (5.123) we have for $a = p = 2$

$$e^{2V(t)} \left|Y_t^\varepsilon - u_0\right|^2 \leq C_{a,p} \Big[\mathbb{E}^{\mathcal{F}_t} e^{2V((T)} \left|\eta - u_0\right|^2 + \mathbb{E}^{\mathcal{F}_t} \Big(\int_t^T e^{V(s)} \left[|\hat{u}_0| + |F(s, u_0, 0)|\right] ds\Big)^2\Big],$$

and consequently if $|\eta| + \int_0^T |F(s, u_0, 0)|\, ds \leq M_0$, then a.s. for all $t \in [0, T]$,

$$|Y_t^\varepsilon| \leq R_0 = |u_0| + C\, e^{2\|V\|_T} \left(|u_0| + T\,|\hat{u}_0| + M_0\right). \tag{5.128}$$

Corollary 5.54. *If in Theorem 5.52 we replace the assumption (A_4) by*

(A_5) There exist $M_0, L > 0$ such that:

$$\begin{array}{ll} (i) & 0 \leq \ell_t \leq L, \ a.e., \ t \in [0, T], \\ (ii) & |\eta| + \int_0^T |F(s, u_0, 0)|\, ds \leq M_0, \ a.s., \omega \in \Omega, \\ (iii) & \exists R_0 > 0 \text{ sufficient large such that} \end{array}$$

$$\mathbb{E} \int_0^T \left(F_{R_0}^{\#}(s)\right)^2 ds < \infty,$$

(in the proof R_0 is defined by (5.128)) the conclusions of Theorem 5.52 hold.

Proof. Let R_0 be defined by (5.128). The proof follows the same steps and computations as in Theorem 5.52 with the modification of Step 3: the estimate (5.126) now takes the following form (considering (5.128)),

$$\begin{aligned} \left\langle \nabla\varphi_\varepsilon(Y_s^\varepsilon), F\left(s, Y_s^\varepsilon, Z_s^\varepsilon\right)\right\rangle &\leq |\nabla\varphi_\varepsilon(Y_s^\varepsilon)| \left|F\left(s, Y_s^\varepsilon, 0\right)\right| + |\nabla\varphi_\varepsilon(Y_s^\varepsilon)|\, L\, |Z_s^\varepsilon| \\ &\leq \frac{1}{2} |\nabla\varphi_\varepsilon(Y_s^\varepsilon)|^2 + \left(F_{R_0}^{\#}(s)\right)^2 + L^2 |Z_s^\varepsilon|^2. \end{aligned}$$

Using this inequality in (5.125) we directly obtain (5.127). ■

Remark 5.55. We note that if $F(\omega, t, y, z) = F(y, z)$, then the assumption ($A_5$) becomes $|\eta| \leq M_0$, a.s., $\omega \in \Omega$.

Remark 5.56. In the particular case where φ is the convex indicator of a convex subset $D \subset \mathbb{R}^m$, the BSDE (5.116) is a reflected BSDE. As first noted in [34], the process K which maintains the solution inside D is absolutely continuous, unlike in the case of forward SDEs. The intuitive reason for this is that K does not need to fight against the martingale term. The situation is probably quite different in the case of nonconvex sets, but reflecting BSDEs at the boundary of nonconvex sets remains an open problem. The theory of reflected BSDEs was initiated in [25], where reflection in $\mathbb{R}$ above a given continuous adapted process was considered.

5.6 BSDEs with Random Final Time

5.6.1 BSDEs with a Monotone Coefficient

Let us now discuss the existence and uniqueness of a solution to an equation which we would like to write as

$$Y_t = \eta + \int_t^\infty \Phi(s, Y_s, Z_s)\, dQ_s - \int_t^\infty Z_s dB_s, \quad a.s., \ \forall\, t \geq 0. \tag{5.129}$$

In most cases the above integrals will not make sense. For this reason we shall give below a weaker formulation of the above BSDE.

We formulate the following assumptions:

$$(\mathbf{BSDE\text{-}H_\infty}) \tag{5.130}$$

(i) $p, a > 1, n_p = 1 \wedge (p-1)$,

(ii) $\eta \in L^p(\Omega, \mathcal{F}_\infty, \mathbb{P}; \mathbb{R}^m)$ *and* $(\xi, \zeta) \in S_d^p \times \Lambda_{d\times k}^p(0,\infty)$ *is the unique pair such that*

$$\xi_t = \eta - \int_t^\infty \zeta_s dB_s, \quad t \geq 0,\ a.s.,$$

(in particular $(\xi_t)_{t\geq 0}$ *is given by* $\xi_t = \mathbb{E}^{\mathcal{F}_t}\eta$*).*

(iii) $(\omega, t) \longmapsto Q_t(\omega) : \Omega \times [0,\infty[\to \mathbb{R}$ is a $\mathcal{P}$-m.i.c.s.p. such that $Q_0 = 0$.

- ♦ $\forall\, y \in \mathbb{R}^m$, $z \in \mathbb{R}^{m\times k}$, *the function* $\Phi(\cdot,\cdot,y,z) : \Omega \times [0,\infty[\to \mathbb{R}^m$ *is* $\mathcal{P}$*-measurable*;
- ♦ *there exist* $\ell \in L^2_{loc}(\mathbb{R}_+; \mathbb{R}_+)$ *(a deterministic function) and two* $\mathcal{P}$*-m.s.p* $\mu, \alpha : \Omega \times [0,\infty[\to \mathbb{R}$, $\alpha \geq 0$, *such that* $\alpha_t dQ_t = dt$ *and*

$$\int_0^T |\mu_t| dQ_t < \infty, \quad \textit{for all } T > 0, \mathbb{P}\text{-a.s.};$$

♦ *for all* $y, y' \in \mathbb{R}^m$ and $z, z' \in \mathbb{R}^{m\times k}$, $d\mathbb{P} \otimes dQ_t$*-a.e.:*

$$\begin{array}{ll} & \textit{Continuity:} \\ (C_y) & y \longrightarrow \Phi(t,y,z) : \mathbb{R}^m \to \mathbb{R}^m \textit{ is continuous;} \\ & \textit{Monotonicity condition:} \\ (M_y) & \langle y'-y, \Phi(t,y',z) - \Phi(t,y,z)\rangle \le \mu_t |y'-y|^2\,; \\ & \textit{Lipschitz condition:} \\ (L_z) & |\Phi(t,y,z') - \Phi(t,y,z)| \le \alpha_t\, \ell(t)\, |z'-z|; \\ & \textit{Boundedness condition:} \\ (B_y) & \displaystyle\int_0^T \Phi_\rho^{\#}(s)\, dQ_s < \infty, \quad \forall\, \rho, T \ge 0, \end{array} \tag{5.131}$$

□

where $\Phi_\rho^{\#}(t) = \sup\{|\Phi(t,y,0)| : |y| \le \rho\}$.

Define

$$\bar{\mu}_t = \int_0^t \mu_s dQ_s$$

and

$$S_m^p\left(e^{\bar{\mu}}\right) = \left\{ Y \in S_m^0 : \mathbb{E} \sup_{s\in[0,T]} \left|e^{\bar{\mu}_s} Y_s\right|^p < \infty \text{ for all } T \ge 0 \right\}.$$

Note that

$$\bar{\mu}_t \le \bar{\mu}_t^+ \le \left\|\bar{\mu}^+\right\|_t = \sup_{s\in[0,t]} \left(\int_0^s \mu_r dQ_r\right)^+ \le \int_0^t \mu_r^+ dQ_r.$$

Finally we recall the usual notation

$$V_t = V_t^{a,p} \overset{def}{=} \int_0^t \mu_s dQ_s + \frac{a}{2n_p} \int_0^t \ell^2(s)\, ds. \tag{5.132}$$

Theorem 5.57. *Let* $p, a > 1$ *and* V *be defined by (5.132). Let the assumptions* (**BSDE-H**$_\infty$) *be satisfied and*

$$\begin{array}{ll} (i) & \mathbb{E}\left(\sup\limits_{t\in[0,T]} e^{p\bar{\mu}_t} |\eta|^p\right) < \infty < \infty, \textit{ for all } T \ge 0, \\ (ii) & \mathbb{E}\left(\displaystyle\int_0^\infty e^{V_t} |\Phi(t,\xi_t,\zeta_t)|\, dQ_t\right)^p < \infty. \end{array}$$

If, moreover, for all $\rho \geq 0$

$$(h_1) \quad \mathbb{E}\left(\int_0^T \sup_{|y|\leq\rho} \left|e^{\bar{\mu}_t}\Phi\left(s, e^{-\bar{\mu}_t}y, 0\right) - \mu_t y\right| dQ_t\right)^p < \infty, \text{ or}$$

$$(h_2) \quad \mu \geq 0 \text{ and } \mathbb{E}\left(\int_0^T e^{\bar{\mu}_t} \sup_{|y|\leq\rho} |\Phi(t, y, 0)| \, dQ_s\right)^p < \infty,$$

then there exists a unique solution $(Y_t, Z_t)_{t\geq 0} \in S_m^0 \times \Lambda_{m\times k}^0$ *of the BSDE (5.129) in the sense that (here* $\forall 0 \leq t \leq T$ *means for all* t *and all* T *such that* $0 \leq t \leq T$*)*

$$\begin{cases} (j) & Y_t = Y_T + \int_t^T \Phi(s, Y_s, Z_s) \, dQ_s - \int_t^T Z_s dB_s, \quad a.s., \ \forall \, 0 \leq t \leq T, \\ (jj) & \mathbb{E} \sup_{0\leq t\leq T} e^{pV_s} |Y_s|^p < \infty, \quad \text{for all } T \geq 0, \\ (jjj) & \lim_{T\to\infty} \mathbb{E}\, e^{pV_T} |Y_T - \xi_T|^p = 0. \end{cases} \tag{5.133}$$

Moreover

$$\mathbb{E}\left(\int_0^T e^{2V_s} |Z_s|^2 ds\right)^{p/2} < \infty, \quad \text{for all } T \geq 0,$$

and there exists a constant $C_{a,p}$ *depending only on* (a, p) *such that for all* $t \geq 0$,

$$\begin{aligned} &\mathbb{E}^{\mathcal{F}_t} \sup_{s\geq t} \left|e^{V_s}(Y_s - \xi_s)\right|^p + \mathbb{E}^{\mathcal{F}_t}\left(\int_t^\infty e^{2V_s} |Z_s - \zeta_s|^2 ds\right)^{p/2} \\ &\quad \leq C_{a,p}\mathbb{E}^{\mathcal{F}_t}\left(\int_t^\infty e^{V_s} |\Phi(s, \xi_s, \zeta_s)| \, dQ_s\right)^p, \quad a.s. \end{aligned} \tag{5.134}$$

Proof. Uniqueness. If (Y, Z) and $\left(\hat{Y}, \hat{Z}\right)$ are two solutions of (5.133) in the space $S_m^p\left(e^{\bar{\mu}}\right) \times \Lambda_{m\times k}^0 = S_m^p\left(e^V\right) \times \Lambda_{m\times k}^0$. Then from (5.24) there exists a positive constant $C_{a,p}$ depending only on (a, p), such that

$$\begin{aligned} &\mathbb{E}\left(\sup_{t\in[0,T]} e^{pV_t}\left|Y_t - \hat{Y}_t\right|^p\right) + \mathbb{E}\left(\int_0^T e^{2V_s}\left|Z_s - \hat{Z}_s\right|^2 ds\right)^{p/2} \\ &\quad \leq C_{a,p}\mathbb{E}\, e^{pV_T}\left|Y_T - \hat{Y}_T\right|^p \to 0, \quad \text{as } T \to \infty, \end{aligned}$$

where we have used $((5.133)(jj))$. Uniqueness follows.

Existence. Note that

$$\xi_t = \mathbb{E}^{\mathcal{F}_n}\eta - \int_t^n \zeta_s dB_s, \quad t \in [0, n], \ a.s.,$$

and since $\mathbb{E}\left(\sup_{t\in[0,T]} e^{p\bar{\mu}_t}|\eta|^p\right) < \infty$, by Corollary 6.83

$$\mathbb{E}\sup_{t\in[0,T]} e^{p\bar{\mu}_t}|\xi_t|^p + \mathbb{E}\left(\int_0^T e^{2\bar{\mu}_t}|\zeta_t|^2 dt\right)^{p/2} \le C_p\,\mathbb{E}\left(\sup_{t\in[0,T]} e^{p\bar{\mu}_t}|\eta|^p\right) < \infty. \tag{5.135}$$

Hence

$$\begin{aligned}(\xi,\zeta) &\in S_m^p\left([0,T];e^{\bar{\mu}}\right)\times\Lambda_{m\times k}^p\left(0,T;e^{\bar{\mu}}\right)\\ &\equiv S_m^p\left([0,T];e^{V}\right)\times\Lambda_{m\times k}^p\left(0,T;e^{V}\right).\end{aligned}$$

For any fixed $n\in\mathbb{N}^*$, we consider the approximating equation

$$Y_t^n = \mathbb{E}^{\mathcal{F}_n}\eta + \int_t^n \Phi\left(s,Y_s^n,Z_s^n\right)dQ_s - \int_t^n Z_s^n dB_s,\quad t\in[0,n],\ a.s.$$

By Lemma 5.29, this equation has a unique solution $(Y^n,Z^n)\in S_m^p\left([0,n];e^{\bar{\mu}}\right)\times\Lambda_{m\times k}^p\left(0,n;e^{\bar{\mu}}\right)$. We set $Y_s^n=\xi_s$ and $Z_s^n=\zeta_s$ for $s>n$.

Since the approximating equation can be written in the form: $\mathbb{P}$-a.s., for all $t\in[0,n]$,

$$Y_t^n-\xi_t = \int_t^n \Phi\left(s,\xi_s+\left(Y_s^n-\xi_s\right),\zeta_s+\left(Z_s^n-\zeta_s\right)\right)dQ_s - \int_t^n\left(Z_s^n-\zeta_s\right)dB_s,$$

we deduce from (5.19) that $\mathbb{P}$-a.s., for all $0\le t\le n$,

$$\begin{aligned}\mathbb{E}^{\mathcal{F}_t}\sup_{s\ge t} e^{pV_s}|Y_s^n-\xi_s|^p &+ \mathbb{E}^{\mathcal{F}_t}\left(\int_t^\infty e^{2V_s}|Z_s^n-\zeta_s|^2 ds\right)^{p/2}\\ &\le C_{a,p}\mathbb{E}^{\mathcal{F}_t}\left(\int_t^n e^{V_s}|\Phi(s,\xi_s,\zeta_s)|\,dQ_s\right)^p\\ &\le C_{a,p}\mathbb{E}^{\mathcal{F}_t}\left(\int_t^\infty e^{V_s}|\Phi(s,\xi_s,\zeta_s)|\,dQ_s\right)^p\end{aligned} \tag{5.136}$$

where $C_{a,p}$ is a constant depending only upon (a,p). In particular for $i\in\mathbb{N}^*$:

$$\begin{aligned}\mathbb{E}\sup_{s\ge n}\left\|e^{V_s}\left(Y_s^{n+i}-\xi_s\right)\right\|^p &+ \mathbb{E}\left(\int_n^\infty e^{2V_s}\left|Z_s^{n+i}-\zeta_s\right|^2 ds\right)^{p/2}\\ &\le C_{a,p}\mathbb{E}\left(\int_n^\infty e^{V_s}|\Phi(s,\xi_s,\zeta_s)|\,dQ_s\right)^p \to 0,\quad \text{as } n\to\infty.\end{aligned} \tag{5.137}$$

Note that by uniqueness

$$Y_t^{n+i} = Y_n^{n+i} + \int_t^n \Phi\left(s,Y_s^{n+i},Z_s^{n+i}\right)dQ_s - \int_t^n Z_s^{n+i}dB_s,\ t\in[0,n],\ a.s.$$

Using the inequality (5.24) in this context we infer that

$$\mathbb{E}\left(\sup_{t\in[0,n]} e^{pV_t}\left|Y_t^{n+i}-Y_t^n\right|^p\right)+\mathbb{E}\left(\int_0^n e^{2V_s}\left|Z_s^{n+i}-Z_s^n\right|^2 ds\right)^{p/2} \leq C_{a,p}\mathbb{E}\, e^{pV_n}\left|Y_n^{n+i}-\xi_n\right|^p \to 0, \quad \text{as } n\to\infty.$$

Hence

$$\begin{aligned}
&\mathbb{E}\left(\sup_{s\geq 0} e^{pV_s}\left|Y_s^{n+i}-Y_s^n\right|^p\right)+\mathbb{E}\left(\int_0^\infty e^{2V_s}\left|Z_s^{n+i}-Z_s^n\right|^2 ds\right)^{p/2}\\
&\leq \mathbb{E}\left(\sup_{s\in[0,n]} e^{pV_s}\left|Y_s^{n+i}-Y_s^n\right|^p\right)+2^{p/2}\mathbb{E}\left(\int_0^n e^{2V_s}\left|Z_s^{n+i}-Z_s^n\right|^2 ds\right)^{p/2}\\
&+\mathbb{E}\left(\sup_{s>n} e^{pV_s}\left|Y_s^{n+i}-\xi_s\right|^p\right)+2^{p/2}\mathbb{E}\left(\int_n^\infty e^{2V_s}\left|Z_s^{n+i}-\zeta_s\right|^2 ds\right)^{p/2}\\
&\to 0, \quad \text{as} \quad n\to\infty.
\end{aligned}$$

This shows there exist progressively measurable stochastic processes $Y:\mathbb{R}_+\times\Omega\to\mathbb{R}^m$ and $Z:\mathbb{R}_+\times\Omega\to\mathbb{R}^{m\times k}$ such that

$$\lim_{n\to\infty}\mathbb{E}\left(\sup_{s\geq 0} e^{pV_s}\left|Y_s^n-Y_s\right|^p\right)+\mathbb{E}\left(\int_0^\infty e^{2V_s}\left|Z_s^n-Z_s\right|^2 ds\right)^{p/2}=0. \quad (5.138)$$

From (5.136) we deduce by letting $n\to\infty$ that for all $t\geq 0$, $\mathbb{P}$-a.s.,

$$\begin{aligned}
&\mathbb{E}^{\mathcal{F}_t}\sup_{s\geq t}\left|e^{V_s}\left(Y_s-\xi_s\right)\right|^p+\mathbb{E}^{\mathcal{F}_t}\left(\int_t^\infty e^{2V_s}\left|Z_s-\zeta_s\right|^2 ds\right)^{p/2}\\
&\qquad\leq C_{a,p}\mathbb{E}^{\mathcal{F}_t}\left(\int_t^\infty e^{V_s}\left|\Phi\left(s,\xi_s,\zeta_s\right)\right| dQ_s\right)^p.
\end{aligned} \quad (5.139)$$

Since $(\xi,\zeta)\in S_m^p\left([0,T];e^V\right)\times\Lambda_{m\times k}^p\left(0,T;e^V\right)$, for all $T>0$, it clearly follows from (5.139) that

$$\mathbb{E}\sup_{s\in[0,T]}\left|e^{V_s}Y_s\right|^p+\mathbb{E}\left(\int_0^T e^{2V_s}\left|Z_s\right|^2 ds\right)^{p/2}<\infty.$$

Let $0\leq t\leq T\leq n$. Now by Lemma 5.5 we can pass to the limit in

$$Y_t^n=Y_T^n+\int_t^T\Phi\left(s,Y_s^n,Z_s^n\right)dQ_s-\int_t^T Z_s^n dB_s, \quad a.s.,\ t\in[0,T] \quad (5.140)$$

and taking into account (5.139) we deduce that (Y, Z) satisfies (5.133). The proof is complete.

Remark 5.58. If, moreover, there exists a constant b such that $\sup_{t\geq 0} V_t \leq b$, $\mathbb{P}$-a.s., then the conditions (5.143-(*jjj*)) can be replaced by the stronger statement than (5.133):

$$(jjj') \quad \lim_{T\to\infty} \mathbb{E}\, e^{pV_T} |Y_T - \eta|^p = 0. \tag{5.141}$$

Indeed using the backward Burkholder–Davis–Gundy inequality (2.51) we have

$$c_p \mathbb{E} \left(\int_t^\infty |\zeta_r|^2 dr\right)^{p/2} \leq \mathbb{E} \sup_{s\geq t} |\eta - \xi_s|^p \leq C_p \mathbb{E} \left(\int_t^\infty |\zeta_r|^2 dr\right)^{p/2}.$$

□

Let $\tau : \Omega \to [0,\infty]$ be a stopping time and $\eta \in L^p(\Omega, \mathcal{F}_\tau, \mathbb{P}; \mathbb{R}^m)$, $p > 1$. We now consider the BSDE

$$Y_t = \eta + \int_{t\wedge\tau}^{\tau} \Phi(s, Y_s, Z_s)\, dQ_s - \int_{t\wedge\tau}^{\tau} Z_s dB_s, \quad a.s., \ \forall\, t \geq 0, \tag{5.142}$$

in the sense which will be made precise in the next theorem. Plainly the BSDE (5.142) a particular case of Eq. (5.129) where Φ is of the form $\mathbf{1}_{[0,\tau]}\Phi$, since by Lemma 2.43 $Z_t = 0$ for all $t > \tau$.

Recall that the unique pair $(\xi, \zeta) \in S_d^p \times \Lambda_{d\times k}^p(0,\infty)$ such that

$$\xi_t = \eta - \int_t^\infty \zeta_s dB_s, \quad t \geq 0, \ a.s.,$$

satisfies $\xi_t = \mathbb{E}^{\mathcal{F}_{t\wedge\tau}} \eta$ and $\zeta_t = \mathbf{1}_{[0,\tau]}(t)\, \zeta_t$.

Define

$$V_t \overset{def}{=} \int_0^{t\wedge\tau} \mu_s dQ_s + \frac{a}{2n_p} \int_0^{t\wedge\tau} \ell^2(s)\, ds \quad \text{and} \quad \bar{\mu}_t = \int_0^{t\wedge\tau} \mu_s dQ_s.$$

We deduce from Theorem 5.57:

Corollary 5.59. *Let $a, p > 1$ and $\tau : \Omega \to [0,\infty]$ be a stopping time. Let the assumptions* (**BSDE-H**$_\infty$) *with $\Phi(s, y, z) = \mathbf{1}_{[0,\tau]}(s)\, \Phi(s, y, z)$ be satisfied and $\eta \in L^p(\Omega, \mathcal{F}_\tau, \mathbb{P}; \mathbb{R}^m)$. Assume moreover*

$$(i) \quad \mathbb{E}\left(e^{p\|\bar{\mu}^+\|_{T\wedge\tau}} |\eta|^p\right) < \infty, \ \textit{for all } T \geq 0,$$

$$(ii) \quad \mathbb{E}\left(\int_0^\tau e^{V_t} |\Phi(t, \xi_t, \zeta_t)|\, dQ_t\right)^p < \infty,$$

and for all $\rho \geq 0$

(h_1) $\mathbb{E}\left(\int_0^{T\wedge\tau} \sup_{|y|\leq\rho} \left|e^{\bar{\mu}_t}\Phi\left(s, e^{-\bar{\mu}_t}y, 0\right) - \mu_t y\right| dQ_t\right)^p < \infty$, *or*

(h_1) $\mu \geq 0$ *and* $\mathbb{E}\left(\int_0^{T\wedge\tau} e^{\bar{\mu}_t} \sup_{|y|\leq\rho} |\Phi(t, y, 0)|\, dQ_s\right)^p < \infty.$

(A) *Then there exists a unique solution* $(Y_t, Z_t)_{t\geq 0} \in S_m^0 \times \Lambda_{m\times k}^0$, $(Y_t, Z_t) = (\eta, 0)$ *if* $t > \tau$, *of the BSDE (5.142) in the sense that*

$$\begin{cases} (j) \quad Y_t = Y_T + \int_t^T \Phi(s, Y_s, Z_s)\, dQ_s - \int_t^T Z_s dB_s, \quad a.s., \\ \qquad\qquad\qquad\qquad\qquad\qquad \text{for all } 0 \leq t \leq T, \\ (jj)\ \mathbb{E} \sup_{s\in[0,T]} e^{pV_s} |Y_s|^p < \infty, \quad \text{for all } T \geq 0, \\ (jjj)\ \lim_{T\to\infty} \mathbb{E}\, e^{pV_{T\wedge\tau}} |Y_{T\wedge\tau} - \xi_{T\wedge\tau}|^p = 0. \end{cases} \tag{5.143}$$

Moreover

$$\mathbb{E}\left(\int_0^\tau e^{2V_s} |Z_s|^2 ds\right)^{p/2} < \infty$$

and there exists a constant $C_{a,p}$ *depending only on* (a, p) *such that for all* $t \geq 0$,

$$\begin{aligned} \mathbb{E} \sup_{t\wedge\tau\leq s\leq\tau} \left|e^{V_s}(Y_s - \xi_s)\right|^p + \mathbb{E}\left(\int_{t\wedge\tau}^\tau e^{2V_s} |Z_s - \zeta_s|^2 ds\right)^{p/2} \\ \leq C_{a,p}\mathbb{E}\left(\int_{t\wedge\tau}^\tau e^{V_s} |\Phi(s, \xi_s, \zeta_s)|\, dQ_s\right)^p. \end{aligned} \tag{5.144}$$

(B) *If, moreover, there exists a constant* b *such that* $\sup_{0\leq t\leq\tau} V_t \leq b$, $\mathbb{P}$-*a.s., then the conditions (5.143-(jjj)) can be replaced by*

$$(jjj') \quad \lim_{T\to\infty} \mathbb{E}\, e^{pV_{T\wedge\tau}} |Y_{T\wedge\tau} - \eta|^p = 0. \tag{5.145}$$

□

5.6.2 BSVIs with Random Final Time

In this section we are interested in the following generalized backward stochastic variational inequality (BSVI for short):

$$\begin{cases} Y_t + \int_{t\wedge\tau}^{\tau} dK_s = \eta + \int_{t\wedge\tau}^{\tau} [F(s, Y_s, Z_s)\, ds + G(s, Y_s)\, dA_s] \\ \qquad\qquad\qquad\qquad\qquad\qquad -\int_{t\wedge\tau}^{\tau} Z_s dB_s,\ t \geq 0, \\ dK_t \in \partial\varphi(Y_t)\, dt + \partial\psi(Y_t)\, dA_t \quad \text{on } \mathbb{R}_+, \end{cases} \tag{5.146}$$

where $\partial\varphi$, $\partial\psi$ are the subdifferentials of the convex lower semicontinuous functions φ, ψ, $\{A_t : t \geq 0\}$ is a progressively measurable increasing continuous stochastic process, and τ is a stopping time.

In fact we will define and prove the existence of the solution for an equivalent form of (5.146):

$$\begin{cases} Y_t + \int_t^{\infty} dK_s = \eta + \int_t^{\infty} \Phi(s, Y_s, Z_s)\, dQ_s - \int_t^{\infty} Z_s dB_s,\ \ t \geq 0, \\ dK_t \in \partial_y \Psi(t, Y_t)\, dQ_t,\ \text{on } \mathbb{R}_+, \end{cases} \tag{5.147}$$

with Q, Φ and Ψ adequately defined.

We mention that the presence of the process A is justified by the possible applications of Eq. (5.146) in obtaining a probabilistic interpretation for the solution of PDEs with Neumann boundary conditions; since τ is a stopping time the BSVI (5.146) can be used for elliptic PDEs.

Because (5.146) is quite a complicated equation, in order to simplify the presentation we shall restrict ourselves to $p = 2$. The case $p \geq 2$ can be found in [47].

We begin to give the main assumptions for this section.

(A_1) *The random variable $\tau : \Omega \to [0, \infty]$ is a stopping time.*

(A_2) *The random variable $\eta : \Omega \to \mathbb{R}^m$ is $\mathcal{F}_\tau$-measurable, $\mathbb{E}\,|\eta|^2 < \infty$ and the stochastic process $(\xi, \zeta) \in S_m^2 \times \Lambda_{m\times k}^2(0, \infty)$ is the unique pair associated to η given by the martingale representation formula (Corollary 2.44)*

$$\begin{cases} \xi_t = \eta - \int_t^{\infty} \zeta_s dB_s,\ t \geq 0,\ \textit{a.s.}, \\ \xi_t = \mathbb{E}^{\mathcal{F}_t}\eta = \mathbb{E}^{\mathcal{F}_{t\wedge\tau}}\eta \quad \textit{and} \quad \zeta_t = \mathbf{1}_{[0,\tau]}(t)\, \zeta_t. \end{cases}$$

(A_3) *The process $\{A_t : t \geq 0\}$ is a progressively measurable increasing continuous stochastic process such that $A_0 = 0$,*

$$Q_t(\omega) = t + A_t(\omega),$$

and $\{\alpha_t : t \geq 0\}$ is a real positive p.m.s.p. (given by the Radon–Nikodym representation theorem) such that $\alpha \in [0, 1]$ and

$$dt = \alpha_t dQ_t \quad \textit{and} \quad dA_t = (1 - \alpha_t)\, dQ_t.$$

(A_4) *The functions* $F : \Omega\times\mathbb{R}_+\times\mathbb{R}^m\times\mathbb{R}^{m\times k}\to\mathbb{R}^m$ *and* $G : \Omega\times\mathbb{R}_+\times\mathbb{R}^m\to\mathbb{R}^m$ *are such that*

$$\begin{cases} F(\cdot,\cdot,y,z)\,,\ G(\cdot,\cdot,y) \text{ are p.m.s.p., for all } (y,z)\in\mathbb{R}^m\times\mathbb{R}^{m\times k}\,,\\ F(\omega,t,\cdot,\cdot)\,,\ G(\omega,t,\cdot) \text{ are continuous functions, } d\mathbb{P}\otimes dt\text{-a.e.,}\end{cases}$$

and $\mathbb{P}$*-a.s.,*

$$\int_0^T F_\rho^\#(s)\,ds+\int_0^T G_\rho^\#(s)\,dA_s<\infty,\quad \forall\,\rho,T\geq 0,$$

where

$$F_\rho^\#(\omega,s):=\sup_{|y|\leq\rho}|F(\omega,s,y,0)|\,,\quad G_\rho^\#(\omega,s):=\sup_{|y|\leq\rho}|G(\omega,s,y)|\,.$$

(A_5) *Assume that there exist three progressively measurable positive stochastic processes* $\mu,\nu,\ell:\Omega\times\mathbb{R}_+\to\mathbb{R}_+$ *such that*

$$\int_0^T\left(\mu_s+(\ell_s)^2\right)ds+\int_0^T\nu_s dA_s<\infty,\ \text{for all } T>0,\ \mathbb{P}\text{-a.s.,}$$

and $\mathbb{P}$-a.s. $\omega\in\Omega$, *for all* $t\in[0,\tau(\omega)]$, $y,y'\in\mathbb{R}^m$, $z,z'\in\mathbb{R}^{m\times k}$,

$$\begin{array}{ll}(i) & \langle y'-y,F(t,y',z)-F(t,y,z)\rangle\leq\mu_t\,|y'-y|^2\,,\\ (ii) & \langle y'-y,G(t,y')-G(t,y)\rangle\leq\nu_t\,|y'-y|^2\,,\\ (iii) & |F(t,y,z')-F(t,y,z)|\leq\ell_t\,|z'-z|\,.\end{array}\tag{5.148}$$

Let us introduce the functions

$$\begin{aligned}&H(\omega,t,y,z):=\mathbf{1}_{[0,\tau(\omega)]}(t)\left[\alpha_t(\omega)F(\omega,t,y,z)+(1-\alpha_t(\omega))G(\omega,t,y)\right],\\&\bar{\mu}_t:=\int_0^t\mathbf{1}_{[0,\tau]}(s)\,\mu_s ds,\quad \bar{\nu}_t:=\int_0^t\mathbf{1}_{[0,\tau]}(s)\,\nu_s dA_s,\\&\sigma_t:=\mathbf{1}_{[0,\tau]}(t)\left[\mu_t\alpha_t+\nu_t(1-\alpha_t)\right],\quad \bar{\sigma}_t:=\int_0^t\mathbf{1}_{[0,\tau]}(s)\,\sigma_s dQ_s=\bar{\mu}_t+\bar{\nu}_t,\\&\lambda_t=\mathbf{1}_{[0,\tau]}(t)\,\alpha_t\ell_t,\quad \hat{\lambda}_t=\int_0^t\mathbf{1}_{[0,\tau]}(s)\,(\ell_s)^2\alpha_s dQ_s=\int_0^{t\wedge\tau}(\ell_s)^2ds.\end{aligned}\tag{5.149}$$

The relations (5.148) yield

$$\begin{array}{ll}(a) & \langle y'-y,H(t,y',z)-H(t,y,z)\rangle\leq\sigma_t\,|y'-y|^2\,,\\ (b) & |H(t,y,z')-H(t,y,z)|\leq\boldsymbol{\lambda}_t\,|z'-z|\,.\end{array}\tag{5.150}$$

Let

$$V_t=\int_0^t\mathbf{1}_{[0,\tau]}(s)\left[\left(\mu_s+(\ell_s)^2\right)\alpha_s+\nu_s(1-\alpha_s)\right]dQ_s=\bar{\sigma}_t+\hat{\lambda}_t.\tag{5.151}$$

Concerning φ and ψ we shall assume:

(A_6) $\varphi, \psi : \mathbb{R}^m \to [0, +\infty]$ *are proper convex lower semicontinuous (l.s.c.) functions,* $\partial\varphi$ *and* $\partial\psi$ *are the subdifferentials of* φ *and* ψ, *respectively, and there exists a* $u_0 \in \mathbb{R}^m$ *such that* $0 \in \partial\varphi(u_0) \cap \partial\psi(u_0)$ *(which is equivalent to* $\varphi(u_0) \leq \varphi(y)$ *and* $\psi(u_0) \leq \psi(y)$ for all $y \in \mathbb{R}^m$). *Define*

$$\Psi(\omega, t, y) = \mathbf{1}_{[0,\tau(\omega)]}(t) \left[\alpha_t(\omega) \varphi(y) + (1 - \alpha_t(\omega)) \psi(y)\right].$$

(A_7) *If* $\mathbb{P}(\tau > N) > 0$, *for all* $N \in \mathbb{N}^*$, *then for every* $\bar{\eta} \in \overline{\eta, u_0} = conv\{\eta, u_0\}$ *there exist two progressively measurable stochastic processes* $\bar{\xi}^{(1)}, \bar{\xi}^{(2)}$ *such that* $\bar{\xi}_t^{(1)} \in \partial\varphi\left(\mathbb{E}^{\mathcal{F}_t}\bar{\eta}\right)$, $\bar{\xi}_t^{(2)} \in \partial\psi\left(\mathbb{E}^{\mathcal{F}_t}\bar{\eta}\right)$ *a.e.* $t \geq 0$ *and*

$$\mathbb{E}\left(\int_0^\tau e^{V_s}|\bar{\xi}_s^{(1)}|ds\right)^2 + \mathbb{E}\left(\int_0^\tau e^{V_s}|\bar{\xi}_s^{(2)}|dA_s\right)^2 < \infty;$$

if $\bar{\xi}_s = \mathbf{1}_{[0,\tau]}(s)\,[\bar{\xi}_s^{(1)}\alpha_s + \bar{\xi}_s^{(2)}(1 - \alpha_s)]$, *then* $\bar{\xi}_s(\omega) \in \partial\Psi\left(\omega, s, \mathbb{E}^{\mathcal{F}_s}\bar{\eta}\right)$ *and*

$$\mathbb{E}\left(\int_0^\tau e^{V_s}\left|\bar{\xi}_s\right| dQ_s\right)^2 < \infty$$

(in the case $\bar{\eta} = \eta$ *we define* $\xi_t = \mathbb{E}^{\mathcal{F}_t}\eta$ *and in the place of* $(\bar{\xi}^{(1)}, \bar{\xi}^{(2)}, \bar{\xi})$ *we shall use the notation* $(\hat{\xi}^{(1)}, \hat{\xi}^{(2)}, \hat{\xi})$).

Remark 5.60. In place of the assumption (A_7) we can consider two particular cases:

(A_7') $\eta : \Omega \to \mathbb{O}$, where $\mathbb{O}$ is the closed convex set defined by

$$\mathbb{O} \stackrel{def}{=} \{y \in \mathbb{R}^m : \varphi(y) = \varphi(u_0) \text{ and } \psi(y) = \psi(u_0)\};$$

or

(A_7'') there exist $r_0 > 0$ and $v_0 \in \text{Dom}(\varphi) \cap \text{Dom}(\psi)$ such that

$$\begin{array}{ll} (i) & \eta : \Omega \to \overline{B(v_0, r_0)} \subset \text{int}(\text{Dom}(\varphi)) \cap \text{int}(\text{Dom}(\psi)), \\ (ii) & \mathbb{E}\left(e^{V_\tau}(\tau + A_\tau)\right) < \infty. \end{array}$$

Indeed:

(A_7') $\Rightarrow$(A_7): since $\bar{\xi}_t = \mathbb{E}^{\mathcal{F}_t}\bar{\eta} \in \mathbb{O}$ for all $t \geq 0$ we can set $\bar{\xi}^{(1)} = \bar{\xi}^{(2)} = \bar{\xi} = 0$ for every $\bar{\eta} \in \overline{\eta, u_0}$;

(A_7'') $\Rightarrow$ (A_7): by Proposition 6.2-(d) there exists an $M_0 > 0$ such that $\partial\varphi(u) \subset B(0, M_0)$ and $\partial\psi(u) \subset B(0, M_0)$ for all $u \in \overline{B(v_0, r_0)}$ and consequently $\left|\bar{\xi}_t^{(1)}\right| + \left|\bar{\xi}_t^{(2)}\right| \leq 2M_0$ for all $\bar{\eta} \in \overline{\eta, u_0}$ because $\mathbb{E}^{\mathcal{F}_t}\bar{\eta} \in \overline{B(v_0, r_0)}$ for all $t \geq 0$. Observe that from $\left(A_7''\text{-ii}\right)$ it follows that $\mathbb{P}(\tau = \infty) = 0$.

Definition 5.61. By the notation $dK_t \in \partial\psi(Y_t)\,dA_t$ we shall understand that:

- K is an $\mathbb{R}^m$-valued locally bounded variation stochastic process;
- Y is an $\mathbb{R}^m$-valued continuous stochastic process such that $\int_0^T \psi(Y_t)\,dA_t < \infty$, a.s. $\forall\, T \geq 0$; and
- $\mathbb{P}$-a.s., for all $0 \leq t \leq s$

$$\int_t^s \langle y(r) - Y_r, dK_r\rangle + \int_t^s \psi(Y_r)\,dA_r \leq \int_t^s \varphi(y(r))\,dA_r,\ \forall y \in C(\mathbb{R}_+;\mathbb{R}^m),$$

(we have an analogous definition for $dK_t \in \partial\varphi(Y_t)\,dt$).

Remark 5.62. The condition $0 \in \partial\varphi(u_0) \cap \partial\psi(u_0)$ does not restrict the generality of the problem because from $\mathrm{Dom}(\partial\varphi) \cap \mathrm{Dom}(\partial\psi) \neq \emptyset$ it follows that *there exists* $u_0 \in \mathrm{Dom}(\partial\varphi) \cap \mathrm{Dom}(\partial\psi)$ and $\hat{u}_{01} \in \partial\varphi(u_0)$, $\hat{u}_{02} \in \partial\psi(u_0)$; in this case equation (5.146) is equivalent to

$$\begin{cases} Y_t + \displaystyle\int_{t\wedge\tau}^{\tau} d\hat{K}_s = \eta + \int_{t\wedge\tau}^{\tau} \Big[\hat{F}(s, Y_s, Z_s)\,ds + \hat{G}(s, Y_s)\,dA_s\Big] \\ \qquad\qquad\qquad\qquad -\displaystyle\int_{t\wedge\tau}^{\tau} Z_s dB_s, \quad t \geq 0, \\ d\hat{K}_t \in \partial\hat{\varphi}(Y_t)\,dt + \partial\hat{\psi}(Y_t)\,dA_t, \text{ on } \mathbb{R}_+, \end{cases}$$

where $\hat{F}(s, y, z) = F(t, y, z) - \hat{u}_{01}$, $\hat{G}(s, y, z) = G(t, y) - \hat{u}_{02}$, $\hat{\varphi}(y) = \varphi(y) - \langle \hat{u}_{01}, y - u_0\rangle$, $\hat{\psi}(y) = \psi(y) - \langle \hat{u}_{02}, y - u_0\rangle$, $\partial\hat{\varphi}(y) = \partial\varphi(y) - \hat{u}_{01}$, $\partial\hat{\psi}(y) = \partial\psi(y) - \hat{u}_{02}$ and $d\hat{K}_t = dK_t - \hat{u}_{01}dt - \hat{u}_{02}dA_t$.

Let $\varepsilon > 0$ and define the Moreau–Yosida regularization of φ by

$$\varphi_\varepsilon(y) := \inf\left\{\frac{1}{2\varepsilon}|y - v|^2 + \varphi(v) : v \in \mathbb{R}^m\right\},$$

which is a C^1 convex function and $\nabla\varphi_\varepsilon(x) = \partial\varphi_\varepsilon(x) \in \partial\varphi(J_\varepsilon x)$, where $J_\varepsilon x = x - \varepsilon\nabla\varphi_\varepsilon(x)$. (For further properties see Annex B, Section "Convex Functions".) Since $0 \in \partial\varphi(u_0)$ we deduce that $\varphi(u_0) = \varphi_\varepsilon(u_0) \leq \varphi_\varepsilon(u) \leq \varphi(u)$, $J_\varepsilon(u_0) = u_0$ and $\nabla\varphi_\varepsilon(u_0) = 0$.

We introduce the *compatibility conditions* between φ, ψ and F, G:

(A_8) *There exists a* $c > 0$ *and two progressively measurable stochastic processes* $f, g \colon \Omega \times \mathbb{R}_+ \to \mathbb{R}_+$ *satisfying*

$$\mathbb{E}\int_0^\tau e^{2V_s}|f_s|^2\,ds + \mathbb{E}\int_0^\tau e^{2V_s}|g_s|^2\,dA_s < \infty,$$

such that for all $\varepsilon > 0$, $t \geq 0$, $y \in \mathbb{R}^m$, $z \in \mathbb{R}^{m\times k}$, $\mathbb{P}$-a.s.

$$\begin{array}{ll} (i) & \langle \nabla\varphi_\varepsilon(y), \nabla\psi_\varepsilon(y)\rangle \geq 0, \\ (ii) & \langle \nabla\varphi_\varepsilon(y), G(t,y)\rangle \leq c\,|\nabla\psi_\varepsilon(y)|\,[|G(t,y)| + g_t], \\ (iii) & \langle \nabla\psi_\varepsilon(y), F(t,y,z)\rangle \leq c\,|\nabla\varphi_\varepsilon(y)|\,[|F(t,y,z)| + f_t], \end{array} \tag{5.152}$$

and

$$\begin{array}{ll} (iv) & \langle \nabla\varphi_\varepsilon(y), -G(t,u_0)\rangle \leq c\,|\nabla\psi_\varepsilon(y)|\,[|G(t,u_0)| + g_t], \\ (v) & \langle \nabla\psi_\varepsilon(y), -F(t,u_0,0)\rangle \leq c\,|\nabla\varphi_\varepsilon(y)|\,[|F(t,u_0,0)| + f_t]. \end{array} \tag{5.153}$$

Example 5.63. (e_1) If $\varphi = \psi$ then the compatibility assumptions (5.152) and (5.153) are clearly satisfied.

(e_2) Let $m = 1$. Since $\nabla\varphi_\varepsilon$ and $\nabla\psi_\varepsilon$ are increasing monotone functions on $\mathbb{R}$, we see that, if $G(t,u_0) = F(t,u_0,0) = 0$ and

$$(y - u_0)\,G(t,y) \leq 0 \quad \text{and} \quad (y - u_0)\,F(t,y,z) \leq 0, \ \forall\, t,y,z,$$

then the compatibility assumptions (5.152) and (5.153) are satisfied.

(e_3) Let $m = 1$. If $\varphi, \psi : \mathbb{R} \to (-\infty, +\infty]$ are the convex indicator functions

$$\varphi(y) = \begin{cases} 0, \text{ if } y \in [a,b], \\ +\infty, \text{ if } y \notin [a,b], \end{cases} \quad \text{and} \quad \psi(y) = \begin{cases} 0, \text{ if } y \in [c,d], \\ +\infty, \text{ if } y \notin [c,d], \end{cases}$$

where $-\infty \leq a \leq b \leq +\infty$ and $-\infty \leq c \leq d \leq +\infty$ are such that $[a,b] \cap [c,d] \neq \emptyset$ (see the assumption (A_6)), then

$$\nabla\varphi_\varepsilon(y) = \frac{1}{\varepsilon}\left[(y-b)^+ - (a-y)^+\right], \quad \text{and}$$

$$\nabla\psi_\varepsilon(y) = \frac{1}{\varepsilon}\left[(y-d)^+ - (c-y)^+\right].$$

The assumption (A_8-i) is clearly fulfilled; the next *compatibility assumptions* (A_8-ii, iii, iv, v) *are satisfied if for example* $G(t,u_0) = F(t,u_0,0) = 0$ and

$$G(t,y) \geq 0, \textit{ for } y \leq a, \quad G(t,y) \leq 0, \textit{ for } y \geq b,$$

and, respectively,

$$F(t,y,z) \geq 0, \textit{ for } y \leq c, \quad F(t,y,z) \leq 0, \textit{ for } y \geq d.$$

We complete the assumptions with some general boundedness conditions

(A_9) *For all* $\rho > 0$

$$\begin{aligned}
&(i)\ \mathbb{E}\, e^{2V_\tau}\Big(|\eta-u_0|^2+\varphi(\eta)-\varphi(u_0)+\psi(\eta)-\psi(u_0)\Big)<\infty,\\
&(ii)\ \mathbb{E}\left(\int_0^\tau e^{V_s}F_\rho^{\#}(s)\,ds\right)^2+\mathbb{E}\left(\int_0^\tau e^{V_s}G_\rho^{\#}(s)\,dA_s\right)^2<\infty,\\
&(iii)\ \mathbb{E}\left[\int_0^\tau e^{2V_s}\left|F_\rho^{\#}(s)\right|^2ds+\int_0^\tau e^{2V_s}\left|G_\rho^{\#}(s)\right|^2dA_s\right]<\infty,\\
&(iv)\ \mathbb{E}\left(\int_0^T e^{V_s}dQ_s\right)^2<\infty,\text{ for all } T\geq 0;
\end{aligned}\tag{5.154}$$

and some special boundedness conditions

(A_{10}) *There exist* $L, b > 0$ *such that for all* $0 \le t \le \tau$, $\mathbb{P}$-a.s.

$$\begin{aligned}
&(a)\quad \ell_t+\int_0^\tau(\ell_s)^2\,ds\leq L,\\
&(b)\ e^{V_\tau}|\eta-u_0|+|H(t,u_0,0)|+\int_0^\tau e^{V_s}|H(s,u_0,0)|\,dQ_s\leq b,
\end{aligned}\tag{5.155}$$

where again H is defined by

$$H(t,y,z)=\alpha_tF(t,y,z)+(1-\alpha_t)G(t,y).$$

We also recall the definition of

$$\Psi(t,y)=\alpha_t\varphi(y)+(1-\alpha_t)\psi(y).$$

Since $V \ge 0$, we remark that under (A_{10}) we have

$$|\eta-u_0|\leq\left|e^{V_\tau}(\eta-u_0)\right|\leq b$$

and for all $t \ge 0$,

$$|\xi_t-u_0|\leq e^{V_t}|\xi_t-u_0|=\mathbb{E}^{\mathcal{F}_t}\left(e^{V_t}|\xi_t-u_0|\right)\leq b.$$

Therefore by Proposition 6.80-A, for all $q > 0$

$$\mathbb{E}\left(\int_0^\tau e^{2V_s}|\zeta_s|^2\,ds\right)^{q/2}\leq C_{b,p}.\tag{5.156}$$

Using the definition of Q, H and Ψ we can rewrite (5.146) in the form

$$\begin{cases}Y_t+\displaystyle\int_t^\infty dK_s=\eta+\int_t^\infty H(s,Y_s,Z_s)\,dQ_s-\int_t^\infty Z_sdB_s,\ t\geq 0,\\ dK_s\in\partial_y\Psi(s,Y_s)\,dQ_s \text{ on } \mathbb{R}_+.\end{cases}\tag{5.157}$$

Definition 5.64. We call $(Y_t, Z_t)_{t\geq 0}$ a solution of (5.157) if

(d_1) $(Y, Z) \in S_m^0 \times \Lambda_{m\times k}^0$,
(d_2) $(Y_t, Z_t) = (\xi_t, \zeta_t) = (\eta, 0)$, if $t > \tau$,
(d_3) $\mathbb{P}$-a.s., for all $T \geq 0$,

$$\int_0^T [|F(s, Y_s, Z_s)| + |\varphi(Y_s)|] \, ds + \int_0^T [|G(s, Y_s)| + |\psi(Y_s)|] \, dA_s < \infty,$$

(d_4) there exists a $K \in S_m^0$ such that $\mathbb{P}$-a.s.

$$\begin{array}{ll} (i) & \updownarrow K \updownarrow_T < \infty, \quad \forall T \geq 0, \\ (ii) & dK_t \in \partial_y \Psi(t, Y_t) \, dQ_t, \end{array}$$

(d_5) $e^{2V_T} |Y_T - \xi_T|^2 + \int_T^\infty e^{2V_s} |Z_s - \zeta_s|^2 \, ds \xrightarrow{prob.} 0$, as $T \to \infty$,
(d_6) $\mathbb{P}$-a.s., for all $0 \leq t \leq T$,

$$Y_t + K_T - K_t = Y_T + \int_t^T H(s, Y_s, Z_s) \, dQ_s - \int_t^T Z_s dB_s \tag{5.158}$$

(we also say that the triplet (Y, Z, K) is a solution of (5.157)).

Remark 5.65. If there exists a constant C such that $\sup_{t\in[0,\tau]} |V_t(\omega)| \leq C$, $\mathbb{P}$-a.s. $\omega \in \Omega$, then the condition (d_5) from Definition 5.64 is equivalent to

$$|Y_T - \eta|^2 + \int_T^\infty |Z_s|^2 \, ds \xrightarrow{prob.} 0, \text{ as } T \to \infty. \tag{5.159}$$

In the rest of this book, a constant depending upon $p > 0$ is denoted by C_p; in this section since we are only considering the case $p = 2$ we will denote the corresponding constant by C_2.

We now give the main result.

Theorem 5.66. *Let the assumptions (A_1–A_{10}) be satisfied. Then the backward stochastic variational inequality (5.157) has a unique solution $(Y, Z, K) \in S_m^0 \times \Lambda_{m\times k}^0 \times S_m^0$ such that*

$$\begin{array}{ll} (j) & \mathbb{E} \sup_{s\geq 0} e^{2V_s} |Y_s - u_0|^2 + \mathbb{E} \int_0^\infty e^{2V_s} |Z_s|^2 \, ds < \infty, \\ (jj) & \lim_{T\to\infty} \mathbb{E}\Big[e^{2V_T} |Y_T - \xi_T|^2 + \int_T^\infty e^{2V_s} |Z_s - \zeta_s|^2 \, ds \Big] = 0. \end{array} \tag{5.160}$$

Moreover there exists $U^{(1)}, U^{(2)} \in \Lambda_m^0$, with $U^{(1)} \in \partial\varphi(Y_t) \, d\mathbb{P} \otimes dt$ a.e. and $U_t^{(2)} \in \partial\psi(Y_t) \, d\mathbb{P} \otimes dA_t$ a.e., so that with $U_t = \mathbf{1}_{[0,\tau]}(t)[\alpha_t U_t^{(1)} + (1 - \alpha_t) U_t^{(2)}]$, $dK_t = U_t dQ_t \in \partial_y \Psi(t, Y_t) dQ_t$,

$$U_t = \mathbf{1}_{[0,\tau]}(t)\left[\alpha_t U_t^{(1)} + (1-\alpha_t)\, U_t^{(2)}\right]$$

and for all $0 \le t \le T$,

$$Y_t + \int_t^T U_s dQ_s = Y_T + \int_t^T H(s, Y_s, Z_s)\, dQ_s - \int_t^T Z_s dB_s.$$

The solution also satisfies for some positive constants:

(A) *for all* $t \ge 0$ *and all* $q > 0$,

$$\begin{aligned} &(j) \quad |Y_t - u_0| \le e^{V_t}|Y_t - u_0| \le C_b, \\ &(jj) \quad \mathbb{E}\left(\int_0^\infty e^{2V_r}|Z_r|^2 dr\right)^{q/2} \le C_{q,b}; \end{aligned} \tag{5.161}$$

(B) *for all* $t \ge 0$,

$$\begin{aligned} &\mathbb{E}^{\mathcal{F}_t}\sup_{s\ge t}\left|e^{V_s}(Y_s - u_0)\right|^2 + \mathbb{E}^{\mathcal{F}_t}\left(\int_t^\infty e^{2V_s}|Z_s|^2 ds\right) \\ &\quad + \mathbb{E}^{\mathcal{F}_t}\int_t^\infty e^{2V_s}\mathbf{1}_{[0,\tau]}(s)\left[|\varphi(Y_s) - \varphi(u_0)|\, ds + |\psi(Y_s) - \psi(u_0)|\, dA_s\right] \\ &\le C_2\,\mathbb{E}^{\mathcal{F}_t}\left[e^{2V_\tau}|\eta - u_0|^2 + \left(\int_t^\tau e^{V_s}\left(|F(s,u_0,0)|\, ds + |G(s,u_0)|\, dA_s\right)\right)^2\right]; \end{aligned} \tag{5.162}$$

(C) *for all* $t \ge 0$,

$$\begin{aligned} &\mathbb{E}\sup_{s\ge t} e^{2V_s}|Y_s - \xi_s|^2 + \mathbb{E}\int_t^\infty e^{2V_s}|Z_s - \zeta_s|^2 ds \\ &\qquad + \mathbb{E}\int_t^\infty e^{2V_s}|\Psi(s,Y_s) - \Psi(s,\xi_s)|\, dQ_s \\ &\le C_2\mathbb{E}\left(\int_t^\infty \mathbf{1}_{[0,\tau]}(s)\, e^{V_s}\left[|\hat{\xi}_s| dQ_s + (|F(s,\xi_s,0)| + \ell_s|\zeta_s|)\, dsx\right.\right. \\ &\qquad \left.\left. + |G(s,\xi_s)|\, dA_s\right]\right)^2; \end{aligned} \tag{5.163}$$

(D) *for all* $t \ge 0$

$$\begin{aligned} &\mathbb{E}\left[e^{2V_t}(\varphi(Y_t) - \varphi(u_0) + \psi(Y_t) - \psi(u_0))\right] \\ &\quad + \frac{1}{2}\mathbb{E}\int_t^\infty \mathbf{1}_{[0,\tau]}(s)\, e^{2V_s}\left(\left|U_s^{(1)}\right|^2 ds + \left|U_s^{(2)}\right|^2 dA_s\right) \\ &\le \mathbb{E}\left[e^{2V_\tau}(\varphi(\eta) - \varphi(u_0) + \psi(\eta) - \psi(u_0))\right] \\ &\quad + (1+c)^2\,\mathbb{E}\int_t^\infty \mathbf{1}_{[0,\tau]}(s)\, e^{2V_s}\left(|F(s,Y_s,Z_s)|^2 + |f_s|^2\right) ds \\ &\quad + (1+c)^2\,\mathbb{E}\int_t^\infty \mathbf{1}_{[0,\tau]}(s)\, e^{2V_s}\left(|G(s,Y_s)|^2 + |g_s|^2\right) dA_s. \end{aligned} \tag{5.164}$$

Proof. Uniqueness. If (Y, Z, K), (Y', Z', K') are two solutions, in the sense of Definition 5.64, that satisfy (5.160), then

$$\mathbb{E} \sup_{t\in[0,T]} e^{2V_s}|Y_s - Y_s'|^2 < \infty.$$

Applying the monotonicity and Lipschitz property of the function H and taking into account that

$$\left\langle Y_s - Y_s', dK_s - dK_s' \right\rangle \geq 0$$

for $dK_s \in \partial_y \Psi(s, Y_s)dQ_s$ and $dK_s' \in \partial_y \Psi(s, Y_s')dQ_s$, then

$$\begin{aligned}
&\langle Y_s - Y_s', \left[H\left(s, Y_s, Z_s\right) - H(s, Y_s', Z_s')\right] dQ_s - dK_s + dK_s' \rangle \\
&\leq |Y_s - Y_s'|^2 dV_s + \frac{1}{4}|Z_s - Z_s'|^2 ds.
\end{aligned}$$

Using Corollary 6.82 from Annex C, it follows that

$$\begin{aligned}
&\mathbb{E} \sup_{s\in[0,T]} e^{2V_s}\left|Y_s - Y_s'\right|^2 + \mathbb{E}\int_0^T e^{2V_s}\left|Z_s - Z_s'\right|^2 ds \\
&\leq C_2\,\mathbb{E}\left(e^{2V_T}\left|Y_T - Y_T'\right|^2\right) \xrightarrow[T\to\infty]{} 0,
\end{aligned}$$

which yields the uniqueness.

The proof of the existence will be split into several steps.

A. *Approximating problem.* Let $n \in \mathbb{N}^*$ and $\varepsilon = \dfrac{1}{n}$.

Let

$$\begin{aligned}
\Psi^n\left(\omega, t, y\right) &= \mathbf{1}_{[0,n\wedge\tau(\omega)]}\left(t\right)\left[\alpha_t\left(\omega\right)\varphi_\varepsilon\left(y\right) + \left(1 - \alpha_t\left(\omega\right)\right)\psi_\varepsilon\left(y\right)\right] \\
\nabla_y\Psi^n\left(\omega, t, y\right) &= \mathbf{1}_{[0,n\wedge\tau(\omega)]}\left(t\right)\left[\alpha_t\left(\omega\right)\nabla_y\varphi_\varepsilon\left(y\right) + \left(1 - \alpha_t\left(\omega\right)\right)\nabla_y\psi_\varepsilon\left(y\right)\right] \\
H_n\left(\omega, t, y, z\right) &= \mathbf{1}_{[0,n]}\left(t\right) H\left(\omega, t, y, z\right) \\
&= \mathbf{1}_{[0,n\wedge\tau(\omega)]}\left(t\right)\left[\alpha_t\left(\omega\right) F\left(\omega, t, y, z\right) + \left(1 - \alpha_t\left(\omega\right)\right) G\left(\omega, t, y\right)\right]
\end{aligned}$$

and

$$\Phi_n\left(\omega, t, y, z\right) = H_n\left(\omega, t, y, z\right) - \nabla_y\Psi^n\left(\omega, t, y\right).$$

We note that

$$\begin{aligned}
&\left|\Phi_n\left(t, u_0, 0\right)\right| dQ_t \\
&= \mathbf{1}_{[0,n\wedge\tau]}\left(t\right)\left[\left|H\left(t, u_0, 0\right)\right| dQ_t + \left|\nabla_y\Psi^n\left(t, u_0\right)\right| dQ_t\right]
\end{aligned}$$

$$= \mathbf{1}_{[0,n\wedge\tau]}(t)\,|H(t,u_0,0)|\,dQ_t$$
$$\leq \mathbf{1}_{[0,n\wedge\tau]}(t)\,[|F(t,u_0,0)|\,dt + |G(t,u_0)|\,dA_t]$$

and

$$\left|\nabla_y\Psi^n(s,y) - \nabla_y\Psi^n(s,y')\right| \leq n\mathbf{1}_{[0,n\wedge\tau]}(t)\left|y-y'\right|.$$

We consider the approximating stochastic equation: for all $t \geq 0$,

$$Y_t^n + \int_t^\infty \nabla_y\Psi^n(s,Y_s^n)dQ_s = \eta + \int_t^\infty H_n(s,Y_s^n,Z_s^n)dQ_s - \int_t^\infty Z_s^n dB_s, \quad (5.165)$$

or equivalently

$$\begin{cases} Y_t^n - u_0 = \left(\mathbb{E}^{\mathcal{F}_n}\eta - u_0\right) + \displaystyle\int_t^n \Phi_n(s,u_0+\left(Y_s^n-u_0\right),Z_s^n)dQ_s \\ \qquad\qquad\qquad\qquad - \displaystyle\int_t^n Z_s^n dB_s, \quad \forall\, t\in[0,n], \\ (Y_t^n,Z_t^n) = (\xi_t,\zeta_t),\ \forall\, t>n. \end{cases} \quad (5.166)$$

To show the existence of a solution (Y^n,Z^n) of (5.166) we intend to use Lemma 5.29-(h_2).

Since $\langle y'-y, \nabla\varphi_\varepsilon(y') - \nabla\varphi_\varepsilon(y)\rangle \geq 0$ (and similarly for $\nabla\psi_\varepsilon$) we notice that Φ_n satisfies the inequalities

$$\begin{array}{ll} (a) & \langle y'-y, \Phi_n(t,y',z) - \Phi_n(t,y,z)\rangle \\ & \quad \leq \mathbf{1}_{[0,n\wedge\tau]}(t)\,[\mu_t\alpha_t + \nu_t(1-\alpha_t)]\,|y'-y|^2 \leq \sigma_t\,|y'-y|^2 \\ (b) & |\Phi_n(t,y,z') - \Phi_n(t,y,z)| \leq \mathbf{1}_{[0,n\wedge\tau]}(t)\,\ell_t\alpha_t\,|z'-z| \leq \alpha_t L\,|z'-z|\,. \end{array} \quad (5.167)$$

Consequently the corresponding assumptions (5.13-**BSDE-H**$_\Phi$) for Φ_n are satisfied.

We have

$$\mathbb{E}\left(e^{2\bar{\sigma}_n}\left|\mathbb{E}^{\mathcal{F}_n}\eta - u_0\right|^2\right) \leq \mathbb{E}\left(e^{2\bar{\sigma}_n}|\eta-u_0|^2\right) \leq b^2 < \infty.$$

For the assumption (h_2) from Lemma 5.29 we have for all $\rho > 0$,

$$\mathbb{E}\left(\int_0^n e^{\bar{\sigma}_s}\sup_{|y|\leq\rho}|\Phi_n(s,y,0)|\,dQ_s\right)^2$$
$$\leq \mathbb{E}\left(\int_0^{n\wedge\tau} e^{\bar{\sigma}_s}\left|F_\rho^\#(s)\right|ds + \int_0^{n\wedge\tau} e^{\bar{\sigma}_s}\left|G_\rho^\#(s)\right|dA_s\right.$$
$$\left.+ \int_0^{n\wedge\tau} e^{\bar{\sigma}_s}\sup_{|y|\leq\rho}\left|\nabla_y\Psi^n(s,y)\right|dQ_s\right)^2$$

$$\leq 3\mathbb{E}\left(\int_0^{n\wedge\tau} e^{\bar{\sigma}_s}\left|F_\rho^{\#}(s)\right| ds\right)^2 + 3\mathbb{E}\left(\int_0^{n\wedge\tau} e^{\bar{\sigma}_s}\left|G_\rho^{\#}(s)\right| dA_s\right)^2$$
$$+ 3\mathbb{E}\left(\int_0^{n\wedge\tau} e^{\bar{\sigma}_s} n\left(\rho + |u_0|\right) dQ_s\right)^2 < \infty$$

because $\nabla_y \Psi^n(s, u_0) = 0$ and

$$\sup_{|y|\leq\rho}\left|\nabla_y \Psi^n(s,y)\right| = \sup_{|y|\leq\rho}\left|\nabla_y \Psi^n(s,y) - \nabla_y \Psi^n(s,u_0)\right| \leq n \sup_{|y|\leq\rho}|y-u_0|.$$

By Lemma 5.29-(h_2) Eq. (5.166) has a unique solution $(Y^n, Z^n) \in S_m^0 \times \Lambda_{m\times k}^0$ such that

$$\mathbb{E}\sup_{s\in[0,n]}\left|e^{V_s}\left(Y_s^n - u_0\right)\right|^2 + \mathbb{E}\int_0^n e^{2V_s}|Z_s^n|^2 ds < \infty.$$

Consequently for all $T > n$,

$$\begin{aligned}
&\mathbb{E}\sup_{s\in[0,T]}\left|e^{V_s}\left(Y_s^n - u_0\right)\right|^2 \\
&\leq \mathbb{E}\sup_{s\in[0,n]}\left|e^{V_s}\left(Y_s^n - u_0\right)\right|^2 + \mathbb{E}\sup_{s\in[n,T]}\left|e^{V_s}\left(\mathbb{E}^{\mathcal{F}_s}\eta - u_0\right)\right|^2 \\
&\leq \mathbb{E}\sup_{s\in[0,n]}\left|e^{V_s}\left(Y_s^n - u_0\right)\right|^2 + b^2 < \infty.
\end{aligned}$$

Now we remark that

$$\begin{aligned}
|\Psi^n(s,y) - \Psi^n(s,u_0)|\, dQ_s &= (\Psi^n(s,y) - \Psi^n(s,u_0))\, dQ_s \\
&\leq \left\langle y - u_0, \nabla_y \Psi^n(s,y)\right\rangle,
\end{aligned}$$

and therefore

$$\begin{aligned}
&|\Psi^n(s,Y_s^n) - \Psi^n(s,u_0)|\, dQ_s + \left\langle Y_s^n - u_0, \Phi_n\left(s, Y_s^n, Z_s^n\right) dQ_s\right\rangle \\
&\leq |Y_s^n - u_0|\,|H(s,u_0,0)|\, dQ_s + |Y_s^n - u_0|^2\, dV_s + \frac{1}{4}|Z_s^n|^2 ds.
\end{aligned}$$

By Proposition 6.80 we have for all $q, T > 0$

$$\begin{aligned}
&\mathbb{E}\left(\int_0^T e^{2V_s}|\Psi^n(s,Y_s^n) - \Psi^n(s,u_0)|\, dQ_s\right)^{q/2} + \mathbb{E}\left(\int_0^T e^{2V_s}|Z_s^n|^2 ds\right)^{q/2} \\
&\leq C_q \mathbb{E}\left[\sup_{s\in[0,T]}\left|e^{V_s}\left(Y_s^n - u_0\right)\right|^q + \left(\int_0^T e^{V_s}|H(s,u_0,0)|\, dQ_s\right)^q\right],
\end{aligned} \tag{5.168}$$

and

$$\begin{aligned}\mathbb{E}^{\mathcal{F}_t}\sup_{s\in[t,T]}\left|e^{V_s}\left(Y_s^n-u_0\right)\right|^2+\mathbb{E}^{\mathcal{F}_t}\left(\int_t^T e^{2V_s}\left|Z_s^n\right|^2ds\right)\\ +\mathbb{E}^{\mathcal{F}_t}\left(\int_t^T e^{2V_s}\left|\Psi^n(s,Y_s^n)-\Psi^n(s,u_0)\right|dQ_s\right)\\ \leq C_2\,\mathbb{E}^{\mathcal{F}_t}\Big[\left|e^{V_T}\left(Y_T^n-u_0\right)\right|^2+\left(\int_t^T e^{V_s}\left|H\left(s,u_0,0\right)\right|dQ_s\right)^2\Big].\end{aligned} \tag{5.169}$$

B. *Boundedness of* Y^n *and* Z^n.

If $n\leq T$, then

$$\mathbb{E}^{\mathcal{F}_t}\left|e^{V_T}\left(Y_T^n-u_0\right)\right|^2=\mathbb{E}^{\mathcal{F}_t}\left|e^{V_T}\mathbb{E}^{\mathcal{F}_T}\left(\eta-u_0\right)\right|^2\leq b^2.$$

Passing to the limit as $T\to\infty$ in (5.169) we infer (by the Beppo Levi monotone convergence theorem) that for all $t\geq 0$

$$\begin{aligned}\mathbb{E}^{\mathcal{F}_t}\sup_{s\geq t}\left|e^{V_s}\left(Y_s^n-u_0\right)\right|^2+\mathbb{E}^{\mathcal{F}_t}\left(\int_t^\infty e^{2V_s}\left|Z_s^n\right|^2ds\right)\\ +\mathbb{E}\left(\int_0^\infty e^{2V_s}\left|\Psi^n(s,Y_s^n)-\Psi^n(s,u_0)\right|dQ_s\right)\\ \leq C_2\,\mathbb{E}^{\mathcal{F}_t}\Big[\left|e^{V_\tau}\left(\eta-u_0\right)\right|^2+\left(\int_t^\infty e^{V_s}\left|H\left(s,u_0,0\right)\right|dQ_s\right)^2\Big].\end{aligned} \tag{5.170}$$

In particular, using the assumption (5.155), we deduce that for all $t\geq 0$

$$\left|Y_t^n-u_0\right|\leq e^{V_t}\left|Y_t^n-u_0\right|\leq C_2b^2\overset{def}{=}R_0. \tag{5.171}$$

Moreover from (5.168) for all $q>0$

$$\mathbb{E}\left(\int_0^\infty e^{2V_r}\left|Z_r^n\right|^2dr\right)^{q/2}\leq C_{q,b}. \tag{5.172}$$

C. *Estimates on* $|Y_t^n-\xi_t|$ *and* $|Z_t^n-\zeta_t|$ *for large* $t\geq 0$.

If there exists an $N_0>0$ such that $\tau(\omega)\leq N_0$, $\mathbb{P}$-a.s. $\omega\in\Omega$, then $Y_t^n=\xi_t=\eta$ and $Z_t^n=\zeta_t=0$ for all $t\geq N_0$.

We next consider the case where $\mathbb{P}(\tau>N)>0$ for all $N\in\mathbb{N}^*$.

Since

$$\xi_t=\xi_n-\int_t^n\zeta_s dB_s,\ \forall\, t\in[0,n],$$

we infer, from (5.166), that (Y^n, Z^n) satisfies for all $t \in [0, n]$ the equality

$$Y_t^n - \xi_t = \int_t^n \Phi_n\left(s, \xi_s + \left(Y_s^n - \xi_s\right), \zeta_s + \left(Z_s^n - \zeta_s\right)\right) dQ_s - \int_t^n (Z_s^n - \zeta_s) dB_s.$$

We have

$$\Psi^n(t, u_0) \le \Psi^n(t, \xi_t) \le \Psi(t, \xi_t) = \Psi(t, \mathbb{E}^{\mathcal{F}_t}\eta) \le \mathbb{E}^{\mathcal{F}_t}\Psi(t, \eta).$$

From $\left\langle \nabla_y \Psi^n(t, \xi_t), y - \xi_t \right\rangle \le \Psi^n(t, y) - \Psi^n(t, \xi_t) \le \langle y - \xi_t, \nabla_y \Psi^n(t, y)\rangle$ we infer that

$$\begin{aligned}
|\Psi^n(t, y) - \Psi^n(t, \xi_t)| &\le \Psi^n(t, y) - \Psi^n(t, \xi_t) + 2\left|\nabla_y \Psi^n(t, \xi_t)\right| |y - \xi_t| \\
&\le \langle y - \xi_t, \nabla_y \Psi^n(t, y)\rangle + 2\left|\nabla_y \Psi^n(t, \xi_t)\right| |y - \xi_t| \\
&\le \langle y - \xi_t, \nabla_y \Psi^n(t, y)\rangle + 2|\hat{\xi}_t| \, |y - \xi_t|
\end{aligned}$$

where $\hat{\xi}_t \in \partial_y \Psi(t, \xi_t)$ is given by the assumption (A_7). Using the inequality (5.167)) it follows that, as signed measures on $\mathbb{R}_+$,

$$\begin{aligned}
&\left|\Psi^n\left(t, Y_t^n\right) - \Psi^n(t, \xi_t)\right| dQ_t + \langle Y_t^n - \xi_t, \Phi_n(t, Y_t^n, Z_t^n)\rangle dQ_t \\
&\quad \le |Y_t^n - \xi_t| \left[2|\hat{\xi}_t| + |H(t, \xi_t, \zeta_t)|\right] dQ_t + |Y_t^n - \xi_t|^2 dV_t + \frac{1}{4}|Z_t^n - \zeta_t|^2 dt.
\end{aligned} \tag{5.173}$$

Since

$$\begin{aligned}
\mathbb{E} \sup_{t\in[0,T]} e^{2V_t} |Y_t^n - \xi_t|^2 &\le 2\mathbb{E}\left[\sup_{t\in[0,T]} e^{2V_t} |Y_t^n - u_0|^2 + \sup_{t\in[0,T]} e^{2V_t} |\xi_t - u_0|^2\right] \\
&\le 2R_0^2 + \mathbb{E} \sup_{t\in[0,T]} e^{2V_t} \left|\mathbb{E}^{\mathcal{F}_t}\eta - u_0\right|^2 \le 2R_0^2 + \mathbb{E} \sup_{t\in[0,T]} \mathbb{E}^{\mathcal{F}_t}\left(e^{2V_t} |\eta - u_0|^2\right) \\
&\le 2R_0^2 + b^2
\end{aligned}$$

by Proposition 5.2 we deduce that for $0 \le t \le T$,

$$\begin{aligned}
&\mathbb{E} \sup_{s\in[t,T]} e^{2V_s} |Y_s^n - \xi_s|^2 + \mathbb{E}\int_t^T e^{2V_s}|Z_s^n - \zeta_s|^2 ds \\
&\qquad\qquad + \mathbb{E}\int_t^T e^{2V_s} \left|\Psi^n\left(s, Y_s^n\right) - \Psi^n(s, \xi_s)\right| dQ_s \\
&\le C_p\left[\mathbb{E}\left(e^{pV_T} |Y_T^n - \xi_T|^2\right) + \mathbb{E}\left(\int_t^T e^{V_s}\left[|\hat{\xi}_s| + |H(s, \xi_s, \zeta_s)|\right] dQ_s\right)^2\right].
\end{aligned} \tag{5.174}$$

Recall that $(Y_s^n, Z_s^n) = (\xi_s, \zeta_s)$, $\forall\, s > n$. Passing to the limit $T \to \infty$ in (5.174), we obtain by the Beppo Levi monotone convergence theorem

$$\begin{aligned}
&\mathbb{E}\sup_{s\geq t} e^{pV_s}\left|Y_s^n-\xi_s\right|^2+\mathbb{E}\int_t^{\infty} e^{2V_s}|Z_s^n-\zeta_s|^2 ds\\
&\qquad\qquad +\mathbb{E}\int_t^{\infty} e^{2V_s}\left|\Psi^n\left(s,Y_s^n\right)-\Psi^n\left(s,\xi_s\right)\right| dQ_s \qquad (5.175)\\
&\leq C_2\mathbb{E}\left(\int_t^{\infty} e^{V_s}\left[|\hat{\xi}_s|+|H(s,\xi_s,\zeta_s)|\right] dQ_s\right)^2.
\end{aligned}$$

D. *Boundedness of* $\nabla\varphi_\varepsilon(Y_t^n)$ *and* $\nabla\psi_\varepsilon(Y_t^n)$.

By the stochastic subdifferential inequality from Lemma 2.38 and Remark 2.39, for all $0\leq t\leq T$

$$\begin{aligned}
&e^{2V_t}\left[\varphi_\varepsilon(Y_t^n)-\varphi_\varepsilon\left(u_0\right)\right]\leq e^{2V_T}\left[\varphi_\varepsilon(Y_T^n)-\varphi_\varepsilon\left(u_0\right)\right]\\
&\quad+\int_t^T e^{2V_s}\left\langle\nabla\varphi_\varepsilon(Y_s^n),\Phi_n(s,Y_s^n,Z_s^n)\right\rangle dQ_s-\int_t^T e^{2V_s}\left\langle\nabla\varphi_\varepsilon(Y_s^n),Z_s^n dB_s\right\rangle
\end{aligned}$$

(and a similar inequality for ψ_ε). We infer that

$$\begin{aligned}
&e^{2V_t}\left[\varphi_\varepsilon(Y_t^n)-\varphi_\varepsilon\left(u_0\right)+\psi_\varepsilon(Y_t^n)-\varphi_\varepsilon\left(u_0\right)\right]+\int_t^T \mathbf{1}_{s\leq n\wedge\tau}e^{2V_s}\Big[\alpha_s|\nabla\varphi_\varepsilon(Y_s^n)|^2\\
&\qquad+\left\langle\nabla\varphi_\varepsilon(Y_s^n),\nabla\psi_\varepsilon(Y_s^n)\right\rangle+(1-\alpha_s)\left|\nabla\psi_\varepsilon(Y_s^n)\right|^2\Big]dQ_s\\
&\quad\leq e^{2V_T}\left[\varphi_\varepsilon(Y_T^n)-\varphi_\varepsilon\left(u_0\right)+\psi_\varepsilon(Y_T^n)-\varphi_\varepsilon\left(u_0\right)\right]\\
&\qquad+\int_t^T \mathbf{1}_{[0,n]}\left(s\right)e^{2V_s}\left\langle\nabla\varphi_\varepsilon(Y_s^n)+\nabla\psi_\varepsilon(Y_s^n),H(s,Y_s^n,Z_s^n)\right\rangle dQ_s\\
&\qquad-\int_t^T e^{2V_s}\left\langle\nabla\varphi_\varepsilon(Y_s^n)+\nabla\psi_\varepsilon(Y_s^n),Z_s^n dB_s\right\rangle.
\end{aligned} \tag{5.176}$$

Using the definition of the function $H(t,y,z)$ given in (5.149), the compatibility assumptions (5.152) yield

$$\begin{aligned}
&\left\langle\nabla\varphi_\varepsilon(y),H(t,y,z)\right\rangle=\mathbf{1}_{[0,\tau]}\left(t\right)\left\langle\nabla\varphi_\varepsilon(y),\alpha_t F(t,y,z)+\left(1-\alpha_t\right)G(t,y)\right\rangle\\
&\leq\mathbf{1}_{[0,\tau]}\left(t\right)\left[\alpha_t|\nabla\varphi_\varepsilon(y)||F(t,y,z)|+c\left(1-\alpha_t\right)|\nabla\psi_\varepsilon(y)|\left(|G(t,y)|+g_s\right)\right]
\end{aligned} \tag{5.177}$$

and respectively

$$\begin{aligned}
&\left\langle\nabla\psi_\varepsilon(y),H(t,y,z)\right\rangle=\mathbf{1}_{[0,\tau]}\left(t\right)\left\langle\nabla\psi_\varepsilon(y),\alpha_t F(t,y,z)+\left(1-\alpha_t\right)G(t,y)\right\rangle\\
&\leq\mathbf{1}_{[0,\tau]}\left(t\right)\left[c\,\alpha_t|\nabla\varphi_\varepsilon(y)|\left(|F(t,y,z)|+f_s\right)+\left(1-\alpha_t\right)|G(t,y)||\nabla\psi_\varepsilon(y)|\right].
\end{aligned} \tag{5.178}$$

Recall that $\varphi\left(y\right)\geq\varphi_\varepsilon\left(u_0\right)=\varphi\left(u_0\right)$ and $\psi\left(y\right)\geq\psi_\varepsilon\left(u_0\right)=\psi\left(u_0\right)$. From (5.152), (5.176–5.178) and the inequality $a\left(x+y\right)\leq\frac{1}{2}a^2+x^2+y^2$ we obtain

$$
\begin{aligned}
& e^{2V_t}\left(\varphi_\varepsilon(Y_t^n)-\varphi(u_0)+\psi_\varepsilon(Y_t^n)-\psi(u_0)\right) \\
& +\frac{1}{2}\int_t^T \mathbf{1}_{[0,n\wedge\tau]}(s)\, e^{2V_s}\Big[|\nabla\varphi_\varepsilon(Y_s^n)|^2 ds+|\nabla\psi_\varepsilon(Y_s^n)|^2 dA_s\Big] \\
& \leq e^{2V_T}\left[\varphi_\varepsilon(Y_T^n)-\varphi(u_0)+\psi_\varepsilon(Y_T^n)-\psi(u_0)\right] \\
& \quad +(1+c)^2\int_t^T \mathbf{1}_{[0,n\wedge\tau]}(s)\, e^{2V_s}\left(|F(s,Y_s^n,Z_s^n)|^2+|f_s|^2\right) ds \\
& \quad +(1+c)^2\int_t^T \mathbf{1}_{[0,n\wedge\tau]}(s)\, e^{2V_s}\left(|G(s,Y_s^n)|^2+|g_s|^2\right) dA_s \\
& \quad -\int_t^T e^{2V_s}\left\langle\nabla\varphi_\varepsilon(Y_s^n)+\nabla\psi_\varepsilon(Y_s^n), Z_s^n dB_s\right\rangle .
\end{aligned}
\tag{5.179}
$$

The stochastic integral from this last inequality has the property

$$
\mathbb{E}^{\mathcal{F}_t}\int_t^T e^{2V_r}\left\langle\nabla\varphi_\varepsilon(Y_r^n)+\nabla\psi_\varepsilon(Y_r^n), Z_r^n dB_r\right\rangle=0,
$$

because by $\nabla\varphi_\varepsilon(u_0)=\nabla\psi_\varepsilon(u_0)=0$ we have

$$
|\nabla\varphi_\varepsilon(Y_s^n)+\nabla\psi_\varepsilon(Y_r^n)|\leq 2n\,|Y_s^n-u_0|
$$

and by (5.171) and (5.172)

$$
\begin{aligned}
& \mathbb{E}\left(\int_t^T e^{4V_s}\,|\nabla\varphi_\varepsilon(Y_s^n)+\nabla\psi_\varepsilon(Y_s^n)|^2\,|Z_s^n|^2 ds\right)^{1/2} \\
& \leq 2nR_0\,\mathbb{E}\left(\int_0^T e^{2V_s}|Z_s^n|^2 ds\right)^{1/2}<\infty .
\end{aligned}
$$

Let $T\geq n$. By Jensen's inequality it follows that

$$
\begin{aligned}
\mathbb{E}\left[e^{2V_T}\left(\varphi_\varepsilon(Y_T^n)+\psi_\varepsilon(Y_T^n)\right)\right] & \leq \mathbb{E}\left[e^{2V_T}\left(\varphi(\xi_T)+\psi(\xi_T)\right)\right] \\
& \leq \mathbb{E}\left[e^{2V_T}\left(\varphi(\eta)+\psi(\eta)\right)\right].
\end{aligned}
$$

Now from inequality (5.179) we infer by Beppo Levi's monotone convergence theorem for $T\to\infty$

$$
\begin{aligned}
& \mathbb{E}\left[e^{2V_t}\left(\varphi_\varepsilon(Y_t^n)-\varphi(u_0)+\psi_\varepsilon(Y_t^n)-\psi(u_0)\right)\right] \\
& +\frac{1}{2}\mathbb{E}\int_t^\infty \mathbf{1}_{[0,n\wedge\tau]}(s)\, e^{2V_s}\Big[|\nabla\varphi_\varepsilon(Y_s^n)|^2 ds+|\nabla\psi_\varepsilon(Y_s^n)|^2 dA_s\Big] \\
& \leq \mathbb{E}\left[e^{2V_\tau}\left(\varphi(\eta)-\varphi(u_0)+\psi(\eta)-\psi(u_0)\right)\right] \\
& \quad +(1+c)^2\,\mathbb{E}\int_t^\infty \mathbf{1}_{[0,\tau]}(s)\, e^{2V_s}\left(|F(s,Y_s^n,Z_s^n)|^2+|f_s|^2\right) ds \\
& \quad +(1+c)^2\,\mathbb{E}\int_t^\infty \mathbf{1}_{[0,\tau]}(s)\, e^{2V_s}\left(|G(s,Y_s^n)|^2+|g_s|^2\right) dA_s .
\end{aligned}
\tag{5.180}
$$

By (5.171), (5.168) and the assumption (5.154(iii)) we deduce that there exists a constant C independent of n such that

$$\begin{aligned}
&\mathbb{E}\int_t^{\infty} \mathbf{1}_{[0,\tau]}(s)\, e^{2V_s}\left(|F(s,Y_s^n,Z_s^n)|^2+|f_s|^2\right)ds \\
&\leq \mathbb{E}\int_t^{\infty} \mathbf{1}_{[0,\tau]}(s)\, e^{2V_s}\left[2\left|F^{\#}_{R_0+|u_0|}(s)\right|^2+2L^2|Z_s^n|^2+|f_s|^2\right]ds \leq C
\end{aligned}$$

and

$$\begin{aligned}
&\mathbb{E}\int_t^{\infty} \mathbf{1}_{[0,\tau]}(s)\, e^{2V_s}\left(|G(s,Y_s^n)|^2+|g_s|^2\right)dA_s \\
&\leq \mathbb{E}\int_t^{\infty} \mathbf{1}_{[0,\tau]}(s)\, e^{2V_s}\left[\left|G^{\#}_{R_0+|u_0|}(s)\right|^2+|g_s|^2\right]dA_s \leq C.
\end{aligned}$$

Therefore from (5.180) we have

$$\mathbb{E}\left[e^{2V_t}\left(\varphi_\varepsilon(Y_t^n)-\varphi(u_0)+\psi_\varepsilon(Y_t^n)-\psi(u_0)\right)\right]\leq C,\ \text{ for all } t\geq 0 \tag{5.181}$$

and

$$\mathbb{E}\int_0^{\infty} \mathbf{1}_{[0,n\wedge\tau]}(r)\left[e^{2V_r}|\nabla\varphi_\varepsilon(Y_r^n)|^2dr+e^{2V_r}|\nabla\psi_\varepsilon(Y_r^n)|^2dA_r\right]\leq C. \tag{5.182}$$

Since

$$\varphi_\varepsilon(y)-\varphi(u_0)=\frac{\varepsilon}{2}|\nabla\varphi_\varepsilon(y)|^2+\left[\varphi(y-\varepsilon\nabla\varphi_\varepsilon(y))-\varphi(u_0)\right]$$

we see from (5.181) that, for all $t\geq 0$,

$$\mathbb{E}\left[e^{2V_t}\left(|\varepsilon\nabla\varphi_\varepsilon(Y_t^n)|^2+|\varepsilon\nabla\psi_\varepsilon(Y_t^n)|^2\right)\right]\leq 2C\varepsilon \tag{5.183}$$

(recall that $\varepsilon=1/n$).

E. *Cauchy sequences and convergence.*

Note that by assumption (A_{10}) and (5.156) we have $|\xi_s-u_0|\leq b$ and

$$\begin{aligned}
&\mathbb{E}\left(\int_n^{\infty} \mathbf{1}_{[0,\tau]}(s)\, e^{V_s}\ell_s\,|\zeta_s|\,ds\right)^2 \\
&\leq \mathbb{E}\left[\left(\int_n^{\infty} \mathbf{1}_{[0,\tau]}(s)\,(\ell_s)^2\,ds\right)\left(\int_0^{\infty} \mathbf{1}_{[0,\tau]}(s)\, e^{2V_s}\,|\zeta_s|^2\,ds\right)\right]\to 0,\quad \text{as } n\to\infty.
\end{aligned}$$

Hence by assumption (5.154ii)

$$\begin{aligned}
&\mathbb{E}\left(\int_n^\infty e^{V_s}|H(s,\xi_s,\zeta_s)|dQ_s\right)^2\\
&\leq 3\mathbb{E}\left(\int_n^\infty \mathbf{1}_{[0,\tau]}(s)\,e^{V_s}F^\#_{b+|u_0|}(s)\,ds\right)^2+3\mathbb{E}\left(\int_n^\infty \mathbf{1}_{[0,\tau]}(s)\,e^{V_s}G^\#_{b+|u_0|}(s)\,dA_s\right)^2\\
&\quad+3\mathbb{E}\left(\int_n^\infty e^{V_s}\mathbf{1}_{[0,\tau]}(s)\,\ell_s\,|\zeta_s|\,ds\right)^2\rightarrow 0,\quad \text{as } n\rightarrow\infty,
\end{aligned}$$

and by (5.175),

$$\begin{aligned}
&\mathbb{E}\sup_{s\geq n}e^{2V_s}|Y_s^{n+i}-\xi_s|^2+\mathbb{E}\int_n^\infty e^{2V_s}|Z_s^{n+i}-\zeta_s|^2ds\\
&\qquad+\mathbb{E}\int_n^\infty e^{2V_s}\left|\Psi^{n+i}\left(s,Y_s^{n+i}\right)-\Psi^{n+i}\left(s,\xi_s\right)\right|dQ_s\\
&\leq C_p\mathbb{E}\left(\int_n^\infty e^{V_s}\left[\left|\hat{\xi}_s\right|+|H(s,\xi_s,\zeta_s)|\right]dQ_s\right)^2\longrightarrow 0,\ n\rightarrow\infty.
\end{aligned}\tag{5.184}$$

By uniqueness it follows that, for all $t\in[0,n]$,

$$Y_t^{n+i}-Y_t^n=Y_n^{n+i}-\xi_n+\int_t^n dK_s^{n,i}-\int_t^n(Z_s^{n+i}-Z_s^n)dB_s,\quad a.s.,$$

where on $[0,n]$

$$\begin{aligned}
&dK_s^{n,i}\\
&=\left[H_{n+i}(s,Y_s^{n+i},Z_s^{n+i})-H_n(s,Y_s^n,Z_s^n)-\nabla_y\Psi^{n+i}(s,Y_s^{n+i})+\nabla_y\Psi^n(s,Y_s^n)\right]dQ_s\\
&=\left[H(s,Y_s^{n+i},Z_s^{n+i})-H(s,Y_s^n,Z_s^n)-\nabla_y\Psi(s,Y_s^{n+i})+\nabla_y\Psi(s,Y_s^n)\right]dQ_s.
\end{aligned}$$

By (6.28, with $a=0$, $\varepsilon=1/n$ and $\delta=1/(n+i)$)

$$\begin{aligned}
&-\left\langle Y_s^{n+i}-Y_s^n,\left(\nabla_y\Psi(s,Y_s^{n+i})-\nabla_y\Psi(s,Y_s^n)\right)dQ_s\right\rangle\\
&\leq(\varepsilon+\delta)\mathbf{1}_{[0,\tau]}(s)\left(\langle\nabla\varphi_\varepsilon(Y_s^{n+i}),\nabla\varphi_\delta(Y_s^n)\rangle ds+\langle\nabla\psi_\varepsilon(Y_s^{n+i}),\nabla\psi_\delta(Y_s^n)\rangle dA_s\right),
\end{aligned}$$

and using (5.150) we have on $[0,n]$

$$\begin{aligned}
&\langle Y_s^{n+i}-Y_s^n,dK_s^{n,i}\rangle\\
&\leq\frac{\varepsilon+\delta}{2}\mathbf{1}_{[0,\tau]}(s)\Big[\left(|\nabla\varphi_\varepsilon(Y_s^n)|^2+|\nabla\varphi_\delta(Y_s^{n+i})|^2\right)ds\\
&\qquad+\left(|\nabla\psi_\varepsilon(Y_s^n)|^2+|\nabla\psi_\delta(Y_s^{n+i})|^2\right)dA_s\Big]\\
&\qquad+|Y_s^{n+i}-Y_s^n|^2dV_s+\frac{1}{4}|Z_s^{n+i}-Z_s^n|^2ds.
\end{aligned}$$

Since by (5.171),

$$\mathbb{E}\sup_{s\in[0,n]} e^{2V_s}|Y_s^{n+i}-Y_s^n|^2 \le 2\mathbb{E}\sup_{s\in[0,n]} e^{2V_s}\left[\left|Y_s^{n+i}-u_0\right|^2+|Y_s^n-u_0|^2\right]$$
$$\le 2R_0^2<\infty,$$

we obtain by Proposition 5.2 that

$$\begin{aligned}
&\mathbb{E}\sup\nolimits_{s\in[0,n]} e^{2V_s}|Y_s^{n+i}-Y_s^n|^2+\mathbb{E}\int_0^n e^{2V_s}|Z_s^{n+i}-Z_s^n|^2ds\\
&\le C\,\mathbb{E}e^{2V_n}|Y_n^{n+i}-\xi_n|^2\\
&\quad+(\varepsilon+\delta)\,C\,\mathbb{E}\int_0^{n\wedge\tau} e^{2V_s}\left(|\nabla\varphi_\varepsilon(Y_s^n)|^2+|\nabla\varphi_\delta(Y_s^{n+i})|^2\right)ds\\
&\quad+(\varepsilon+\delta)\,C\,\mathbb{E}\int_0^{n\wedge\tau} e^{2V_s}\left(|\nabla\psi_\varepsilon(Y_s^n)|^2+|\nabla\psi_\delta(Y_s^{n+i})|^2\right)dA_s.
\end{aligned} \tag{5.185}$$

The estimates (5.182) and (5.184) give us

$$\mathbb{E}\sup_{s\in[0,n]} e^{2V_s}|Y_s^{n+i}-Y_s^n|^2+\mathbb{E}\int_0^n e^{2V_s}|Z_s^{n+i}-Z_s^n|^2ds$$
$$\le \mathbb{E}\sup_{s\ge n} e^{2V_s}|Y_s^{n+i}-\xi_s|^2+\frac{C}{n}\longrightarrow 0,\ \text{as } n\to\infty.$$

Hence

$$\begin{aligned}
&\mathbb{E}\sup_{s\ge 0} e^{2V_s}|Y_s^{n+i}-Y_s^n|^2\\
&\quad\le \mathbb{E}\sup_{s\in[0,n]} e^{2V_s}|Y_s^{n+i}-Y_s^n|^2+\mathbb{E}\sup_{s\ge n} e^{2V_s}|Y_s^{n+i}-\xi_s|^2\longrightarrow 0,\ \text{as } n\to\infty
\end{aligned}$$

and

$$\begin{aligned}
&\mathbb{E}\int_0^\infty e^{2V_s}|Z_s^{n+i}-Z_s^n|^2ds\\
&\quad\le \mathbb{E}\int_0^n e^{2V_s}|Z_s^{n+i}-Z_s^n|^2ds+\mathbb{E}\int_n^\infty e^{2V_s}|Z_s^{n+i}-\zeta_s|^2ds\longrightarrow 0,\ \text{as } n\to\infty.
\end{aligned}$$

F. *Passage to the limit.*

Consequently there exists $(Y,Z)\in S_m^0\times\Lambda_{m\times k}^0$ such that

$$\mathbb{E}\sup_{s\ge 0} e^{2V_s}|Y_s^n-Y_s|^2+\mathbb{E}\int_0^\infty e^{2V_s}|Z_s^n-Z_s|^2ds\longrightarrow 0,\ \text{as } n\to\infty.$$

We have that $(Y_t,Z_t)=(\eta,0)$ for $t>\tau$, since $Y_t^n=\xi_t=\eta$ and $Z_t^n=\zeta_t=0$ for $t>\tau$.

Taking into account (5.183) and

$$\begin{aligned}
&|\Psi^n(s,y)-\Psi^n(s,u_0)|\\
&=\mathbf{1}_{[0,n\wedge\tau]}(s)\left[\alpha_s\left(\varphi_\varepsilon(y)-\varphi_\varepsilon(u_0)\right)+(1-\alpha_s)\left(\psi_\varepsilon(y)-\psi_\varepsilon(u_0)\right)\right]\\
&\geq \mathbf{1}_{[0,n\wedge\tau]}(s)\,\alpha_s\left[\varphi\left(y-\varepsilon\nabla\varphi_\varepsilon(y)\right)-\varphi(u_0)\right]\\
&\quad+\mathbf{1}_{[0,n\wedge\tau]}(s)\,(1-\alpha_s)\left|\psi\left(y-\varepsilon\nabla\psi_\varepsilon(y)\right)-\psi(u_0)\right|,
\end{aligned}$$

the inequality (5.162) follows from (5.170) by Fatou's Lemma.

Also by Fatou's Lemma from (5.175) we obtain (5.163) and from (5.171) and (5.172) we deduce (5.161).

From (5.182) there exist two p.m.s.p. $U^{(1)}$ and $U^{(2)}$, such that along a subsequence still indexed by n, we have for $\varepsilon=\frac{1}{n}\to 0$

$$\begin{aligned}
&e^V\nabla\varphi_\varepsilon(Y^n)\mathbf{1}_{[0,\tau\wedge n]}\rightharpoonup e^VU^{(1)}\mathbf{1}_{[0,\tau)},\quad \text{weakly in } L^2\left(\Omega\times\mathbb{R}_+,d\mathbb{P}\otimes dt;\mathbb{R}^m\right),\\
&e^V\nabla\psi_\varepsilon(Y^n)\mathbf{1}_{[0,\tau\wedge n]}\rightharpoonup e^VU^{(2)}\mathbf{1}_{[0,\tau)},\quad \text{weakly in } L^2\left(\Omega\times\mathbb{R}_+,d\mathbb{P}\otimes dA_t;\mathbb{R}^m\right).
\end{aligned}$$

Using (5.183) and applying Fatou's Lemma we have

$$\begin{aligned}
\mathbb{E}\Big(e^{2V_t}\left[\varphi(Y_t)-\varphi(u_0)\right]\Big)&\leq\liminf_{n\to+\infty}\mathbb{E}\Big(e^{2V_t}\left[\varphi\left(Y_t^n-\varepsilon\nabla\varphi_\varepsilon(Y_t^n)\right)-\varphi(u_0)\right]\Big)\\
&\leq\liminf_{n\to+\infty}\mathbb{E}\Big(e^{2V_t}\left[\varphi_\varepsilon\left(Y_t^n\right)-\varphi(u_0)\right]\Big),
\end{aligned}$$

and similarly for ψ. Passing to $\liminf_{n\to+\infty}$ in (5.180) we obtain (5.164).

From (5.165) we have for all $0\leq t\leq T\leq n$, $\mathbb{P}$-a.s.

$$Y_t^n+\int_t^T\nabla_y\Psi^n(s,Y_s^n)dQ_s=Y_T^n+\int_t^T H(s,Y_s^n,Z_s^n)dQ_s-\int_t^T Z_s^n dB_s,$$

and passing to the limit we conclude that

$$Y_t+\int_t^T U_s dQ_s=Y_T+\int_t^T H(s,Y_s,Z_s)dQ_s-\int_t^T Z_s dB_s,\ \text{a.s.}\tag{5.186}$$

with

$$U_s=\mathbf{1}_{[0,\tau]}(s)\left[\alpha_s U_s^1+(1-\alpha_s)U_s^2\right],\ \text{for } s\geq 0.\tag{5.187}$$

By (5.118–b), we see that, for all $E\in\mathcal{F}$, $0\leq s\leq t$ and $X\in S_m^2$,

$$\begin{aligned}
\mathbb{E}\int_s^t\langle e^{2V_r}\nabla\varphi_\varepsilon(Y_r^n),X_r-Y_r^n\rangle\mathbf{1}_E dr+\mathbb{E}\int_s^t e^{2V_r}\varphi(Y_r^n-\varepsilon\nabla\varphi_\varepsilon(Y_s^n))\mathbf{1}_E dr\\
\leq\mathbb{E}\int_s^t e^{2V_r}\varphi(X_r)\mathbf{1}_E dr.
\end{aligned}$$

Passing to lim inf for $n \to \infty$ in the above inequality we obtain $U_s^{(1)} \in \partial\varphi(Y_s), d\mathbb{P}\otimes ds$-a.e. and, with similar arguments, $U_s^{(2)} \in \partial\psi(Y_s), d\mathbb{P}\otimes dA_s$-a.e. Summarizing the above conclusions we conclude that (Y, Z, U) is a solution of the BSVI (5.157).

We want to highlight the fact that the assumption (A_{10}-b) is too strong for many applications. The next two results are concerned with the existence of a solution for the backward stochastic variational inequality (5.146) recalled here for convenience:

$$\begin{cases} Y_t + \displaystyle\int_{t\wedge\tau}^{\tau} dK_s = \eta + \int_{t\wedge\tau}^{\tau} \left[F\left(s, Y_s, Z_s\right) ds + G\left(s, Y_s\right) dA_s\right] \\ \qquad\qquad\qquad\qquad - \displaystyle\int_{t\wedge\tau}^{\tau} Z_s dB_s, \quad \text{for } t \geq 0, \\ dK_t \in \partial\varphi\left(Y_t\right) dt + \partial\psi\left(Y_t\right) dA_t, \text{ on } \mathbb{R}_+, \end{cases} \tag{5.188}$$

without the boundedness conditions from (A_{10}).

Consider the closed convex sets

$$\begin{aligned} \mathbb{O}_\varphi &= \left\{y \in \mathbb{R}^m : \varphi\left(y\right) = \varphi\left(u_0\right)\right\}, \\ \mathbb{O}_\psi &= \left\{y \in \mathbb{R}^m : \psi\left(y\right) = \psi\left(u_0\right)\right\}, \text{ and} \\ \mathbb{O} &= \mathbb{O}_\varphi \cap \mathbb{O}_\psi. \end{aligned}$$

Since every point of $\mathbb{O}$ is a minimum point for φ and ψ, it follows that $\nabla\varphi_\varepsilon\left(u\right) = \nabla\psi_\varepsilon\left(u\right) = 0$ for all $u \in \mathbb{O}$.

Theorem 5.67. *Let the assumptions* $(\mathrm{A}_1), \ldots, (\mathrm{A}_9)$ *be satisfied and assume that there exists a* $\delta_0 > 0$ *such that*

$$\overline{B\left(u_0, \delta_0\right)} \subset \operatorname{int}\left(\mathbb{O}\right). \tag{5.189}$$

Assume moreover there exists $\delta \in (0, \infty]$ *and* $q = 1 + \frac{\delta}{2+\delta}$ *(with* $q = 2$ *if* $\delta = \infty$*) such that*

$$\begin{aligned} &(i) \quad \textit{for } 0 < \delta < \infty\text{: } \mathbb{E}\left(\left(\int_0^\infty \left(\ell_s\right)^2 ds\right)^{1+\delta}\right) < \infty, \\ &(i') \quad \textit{for } \delta = \infty\text{: } \left(\ell_s\right)_{s\geq 0} \textit{ is a deterministic process and } \int_0^\infty \left(\ell_s\right)^2 ds < \infty, \\ &(ii) \quad \lim_{t\to\infty} \mathbb{E}\left(\int_t^\infty \mathbf{1}_{[0,\tau]}\left(s\right) e^{V_s} \left|\left|F(s, \xi_s, 0)\right|ds + \left|G(s, \xi_s,)\right| dA_s\right)^q = 0. \end{aligned} \tag{5.190}$$

Then the BSVI (5.188) has a unique solution $(Y, Z, K) \in S_m^0 \times \Lambda_{m\times k}^0 \times S_m^0$ *in the sense of Definition 5.64 such that for* $q = 1 + \frac{\delta}{2+\delta}$ *and* $q = 2$ *if* $\delta = \infty$,

$$
\begin{aligned}
&(j) \quad \mathbb{E}\sup_{s\geq 0} e^{2V_s}\left|Y_s-u_0\right|^2+\mathbb{E}\int_0^{\infty} e^{2V_s}\left|Z_s\right|^2 ds<\infty, \\
&(jj) \quad \lim_{T\to\infty}\left[\mathbb{E}\, e^{qV_T}\left|Y_T-\xi_T\right|^q+\mathbb{E}\left(\int_T^{\infty} e^{2V_s}\left|Z_s-\zeta_s\right|^2 ds\right)^{q/2}\right]=0.
\end{aligned}
\tag{5.191}
$$

Moreover the inequalities (5.162) and (5.163) hold.

Proof. ***(I) Uniqueness***. The proof of uniqueness is similar to that given for Theorem 5.66 except that now by Corollary 6.82 from Annex C, we have

$$
\begin{aligned}
&\mathbb{E}\sup_{s\in[0,T]} e^{qV_s}\left|Y_s-Y_s'\right|^q+\mathbb{E}\left(\int_0^T e^{2V_s}\left|Z_s-Z_s'\right|^2 ds\right)^{q/2} \\
&\leq C_q\,\mathbb{E}\left(e^{qV_T}\left|Y_T-Y_T'\right|^q\right)\xrightarrow[T\to\infty]{} 0.
\end{aligned}
$$

(II) Existence. *Step 1. Approximation of the problem's data to satisfy* (A_{10}). Let

$$
\begin{aligned}
\sigma_s &= \mathbf{1}_{[0,\tau]}(s)\left[\mu_s\alpha_s+\nu_s\left(1-\alpha_s\right)\right], \quad dQ_s=ds+dA_s, \\
\beta_t &= Q_{t\wedge\tau}+\int_0^{t\wedge\tau}\sigma_s dQ_s+\int_0^{t\wedge\tau}\left|F\left(s,u_0,0\right)\right|ds+\int_0^{t\wedge\tau}\left|G\left(s,u_0\right)\right|dA_s, \\
\gamma_t &= t+\beta_t+\ell_t+\left|F\left(t,u_0,0\right)\right|+\left|G\left(t,u_0\right)\right| \text{ and} \\
\lambda_t &= t+\ell_t.
\end{aligned}
$$

Define, for $n\in\mathbb{N}^*$,

$$
\begin{aligned}
\ell_t^n &= \ell_t\mathbf{1}_{[0,n]}\left(\lambda_t\right), \\
\eta_n &= \left(\eta-u_0\right)\mathbf{1}_{[0,n]}\left(\beta_\tau+\left|\eta-u_0\right|\right)+u_0\in\overline{\eta,u_0}, \\
\hat{F}_n(t,y,z) &= F\left(t,y,z\mathbf{1}_{[0,n]}\left(\lambda_t\right)\right)-F\left(t,u_0,0\right)\,\mathbf{1}_{(n,\infty)}\left(\gamma_t\right), \\
\hat{G}_n(t,y) &= G(t,y)-G\left(t,u_0\right)\,\mathbf{1}_{(n,\infty)}\left(\gamma_t\right), \\
\hat{H}_n(t,y,z) &= \left[\alpha_t\hat{F}_n(t,y,z)+\left(1-\alpha_t\right)\hat{G}_n(t,y)\right]\mathbf{1}_{[0,\tau]}(t),
\end{aligned}
$$

and

$$
V_t^n=\int_0^{t\wedge\tau}\left[\mu_s ds+\left(\ell_s^n\right)^2 ds+\nu_s dA_s\right]=V_t-\int_0^{t\wedge\tau}\left(\ell_s\right)^2\mathbf{1}_{(n,\infty)}\left(\lambda_s\right)ds.
$$

Let (ξ^n,ζ^n) be given by the martingale representation theorem (Corollary 2.44): for all $t\geq 0$, $\xi_t^n=\mathbb{E}^{\mathcal{F}_t}\eta_n$ and

$$\xi_t^n = \eta_n - \int_t^\infty \zeta_s^n dB_s,$$

or equivalently, for all $T > 0$

$$\xi_t^n = \mathbb{E}^{\mathcal{F}_T} \eta_n - \int_t^T \zeta_s^n dB_s, \quad t \in [0, T].$$

It is easy to verify that

$$\begin{aligned}
&\mathbb{E} \sup_{t \geq 0} e^{2V_t} |\eta_n - u_0|^2 \leq \mathbb{E} \sup_{t \geq 0} e^{2V_t} |\eta - u_0|^2 \leq \mathbb{E} \left(e^{2V_\tau} |\eta - u_0|^2 \right) < \infty, \\
&\mathbb{E} \sup_{t \geq 0} e^{2V_t} |\eta_n - \eta|^2 \leq \mathbb{E} \left[e^{2V_\tau} |\eta - u_0|^2 \mathbf{1}_{\beta_T + |\eta - u_0| > n} \right].
\end{aligned}$$

Applying Corollary 6.83, first on $[t, T]$ and then letting $T \to \infty$, we infer that for all $t \geq 0$

$$\begin{aligned}
&(a)\ \mathbb{E}^{\mathcal{F}_t} \sup_{s \geq t} e^{2V_s} |\xi_s - u_0|^2 + \mathbb{E}^{\mathcal{F}_t} \int_t^\infty e^{2V_s} |\zeta_s|^2 ds \leq C_2\, \mathbb{E}^{\mathcal{F}_t} \left(e^{2V_\tau} |\eta - u_0|^2 \right), \\
&(b)\ \mathbb{E}^{\mathcal{F}_t} \sup_{s \geq t} e^{2V_s} |\xi_s^n - u_0|^2 + \mathbb{E}^{\mathcal{F}_t} \int_t^\infty e^{2V_s} |\zeta_s^n|^2 ds \leq C_2\, \mathbb{E}^{\mathcal{F}_t} \left(e^{2V_\tau} |\eta - u_0|^2 \right), \\
&(c)\ \mathbb{E}^{\mathcal{F}_t} \sup_{s \geq t} e^{2V_s} |\xi_s^n - \xi_s|^2 + \mathbb{E}^{\mathcal{F}_t} \int_t^\infty e^{2V_s} |\zeta_s^n - \zeta_s|^2 ds \\
&\qquad\qquad \leq C_2\, \mathbb{E}^{\mathcal{F}_t} \left[e^{2V_\tau} |\eta - u_0|^2 \mathbf{1}_{\beta_T + |\eta - u_0| > n} \right].
\end{aligned} \tag{5.192}$$

Since the assumptions (A_1, ..., A_9) are satisfied by $(\eta, F, G, \varphi, \psi, V, \mu, \nu, \ell)$ it follows that the same assumptions are satisfied replacing $(\eta, F, G, \varphi, \psi, V, \mu, \nu, \ell)$ by $(\eta_n, \hat{F}_n, \hat{G}_n, \varphi, \psi, V^n, \mu, \nu, \ell^n)$.

With respect to (A_{10}) we have

$$\ell_t^n + \int_0^\infty (\ell_s^n)^2 ds \leq n + \int_0^n n^2 ds = n + n^3$$

and

$$\begin{aligned}
&e^{V_\tau^n} |\eta_n - u_0| + \left| \hat{H}_n(t, u_0, 0) \right| + \int_0^\tau e^{V_s^n} \left| \hat{H}_n(s, u_0, 0) \right| dQ_s \\
&\leq e^{V_\tau} |\eta_n - u_0| + |H(t, u_0, 0)| \mathbf{1}_{[0,n]}(\gamma_t) + \int_0^\tau e^{V_s} |H(s, u_0, 0)| \mathbf{1}_{[0,n]}(\gamma_s) dQ_s \\
&\leq n + e^n n + e^n n^2 = b_n.
\end{aligned}$$

Hence $\left(\eta_n, \hat{F}_n, \hat{G}_n, \mu, \nu, \ell^n, V^n \right)$ satisfies (A_{10}).

Step 2. Approximating equation and estimates. By *Step 1* we are in the conditions of Theorem 5.66 and therefore the approximating equation

$$\begin{cases} Y_t^n + \int_t^\infty U_s^n dQ_s = \eta_n + \int_t^\infty \hat{H}_n\left(s, Y_s^n, Z_s^n\right) dQ_s - \int_t^\infty Z_s^n dB_s, \\ dK_s^n = U_s^n dQ_s \in \partial_y \Psi\left(s, Y_s^n\right) dQ_s \end{cases}$$

has a unique solution $(Y^n, Z^n, K^n) \in S_m^0 \times \Lambda_{m\times k}^0 \times S_m^0$, $\left(Y_s^n, Z_s^n\right) = \left(\xi_s^n, 0\right)$ for $s > \tau$ and $U_s^n = \alpha_s U_s^{1,n} + (1-\alpha_s) U_s^{2,n}$ such that

$$\begin{aligned} &(j) \quad \mathbb{E}\sup_{s\ge 0} e^{2V_s^n}\left|Y_s^n - u_0\right|^2 + \mathbb{E}\int_0^\infty e^{2V_s^n}\left|Z_s^n\right|^2 ds < \infty, \\ &(jj) \quad \lim_{T\to\infty} \mathbb{E}\Big[e^{2V_T^n}\left|Y_T^n - \xi_T^n\right|^2 + \int_T^\infty e^{2V_s^n}\left|Z_s^n - \zeta_s^n\right|^2 ds\Big] = 0. \end{aligned} \tag{5.193}$$

Moreover the inequalities (5.161), (5.162), (5.163) and (5.164) hold with $(\eta, F, G, \varphi, \psi, V, \mu, \nu, \ell, C_b, C_{q,b})$ by $(\eta_n, \hat{F}_n, \hat{G}_n, \varphi, \psi, V^n, \mu, \nu, \ell^n, C_n, C_{q,n})$.

Using in (5.193) $V_t^{n+i} - (n+i)^3 \le V_t^n$ we get for all $i \in \mathbb{N}$

$$\begin{aligned} &(j') \quad \mathbb{E}\sup_{s\ge 0} e^{2V_s^{n+i}}\left|Y_s^n - u_0\right|^2 + \mathbb{E}\int_0^\infty e^{2V_s^{n+i}}\left|Z_s^n\right|^2 ds < \infty, \\ &(jj') \quad \lim_{T\to\infty} \mathbb{E}\Big[e^{2V_T^{n+i}}\left|Y_T^n - \xi_T^n\right|^2 + \int_T^\infty e^{2V_s^{n+i}}\left|Z_s^n - \zeta_s^n\right|^2 ds\Big] = 0. \end{aligned} \tag{5.194}$$

Since $\langle Y_s^n - u_0, U_s^n - 0\rangle \ge 0$ and

$$\begin{aligned} &|\Psi(s, Y_s^n) - \Psi(s, u_0)|\, dQ_s + \left\langle Y_s^n - u_0, \hat{H}_n\left(s, Y_s^n, Z_s^n\right) - U_s^n\right\rangle dQ_s \\ &\le |Y_s^n - u_0|\left|\hat{H}_n\left(s, u_0, 0\right)\right| dQ_s + |Y_s^n - u_0|^2\, dV_s^n + \frac{1}{4}|Z_s^n|^2 ds \\ &\le |Y_s^n - u_0|\, \mathbf{1}_{[0,\tau]}(s)\left(|F(s, u_0, 0)|\, ds + |G(s, u_0)|\, dA_s\right) + |Y_s^n - u_0|^2\, dV_s^{n+i} \\ &\quad + \frac{1}{4}|Z_s^n|^2 ds, \end{aligned}$$

we infer by Corollary 6.82 for $p = 2$ and $0 \le t \le T$,

$$\begin{aligned} &\mathbb{E}^{\mathcal{F}_t}\sup_{s\in[t,T]}\left|e^{V_s^{n+i}}\left(Y_s^n - u_0\right)\right|^2 + \mathbb{E}^{\mathcal{F}_t}\left(\int_t^T e^{2V_s^{n+i}}\left|Z_s^n\right|^2 ds\right) \\ &\qquad + \mathbb{E}^{\mathcal{F}_t}\left(\int_t^T e^{2V_s^{n+i}}\left|\Psi(s, Y_s^n) - \Psi(s, u_0)\right| dQ_s\right) \\ &\le C_2\, \mathbb{E}^{\mathcal{F}_t}\Bigg[\left|e^{V_T^{n+i}}\left(Y_T^n - u_0\right)\right|^2 \\ &\qquad + \left(\int_t^T \mathbf{1}_{[0,\tau]}(s)\, e^{V_s^{n+i}}\left(|F(s, u_0, 0)|\, ds + |G(s, u_0)|\, dA_s\right)\right)^2\Bigg]. \end{aligned} \tag{5.195}$$

But by (5.192-b) and $V_T^{n+i} \leq V_T$ we have

$$\begin{aligned}
&\left[\mathbb{E}^{\mathcal{F}_t}\left(\left|e^{V_T^{n+i}}\left(Y_T^n-u_0\right)\right|^2\right)\right]^{1/2}\\
&\leq \left[\mathbb{E}^{\mathcal{F}_t}\left(e^{2V_T^{n+i}}\left|Y_T^n-\xi_T^n\right|^2\right)\right]^{1/2}+\left[\mathbb{E}^{\mathcal{F}_t}\left(e^{2V_T}\left|\xi_T^n-u_0\right|^2\right)\right]^{1/2}\\
&\leq \left[\mathbb{E}^{\mathcal{F}_t}\left(e^{2V_T^{n+i}}\left|Y_T^n-\xi_T^n\right|^2\right)\right]^{1/2}+\left[\mathbb{E}^{\mathcal{F}_t}\left(e^{2V_\tau}\left|\eta-u_0\right|^2\right)\right]^{1/2}.
\end{aligned}$$

Using Beppo Levi's monotone convergence theorem and (5.194-*jj*$''$) we can pass to the limit in (5.195), first $\limsup_{T\to\infty}$ and then $\lim_{i\to\infty}$. We obtain

$$\begin{aligned}
&\mathbb{E}^{\mathcal{F}_t}\sup_{s\geq t}\left|e^{V_s}\left(Y_s^n-u_0\right)\right|^2+\mathbb{E}^{\mathcal{F}_t}\left(\int_t^\infty e^{2V_s}\left|Z_s^n\right|^2 ds\right)\\
&\qquad+\mathbb{E}^{\mathcal{F}_t}\left(\int_t^\infty e^{2V_s}\left|\Psi(s,Y_s^n)-\Psi(s,u_0)\right| dQ_s\right)\\
&\leq C_2\,\mathbb{E}^{\mathcal{F}_t}\Big[e^{2V_\tau}\left|\eta-u_0\right|^2\\
&\qquad+\Big(\int_t^\infty \mathbf{1}_{[0,\tau]}(s)\,e^{V_s}\left(\left|F(s,u_0,0)\right| ds+\left|G(s,u_0)\right| dA_s\right)\Big)^2\Big],
\end{aligned}\tag{5.196}$$

and, in particular,

$$\mathbb{E}\,\sup_{s\geq 0} e^{2V_s}\left|Y_s^n-u_0\right|^2<\infty. \tag{5.197}$$

Step 3. Cauchy sequence and convergences. Let

$$\mathcal{K}_t^n=\int_0^t\left[\hat{H}_n\left(s,Y_s^n,Z_s^n\right)-U_s^n\right]dQ_s.$$

We have for any $j>i>1$

$$\begin{aligned}
&\left\langle Y_s^n-Y_s^{n+i},d\left(\mathcal{K}_s^n-\mathcal{K}_s^{n+i}\right)\right\rangle\\
&\leq\left\langle Y_s^n-Y_s^{n+i},H\left(s,Y_s^n,Z_s^n\mathbf{1}_{[0,n]}(\lambda_s)\right)-H\left(s,Y_s^{n+i},Z_s^{n+i}\mathbf{1}_{[0,n+i]}(\lambda_s)\right)\right\rangle dQ_s\\
&\quad-\left\langle Y_s^n-Y_s^{n+i},H(s,u_0,0)\right\rangle\left[\mathbf{1}_{[n,\infty[}(\gamma_s)-\mathbf{1}_{[n+i,\infty[}(\gamma_s)\right]dQ_s\\
&\leq\left|Y_s^n-Y_s^{n+i}\right|\Big[\left|H(s,u_0,0)\right|\,\mathbf{1}_{[n,\infty[}(\gamma_s)\,dQ_s\\
&\quad+\mathbf{1}_{[0,\tau]}(s)\,\ell_s\left|\mathbf{1}_{[0,n]}(\lambda_s)-\mathbf{1}_{[0,n+i]}(\lambda_s)\right|\left|Z_s^n\right|ds\Big]\\
&\quad+\left|Y_s^n-Y_s^{n+i}\right|^2\left(\sigma_s dQ_s+\mathbf{1}_{[0,\tau]}(s)\,\mathbf{1}_{[0,n+j]}(\lambda_s)\,(\ell_s)^2 ds\right)+\frac{1}{4}\left|Z_s^n-Z_s^{n+i}\right|^2 ds,
\end{aligned}$$

then for all $T > 0$, by Proposition 5.2 with $q = 1 + \frac{\delta}{2+\delta} \in (1,2)$ and $q = 2$ if $\delta = \infty$,

$$\mathbb{E}\left(\sup_{s\in[0,T]} e^{qV_s^{n+j}} \left|Y_s^n - Y_s^{n+i}\right|^q\right) + \mathbb{E}\left(\int_0^T e^{2V_s^{n+j}} \left|Z_s^n - Z_s^{n+i}\right|^2 ds\right)^{q/2}$$

$$\leq C_q\, \mathbb{E}\, e^{qV_T^{n+j}} \left|Y_T^n - Y_T^{n+i}\right|^q$$

$$+C_q\, \mathbb{E}\left(\int_0^T \mathbf{1}_{(n,\infty)}(\gamma_s)\, e^{V_s^{n+j}}\Big[\left|H(t,u_0,0)\right|\, dQ_s + \mathbf{1}_{[0,\tau]}(s)\,\ell_s\left|Z_s^n\right| ds\Big]\right)^q.$$

But

$$\mathbb{E}\left(\int_0^T \mathbf{1}_{(n,\infty)}(\gamma_s)\, e^{V_s^{n+j}} \mathbf{1}_{[0,\tau]}(s)\,\ell_s\left|Z_s^n\right| ds]\right)^q$$

$$\leq \mathbb{E}\left[\left(\int_0^T \mathbf{1}_{[0,\tau]}(s)\,(\ell_s)^2\,\mathbf{1}_{(n,\infty)}(\lambda_s)\, ds\right)^{q/2}\left(\int_0^T e^{2V_s^{n+j}}\left|Z_s^n\right|^2 ds\right)^{q/2}\right]$$

$$\leq \Lambda_{n,\delta} \times \left[\mathbb{E}\left(\int_0^T e^{2V_s}\left|Z_s^n\right|^2 ds\right)\right]^{\frac{q}{2}},$$

with

$$\Lambda_{n,\delta} = \begin{cases} \left[\mathbb{E}\left(\int_0^T \mathbf{1}_{[0,\tau]}(s)\,(\ell_s)^2\,\mathbf{1}_{(n,\infty)}(\lambda_s)\, ds\right)^{1+\delta}\right]^{\frac{1}{2+\delta}}, & \text{if } 0 < \delta < \infty, \\ \displaystyle\int_0^\infty (\ell_s)^2\,\mathbf{1}_{(n,\infty)}(\lambda_s)\, ds, & \text{if } q = 2\ (\delta = \infty,\ \ell \text{ is deterministic}); \end{cases}$$

the last inequality is obtained by Hölder's inequality since $\frac{1}{2+\delta} + \frac{1}{2/q} = 1$. Thus

$$\begin{aligned} &\mathbb{E}\left(\sup_{s\in[0,T]} e^{qV_s^{n+j}} \left|Y_s^n - Y_s^{n+i}\right|^q\right) + \mathbb{E}\left(\int_0^T e^{2V_s^{n+j}} \left|Z_s^n - Z_s^{n+i}\right|^2 ds\right)^{q/2} \\ &\leq C_q\, \mathbb{E}\, e^{qV_T^{n+j}} \left|Y_T^n - Y_T^{n+i}\right|^q + C_q\, \mathbb{E}\left(\int_0^T e^{V_s}\Big[\left|H(t,u_0,0)\right|\, \mathbf{1}_{(n,\infty)}(\gamma_s)\, dQ_s\right)^q \\ &\qquad + C_q\, \Lambda_{n,\delta} \times \left[\mathbb{E}\left(\int_0^T e^{2V_s}\left|Z_s^n\right|^2 ds\right)\right]^{\frac{q}{2}}. \end{aligned} \tag{5.198}$$

Here we have

$$\begin{aligned}
&\left(\mathbb{E}\, e^{2V_T^{n+j}} \left|Y_T^n - Y_T^{n+i}\right|^2\right)^{1/2} \\
&\leq \left(\mathbb{E}\, e^{2V_T^{n+j}} \left|Y_T^n - \xi_T^n\right|^2\right)^{1/2} + \left(\mathbb{E}\, e^{2V_T^{n+j}} \left|\xi_T^n - \xi_T^{n+i}\right|^2\right)^{1/2} \\
&\quad + \left(\mathbb{E}\, e^{2V_T^{n+j}} \left|\xi_T^{n+i} - Y_T^{n+i}\right|^2\right)^{1/2} \\
&\leq \left(\mathbb{E}\, e^{2V_T^{n+j}} \left|Y_T^n - \xi_T^n\right|^2\right)^{1/2} + \left(\mathbb{E}\, e^{2V_T^{n+j}} \left|\xi_T^{n+i} - Y_T^{n+i}\right|^2\right)^{1/2} \\
&\quad + \left[\mathbb{E}\left(e^{2V_\tau} \left|\eta - u_0\right|^2 \mathbf{1}_{\beta_\tau + |\eta - u_0| > n}\right)\right]^{1/2},
\end{aligned}$$

and as $T \to \infty$ we infer

$$\limsup_{T\to\infty} \mathbb{E}\, e^{2V_T^{n+j}} \left|Y_T^n - Y_T^{n+i}\right|^2 \leq \mathbb{E}\left(e^{2V_\tau} \left|\eta - u_0\right|^2 \mathbf{1}_{\beta_\tau + |\eta - u_0| > n}\right).$$

Using (5.196) for the boundedness of $\mathbb{E}\int_0^\infty e^{2V_s} |Z_s^n|^2\, ds$ we get from (5.198) as $T \to \infty$ and then passing to the limit as $j \to \infty$:

$$\begin{aligned}
&\mathbb{E}\left(\sup_{s\geq 0} e^{qV_s} \left|Y_s^n - Y_s^{n+i}\right|^q\right) + \mathbb{E}\left(\int_0^\infty e^{2V_s} \left|Z_s^n - Z_s^{n+i}\right|^2 ds\right)^{q/2} \\
&\leq C\left[\mathbb{E}\left(e^{2V_\tau} \left|\eta - u_0\right|^2 \mathbf{1}_{|\eta - u_0| > n}\right)\right]^{q/2} \\
&\quad + C\, \mathbb{E}\left(\int_0^\tau e^{V_s}\left[\left|H\left(t, u_0, 0\right)\right| \mathbf{1}_{(n,\infty)}\left(\gamma_s\right) dQ_s\right)^q + C\, \Lambda_{n,\delta},
\end{aligned}$$

which yields by (5.190) the existence of a pair $(Y, Z) \in S_m^0 \times \Lambda_{m\times k}^0$ such that

$$\lim_{n\to\infty}\left[\mathbb{E}\sup_{s\geq 0} e^{qV_s} \left|Y_s^n - Y_s\right|^q + \mathbb{E}\left(\int_0^\infty e^{2V_s} \left|Z_s^n - Z_s\right|^2 ds\right)^{q/2}\right] = 0. \quad (5.199)$$

Now by Fatou's lemma from (5.196) we obtain (5.162) and consequently (5.191-j).

To verify (5.191-*jj*), following the proof of Theorem 5.66, we have

$$\begin{aligned}
&\left|\Psi\left(s, Y_s^n\right) - \Psi\left(s, \xi_s\right)\right| dQ_s + \left\langle Y_s^n - \xi_s, \hat{H}_n\left(s, Y_s^n, Z_s^n\right) - U_s^n\right\rangle dQ_s \\
&\leq \left|Y_s^n - \xi_s\right|\left[|\hat{\xi}_s| + |\hat{H}_n(s, \xi_s, \zeta_s)|\right] dQ_s \\
&\quad + |Y_s^n - \xi_s|^2 dV_s^{n+i} + \frac{1}{4}|Z_s^n - \zeta_s|^2 ds.
\end{aligned}$$

By (5.192-a) and (5.197) we have

$$\mathbb{E}\ \sup_{s\geq 0} e^{2V_s^{n+i}}\ |Y_s^n-\xi_s|^2\leq \mathbb{E}\ \sup_{s\geq 0} e^{2V_s}\ |Y_s^n-\xi_s|^2<\infty .$$

Furthermore, using (5.194-*jj''*) and (5.192-c), we have

$$\limsup_{T\to\infty} \mathbb{E}\ e^{2V_T^{n+i}}\ |Y_T^n-\xi_T|^2\leq \mathbb{E}\ \left(e^{2V_\tau}\ |\eta-u_0|^2\ \mathbf{1}_{\beta_\tau+|\eta-u_0|>n}\right).$$

In the same manner as above when we proved (5.196) we obtain, by Corollary 6.82 for $p\in(1,2]$, similar inequalities with (ξ_s,ζ_s) in place of $(u_0,0)$ and passing successively to the limit $T\to\infty$ and $i\to\infty$ we get that for all $t\geq 0$

$$\begin{aligned}&\mathbb{E}\sup_{s\geq t} e^{pV_s}\ |Y_s^n-\xi_s|^p+\mathbb{E}\ \left(\int_t^\infty e^{2V_s}|Z_s^n-\zeta_s|^2ds\right)^{p/2}\\&\qquad\qquad+\mathbb{E}\ \left(\int_t^\infty e^{2V_s}\left|\Psi\left(s,Y_s^n\right)-\Psi\left(s,\xi_s\right)\right|dQ_s\right)^{p/2}\\&\leq C_p\Big[\mathbb{E}\ (e^{pV_\tau}\ |\eta-u_0|^p\ \mathbf{1}_{\beta_\tau+|\eta-u_0|>n})\\&\quad+\mathbb{E}\ \Big(\int_t^\infty \mathbf{1}_{[0,\tau]}(s)\,e^{V_s}\big[|\hat{\xi}_s|+|\hat{H}_n(s,\xi_s,\zeta_s)|\big]dQ_s\Big)^p\Big].\end{aligned}\tag{5.200}$$

Since $|\hat{H}_n(s,\xi_s,\zeta_s)|\leq|H(s,\xi_s,0)|+\ell_s\,|\zeta_s|+|H(s,u_0,0)|\mathbf{1}_{(n,\infty)}(\gamma_s)$, from (5.200) with Fatou's Lemma applied to the left-hand side and the Lebesgue dominated convergence theorem applied to the right-hand side we infer by taking the limit as $n\to\infty$

$$\begin{aligned}&\mathbb{E}\sup_{s\geq t} e^{pV_s}\ |Y_s-\xi_s|^p+\mathbb{E}\ \left(\int_t^\infty e^{2V_s}|Z_s-\zeta_s|^2ds\right)^{p/2}\\&\qquad\qquad+\mathbb{E}\ \left(\int_t^\infty e^{2V_s}\left|\Psi\left(s,Y_s\right)-\Psi\left(s,\xi_s\right)\right|dQ_s\right)^{p/2}\\&\leq C_p\,\mathbb{E}\ \left(\int_t^\infty \mathbf{1}_{[0,\tau]}(s)\,e^{V_s}\left[|\hat{\xi}_s|dQ_s+|H(s,\xi_s,0)|\,dQ_s+\ell_s\,|\zeta_s|\,ds\right]\right)^p,\end{aligned}\tag{5.201}$$

which yields (5.163) if we choose $p=2$.

In the case $p=q=1+\frac{\delta}{2+\delta}\in(1,2)$, by Hölder's inequality, we have

$$\begin{aligned}&\mathbb{E}\ \left(\int_t^\infty \mathbf{1}_{[0,\tau]}(s)\,e^{V_s}\ell_s|\zeta_s|ds\right)^q\\&\quad\leq \mathbb{E}\Big[\left(\int_t^\infty \mathbf{1}_{[0,\tau]}(s)\,(\ell_s)^2\,ds\right)^{q/2}\left(\int_t^\infty e^{2V_s}|\zeta_s|^2ds\right)^{q/2}\Big]\\&\quad\leq\left[\mathbb{E}\ \left(\int_t^\infty \mathbf{1}_{[0,\tau]}(s)\,(\ell_s)^2\,ds\right)^{1+\delta}\right]^{\frac{1}{2+\delta}}\left[\mathbb{E}\int_t^\infty e^{2V_s}|\zeta_s|^2ds\right]^{\frac{q}{2}}.\end{aligned}$$

In the case $p = q = 2$, $\delta = \infty$ and ℓ is a deterministic process

$$\mathbb{E}\left(\int_t^\infty \mathbf{1}_{[0,\tau]}(s)\, e^{V_s}\ell_s|\zeta_s|ds\right)^2 \le \left(\int_t^\infty (\ell_s)^2\, ds\right)\mathbb{E}\left(\int_t^\infty e^{2V_s}|\zeta_s|^2 ds\right).$$

Using the assumptions (5.190) and (5.154-ii), from (5.201) we infer (5.191-*jj*).
Step 4. Estimates on the subdifferential term $dK_s^n = U_s^n dQ_s \in \partial_y \Psi\left(s, Y_s^n\right) dQ_s$. We now use the assumption (5.189) on the interior of Dom (φ). From the proof of Corollary 5.49 we have

$$\begin{aligned}
&\delta_0\, d \updownarrow K^n \updownarrow_t + \left\langle Y_t - u_0, \hat{H}_n\left(t, Y_t^n, Z_t^n\right) dQ_t - dK_t^n\right\rangle \\
&\quad \le \left[\Psi^\#_{u_0,\delta_0}(t) - \Psi(t,u_0)\right] dQ_t + |Y_t^n - u_0|\left[|\hat{u}_t| + \left|\hat{H}_n(t,u_0,0)\right|\right] dQ_t \\
&\quad\quad + |Y_t^n - u_0|^2\, dV_t + \frac{1}{4}|Z_t^n|^2\, dt,
\end{aligned}$$

where

$$\begin{aligned}
\Psi^\#_{u_0,\delta_0}(t) &= \sup\left\{\mathbf{1}_{[0,\tau]}(t)\left[\alpha_t\varphi(u_0 + \delta_0 v) + (1-\alpha_t)\psi(u_0+\delta_0 v)\right] : |v| \le 1\right\} \\
&= \sup\left\{\mathbf{1}_{[0,\tau]}(t)\left|\alpha_t\varphi(u_0) + (1-\alpha_t)\psi(u_0)\right|\right\} \\
&= \Psi(t,u_0),
\end{aligned}$$

and $\hat{u}_t = 0 \in \partial_y\Psi(\omega, t, u_0)$. Hence

$$\begin{aligned}
&\delta_0 d \updownarrow K^n \updownarrow_t + \left\langle Y_t^n - u_0, \hat{H}_n\left(t, Y_t^n, Z_t^n\right) dQ_t - dK_t^n\right\rangle \\
&\quad \le |Y_t^n - u_0|\,|H(t,u_0,0)|\, dQ_t + |Y_t^n - u_0|^2\, dV_t + \frac{1}{4}|Z_t^n|^2\, dt.
\end{aligned}$$

By Proposition 6.80-B we obtain

$$\begin{aligned}
&\delta_0\mathbb{E}\int_0^T e^{2V_s} d \updownarrow K^n \updownarrow_s \\
&\quad \le C_2\left[\mathbb{E}\, e^{2V_T}|Y_T^n - u_0|^2 + \mathbb{E}\left(\int_0^T e^{V_s}|H(s,u_0,0)|\, dQ_s\right)^2\right] \le C.
\end{aligned}$$

From the convergence (5.199) and the equality

$$Y_0^n + K_t^n = Y_t^n + \int_0^t \hat{H}_n\left(s, Y_s^n, Z_s^n\right) dQ_s - \int_0^t Z_s^n dB_s,\ \forall\, t \ge 0,$$

it follows, via Lemma 5.16, that there exists a $K \in S_m^0$ such that

$$\|K^n - K\|_T \xrightarrow{prob.} 0, \quad \text{as } n \to \infty.$$

As in Proposition 1.20 and Corollary 1.22 we obtain

$$\mathbb{E}\int_0^{\tau} e^{2V_s} d \updownarrow K \updownarrow_s \leq \liminf_{n\to+\infty} \mathbb{E}\int_0^{\tau} e^{2V_s} d \updownarrow K^n \updownarrow_s \leq C$$

and

$$dK_t \in \partial_y \Psi(t, Y_t)\, dQ_t \text{ on } \mathbb{R}_+.$$

Finally passing to the limit in

$$Y_t^n + K_T^n - K_t^n = Y_T^n + \int_t^T \hat{H}_n\left(s, Y_s^n, Z_s^n\right) dQ_s - \int_t^T Z_s^n dB_s,$$

we complete the proof. ∎

Remark 5.68. In this last theorem, in contrast to the results in Theorem 5.66, we have not been able to show that the process K is absolutely continuous.

To end this section we discuss a particular case of BSVI (5.146) that we recall here for the convenience of the reader:

$$\begin{cases} Y_t + \displaystyle\int_{t\wedge\tau}^{\tau} dK_s = \eta + \int_{t\wedge\tau}^{\tau} \left[F(s, Y_s, Z_s)\, ds + G(s, Y_s)\, dA_s\right] \\ \qquad\qquad\qquad\qquad\qquad - \displaystyle\int_{t\wedge\tau}^{\tau} Z_s dB_s, \quad t \geq 0, \\ dK_t \in \partial\varphi(Y_t)\, dt + \partial\psi(Y_t)\, dA_t, \text{ on } \mathbb{R}_+, \end{cases} \tag{5.202}$$

where the assumptions $(A_1), \ldots, (A_{10})$ from the beginning of this section will to be replaced by

(L_1): $(A_1) + (A_2) + (A_3)$ *are satisfied;*

(L_2): *the functions* $F : \Omega\times\mathbb{R}_+\times\mathbb{R}^m\times\mathbb{R}^{m\times k} \to \mathbb{R}^m$ *and* $G : \Omega\times\mathbb{R}_+\times\mathbb{R}^m \to \mathbb{R}^m$ *satisfy*

$$F(\cdot,\cdot, y, z),\ G(\cdot,\cdot, y) \text{ is p.m.s.p., for each } (y, z) \in \mathbb{R}^m \times \mathbb{R}^{m\times k},$$

and there exists an $L > 0$ *such that,* $\mathbb{P}$-a.s. $\omega \in \Omega$, *a.e.* $t \geq 0$, for all y, y', z, z'

$$\begin{array}{rl} (i) & \langle y' - y, F(t, y', z) - F(t, y, z)\rangle \leq \frac{L}{2}\left|y' - y\right|^2, \\ (ii) & \left|F(t, y, z') - F(t, y, z)\right| \leq \sqrt{\frac{L}{2}}\left|z' - z\right|, \\ (iii) & \left|F(\omega, t, y, 0)\right| \leq \frac{L}{2}(1 + |y|), \\ (iv) & \langle y' - y, G(t, y') - G(t, y)\rangle \leq L\left|y' - y\right|^2, \\ (v) & \left|G(\omega, t, y)\right| \leq L(1 + |y|). \end{array} \tag{5.203}$$

We remark that in this case $\mu_t = \frac{L}{2}\mathbf{1}_{[0,\tau]}(t)$, $\ell_t = \sqrt{\frac{L}{2}}\mathbf{1}_{[0,\tau]}(t)$, $\nu_t = L\mathbf{1}_{[0,\tau]}(t)$ and

$$V_t = \int_0^{t\wedge\tau} \left[\mu_s ds + \nu_s dA_s + (\ell_s)^2\, ds\right] = LQ_{t\wedge\tau},$$

$$F_\rho^{\#}(s) = \sup_{|y|\le\rho} |F(t,y,0)| \le \frac{L}{2}(1+\rho),$$

$$G_\rho^{\#}(s) = \sup_{|y|\le\rho} |G(t,y)| \le L(1+\rho).$$

(L_3): (A_6)+(A_7)+(A_8) *are satisfied;*
(L_5): *assume that*

$$\mathbb{E}\, e^{2LQ_\tau}\left(1+|\eta|^2+|\varphi(\eta)|+|\psi(\eta)|\right) < \infty. \tag{5.204}$$

We highlight that under $(L_1),\ldots,(L_5)$, the assumptions $(A_1),\ldots,(A_9)$ are satisfied. Also from (L_5) we have

$$\frac{(2L)^j}{j!}\,\mathbb{E}\left[(\tau+A_\tau)^j\right] \le \mathbb{E}\, e^{2LQ_\tau} < \infty, \quad \text{for all } j\in\mathbb{N}^*,$$

and consequently $\tau<\infty$, $\mathbb{P}$-a.s. Moreover it is not difficult to verify that by (L_2), (L_5) and (5.192-a) the condition (5.190) is satisfied for all $\delta\in(0,\infty)$ and for all $q = 1+\frac{\delta}{2+\delta}\in(1,2)$. Hence with the exception of assumption (5.189) on the interior of Dom (φ) all other assumptions of Theorem 5.67 are satisfied.

Theorem 5.69. *Under the assumptions $(L_1),\ldots,(L_5)$ the BSVI (5.202) has a unique solution $(Y,Z,K)\in S_m^0\times\Lambda_{m\times k}^0\times S_m^0$ in the sense of Definition 5.64 which satisfies for all $q\in(1,2)$:*

$$(j)\quad \mathbb{E}\sup_{s\ge0} e^{2LQ_s}|Y_s-u_0|^2+\mathbb{E}\int_0^\infty e^{2LQ_s}|Z_s|^2 ds<\infty,$$

$$(jj)\quad \lim_{T\to\infty}\left[\mathbb{E}\, e^{qLQ_T}|Y_T-\xi_T|^q+\mathbb{E}\left(\int_T^\infty e^{2LQ_s}|Z_s-\zeta_s|^2 ds\right)^{q/2}\right]=0.$$

Moreover there exist $U^{(1)},U^{(2)}\in\Lambda_m^0$, $U_t^{(1)}\in\partial\varphi(Y_t)$ and $U_t^{(2)}\in\partial\psi(Y_t)$, $d\mathbb{P}\otimes dt$-a.e. such that $dK_t = U_t dQ_t \in \partial_y\Psi(t,Y_t)\,dQ_t$, where

$$U_t = \mathbf{1}_{[0,\tau]}(t)\left[\alpha_t U_t^{(1)}+(1-\alpha_t)\,U_t^{(2)}\right]$$

and

$$Y_t + \int_t^T U_s dQ_s = Y_T + \int_t^T H(s, Y_s, Z_s) dQ_s - \int_t^T Z_s dB_s, \ a.s.$$

The inequalities (5.162), (5.163) and (5.164) hold with $V_t = LQ_t$.

Proof. The proof is similar to that of Theorem 5.67: the *Steps 1–3* are exactly the same. To pass to the limit in the approximating equation

$$\begin{cases} Y_t^n + \int_t^\infty U_s^n dQ_s = \eta_n + \int_t^\infty \hat{H}_n\left(s, Y_s^n, Z_s^n\right) dQ_s - \int_t^\infty Z_s^n dB_s, \\ dK_s^n = U_s^n dQ_s = U_s^{1,n} ds + U_s^{2,n} dA_s, \\ \qquad \text{with } U_s^{1,n} ds \in \partial\varphi\left(Y_s^n\right) ds, \ U_s^{2,n} dA_s \in \partial\psi\left(Y_s^n\right) dA_s, \end{cases} \tag{5.205}$$

we need a new argument for *Step 4* since now the interior condition (5.189) is not satisfied.

Step 4′. Estimates on subdifferential terms $U^{1,n}$ and $U^{2,n}$. By Theorem 5.66 we have

$$\begin{aligned} &\mathbb{E}\left[e^{2V_t^n}\left(\varphi(Y_t^n) - \varphi(u_0) + \psi(Y_t^n) - \psi(u_0)\right)\right] \\ &\quad + \frac{1}{2}\mathbb{E}\int_{t\wedge\tau}^{\tau} e^{2V_s^n}\left[\left|U_s^{1,n}\right|^2 ds + \left|U_s^{2,n}\right|^2 dA_s\right] \\ &\le \mathbb{E}\left[e^{2V_\tau^n}\left(\varphi(\eta_n) - \varphi(u_0) + \psi(\eta_n) - \psi(u_0)\right)\right] \\ &\quad + 4\mathbb{E}\int_t^\infty \mathbf{1}_{[0,\tau]}(s)\, e^{2V_s^n}\left[|\hat{F}_n(s, Y_s^n, Z_s^n)|^2 ds + |\hat{G}_n(s, Y_s^n)|^2 dA_s\right]. \end{aligned}$$

Note that $V_t^n = V_t - \theta_t^n$, where $\theta_t^n = \int_0^{t\wedge\tau} \frac{L}{2} \mathbf{1}_{(n,\infty)}(\lambda_s)\, ds$. Since

$$\begin{aligned} \varphi(\eta_n) - \varphi(u_0) &= (\varphi(\eta) - \varphi(u_0))\, \mathbf{1}_{[0,n]}(\beta_\tau + |\eta - u_0|), \\ \psi(\eta_n) - \psi(u_0) &= (\psi(\eta) - \psi(u_0))\, \mathbf{1}_{[0,n]}(\beta_\tau + |\eta - u_0|), \end{aligned}$$

$$|\hat{F}_n(s, Y_s^n, Z_s^n)| \le \frac{L}{2}\left(1 + |Y_s^n|\right) + \sqrt{\frac{L}{2}}\, |Z_s^n|\, \mathbf{1}_{[0,n]}(\lambda_s) + |F(s, u_0, 0)|\, \mathbf{1}_{[0,n]}(\gamma_s),$$

$$|\hat{G}_n(s, Y_s^n)| \le L\left(1 + |Y_s^n|\right) + |G(s, u_0)|\, \mathbf{1}_{[0,n]}(\gamma_s),$$

we obtain

$$\mathbb{E}\left[e^{2V_t^n}\left(\varphi(Y_t^n) - \varphi(u_0) + \psi(Y_t^n) - \psi(u_0)\right)\right] \le C,$$

and

$$\mathbb{E}\int_0^\infty \mathbf{1}_{[0,\tau]}(s)\, e^{2V_s^n}\left[\left|U_s^{1,n}\right|^2 ds + \left|U_s^{2,n}\right|^2 dA_s\right] \le C.$$

Consequently there exist two p.m.s.p. $U^{(1)}$ and $U^{(2)}$, such that along a subsequence still indexed by n,

$$e^{V}U^{1,n}e^{\theta^n}\mathbf{1}_{[0,\tau]} \rightharpoonup e^{V}U^{(1)}\mathbf{1}_{[0,\tau]}, \quad \text{weakly in } L^2\left(\Omega\times\mathbb{R}_+, d\mathbb{P}\otimes dt;\mathbb{R}^m\right),$$
$$e^{V}U^{2,n}e^{\theta^n}\mathbf{1}_{[0,\tau]} \rightharpoonup e^{V}U^{(2)}\mathbf{1}_{[0,\tau]}, \quad \text{weakly in } L^2\left(\Omega\times\mathbb{R}_+, d\mathbb{P}\otimes dA_t;\mathbb{R}^m\right).$$

Passing to the limit in (5.205) the result follows in a standard manner (see the proof of Theorem 5.66). ■

Remark 5.70. If $\tau = T < \infty$ is a deterministic final time, then the assertions of Theorem 5.69 are also true with $q = 2$ (and $\delta = \infty$) by setting $\ell_s = \sqrt{\frac{L}{2}}\mathbf{1}_{[0,T]}(s)$.

5.6.3 *Weak Variational Solutions*

In this subsection we discuss again the existence and the uniqueness of a solution (Y, Z) of BSVI (5.146) that we recall here:

$$\begin{cases} Y_t + \int_{t\wedge\tau}^{\tau} dK_s = \eta + \int_{t\wedge\tau}^{\tau} \left[F\left(s, Y_s, Z_s\right)ds + G\left(s, Y_s\right)dA_s\right] - \int_{t\wedge\tau}^{\tau} Z_s dB_s, \ t \geq 0, \\ dK_t \in \partial\varphi\left(Y_t\right)dt + \partial\psi\left(Y_t\right)dA_t, \ \text{on } \mathbb{R}_+, \end{cases} \tag{5.206}$$

under the assumptions $(A_1), \ldots, (A_9)$ presented in Sect. 5.6.2. Adding the assumption (A_{10}) we have Theorem 5.66. Replacing the assumption (A_{10}) by (5.190) and the interior condition (5.189) we have Theorem 5.67. Furthermore if the stochastic processes $(\mu_t)_{t\geq 0}$, $(\nu_t)_{t\varepsilon 0}$, $(\ell_t)_{t\geq 0}$ are constants and some boundedness assumptions (5.203-*iii*, *v*) and (5.204) are satisfied then we can renounce assumption (A_{10}), and obtain the existence and uniqueness of a solution (Y, Z) for (5.206): see Theorem 5.69.

The aim of this subsection is to obtain existence and uniqueness under the assumptions $(A_1), \ldots, (A_9)$ and (5.190), i.e. to see what happens in Theorem 5.67 without the interior condition (5.189). It is not clear how we can obtain some estimates on the subdifferential term $dK_s^n = U_s^n dQ_s \in \partial_y\Psi\left(s, Y_s^n\right)dQ_s$, except for the particular case treated in Theorem 5.69. For this reason we shall give a weak variational formulation for the solution as in [47]. The stochastic variational formulation for forward SDEs was introduced by Răşcanu in [62].

Let us define the space $\mathcal{L}_m^p$, $p \geq 0$, of continuous semimartingales M of the form

$$M_t = \gamma - \int_0^t \Lambda_r dQ_r + \int_0^t \Theta_r dB_r,$$

where $\gamma \in \mathbb{R}^m$, Λ and Θ are two p.m.s.p. such that on every interval $[0, T] \subset \mathbb{R}_+$, $\Lambda \in L^p\left(\Omega; L^1\left(0, T;\mathbb{R}^m\right)\right)$, $\Theta \in L^p\left(\Omega; L^2\left(0, T;\mathbb{R}^{m\times k}\right)\right)$.

For an intuitive introduction, let $M \in \mathcal{L}_m^0$ and (Y, Z, K) be a solution of (5.157), in the sense of Definition 5.64. By Itô's formula for $\frac{1}{2}|M_t - Y_t|^2$ and the subdifferential inequality

$$\int_t^T \langle M_r - Y_r, dK_r\rangle + \int_t^T \Psi(r, Y_r)\, dQ_r \le \int_t^T \Psi(r, M_r)\, dQ_r$$

we obtain the inequality

$$\begin{aligned}
&\frac{1}{2}|M_t - Y_t|^2 + \frac{1}{2}\int_t^T |\Theta_r - Z_r|^2 dr + \int_t^T \Psi(r, Y_r)\, dQ_r \\
&\le \frac{1}{2}|M_T - Y_T|^2 + \int_t^T \Psi(r, M_r)\, dQ_r + \int_t^T \langle M_r - Y_r, \Lambda_r - H(r, Y_r, Z_r)\rangle dQ_r \\
&\qquad - \int_t^T \langle M_r - Y_r, (\Theta_r - Z_r) dB_r\rangle.
\end{aligned}$$

Therefore, we propose the following weak formulation for the solution.

Definition 5.71. We call $(Y_t, Z_t)_{t\ge 0}$ a weak variational solution of (5.206) if $(Y, Z) \in S_m^0 \times \Lambda_{m\times k}^0$, $(Y_t, Z_t) = (\xi_t, \zeta_t) = (\eta, 0)$ for $t > \tau$ and

$$\begin{aligned}
(i)\quad & \int_0^T (|H(r, Y_r, Z_r)| + \Psi(r, Y_r))\, dQ_r < \infty, \ \mathbb{P}\text{-a.s., for all } T \ge 0, \\
(ii)\quad & \frac{1}{2}|M_t - Y_t|^2 + \frac{1}{2}\int_t^s |\Theta_r - Z_r|^2 dr + \int_t^s \Psi(r, Y_r)\, dQ_r \\
& \le \frac{1}{2}|M_s - Y_s|^2 + \int_t^s \Psi(r, M_r)\, dQ_r \\
& + \int_t^s \langle M_r - Y_r, \Lambda_r - H(r, Y_r, Z_r)\rangle dQ_r - \int_t^s \langle M_r - Y_r, (\Theta_r - Z_r) dB_r\rangle, \\
& \qquad \forall\, 0 \le t \le s \le \tau,\ \forall\, M_\cdot = \gamma - \textstyle\int_0^\cdot \Lambda_r dQ_r + \int_0^\cdot \Theta_r dB_r \in \mathcal{L}_m^0, \\
(iii)\quad & e^{2V_T}|Y_T - \xi_T|^2 + \int_T^\infty e^{2V_s}|Z_s - \zeta_s|^2 ds \xrightarrow{prob.} 0, \ \text{as } T \to \infty.
\end{aligned} \tag{5.207}$$

Theorem 5.72. *Let the assumptions* $(A_1, \ldots, A_9)$ *and* $(5.190\text{-}(i')$ *and* $(ii,$ *with* $q = 2))$ *be satisfied. Then the BSVI (5.206) has a unique weak variational solution* $(Y, Z) \in S_m^0 \times \Lambda_{m\times k}^0$ *in the sense of Definition 5.71 such that*

$$\begin{aligned}
(j)\quad & \mathbb{E}\sup_{s\ge 0} e^{2V_s}|Y_s - u_0|^2 + \mathbb{E}\int_0^\infty e^{2V_s}|Z_s|^2 ds < \infty, \\
(jj)\quad & \lim_{T\to\infty}\left[\mathbb{E}\, e^{2V_T}|Y_T - \xi_T|^2 + \mathbb{E}\int_T^\infty e^{2V_s}|Z_s - \zeta_s|^2 ds\right] = 0.
\end{aligned} \tag{5.208}$$

Moreover the inequalities (5.162) and (5.163) hold.

Proof. Existence We remark that we are in the conditions of Theorem 5.67 without the interior condition (5.189). Therefore we start with the same approximating equation as in the proof of Theorem 5.67

$$\begin{cases} Y_t^n + \int_t^\infty U_s^n dQ_s = \eta_n + \int_t^\infty \hat{H}_n\left(s, Y_s^n, Z_s^n\right) dQ_s - \int_t^\infty Z_s^n dB_s, \\ dK_s^n = U_s^n dQ_s \in \partial_y \Psi\left(s, Y_s^n\right) dQ_s \equiv \mathbf{1}_{[0,\tau]}(s)\, \partial\varphi\left(Y_s^n\right) dQ_s \end{cases} \tag{5.209}$$

and we follow exactly the same *Steps 1–3* as there.

We obtain the existence of $(Y, Z) \in S_m^0 \times \Lambda_{m\times k}^0$ such that

$$\mathbb{E}\left[\sup_{s\geq 0} e^{2V_s}|Y_s^n - Y_s|^2 + \int_0^\infty e^{2V_s}|Z_s^n - Z_s|^2 ds\right] \longrightarrow 0, \text{ as } n \to \infty,$$

$(Y_t, Z_t) = (\eta, 0)$ for $t > \tau$ and (Y, Z) satisfies (5.208), the inequalities (5.162) and (5.163), and (5.207-i, iii).

Let $M_\cdot = \gamma - \int_0^\cdot \Lambda_r dQ_r + \int_0^\cdot \Theta_r dB_r \in \mathcal{L}_m^0$. By Itô's formula for $\frac{1}{2}|M_t - Y_t^n|^2$ we deduce that, for all $0 \leq t \leq s$,

$$\begin{aligned} \frac{1}{2}|M_t - Y_t^n|^2 + \frac{1}{2}\int_t^s |\Theta_r - Z_r^n|^2 dr + \int_t^s \Psi\left(r, Y_r^n\right) dQ_r \leq \frac{1}{2}\mathbb{E}|M_s - Y_s^n|^2 \\ + \int_t^s \Psi\left(r, M_r^n\right) dQ_r + \int_t^s \langle M_r - Y_r^n, \Lambda_r - \hat{H}_n\left(r, Y_r^n, Z_r^n\right)\rangle dQ_r \\ - \int_t^s \langle M_r - Y_r^n, (\Theta_r - Z_r^n) dB_r\rangle. \end{aligned}$$

Passing to the lim inf it follows that the pair (Y, Z) satisfies the inequality (5.207-ii).

Uniqueness. In order to prove the uniqueness of the solution, let $\left(\hat{Y}, \hat{Z}\right) \in S_m^0 \times \Lambda_{m\times k}^0$ and $\left(\tilde{Y}, \tilde{Z}\right) \in S_m^0 \times \Lambda_{m\times k}^0$ be two weak variational solutions of (5.206) corresponding to $\hat{\eta}$ and $\tilde{\eta}$, respectively. Therefore for all $M_\cdot = \gamma - \int_0^\cdot \Lambda_r dQ_r + \int_0^\cdot \Theta_r dB_r \in \mathcal{L}_m^0$,

$$\begin{aligned} &\frac{1}{2}\left(|M_t - \hat{Y}_t|^2 + |M_t - \tilde{Y}_t|^2\right) + \frac{1}{2}\int_t^s \left(|\Theta_r - \hat{Z}_r|^2 + |\Theta_r - \tilde{Z}_r|^2\right) dr \\ &\quad + \int_t^s \left(\Psi(r, \hat{Y}_r) + \Psi(r, \tilde{Y}_r)\right) dQ_r \\ &\leq \frac{1}{2}\left(|M_s - \hat{Y}_s|^2 + |M_s - \tilde{Y}_s|^2\right) + 2\int_t^s \Psi\left(r, M_r\right) dQ_r \\ &\quad + \int_t^s \left(\langle M_r - \hat{Y}_r, \Lambda_r - H(r, \hat{Y}_r, Z_r)\rangle + \langle M_r - \tilde{Y}_r, \Lambda_r - H(r, \tilde{Y}_r, \tilde{Z}_r)\rangle\right) dQ_r \\ &\quad - \int_t^s \left(\langle M_r - \hat{Y}_r, (\Theta_r - \hat{Z}_r) dB_r\rangle + \langle M_r - \tilde{Y}_r, (\Theta_r - \tilde{Z}_r) dB_r\rangle\right), \ \forall\, 0 \leq t \leq s. \end{aligned}$$

Let $Y = \frac{\hat{Y}+\tilde{Y}}{2}$, $Z = \frac{\hat{Z}+\tilde{Z}}{2}$ and $h_r = \frac{1}{2}\Big[H(r,\hat{Y}_r,\hat{Z}_r) + H(r,\tilde{Y}_r,\tilde{Z}_r)\Big]$.

From the convexity of φ we see that

$$2\varphi(Y_r) \le \varphi(\hat{Y}_r) + \varphi(\tilde{Y}_r),$$

and using the identity

$$2\left\langle \frac{u+v}{2}, \frac{f+g}{2}\right\rangle + \frac{1}{2}\langle u-v, f-g\rangle = \langle u, f\rangle + \langle v, g\rangle,$$

we obtain

$$\begin{aligned}
&\langle M_r - \hat{Y}_r, \Lambda_r - H(r,\hat{Y}_r,\hat{Z}_r)\rangle + \langle M_r - \tilde{Y}_r, \Lambda_r - H(r,\tilde{Y}_r,\tilde{Z}_r)\rangle \\
&= 2\langle M_r - Y_r, \Lambda_r - h_r\rangle + \frac{1}{2}\langle \hat{Y}_r - \tilde{Y}_r, H(r,\hat{Y}_r,\hat{Z}_r) - H(r,\tilde{Y}_r,\tilde{Z}_r)\rangle,
\end{aligned}$$

and

$$\begin{aligned}
&\int_t^s \langle M_r - \hat{Y}_r, (\Theta_r - \hat{Z}_r)dB_r\rangle + \int_t^s \langle M_r - \tilde{Y}_r, (\Theta_r - \tilde{Z}_r)dB_r\rangle \\
&= 2\int_t^s \langle M_r - Y_r, (\Theta_r - Z_r)\rangle\, dB_r + \frac{1}{2}\int_t^s \langle \hat{Y}_r - \tilde{Y}_r, (\hat{Z}_r - \tilde{Z}_r)dB_r\rangle.
\end{aligned}$$

Therefore, since

$$\frac{1}{2}\left(|m-u|^2 + |m-v|^2\right) = \left|m - \frac{u+v}{2}\right|^2 + \frac{1}{4}|u-v|^2,$$

we have for all $M_\cdot = \gamma - \int_0^\cdot \Lambda_r dQ_r + \int_0^\cdot \Theta_r dB_r \in \mathcal{L}_m^0$

$$\begin{aligned}
\left|\hat{Y}_t - \tilde{Y}_t\right|^2 + \int_t^s \left|\hat{Z}_r - \tilde{Z}_r\right|^2 dr \le 8B_{t,s}(M) + \left|\hat{Y}_s - \tilde{Y}_s\right|^2 \\
+2\int_t^s \langle \hat{Y}_r - \tilde{Y}_r, H(r,\hat{Y}_r,\hat{Z}_r) - H(r,\tilde{Y}_r,\tilde{Z}_r)\rangle dQ_r \\
-2\int_t^s \langle \hat{Y}_r - \tilde{Y}_r, (\hat{Z}_r - \tilde{Z}_r)dB_r\rangle, \quad \forall\, 0 \le t \le s,
\end{aligned} \tag{5.210}$$

where

$$\begin{aligned}
B_{t,s}(M) &= \frac{1}{2}|M_s - Y_s|^2 + \int_t^s \Psi(r, M_r)\, dQ_r \\
&+ \int_t^s \langle M_r - Y_r, \Lambda_r - h_r\rangle dQ_r - \frac{1}{2}|M_t - Y_t|^2 \\
&- \frac{1}{2}\textstyle\int_t^s |\Theta_r - Z_r|^2 dr - \displaystyle\int_t^s \Psi(r, Y_r)\, dQ_r - \int_t^s \langle M_r - Y_r, (\Theta_r - Z_r)\, dB_r\rangle.
\end{aligned}$$

Let

$$M_t^\varepsilon = e^{-\frac{Q_t}{Q_\varepsilon}} \left[Y_0 + \frac{1}{Q_\varepsilon} \int_0^t e^{\frac{Q_r}{Q_\varepsilon}} Y_r dQ_r \right]. \tag{5.211}$$

Clearly, $M^\varepsilon \in \mathcal{L}_m^0$ since $M_t^\varepsilon = M_0^\varepsilon + \int_0^t dM_r^\varepsilon = Y_0 + \int_0^t \frac{Y_r - M_r^\varepsilon}{Q_\varepsilon} dQ_r + \int_0^t 0 dB_r$. By Lemma 6.21 it follows that for all $0 \le t \le s \le T$

$$\begin{aligned}
(a) \quad & \lim_{\varepsilon \to 0_+} \left[\sup_{r \in [0,T]} |M_r^\varepsilon - Y_r| \right] = 0, \\
(b) \quad & \lim_{\varepsilon \to 0_+} \int_t^s \mathbf{1}_{[0,\tau]}(r)\, \varphi(M_r^\varepsilon) dr = \int_t^s \mathbf{1}_{[0,\tau]}(r)\, \varphi(Y_r) dr, \\
(c) \quad & \lim_{\varepsilon \to 0_+} \int_t^s \mathbf{1}_{[0,\tau]}(r)\, \psi(M_r^\varepsilon) dA_r = \int_t^s \mathbf{1}_{[0,\tau]}(r)\, \psi(Y_r) dA_r
\end{aligned}$$

and consequently

$$\limsup_{\varepsilon \to 0_+} B_{t,s}(M^\varepsilon) \le 0,$$

because $\Psi(r, M_r^\varepsilon)\, dQ_r = \mathbf{1}_{[0,\tau]}(r) \left[\varphi(M_r^\varepsilon) dr + \psi(M_r^\varepsilon) dA_r \right]$.

Using the inequality

$$\langle \hat{Y}_r - \tilde{Y}_r, H(r, \hat{Y}_r, \hat{Z}_r) - H(r, \tilde{Y}_r, \tilde{Z}_r) \rangle dQ_r \le |\hat{Y}_r - \tilde{Y}_r|^2 dV_r + \frac{1}{4} |\hat{Z}_r - \tilde{Z}_r|^2 dr$$

from (5.210) with $M = M^\varepsilon$, $\varepsilon \to 0_+$, we obtain that for all $0 \le t \le s$,

$$\begin{aligned}
|\hat{Y}_t - \tilde{Y}_t|^2 + \frac{1}{2} \int_t^s |\hat{Z}_r - \tilde{Z}_r|^2 dr \le |\hat{Y}_s - \tilde{Y}_s|^2 + 2 \int_t^s |\hat{Y}_r - \tilde{Y}_r|^2 dV_r \\
-2 \int_t^s \langle \hat{Y}_r - \tilde{Y}_r, (\hat{Z}_r - \tilde{Z}_r) dB_r \rangle,
\end{aligned}$$

which yields, by Proposition 6.69

$$\begin{aligned}
e^{2V_t} |\hat{Y}_t - \tilde{Y}_t|^2 + \frac{1}{2} \int_t^s e^{2V_r} |\hat{Z}_r - \tilde{Z}_r|^2 dr \le e^{2V_s} |\hat{Y}_s - \tilde{Y}_s|^2 \\
- 2 \int_t^s e^{2V_r} \langle \hat{Y}_r - \tilde{Y}_r, (\hat{Z}_r - \tilde{Z}_r) dB_r \rangle.
\end{aligned}$$

Taking the expectation and then passing to the limit as $s \to \infty$ uniqueness follows (see the properties of the solutions given in (5.208)).

5.7 Semilinear Elliptic PDEs

5.7.1 Elliptic Equations in the Whole Space

We will first consider elliptic PDEs in $\mathbb{R}^d$, and then in a bounded open subset of $\mathbb{R}^d$, with Dirichlet boundary condition.

Let $\{X_t^x;\ t \geq 0\}$ denote the solution of the forward SDE:

$$X_t^x = x + \int_0^t f(X_s^x)\,ds + \int_0^t g(X_s^x)\,dB_s,\ t \geq 0, \tag{5.212}$$

where $f : \mathbb{R}^d \to \mathbb{R}^d$ is continuous and globally monotone, $g : \mathbb{R}^d \to \mathbb{R}^{d \times d}$ is globally Lipschitz, and consider the backward SDE

$$Y_t^x = Y_T^x + \int_t^T F(X_s^x, Y_s^x, Z_s^x)\,ds - \int_t^T Z_s^x\,dB_s, \text{ for all } t, T \text{ s.t. } 0 \leq t \leq T, \tag{5.213}$$

where $F : \mathbb{R}^d \times \mathbb{R}^k \times \mathbb{R}^{k \times d} \to \mathbb{R}^k$ is continuous and such that for some K, K', $\mu < 0$, $p > 0$,

$$\begin{aligned}
|F(x, y, z)| &\leq K'(1 + |x|^p + |y| + |z|),\\
\langle y - y', F(x, y, z) - F(x, y', z)\rangle &\leq \mu |y - y'|^2,\\
|F(x, y, z) - F(x, y, z')| &\leq K\|z - z'\|.
\end{aligned}$$

We assume moreover that for some $\lambda > 2\mu + K^2$, and all $x \in \mathbb{R}^d$,

$$\mathbb{E}\int_0^\infty e^{\lambda t}|F(X_t^x, 0, 0)|^2 dt < \infty, \tag{5.214}$$

which essentially implies that $\lambda < 0$.

Under these assumptions, the BSDE (5.213) has a unique solution, in the sense of Theorem 5.27.

It is not hard to see, using uniqueness for BSDEs, that

$$Y_t^x = Y_0^{X_t^x}, \quad t > 0. \tag{5.215}$$

Denote by

$$\mathcal{A} = \frac{1}{2}\sum_{i,j}(gg^*)_{ij}(x)\frac{\partial^2}{\partial x_i \partial x_j} + \sum_i f_i(x)\frac{\partial}{\partial x_i}$$

the infinitesimal generator of the Markov process $\{X_t^x; t \geq 0\}$, and consider the following system of semilinear elliptic PDEs in $\mathbb{R}^d$

$$\mathcal{A}u_i(x) + F_i(x, u(x), (\nabla ug)(x)) = 0, x \in \mathbb{R}^d,\ 0 \leq i \leq k. \tag{5.216}$$

As in Sect. 5.4, one easily establishes the following:

Theorem 5.73. *Let $u \in C^2(\mathbb{R}^d; \mathbb{R}^m)$ be a classical solution of (5.216) such that for some $M, q > 0$,*

$$|u(x)| \leq M(1 + |x|^q),\ \forall x \in \mathbb{R}^d.$$

Then for each $x \in \mathbb{R}^d$, $\{(u(X_t^x), (\nabla ug)(X_t^x));\ t \geq 0\}$ is the solution of the BSDE (5.213). In particular $u(x) = Y_0^x$.

We now want to prove that (5.212)–(5.213) provide a viscosity solution to (5.216) Again, for the notion of a viscosity solution of the system of PDEs we need (5.216) to make sense, therefore we need to make the following restriction: *for $0 \leq i \leq k$, the i-th coordinate of F depends only on the i-th row of the matrix z.*

Define the mapping

$$\Phi : \mathbb{R}^d \times \mathbb{R}^m \times \mathbb{R}^d \times \mathbb{S}^d \to \mathbb{R}^m$$

by

$$\Phi_i(x, r, p, X) = -\frac{1}{2}\mathbf{Tr}[g(x)g^*(x)X] - \langle f(x), p\rangle - F_i(x, r, pg(x)),\ 1 \leq i \leq m.$$

Then the system (5.216) reads

$$\Phi_i(x, u(x), Du_i(x), D^2u_i(x)) = 0,\ x \in \mathbb{R}^d,\ 0 \leq i \leq m.$$

All the assumptions from Theorem 5.37 are assumed to hold below (with of course f, g and F independent of the time variable t). The notion of a viscosity solution of (5.216) is defined by Definition 6.94 in Annex D.

We can now prove the following:

Theorem 5.74. *Under the above assumptions, $u(x) \stackrel{def}{=} Y_0^x$ is a continuous function which satisfies*

$$|Y_0^x| \leq c\sqrt{\mathbb{E}\int_0^\infty e^{\lambda t}|F(X_t^x, 0, 0)|^2 dt}, \tag{5.217}$$

for any $\lambda > 2\mu + K^2$, and it is a viscosity solution of (5.216).

Proof. The continuity follows from the mean-square continuity of $\{Y_\cdot^x,\ x \in \mathbb{R}^d\}$. The inequality (5.217) follows from (5.134) with $\eta = 0$ (hence $\xi = 0$ and $\zeta = 0$).

To prove that u is a viscosity sub-solution, we take any $1 \leq i \leq m$, $\varphi \in C^2(\mathbb{R}^d)$ and $x \in \mathbb{R}^d$ such that $u_i - \varphi$ has a local maximum at x. We assume without loss of generality that

$$u_i(x) = \varphi(x).$$

We suppose that

$$\Phi_i(x, u(x), D\varphi_i(x), D^2\varphi_i(x)) > 0,$$

and we will find a contradiction.

Let $\alpha > 0$ be such that whenever $|y - x| \leq \alpha$,

$$u_i(y) \leq \varphi(y),$$
$$\Phi_i(y, u(y), D\varphi_i(y), D^2\varphi_i(y)) > 0,$$

and define, for some $T > 0$,

$$\tau = \inf\{t > 0;\ |X_t^x - x| \geq \alpha\} \wedge T.$$

Let now

$$(\overline{Y}_t, \overline{Z}_t) = ((Y_{t\wedge\tau}^x)^i, \mathbf{1}_{[0,\tau]}(t)(Z_t^x)^i),\ 0 \leq t \leq T.$$

$(\overline{Y}, \overline{Z})$ solves the one-dimensional BSDE

$$\overline{Y}_t = u_i(X_\tau^x) + \int_t^T \mathbf{1}_{[0,\tau]}(s) F_i(X_s^x, u(X_s^x), \overline{Z}_s)ds - \int_t^T \overline{Z}_s dB_s,\ 0 \leq t \leq T.$$

On the other hand, from Itô's formula,

$$(\hat{Y}_t, \hat{Z}_t) = (\varphi(X_{t\wedge\tau}^x), \mathbf{1}_{[0,\tau]}(t)(\nabla\varphi g)(X_t^x)),\ 0 \leq t \leq T$$

solves the BSDE

$$\hat{Y}_t = \varphi(X_\tau^x) - \int_t^T \mathbf{1}_{[0,\tau]}(s)\mathcal{A}\varphi(X_s^x)ds - \int_t^T \hat{Z}_s dB_s,\ 0 \leq t \leq T.$$

From $u_i \leq \varphi$, and the choice of α and τ, we deduce with the help of Proposition 5.34 that $\overline{Y}_0 < \hat{Y}_0$, i.e. $u_i(x) < \varphi(x)$, which is a contradiction. ■

5.7.2 Elliptic Dirichlet Problem

We now give a similar result for a system of elliptic PDEs in an open bounded subset of $\mathbb{R}^d$, with Dirichlet boundary condition, following [20]. Let $D \subset \mathbb{R}^d$ be a bounded domain (i.e. D is an open bounded subset of $\mathbb{R}^d$), whose boundary ∂D is of class C^1. We are given a function $\chi \in C(\mathbb{R}^d)$ and we consider the system of elliptic PDEs

$$\begin{cases} \Phi_i(x, u(x), Du(x), D^2u(x)) = 0, & 1 \le i \le m,\ x \in D; \\ u_i(x) = \chi_i(x), & 1 \le i \le m,\ x \in \partial D. \end{cases} \tag{5.218}$$

The process $\{X_t^x;\ t \ge 0\}$ is defined as in the preceding subsection. For each $x \in \overline{D}$, we define the stopping time

$$\tau_x = \inf\{t \ge 0;\ X_t^x \notin \overline{D}\}.$$

Let $\{(Y_t^x, Z_t^x);\ 0 \le t \le \tau_x\}$ be the solution, in the sense of Corollary 5.59, of the BSDE

$$Y_t^x = \chi(X_{\tau_x}^x) + \int_{t\wedge\tau_x}^{\tau_x} F(X_s^x, Y_s^x, Z_s^x)\,ds - \int_{t\wedge\tau_x}^{\tau_x} Z_s^x dB_s,\ t \ge 0. \tag{5.219}$$

Using Itô's formula, it is not hard to establish the following:

Theorem 5.75. *Let $u \in C^2(D;\mathbb{R}^m)\cap C^0(\overline{D};\mathbb{R}^m)$ be a classical solution of (5.218). Then for each $x \in \mathbb{R}^d$, $\{(u(X_t^x), (\nabla u g)(X_t^x));\ t \ge 0\}$ is the solution of the BSDE (5.219). In particular $u(x) = Y_0^x$.*

We now assume that $P(\tau_x < \infty) = 1$, for all $x \in \overline{D}$, that the set

$$\Lambda = \{x \in \partial D;\ P(\tau_x > 0) = 0\} \quad \text{is closed,} \tag{5.220}$$

and that for some $\lambda > 2\mu + K^2$, and all $x \in \overline{D}$,

$$\mathbb{E}e^{\lambda\tau_x} < \infty.$$

We again define $u(x) = Y_0^x$. Besides some arguments which we have already used, the continuity of u also relies on the following:

Proposition 5.76. *Under the condition (5.220), the mapping $x \to \tau_x$ is a.s. continuous on $\overline{D}$.*

Proof. Let $\{x_n,\ n \in \mathbb{N}\}$ be a sequence in $\overline{D}$ such that $x_n \to x$, as $n \to \infty$. We first show that

$$\limsup_{n\to\infty} \tau_{x_n} \le \tau_x \quad \text{a.s.} \tag{5.221}$$

Suppose that (5.221) is false. Then

$$P(\tau_x < \limsup_{n\to\infty} \tau_{x_n}) > 0. \tag{5.222}$$

For each $\varepsilon > 0$, let

$$\tau_x^\varepsilon = \inf\{t \geq 0;\ d(X_t^x, D) \geq \varepsilon\}.$$

From (5.222), there exists ε and T such that

$$P(\tau_x^\varepsilon < \limsup_{n\to\infty} \tau_{x_n} \leq T) > 0.$$

But since $X_\cdot^{x_n} \to X_\cdot^x$ uniformly on $[0, T]$ a.s., this implies that

$$P(\limsup_{n\to\infty} \tau_{x_n}^{\varepsilon/2} \leq \tau_x^\varepsilon < \limsup_{n\to\infty} \tau_{x_n} \leq T) > 0,$$

which would mean that for some n, X^{x_n} exits the $\varepsilon/2$-neighbourhood of D before exiting D, which is impossible.

We next prove that

$$\liminf_{n\to\infty} \tau_{x_n} \geq \tau_x \quad \text{a.s.} \tag{5.223}$$

For this part of the proof, we will need the assumption (5.220) that Λ is closed.

It suffices to prove that (5.223) holds a.s. on $\Omega_M = \{\tau_x \leq M\}$, with M arbitrary. From the result of the first step, for almost all $\omega \in \Omega_M$, there exists an $n(\omega)$ such that $n \geq n(\omega)$ implies $\tau_{x_n} \leq M + 1$. From the a.s. (on Ω_M) uniform convergence of $X_\cdot^{x_n} \to X_\cdot^x$ on the interval $[0, M+1]$, $X_\cdot^x$ hits the set

$$\overline{\{X_{\tau_{x_n}}^{x_n};\ n \in \mathbb{N}\}} \subset \overline{\Lambda} = \Lambda$$

on the random interval $[0, \liminf_n \tau_{x_n}]$ a.s. on Ω_M. The result follows, since $X_\cdot^x$ exits $\overline{D}$ when it hits Λ. ■

We now prove the following:

Theorem 5.77. *Under the assumptions of Theorem 5.74, the above conditions on D and the condition (5.220), $u(x) \stackrel{def}{=} Y_0^x$ is continuous on $\overline{D}$ and it is a viscosity solution of the system of Eq. (5.218).*

Proof. We only prove that u is a sub-solution. Let $1 \leq i \leq m$, $\varphi \in C^2(\mathbb{R}^d)$ $u_i - \varphi$ have a local maximum at $x \in \overline{D}$, such that $u_i(x) = \varphi(x)$. If $x \in \Lambda$, then $\tau_x = 0$, and hence $u(x) = \chi(x)$. If however $x \in D \cup (\partial D \backslash \Lambda)$, the result follows by the same argument as in the proof of Theorem 5.74.

5.7.3 Elliptic Equations with Neumann Boundary Conditions

The data and assumptions are the same as in Sect. 5.4.3, except that we suppress the dependence of all coefficients upon the time variable. Moreover we also assume that all assumptions of Section 5.4.1 are satisfied.

Consider the following system of semilinear elliptic PDEs with nonlinear Neumann boundary condition

$$\begin{cases} \Phi_i(x, u(x), Du_i(x), D^2u_i(x)) = 0, \ x \in D, \quad 0 \le i \le m; \\ \dfrac{\partial u_i}{\partial n}(x) - G_i(x, u(x)) = 0, \ x \in \partial D, \ 1 \le i \le m. \end{cases} \tag{5.224}$$

Let X^x be the process solution of the reflected stochastic differential equation, for all $t \geq 0$, $\mathbb{P}$ a.s.

$$\begin{cases} X_t^x + K_t^x = x + \displaystyle\int_0^t f(r, X_r^x)dr + \int_0^t g(r, X_r^x)dB_r, \\ X_t^x \in \overline{D}, \ K_t^x = \displaystyle\int_0^t n(X_r^x)\mathbf{1}_{\partial D}\left(X_r^x\right) d\updownarrow K^x \updownarrow_r . \end{cases} \tag{5.225}$$

To each $x \in \overline{D}$ we associate the BSDE

$$Y_t^x = Y_T^x + \int_t^T F(r, X_r^x, Y_r^x, Z_r^x)dr + \int_t^T G(r, X_r^x, Y_r^x)d\updownarrow K^x \updownarrow_r \tag{5.226}$$

$$- \int_t^T Z_r^x dB_r, \text{ for all pairs } 0 \le t < T.$$

Itô's formula again allows us to establish the following:

Theorem 5.78. *Let $u \in C^2(D; \mathbb{R}^m) \cap C^1(\overline{D}; \mathbb{R}^m)$ be a classical solution of (5.224). Then for each $x \in \mathbb{R}^d$, $\{(u(X_t^x), (\nabla ug)(X_t^x)); \ t \geq 0\}$ is the solution of the BSDE (5.226). In particular $u(x) = Y_0^x$.*

We now have:

Theorem 5.79. *Under the above conditions and those of Theorem 5.43, $u(x) := Y_0^x$ is a continuous function of x, and it is a viscosity solution of (5.224).*

The proof of this Theorem is easily done by combining the arguments in the proofs of Theorems 5.74 and 5.43.

5.8 Parabolic Variational Inequality

The aim of this section is to prove the existence of a viscosity solution for the following parabolic variational inequality (PVI) with a mixed nonlinear multivalued Neumann–Dirichlet boundary condition:

$$\begin{cases} \dfrac{\partial u(t,x)}{\partial t} + \mathcal{A}_t u\,(t,x) + F\,(t,x,u(t,x),(\nabla u g)(t,x)) \in \partial\varphi\,(u(t,x))\,, \\ \hfill t \in (0,T)\,,\ x \in D, \\ -\dfrac{\partial u(t,x)}{\partial n} + G\,(t,x,u(t,x)) \in \partial\psi\,(u(t,x))\,,\ \ t \in (0,T)\,,\ x \in \mathrm{Bd}\,(D)\,, \\ u(T,x) = \kappa(x),\ \ x \in \overline{D}, \end{cases} \tag{5.227}$$

where the operator $\mathcal{A}_t$ is given by

$$\mathcal{A}_t v(x) = \frac{1}{2}\mathrm{Tr}\big[g(t,x)g^*(t,x)D^2v(x)\big] + \big\langle f(t,x),\nabla v(x)\big\rangle,$$

and D is an open connected bounded subset of $\mathbb{R}^d$ of the form

$$D = \left\{x \in \mathbb{R}^d : \phi\,(x) < 0\right\},\ \ \mathrm{Bd}\,(D) = \left\{x \in \mathbb{R}^d : \phi\,(x) = 0\right\},$$

where $\phi \in C_b^3\left(\mathbb{R}^d\right)$, $|\nabla\phi\,(x)| = 1$, for all $x \in \mathrm{Bd}\,(D)$. The outward normal derivative of v at the point $x \in \mathrm{Bd}\,(D)$ is given by

$$\frac{\partial v\,(x)}{\partial n} = \sum_{j=1}^{d} \frac{\partial\phi\,(x)}{\partial x_j}\frac{\partial v\,(x)}{\partial x_j} = \langle\nabla\phi\,(x)\,,\nabla v\,(x)\rangle\,.$$

The functions $f : \mathbb{R}_+ \times \mathbb{R}^d \to \mathbb{R}^d$, $g : \mathbb{R}_+ \times \mathbb{R}^d \to \mathbb{R}^{d\times d}$, $F : \mathbb{R}_+ \times \overline{D} \times \mathbb{R} \times \mathbb{R}^d \to \mathbb{R}$, $G : \mathbb{R}_+ \times \mathrm{Bd}\,(D) \times \mathbb{R} \to \mathbb{R}$ and $\kappa : \overline{D} \to \mathbb{R}$ are continuous. Assume that for all $T > 0$, there exist $a, L > 0$ (which can depend on T) such that $\forall t \in [0,T]\,,\ \forall x,\tilde{x} \in \mathbb{R}^d$:

$$|f\,(t,x) - f\,(t,\tilde{x})| + |g\,(t,x) - g\,(t,\tilde{x})| \le a\,|x - \tilde{x}|\,, \tag{5.228}$$

and $\forall t \in [0,T]$, $\forall x \in \overline{D}$, $x' \in \mathrm{Bd}\,(D)$, $y,\tilde{y} \in \mathbb{R}$, $z,\tilde{z} \in \mathbb{R}^d$:

$$\begin{array}{rl} (i) & (y-\tilde{y})\,[F(t,x,y,z) - F(t,x,\tilde{y},z)] \le \frac{L}{2}|y-\tilde{y}|^2, \\ (ii) & |F(t,x,y,z) - F(t,x,y,\tilde{z})| \le \sqrt{\frac{L}{2}}|z-\tilde{z}|, \\ (iii) & |F(t,x,y,0)| \le \frac{L}{2}\,(1+|y|)\,, \\ (iv) & (y-\tilde{y})\,[G(t,x',y) - G(t,x',\tilde{y})] \le L|y-\tilde{y}|^2, \\ (v) & |G(t,x',y)| \le L\,(1+|y|)\,. \end{array} \tag{5.229}$$

We also assume that

$$\begin{array}{rl} (i) & \varphi,\psi : \mathbb{R} \to (-\infty,+\infty] \text{ are proper convex l.s.c. functions,} \\ (ii) & 0 = \varphi\,(0) \le \varphi\,(y) \text{ and } 0 = \psi\,(0) \le \psi\,(y)\,,\ \forall\ y \in \mathbb{R}, \\ (iii) & \kappa(x) \in \mathrm{int}\,(\mathrm{Dom}\,(\varphi)) \cap \mathrm{int}\,(\mathrm{Dom}\,(\psi)) \text{ for all } x \in \overline{D}, \end{array} \tag{5.230}$$

and the *compatibility conditions*:
for all $\varepsilon > 0$, $t \geq 0$, $x \in \operatorname{Bd}(D)$, $\tilde{x} \in \overline{D}$, $y \in \mathbb{R}$ and $z \in \mathbb{R}^d$

$$\begin{aligned}&(i)\ \nabla\varphi_\varepsilon(y)\, G(t,x,y) \leq |\nabla\psi_\varepsilon(y)|\, |G(t,x,y)|,\\&(ii)\ \nabla\psi_\varepsilon(y)\, F(t,\tilde{x},y,z) \leq |\nabla\varphi_\varepsilon(y)|\, |F(t,\tilde{x},y,z)|,\end{aligned} \tag{5.231}$$

where $a^+ = \max\{0,a\}$ and $\nabla\varphi_\varepsilon(y)$, $\nabla\psi_\varepsilon(y)$ are the unique solutions U and V, respectively, of equations

$$\partial\varphi(y-\varepsilon U) \ni U \quad \text{and} \quad \partial\psi(y-\varepsilon V) \ni V$$

(for the Moreau–Yosida approximations $\nabla\varphi_\varepsilon$, $\nabla\psi_\varepsilon$ see section "Convex function" from Annex B and for the compatibility conditions see Example 5.63). We mention that in the one dimensional case (which is our case here)

$$\partial\varphi(y) = \left[\varphi'_-(y), \varphi'_+(y)\right] \quad \text{and} \quad \partial\psi(y) = \left[\psi'_-(y), \psi'_+(y)\right].$$

Since $\overline{D}$ is bounded and κ is continuous it follows from (5.230-*iii*) that there exists an $M_0 > 0$ such that

$$\sup_{x\in\overline{D}} |\kappa(x)| + \sup_{x\in\overline{D}} \varphi(\kappa(x)) + \sup_{x\in\overline{D}} \psi(\kappa(x)) \leq M_0.$$

Let $(t,x) \in [0,T] \times \overline{D}$ be arbitrarily fixed. Consider the stochastic basis $\left(\Omega, \mathcal{F}, \mathbb{P}, (\mathcal{F}^t_s)_{s\geq 0}\right)$, where the filtration is generated by a d-dimensional Brownian motion as follows: $\mathcal{F}^t_s = \mathcal{N}$ if $0 \leq s \leq t$ and

$$\mathcal{F}^t_s = \sigma\{B_r - B_t : t \leq r \leq s\} \vee \mathcal{N}, \quad \text{if } s > t.$$

From Theorem 4.54 and Theorem 4.47 we infer that there exists a unique pair $(X^{t,x}, A^{t,x}) : \Omega \times [0,\infty[\to \mathbb{R}^d \times \mathbb{R}^d$ of continuous progressively measurable stochastic processes such that, $\mathbb{P}$-a.s.:

$$\begin{cases}(j)\ X^{t,x}_s \in \overline{D} \text{ and } X^{t,x}_{s\wedge t} = x \text{ for all } s \geq 0,\\(jj)\ 0 = A^{t,x}_u \leq A^{t,x}_s \leq A^{t,x}_v \text{ for all } 0 \leq u \leq t \leq s \leq v,\\(jjj)\ X^{t,x}_s + \displaystyle\int_t^s \nabla\phi\left(X^{t,x}_r\right) dA^{t,x}_r = x + \int_t^s f\left(r, X^{t,x}_r\right) dr\\ \qquad\qquad\qquad\qquad + \displaystyle\int_t^s g\left(r, X^{t,x}_r\right) dB_r, \quad \forall\, s \geq t,\\(jv)\ A^{t,x}_s = \displaystyle\int_t^s \mathbf{1}_{\operatorname{Bd}(\overline{D})}\left(X^{t,x}_r\right) dA^{t,x}_r, \quad \forall\, s \geq t.\end{cases} \tag{5.232}$$

Then by Proposition 4.55 and Corollary 4.56 we have for all $p \geq 1$, $\lambda > 0$ and $s \geq t$,

$$
\begin{aligned}
&(j)\quad \mathbb{E}\sup_{r\in[t,s]}|X_r^{t,x}-X^{t,y}|^p+\mathbb{E}\sup_{r\in[t,s]}\left|A_r^{t,x}-A_r^{t,y}\right|^p\le Ce^{C(s-t)}\,|x-y|^p\,,\\
&(jj)\quad \mathbb{E}e^{\lambda A_s^{t,x}}\le\exp\left(C\lambda+C\lambda t+\frac{C^2\lambda^2}{2}t\right),
\end{aligned}
\tag{5.233}
$$

and for every $T>0$, $p\ge 1$ and continuous functions $h_1,h_2:[0,T]\times\overline{D}\to\mathbb{R}$, the mappings

$$(t,x)\mapsto\left(X_\cdot^{t,x},A_\cdot^{t,x}\right):[0,T]\times\overline{D}\to S_d^p\left[0,T\right]\times S_1^p\left[0,T\right]$$

and

$$(t,x)\mapsto\mathbb{E}\int_t^T h_1(s,X_s^{t,x})ds+\mathbb{E}\int_t^T h_2(s,X_s^{t,x})\,dA_s^{t,x}:[0,T]\times\overline{D}\to\mathbb{R}$$

are continuous.

We consider the backward stochastic variational inequality (BSVI):

$$
\begin{cases}
Y_s^{t,x}+\displaystyle\int_s^T dK_r^{t,x}=\kappa\left(X_T^{t,x}\right)+\int_s^T 1_{[t,T]}(r)\;F\left(r,X_r^{t,x},Y_r^{t,x},Z_r^{t,x}\right)dr,\\
\qquad+\displaystyle\int_s^T 1_{[t,T]}(r)\,G\left(r,X_r^{t,x},Y_r^{t,x}\right)dA_r^{t,x}-\int_s^T\langle Z_r^{t,x},dB_r\rangle\,,\ \forall\, s\in[0,T]\,,\\
Y_s^{t,x}=Y_t^{t,x},\ Z_s^{t,x}=0,\ K_s^{t,x}=U_s^{t,x}=V_s^{t,x}=0,\ \forall\, s\in[0,t]\,,\\
K_s^{t,x}=\displaystyle\int_0^s\left(U_r^{t,x}dr+V_r^{t,x}dA_r^{t,x}\right),\quad\forall\, s\in[0,T]\,,\\
U_s^{t,x}\in\partial\varphi\left(Y_s^{t,x}\right)\text{ and }V_s^{t,x}\in\partial\psi\left(Y_s^{t,x}\right)\quad a.e.\text{ on }\ \Omega\times[t,T]\,.
\end{cases}
\tag{5.234}
$$

Note that the backward stochastic variational inequality (5.234) satisfies the assumptions of Theorem 5.69 and Remark 5.70 with $\tau=T$, $\eta=\kappa\left(X_T^{t,x}\right)$ satisfying (A_7''), $\mu_s=\frac{L}{2}\mathbf{1}_{[0,T]}(s)$, $\ell_s=\sqrt{\frac{L}{2}\mathbf{1}_{[0,T]}(s)}$, $\nu_s=L\mathbf{1}_{[0,T]}(s)$, $V_s=LQ_{s\wedge T}^{t,x}$, $u_0=0$, where

$$Q_s^{t,x}=s+A_s^{t,x}\quad\text{and}\quad\mathbb{E}\left(e^{\lambda Q_T^{t,x}}\right)<\infty,\text{ for all }\lambda>0.$$

Therefore (5.234) has a unique solution $(Y^{t,x},Z^{t,x},K^{t,x})$ of continuous progressively measurable stochastic processes such that

$$\mathbb{E}\sup_{r\in[t,T]}e^{2LQ_r^{t,x}}\left|Y_r^{t,x}\right|^2+\mathbb{E}\left(\int_t^T e^{2LQ_r^{t,x}}\left|Z_r^{t,x}\right|^2dr\right)<\infty,$$

and $dK_s^{t,x}=U_s^{t,x}ds+V_s^{t,x}dA_s^{t,x}$, where $U^{t,x},V^{t,x}$ are progressively measurable stochastic processes and $U_s^{t,x}\in\partial\varphi\left(Y_s^{t,x}\right)$, $V_s^{t,x}\in\partial\varphi\left(Y_s^{t,x}\right)$ $d\mathbb{P}\otimes dt-a.e.$ on $\Omega\times[t,T]$.

Moreover by (5.162) and (5.164) the solution satisfies for all $s \in [t, T]$:

$$\begin{aligned}
&\mathbb{E}^{\mathcal{F}_s} \sup_{r\in[s,T]} e^{2LQ_r^{t,x}} |Y_r^{t,x}|^2 + \mathbb{E}^{\mathcal{F}_s} \left(\int_s^T e^{2LQ_r^{t,x}} |Z_r^{t,x}|^2 dr \right) \\
&\quad + \mathbb{E}^{\mathcal{F}_s} \int_s^T e^{2LQ_r^{t,x}} \left[\varphi\left(Y_r^{t,x}\right) dr + \left|\psi\left(Y_r^{t,x}\right)\right| dA_r^{t,x} \right] \\
&\leq C_2 \mathbb{E}^{\mathcal{F}_s} \Big[e^{2LQ_T^{t,x}} \left|\kappa\left(X_T^{t,x}\right)\right|^2 \\
&\quad + \Big(\int_s^T e^{LQ_r^{t,x}} \left(\left|F\left(r, X_r^{t,x}, 0, 0\right)\right| dr + \left|G\left(r, X_r^{t,x}, 0, \right)\right| dA_r^{t,x} \right) \Big)^2 \Big] \\
&\leq C_2\, \mathbb{E}^{\mathcal{F}_s} \left[e^{2LQ_T^{t,x}} M_0 + \left(e^{LQ_T^{t,x}} - e^{LQ_s^{t,x}} \right)^2 \right] \\
&\leq C_{M_0} \mathbb{E}^{\mathcal{F}_s} \left(e^{2LQ_T^{t,x}} \right),
\end{aligned} \tag{5.235}$$

and

$$\begin{aligned}
&\mathbb{E}\left[e^{2LQ_s^{t,x}} \varphi(Y_s^{t,x}) + \psi(Y_s^{t,x}) \right] + \frac{1}{2}\mathbb{E} \int_s^T e^{2LQ_r^{t,x}} \left(|U_r^{t,x}|^2 dr + |V_r^{t,x}|^2 dA_r \right) \\
&\leq \mathbb{E}\left[e^{2LQ_T^{t,x}} \left(\varphi(\kappa\left(X_T^{t,x}\right)) + \psi(\kappa\left(X_T^{t,x}\right)) \right) \right] \\
&\quad + 4\mathbb{E} \int_s^T e^{2LQ_r^{t,x}} \left(|F(r, Y_r^{t,x}, Z_r^{t,x})|^2 \right) dr + 4\mathbb{E} \int_s^T e^{2LQ_r^{t,x}} \left(|G(r, Y_r^{t,x})|^2 \right) dA_r \\
&\leq C_{M_0, L} \mathbb{E} e^{2LQ_T^{t,x}}.
\end{aligned} \tag{5.236}$$

We define

$$u(t, x) = Y_t^{t,x}, \quad (t, x) \in [0, T] \times \overline{D}, \tag{5.237}$$

which is a deterministic quantity since $Y_t^{t,x}$ is $\mathcal{F}_t^t \equiv \mathcal{N}$-measurable.

From the Markov property, we have

$$u(s, X_s^{t,x}) = Y_s^{t,x}.$$

By (5.236) we infer that

$$u(t, x) \in \mathrm{Dom}\,(\varphi) \cap \mathrm{Dom}\,(\psi) \ \text{ for all } (t, x) \in [0, T] \times \overline{D}. \tag{5.238}$$

In the sequel we shall prove that u defined by (5.237) is a viscosity solution of (5.227). Reversing the time by $\tilde{u}(t, x) = u(T - t, x)$, the PVI (5.227) becomes (6.137) and the uniqueness of the viscosity solution follows from Theorem 6.112.

We now give the definition of the viscosity solution of the PVI (5.227).

A triple $(p, q, X) \in \mathbb{R} \times \mathbb{R}^d \times \mathbb{S}^d$ is a parabolic super-jet to u at $(t, x) \in (0, T) \times \overline{D}$ if for all $(s, y) \in (0, T) \times \overline{D}$

$$u(s,y) \le u(t,x) + p(s-t) + \langle q, y-x\rangle + \tfrac{1}{2}\langle X(y-x), y-x\rangle \\ +o(|s-t| + |y-x|^2);$$

the set of parabolic super-jets is denoted $\mathcal{P}^{2,+}u(t,x)$. The set of parabolic sub-jets is defined by $\mathcal{P}_{\mathcal{O}}^{2,-}u = -\mathcal{P}_{\mathcal{O}}^{2,+}(-u)$.

Let

$$\Phi(t,x,y,q,X) = -\frac{1}{2}\mathrm{Tr}\left((gg^*)(t,x)X\right) - \langle f(t,x), q\rangle - F(t,x,y,qg(t,x)),$$

$$\Gamma(t,x,y,q) = \langle \nabla\phi(x), q\rangle - G(t,x,y).$$

We clearly have

$$\Phi\left(s,y,r,\nabla v(y), D^2v(y)\right) = -\mathcal{A}_s v(y) - F(s,y,r,\nabla v(y)g(s,y)). \tag{5.239}$$

Definition 5.80. Let $u : [0,T]\times\overline{D} \to \mathbb{R}$ be a continuous function, which satisfies $u(T,x) = \kappa(x)$, $\forall\, x \in \overline{D}$.

(a) u is a viscosity sub-solution of (5.227) if:

$$\left|\begin{array}{l} u(t,x) \in \mathrm{Dom}(\varphi), \quad \forall (t,x) \in (0,T)\times\overline{D}, \\ u(t,x) \in \mathrm{Dom}(\psi), \quad \forall (t,x) \in (0,T)\times \mathrm{Bd}(D), \end{array}\right.$$

and for any $(t,x) \in (0,T)\times\overline{D}$ and any $(p,q,X) \in \mathcal{P}^{2,+}u(t,x)$:

$$\begin{cases} (d_1) & p + \Phi(t,x,u(t,x),q,X) + \varphi'_-(u(t,x)) \le 0 \text{ if } x \in D, \\ (d_2) & \min\Big\{p + \Phi(t,x,u(t,x),q,X) + \varphi'_-(u(t,x)), \\ & \quad \Gamma(t,x,u(t,x),q) + \psi'_-(u(t,x))\Big\} \le 0 \text{ if } x \in \mathrm{Bd}(D). \end{cases} \tag{5.240}$$

(b) The viscosity super-solution of (5.227) is defined in a similar manner as above, with $\mathcal{P}^{2,+}$ replaced by $\mathcal{P}^{2,-}$, the left derivative replaced by the right derivative, min by max, and the inequalities $\le$ by $\ge$.

(c) A continuous function $u : [0,\infty)\times\overline{D}$ is a viscosity solution of (6.137) if it is both a viscosity sub- and super-solution.

Theorem 5.81. *Let the assumptions (5.228), (5.229), (5.230) and (5.231) be satisfied. If u defined by (5.237) is continuous on $[0,T]\times\overline{D}$, then u is a viscosity solution of PVI (5.227).*

Proof. We show only that u is a viscosity sub-solution of (5.227) (the proof of the super-solution property is similar).

Let $(t,x) \in [0,T]\times\overline{D}$ and $(p,q,X) \in \mathcal{P}^{2,+}u(t,x)$.

Cases. $(t,x) \in [0,T] \times \mathrm{Bd}(D)$.

Aiming to deduce a contradiction we suppose that

$$\min\Big\{-p+\Phi(t,x,u(t,x),q,X)+\varphi'_-\big(u(t,x)\big),\Gamma(t,x,u(t,x),q)+\psi'_-\big(u(t,x)\big)\Big\}>0.$$

It follows by continuity of F, G, u, f, g, ϕ, Φ, Γ left continuity and nondecreasing monotonicity of φ'_- and ψ'_- that there exists $\varepsilon > 0$, $\delta > 0$ such that for all $(s,x') \in [0,T] \times \overline{D}$, $|s-t| \le \delta$, $|x'-x| \le \delta$,

$$-(p+\varepsilon)+\Phi\big(s,x',u(s,x'),q+(X+\varepsilon I)\left(x'-x\right),X+\varepsilon I\big)+\varphi'_-\big((u(s,x')\big)>0 \tag{5.241}$$

and

$$\Gamma(s,x',u(s,x'),q+(X+\varepsilon I)\left(x'-x\right))+\psi'_-\big(u(s,x')\big)>0. \tag{5.242}$$

Now since $(p,q,X) \in \mathcal{P}^{2,+}u(t,x)$ there exists $0 < \delta' \le \delta$ such that for all $s \in [0,T]$, $s \ne t$, $x' \in \overline{D}$, $x' \ne x$, $|s-t| \le \delta'$, $|x'-x| \le \delta'$ we have

$$\begin{aligned} u(s,x') < \hat{u}(s,x') \overset{def}{=} u(t,x) &+ (p+\varepsilon)(s-t) + \langle q, x'-x\rangle \\ &+ \frac{1}{2}\big\langle (X+\varepsilon I)(x'-x), x'-x\big\rangle. \end{aligned}$$

By (5.239) the condition (5.241) becomes

$$-\frac{\partial \hat{u}(r,x')}{\partial t}-\mathcal{A}_s\hat{u}(s,x')-F\left(s,x',u(s,x'),\nabla\hat{u}(s,x')g(s,x')\right)+\varphi'_-\big((u(s,x')\big)>0. \tag{5.243}$$

The condition (5.242) can be written as follows

$$\big\langle\nabla\hat{u}(s,x'),\nabla\phi(x')\big\rangle-G(s,x',u(s,x'))+\psi'_-\big(u(s,x')\big)>0. \tag{5.244}$$

Let

$$\theta \overset{def}{=} \left(t+\delta'\right)\wedge\inf\left\{s>t:\ |X_s^{t,x}-x|\ge\delta'\right\}.$$

We note that $(Y_s^{t,x}, Z_s^{t,x})$, $t \le s \le \theta$, solves the BSDE

$$\begin{cases} Y_s^{t,x} = u\left(\theta, X_\theta^{t,x}\right) + \displaystyle\int_s^\theta \left(F(r,X_r^{t,x},Y_r^{t,x},Z_r^{t,x}) - U_r^{t,x}\right)dr - \int_s^\theta Z_r^{t,x}dB_r \\ \qquad\qquad + \displaystyle\int_s^\theta \left(G(r,X_r^{t,x},Y_r^{t,x}) - V_r^{t,x}\right)dA_r^{t,x}, \\ U_s^{t,x} \in \partial\varphi\left(Y_s^{t,x}\right) \text{ and } V_s^{t,x} \in \partial\psi\left(Y_s^{t,x}\right) \quad d\mathbb{P}\otimes dt\text{-}a.e. \end{cases}$$

Moreover, it follows from Itô's formula that

$$(\hat{Y}_s^{t,x}, \hat{Z}_s^{t,x}) = \left(\hat{u}(s, X_s^{t,x}), (\nabla\hat{u}g)(s, X_s^{t,x})\right), \ t \le s \le \theta$$

satisfies

$$\hat{Y}_s^{t,x} = \hat{u}(\theta, X_\theta^{t,x}) - \int_s^\theta \Big[\frac{\partial\hat{u}(r, X_r^{t,x})}{\partial t} + \mathcal{A}_r\hat{u}(r, X_r^{t,x})\Big]dr - \int_s^\theta \hat{Z}_r^{t,x}dB_r$$
$$+ \int_s^\theta \left\langle \nabla_x\hat{u}\left(r, X_r^{t,x}\right), \nabla\phi(X_r^{t,x})\right\rangle dA_r^{t,x}.$$

Let $(\tilde{Y}_s^{t,x}, \tilde{Z}_s^{t,x}) = (\hat{Y}_s^{t,x} - Y_s^{t,x}, \hat{Z}_s^{t,x} - Z_s^{t,x})$. We have

$$\tilde{Y}_s^{t,x} = \left[\hat{u}(\theta, X_\theta^{t,x}) - u\left(\theta, X_\theta^{t,x}\right)\right] + \int_s^\theta \Big[- \frac{\partial\hat{u}(r, X_r^{t,x})}{\partial t} - \mathcal{A}_r\hat{u}(r, X_r^{t,x})$$
$$-F(r, X_r^{t,x}, Y_r^{t,x}, Z_r^{t,x}) + U_r^{t,x}\Big]dr - \int_s^\theta \tilde{Z}_r^{t,x}dB_r$$
$$+ \int_s^\theta \Big[\left\langle\nabla_x\hat{u}(r, X_r^{t,x}), \nabla\phi(X_r^{t,x})\right\rangle - G(r, X_r^{t,x}, Y_r^{t,x}) + V_r^{t,x}\Big]dA_r^{t,x}.$$

Let

$$\beta_s = \mathcal{A}_s\hat{u}(s, X_s^{t,x}) + F(s, X_s^{t,x}, Y_s^{t,x}, Z_s^{t,x}) \ \text{ and}$$
$$\hat{\beta}_s = \mathcal{A}_s\hat{u}(s, X_s^{t,x}) + F(s, X_s^{t,x}, Y_s^{t,x}, \hat{Z}_s^{t,x}).$$

Since $|\hat{\beta}_s - \beta_s| \le \sqrt{\frac{L}{2}}\,|\hat{Z}_s^{t,x} - Z_s^{t,x}|$, there exists a bounded d-dimensional p.m.s.p. $\{\zeta_s; t \le s \le \theta\}$ such that $\hat{\beta}_s - \beta_s = \langle\zeta_s, \tilde{Z}_s^{t,x}\rangle$ and therefore

$$\tilde{Y}_s^{t,x} = \hat{u}(\theta, X_\theta^{t,x}) - u(\theta, X_\theta^{t,x}) + \int_s^\theta \Big[- \frac{\partial\hat{u}(r, X_r^{t,x})}{\partial t} + \langle\zeta_r, \tilde{Z}_r^{t,x}\rangle - \hat{\beta}_r + U_r^{t,x}\Big]dr$$
$$+ \int_s^\theta \Big[\left\langle\nabla_x\hat{u}(r, X_r^{t,x}), \nabla\phi(X_r^{t,x})\right\rangle - g(r, X_r^{t,x}, Y_r^{t,x}) + V_r^{t,x}\Big]dA_r^{t,x} - \int_s^\theta \tilde{Z}_r^{t,x}dB_r.$$

Let

$$\Lambda_s = \exp\left(\int_t^s \langle\zeta_r, dB_r\rangle - \frac{1}{2}\int_t^s |\zeta_r|^2\,dr\right), \ t \le s \le \theta.$$

Then by Itô's formula,

$$\Lambda_s = 1 + \int_t^s \Lambda_r \ \langle\zeta_r, dB_r\rangle, \ t \le s \le \theta,$$

and so

$$d(\tilde{Y}_r^{t,x}\,\Lambda_r) = \Lambda_r\Big[\frac{\partial \hat{u}(r,X_r^{t,x})}{\partial t} + \hat{\beta}_r - U_r^{t,x}\Big]dr + \Lambda_r\langle \tilde{Z}_r^{t,x} + \tilde{Y}_r^{t,x}\zeta_r, dB_r\rangle$$
$$+ \Lambda_r\Big[-\big\langle \nabla_x \hat{u}(r,X_r^{t,x}), \nabla\phi(X_r^{t,x})\big\rangle + g(r,X_r^{t,x},Y_r^{t,x}) - V_r^{t,x}\Big]dA_r^{t,x}.$$

Then

$$\tilde{Y}_t^{t,x} = \mathbb{E}\Big[\Lambda_\theta\ \big(\hat{u}(\theta,X_\theta^{t,x}) - u(\theta,X_\theta^{t,x})\big)\Big] + \mathbb{E}\int_t^\theta \Lambda_r\Big[-\tfrac{\partial \hat{u}(r,X_r^{t,x})}{\partial t} - \hat{\beta}_r + U_r^{t,x}\Big]dr$$
$$+\mathbb{E}\int_t^\theta \Lambda_r\Big[\big\langle \nabla_x \hat{u}(r,X_r^{t,x}), \nabla\phi(X_r^{t,x})\big\rangle - g(r,X_r^{t,x},Y_r^{t,x}) + V_r^{t,x}\Big]dA_r^{t,x}. \tag{5.245}$$

Since $U_r^{t,x} \in \partial\varphi\left(Y_r^{t,x}\right)$ and $V_r^{t,x} \in \partial\psi\left(Y_r^{t,x}\right)$, we have

$$U_r^{t,x}dr \geq \varphi'_-\big(u(r,X_r^{t,x})\big)dr, \quad V_r^{t,x}dA_r^{t,x} \geq \psi'_-\big(u(r,X_r^{t,x})\big)dA_r^{t,x},$$

and therefore by (5.243) and (5.244)

$$-\frac{\partial \hat{u}(r,X_r^{t,x})}{\partial t} - \hat{\beta}_r + U_r^{t,x} > 0,$$

and

$$\Big[\big\langle \nabla_x \hat{u}(r,X_r^{t,x}), \nabla\phi(X_r^{t,x})\big\rangle - g(r,X_r^{t,x},Y_r^{t,x}) + V_r^{t,x}\Big]dA_r^{t,x} \geq 0.$$

Moreover, the choice of δ' and θ implies that $u(\theta,X_\theta^{t,x}) < \hat{u}(\theta,X_\theta^{t,x})$. Hence

$$0 = \hat{u}\,(t,x) - u\,(t,x) = \tilde{Y}_t^{t,x} \geq \mathbb{E}\Big[\Lambda_\theta\ \big(\hat{u}(\theta,X_\theta^{t,x}) - u(\theta,X_\theta^{t,x})\big)\Big] > 0,$$

which is a contradiction. It follows that (5.240-d_2) holds.

Cases. $(t,x) \in [0,T] \times D$.

The proof follows the same steps from *Case 5.8,* where we now choose δ and δ' such that $\overline{B\,(x,\delta')} \subset \overline{B\,(x,\delta)} \subset D$ and, by condition (5.232-*iv*), $A_r^{t,x} = 0$ for all $t \leq r \leq \theta$.

This proves that u is a viscosity sub-solution of PVI (5.227). Symmetric arguments show that u is also a super-solution; hence u is a viscosity solution of PVI (5.227).

Corollary 5.82. *We have*

$$u(t,x) \in \text{Dom}\,(\partial\varphi)\,, \quad \forall (t,x) \in [0,T] \times D.$$

Proof. Let $(t,x) \in [0,T] \times D$ be fixed. We have two cases:

(1) $\mathrm{Dom}\,(\partial\varphi) = \mathrm{Dom}\,(\varphi)$, and so, from (5.238), $u(t,x) \in \mathrm{Dom}\,(\partial\varphi)$.
(2) $\mathrm{Dom}\,(\partial\varphi) \neq \mathrm{Dom}\,(\varphi)$. Let $b \in \mathrm{Dom}\,\varphi \setminus \mathrm{Dom}\,(\partial\varphi)$. Then $b = \sup(\mathrm{Dom}\,\varphi)$ or $b = \inf \mathrm{Dom}\,\varphi$. If $b = \sup(\mathrm{Dom}\,\varphi)$ and $u(t,x) = b$, then $(0,0,0) \in \mathcal{P}^{2,+}u(t,x)$ since

$$u(s,y) \leq u(t,x) + o\big(|s-t| + |y-x|^2\big)$$

and from (6.143-d_1) it follows that $\varphi'_-(b) = \varphi'_-\big(u(t,x)\big) < \infty$ and consequently $b \in \mathrm{Dom}\,(\partial\varphi)$; a contradiction which shows that $u(t,x) < b$. Similarly for $b = \inf(\mathrm{Dom}\,\varphi)$. ■

The problem now is to see when $(t,x) \longmapsto u(t,x) = Y_t^{t x} : [0,T] \times \overline{D} \to \mathbb{R}$ is continuous. A recent result of Maticiuc and Răşcanu [46] gives a sufficient condition for u to be continuous. The idea is to show that if $(t_n, x_n) \to (t,x)$ then $(Y_\cdot^{t_n,x_n})_{n\in\mathbb{N}^*}$ is tight in the Skorohod space $\mathbb{D}\left([0,T], \mathbb{R}\right)$ of càdlàg functions endowed with the S-topology (introduced by Jakubowski in [41]). This topology makes the mapping $y \longmapsto \int_0^s G\left(r, y\left(r\right)\right) dA_r$ continuous from $\mathbb{D}\left([0,T], \mathbb{R}\right)$ into $\mathbb{R}$. The result is the following:

Proposition 5.83. *Let the assumptions (5.228), ..., (5.231) be satisfied. If moreover there exists an $L_0 > 0$ such that*

$$\begin{array}{ll} (i) & F \text{ is independent of } z,\\ (ii) & g\left(t,x\right) \text{ is an invertible matrix, for all } (t,x) \in [0,T] \times \overline{D},\\ (iii) & |G\left(t,x,y\right) - G\left(t',x',y'\right)| \leq L_0\left(|t-t'| + |x-x'| + |y-y'|\right)\\ & \qquad \text{for all } t,t' \in [0,T],\ x,x' \in \mathrm{Bd}\left(\mathcal{D}\right),\ y,y' \in \mathbb{R} \end{array} \tag{5.246}$$

then the function

$$(t,x) \mapsto u\left(t,x\right) = Y_t^{t,x} : [0,T] \times \overline{\mathcal{D}} \to \mathbb{R}$$

is continuous.

Finally let f, g, F, G be independent of t and $(X_s^x,\ A_s^x,\ Y_s^{x;t},\ Z_s^{x;t},\ U_s^{x;t},\ V_s^{x;t})_{0\leq s\leq t}$ be defined by

$$\begin{cases} (j)\ X_s^x \in \overline{D} \text{ for all } s \geq 0,\\ (jj)\ 0 = A_0^x \leq A_u^x \leq A_s^x \text{ for all } 0 \leq u \leq s,\\ (jjj)\ X_s^x + \displaystyle\int_0^s \nabla\phi(X_r^x) dA_r^x = x + \int_0^s f(X_r^x) dr + \int_0^s g(X_r^x) dB_r,\\ \qquad\qquad\qquad\qquad\qquad\qquad\qquad\qquad \forall\ s \geq 0,\\ (jv)\ A_s^x = \displaystyle\int_0^s \mathbf{1}_{\mathrm{Bd}(\overline{D})}\left(X_r^x\right) dA_r^x, \quad \forall\ s \geq 0, \end{cases}$$

and

$$\begin{cases} Y_s^{x;t} + \int_s^t (U_r^{x;t} dr + V_r^{x;t} dA_r^x) = \kappa(X_t^x) + \int_s^t F(X_r^x, Y_r^{x;t}, Z_r^{x;t}) dr, \\ \qquad + \int_s^t G(X_r^x, Y_r^{x;t}) dA_r^x - \int_s^t \langle Z_r^{x;t}, dB_r \rangle, \ \forall\, s \in [0,t], \\ U_s^{x;t} \in \partial\varphi(Y_s^{x;t}) \text{ and } V_s^{x;t} \in \partial\psi(Y_s^{x;t}) \quad a.e. \text{ on } \Omega \times [0,t]. \end{cases}$$

Summarizing Theorem 5.81 and Theorem 6.112 we have:

Theorem 5.84. *Let the assumptions (5.228),…,(5.231) be satisfied. Assume there exists a continuous function* $\mathbf{m} : [0,\infty) \to [0,\infty)$*,* $\mathbf{m}(0) = 0$*, such that*

$$\begin{array}{ll} (i) & yG(x,y) \leq 0, \ \forall x \in \mathrm{Bd}(D) \text{ and } y \in \mathbb{R}, \\ (ii) & \left|F(x,y) - F(x',y)\right| \leq \mathbf{m}(|x - x'|) \ \forall\, x, x' \in \overline{D} \text{ and } y \in \mathbb{R}. \end{array} \tag{5.247}$$

If $(t,x) \longmapsto u(t,x) \overset{def}{=} Y_0^{x;t} : [0,T] \times \overline{D} \to \mathbb{R}$ *is continuous (this is true in particular under the assumptions of Proposition 5.83), then u is the unique viscosity solution of the parabolic variational inequality*

$$\begin{cases} \dfrac{\partial u(t,x)}{\partial t} - \mathcal{A}u(t,x) + \partial\varphi(u(t,x)) \ni F(x, u(t,x), (\nabla u g)(t,x)), \ t > 0, \ x \in D, \\ \dfrac{\partial u(t,x)}{\partial n} + \partial\psi(u(t,x)) \ni G(x, u(t,x)), \ t > 0, \ x \in \mathrm{Bd}(D), \\ u(0,x) = \kappa(x), \ \ x \in \overline{D}, \end{cases}$$

where the operator $\mathcal{A}$ *is given by*

$$\mathcal{A}v(x) = \frac{1}{2}\mathrm{Tr}\big[g(x)g^*(x)D^2v(x)\big] + \big\langle f(x), \nabla v(x)\big\rangle.$$

5.9 Invariant Sets of BSDEs

Let $\{B_t : t \geq 0\}$ be a k-dimensional standard Brownian motion defined on some complete probability space $(\Omega, \mathcal{F}, \mathbb{P})$. We denote by $\{\mathcal{F}_t : t \geq 0\}$ the natural filtration generated by $\{B_t, t \geq 0\}$ and augmented by the $\mathbb{P}$-null sets of $\mathcal{F}$.

Let $x \in \mathbb{R}^d$, $0 \leq t \leq \tilde{T} \leq T$. Consider the SDE

$$\begin{cases} X_s^{t,x} = x + \int_t^s b(r, X_r^{t,x}) dr + \int_t^s \sigma(r, X_r^{t,x}) dB_r, \ t \leq s \leq T, \\ X_s^{t,x} = x, \quad 0 \leq s \leq t, \end{cases} \tag{5.248}$$

and the BSDE

$$\begin{cases} Y_s^{t,x} = \kappa(X_{\tilde{T}}^{t,x}) + \int_s^{\tilde{T}} F(r, X_r^{t,x}, Y_r^{t,x}, Z_r^{t,x})dr - \int_s^{\tilde{T}} Z_r^{t,x} dB_r, \\ \qquad\qquad\qquad\qquad\qquad\qquad\qquad\qquad t \leq s \leq \tilde{T}, \\ Y_s^{t,x} = \kappa(X_{\tilde{T}}^{t,x}), \quad \tilde{T} \leq s \leq T, \\ Y_s^{t,x} = Y_t^{t,x}, \quad 0 \leq s \leq t. \end{cases} \tag{5.249}$$

The aim of this section is to state necessary and sufficient conditions which guarantee that the solution of the BSDE (5.249) does not leave a given set

$$\mathcal{E} = \{E(t,x) \subset \mathbb{R}^m : (t,x) \in [0,T] \times \mathbb{R}^d\},$$

i.e., under which we have that for all $0 \leq t \leq \tilde{T} \leq T$, $x \in \mathbb{R}^d$ and $\kappa(X_{\tilde{T}}^{t,x}) \in E(\tilde{T}, X_{\tilde{T}}^{t,x})$ $a.s.\ \omega \in \Omega$:

$$Y_s^{t,x} \in E(s, X_s^{t,x}) \quad a.s.\ \omega \in \Omega, \quad \forall s \in [t, \tilde{T}].$$

As a by-product, we will derive a result on the existence of constrained viscosity solutions to some PDEs. Together with the Eqs. (5.248) and (5.249), we consider the following system of semilinear parabolic PDEs

$$\begin{cases} \dfrac{\partial u_i(t,x)}{\partial t} + \mathcal{A}(t)u_i(t,x) + f_i(t,x,u(t,x),\sigma^*(t,x)\nabla_x u_i(t,x)) = 0, \\ u(T,x) = \kappa(x), \quad (t,x) \in [0,T] \times \mathbb{R}^d,\ 1 \leq i \leq n, \end{cases} \tag{5.250}$$

with the second order differential operator

$$\begin{aligned} \mathcal{A}(t)\varphi(x) &= \frac{1}{2}\mathrm{Tr}[\sigma\sigma^*(t,x)D_x^2\varphi(x)] + < b(t,x), \nabla_x\varphi(x) > \\ &= \frac{1}{2}\sum_{j,\ell=1}^m (\sigma\sigma^*)_{j\ell}(t,x)\frac{\partial^2\varphi(x)}{\partial x_j \partial x_\ell} + \sum_{j=1}^m b_j(t,x)\frac{\partial\varphi(x)}{\partial x_j}, \quad \varphi \in C^2\left(\mathbb{R}^d\right), \end{aligned}$$

where $b : [0,T] \times \mathbb{R}^d \to \mathbb{R}^d$, $\sigma : [0,T] \times \mathbb{R}^d \to \mathbb{R}^{d \times k}$ and $f_i : [0,T] \times \mathbb{R}^d \times \mathbb{R}^m \times \mathbb{R}^k \to \mathbb{R}$, $1 \leq i \leq n$.

We make the following standard assumptions:

(AV_1) We assume that the functions $b : [0,T] \times \mathbb{R}^d \to \mathbb{R}^d$, $\sigma : [0,T] \times \mathbb{R}^d \to \mathbb{R}^{d \times k}$, $F : [0,T] \times \mathbb{R}^d \times \mathbb{R}^m \times \mathbb{R}^{m \times k} \to \mathbb{R}^m$ and $f : [0,T] \times \mathbb{R}^d \times \mathbb{R}^m \times \mathbb{R}^k \to \mathbb{R}^m$ are continuous and such that, for some constants $L > 0$ and $q \geq 2$,

$$|b(t,x) - b(t,\tilde{x})| + \|\sigma(t,x) - \sigma(t,\tilde{x})\| \leq L|x - \tilde{x}|,$$

$$\begin{aligned} &i)\ \left|F\left(t,x,y,z\right)\right| \leq L\left(1+|x|^{q}+|y|+\|z\|\right),\\ &ii)\ \left|F\left(t,x,y,z\right)-F\left(t,x,y,\tilde{z}\right)\right| \leq L\left\|z-\tilde{z}\right\|,\\ &iii)\ \left\langle F\left(t,x,y,z\right)-F\left(t,x,\tilde{y},z\right),y-\tilde{y}\right\rangle \leq L\left|y-\tilde{y}\right|^{2} \end{aligned}$$

and

$$\begin{aligned} &j)\ \left|f\left(t,x,y,u\right)\right| \leq L\left(1+|x|^{q}+|y|+|u|\right),\\ &jj)\ \left|f\left(t,x,y,u\right)-f\left(t,x,y,\tilde{u}\right)\right| \leq L\left|u-\tilde{u}\right|,\\ &jjj)\ \left\langle f\left(t,x,y,u\right)-f\left(t,x,\tilde{y},u\right),y-\tilde{y}\right\rangle \leq L\left|y-\tilde{y}\right|^{2}, \end{aligned}$$

for all $t \in [0,T]$, $x,\tilde{x} \in \mathbb{R}^d$, $y,\tilde{y} \in \mathbb{R}^m$, and $z,\tilde{z} \in \mathbb{R}^{m\times k}$, $u,\tilde{u} \in \mathbb{R}^k$.

(AV_2) We assume that $\kappa : \mathbb{R}^d \to \mathbb{R}^m$ is a Borel measurable function of at most polynomial growth, i.e., there are some $a > 0, q \geq 1$ such that, for all $x \in \mathbb{R}^d$,

$$|\kappa(x)| \leq a(1+|x|^q),\ \forall x \in \mathbb{R}^d.$$

We shall now recall some basic properties of forward and backward stochastic differential equations.

Proposition 5.85. *Under the assumptions (AV_1) and (AV_2):*

I. *Equations (1.1) and (1.2) have unique solutions $X^{t,x} \in S_d^2[0,T]$ and*

$$\left(Y^{t,x},Z^{t,x}\right) \in S_m^2[0,T] \times \Lambda_{m\times k}^2[0,T]$$

with $Z_s^{t,x} = 0$ for $s \in [0,t] \cup [\tilde{T},T]$ and the solutions satisfy:

II. *For all $p \geq 2$, there exist some constants $C_p > 0$, $q \in \mathbb{N}^*$, which don't depend on $t,t' \in [0,T]$ and $x,x' \in \mathbb{R}^m$, such that*

$$\begin{aligned} &a)\quad \mathbb{E}\left(\sup\nolimits_{s\in[0,T]}|X_s^{t,x}|^p\right) \leq C_p(1+|x|^p),\\ &b)\quad \mathbb{E}\left(\sup\nolimits_{s\in[0,T]}|X_s^{t,x}-X_s^{t',x'}|^p\right) \leq\\ &\leq C_p(1+|x|^p+|x'|^{pq})(|t-t'|^{p/2}+|x-x'|^p), \end{aligned} \tag{5.251}$$

and

$$\begin{aligned} &c)\quad \mathbb{E}\left(\sup\nolimits_{s\in[0,T]}|Y_s^{t,x}|^p\right) \leq C_p(1+|x|^{pq}),\\ &d)\quad \mathbb{E}\left(\sup\nolimits_{s\in[0,T]}|Y_s^{t,x}-Y_s^{t',x'}|^2\right) \leq C_2[\mathbb{E}|\kappa(X_T^{t,x})-\kappa(X_T^{t',x'})|^2,\\ &+\mathbb{E}\int_0^T|1_{[t,T]}(r)F(r,X_r^{t,x},Y_r^{t,x},Z_r^{t,x})\\ &\qquad -1_{[t',T]}(r)F(r,X_r^{t',x'},Y_r^{t,x},Z_r^{t,x})|^2dr]. \end{aligned}$$

III. *There are some Borel measurable functions $u : [0,T]\times\mathbb{R}^d \to \mathbb{R}^m$, and $v : [0,T]\times\mathbb{R}^d \to \mathbb{R}^{m\times d}$ such that for all $0 \leq t \leq s \leq \tilde{T} \leq T$*

$$Y_s^{t,x} = u\left(s \wedge \tilde{T}, X_{s\wedge\tilde{T}}^{t,x}\right), \quad Z_s^{t,x} = (\upsilon\sigma)\left(s \wedge \tilde{T}, X_{s\wedge\tilde{T}}^{t,x}\right), \quad d\mathbb{P}\otimes ds - a.e.$$

(see [30]).

For the convenience of the reader we recall the definition of a viscosity solution corresponding to the PDE (5.250).

Definition 5.86. a) A lower semicontinuous function $u : [0,T]\times\mathbb{R}^d \to \mathbb{R}^m$ is a *viscosity super-solution* of (5.250), if, firstly, $u_i(T,x) \ge \kappa_i(x)$, for all $x \in \mathbb{R}^d$, $1 \le i \le n$, and secondly, for any $1 \le i \le n$, $\varphi \in C^{1,2}((0,T)\times\mathbb{R}^d)$ and $(t,x) \in [0,T]\times\mathbb{R}^d$ such that $u_i - \varphi$ achieves a local minimum at (t,x), it holds that

$$\frac{\partial}{\partial t}\varphi(t,x) + \mathcal{A}(t)\varphi(t,x) + f_i(t,x,u(t,x),(\sigma^*\nabla\varphi)(t,x)) \le 0.$$

b) An upper semicontinuous function $u : [0,T]\times\mathbb{R}^d \to \mathbb{R}^m$ is a *viscosity sub-solution* of (5.250), if, firstly, $u_i(T,x) \le \kappa_i(x)$, for all $x \in \mathbb{R}^d$, $1 \le i \le n$, and secondly, for any $1 \le i \le n$, $\varphi \in C^{1,2}((0,T)\times\mathbb{R}^d)$ and $(t,x) \in [0,T]\times\mathbb{R}^d$ such that $u_i - \varphi$ attains a local maximum at (t,x), we have that

$$\frac{\partial}{\partial t}\varphi(t,x) + \mathcal{A}(t)\varphi(t,x) + f_i(t,x,u(t,x),(\sigma^*\nabla\varphi)(t,x)) \ge 0.$$

c) Finally, a continuous function $u : [0,T]\times\mathbb{R}^d \to \mathbb{R}^m$ is a *viscosity solution* of (5.250) if it is both a viscosity super-solution and a viscosity sub-solution of this equation.

From Sect. 5.4.1 of this chapter we have:

Proposition 5.87. *We suppose that the function f satisfies hypothesis (AV_1) and that $\kappa : \mathbb{R}^d \to \mathbb{R}^m$ is a continuous function satisfying (AV_2). Let $X^{t,x}$ and $(Y^{t,x}, Z^{t,x})$ be the solutions to (5.248) and (5.249), respectively, where the driver F of BSDE (1.2) is of the form*

$$F(t,x,y,z) = (f_1(t,x,y,z^*e_1),\ldots,f_n(t,x,y,z^*e_n)),$$

and e_i denotes the unit vector pointing in the i-th coordinate direction of $\mathbb{R}^m$. Then $u(t,x) = Y_t^{t,x}$, $(t,x) \in [0,T]\times\mathbb{R}^d$, is a deterministic continuous function of at most polynomial growth. This function is a viscosity solution to (5.250). Moreover if, for each $R > 0$, there exists a continuous function $\alpha_R : \mathbb{R}_+ \to \mathbb{R}$, $\alpha_R(0) = 0$, such that, for all t,y,z,x,x' with $|x| \le R$, $|x'| \le R$,

$$|f(t,x,y,z) - f(t,x',y,z)| \le \alpha_R(|x-x'|(1+\|z\|)), \tag{5.252}$$

then u is the unique viscosity solution in the class $C_{pol}([0,T]\times\mathbb{R}^d,\mathbb{R}^m)$.

We now give the notion of the viability property for BSDEs and PDEs. We recall some notations. For any closed set $S \subset \mathbb{R}^d$ we denote by $x \to d_S(x) = \min\{|x-y : y \in S\}$ the distance function to S, and for $x \in \mathbb{R}^d$, we denote by $\Pi_S(x) := \{z \in S : d_S(x) = |x-z|\}$ the set of projections of x on S.

For all $t \in [0,T]$, $x \in \mathbb{R}^d$, let $E(t,x)$ be a non-empty and closed subset of $\mathbb{R}^m$. We consider the following set of *moving* constraints

$$\mathcal{E} = \{E(t,x) : (t,x) \in [0,T] \times \mathbb{R}^d\}.$$

Definition 5.88 (Viability for BSDEs). The moving set $E(t,x)$, $(t,x) \in [0,T] \times \mathbb{R}^d$, is viable (invariant) for the BSDE (5.249) (or Eq. (5.249) is said to be $\mathcal{E}$-viable on $[0,T]$) if, for all $(t,x) \in [0,T] \times \mathbb{R}^d$, $\tilde{T} \in [t,T]$, and all Borel measurable functions $\kappa : \mathbb{R}^d \to \mathbb{R}^m$ of at most polynomial growth, such that $\kappa(\tilde{x}) \in E(\tilde{T},\tilde{x})$, $\mathbb{P} \circ \left[X_{\tilde{T}}^{t,s}\right]^{-1}(d\tilde{x})$-a.s., it holds that the solution of (1.2) satisfies

$$Y_s^{t,x} \in E(s, X_s^{t,x}), \quad \forall s \in [t,\tilde{T}], \ \mathbb{P}\text{-a.s.}$$

Viability for PDEs: Equation (5.250) is said to be $\mathcal{E}$-viable ($\mathcal{E}$-invariant) on $[0,T]$ if, for all $\tilde{T} \in [0,T]$ and $\kappa \in C_{pol}(\mathbb{R}^d,\mathbb{R}^m)$ such that $\kappa(\tilde{x}) \in E(\tilde{T},\tilde{x})$, for all $\tilde{x} \in \mathbb{R}^d$, it holds that there exists a viscosity solution $u \in C_{pol}([0,\tilde{T}] \times \mathbb{R}^d, \mathbb{R}^m)$ of (5.250) with time horizon $\tilde{T}$ and terminal condition $u(\tilde{T},x) = \kappa(x)$, $x \in R^m$, such that

$$u(t,x) \in E(t,x), \ \forall (t,x) \in [0,\tilde{T}] \times \mathbb{R}^d.$$

From Proposition 5.87 we see immediately that:

Remark 5.89. If BSDE (5.249) is $\mathcal{E}$-viable then PDE (5.250) is also $\mathcal{E}$-viable.

Therefore the next result also concerns constrained the BSDEs and the PDEs.

Theorem 5.90 (Viability Criterion for BSDEs). *Assume that (AV_1) and (AV_2) are satisfied and moreover*

(i) *the function $(t,x) \mapsto d^2_{E(t,x)}(y) : [0,T] \times \mathbb{R}^d \to \mathbb{R}$ is jointly upper semicontinuous,*

(ii) *there exist some constants $M > 0$, $p \geq 1$ such that*

$$d^2_{E(t,x)}(0) \leq M(1 + |x|^p), \ \forall (t,x) \in [0,T] \times \mathbb{R}^d.$$

Then the following assertions (c) and (cc) are equivalent:

(c) *Equation (5.249) is $\mathcal{E}$-viable on $[0,T]$.*

(cc) *For any sufficiently large $C > 0$ and for every $z \in \mathbb{R}^{m\times d}$, the function $h(t,x,y) = d^2_{E(t,x)}(y)$ is an upper semicontinuous viscosity sub-solution of the PDE*

$$\frac{\partial V(t,x,y)}{\partial t} + \mathcal{L}_z(t)V(t,x,y) + Cd^2_{E(t,x)}(y) = 0, \qquad (t,x,y) \in [0,T] \in \mathbb{R}^d \times \mathbb{R}^m. \tag{5.253}$$

In the above relation, $\mathcal{L}_z(t)$ denotes the following second order differential operator

$$\mathcal{L}_z(t)\varphi(x,y) = \frac{1}{2}\text{Tr}[\Gamma_z\Gamma_z^*(t,x,y)D^2_{(x,y)}\varphi(x,y)] + \left\langle B_z(t,x,y), \nabla_{(x,y)}\varphi(x,y)\right\rangle, \tag{5.254}$$

where

$$\Gamma_z(t,x,y) = \begin{pmatrix} \sigma(t,x) \\ z\sigma(t,x) \end{pmatrix}, \quad B_z(t,x,y) = \begin{pmatrix} b(t,x) \\ -F(t,x,y,z\sigma(t,x)) \end{pmatrix}.$$

This theorem yields:

Corollary 5.91 (Viability Criterion for BSDEs). *We assume that the moving sets of Theorem 5.90 are independent of the spatial variable, $E(t,x) \equiv E(t)$, $(t,x) \in [0,T] \times \mathbb{R}^m$. Then the following assertions (j) and (jj) are equivalent:*

- (j) *Equation (5.249) is $\mathcal{E}$-viable on $[0,T]$.*
- (jj) *The function $h(t,y) = d^2_{E(t)}(y)$ is an upper semicontinuous viscosity subsolution of the PDE:*

$$\frac{\partial V(t,y)}{\partial t} + \mathcal{A}_z(t;x)V(t,y) + Cd^2_{E(t)}(y) = 0, \quad (t,y) \in [0,T] \times \mathbb{R}^m,$$

for all $x \in \mathbb{R}^d$, $z \in \mathbb{R}^{m\times d}$, where

$$\mathcal{A}_z(t;x)\psi(y) = \frac{1}{2}\text{Tr}[z\sigma\sigma^*(t,x)z^* D^2_y\psi(y)] \ \left\langle F(t,x,y,z\sigma(t,x)), \nabla_y\psi(y)\right\rangle,$$

and $C > 0$ is any sufficiently large constant.

Before proving the main results stated above, we shall present some clarifying examples. In the first example we find a criterion such that a family of moving balls has the viability property for a given BSDE.

Example 5.92 (Control Security Tube). We consider an arbitrary function $r \in C^1([0,T];\mathbb{R}_+)$ with $r(t) > 0$ for all $t \in [0,T]$, and we put

$$E(t) = \{y \in \mathbb{R}^m : |y| \le r(t)\}, \quad t \in [0,T].$$

Then the square-distance function is

$$d^2_{E(t)}(y) = h_0(t,y) = ((|y| - r(t))^+)^2,$$

and, for $|y| > r(t)$, the operator $\mathcal{A}_z(t)$ applied to h_0 at (t, y) takes the form

$$\mathcal{A}_z(t)h_0(t,y) = \frac{|y|-r(t)}{|y|}|z\sigma(t,x)|^2 + \frac{r(t)}{|y|^3}|(z\sigma(t,x))^* y|^2 - 2\frac{|y|-r(t)}{|y|}\langle F(t,x,y,z\sigma(t,x)), y\rangle .$$

Hence, the inequality in Corollary 5.91(jj) yields that, for all $(t, x, y, z) \in [0, T] \times \mathbb{R}^d \times \mathbb{R}^m \times \mathbb{R}^{m\times d}$ with $|y| > r(t)$,

$$2\frac{|y|-r(t)}{|y|}\left[\langle F(t,x,y,z\sigma(t,x)), y\rangle + |y|\,r'(t)\right] \le \frac{|y|-r(t)}{|y|}\|z\sigma(t,x)\|^2 + \frac{r(t)}{|y|^3}|(z\sigma(t,x))^* y|^2 + C\,(|y|-r(t))^2,$$

from where we easily deduce the following necessary condition for the $\mathcal{E}$-viability of BSDE (5.249):

For all (t, x, y, z) with $|y| = r(t)$ and $(z\sigma(t,x))^* y = 0$,

$$2r(t)\,r'(t) + 2\langle F(t,x,y,z\sigma(t,x)), y\rangle \le \|z\sigma(t,x)\|^2 . \tag{5.255}$$

If the assumption $(AV_1\text{-}i)$ is replaced by

$$i') \qquad |F(t,x,y,z)| \le L\,(1+|y|),$$

for all (t, x, y, z), then this condition is not only necessary but also sufficient as the following argument proves. We fix any $(t, x, y, z) \in [0, T] \times \mathbb{R}^d \times \mathbb{R}^m \times \mathbb{R}^{m\times d}$ with $|y| > r(t)$, and for simplicity of notation we put $\overline{y} = |y|^{-1} r(t)\,y$ and, for $1 \le j \le m$

- $u_j = (z\sigma(t,x))_j = \sum_{i=1}^d z_{\cdot,i}\sigma_{i,j}(t,x)$,
- $\hat{u}_j = |y|^{-2}\langle u_j, y\rangle y, \quad u_j^\perp = u_j - \hat{u}_j$,
- $\hat{u} = (\hat{u}_1, \ldots, \hat{u}_m), \quad u^\perp = (u_1^\perp, \ldots, u_m^\perp) = u - \hat{u}$.

From the assumptions (AV_1) and $(AV_1\text{-}i')$ we get that, for some generic constant C which can change from line to line but does not depend on (t, x, y, z),

$$\begin{aligned}
&2\frac{|y|-r(t)}{|y|}\left(\langle F(t,x,y,u), y\rangle + |y|\,r'(t)\right) \\
&= 2\langle F(t,x,y,u), y-\overline{y}\rangle + 2\,(|y|-r(t))\,r'(t) \\
&\le 2\langle F(t,x,\overline{y},u), y-\overline{y}\rangle + 2\,(|y|-r(t))\,r'(t) + C\,(|y|-r(t))^2 \\
&\le 2\left\langle F(t,x,\overline{y},u^\perp), y-\overline{y}\right\rangle + 2\,(|y|-r(t))\,r'(t) + C\,(|y|-r(t))^2
\end{aligned}$$

$$+C\left(|y|-r\left(t\right)\right)|\hat{u}|$$

$$=2\frac{|y|-r(t)}{|y|}\left(\left\langle F(t,x,\overline{y},u^{\perp}),y\right\rangle+C\,|y|\,|\hat{u}|+|y|\,r'(t)\right)$$

$$+C\left(|y|-r\left(t\right)\right)^2$$

$$\leq 2\frac{|y|-r(t)}{|y|}\left(\left\langle F(t,x,\overline{y},u^{\perp}),\overline{y}\right\rangle+|y|\,r'(t)+C\left(|y|-r\left(t\right)\right)\right)$$

$$+C\left(|y|-r(t)\right)\|\hat{u}\|+C\left(|y|-r\left(t\right)\right)^2.$$

Thus, since $|y|\,r'(t)\leq r\left(t\right)r'(t)+C\left(|y|-r\left(t\right)\right)$, for all $(t,y)\in[0,T]\times\mathbb{R}^m$, we can deduce from (5.255) that

$$2\frac{|y|-r(t)}{|y|}\left(\langle F(t,x,y,u),y\rangle+|y|\,r'(t)\right)$$

$$\leq\frac{|y|-r(t)}{|y|}\left\|u^{\perp}\right\|^2+C\left(|y|-r(t)\right)\|\hat{u}\|+C\left(|y|-r\left(t\right)\right)^2$$

$$\leq\frac{|y|-r(t)}{|y|}\|u\|^2+\frac{r(t)}{|y|}\|\hat{u}\|^2+C\left(|y|-r\left(t\right)\right)^2$$

$$\leq\frac{|y|-r(t)}{|y|}\|u\|^2+\frac{r(t)}{|y|^3}\left|u^*y\right|^2+C\left(|y|-r\left(t\right)\right)^2.$$

This proves the sufficiency of (5.255).

The next example shows that, in the general case, there is no possibility of null-controllability of BSDEs; although we don't consider controlled equations, we can interpret the choice of the coefficients as controls.

Example 5.93. For any given $(t_0,y_0)\in\,]0,T[\,\times\mathbb{R}^m$, we introduce the family of moving constraints

$$E\left(t\right)=\begin{cases}\mathbb{R}^m, & \text{if } t\neq t_0,\\ \{y_0\}, & \text{if } t=t_0.\end{cases}$$

The associated square-distance function is of the form:

$$h\left(t,y\right)=d^2_{E(t)}\left(y\right)=\begin{cases}0, & \text{if } t\neq t_0,\\ |y-y_0|^2, & \text{if } t=t_0.\end{cases}$$

This function is upper semicontinuous in $(t,y)\in[0,T]\times\mathbb{R}^m$, and if $t=t_0, y\neq y_0$, then, for every $a\in\mathbb{R}$, there is some $\varphi_a\in C^{1,2}\left([0,T]\times\mathbb{R}^m\right)$ with

$$\left(\frac{\partial}{\partial t},\nabla_y,D^2_y\right)\varphi_a\left(t_0,y\right)=\left(-a,2\left(y-y_0\right),2I\right)$$

such that $h - \varphi_a$ achieves a local maximum at (t_0, y). Since

$$\mathcal{A}_z(t_0; x)\, \varphi_a(t_0, y) = |z\sigma(t, x)|^2 - 2\langle F(t, x, y, z\sigma(t, x)), y - y_0\rangle$$

does not depend on $a \in \mathbb{R}$, we can choose $a > 0$ sufficiently large in order to guarantee that the inequality in Corollary 5.91(jj) is not satisfied. This shows that Eq. (5.249) cannot be $\mathcal{E}$-viable.

The proof of Theorem 5.90 reduces to that of the following two lemmas, see [15].

Lemma 5.94. *Under our standard assumptions we have the equivalence between the following statements:*

i) Equation (5.249) is $\mathcal{E}$-viable on $[0, T]$.
ii) There exists a $C > 0$ such that, for all $t, \tilde{T}$ with $0 \leq t \leq \tilde{T} \leq T$, and for all $x \in \mathbb{R}^d$, the solution of BSDE (5.249) with time horizon $\tilde{T}$ and arbitrary Borel measurable terminal function $\kappa : \mathbb{R}^d \to \mathbb{R}^m$ of at most polynomial growth satisfies:

$$d^2_{E(t,x)}(Y_t^{t,x}) \leq e^{C(\tilde{T}-t)} \mathbb{E} d^2_{E(\tilde{T}, X_{\tilde{T}}^{t,x})}(Y_{\tilde{T}}^{t,x}).$$

Lemma 5.95. *Let $Y^{t,x}$ be the solution of BSDE (5.249) with time horizon $\tilde{T}$ and arbitrary terminal function $\kappa \in C_{pol}\left([0, \tilde{T}] \times \mathbb{R}^d\right)$.*

Let C be a positive constant and $h : [0, T] \times \mathbb{R}^d \times \mathbb{R}^m \to \mathbb{R}$ be an upper semicontinuous function of at most polynomial growth such that, for some positive constants $M, p > 0$,

$$|h(t, x, y') - h(t, x, y)| \leq M|y - y'|(1 + |x|^p + |y|^p + |y'|^p) \tag{5.256}$$

for all $(t, x) \in [0, T] \times \mathbb{R}^d$ and all $y, y' \in \mathbb{R}^m$. Then the following assertions are equivalent:

i) For all $x \in \mathbb{R}^d$ and $t, \tilde{T}$ with $0 \leq t \leq \tilde{T} \leq T$, it holds that

$$h(t, x, Y_t^{t,x}) \leq e^{C(\tilde{T}-t)} \mathbb{E} h(\tilde{T}, X_{\tilde{T}}^{t,x}, Y_{\tilde{T}}^{t,x}).$$

ii) For every $z \in \mathbb{R}^{m \times d}$, the function h is a viscosity sub-solution of the equation

$$\frac{\partial V(t, x, y)}{\partial t} + \mathcal{L}_z(t) V(t, x, y) + C h(t, x, y) = 0 \text{ on } [0, \tilde{T}] \times \mathbb{R}^d \times \mathbb{R}^m. \tag{5.257}$$

Recall that $\mathcal{L}_z(t)$ is defined in (5.254).

Proof of Lemma 5.94. We first remark that (ii) obviously implies (i). Thus, it only remains to show that (ii) can be deduced from (i). Let $\tilde{T} \in [0, T]$, $(t, x) \in [0, \tilde{T}] \times \mathbb{R}^d$. For simplicity of notation we put $u(t, x) = Y_t^{t,x}$, and we select a Borel measurable mapping $\hat{u} : [t,T] \times \mathbb{R}^d \to \mathbb{R}^m$ such that $\hat{u}(s,x') \in \prod_{E(s,x')}(u(s,x'))$, for all $(s,x') \in [t,T] \times \mathbb{R}^d$. Recall that $\prod_{E(s,x')}(z) = \{y \in E(s,x') : |z-y| = d_{E(s,x')}(z)\}$. Then, since Eq. (1.2) is $\mathcal{E}$-viable, the unique square integrable adapted solution $(\tilde{Y}^{t,x}, \tilde{Z}^{t,x})$ of the BSDE

$$\tilde{Y}_s^{t,x} = \hat{u}(\tilde{T}, X_{\tilde{T}}^{t,x}) + \int_s^{\tilde{T}} F(r, X_r^{t,x}, \tilde{Y}_r^{t,x}, \tilde{Z}_r^{t,x})dr - \int_s^{\tilde{T}} \tilde{Z}_r^{t,x} dW_r, \quad s \in [t,T],$$

is such that $\tilde{Y}_s^{t,x} \in E(s, X_s^{t,x}), t \le s \le T, \mathbb{P}\text{-}a.s.$

Consequently, $\mathbb{E}d^2_{E(s,X_s^{t,x})}(Y_s^{t,x}) \le \mathbb{E}|Y_s^{t,x} - \tilde{Y}_s^{t,x}|^2$, and a standard estimate of $\mathbb{E}|Y_s^{tx} - \tilde{Y}_s^{tx}|^2$ involving Itô's formula and Gronwall's formula, yields the desired result:

$$\begin{aligned}
&\mathbb{E}d^2_{E(s,X_s^{t,x})}(Y_s^{t,x}) \\
&\le \mathbb{E}|Y_s^{t,x} - \tilde{Y}_s^{t,x}|^2 \le e^{C(\tilde{T}-s)}\mathbb{E}|Y_{\tilde{T}}^{t,x} - \tilde{Y}_{\tilde{T}}^{t,x}|^2 \\
&= e^{C(\tilde{T}-s)}\mathbb{E}|Y_{\tilde{T}}^{t,x} - \hat{u}(\tilde{T}, X_{\tilde{T}}^{t,x})|^2 = e^{C(\tilde{T}-s)}\mathbb{E}d^2_{E(\tilde{T},X_{\tilde{T}}^{t,x})}(Y_{\tilde{T}}^{t,x}),
\end{aligned}$$

$0 \le t \le s \le \tilde{T} \le T, x \in \mathbb{R}^d$. This completes the proof of Lemma 5.94.

We now come to the proof of Lemma 5.95.

Proof of Lemma 5.95. We first show that, under the assumption (i), we have (ii). To this end we fix an arbitrary function $\varphi : [0,T] \times \mathbb{R}^d \times \mathbb{R}^m \to \mathbb{R}$ of class $C^{1,2,2}_{pol}$ and a point $(t,x,y) \in (0,T) \times \mathbb{R}^d \times \mathbb{R}^m$ such that the mapping $h - \varphi$ achieves a global maximum at (t,x,y). For an arbitrary but fixed $z \in \mathbb{R}^{m\times d}$ we denote by $(Y^\varepsilon, Z^\varepsilon) \in S^2_m[t,t+\varepsilon] \times \Lambda^2_{m\times k}(t,t+\varepsilon)$ the unique solution of the BSDE

$$Y_s^\varepsilon = \kappa_\varepsilon(X_{t+\varepsilon}^{t,x}) + \int_s^{t+\varepsilon} F(r, X_r^{t,x}, Y_r^\varepsilon, Z_r^\varepsilon)dr - \int_s^{t+\varepsilon} Z_r^\varepsilon dW_r, \ t \le s \le t+\varepsilon,$$

where

$$\kappa_\varepsilon(x') = y + z(x'-x) - \varepsilon z b(t,x) - \varepsilon F(t,x,y,z\sigma(t,x)).$$

From the assumption made on h in assertion (i), we obtain

$$\begin{aligned}
&h(t,x,Y_t^\varepsilon) - h(t,x,y) \\
&\quad \le e^{C\varepsilon}[\mathbb{E}h(t+\varepsilon, X_{t+\varepsilon}^{t,x}, Y_{t+\varepsilon}^\varepsilon) - h(t,x,y))] + (e^{C\varepsilon}-1)h(t,x,y) \\
&\quad \le e^{C\varepsilon}[\mathbb{E}\varphi(t+\varepsilon, X_{t+\varepsilon}^{t,x}, Y_{t+\varepsilon}^\varepsilon) - \varphi(t,x,y)] + (e^{C\varepsilon}-1)h(t,x,y).
\end{aligned}$$

Then, with the help of a Taylor expansion of φ, we get

$$\begin{aligned}
&\frac{1}{\varepsilon}\left(h(t,x,Y_t^\varepsilon)-h(t,x,y)\right) \\
&\le e^{C\varepsilon}[\frac{\partial\varphi}{\partial t}(t,x,y)+\frac{1}{\varepsilon}\mathbb{E}\left\langle\nabla_{(x,y)}\varphi(t,x,y),\begin{pmatrix}X_{t+\varepsilon}^{t,x}-x\\ Y_{t+\varepsilon}^\varepsilon-y\end{pmatrix}\right\rangle+ \\
&+\frac{1}{2\varepsilon}\mathbb{E}\left\langle D^2_{(x,y)}\varphi(t,x,y)\begin{pmatrix}X_{t+\varepsilon}^{t,x}-x\\ Y_{t+\varepsilon}^\varepsilon-y\end{pmatrix},\begin{pmatrix}X_{t+\varepsilon}^{t,x}-x\\ Y_{t+\varepsilon}^\varepsilon-y\end{pmatrix}\right\rangle+ \\
&+\frac{1}{\varepsilon}\mathbb{E}\gamma^{t,x,y}(t+\varepsilon,X_{t+\varepsilon}^{t,x},Y_{t+\varepsilon}^\varepsilon)]+\frac{e^{C\varepsilon}-1}{\varepsilon}h(t,x,y),
\end{aligned} \tag{5.258}$$

where,

$$\begin{aligned}
&\gamma^{t,x,y}(t',x',y') \\
&=\int_0^1\left(\frac{\partial}{\partial t}\varphi\left(t+\theta\left(t'-t\right),x',y'\right)-\frac{\partial}{\partial t}\varphi(t,x,y)\right)\left(t'-t\right)d\theta \\
&+\int_0^1\int_0^\theta\Big\langle\Big(D^2_{(x,y)}\varphi\left(t,x+\vartheta\left(x'-x\right),y+\vartheta\left(y'-y\right)\right)-D^2_{(x,y)}\varphi(t,x,y)\Big) \\
&\qquad\qquad\begin{pmatrix}x'-x\\ y'-y\end{pmatrix},\begin{pmatrix}x'-x\\ y'-y\end{pmatrix}\Big\rangle d\vartheta d\theta.
\end{aligned}$$

Note that

$$\begin{aligned}
\begin{pmatrix}X_{t+\varepsilon}^{t,x}-x\\ Y_{t+\varepsilon}^\varepsilon-y\end{pmatrix}&=\int_t^{t+\varepsilon}\begin{pmatrix}b(r,X_r^{t,x})\\ z(b(r,X_r^{t,x})-b(t,x))-F(t,x,y,z\sigma(t,x))\end{pmatrix}dr \\
&+\int_t^{t+\varepsilon}\begin{pmatrix}\sigma(r,X_r^{t,x})\\ z\sigma(r,X_r^{t,x})\end{pmatrix}dW_r.
\end{aligned}$$

Hence,

$$\lim_{\varepsilon\to0}\frac{1}{\varepsilon}\mathbb{E}\begin{pmatrix}X_{t+\varepsilon}^{t,x}-x\\ Y_{t+\varepsilon}^\varepsilon-y\end{pmatrix}=\begin{pmatrix}b\left(t,x\right)\\ -F(t,x,y,z\sigma(t,x))\end{pmatrix}$$

and

$$\begin{aligned}
&\lim_{\varepsilon\to0}\frac{1}{\varepsilon}\mathbb{E}\left\langle D^2_{(x,y)}\varphi(t,x,y)\begin{pmatrix}X_{t+\varepsilon}^{t,x}-x\\ Y_{t+\varepsilon}^\varepsilon-y\end{pmatrix},\begin{pmatrix}X_{t+\varepsilon}^{t,x}-x\\ Y_{t+\varepsilon}^\varepsilon-y\end{pmatrix}\right\rangle \\
&=\frac{1}{2}\mathrm{Tr}\left(\left(\sigma,z\sigma\right)\left(\sigma,z\sigma\right)^*\left(t,x\right)D^2_{(x,y)}\varphi(t,x,y)\right).
\end{aligned}$$

Moreover, from the assumptions on h,

$$\frac{1}{\varepsilon}|h(t,x,Y_t^\varepsilon)-h(t,x,y)| \leq \frac{M}{\varepsilon}|Y_t^\varepsilon - y|(1+|x|^p+|y|^p+|Y_t^\varepsilon|^p).$$

Therefore, applying the following auxiliary lemma, the proof of which will be given at the end of this section, we can take the limit as $\varepsilon \to 0$ in (5.258) and obtain assertion (*ii*).

Lemma 5.96. *Under the assumptions of Lemma 5.95, and with the notations introduced above, we have*

$$a)\ \lim_{\varepsilon \searrow 0} \frac{1}{\varepsilon^2}|Y_t^\varepsilon - y|^2 = 0,$$
$$b)\ \lim_{\varepsilon \searrow 0} \frac{1}{\varepsilon}\mathbb{E}|\gamma^{t,x,y}(\varepsilon, X_{t+\varepsilon}^{t,x}, Y_{t+\varepsilon}^\varepsilon)| = 0,$$

for all $(t,x,y) \in [0,T] \times \mathbb{R}^d \times \mathbb{R}^m$.

We shall now prove the reverse implication: Under the assumption that (*ii*) holds we have to show the validity of (*i*). For this we first remark that, for any continuous function $\Phi : [0,T] \times \mathbb{R}^d \times \mathbb{R}^m \times \mathbb{R} \times \mathbb{R}^{m+n} \times \mathbb{S}^{m+n} \times \mathbb{R}^{m\times d} \to \mathbb{R}$ satisfying the standard assumptions of monotonicity with respect to the $\mathbb{R}^{m+n}$-variable and of degenerate ellipticity with respect to the $\mathbb{S}^{m+n}$-variable (see Annex D),

(α) $h(t,x,y)$ is a viscosity sub-solution of the PDE
$$\Phi(t,x,y,\partial_t h(t,x,y), \nabla_{(x,y)}h(t,x,y), D^2_{(x,y)}h(t,x,y); z) = 0, \quad \text{for all } z \in \mathbb{R}^{m\times d}$$

if and only if

(β) $h(t,x,y)$ is a viscosity sub-solution of the PDE
$$\Phi(t,x,y,\partial_t h(t,x,y), \nabla_{(x,y)}h(t,x,y), D^2_{(x,y)}h(t,x,y); g(t,x)) = 0, \quad \text{for all } g \in C_{pol}([0,T]\times\mathbb{R}^d;\mathbb{R}^{m\times d}).$$

Indeed, in order to see that (β) implies (α), it suffices to choose $g \in C_{pol}([0,T] \times \mathbb{R}^d;\mathbb{R}^{m\times d})$ with $g(t,x) = z \in \mathbb{R}^{m\times d}$. On the other hand, to get the necessity of (β) under (α), we remark that, for all test functions $\varphi \in C^{1,2,2}$ for which $h-\varphi$ achieves a local maximum at (t,x,y), and with the notation

$$(a,p,S) = \left(\frac{\partial}{\partial t}\varphi, \nabla_{(x,y)}\varphi, D^2_{(x,y)}\varphi\right)(t,x,y),$$

we have that $\Phi(t,x,y,a,p,S;z) \geq 0$, for all $z \in \mathbb{R}^{m\times d}$, and hence also for $z = g(t,x)$, where g runs over the set of functions belonging to $C_{pol}([0,T] \times \mathbb{R}^d;$

$\mathbb{R}^{m\times d}$). We now fix any $g \in C_{pol}([0,T]\times\mathbb{R}^d;\mathbb{R}^{m\times d})$ and consider the unique square integrable adapted solution $(X, \overline{Y}^{t,x,y})$ of the (forward) SDE

$$\begin{pmatrix} X_s^{t,x} \\ \overline{Y}_s^{t,x,y} \end{pmatrix} = \begin{pmatrix} x \\ y \end{pmatrix} + \int_t^s \begin{pmatrix} b(r, X_r^{t,x}) \\ -F(r, X_r^{t,x}, \overline{Y}_r^{t,x,y}, g(r, X_r^{t,x})\sigma(r, X_r^{t,x})) \end{pmatrix} dr$$
$$+ \int_t^s \begin{pmatrix} \sigma(r, X_r^{t,x}) \\ g(r, X_r^{t,x})\sigma(r, X_r^{t,x}) \end{pmatrix} dW_r, \quad s \in [t,T].$$

Of course, here the process $X^{t,x}$ is nothing else than the unique solution of SDE (5.248). Moreover, we denote by $\left(\tilde{Y}_{k,\cdot}^{t,x,y}, \tilde{Z}_{k,\cdot}^{t,x,y}\right) \in S_m^2\left[t,\tilde{T}\right] \times \Lambda_{m\times k}^2\left(t,\tilde{T}\right)$ the unique solution of the BSDE

$$\tilde{Y}_{k,s}^{t,x,y} = h_k(\tilde{T}, X_{\tilde{T}}^{t,x}, \overline{Y}_{\tilde{T}}^{t,x,y}) + C\int_s^{\tilde{T}} h_k(r, X_r^{t,x}, \overline{Y}_r^{t,x,y})dr$$
$$-\int_s^{\tilde{T}} \tilde{Z}_{k,r}^{t,x,y} dW_r, \quad s \in [t,\tilde{T}],$$

where $\tilde{T} \in [0,T]$ and $(h_k)_{k\geq 1} \subset C_{pol}([0,T] \times \mathbb{R}^d \times \mathbb{R}^m)$ is a monotonically decreasing sequence of continuous functions with at most polynomial growth, such that its pointwise limit is equal to h. Then the function

$$V_k(t,x,y) = \tilde{Y}_{k,t}^{t,x,y}, \quad (t,x,y) \in [0,\tilde{T}] \times \mathbb{R}^d \times \mathbb{R}^m,$$

is a continuous viscosity solution of the equation

$$\begin{cases} \dfrac{\partial V_k(t,x,y)}{\partial t} + \mathcal{L}_{g(t,x)}(t)V_k(t,x,y) + Ch_k(t,x,y) = 0, \\ V_k(\tilde{T},x,y) = h_k(\tilde{T},x,y), \quad (x,y) \in \mathbb{R}^d \times \mathbb{R}^m, \end{cases}$$

and it is the unique solution in the class of continuous functions of at most polynomial growth. We also note that, by the Markov property,

$$\tilde{Y}_{k,s}^{t,x,y} = V_k(t, X_s^{t,x}, \overline{Y}_s^{t,x,y}), s \in [t,\tilde{T}].$$

Since, due to assumption (ii), h is an upper semicontinuous viscosity sub-solution of at most polynomial growth of the above PDE, we know that h must be smaller than or equal to the viscosity solution V_k. Thus,

$$\begin{aligned} &\mathbb{E}h(s, X_s^{t,x}, \overline{Y}_s^{t,x,y}) \\ &\quad \leq \mathbb{E}V_k(s, X^{t,x}, \overline{Y}_s^{t,x,y}) \\ &\quad = \mathbb{E}h_k(\tilde{T}, X_{\tilde{T}}^{t,x}, \overline{Y}_{\tilde{T}}^{t,x,y}) + C\int_s^{\tilde{T}} \mathbb{E}h_k(r, X_r^{t,x}, \overline{Y}_r^{t,x,y})dr, \ s \in [t,\tilde{T}], \end{aligned}$$

then, by passing to the limit as $k \to \infty$ and applying Gronwall's inequality, we obtain the following estimate

$$\mathbb{E}h(s, X_s^{t,x}, \overline{Y}_s^{t,x,y}) \leq e^{C(\tilde{T}-s)}\mathbb{E}h(\tilde{T}, X_{\tilde{T}}^{t,x}, \overline{Y}_{\tilde{T}}^{t,x,y}).$$

Setting $s = t$ and $y = u(t,x) = Y_t^{t,x}$ and using the assumption (5.256) we obtain for some positive constant C_1,

$$\begin{aligned}
&h(t,x,u(t,x)) \\
&\quad \leq e^{C(\tilde{T}-t)}\mathbb{E}h(\tilde{T}, X_{\tilde{T}}^{t,x}, \overline{Y}_{\tilde{T}}^{t,x,y}) \\
&\quad \leq e^{C(\tilde{T}-t)}\Big[\mathbb{E}h(\tilde{T}, X_{\tilde{T}}^{t,x}, Y_{\tilde{T}}^{t,x}) \\
&\qquad +M\,\mathbb{E}\Big(|\overline{Y}_{\tilde{T}}^{t,x,y} - Y_{\tilde{T}}^{t,x}|(1 + |X_{\tilde{T}}^{t,x}|^p + |\overline{Y}_{\tilde{T}}^{t,x,y}|^p + |Y_{\tilde{T}}^{t,x}|^p)\Big)\Big] \\
&\leq e^{C(\tilde{T}-t)}\Bigg[\mathbb{E}h(\tilde{T}, X_{\tilde{T}}^{t,x}, Y_{\tilde{T}}^{t,x}) \\
&\qquad +C_1(1 + |x|^{pq} + |y|^p)\left(\mathbb{E}\int_t^{\tilde{T}} \left|Z_r^{t,x} - (g\sigma)(r, X_r^{t,x})\right|^2 dr\right)^{1/2}\Bigg]
\end{aligned}$$

for all $g \in C_{pol}([0,T]\times\mathbb{R}^d;\mathbb{R}^{m\times d})$. Since by a result from [30] (Theorem 4.1) there is a Borel measurable function $v : [0,\tilde{T}]\times\mathbb{R}^d \to \mathbb{R}^{m\times d}$ such that

$$Z_s^{t,x} = (v\sigma)\left(s, X_s^{t,x}\right),\ s \in [t,\tilde{T}],\ ds\,d\mathbb{P} - a.e.,$$

we deduce that by density (and Lebesgue's dominated convergence theorem)

$$h(t,x,u(t,x)) \leq e^{C(\tilde{T}-t)}\mathbb{E}h(\tilde{T}, X_{\tilde{T}}^{t,x}, Y_{\tilde{T}}^{t,x}).$$

Since this result holds true for all $x \in \mathbb{R}^d$, $0 \leq t \leq \tilde{T} \leq T$, we have proved (i).

Let us now prove Lemma 5.96.

Proof of Lemma 5.96. We first prove part a) of the lemma. Obviously, we have that

$$\begin{aligned}
Y_t^{\varepsilon} &= \kappa_{\varepsilon}(X_{t+\varepsilon}^{t,x}) + \int_t^{t+\varepsilon} F(r, X_r^{t,x}, Y_r^{\varepsilon}, Z_r^{\varepsilon})dr - \int_t^{t+\varepsilon} Z_r^{\varepsilon} dW_r \\
&= y + \int_t^{t+\varepsilon} z\left(b\left(r, X_r^{t,x}\right) - b(t,x)\right) dr \\
&\quad + \int_t^{t+\varepsilon} \left(F(r, X_r^{t,x}, Y_r^{\varepsilon}, Z_r^{\varepsilon}) - F(t,x,y,z\sigma(t,x))\right) dr \\
&\quad - \int_t^{t+\varepsilon} \left[Z_r^{\varepsilon} - z\sigma(r, X_r^{t,x})\right] dW_r.
\end{aligned}$$

Thus for $0 < \varepsilon < \dfrac{1}{6L^2}$,

$$
\begin{aligned}
&|Y_t^\varepsilon - y|^2 + \mathbb{E}\int_t^{t+\varepsilon} \left|Z_r^\varepsilon - z\sigma(r, X_r^{t,x})\right|^2 dr \\
&\leq 3\varepsilon\, |z|^2 \int_t^{t+\varepsilon} \mathbb{E}\left|b(r, X_r^{t,x}) - b(t,x)\right|^2 dr \\
&\quad +3\varepsilon\mathbb{E}\int_t^{t+\varepsilon} \left|F(r, X_r^{t,x}, Y_r^\varepsilon, Z_r^\varepsilon) - F(r, X_r^{t,x}, Y_r^\varepsilon, z\sigma(t, X_r^{t,x})\right|^2 dr \\
&\quad +3\varepsilon\mathbb{E}\int_t^{t+\varepsilon} \left|F(r, X_r^{t,x}, Y_r^\varepsilon, z\sigma(t, X_r^{t,x})) - F(t,x,y,z\sigma(t,x)\right|^2 dr \\
&\leq 3\varepsilon\, |z|^2 \int_t^{t+\varepsilon} \mathbb{E}\left|b(r, X_r^{t,x}) - b(t,x)\right|^2 dr + \frac{1}{2}\mathbb{E}\int_t^{t+\varepsilon} \left|Z_r^\varepsilon - z\sigma(t, X_r^{t,x})\right|^2 dr \\
&\quad +3\varepsilon\mathbb{E}\int_t^{t+\varepsilon} \left|F(r, X_r^{t,x}, Y_r^\varepsilon, z\sigma(t, X_r^{t,x})) - F(t,x,y,z\sigma(t,x)\right|^2 dr,
\end{aligned}
$$

which yields

$$
\limsup_{\varepsilon\searrow 0} \frac{1}{\varepsilon^2}\mathbb{E}\,|Y_t^\varepsilon - y|^2 + \limsup_{\varepsilon\searrow 0} \frac{1}{2\varepsilon^2}\mathbb{E}\int_t^{t+\varepsilon} \left|Z_r^\varepsilon - z\sigma(r, X_r^{tx})\right|^2 dr \leq 0.
$$

Finally, the proof of part b) of Lemma 5.96 uses the same argument as that of Lemma 4.82. The only difference is that the role of the diffusion process $X^{t,x}$ in the proof of Lemma 4.82 is now replaced by that of the pair $(X^{t,x}, Y^\varepsilon)$.

5.10 Exercises

Without further mention, $(\Omega, \mathcal{F}, \mathbb{P}, \{\mathcal{F}_t\}_{t\geq 0})$ will be a stochastic basis, $\{B_t : t \geq 0\}$ will be a k-dimensional Brownian motion with respect to this basis and $\mathcal{F}_t = \mathcal{F}_t^B$ for all $t \geq 0$.

Exercise 5.1. Consider the BSDE

$$
Y_t = \eta + \int_t^T \Phi\left(s, Y_s, Z_s\right) dQ_s - \int_t^T Z_s dB_s, \tag{5.259}
$$

under the assumptions (5.41). Let

$$
V_t = \int_0^t L_s dQ_s + \frac{1}{n_p}\int_0^t (\ell_s)^2\, ds.
$$

Show that if $p \geq 2$ and for all $\delta \geq 0$

$$\mathbb{E}\ \left|e^{\delta V_T}\eta\right|^p + \mathbb{E}\Big(\int_0^T e^{\delta V_t}\left|\Phi\left(t,0,0\right)\right| dQ_t\Big)^p < \infty,$$

then the BSDE (5.259) has a unique solution $(Y,Z) \in S_m^0\,[0,T] \times \Lambda_{m\times k}^0\,(0,T)$ such that

$$\mathbb{E}\ \sup_{s\in[0,T]} e^{\delta p V_s}\left|Y_s\right|^p + +\mathbb{E}\left(\int_0^T e^{2\delta V_s}\left|Z_s\right|^2 ds\right)^{p/2} < \infty, \text{ for all } \delta \geq 0.$$

Remark. Note that our assumptions hold in particular if both V_t has exponential moments of all orders (e.g. the tail of its law behaves like that of a Gaussian random variable) and $|\eta| + \mathbb{E}\int_0^T |\Phi(t,0,0)|dQ_t$ has a finite moment of some order $p > 1$.

Exercise 5.2 (g-Expectation). Consider the BSDE: $\mathbb{P}$-a.s., for all $t \in [0,T]$

$$Y_t = \eta + \int_t^T g\left(s, Y_s, Z_s\right) ds - \int_t^T \left\langle Z_s, dB_s\right\rangle, \tag{5.260}$$

where we assume:

(*i*) $\eta \in L^p\left(\Omega, \mathcal{F}_T, \mathbb{P}; \mathbb{R}\right)$, $p > 1$;
(*ii*) *for every* $(y,z) \in \mathbb{R}\times\mathbb{R}^k$, *the function* $g\left(\cdot,\cdot,y,z\right) : \Omega\times[0,T] \to \mathbb{R}$ *is* $\mathcal{P}$-*measurable*;
(*iii*) g satisfies the assumptions of Theorem 5.27 (F replaced by g) and $g\left(t,y,0\right) = 0$ for all $y \in \mathbb{R}$, a.e. $t \in [0,T]$.

Then by Theorem 5.17 the BSDE (5.260) has a unique solution $(Y,Z) \in S_1^p\,[0,T] \times \Lambda_k^p\,(0,T)$. Moreover if $\tau : \Omega \to [0,T]$ is a stopping time and $\eta \in L^p\left(\Omega, \mathcal{F}_\tau, \mathbb{P}; \mathbb{R}\right)$ then $(Y_t, Z_t) = (\eta, 0)$ for all $t \geq \tau$.

Define the g-expectation of η by $\mathbf{E}_g\left(\eta\right) \overset{def}{=} Y_0$ and the conditional g-expectation of η with respect to $\mathcal{F}_t$ by $\mathbf{E}_g\left(\eta|\mathcal{F}_t\right) \overset{def}{=} Y_t$. Clearly $\mathbf{E}_0\left(\eta\right) = \mathbb{E}\eta$ and $\mathbf{E}_0\left(\eta|\mathcal{F}_t\right) = \mathbb{E}\left(\eta|\mathcal{F}_t\right)$.

Show that:

1. $\mathbf{E}_g\left(a\right) = a$, for all $a \in \mathbb{R}$.
2. $\eta_1 \leq \eta_2$, $\mathbb{P}$-a.s. $\implies$ $\mathbf{E}_g\left(\eta_1\right) \leq \mathbf{E}_g\left(\eta_2\right)$.
3. $\eta_1 \leq \eta_2$, $\mathbb{P}$-a.s. and $\mathbf{E}_g\left(\eta_1\right) = \mathbf{E}_g\left(\eta_2\right)$ $\implies$ $\eta_1 = \eta_2$, $\mathbb{P}$-a.s.
4. If $g\left(t,\cdot,\cdot\right) : \mathbb{R}\times\mathbb{R} \to \mathbb{R}$ is a convex function, a.e. $t \in [0,T]$, then $\mathbf{E}_g : L^p\left(\Omega, \mathcal{F}_T, \mathbb{P}\right) \to \mathbb{R}$ is convex, too.
5. Let $U \in L^p\left(\Omega, \mathcal{F}_t, \mathbb{P}\right)$. Then $\mathbf{E}_g\left(1_A\eta\right) = \mathbf{E}_g\left(1_A U\right)$, for all $A \in \mathcal{F}_t$, if and only if $U = Y_t$.
6. $\mathbf{E}_g\left(a|\mathcal{F}_t\right) = a$, for all $a \in \mathbb{R}$.
7. $\mathbf{E}_g\left(\eta|\mathcal{F}_t\right) = \eta$, for all $\eta \in L^p\left(\Omega, \mathcal{F}_t, \mathbb{P}\right)$.

8. $\eta_1 \leq \eta_2$, $\mathbb{P}$-a.s. $\implies$ $\mathbf{E}_g(\eta_1|\mathcal{F}_t) \leq \mathbf{E}_g(\eta_2|\mathcal{F}_t)$, $\mathbb{P}$-a.s.
9. $\mathbf{E}_g(1_A\eta|\mathcal{F}_t) = 1_A\mathbf{E}_g(\eta|\mathcal{F}_t)$, for all $A \in \mathcal{F}_t$.

Exercise 5.3 (Peano Type Result). Consider the BSDE

$$Y_t = \eta + \int_t^T G(s, Y_s, Z_s)\, ds - \int_t^T \langle Z_s, dB_s\rangle,$$

where $\eta \in L^p(\Omega, \mathcal{F}_T, \mathbb{P}; \mathbb{R})$, $p \geq 2$, and $G : [0,T] \times \mathbb{R} \times \mathbb{R}^k \to \mathbb{R}$ is a function such that

- $G(\cdot, x, z) : [0,T] \to \mathbb{R}$ is measurable for all $x \in \mathbb{R}$ and $z \in \mathbb{R}^k$,
- $G(t, \cdot, \cdot) : \mathbb{R} \times \mathbb{R}^k \to \mathbb{R}$ is continuous for all $t \in [0,T]$,
- there exists an $L > 0$ such that for all $(t, y, z) \in [0,T] \times \mathbb{R} \times \mathbb{R}^k$,

$$|G(t, y, z)| \leq L(1 + |y| + |z|).$$

Under these conditions we shall prove that the BSDE (5.260) has at least one solution $(Y, Z) \in S^p[0,T] \times \Lambda_k^p(0,T)$.

Let $0 < \varepsilon \leq \varepsilon_0 = 1 \wedge (1/L)$ and $G_\varepsilon : [0,T] \times \mathbb{R} \times \mathbb{R}^k \to \mathbb{R}$,

$$G_\varepsilon(t, y, z) = \inf\left\{G(t, u, v) + \frac{1}{\varepsilon}|y - u| + \frac{1}{\varepsilon}|z - v| : (u, v) \in \mathbb{R} \times \mathbb{R}^k\right\}.$$

Prove that:

1. For all $t \in [0,T]$, $y, y' \in \mathbb{R}$ and $z, z' \in \mathbb{R}^k$:
   (a) $|G_\varepsilon(t, y, z)| \leq L(1 + |y| + |z|)$;
   (b) $|G_\varepsilon(t, y, z) - G_\varepsilon(t, y', z')| \leq \frac{1}{\varepsilon}(|y - y'| + |z - z'|)$;
   (c) $yG_\varepsilon(t, y, z) \leq L|y| + (L + L^2)|y|^2 + \frac{1}{4}|z|^2$;
   (d) $0 < \delta < \varepsilon \implies G_\delta(t, y, z) \geq G_\varepsilon(t, y, z)$;
   (e) if $\lim_{\varepsilon\to 0}(y_\varepsilon, z_\varepsilon) = (y, z)$, then $\lim_{\varepsilon\to 0} G_\varepsilon(t, y_\varepsilon, z_\varepsilon) = G(t, y, z)$.
2. The BSDEs

$$Y_t^\varepsilon = \eta + \int_t^T G_\varepsilon\left(s, Y_s^\varepsilon, Z_s^\varepsilon\right) ds - \int_t^T Z_s^\varepsilon dB_s,$$

$$U_t = \eta + \int_t^T L(1 + |U_s| + |V_s|)\, ds - \int_t^T Z_s dB_s$$

have unique solutions $(Y^\varepsilon, Z^\varepsilon)$, $(U, V) \in S^p[0,T] \times \Lambda_{m\times k}^p(0,T)$ and:

(a)

$$\mathbb{E}^{\mathcal{F}_t}\left(\sup_{s\in[t,T]}|Y_s^{\varepsilon}|^p\right)+\mathbb{E}^{\mathcal{F}_t}\left(\int_t^T|Z_s^{\varepsilon}|^2\,ds\right)^{p/2}$$

$$\leq C_p\exp\left[\left(L+L^2\right)(T-t)\right]\left[\mathbb{E}^{\mathcal{F}_t}|\eta|^p+L^p(T-t)^p\right]$$

where C_p is a constant depending only on p.

(b) For all $0<\delta<\varepsilon\leq\varepsilon_0=1\wedge(1/L)$, $\mathbb{P}$-a.s.,

$$Y_t^{\varepsilon_0}\leq Y_t^{\varepsilon}\leq Y_t^{\delta}\leq U_t,\quad\text{for all } t\in[0,T],$$

and there exists a $Y\in S^p[0,T]$ such that

$$\lim_{\varepsilon\to0}\mathbb{E}\left(\sup_{s\in[0,T]}|Y_s^{\varepsilon}-Y_s|^p\right)=0.$$

(c) There exists a $Z\in\Lambda^p_{m\times k}(0,T)$ such that

$$\lim_{\varepsilon\to0}\mathbb{E}\left(\int_0^T|Z_s^{\varepsilon}-Z_s|^2\,ds\right)^{p/2}=0.$$

Exercise 5.4 (BSDE Reflected Above 0). Let $\xi\in L^2\left(\Omega,\mathcal{F}_T^B,\mathbb{P};\mathbb{R}\right)$, where $\{B_t,\ 0\leq t\leq T\}$ is a k-dimensional BM, and $F:\mathbb{R}\times\mathbb{R}^k\to\mathbb{R}$ be a Lipschitz continuous mapping. Consider for each $n\in\mathbb{N}$ the solution $\{(Y_t^n,Z_t^n),\ 0\leq t\leq T\}$ of the BSDE

$$Y_t^n=\xi+\int_t^T F(Y_s^n,Z_s^n)ds+n\int_t^T(Y_s^n)^-ds-\int_0^t\langle Z_s^n,dB_s\rangle,$$

and let $K_t^n=n\int_0^t(Y_s^n)^-ds$.

1. Show that $Y_t^{n+1}\geq Y_t^n$, $0\leq t\leq T$.
2. Show that

$$\sup_n\mathbb{E}\left(\sup_{0\leq t\leq T}|Y_t^n|^2\right)<\infty.$$

3. Deduce that there exists a progressively measurable process $\{Y_t,\ 0\leq t\leq T\}$ such that $Y_t^n\to Y_t$ a.s. for all $t\in[0,T]$, and

$$\mathbb{E}\left(\sup_{0\leq t\leq T}|Y_t|^2\right)<\infty.$$

4. Show that $Y_t^n \geq \tilde{Y}_t^n$, where $\{\tilde{Y}_t^n,\ 0 \leq t \leq T\}$ solves the BSDE

$$\tilde{Y}_t^n = \xi + \int_t^T F(\tilde{Y}_s^n, \tilde{Z}_s^n)ds - n\int_t^T \tilde{Y}_s^n ds - \int_0^t \langle \tilde{Z}_s^n, dB_s\rangle .$$

5. Identify $\lim_{n\to\infty} \tilde{Y}_t^n$ and deduce that $Y_t \geq 0,\ 0 \leq t \leq T$, a.s., and (with the help of Dini's theorem) that $\sup_{0\leq t\leq T}(Y_t^n)^- \to 0$ in mean square.
6. Show that $\{Z_t^n,\ 0 \leq t \leq T\}$ is a Cauchy sequence in $\Lambda_k^2(0,T)$. Hint: check that

$$\int_t^T (Y_s^n - Y_s^p)(dK_s^n - dK_s^p) \leq \int_t^T \left[(Y_s^p)^- dK_s^n + (Y_s^n)^- dK_s^p\right] \to 0.$$

7. Deduce that K_t^n converges in probability to a progressively measurable increasing continuous stochastic process K_t.
8. Show that the just constructed triple $\{(X_t, Z_t, K_t),\ 0 \leq t \leq T\}$ is a unique progressively measurable solution of the reflected BSDE: for all $t \in [0,T]$, $\mathbb{P}$-a.s.

$$\begin{cases} (i) & Y \text{ is a continuous stochastic process, } Y_t \geq 0, \\ (ii) & K \text{ is c.i.s.p., } \displaystyle\int_0^T Y_s dK_s = 0, \\ (iii) & \mathbb{E}\displaystyle\int_0^T |Z_s|^2 dt < \infty, \\ (iv) & Y_t = \xi + \displaystyle\int_t^T F(Y_s, Z_s)ds + K_T - K_t - \int_t^T \langle Z_s, dB_s\rangle . \end{cases}$$

9. With the help of Tanaka's formula applied to $(Y_t)^+ = Y_t$, show that in the sense of inequality between measures,

$$0 \leq dK_t \leq \mathbf{1}_{\{Y_t=0\}}\left[F(Y_t, Z_t)\right]^- dt.$$

Deduce that K is absolutely continuous.
10. Show that the points 2–9 constitute a particular case of Theorem 5.52.

Exercise 5.5. Let $\eta \in L^0(\Omega, \mathcal{F}_T, \mathbb{P}; \mathbb{R})$ be such that $0 \leq \eta \leq 1$, $\mathbb{P}$-a.s. Prove that the BSDE

$$Y_t = \eta + \int_t^T Y_s(1 - Y_s)\, ds - \int_t^T \langle Z_s, dB_s\rangle$$

has a unique solution $(Y, Z) \in S_1^2[0,T] \times \Lambda_k^2(0,T)$. Moreover

$$\mathbb{E}\left(\int_0^T |Z_s|^2\, ds\right)^{p/2} < \infty, \quad \text{for all } p > 0,$$

$0 \leq Y_t \leq 1, \quad \mathbb{P} - a.s.$

Exercise 5.6. Let $\varepsilon > 0$, $\kappa : \mathbb{R} \to \mathbb{R}$ be a continuous bounded function and $g : \mathbb{R} \to \mathbb{R}$ be a bounded Lipschitz continuous function. Consider the PDEs

$$\begin{cases} u'_t(t,x) + \frac{1}{2}u''_{xx}(t,x) = 0, & (t,x) \in]0,T[\times\mathbb{R}, \\ u(T,x) = \kappa(x) & x \in \mathbb{R}, \end{cases} \tag{5.261}$$

and

$$\begin{cases} (u^\varepsilon)'_t(t,x) + \frac{1}{2}(u^\varepsilon)''_{xx}(t,x) + \sin\left(\frac{x}{\varepsilon}\right) g\left(u^\varepsilon(t,x), (u^\varepsilon)'_x(t,x)\right) = 0, \\ \qquad\qquad\qquad\qquad\qquad\qquad\qquad\qquad (t,x) \in]0,T[\times\mathbb{R}, \\ u(T,x) = \kappa(x), \qquad x \in \mathbb{R}. \end{cases} \tag{5.262}$$

1. Write the BSDEs in $(Y^{t,x}, Z^{t,x})$ and respectively in $(Y^{\varepsilon;t,x}, Z^{\varepsilon;t,x})$ such that $u(t,x) = Y_t^{t,x}$ and $u^\varepsilon(t,x) = Y_t^{\varepsilon;t,x}$ are viscosity solutions of the PDEs (5.261) and, respectively, (5.262). Are the corresponding viscosity solutions unique?
2. Prove that

$$\lim_{\varepsilon \to 0} u^\varepsilon(0,x) = u(0,x), \quad \text{for all } x \in \mathbb{R}.$$

Exercise 5.7. Let E be a non-empty closed subset of $\mathbb{R}^m$, $g : \mathbb{R}^k \to E$ be a bounded Borel measurable function and $F : \Omega \times [0,T] \to \mathbb{R}^m$ be a bounded progressively measurable stochastic process. Let $(Y,Z) \in S_m^1[0,T] \times \Lambda_{m\times k}^1(0,T)$ be such that

$$Y_t = g(B_T) + \int_t^T F_s ds - \int_t^T Z_s dB_s, \quad a.s., \quad t \in [0,T].$$

Show that $(i) \Rightarrow (ii)$, where:

(i) $\mathbb{P}$-a.s., $\{Y_t : t \in [0,T]\} \subset E$, for all bounded Borel measurable function $g : \mathbb{R}^k \to E$;

(ii) E is a convex set.

Chapter 6
Annexes

6.1 Introduction

In this chapter, we collect several results which are used in the book, but whose presentation we have preferred to postpone until now. A first section presents notations and elementary results on matrices. The second section presents some elements of nonlinear and convex analysis. It is mainly used in Chap. 4. The third section presents Gronwall's inequality, both in the forward and in the backward time direction, together with various original extensions of this inequality to stochastic processes. The most important stochastic inequalities are given in Propositions 6.71, 6.74, 6.80. Section four presents the notion of viscosity solutions of nonlinear PDEs, and establishes three different uniqueness results for viscosity solutions of PDEs which appear in previous chapters of this book. These are variants of more or less known results scattered in the literature. We could not possibly cover all types of elliptic and parabolic equations (and systems of equations) with various types of boundary conditions. But we believe that the reader can adapt our arguments to all situations considered in Chaps. 3–5 of the book.

Finally a last section is devoted to giving hints for the solutions to some of the exercises from the book.

6.2 Annex A: Vectors and Matrices

Denote by $\mathbb{R}^{d\times k}$ the linear space of matrices $A=\left(a_{i,j}\right)_{d\times k}$, where $a_{i,j}\in\mathbb{R}$. If $k=1$ then $\mathbb{R}^{d\times 1}$ is the Euclidean space $\mathbb{R}^d$. Denote by $A^*=\left(a_{j,i}\right)_{k\times d}$ the transposed matrix of A.

Let $x=(x_i)_{i=\overline{1,d}}\in\mathbb{R}^d$ and $y=(y_i)_{i=\overline{1,d}}\in\mathbb{R}^d$. The usual inner product on $\mathbb{R}^d$ is given by

E. Pardoux and A. Răşcanu, *Stochastic Differential Equations, Backward SDEs, Partial Differential Equations*, Stochastic Modelling and Applied Probability 69, DOI 10.1007/978-3-319-05714-9_6,

$$\langle x, y\rangle = x_1 y_1 + x_2 y_2 + \cdots + x_d y_d = x^* y$$

and the norm

$$|x| = \sqrt{\langle x, x\rangle} = \left(x_1^2 + x_2^2 + \cdots + x_d^2\right)^{1/2} = \sqrt{x^* x}.$$

We also introduce the notation $x^+ := \left(x_i^+\right)_{d\times 1}$.

The tensor product of the two vectors x and y is the linear operator $x \otimes y : \mathbb{R}^d \times \mathbb{R}^d \to \mathbb{R}$ defined by

$$(x \otimes y)(u, v) = \langle x, u\rangle \langle y, v\rangle = u^* \left(xy^*\right) v.$$

Hence one can identify

$$x \otimes y = \left(x_i y_j\right)_{d\times d} = xy^*.$$

If $A = \left(a_{i,j}\right)_{d\times d}$ and $\{\mathbf{u}_1, \ldots, \mathbf{u}_d\}$ is an orthonormal basis of $\mathbb{R}^d$, that is

$$\left\langle \mathbf{u}_i, \mathbf{u}_j\right\rangle = \delta_{i,j} = \begin{cases} 1 & \text{if } i = j, \\ 0 & \text{if } i \neq j, \end{cases}$$

we define

$$\mathbf{Tr}\, A = \text{Trace}\,(A) = \sum_{i=1}^{d} \langle A\mathbf{u}_i, \mathbf{u}_i\rangle\,.$$

The "Trace" is independent of the basis $\{\mathbf{u}_1, \ldots, \mathbf{u}_d\}$ and

$$\mathbf{Tr} A = \sum_{i=1}^{d} a_{ii} = \mathbf{Tr} A^*.$$

Moreover if $A, B \in \mathbb{R}^{d\times d}$ then one verifies that

$$\mathbf{Tr}\,(AB) = \mathbf{Tr}\,(BA) = \mathbf{Tr}\left(A^* B^*\right) = \mathbf{Tr}\left(B^* A^*\right).$$

Let $A = \left(a_{i,j}\right)_{d\times k} \in \mathbb{R}^{d\times k}$, $B = \left(b_{i,j}\right)_{d\times k} \in \mathbb{R}^{d\times k}$. We define the inner product on $\mathbb{R}^{d\times k}$ by

$$\begin{aligned} \langle A, B\rangle &= \mathbf{Tr}\left(A^* B\right) = \mathbf{Tr}\left(AB^*\right) \\ &= \sum_{i=1}^{d} \sum_{j=1}^{k} a_{ij} b_{ij} \end{aligned}$$

and the norm

$$|A| = \sqrt{\mathbf{Tr}\,(A^*A)} = \sqrt{\mathbf{Tr}\,(AA^*)} = \left(\sum_{i=1}^{d}\sum_{j=1}^{k} a_{ij}^2\right)^{1/2}.$$

We have

$$\begin{array}{ll} a) & |AB| \le |A|\,|B|\,, \\ b) & |Ax| \le |A|\,|x|\,, \\ c) & |A| = |A^*|\,, \\ d) & \mathbf{Tr}\,(x \otimes y) = \langle x, y\rangle\,, \\ e) & \mathbf{Tr}\,[(x \otimes y)\,AB^*] = \langle x, BA^* y\rangle\,, \\ f) & \mathbf{Tr}\,[(x \otimes x)\,AA^*] = |A^* x|^2\,, \\ g) & |x \otimes y| = |x|\,|y|\,. \end{array}$$

We note that the above matrix norm is not the operator norm

$$\|A\| = \sup\{|Ax| : |x| \le 1\} \le |A|\,,$$

since $\|I_d\| = 1 \ne \sqrt{d} = |I_d|$. Note that

$$\|A\| \le |A|\,.$$

We denote by $\mathbb{S}^{d\times d} \subset \mathbb{R}^{d\times d}$ the set of symmetric matrices. If $Q, P \in \mathbb{S}^{d\times d}$, we say that $Q \le P$ if $\langle Qx, x\rangle \le \langle Px, x\rangle$, for all $x \in \mathbb{R}^d$; Q is semipositive definite if $Q \ge 0$.

$Q \in \mathbb{S}^{d\times d}$ is semipositive definite if and only if there exists an orthonormal basis $\{\mathbf{v}_1, \ldots, \mathbf{v}_d\}$ of $\mathbb{R}^d$ and $\{\lambda_1, \ldots, \lambda_d\} \subset [0, \infty[$, such that

$$Q\mathbf{v}_i = \lambda_i \mathbf{v}_i, \ \forall\, i \in \overline{1, n}.$$

Then $\mathbf{Tr}\ Q = \sum\limits_{i=1}^{d} \lambda_i$ and for all $A \in \mathbb{R}^{d\times d}$ we have

$$\mathbf{Tr}\,(AQ) = \sum_{i=1}^{d} \lambda_i\,\langle A\mathbf{v}_i, \mathbf{v}_i\rangle \le \|A\|\,\mathbf{Tr}Q \le |A|\,\mathbf{Tr}Q. \tag{6.1}$$

6.3 Annex B: Elements of Nonlinear Analysis

6.3.1 Notations

As references for this Annex, see e.g. [2] or [12]. Throughout in this Annex $\mathbb{H}$ is a real separable Hilbert space with norm $|\cdot|$ and scalar product $\langle \cdot, \cdot\rangle$.

Let $(\mathbb{X}, \|\cdot\|_{\mathbb{X}})$ be a real Banach space with dual $(\mathbb{X}^*, \|\cdot\|_{\mathbb{X}^*})$. The duality paring $(\mathbb{X}^*, \mathbb{X})$ is also denoted $\langle\cdot,\cdot\rangle$; hence if $x \in \mathbb{X}$ and $\hat{x} \in \mathbb{X}^*$, then by $\langle \hat{x}, x\rangle$ and $\langle x, \hat{x}\rangle$ we understand the value, $\hat{x}(x)$, of $\hat{x}$ in x.

Given $x \in \mathbb{X}$, $\hat{x} \in \mathbb{X}^*$ and the sequences $x_n \in \mathbb{X}$, $\hat{x}_n \in \mathbb{X}^*$ we say that as $n \to \infty$

$$x_n \to x \text{ (strongly) in } \mathbb{X} \text{ if } \|x_n - x\|_{\mathbb{X}} \to 0,$$

$$x_n \overset{w}{\rightharpoonup} x \text{ (weakly) in } \mathbb{X} \text{ if } \langle \hat{y}, x_n\rangle \to \langle \hat{y}, x\rangle, \text{ for all } \hat{y} \in \mathbb{X}^*,$$

$$\hat{x}_n \overset{w^*}{\rightharpoonup} \hat{x} \text{ (weak star) in } \mathbb{X}^* \text{ if } \langle \hat{x}_n, y\rangle \to \langle \hat{x}, y\rangle, \text{ for all } y \in \mathbb{X}.$$

6.3.2 Maximal Monotone Operators

Let $\mathbb{X}$ and $\mathbb{Y}$ be Banach spaces. A multivalued operator $A : \mathbb{X} \rightrightarrows \mathbb{Y}$ (a point-to-set operator $A : \mathbb{X} \to 2^{\mathbb{Y}}$) will also be regarded as a subset of $\mathbb{X} \times \mathbb{Y}$ setting for $A \subset \mathbb{X} \times \mathbb{Y}$,

$$Ax = \{y \in \mathbb{Y} : (x, y) \in A\}.$$

Define

$$D(A) = \mathrm{Dom}(A) = \{x \in \mathbb{X} : Ax \neq \emptyset\} \text{ – the domain of } A,$$

$$R(A) = \{y \in \mathbb{Y} : \exists\, x \in D(A), \text{ s.t. } y \in Ax\} \text{ – the range of } A,$$

and define $A^{-1} : \mathbb{Y} \rightrightarrows \mathbb{X}$ to be the point-to-set operator defined by $x \in A^{-1}(y)$ if $y \in A(x)$.

We give some definitions:

- $A : \mathbb{X} \rightrightarrows \mathbb{X}^*$ is *monotone* if

$$\langle y_1 - y_2, x_1 - x_2\rangle \geq 0, \text{ for all } (x_1, y_1) \in A, (x_2, y_2) \in A.$$

- $A : \mathbb{X} \rightrightarrows \mathbb{X}^*$ is a *maximal monotone* operator if A is a monotone operator and it is maximal in the set of monotone operators: that is,

$$\langle v - y, u - x\rangle \geq 0 \quad \forall\, (x, y) \in A, \implies (u, v) \in A.$$

- $\mathbf{J}_X : \mathbb{X} \rightrightarrows \mathbb{X}^*$ defined by

$$\mathbf{J}_X(x) = \left\{\hat{x} : \|\hat{x}\|_*^2 = \|x\|^2 = \langle \hat{x}, x\rangle\right\}$$
$$= \left\{\hat{x} : \langle \hat{x}, y - x\rangle + \frac{1}{2}\|x\|^2 \leq \frac{1}{2}\|y\|^2, \quad \forall\, y \in \mathbb{X}\right\}$$

is called the *duality mapping*; if $\mathbb{X} = \mathbb{H}$ is a Hilbert space then $\mathbf{J}_X(x) = \mathbf{I}_{\mathbb{H}}(x) = x$ for all $x \in \mathbb{H}$.

- $A : \mathbb{X} \rightrightarrows \mathbb{Y}$ is *locally bounded* at $x_0 \in D(A)$ if there exists a neighborhood V of x_0 such that $A(V) = \{y \in \mathbb{Y} : \exists\, x \in D(A) \cap V, \text{ s.t. } y \in Ax\}$ is bounded in $\mathbb{Y}$.

We have:

Proposition 6.1 (Rockafellar). *Let $\mathbb{X}$ be a reflexive Banach space. Then $A : \mathbb{X} \rightrightarrows \mathbb{X}^*$ is maximal monotone operator if and only if A is a monotone operator and*

$$R(\mathbf{J}_X + \varepsilon A) = \mathbb{X}^*, \textit{ for all } \varepsilon > 0.$$

Proposition 6.2. *Let $A : \mathbb{H} \rightrightarrows \mathbb{H}$ be a maximal monotone operator. Then:*

(a) A is a closed subset of $\mathbb{H} \times \mathbb{H}$; moreover if $(x_n, y_n) \in A$ and

$$\begin{aligned} &x_n \to x \ \textit{(strongly) in } \mathbb{H} \textit{ and } y_n \overset{w}{\to} y \quad \textit{(weakly) in } \mathbb{H}, \quad \textit{or} \\ &x_n \overset{w}{\to} x, \quad \textit{and} \quad y_n \to y, \ \textit{or} \\ &x_n \overset{w}{\to} x, \quad y_n \overset{w}{\to} y, \quad \textit{and } \overline{\lim}_n \langle x_n, y_n \rangle \le \langle x, y \rangle, \end{aligned}$$

then $(x, y) \in A$;

(b) $\overline{D(A)}$ and $\overline{R(A)}$ are convex subsets of $\mathbb{H}$;

(c) Ax is a convex closed subset of $\mathbb{H}$ for all $x \in D(A)$;

(d) A is locally bounded on $\operatorname{int}(D(A))$ that is: for every $u_0 \in \operatorname{int}(\operatorname{Dom}(A))$ there exists an $r_0 > 0$ such that

$$\bar{B}(u_0, r_0) \overset{def}{=} \{u_0 + r_0 v : |v| \le 1\} \subset \operatorname{Dom}(A)$$

and

$$A^{\#}_{u_0, r_0} \overset{def}{=} \sup\{|\hat{u}| : \hat{u} \in A(u_0 + r_0 v),\ |v| \le 1\} < \infty.$$

Proposition 6.3. *1. If $A : \mathbb{H} \to \mathbb{H}$ is a single-valued monotone hemicontinuous operator then A is maximal monotone ($A : \mathbb{H} \to \mathbb{H}$ is hemicontinuous if the function $t \longrightarrow \langle A(x + tz), y \rangle : \mathbb{R} \to \mathbb{R}$ is continuous for all $x, y, z \in \mathbb{H}$).*

2. If $A, B \subset \mathbb{H} \times \mathbb{H}$ are maximal monotone sets and $\operatorname{int}(D(A)) \cap D(B) \neq \emptyset$, then $A + B \overset{def}{=} \{(x, y + z) : (x, y) \in A,\ (x, z) \in B\}$ is maximal monotone in $\mathbb{H} \times \mathbb{H}$.

Let $A \subset \mathbb{H} \times \mathbb{H}$ be a maximal monotone operator. Then for each $\varepsilon > 0$ the operators

$$J_\varepsilon x = (I + \varepsilon A)^{-1}(x) \text{ and } A_\varepsilon(x) = \frac{1}{\varepsilon}(x - J_\varepsilon x)$$

from $\mathbb{H}$ to $\mathbb{H}$ are single-valued. The operator A_ε is called Yosida's approximation of the operator A. In [2, 12] we can find the proof of the following properties:

Proposition 6.4. *Let $A : \mathbb{H} \rightrightarrows \mathbb{H}$ be a maximal monotone operator. Then:*

(j) *For all $\varepsilon, \delta > 0$ and for all $x, y \in \mathbb{H}$*

$$\begin{array}{ll} i) & (J_\varepsilon x, A_\varepsilon x) \in A, \\ ii) & |J_\varepsilon x - J_\varepsilon y| \le |x - y|, \\ iii) & |A_\varepsilon x - A_\varepsilon y| \le \dfrac{1}{\varepsilon} |x - y|, \\ iv) & |J_\varepsilon x - J_\delta x| \le |\varepsilon - \delta| \, |A_\delta x|, \\ v) & |J_\varepsilon x| \le |x| + (1 + |\varepsilon - 1|) \, |J_1 0|, \\ vi) & A_\varepsilon : \mathbb{H} \to \mathbb{H} \quad \text{is a maximal monotone operator.} \end{array}$$

(jj) *If $\varepsilon_n \to 0$, $x_n \xrightarrow{w} x$, $A_{\varepsilon_n} x_n \xrightarrow{w} y$ and*

$$\limsup_{n,m\to\infty} \langle x_n - x_m, A_{\varepsilon_n} x_n - A_{\varepsilon_m} x_m \rangle \le 0,$$

then $(x, y) \in A$ and $\lim\limits_{n,m\to\infty} \langle x_n - x_m, A_{\varepsilon_n} x_n - A_{\varepsilon_m} x_m \rangle = 0$.

(jjj) $\lim\limits_{\varepsilon \searrow 0} J_\varepsilon x = \mathrm{Pr}_{\overline{D(A)}} x, \ \forall x \in \mathbb{H}$ *and*

$$\lim_{\varepsilon \searrow 0} x_\varepsilon = x \in D(A) \quad \Rightarrow \quad \lim_{\varepsilon \searrow 0} J_\varepsilon x_\varepsilon = x.$$

($\mathrm{Pr}_{\overline{D(A)}} x$ *is the orthogonal projection of x on $\overline{D(A)}$.)*

(jv) $\lim\limits_{\varepsilon \searrow 0} A_\varepsilon x = \mathrm{Pr}_{Ax} \{0\} \overset{def}{=} A^0 x \in Ax$, *for all $x \in D(A)$.*

(v) $|A_\varepsilon x|$ *is monotone decreasing in $\varepsilon > 0$, and when $\varepsilon \searrow 0$*

$$|A_\varepsilon (x)| \nearrow \begin{cases} \left|A^0 (x)\right|, & \text{if} \quad x \in D(A), \\ +\infty, & \text{if} \quad x \notin D(A). \end{cases}$$

(vj) $|J_\varepsilon x - x| = \varepsilon \, |A_\varepsilon x| \le \varepsilon \left|A^0 x\right| \le \varepsilon \, |z|$, *for all $(x, z) \in A$.*

(vjj) *For all $x \in \mathbb{H}$,*

$$\begin{aligned} |J_\varepsilon x - x| &\le |J_\varepsilon x - J_\varepsilon (J_1 x)| + |J_\varepsilon (J_1 x) - J_1 x| + |J_1 x - x| \\ &\le 2 \, |J_1 x - x| + \varepsilon \left|A^0 (J_1 x)\right|. \end{aligned}$$

(vjjj) *For all $x \in \mathbb{H}$ and $y \in \mathrm{Dom}\,(A)$*

$$\begin{aligned} |J_\varepsilon x - J_\delta y| &\le |x - y| + |\varepsilon - \delta| \, |A_\delta y| \\ &\le |x - y| + |\varepsilon - \delta| \left|A^0 y\right|. \end{aligned}$$

The operator A is uniquely defined by its principal section $A^0 x \overset{def}{=} \Pr_{Ax}\{0\}$ in the following sense: if $(x, y) \in \overline{D(A)} \times \mathbb{H}$ such that

$$\left\langle y - A^0 u, x - u \right\rangle \geq 0, \text{ for all } u \in D(A)$$

then $(x, y) \in A$.

Proposition 6.5. *Let $A : \mathbb{H} \rightrightarrows \mathbb{H}$ be a maximal monotone operator.*

I. If $\bar{B}(x_0, r_0) \subset \mathrm{Dom}(A)$ and

$$A^{\#}_{x_0,r_0} \overset{def}{=} \sup\{|\hat{u}| : \hat{u} \in A(x_0 + r_0 v),\ |v| \leq 1\},$$

then

$$r_0 |\hat{x}| \leq \langle \hat{x}, x - x_0 \rangle + A^{\#}_{x_0,r_0} |x - x_0| + r_0 A^{\#}_{x_0,r_0}, \quad \forall\ (x, \hat{x}) \in A. \tag{6.2}$$

II. If there exist $x_0 \in \mathbb{H}$ and $a_0, \hat{a}_0 \geq 0$ such that

$$r_0 |\hat{x}| \leq \langle \hat{x}, x - x_0 \rangle + a_0 |x - x_0| + \hat{a}_0, \quad \forall\ (x, \hat{x}) \in A,$$

then there exists a $b_0 \geq 0$ such that for all $x \in \mathbb{H}$, for all $\varepsilon \in]0, 1]$:

$$r_0 |A_\varepsilon x| \leq \langle A_\varepsilon x, x - x_0 \rangle + a_0 |x - x_0| + b_0. \tag{6.3}$$

If $x_0 \in \mathrm{Dom}(A)$ and $0 \in Ax_0$, then $b_0 = \hat{a}_0$.

Proof. I. By monotonicity of A we have $\forall\ (x, \hat{x}) \in A$, $\forall\ |v| \leq 1$:

$$\begin{aligned} r_0 \langle \hat{x}, v \rangle &\leq r_0 \langle \hat{x}, v \rangle + \langle \hat{x} - \hat{y}, x - (x_0 + r_0 v) \rangle \\ &= \langle \hat{x}, x - x_0 \rangle - \langle \hat{y}, x - x_0 \rangle + r_0 \langle \hat{y}, v \rangle \\ &\leq \langle \hat{x}, x - x_0 \rangle + A^{\#}_{x_0,r_0} |x - x_0| + r_0 A^{\#}_{x_0,r_0}, \end{aligned}$$

which yields (6.2).

II. Since $A_\varepsilon(x) \in A(J_\varepsilon(x))$, it follows that

$$\begin{aligned} r_0 |A_\varepsilon x| &\leq \langle A_\varepsilon x, J_\varepsilon(x) - x_0 \rangle + a_0 |J_\varepsilon(x) - x_0| + \hat{a}_0 \\ &\leq \langle A_\varepsilon x, x - x_0 \rangle + a_0 [|J_\varepsilon(x) - J_\varepsilon(x_0)| + |J_\varepsilon(x_0) - x_0|] + \hat{a}_0 \\ &\leq \langle A_\varepsilon x, x - x_0 \rangle + a_0 |x - x_0| + a_0 |J_\varepsilon(x_0) - x_0| + \hat{a}_0. \end{aligned}$$

Hence the inequality (6.3) holds for $b_0 = a_0 \left[2 |J_1 x_0 - x_0| + \left|A^0(J_1 x_0)\right|\right] + \hat{a}_0$. If $0 \in Ax_0$ then $J_\varepsilon(x_0) = x_0$ and $b_0 = \hat{a}_0$.

■

Proposition 6.6. *If A is a maximal monotone set in $\mathbb{H}\times\mathbb{H}$ and $\mathcal{A}\subset L^2(0,T;\mathbb{H})\times L^2(0,T;\mathbb{H})$ is defined by*

$$\mathcal{A}=\left\{(x,\hat{x})\in L^2(0,T;H)\times L^2(0,T;H):(x(t),\hat{x}(t))\in A,\ a.e.\,t\in\left]0,T\right[\right\},$$

then $\mathcal{A}$ is a maximal monotone set in $L^2(0,T;\mathbb{H})\times L^2(0,T;\mathbb{H})$.

6.3.3 Stochastic Monotone Functions

Let $(\Omega,\mathcal{F},\mathbb{P},\{\mathcal{F}_t\}_{t\geq 0})$ be a complete stochastic basis and

$$F:\Omega\times[0,+\infty[\times\mathbb{R}^d\times\mathbb{R}^{d\times k}\to\mathbb{R}^d$$

such that

- ◊ *$F(\cdot,\cdot,y,z)$ is $\mathcal{P}$-m.s.p. for every $(y,z)\in\mathbb{R}^d\times\mathbb{R}^{d\times k}$;*
- ◊ *for all $y,y'\in\mathbb{R}^d$, $z,z'\in\mathbb{R}^{d\times k}$, $t\geq 0$:*

$$\left\langle y-y',F(t,y,z)-F(t,y',z)\right\rangle\leq 0,\quad \mathbb{P}\text{-a.s.};$$

- ◊ *for all $z,z'\in\mathbb{R}^{d\times k}$, $t\geq 0$:*

$$y\longmapsto F(t,y,z):\mathbb{R}^d\to\mathbb{R}^d \text{ is continuous},\quad \mathbb{P}\text{-a.s.};$$

- ◊ *there exists a $\mathcal{P}$-m.s.p. $\ell:\Omega\times[0,+\infty[\to\mathbb{R}_+$ such that for all $y\in\mathbb{R}^d$, $z,z'\in\mathbb{R}^{d\times k}$, $t\geq 0$:*

$$\left|F(t,y,z)-F(t,y,z')\right|\leq \ell_t\left|z-z'\right|,\quad \mathbb{P}\text{-a.s.}$$

Since $y\mapsto -F(t,y,z):\mathbb{R}^d\to\mathbb{R}^d$ is a monotone continuous operator (hence also maximal monotone operator), it follows that for every $\varepsilon>0$ and $(\omega,t,y,z)\in\Omega\times[0,T]\times\mathbb{R}^d\times\mathbb{R}^{d\times k}$ there exists a unique $J_\varepsilon=J_\varepsilon(\omega,t,y,z)\in\mathbb{R}^d$ such that

$$J_\varepsilon-\varepsilon F(\omega,t,J_\varepsilon,z)=y.$$

The Yosida approximation of F is defined by

$$F_\varepsilon(t,y,z)\overset{def}{=}\frac{1}{\varepsilon}\left(J_\varepsilon(t,y,z)-y\right)=F(t,J_\varepsilon(t,y,z),z).$$

Note that $F_\varepsilon=F_\varepsilon(t,y,z)$ is the unique solution of

$$F(\omega,t,y+\varepsilon F_\varepsilon,z)=F_\varepsilon. \tag{6.4}$$

The functions $J_\varepsilon(\cdot,\cdot,y,z)$, $F_\varepsilon(\cdot,\cdot,y,z) : \Omega \times [0,T] \to \mathbb{R}^d$ are $\mathcal{P}$-m.s.p. for every $(y,z) \in \mathbb{R}^d \times \mathbb{R}^{d\times k}$ and we have:

Proposition 6.7. *For all $\varepsilon, \delta > 0$, $\forall\, t \in [0,T]$, $\forall\, y, y' \in \mathbb{R}^d$, $\forall\, z, z' \in \mathbb{R}^{d\times k}$:*

$$\begin{array}{ll} (a) & |J_\varepsilon(t,y,z) - J_\varepsilon(t,y',z')| \le |y-y'| + \varepsilon \ell_t\,|z-z'|\,, \\ (b) & |J_\varepsilon(t,0,0)| \le \varepsilon\,|F(t,0,0)|\,, \\ (c) & \langle F_\varepsilon(t,y,z) - F_\varepsilon(t,y',z'), y-y'\rangle \le \ell_t\,|z-z'|\,|y-y'|\,, \\ (d) & |F_\varepsilon(t,y,z) - F_\varepsilon(t,y',z')| \le \dfrac{2}{\varepsilon}|y-y'| + \ell_t\,|z-z'|\,, \\ (e) & |J_\varepsilon(t,y,z) - y| \le \varepsilon\,|F_\varepsilon(t,y,z)| \le \varepsilon\,|F(t,y,z)|\,, \\ (f) & \lim\limits_{\varepsilon\to 0} F_\varepsilon(t,y,z) = F(t,y,z)\,, \end{array} \tag{6.5}$$

$$|J_\varepsilon(t,y,z) - J_\delta(t,y',z')| \le \left|y-y'\right| + \delta\,\ell_t\,\left|z-z'\right| + |\varepsilon-\delta|\,|F(t,y,z)| \tag{6.6}$$

and

$$\begin{array}{l} \langle y-y', F_\varepsilon(t,y,z) - F_\delta(t,y',z')\rangle + \varepsilon\,|F_\varepsilon(t,y,z)|^2 + \delta\,|F_\delta(t,y',z')|^2 \\ \quad \le (\varepsilon+\delta)\,\langle F_\varepsilon(t,y,z), F_\delta(t,y',z')\rangle \\ \qquad +\ell_t\,[|y-y'| + \varepsilon\,|F(t,y,z)| + \delta\,|F(t,y',z')|]\,|z-z'|\,. \end{array} \tag{6.7}$$

Proof. (a): If $J = J_\varepsilon(t,y,z)$, $J' = J_\varepsilon(t,y',z')$, then

$$\begin{aligned} &\left|J-J'\right|^2 \\ &\quad = \varepsilon\left\langle F(t,J,z) - F\left(t,J',z'\right), J-J'\right\rangle + \left\langle y-y', J-J'\right\rangle \\ &\quad = \varepsilon\left\langle F(t,J,z) - F\left(t,J',z\right), J-J'\right\rangle \\ &\qquad\quad + \varepsilon\left\langle F\left(t,J',z\right) - F\left(t,J',z'\right), J-J'\right\rangle + \left\langle y-y', J-J'\right\rangle \\ &\quad \le \varepsilon\left[\ell_t\left|z-z'\right|\left|J-J'\right|\right] + \left|y-y'\right|\left|J-J'\right| \end{aligned}$$

and (6.5-a) follows.

(b): With the notation $J^0 = J(t,0,0)$,

$$\left|J^0\right|^2 = \varepsilon\left\langle F\left(t,J^0,0\right), J^0\right\rangle \le \varepsilon\left\langle F(t,0,0), J^0\right\rangle \le \varepsilon\,|F(t,0,0)|\,\left|J^0\right|$$

which gives (6.5-b).

(c): We have

$$\begin{aligned} &\left\langle F_\varepsilon(t,y,z) - F_\varepsilon\left(t,y',z'\right), y-y'\right\rangle \\ &= \frac{1}{\varepsilon}\left\langle J_\varepsilon(t,y,z) - J_\varepsilon(t,y',z'), y-y'\right\rangle - \frac{1}{\varepsilon}\left|y-y'\right|^2 \end{aligned}$$

$$\le \frac{1}{\varepsilon}\left[|y-y'|+\varepsilon\ell_t\left|z-z'\right|\right]\left|y-y'\right|-\frac{1}{\varepsilon}\left|y-y'\right|^2$$
$$\le \ell_t\left|z-z'\right|\left|y-y'\right|$$

that is (6.5-b).

(d): From (a) and the definition of F_ε the inequality (d) clearly follows.

(e): The properties follow from those of the Yosida approximation, A_ε, of a maximal operator A (here $A_\varepsilon(y) = -F_\varepsilon(t,y,z)$ for (ω,t,z) fixed.

(6.6): Let $J_\varepsilon = J_\varepsilon(t,y,z)$ and $J'_\delta = J_\delta(t,y',z')$. Then

$$\begin{aligned}\left|J_\varepsilon - J'_\delta\right|^2 &= (\varepsilon-\delta)\left\langle F\left(t,J_\varepsilon,z\right), J_\varepsilon - J'_\delta\right\rangle \\ &\quad +\delta\left\langle F\left(t,J_\varepsilon,z\right)-F\left(t,J'_\delta,z'\right), J_\varepsilon - J'_\delta\right\rangle + \left\langle y-y', J_\varepsilon - J'_\delta\right\rangle \\ &\le |\varepsilon-\delta|\left|F\left(t,J_\varepsilon,z\right)\right|\left|J_\varepsilon - J'_\delta\right| + \delta\ell_t\left|z-z'\right|\left|J_\varepsilon - J'_\delta\right| + \left|y-y'\right|\left|J_\varepsilon - J'_\delta\right|\end{aligned}$$

and (6.6) follows.

(6.7): Now, we have

$$\begin{aligned}&\left\langle J_\varepsilon - J'_\delta, F_\varepsilon(t,y,z) - F_\delta(t,y',z')\right\rangle \\ &\quad= \left\langle J_\varepsilon - J'_\delta, F(t,J_\varepsilon,z) - F(t,J'_\delta,z')\right\rangle \\ &\quad\le 0 + \left\langle J_\varepsilon - J'_\delta, F(t,J'_\delta,z) - F(t,J'_\delta,z')\right\rangle \\ &\quad\le \left|J_\varepsilon - J'_\delta\right|\ell_t\left|z-z'\right| \\ &\quad\le \ell_t\left[\varepsilon\left|F\left(t,y,z\right)\right| + \delta\left|F\left(t,y',z'\right)\right| + \left|y-y'\right|\right]\left|z-z'\right|\end{aligned}$$

and then

$$\begin{aligned}&\left\langle y-y', F_\varepsilon(t,y,z) - F_\delta(t,y',z')\right\rangle \\ &= \left\langle J_\varepsilon - \varepsilon F_\varepsilon\left(t,y,z\right) - J'_\delta + \delta F_\delta\left(t,y',z'\right), F_\varepsilon(t,y,z) - F_\delta(t,y',z')\right\rangle \\ &\le -\varepsilon\left|F_\varepsilon\left(t,y,z\right)\right|^2 - \delta\left|F_\delta\left(t,y',z'\right)\right|^2 + (\varepsilon+\delta)\left\langle F_\varepsilon\left(t,y,z\right), F_\delta(t,y',z')\right\rangle \\ &+ \ell_t\left[\varepsilon\left|F\left(t,y,z\right)\right| + \delta\left|F\left(t,y',z'\right)\right| + \left|y-y'\right|\right]\left|z-z'\right|\end{aligned}$$

that is (6.7). ■

If we define

$$F_R^{\#}(t) \overset{def}{=} \sup_{|y|\le R} |F(t,y,0)|,$$

then we have the following:

Proposition 6.8. *For all* $\varepsilon > 0$, $p, a > 1$, $r_0 \geq 0$, $y \in \mathbb{R}^d$, $z \in \mathbb{R}^{d\times k}$, $t \in [0,T]$*:*

$$\begin{aligned} r_0 |F_\varepsilon(t,y,z)| + \langle F_\varepsilon(t,y,z), y\rangle &\leq r_0 \left(F^\#_{r_0}(t) + r_0 \frac{a}{2n_p}(\ell_t)^2\right) \\ &+ \left(F^\#_{r_0}(t) + r_0 \frac{a}{n_p}(\ell_t)^2\right)|y| + \frac{a}{2n_p}(\ell_t)^2 |y|^2 + \frac{n_p}{2a}|z|^2, \ a.s., \end{aligned} \tag{6.8}$$

where

$$n_p \overset{def}{=} 1 \wedge (p-1).$$

Proof. Let $0 \leq r_0 \leq 1$. The monotonicity property of F_ε implies that for all $|u| \leq 1$:

$$\langle F_\varepsilon(t, r_0 u, z) - F_\varepsilon(t,y,z), r_0 u - y\rangle \leq 0,$$

and, consequently, $\forall\ |u| \leq 1$:

$$\begin{aligned} & r_0 \langle F_\varepsilon(t,y,z), -u\rangle + \langle F_\varepsilon(t,y,z), y\rangle \\ & \leq |F_\varepsilon(t, r_0 u, z)|\,|y - r_0 u| \\ & \leq |F_\varepsilon(t, r_0 u, 0)|\,(|y| + r_0) + \ell_t |z|\,(|y| + r_0) \\ & \leq |F(t, r_0 u, 0)|\,(|y| + r_0) + \frac{a}{2n_p}(\ell_t)^2 (|y| + r_0)^2 + \frac{n_p}{2a}|z|^2. \end{aligned}$$

The inequality (6.8) follows by taking the sup of the left-hand side over all vectors u such that $|u| \leq 1$. ■

Finally we give some convergence results.
Let $F : \Omega \times [0,T] \times \mathbb{R}^d \to \mathbb{R}^{d'}$ be a function satisfying

$$\begin{aligned} &i) \quad F(\cdot,\cdot,x) \quad \text{is } \mathcal{F} \otimes \mathcal{B}_{[0,T]}\text{-}measurable, \ \forall\, x \in \mathbb{R}^d, \\ &ii) \quad F(\omega,t,\cdot) \quad \text{is continuous } d\mathbb{P} \otimes dt\text{-}a.e.\ (\omega,t) \in \Omega \times [0,T], \\ &iii) \quad \exists\, \alpha > 0 \text{ such that} \int_0^T \left(F^\#_R(t)\right)^\alpha dt < +\infty, \quad \mathbb{P}\text{-a.s.}, \ \forall\, R > 0. \end{aligned} \tag{6.9}$$

Proposition 6.9. *Assume that* F *satisfies (6.9). Let*

$$X^\varepsilon, X \in L^0\left(\Omega; C\left([0,T]; \mathbb{R}^d\right)\right)$$

be such that

$$\sup_{t\in[0,T]} |X^\varepsilon_t - X_t| \xrightarrow[\varepsilon\to 0]{prob.} 0.$$

Then

$$\sup_{t\in[0,T]}\left|\int_0^t F\left(s,X_s^\varepsilon\right)ds-\int_0^t F\left(s,X_s\right)ds\right|$$
$$\leq\int_0^T\left|F\left(s,X_s^\varepsilon\right)-F\left(s,X_s\right)\right|ds\xrightarrow[\varepsilon\to 0]{prob.}0.$$

Moreover if for some $p,\alpha>0$:

$$C_{p,\alpha}\overset{def}{=}\sup_{0<\varepsilon\leq\varepsilon_0}\mathbb{E}\left(\int_0^T\left|F\left(t,X_t^\varepsilon\right)\right|^\alpha dt\right)^p<+\infty,\tag{6.10}$$

then

$$\begin{aligned}&c_1)\quad \mathbb{E}\left(\int_0^T\left|F\left(t,X_t\right)\right|^\alpha dt\right)^p\leq C_p,\\&c_2)\quad \mathbb{E}\left(\int_0^T\left|F\left(t,X_t^\varepsilon\right)-F\left(t,X_t\right)\right|^\alpha dt\right)^q\underset{\varepsilon\to 0}{\longrightarrow}0,\quad\forall\, q\in]0,p[.\end{aligned}\tag{6.11}$$

If, in addition, $x\longmapsto -F\,(t,x)$ is a monotone operator and $F_\varepsilon=F_\varepsilon(t,x)$, $\varepsilon>0$, is the Yosida approximation of F (F_ε is the unique solution of $F(\omega,t,x+\varepsilon F_\varepsilon)=F_\varepsilon$) then $\forall\, q\in]0,p[$:

$$\mathbb{E}\left(\int_0^T\left|F_\varepsilon\left(t,X_t^\varepsilon\right)-F\left(t,X_t\right)\right|^\alpha dt\right)^q\underset{\varepsilon\to 0}{\longrightarrow}0.\tag{6.12}$$

Proof. Let $\varepsilon_n\to 0$ such that

$$\lim_{\varepsilon_n\to 0}\sup_{t\in[0,T]}\left|X_t^{\varepsilon_n}-X_t\right|=0,\quad \mathbb{P}\text{-a.s.}$$

Then by the Lebesgue dominated convergence theorem

$$\lim_{\varepsilon_n\to 0}\int_0^T\left|F\left(s,X_s^{\varepsilon_n}\right)-F\left(s,X_s\right)\right|^\alpha ds=0,\quad \mathbb{P}\text{-a.s.}$$

Since the convergence in probability is given by a metric, by *reductio ad absurdum* we infer that

$$\int_0^T\left|F\left(s,X_s^\varepsilon\right)-F\left(s,X_s\right)\right|^\alpha ds\xrightarrow[\varepsilon\to 0]{prob.}0.$$

Also, if $C_p<\infty$, then Fatou's lemma clearly yields (6.11-c_1).

I. Denote by C positive constants independent of ε_n. Let

$$(\Delta_n =)\, \Delta_{\varepsilon_n} \overset{def}{=} \int_0^T \left|F\left(s, X_s^{\varepsilon_n}\right) - F\left(s, X_s\right)\right|^\alpha ds.$$

Then by the Lebesgue dominated convergence theorem $\Delta_n \to 0, \quad \mathbb{P}$-a.s., and

$$\mathbb{E}\Delta_n^p \leq C.$$

Since

$$\begin{aligned}\mathbb{E}\Delta_n^q &\leq \mathbb{E}\left(\Delta_n^q \mathbf{1}_{\Delta_n \leq R}\right) + \mathbb{E}\left(\Delta_n^q \frac{\Delta_n^{p-q}}{R^{p-q}} \mathbf{1}_{\Delta_n > R}\right) \\ &\leq \mathbb{E}\left(\Delta_n^q \mathbf{1}_{\Delta_n \leq R}\right) + \frac{C}{R^{p-q}},\end{aligned}$$

it follows that

$$0 \leq \limsup_{\varepsilon_n \to 0} \mathbb{E}\Delta_n^q \leq \frac{C}{R^{p-q}} \qquad \forall R > 0,$$

that is $\lim_{\varepsilon_n \to 0} \mathbb{E}\Delta_n^q = 0$ and by *reductio ad absurdum* the full sequence Δ_ε has the property (6.11-c_2).

II. Since $\left|F_\varepsilon\left(t, X_t^\varepsilon\right)\right| \leq |F(t, X_t^\varepsilon)|$, on a subsequence

$$\begin{aligned}\lim_{\varepsilon_n \to 0} F_{\varepsilon_n}\left(t, X_t^{\varepsilon_n}\right) &= \lim_{\varepsilon_n \to 0} F\left(t, X_t^\varepsilon + \varepsilon_n F_{\varepsilon_n}\left(t, X_t^{\varepsilon_n}\right)\right) \\ &= F(t, X_t), \quad \mathbb{P}\text{-a.s.}\end{aligned}$$

and then the convergence result, (6.12), follows in exactly the same manner with $\wedge_{\varepsilon_n} := \int_0^T \left|F_{\varepsilon_n}\left(s, X_s^{\varepsilon_n}\right) - F\left(s, X_s\right)\right|^\alpha ds.$

■

6.3.4 Compactness Results

Let $I \subset \mathbb{R}$ be an interval. Denote by $C\left(I; \mathbb{R}^d\right)$ the space of continuous functions $g : I \to \mathbb{R}^d$. If $I = [a, b]$ then $C\left([a, b]; \mathbb{R}^d\right)$ is a separable Banach space with respect to the norm $\|\cdot\|_{[a,b]}$, where if $g : [a, b] \to \mathbb{R}^d$ we define

$$\|g\|_{[a,t]} = \sup\left\{|g(s)| : a \leq s \leq t\right\}.$$

If $[a,b]=[0,t]$ then

$$\|g\|_t \stackrel{def}{=} \|g\|_{[0,t]} = \sup\{|g(s)| : 0\le s\le t\}.$$

For $g\in C([0,T];\mathbb{R}^d)$ we define for $t\in[0,T]$ and $\varepsilon\ge 0$:

$$\mathbf{m}_g(t,\varepsilon)=\sup\{\|g(t)-g(s)\|_{\mathbb{R}^d} : |t-s|\le\varepsilon,\ s\in[0,T]\}$$

the modulus of continuity at t, and

$$\mathbf{m}_g(\varepsilon)=\mathbf{m}(\varepsilon;g)=\sup\{|g(t)-g(s)| : |t-s|\le\varepsilon,\ t,s\in[0,T]\}$$

the modulus of uniformly continuity.

We also introduce the notation

$$\boldsymbol{\mu}_g(\varepsilon)=\boldsymbol{\mu}(\varepsilon;g)=\varepsilon+\mathbf{m}_g(\varepsilon).$$

Note that

$$\begin{array}{ll} m_1) & 0=\mathbf{m}_g(0)\le\mathbf{m}_g(\varepsilon)\le\mathbf{m}_g(\delta)\le 2\|g\|_T, \qquad \forall 0<\varepsilon<\delta,\\ m_2) & 0=\boldsymbol{\mu}_g(0)<\boldsymbol{\mu}_g(\varepsilon)<\boldsymbol{\mu}_g(\delta), \qquad \forall 0<\varepsilon<\delta,\\ m_3) & \mathbf{m}_g(\varepsilon+\delta)\le\mathbf{m}_g(\varepsilon)+\mathbf{m}_g(\delta), \quad \forall\,\varepsilon,\delta\ge 0,\\ m_4) & \lim_{\varepsilon\searrow 0}\mathbf{m}_g(\varepsilon)=\lim_{\varepsilon\searrow 0}\boldsymbol{\mu}_g(\varepsilon)=0 \end{array} \tag{6.13}$$

and

$$\left|\mathbf{m}_g(t,\varepsilon)-\mathbf{m}_h(t,\varepsilon)\right|\le\left|\mathbf{m}_g(\varepsilon)-\mathbf{m}_h(\delta)\right|\le 2\|g-h\|_T+m_g(|\varepsilon-\delta|).$$

If $\mathcal{M}\subset C([0,T];\mathbb{R}^d)$ and $\varepsilon>0$, then

$$\begin{aligned} \mathbf{m}_{\mathcal{M}}(t,\varepsilon) &\stackrel{def}{=} \sup\{\mathbf{m}_g(t,\varepsilon) : g\in\mathcal{M}\},\\ \|\mathcal{M}\|_T &\stackrel{def}{=} \sup\{\|g\|_T : g\in\mathcal{M}\},\\ \mathcal{M}(t) &\stackrel{def}{=} \{g(t) : g\in\mathcal{M}\}, \end{aligned}$$

and

$$\mathbf{m}_{\mathcal{M}}(\varepsilon)\stackrel{def}{=}\sup\{\mathbf{m}_g(\varepsilon) : g\in\mathcal{M}\},$$

$$\boldsymbol{\mu}_{\mathcal{M}}(\varepsilon)\stackrel{def}{=}\varepsilon+\mathbf{m}_{\mathcal{M}}(\varepsilon).$$

Theorem 6.10 (Arzelà–Ascoli). *Let $\mathcal{M}\subset C([0,T];\mathbb{R}^d)$. Then the following three conditions are equivalent:*

(A) $\mathcal{M}$ *is relatively compact in* $C\left([0,T];\mathbb{R}^d\right)$*;*

(B) (B_1) *(equicontinuity):* $\lim_{\varepsilon\to 0} \mathbf{m}_{\mathcal{M}}(t,\varepsilon) = 0,\ \forall\, t \in [0,T]$*;*

(B_2) *(bounded images) for each* $t \in [0,T]$ *the set* $\mathcal{M}(t) = \{g(t) : g \in \mathcal{M}\}$ *is bounded in* $\mathbb{R}^d$*;*

(C) (C_1) *(uniform equicontinuity):* $\lim_{\varepsilon\to 0} \mathbf{m}_{\mathcal{M}}(\varepsilon) = 0$*;*

(C_2) *the set* $\{g(t) : t \in [0,T],\ g \in \mathcal{M}\}$ *is bounded in* $\mathbb{R}^d$*.*

Theorem 6.11 (Kolmogorov–Riesz–Weil). *Let* $p \in [1,\infty[$*. A set* $\mathcal{S} \subset L^p\left(0,T;\mathbb{R}^d\right)$ *is relatively compact in* $L^p\left(0,T;\mathbb{R}^d\right)$ *if and only if:*

(j) (p-equi-integrability)

$$\lim_{\varepsilon\searrow 0}\left[\sup_{g\in\mathcal{S}}\int_{\varepsilon}^{T-\varepsilon}\|g(t+\varepsilon)-g(t)\|_{\mathbb{R}^d}^p\,dt\right]=0,$$

(jj) (boundedness):

$$\sup_{g\in\mathcal{S}}\int_0^T |g(t)|\,dt < \infty.$$

(For the proofs of these two last theorems see as example the book of Vrabie [70].)

Clearly we have:

Corollary 6.12. *Let* $M > 0$ *and* $\gamma_n \searrow 0$*,* $\varepsilon_n \searrow 0$ *be two sequences.*

a) Then the set

$$\mathcal{K}_1 = \left\{ z \in L^2(0,T;\mathbb{R}^d) : \int_0^T |z(t)|^2\,dt \le M,\right.$$
$$\left.\sup_{0\le\theta\le\varepsilon_n}\int_0^{T-\varepsilon_n}|z(t+\theta)-z(t)|^2\,dt \le \gamma_n,\ \forall n \in \mathbb{N}^*\right\}$$

is a compact subset of $L^2(0,T;\mathbb{R}^d)$*.*

b) If $N_n = \left[\frac{T}{\varepsilon_n}\right]$ *and* $t_i = \frac{(i-1)T}{N_n}$*, for* $1 \le i \le N_n, n \ge 1$*, then the set*

$$\mathcal{K}_2 = \left\{ z \in C([0,T];\mathbb{R}^d) : |z(0)| \le M,\right.$$
$$\left.\sup_{1\le i\le N_n}\ \sup_{0<\theta\le\varepsilon_n}|z(t_i+\theta)-z(t_i)| \le \gamma_n, \forall n \in \mathbb{N}^*\right\}$$

is a compact subset of $C([0,T];\mathbb{R}^d)$ *(here* z_t *is extended outside of* $[0,T]$ *by continuity* $z_s = z_T$*, for* $s \ge T$ *and* $z_s = z_0$*, for* $s \le 0$*).*

6.3.5 Bounded Variation Functions

Let $[a, b]$ be a closed interval from $\mathbb{R}$ and $\mathcal{D}_{[a,b]}$ be the set of all partitions

$$\Delta : \ a = t_0 < t_1 < \cdots < t_n = b, \quad n = n_\Delta \in \mathbb{N}^*.$$

Define $\|\Delta\| = \sup\{t_{i+1} - t_i : 0 \leq i \leq n-1\}$.

Let

$$V_\Delta(k) \overset{def.}{=} \sum_{i=0}^{n-1} |k(t_{i+1}) - k(t_i)|$$

be the variation of k corresponding to the partition $\Delta \in \mathcal{D}_{[a,b]}$. We define the total variation of k on $[a, b]$ by

$$\begin{aligned} \updownarrow k \updownarrow_{[a,b]} &= \sup_{\Delta \in \mathcal{D}_{[a,b]}} V_\Delta(k) \\ &= \sup\left\{\sum_{i=0}^{n_\Delta - 1} |k(t_{i+1}) - k(t_i)| : \Delta \in \mathcal{D}_{[a,b]}\right\} \end{aligned}$$

and if $[a, b] = [0, T]$ then

$$\updownarrow k \updownarrow_T := \updownarrow k \updownarrow_{[0,T]}.$$

Proposition 6.13. *If* $k \in C\left([0, T]; \mathbb{R}^d\right)$ *and* $\overline{\Delta}_N \in \mathcal{D}_{[0,T]}$

$$\overline{\Delta}_N : \left\{0 = \frac{0}{2^N}T < \frac{1}{2^N}T < \cdots < \frac{2^N - 1}{2^N}T < \frac{2^N}{2^N}T = T\right\},$$

then

$$V_{\overline{\Delta}_N}(k) \nearrow \updownarrow k \updownarrow_T \quad \text{as } N \nearrow \infty.$$

Proof. Clearly $V_{\overline{\Delta}_N}(k)$ is increasing with respect to N and $V_{\overline{\Delta}_N}(k) \leq \updownarrow k \updownarrow_T$.

Let $\Delta \in \mathcal{D}_{[a,b]}$ be arbitrary

$$\Delta : \ 0 = t_0 < t_1 < \cdots < t_{n_\Delta} = T,$$

and $j_i = \left[\frac{t_i}{T} 2^N\right]$ be the integer part of $\frac{t_i}{T} 2^N$. Then

$$V_\Delta(k) = \sum_{i=1}^{n_\Delta - 1} |k(t_{i+1}) - k(t_i)|$$

$$\leq \sum_{i=1}^{n_\Delta - 1} \left[\left| k(t_{i+1}) - k\left(\frac{j_{i+1}T}{2^N}\right) \right| + \left| k\left(\frac{j_{i+1}T}{2^N}\right) - k\left(\frac{j_i T}{2^N}\right) \right| \right.$$
$$\left. + \left| k\left(\frac{j_i T}{2^N}\right) - k(t_i) \right| \right]$$
$$\leq 2 n_\Delta \mathbf{m}_k \left(\frac{T}{2^N}\right) + V_{\overline{\Delta}_N}(k)$$

and passing to the limit for $N \nearrow \infty$ we obtain

$$V_\Delta(k) \leq \lim_{N \nearrow \infty} V_{\overline{\Delta}_N}(k) \leq \updownarrow k \updownarrow_T, \quad \forall \Delta \in \mathcal{D}_{[a,b]}.$$

Hence $\lim_{N \nearrow \infty} V_{\overline{\Delta}_N}(k) = \updownarrow k \updownarrow_T$. ■

Definition 6.14. A function $k : [a,b] \to \mathbb{R}^d$ has bounded variation on $[a,b]$ if $\updownarrow k \updownarrow_{[a,b]} < \infty$. The space of bounded variation functions on $[a,b]$ will be denoted by $BV\left([a,b];\mathbb{R}^d\right)$.

If $x \in C\left([a,b];\mathbb{R}^d\right)$ and $k \in BV\left([a,b];\mathbb{R}^d\right)$ then the Riemann–Stieltjes integral is defined by

$$\int_a^b \langle x(t), dk(t) \rangle = \lim_{\|\Delta\| \to 0} \sum_{i=0}^{n_\Delta - 1} \langle x(\tau_i), k(t_{i+1}) - k(t_i) \rangle,$$

where the integral is independent of the arbitrary choice of $\tau_i \in [t_i, t_{i+1}]$.

The Riemann–Stieltjes integral satisfies

$$\left| \int_a^b \langle x(t), dk(t) \rangle \right| \leq \|x\|_{[a,b]} \ \updownarrow k \updownarrow_{[a,b]} .$$

Proposition 6.15. *Equipped with the norm*

$$\|k\|_{BV([a,b];\mathbb{R}^d)} := |k(a)| + \updownarrow k \updownarrow_{[a,b]},$$

the space $BV\left([a,b];\mathbb{R}^d\right)$ is a Banach space. An element k of $BV\left([a,b];\mathbb{R}^d\right)$ can be identified with the following linear continuous mapping on $C\left([a,b];\mathbb{R}^d\right)$:

$$x \longmapsto \langle x(a), k(a) \rangle + \int_a^b \langle x(t), dk(t) \rangle .$$

With this identification, $BV\left([a,b];\mathbb{R}^d\right)$ is the dual of the space $C\left([a,b];\mathbb{R}^d\right)$.

Proposition 6.16 (Helly–Bray). *Let $n \in \mathbb{N}^*$, $x_n, x \in C\left([0,T];\mathbb{R}^d\right)$, $k_n \in BV\left([0,T];\mathbb{R}^d\right)$, $k:[0,T]\to\mathbb{R}^d$, such that*

$$\begin{array}{ll} (i) & x_n \to x \quad \text{in } C\left([0,T];\mathbb{R}^d\right), \\ (ii) & k_n(t) \to k(t), \ \forall\, t \in [0,T] \quad \text{and} \\ (iii) & \sup\limits_{n\in\mathbb{N}^*} \updownarrow k_n \updownarrow_T = M < +\infty. \end{array}$$

Then $k \in BV\left([0,T];\mathbb{R}^d\right)$, $\updownarrow k \updownarrow_T \leq M$, and $\forall\, 0 \leq s \leq t \leq T$:

$$\begin{array}{ll} (j) & \displaystyle\int_s^t \langle x_n(r), dk_n(r)\rangle \to \int_s^t \langle x(r), dk(r)\rangle, \text{ as } n\to\infty, \\ (jj) & \displaystyle\int_s^t |x(r)|\, d\updownarrow k\updownarrow_r \leq \liminf_{n\to+\infty} \int_s^t |x_n(r)|\, d\updownarrow k_n\updownarrow_r . \end{array}$$

In particular $k_n \stackrel{w^*}{\to} k$ in $BV\left([0,T];\mathbb{R}^d\right)$, that is for all $y \in C\left([0,T];\mathbb{R}^d\right)$:

$$\int_0^T \langle y(t), dk_n(t)\rangle \to \int_0^T \langle y(t), dk(t)\rangle .$$

Proof. First let $\Delta_N \in \mathcal{D}_{[0,T]}$ be a sequence such that

$$V_{\Delta_N}(k) \nearrow \updownarrow k\updownarrow_T \quad \text{as } N \nearrow \infty.$$

From the definition of $\updownarrow\cdot\updownarrow_T$ we have

$$V_{\Delta_N}(k_n) \leq \updownarrow k_n\updownarrow_T \leq M.$$

Since $k_n(t) \to k(t)$ for all $t \in [0,T]$, it follows that $V_{\Delta_N}(k_n) \to V_{\Delta_N}(k)$. Hence

$$V_{\Delta_N}(k) \leq M \quad \text{for all } N \in \mathbb{N}^*$$

and passing to the limit as $N \nearrow \infty$ we obtain

$$\updownarrow k\updownarrow_T \leq M.$$

Let $\varepsilon > 0$

$$\Delta:\ s = t_0 < t_1 < \cdots < t_N = t, \quad N = N_\Delta \in \mathbb{N}^*,$$

with $t_i \in [0,T]$, $\|\Delta\| = \sup\{t_{i+1} - t_i : 0 \leq i \leq N-1\} \leq \varepsilon$. For $x_i = x(t_i)$, $k_i = k(t_i)$, define

$$S_\Delta(x,k) = \sum_{i=0}^{N-1} \langle x_i, k_{i+1} - k_i\rangle$$

and $\mathbf{m}_x : [0, \infty[\to [0, \infty[$

$$\mathbf{m}_x(\varepsilon) = \sup\{|x(r) - x(s)| : |r - s| \le \varepsilon,\ r, s \in [0, T]\}$$

the modulus of continuity of x on $[0, T]$.

We have

$$\left| \int_s^t \langle x(r), dk(r) \rangle - S_\Delta(x, k) \right| \le \mathbf{m}_x(\varepsilon) \updownarrow k \updownarrow_T . \tag{6.14}$$

Indeed

$$\begin{aligned}
\left| \int_s^t \langle x(r), dk(r) \rangle - S_\Delta(x, k) \right| &= \left| \sum_{i=0}^{N-1} \int_{t_i}^{t_i+1} \langle x(r), dk(r) \rangle - \int_{t_i}^{t_i+1} \langle x_i, dk(r) \rangle \right| \\
&\le \sum_{i=0}^{N-1} \left| \int_{t_i}^{t_i+1} \langle x(r) - x_i, dk(r) \rangle \right| \\
&\le \mathbf{m}_x(\|\Delta\|) \sum_{i=0}^{N-1} \updownarrow k \updownarrow_{[t_i, t_i+1]} \\
&\le \mathbf{m}_x(\varepsilon) \updownarrow k \updownarrow_T .
\end{aligned}$$

Then

$$\begin{aligned}
\left| \int_s^t \langle x(r), d(k_n(r) - k(r)) \rangle - S_\Delta(x, k_n - k) \right| &\le \mathbf{m}_x(\|\Delta\|) \updownarrow k_n - k \updownarrow_T \\
&\le \mathbf{m}_x(\varepsilon) \left[\updownarrow k_n \updownarrow_T + \updownarrow k \updownarrow_T \right].
\end{aligned}$$

Now we obtain the estimate

$$\begin{aligned}
&\left| \int_s^t \langle x_n(r), dk_n(r) \rangle - \int_s^t \langle x(r), dk(r) \rangle \right| \\
&\quad = \left| \int_s^t \langle x_n(r) - x(r), dk_n(r) \rangle + \int_s^t \langle x(r), dk_n(r) - dk(r) \rangle \right| \\
&\quad \le \|x_n - x\|_T \ \updownarrow k_n \updownarrow_T + \left| \int_s^t \langle x(r), dk_n(r) - dk(r) \rangle \right| \\
&\quad \le \|x_n - x\|_T \ \updownarrow k_n \updownarrow_T + \mathbf{m}_x(\varepsilon) \left[\updownarrow k_n \updownarrow_T + \updownarrow k \updownarrow_T \right] + |S_\Delta(x, k_n - k)| .
\end{aligned}$$

Since $k_n(t) \to k(t)$ for all $t \in [0, T]$, it follows that $\lim_{n \to \infty} |S_\Delta(x, k_n - k)| = 0$ and

$$\limsup_{n\to\infty}\left|\int_s^t \langle x_n(r), dk_n(r)\rangle - \int_s^t \langle x(r), dk(r)\rangle\right| \le 2M\,\mathbf{m}_x(\varepsilon), \qquad \forall \varepsilon > 0.$$

Hence the limit $\lim_{n\to\infty}\int_s^t \langle x_n(r), dk_n(r)\rangle$ exists, as does

$$\lim_{n\to\infty}\int_s^t \langle x_n(r), dk_n(r)\rangle = \int_s^t \langle x(r), dk(r)\rangle.$$

Now, let $\alpha \in C\left([0,T];\mathbb{R}^d\right)$, $\|\alpha\|_T \le 1$. Then

$$\begin{aligned}\int_s^t |x(r)|\,\langle \alpha(r), dk(r)\rangle &= \lim_{n\to\infty}\int_s^t |x_n(r)|\,\langle \alpha(r), dk_n(r)\rangle \\ &\le \liminf_{n\to+\infty}\int_s^t |x_n(r)|\,d\updownarrow k_n\updownarrow_r\end{aligned}$$

and passing to $\sup_{\|\alpha\|_T\le 1}$ we obtain

$$\int_s^t |x(r)|\,d\updownarrow k\updownarrow_r \le \liminf_{n\to+\infty}\int_s^t |x_n(r)|\,d\updownarrow k_n\updownarrow_r.$$

■

We now give some other auxiliary results used in the book:

Proposition 6.17. *Let $A:\mathbb{R}^d \rightrightarrows \mathbb{R}^d$ be a maximal monotone operator and $\mathcal{A}: C\left(\mathbb{R}_+;\mathbb{R}^d\right) \rightrightarrows BV_{loc}\left(\mathbb{R}_+;\mathbb{R}^d\right)$ be defined by:*

$$(x,k)\in\mathcal{A} \quad if \quad x \in C\left(\mathbb{R}_+;\overline{D(A)}\right),\ \ k\in BV_{loc}\left(\mathbb{R}_+;\mathbb{R}^d\right) \ and$$

$$\int_s^t \langle x(r) - z, dk(r) - \hat{z}dr\rangle \ge 0, \quad \forall\ (z,\hat{z})\in A,\ \forall\, 0\le s\le t. \tag{6.15}$$

Then the relation (6.15) is equivalent to: for all $u,\hat{u}\in C(\mathbb{R}_+;\mathbb{R}^d)$ such that $(u(r),\hat{u}(r))\in A$, $\forall\, r\ge 0$

$$\int_s^t \langle x(r) - u(r), dk(r) - \hat{u}(r)dr\rangle \ge 0, \quad \forall\, 0\le s\le t, \tag{6.16}$$

and $\mathcal{A}$ is a monotone operator, that is:
for all $(x,k),(y,\ell)\in\mathcal{A}$

$$\int_s^t \langle x(r) - y(r), dk(r) - d\ell(r)\rangle \ge 0, \quad \forall\, 0\le s\le t.$$

Moreover $\mathcal{A}$ is a maximal monotone operator.

Proof. (6.15) $\Longrightarrow$ (6.16):

Let $\forall\, u, \hat{u} \in C\left(\mathbb{R}_+;\mathbb{R}^d\right)$ be such that $(u(r), \hat{u}(r)) \in A,\ \forall\, r \geq 0$. Then

$$\begin{aligned}&\int_s^t \langle x(r) - u(r), dk(r) - \hat{u}(r)dr\rangle \\ &= \lim_{n\to\infty} \int_s^t \left\langle x(r) - u(\frac{\lfloor nr \rfloor}{n}), dk(r) - \hat{u}(\frac{\lfloor nr \rfloor}{n})dr\right\rangle \geq 0.\end{aligned}$$

(6.16) $\Longrightarrow$ (6.15):

The implication is obtained for $u\,(r) = z$ and $\hat{u}\,(r) = \hat{z}$.

Let $(x,k), (y,\ell) \in \mathcal{A}$ be arbitrary. Then for all $u, \hat{u} \in C(\mathbb{R}_+;\mathbb{R}^d)$ such that $(u(r), \hat{u}(r)) \in A,\ \forall\, r \geq 0$ we have for all $0 \leq s \leq t$,

$$\begin{aligned}&\int_s^t \langle y(r) - u(r), d\ell\,(r) - \hat{u}\,(r)\,dr\rangle \geq 0, \\ &\int_s^t \langle x(r) - u(r), dk\,(r) - \hat{u}\,(r)\,dr\rangle \geq 0.\end{aligned}$$

We put here

$$u\,(r) = J_\varepsilon\left(\frac{x\,(r) + y\,(r)}{2}\right) = \frac{x\,(r) + y\,(r)}{2} - \varepsilon A_\varepsilon\left(\frac{x\,(r) + y\,(r)}{2}\right)$$

and

$$\hat{u}\,(r) = A_\varepsilon\left(\frac{x\,(r) + y\,(r)}{2}\right),$$

where $J_\varepsilon(z) = (I + \varepsilon A)^{-1}(z)$, $A_\varepsilon\,(z) = \dfrac{1}{\varepsilon}\,(z - J_\varepsilon\,(z))$. Since A is a maximal operator on $\mathbb{R}^d$ it follows that $\overline{D(A)}$ is convex and $\lim_{\varepsilon\to 0} \varepsilon A_\varepsilon\,(u) \to 0,\ \forall\, u \in \overline{D(A)}$. Also for all $a \in D\,(A)$

$$\varepsilon\,|A_\varepsilon\,(u)| \leq \varepsilon\,|A_\varepsilon\,(u) - A_\varepsilon\,(a)| + \varepsilon\,|A_\varepsilon\,(a)| \leq |u - a| + \varepsilon\left|A^0\,(a)\right|.$$

Adding the inequalities term by term we obtain:

$$0 \leq \frac{1}{2}\int_s^t \langle y\,(r) - x\,(r)\,, d\ell\,(r) - dk\,(r)\rangle + \varepsilon\int_s^t \left\langle A_\varepsilon(\frac{x\,(r) + y\,(r)}{2}), d\ell\,(r) + dk\,(r)\right\rangle.$$

Passing to $\lim_{\varepsilon\searrow 0}$ we obtain $\displaystyle\int_s^t \langle y\,(r) - x\,(r)\,, d\ell\,(r) - dk\,(r)\rangle \geq 0$. $\mathcal{A}$ is a maximal monotone operator since if $(y,\ell) \in C\left(\mathbb{R}_+;\overline{D(A)}\right) \times BV_{loc}\left(\mathbb{R}_+;\mathbb{R}^d\right)$ satisfies

$$\int_s^t \langle y(r) - x(r), d\ell(r) - dk(r)\rangle \geq 0, \quad \forall (x,k) \in \mathcal{A},$$

then this last inequality is satisfied for all (x,k) of the form $(x(t), k(t)) = (z, \hat{z}t)$, where $(z,\hat{z}) \in A$, and consequently (from the definition of $\mathcal{A}$) $(y,\ell) \in \mathcal{A}$. The proof is complete. ■

Remark 6.18. Often we restrict the realization to

$$C\left(\mathbb{R}_+;\mathbb{R}^d\right) \times \left[C\left(\mathbb{R}_+;\mathbb{R}^d\right) \bigcap BV_{0,loc}\left(\mathbb{R}_+;\mathbb{R}^d\right)\right]$$

and we write (for this case) $dk(t) \in A(x(t))(dt)$ if

$$\begin{array}{ll} (a_1) & x \in C\left(\mathbb{R}_+;\overline{\mathrm{Dom}(A)}\right), \\ (a_2) & k \in C\left(\mathbb{R}_+;\mathbb{R}^d\right) \bigcap BV_{loc}\left(\mathbb{R}_+;\mathbb{R}^d\right), \ k(0) = 0, \\ (a_3) & \langle x(t) - u, dk(t) - \hat{u}dt\rangle \geq 0, \quad \text{on } \mathbb{R}_+, \ \forall\ (u,\hat{u}) \in A. \end{array}$$

Proposition 6.19. *Let $A \subset \mathbb{R}^d \times \mathbb{R}^d$ be a maximal subset and $\mathcal{A}$ be the realization of A on $C\left(\mathbb{R}_+;\mathbb{R}^d\right) \times BV_{loc}\left(\mathbb{R}_+;\mathbb{R}^d\right)$ defined by (6.15). Assume that* $\mathrm{int}\,(\mathrm{Dom}\,(A)) \neq \emptyset$. *Let $u_0 \in \mathrm{int}\,(\mathrm{Dom}\,(A))$ and $r_0 > 0$ be such that $\bar{B}(u_0, r_0) \subset$* $\mathrm{Dom}\,(A)$. *Then*

$$A^{\#}_{u_0,r_0} \overset{def}{=} \sup\left\{|\hat{u}| : \hat{u} \in Au,\ u \in \bar{B}(u_0,r_0)\right\} < \infty,$$

and for all $(x,k) \in \mathcal{A}$:

$$r_0 d\, \updownarrow k \updownarrow_t \leq \langle x(t) - u_0, dk(t)\rangle + \left(A^{\#}_{u_0,r_0}|x(t) - u_0| + r_0 A^{\#}_{u_0,r_0}\right) dt \tag{6.17}$$

as signed measures on $\mathbb{R}_+$. Moreover there exists a constant $b_0 > 0$ such that

$$\begin{aligned} r_0 \int_s^t |A_\varepsilon y(r)|\, dr \leq & \int_s^t \langle y(r) - u_0, A_\varepsilon y(r)\rangle\, dr \\ & + A^{\#}_{u_0,r_0} \int_s^t |y(r) - u_0|\, dr + b_0(t-s), \end{aligned} \tag{6.18}$$

for all $0 \leq s \leq t \leq T$, $y \in C\left(\mathbb{R}_+;\mathbb{R}^d\right)$ and $0 < \varepsilon \leq 1$.

Proof. Since A is locally bounded on $\mathrm{int}\,(\mathrm{Dom}\,(A))$, it follows that for $u_0 \in \mathrm{int}\,(\mathrm{Dom}\,(A))$, there exists an $r_0 > 0$ such that $u_0 + r_0 v \in \mathrm{int}\,(\mathrm{Dom}\,(A))$ for all $|v| \leq 1$ and

$$A^{\#}_{u_0,r_0} \overset{def}{=} \sup\left\{|\hat{z}| : \hat{z} \in Az,\ z \in \bar{B}(u_0,r_0)\right\} < \infty.$$

Let $0 \leq s = t_0 < t_1 < \ldots < t_n = t \leq T$, $\max_i (t_{i+1} - t_i) = \delta_n \to 0$.

We put in (6.15) $z = u_0 + r_0 v$. Then

$$\int_{t_i}^{t_i+1} \langle x(r) - (u_0 + r_0 v), dk(r) - \hat{z} dr \rangle \geq 0, \quad \forall\ |v| \leq 1,\ \forall\, 0 \leq s \leq t \leq T,$$

and we obtain

$$\begin{aligned} & r_0 \langle k(t_{i+1}) - k(t_i), v \rangle \\ & \leq \int_{t_i}^{t_i+1} \langle x(r) - u_0, dk(r) \rangle + A^{\#}_{u_0,r_0} \int_{t_i}^{t_i+1} |x(r) - u_0|\, dr + r_0 A^{\#}_{u_0,r_0} (t_{i+1} - t_i), \end{aligned}$$

for all $|v| \leq 1$. Hence

$$\begin{aligned} & r_0 |k(t_{i+1}) - k(t_i)| \\ & \leq \int_{t_i}^{t_i+1} \langle x(r) - u_0, dk(r) \rangle + A^{\#}_{u_0,r_0} \int_{t_i}^{t_i+1} |x(r) - u_0|\, dr + r_0 A^{\#}_{u_0,r_0} (t_{i+1} - t_i) \end{aligned}$$

and adding term by term for $i = 0$ to $i = n - 1$ the inequality

$$\begin{aligned} r_0 \sum_{i=0}^{n-1} |k(t_{i+1}) - k(t_i)| \leq & \int_s^t \langle x(t) - u_0, dk(t) \rangle \\ & + A^{\#}_{u_0,r_0} \int_s^t |x(r) - u_0|\, dr + (t - s)\, r_0 A^{\#}_{u_0,r_0}, \end{aligned}$$

holds and clearly (6.17) follows.

Setting in (6.3) $x = y(r)$, $x_0 = u_0$ and integrating from s to t the inequality (6.18) follows. ■

Often in the book we use some energy type equalities that we describe in the next lemma.

Lemma 6.20. *Let $x, k, m \in C([0, \infty[; \mathbb{R}^d)$, $k \in BV_{loc}([0, \infty[; \mathbb{R}^d)$, $k(0) = m(0) = 0$ such that*

$$x(t) + k(t) = x_0 + m(t), \quad \forall\, t \geq 0.$$

Then

(I): *For all $t \geq 0$ and for all $u \in \mathbb{R}^d$:*

$$\begin{aligned} & |x(t) - m(t) - u|^2 + 2 \int_0^t \langle x(r) - u, dk(r) \rangle \\ & \quad = |x_0 - u|^2 + 2 \int_0^t \langle m(r), dk(r) \rangle. \end{aligned} \tag{6.19}$$

(II): *For all* $0 \le s \le t$*:*

$$|x(t) - x(s) - m(t) + m(s)|^2 + 2\int_s^t \langle x(r) - x(s), dk(r)\rangle = 2\int_s^t \langle m(r) - m(s), dk(r)\rangle . \tag{6.20}$$

Proof. **(I)**: We have

$$\begin{aligned}
&|x(t) - m(t) - u|^2 \\
&= |x_0 - k(t) - u|^2 \\
&= |x_0 - u|^2 + 2\int_0^t \langle x_0 - k(r) - u, d(x_0 - k - u)(r)\rangle \\
&= |x_0 - u|^2 + 2\int_0^t \langle x(r) - m(r) - u, -dk(r)\rangle \\
&= |x_0 - u|^2 + 2\int_0^t \langle m(r), dk(r)\rangle - 2\int_0^t \langle x(r) - u, dk(r)\rangle ,
\end{aligned}$$

that is (6.19).

(II): From (6.19) we have for $u = 0$

$$|x(t) - m(t)|^2 - |x(s) - m(s)|^2 + 2\int_s^t \langle x(r), dk(r)\rangle = 2\int_s^t \langle m(r), dk(r)\rangle .$$

But $k(t) - k(s) = m(t) - x(t) - m(s) + x(s)$,

$$|x(t) - m(t)|^2 = |x(t) - x(s) - m(t) + m(s)|^2 + |x(s) - m(s)|^2 - 2\langle x(s) - m(s), k(t) - k(s)\rangle$$

and

$$2\int_s^t \langle m(r), dk(r)\rangle = 2\int_s^t \langle m(r) - m(s), dk(r)\rangle + 2\langle m(s), k(t) - k(s)\rangle ,$$

$$2\int_s^t \langle x(r), dk(r)\rangle = 2\int_s^t \langle x(r) - x(s), dk(r)\rangle + 2\langle x(s), k(t) - k(s)\rangle .$$

Hence, the equality (6.20) holds. ∎

Finally we give an approximation result via Stieltjes integrals.

Lemma 6.21. *Let*

- $Q : [0,T] \to \mathbb{R}$ *be a strictly increasing continuous function such that* $Q(0) = 0$,
- $f, \gamma : [0,T] \to \mathbb{R}^d$ *be bounded measurable functions,*
- $\varphi : \mathbb{R}^d \to]-\infty, +\infty]$ *be a proper convex lower semicontinuous function.*

If

$$f_\varepsilon(t) = f(0)\, e^{\frac{-Q(t)}{Q(\varepsilon)}} + \frac{1}{Q(\varepsilon)} \int_0^t e^{\frac{Q(r)-Q(t)}{Q(\varepsilon)}} f(r)\, dQ(r), \quad t \in [0,T], \varepsilon > 0$$

then as $\varepsilon \to 0_+$

$$\begin{array}{ll} (j) & f_\varepsilon(r) \longrightarrow f(r), \text{ a.e. } r \in [0,T], \\ (jj) & \displaystyle\int_t^s \varphi(f_\varepsilon(r))\,\gamma(r)\, dQ(r) \longrightarrow \int_t^s \varphi(f(r))\,\gamma(r)\, dQ(r), \ \forall\, [t,s] \subset [0,T]. \end{array}$$

If $f : [0,T] \to \mathbb{R}^d$ *is a continuous function it moreover follows that*

$$\sup_{t\in[0,T]} |f_\varepsilon(t) - f(t)| \longrightarrow 0.$$

Remark 6.22. The same conclusions are true if we replace $f_\varepsilon(t)$ by

$$g_\varepsilon(t) = f(T)\, e^{\frac{Q(t)-Q(T)}{Q(\varepsilon)}} + \frac{1}{Q(\varepsilon)} \int_t^T e^{\frac{Q(t)-Q(r)}{Q(\varepsilon)}} f(r)\, dQ(r), \ t \in [0,T].$$

Proof of Lemma 6.21 (j). Obviously we have

$$\begin{aligned} &\int_0^t \tfrac{1}{Q(\varepsilon)} e^{\frac{Q(r)-Q(t)}{Q(\varepsilon)}} f(r)\, dQ(r) = \int_{\frac{-Q(t)}{Q(\varepsilon)}}^0 e^u f\left(\left(Q^{-1}(uQ(\varepsilon) + Q(t))\right)\right) du \\ &= \int_{\frac{-Q(t)}{Q(\varepsilon)}}^0 e^u \left[f\left(Q^{-1}(uQ(\varepsilon) + Q(t))\right) - f\left(Q^{-1}(Q(t))\right)\right] du + f(t) \int_{\frac{-Q(t)}{Q(\varepsilon)}}^0 e^u du. \end{aligned} \tag{6.21}$$

But

$$\begin{aligned} &\limsup_{\varepsilon\to0} \left| \int_{\frac{-Q(t)}{Q(\varepsilon)}}^0 e^u \left[f\left(Q^{-1}(uQ(\varepsilon) + Q(t))\right) - f\left(Q^{-1}(Q(t))\right)\right] du \right| \\ &\le \limsup_{\varepsilon\to0} \int_{-\infty}^0 e^u \left|f\left(Q^{-1}((uQ(\varepsilon) + Q(t)) \vee 0)\right) - f\left(Q^{-1}(Q(t))\right)\right| du \\ &\le 2C \int_{-\infty}^{-n} e^u du + \int_{-n}^0 e^u \left|f\left(Q^{-1}((uQ(\varepsilon) + Q(t)) \vee 0)\right) - f\left(Q^{-1}(Q(t))\right)\right| du \\ &\le 2Ce^{-n} + \limsup_{\varepsilon\to0} \int_{-n}^0 \left|f\left(Q^{-1}((uQ(\varepsilon) + Q(t)) \vee 0)\right) - f\left(Q^{-1}(Q(t))\right)\right| du \\ &\le 2Ce^{-n}, \text{ for all } n, \end{aligned}$$

since

$$\lim_{\delta\to 0}\int_{\alpha}^{\beta}\left|f\left(Q^{-1}(s+\delta u)\right)-f\left(Q^{-1}(s)\right)\right|du=0,\ \text{a.e.}$$

Therefore the following limit exists

$$\lim_{\varepsilon\to 0}\left|\int_{\frac{-Q(t)}{Q(\varepsilon)}}^{0}e^{u}\left[f\left(Q^{-1}(uQ(\varepsilon)+Q(t))\right)-f(t)\right]du\right|=0,$$

and (j) follows. In the case where f is continuous, it is sufficient to write

$$\begin{aligned}f_{\varepsilon}(t)&=f(0)\,e^{\frac{-Q(t)}{Q(\varepsilon)}}+\tfrac{1}{Q(\varepsilon)}\int_{0}^{t}e^{\frac{Q(r)-Q(t)}{Q(\varepsilon)}}f(r)\,dQ(r)\\&=f(0)\,e^{\frac{-Q(t)}{Q(\varepsilon)}}+\tfrac{1}{Q(\varepsilon)}\int_{0}^{t_{\varepsilon}}e^{\frac{Q(r)-Q(t)}{Q(\varepsilon)}}f(r)\,dQ(r)+\tfrac{1}{Q(\varepsilon)}\int_{t_{\varepsilon}}^{t}e^{\frac{Q(r)-Q(t)}{Q(\varepsilon)}}f(r)\,dQ(r),\end{aligned}$$

where $t_{\varepsilon}:=Q^{-1}\left(Q(t)-\sqrt{Q(\varepsilon)}\right)\longrightarrow t$, as $\varepsilon\to 0$, and $t_{\varepsilon}<t$.
(jj) We have

$$\begin{aligned}&\int_{t}^{s}\varphi\left(f_{\varepsilon}(r)\right)\gamma(r)\,dQ(r)\\&\quad\le\int_{t}^{s}e^{\frac{-Q(r)}{Q(\varepsilon)}}\varphi(f(0))\,\gamma(r)\,dQ(r)\\&\qquad+\int_{t}^{s}\left(\int_{0}^{r}\frac{1}{Q(\varepsilon)}e^{\frac{Q(u)-Q(r)}{Q(\varepsilon)}}\varphi(f(u))\,dQ(u)\right)\gamma(r)\,dQ(r)\\&\quad=\varphi(f(0))\int_{t}^{s}e^{\frac{-Q(r)}{Q(\varepsilon)}}\gamma(r)\,dQ(r)\\&\qquad+\int_{0}^{s}\left(\int_{0}^{s}\frac{1}{Q(\varepsilon)}e^{\frac{Q(u)-Q(r)}{Q(\varepsilon)}}\varphi(f(u))\,\mathbf{1}_{[0,r]}(u)\,dQ(u)\right)\gamma(r)\,dQ(r)\\&\qquad-\int_{0}^{t}\left(\int_{0}^{t}\frac{1}{Q(\varepsilon)}e^{\frac{Q(u)-Q(r)}{Q(\varepsilon)}}\varphi(f(u))\,\mathbf{1}_{[0,r]}(u)\,dQ(u)\right)\gamma(r)\,dQ(r)\\&\quad=\varphi(f(0))\int_{t}^{s}e^{\frac{-Q(r)}{Q(\varepsilon)}}\gamma(r)\,dQ(r)\\&\qquad+\int_{0}^{s}\left(\varphi(f(u))\int_{0}^{s}\frac{1}{Q(\varepsilon)}e^{\frac{Q(u)-Q(r)}{Q(\varepsilon)}}\mathbf{1}_{[u,s]}(r)\,\gamma(r)\,dQ(r)\right)dQ(u)\\&\qquad-\int_{0}^{t}\left(\varphi(f(u))\int_{0}^{t}\frac{1}{Q(\varepsilon)}e^{\frac{Q(u)-Q(r)}{Q(\varepsilon)}}\mathbf{1}_{[u,t]}(r)\,\gamma(r)\,dQ(r)\right)dQ(u).\end{aligned}$$

Using Remark 6.22 we have

$$\lim_{\varepsilon\to 0}\int_u^s \frac{1}{Q(\varepsilon)} e^{\frac{Q(u)-Q(r)}{Q(\varepsilon)}} \gamma(r)\, dQ(r) = \lim_{\varepsilon\to 0}\int_u^t \frac{1}{Q(\varepsilon)} e^{\frac{Q(u)-Q(r)}{Q(\varepsilon)}} \gamma(r)\, dQ(r)\,.$$
$$= \gamma(u)\,,\ \text{a.e.}$$

By Lebesgue's dominated convergence theorem and the lower semicontinuity of φ we conclude that

$$\begin{aligned}\int_t^s \varphi(f(r))\,\gamma(r)\,dQ(r) &\le \liminf_{\varepsilon\to 0}\int_t^s \varphi(f_\varepsilon(r))\,\gamma(r)\,dQ(r)\\ &\le \limsup_{\varepsilon\to 0}\int_t^s \varphi(f_\varepsilon(r))\,\gamma(r)\,dQ(r)\\ &\le \int_t^s \varphi(f(r))\,\gamma(r)\,dQ(r)\,.\end{aligned}$$

■

6.3.6 Semicontinuity

Let $(\mathbb{X},\rho)$ be a metric space.

Definition 6.23. A function $f:\mathbb{X}\to\overline{\mathbb{R}}$ is lower semicontinuous (l.s.c.) at $x\in\mathbb{X}$ if

$$f(x)\le \liminf_{y\to x} f(y)\,,$$

i.e. for all $\varepsilon>0$ there exists a $\delta=\delta(\varepsilon,x)>0$ such that $\rho(x,y)<\delta$ implies $f(y)\ge f(x)-\varepsilon$. The function f is l.s.c. if it is l.s.c. at all $x\in\mathbb{X}$.
A function $g:\mathbb{X}\to\overline{\mathbb{R}}$ is upper semicontinuous (u.s.c.) if $-g$ is l.s.c.

Proposition 6.24. *The following assertions are equivalent:*

(i) $f:\mathbb{X}\to\overline{\mathbb{R}}$ *is lower semicontinuous;*
(ii) the set $\{x\in\mathbb{X}: f(x)\le a\}$ *is closed in* $\mathbb{X}$*, for all* $a\in\mathbb{R}$.

It is easy to prove that:

▲ If $g_n:\mathbb{X}\to\overline{\mathbb{R}}$, $n\in\mathbb{N}$, are l.s.c. functions and

$$g(x)=\sup\{g_n(x): n\in\mathbb{N}\},$$

then $g:\mathbb{X}\to\overline{\mathbb{R}}$ is a l.s.c. function.

▲ If $f:\mathbb{X}\to]-\infty,+\infty]$ is a l.s.c. function, then f is bounded from below on compact subsets of $\mathbb{X}$.

Lemma 6.25. *Let* $(\mathbb{X}, \rho)$ *be a metric space. If* $f : \mathbb{X} \to \overline{\mathbb{R}}$ *is bounded below on bounded subsets of* $\mathbb{X}$*, then there exists a continuous function* $\mu : \mathbb{X} \to \mathbb{R}$ *such that*

$$\mu(x) \le f(x), \quad \text{for all } x \in \mathbb{X}.$$

Proof. Let $n \in \mathbb{N}^*$ and $a \in \mathbb{X}$. Define

$$\mu_n = n \wedge \inf\{f(x) : \rho(x,a) \le n\}.$$

Then $\mu_n \in \mathbb{R}$. Define $\mu : \mathbb{X} \to \mathbb{R}$ such that, if $n - 1 \le \rho(x,a) < n$

$$\mu(x) = \mu_n - 2\left[\rho(x,a) - \left(n - \frac{1}{2}\right)\right]^+ (\mu_n - \mu_{n+1}).$$

The function μ is continuous on $\mathbb{X}$ and

$$\mu(x) \le f(x) \quad \text{for all } x \in \mathbb{X}.$$

■

Proposition 6.26. *Let* $(\mathbb{X}, \rho)$ *be a separable metric space. If* $f : \mathbb{X} \to]-\infty, +\infty]$ *is a l.s.c. function and* $\mu : \mathbb{X} \to \mathbb{R}$ *is a continuous function such that*

$$\mu(x) \le f(x), \quad \text{for all } x \in \mathbb{X}, \tag{6.22}$$

then there exists a sequence of continuous functions $f_n : \mathbb{X} \to \mathbb{R}$*,* $n \in \mathbb{N}^*$*, such that for all* $x \in \mathbb{X}$

$$\mu(x) \le f_1(x) \le \ldots \le f_n(x) \le \ldots \le f(x) \quad \text{and} \quad \lim_{n\to\infty} f_n(x) = f(x).$$

Proof. Using only the boundedness from below (6.22) we shall show that there exists a sequence of continuous functions $f_n : \mathbb{X} \to \mathbb{R}$, $n \in \mathbb{N}^*$, such that

$$\mu(y) \le f_1(y) \le \ldots \le f_n(y) \le \ldots \le f(y), \quad \text{for all } y \in \mathbb{X}, \tag{6.23}$$

and such that for all $x \in \mathbb{X}$ there exists a sequence $y_n \to x$ satisfying

$$n \wedge \left(f(y_n) - \frac{1}{n}\right) \le \sup_{j\in\mathbb{N}^*} f_j(x) \le f(x), \quad \text{for all } n \in \mathbb{N}^*. \tag{6.24}$$

Then the result follows using the lower semicontinuity of f:

$$f(x) \le \liminf_{n\to+\infty} \left[n \wedge \left(f(y_n) - \frac{1}{n}\right)\right] \le \sup_{j\in\mathbb{N}^*} f_j(x) \le f(x).$$

Let us prove (6.23) and (6.24). Let $n, i \in \mathbb{N}^*$ and $a \in \mathbb{X}$. Define

$$\mu_{i,n} = \mu_{i,n}(a) = n \wedge \inf\left\{ f(y) : \rho(y,a) < \frac{i}{n} \right\}.$$

Then $-\infty \leq \mu_{i,n} < +\infty$. Define $\psi_n(\cdot, a) : \mathbb{X} \to \mathbb{R}$ such that, if $\frac{i-1}{n} \leq \rho(y,a) < \frac{i}{n}$, $i \in \mathbb{N}^*$, then

$$\psi_n(y,a) = \mu(y) \vee \mu_{i,n} - 2n \left[\rho(y,a) - \left(i - \frac{1}{2}\right)\frac{1}{n} \right]^+ \times \left[(\mu(y) \vee \mu_{i,n}) - (\mu(y) \vee \mu_{i+1,n}) \right].$$

For each $a \in \mathbb{X}$ the function $\psi_n(\cdot, a)$ is continuous on $\mathbb{X}$ and

$$\psi_n(y,a) \leq f(y) \quad \text{for all } y \in \mathbb{X}.$$

Let $A_1 \subset A_2 \subset \ldots \subset A_n \subset \ldots$ be finite sets such that $A = \bigcup_{n \in \mathbb{N}^*} A_n$ is a dense subset of $\mathbb{X}$. Define $f_n : \mathbb{X} \to \mathbb{R}$

$$f_n(y) = \max_{k \in \overline{1,n}} \left[\max_{a \in A_n} \psi_k(y,a) \right], \quad y \in \mathbb{X}.$$

Clearly f_n, $n \in \mathbb{N}^*$, are continuous functions and

$$f_1(y) \leq \ldots \leq f_n(y) \leq \ldots \leq f(y), \quad \forall\, y \in \mathbb{X}.$$

Let $x \in \mathbb{X}$ be arbitrary. Then there exist $a_n \in A$ and $k_n \geq n$ such that

$$\rho(x, a_n) < \frac{1}{2n} \quad \text{and} \quad a_n \in A_{k_n}.$$

If $\mu_{1,n}(a_n) \in \mathbb{R}$, then from the definition of $\mu_{1,n}(a_n)$, there exists $y_n \in B(a_n, \frac{1}{n})$ such that

$$n \wedge \left(f(y_n) - \frac{1}{n} \right) \leq \mu_{1,n}(a_n) \leq \psi_n(x, a_n) \leq f_{k_n}(x) \leq \sup_{j \in \mathbb{N}^*} f_j(x) \leq f(x).$$

If $\mu_{1,n}(a_n) = -\infty$ then, once again from the definition of $\mu_{1,n}(a_n)$, there exists $y_n \in B(a_n, \frac{1}{n})$ such that

$$n \wedge \left(f(y_n) - \frac{1}{n} \right) \leq f_1(x) \leq \sup_{j \in \mathbb{N}^*} f_j(x) \leq f(x).$$

We remark that

$$\rho(y_n, x) \leq \rho(y_n, a_n) + \rho(a_n, x) \leq \frac{3}{2n}$$

and consequently $y_n \to x$. The proof is complete. ■

We also have the following:

Proposition 6.27. *If $f : \mathbb{X} \to \mathbb{R}$ is a continuous function and $f_n : \mathbb{X} \to \mathbb{R}$, $n \in \mathbb{N}^*$, are lower semicontinuous functions such that for all $x \in \mathbb{X}$:*

$$f_1(x) \leq f_2(x) \leq \ldots \leq f_n(x) \leq \ldots \leq f(x) \quad \text{and} \quad \lim_{n\to\infty} f_n(x) = f(x),$$

then for every compact set $K \subset \mathbb{X}$

$$\lim_{n\to\infty}\left[\sup_{x\in K}|f_n(x) - f(x)|\right] = 0. \tag{6.25}$$

Proof. For each $\varepsilon > 0$, $G_n = \{x \in \mathbb{X} : f(x) - f_n(x) < \varepsilon\}$ is an open subset of $\mathbb{X}$ and

$$K \subset \mathbb{X} = \bigcup_{n\in\mathbb{N}^*} \uparrow G_n.$$

Hence, by the compactness of K, there exists an $n \in \mathbb{N}^*$ such that $K \subset G_n$ and the uniform convergence (6.25) follows. ■

We now give some examples (as exercises for the reader) of lower semicontinuous functions that are used in the book.

Example 6.28. Let $(\mathbb{X}, \rho)$ be a separable metric space and $E \subset \mathbb{X}$. Then E is a closed subset of $\mathbb{X}$ if and only if the function

$$I_E(x) = \begin{cases} 0, & \text{if } x \in E, \\ +\infty, & \text{otherwise,} \end{cases}$$

is a l.s.c. function on $\mathbb{X}$.

Example 6.29. Let $(\mathbb{X}, \rho)$ be a separable metric space. Let $0 \leq s < t \leq T$. If $f : \mathbb{X} \to]-\infty, +\infty]$ is a l.s.c. function bounded below on bounded subsets of $\mathbb{X}$ and $\Phi : C([0,T]; \mathbb{X}) \to]-\infty, +\infty]$ is defined by

$$\Phi(x) = \begin{cases} \displaystyle\int_s^t f(x(r))\,dr, & \text{if } f(x) \in L^1(0,T) \\ +\infty, & \text{otherwise} \end{cases}$$

then Φ is a l.s.c. function.

Let $0 \le s < t \le T$. Let $\mathcal{D}_{[s,t]}$ be the set of partitions $\Delta: \ s = r_0 < r_1 < \cdots < r_n = t$, and

$$V_{\Delta}(k) \stackrel{def.}{=} \sum_{i=0}^{n-1} \rho(k(r_i), k(r_{i+1})).$$

Define the total variation of k on $[s,t]$ by

$$\updownarrow k \updownarrow_{[s,t]} = \sup_{\Delta \in \mathcal{D}_{[a,b]}} V_{\Delta}(k).$$

Then as a sup of continuous functions:

Example 6.30. The mapping $k \longmapsto \updownarrow k \updownarrow_{[s,t]} : C([0,T];\mathbb{X}) \to [0,\infty]$ is a l.s.c. function.

Finally we present Ekeland's principle (see [26], or [4], p. 29, Th. 3.2):

Lemma 6.31 (Ekeland). *Let $(\mathbb{X},\rho)$ be a complete metric space and $J: \mathbb{X} \to]-\infty,+\infty]$ be a proper lower-semicontinuous function bounded from below. Then for any $\varepsilon > 0$ there exists an $x_\varepsilon \in \mathbb{X}$ such that:*

$$\begin{cases} J(x_\varepsilon) \le \inf\limits_{x \in \mathbb{X}} J(x) + \varepsilon \quad \textit{and} \\ J(x_\varepsilon) < J(x) + \sqrt{\varepsilon}\rho(x_\varepsilon, x), \quad \forall x \in \mathbb{X} \setminus \{x_\varepsilon\}. \end{cases}$$

6.3.7 Convex Functions

6.3.7.1 Definitions: Properties

Let $(\mathbb{X}, \|\cdot\|)$ be a real Banach space and $(\mathbb{X}^*, \|\cdot\|_*)$ its dual. A function $\varphi: \mathbb{X} \to]-\infty,+\infty]$ is convex if

$$\varphi((1-\lambda)x + \lambda y) \le (1-\lambda)\varphi(x) + \lambda\varphi(y), \text{ for all } x,y \in \mathbb{X} \text{ and } \lambda \in]0,1[.$$

Denote by

$$\mathrm{Dom}(\varphi) = \{x \in \mathbb{X} : \varphi(x) < +\infty\}$$

the domain of φ and

$$\partial\varphi(x) = \{\hat{x} \in \mathbb{X}^* : \langle \hat{x}, z - x\rangle + \varphi(x) \le \varphi(z), \ \forall z \in \mathbb{X}\}$$

the subdifferential of the function φ at x. We say that φ is proper if $\mathrm{Dom}(\varphi) \ne \emptyset$. Clearly if φ is a convex function then $\mathrm{Dom}(\varphi)$ is a convex subset of $\mathbb{X}$.

Theorem 6.32. *If $\mathbb{X}$ is a Banach space and $\varphi : \mathbb{X} \to]-\infty, +\infty]$ is a proper convex l.s.c. function then*

$$\mathrm{Dom}\,(\partial\varphi) \overset{def}{=} \{x \in \mathbb{X} : \partial\varphi(x) \neq \emptyset\}$$

is non-empty and $\partial\varphi : \mathbb{X} \rightrightarrows \mathbb{X}^*$ is a maximal monotone operator.

If K is a convex subset of $\mathbb{X}$ then the function $I_K : \mathbb{X} \to]-\infty, +\infty]$ defined by

$$I_K(x) = \begin{cases} 0, & \text{if } x \in K, \\ +\infty, & \text{if } x \in \mathbb{X}\backslash K, \end{cases}$$

is a convex function called the *convex indicator function of K.*
Recall, from [71] Chapter 2, the following:

Proposition 6.33. *Let $g : \mathbb{R} \to]-\infty, +\infty]$ be a convex function. Then:*

(a) $\mathrm{Dom}(g)$ *is an interval in* $\mathbb{R}$;
(b) the left derivative $g'_- : \mathrm{Dom}(g) \to [-\infty, +\infty]$ *and the right derivative* $g'_+ : \mathrm{Dom}(g) \to [-\infty, +\infty]$ *are well defined increasing functions and they satisfy:*

$$\begin{array}{rl} (j) & g'_+(r) \le \dfrac{g(s) - g(r)}{s - r} \le g'_-(s), \quad \forall\, r, s \in \mathrm{Dom}(g),\ r < s; \\ (jj) & g'_-(r) \le g'_+(r), \quad \forall\, r \in \mathrm{Dom}(g); \\ (jjj) & g'_- \text{ is left continuous and } g'_+ \text{ is right continuous on } \mathrm{int}(\mathrm{Dom}(g)); \\ (jv) & u \in \left[g'_-(r), g'_+(r)\right] \cap \mathbb{R} \iff u(s - r) \le g(s) - g(r), \quad \forall\, s \in \mathbb{R}; \\ (v) & \left\{r \in \mathrm{Dom}(g) : g'_-(r) \neq g'_+(r)\right\} \text{ is at most countable;} \end{array}$$

(c) g is locally Lipschitz continuous on $\mathrm{int}\,(\mathrm{Dom}(g))$;
(d) $A \subset \mathbb{R} \times \mathbb{R}$ is a maximal monotone operator if and only if there exists a convex function $j : \mathbb{R} \to]-\infty, +\infty]$ such that $\beta = \partial j$.

Note that if φ is a proper convex lower semicontinuous (l.s.c.) function then:

- φ is bounded from below by an affine function, that is $\exists\, v \in \mathbb{X}^*$ and $a \in \mathbb{R}$ such that

$$\varphi(x) \ge \langle v, x\rangle + a, \text{ for all } x \in \mathbb{X},$$

and, moreover, if $\mathbb{X}$ is reflexive and $\lim\limits_{\|x\| \to \infty} \varphi(x) = +\infty$ then there exists an $x_0 \in \mathrm{Dom}(\varphi)$ such that

$$\varphi(x) \ge \varphi(x_0), \text{ for all } x \in \mathbb{X};$$

- (Fenchel–Moreau theorem on biconjugate functions)

$$\varphi(x) = \varphi^{**}(x) = \sup\left\{\langle x, x^* \rangle - \varphi^*\left(x^*\right) : x^* \in \mathbb{X}^*\right\},$$

where $\varphi^* : \mathbb{X}^* \to \bar{\mathbb{R}}$ is the conjugate of the function φ, i.e.

$$\varphi^*\left(x^*\right) = \sup\left\{\langle u, x^* \rangle - \varphi(u) : u \in \mathrm{Dom}(\varphi)\right\};$$

- φ is continuous on int (Dom(φ));
- $\partial\varphi : \mathbb{X} \rightrightarrows \mathbb{X}^*$ is maximal monotone;
- int (Dom (φ)) = int (Dom ($\partial\varphi$)) and $\overline{\mathrm{Dom}\,(\partial\varphi)} = \overline{\mathrm{Dom}\,(\varphi)}$.

We have the following instance of Jensen's inequality.

Lemma 6.34. *Let $\varphi : \mathbb{R}^d \to]-\infty, +\infty]$ be a proper convex lower semicontinuous function. If $a, b \in \mathbb{R}$, $a < b$, $y \in L^\infty\left(a, b; \mathbb{R}^d\right)$ and $\rho \in L^1(a, b; \mathbb{R}_+)$ such that $\int_a^b \rho(r)\, dr = 1$, then*

$$\varphi\left(\int_a^b \rho(r)\, y(r)\, dr\right) \le \int_a^b \rho(r)\, \varphi(y(r))\, dr.$$

Proof. Since there exists a set $\Gamma \subset \mathbb{R}^d \times \mathbb{R}$ such that

$$\varphi(x) = \sup\left\{\langle v, x \rangle + \gamma : (v, \gamma) \in \Gamma\right\},$$

we have

$$\left\langle v, \int_a^b \rho(r)\, y(r)\, dr\right\rangle + \gamma = \int_a^b \rho(r)\left[\langle v, y(r)\rangle + \gamma\right] dr \le \int_a^b \rho(r)\, \varphi(y(r))\, dr$$

and the result follows passing to $\sup_{(v,\gamma)\in\Gamma}$. ■

6.3.7.2 Regularization of Convex Functions

Let $(\mathbb{H}, |\cdot|)$ be a real separable Hilbert space and $\varphi : \mathbb{H} \to]-\infty, +\infty]$ be a proper convex l.s.c. function. The Moreau regularization φ_ε of the convex l.s.c. function φ is defined by

$$\varphi_\varepsilon(x) = \inf\left\{\frac{1}{2\varepsilon}|z - x|^2 + \varphi(z); z \in \mathbb{H}\right\}, \quad \varepsilon > 0.$$

The function φ_ε is a convex function of class C^1 on $\mathbb{H}$; the gradient $\nabla\varphi_\varepsilon$ is a Lipschitz function on $\mathbb{H}$ with the Lipschitz constant equal to ε^{-1}. If we define:

$$J_\varepsilon x = x - \varepsilon \nabla \varphi_\varepsilon(x),$$

then one can easily prove (see e.g. Brezis [12], Barbu [2], Rockafellar [65] or Zălinescu [71]) that for all $x \in \mathbb{H}$ and $\varepsilon > 0$:

1. $\varphi_\varepsilon(x) = \frac{\varepsilon}{2}|\nabla \varphi_\varepsilon(x)|^2 + \varphi(J_\varepsilon x)$,
2. $\varphi(J_\varepsilon x) \le \varphi_\varepsilon(x) \le \varphi(x)$,
3. $\nabla \varphi_\varepsilon(x) = \partial \varphi_\varepsilon(x)$ and

$$\begin{aligned}\varphi(J_\varepsilon x) &\le \varphi_\varepsilon(x)\\ &\le \varphi_\varepsilon(z) + \langle x - z, \nabla \varphi_\varepsilon(x)\rangle\\ &\le \varphi(z) + \langle x - z, \nabla \varphi_\varepsilon(x)\rangle, \ \forall\, z \in \mathbb{H},\end{aligned}$$

4. $\nabla \varphi_\varepsilon(x) \in \partial \varphi(J_\varepsilon x)$ i.e.

$$\langle \nabla \varphi_\varepsilon(x), z - J_\varepsilon x\rangle + \varphi(J_\varepsilon x) \le \varphi(z), \ \forall\, z \in \mathbb{H}.$$

   Hence $J_\varepsilon x = (I + \varepsilon \partial \varphi)^{-1}(x)$ and $\nabla \varphi_\varepsilon(x) = A_\varepsilon(x)$, where A is the maximal monotone operator $\partial \varphi$; $\nabla \varphi_\varepsilon$ is called the Moreau–Yosida approximation of $\partial \varphi$.
5. If $(u_0, \hat{u}_0) \in \partial \varphi$, then for all $y \in \mathbb{H}$

$$\begin{cases} (a) & |\nabla \varphi_\varepsilon(u_0)| \le |\hat{u}_0|,\\ (b) & 0 \le \varphi(u_0) - \varphi_\varepsilon(u_0) \le \varphi(u_0) - \varphi(J_\varepsilon u_0) \le \varepsilon |\hat{u}_0|^2,\\ (c) & |J_\varepsilon(y)| \le |y - u_0| + \varepsilon |\hat{u}_0| + |u_0|,\\ (d) & \varphi(J_\varepsilon y) \ge \varphi(u_0) - |\hat{u}_0|\,|y - u_0| - \varepsilon |\hat{u}_0|^2,\\ (e) & \frac{\varepsilon}{2}|\nabla \varphi_\varepsilon(y)|^2 \le \varphi_\varepsilon(y) - \varphi(u_0) + |\hat{u}_0|\,|y - u_0| + \varepsilon |\hat{u}_0|^2. \end{cases} \tag{6.26}$$

   Indeed $|\nabla \varphi_\varepsilon(u_0)| = |A_\varepsilon(u_0)| \le \left|A^0(x)\right|$ and

$$\begin{aligned}-\varepsilon |\hat{u}_0|^2 &\le -\varepsilon \langle \hat{u}_0, \nabla \varphi_\varepsilon(u_0)\rangle\\ &= \langle \hat{u}_0, J_\varepsilon u_0 - u_0\rangle\\ &\le \varphi(J_\varepsilon u_0) - \varphi(u_0)\\ &\le \varphi_\varepsilon(u_0) - \varphi(u_0)\\ &\le 0.\end{aligned}$$

   For the inequality (c) we have

$$|J_\varepsilon(y)| \le |J_\varepsilon(y) - J_\varepsilon(u_0)| + |J_\varepsilon(u_0) - u_0| + |u_0|,$$

and therefore

$$\begin{aligned}\varphi\left(J_{\varepsilon}y\right) &\geq \varphi\left(u_0\right)+\left\langle \hat{u}_0, J_{\varepsilon}\left(y\right)-u_0\right\rangle \\ &\geq \varphi\left(u_0\right)-\left|\hat{u}_0\right|\left|J_{\varepsilon}\left(y\right)-J_{\varepsilon}\left(u_0\right)\right|-\left|\hat{u}_0\right|\left|J_{\varepsilon}\left(u_0\right)-u_0\right|\end{aligned}$$

which yields (d).
For the last inequality, (d), we have

$$\begin{aligned}\frac{\varepsilon}{2}\left|\nabla\varphi_{\varepsilon}\left(y\right)\right|^2 &= \varphi_{\varepsilon}\left(y\right)-\varphi\left(J_{\varepsilon}y\right) \\ &\leq \varphi_{\varepsilon}\left(y\right)-\varphi\left(u_0\right)+\left|\hat{u}_0\right|\left|y-u_0\right|+\varepsilon\left|\hat{u}_0\right|^2.\end{aligned}$$

6. If $0=\varphi(0)\leq\varphi(x)$, $\forall x\in\mathbb{H}$, it is easy to verify that, moreover

$$\begin{array}{ll} j) & 0\in\partial\varphi\left(0\right),\ 0=\varphi_{\varepsilon}(0)\leq\varphi_{\varepsilon}(x),\ \ J_{\varepsilon}\left(0\right)=\nabla\varphi_{\varepsilon}\left(0\right)=0, \\ jj) & \frac{\varepsilon}{2}|\nabla\varphi_{\varepsilon}(x)|^2\leq\varphi_{\varepsilon}(x)\leq\left\langle\nabla\varphi_{\varepsilon}(x),x\right\rangle, \quad \forall x\in\mathbb{H}, \\ jjj) & |\nabla\varphi_{\varepsilon}(x)|\leq\frac{1}{\varepsilon}|x|,\ \text{and } 0\leq\varphi_{\varepsilon}(x)\leq\frac{1}{2\varepsilon}|x|^2,\ \forall x\in\mathbb{H}, \\ jv) & \left\langle\nabla\varphi_{\varepsilon}(x),x-y\right\rangle\geq-\varphi(J_{\varepsilon}x)-\varepsilon\left\langle\nabla\varphi_{\varepsilon}(x),\nabla\varphi_{\varepsilon}(y)\right\rangle,\ \forall x,y\in\mathbb{H}.\end{array} \tag{6.27}$$

If for a fixed $a\geq 0$

$$\langle\hat{x}-\hat{y},x-y\rangle\geq a|x-y|^2,\ \forall\ (x,\hat{x}),(y,\hat{y})\in\partial\varphi,$$

or equivalently the function

$$\psi\left(x\right)=\varphi\left(x\right)-\frac{a}{2}|x|^2$$

is convex, too, then by the definition of J_{ε} and the monotonicity of the operator $\partial\varphi$ we have $\forall\ r\in\left]0,1\right[$:

$$\begin{aligned}&a\left[(1-r)|x-y|^2-\frac{1-r}{r}|\varepsilon\nabla\varphi_{\varepsilon}(x)-\delta\nabla\varphi_{\delta}(y)|^2\right] \\ &\leq a|J_{\varepsilon}x-J_{\delta}y|^2 \\ &\leq\left\langle\nabla\varphi_{\varepsilon}(x)-\nabla\varphi_{\delta}(y),J_{\varepsilon}x-J_{\delta}y\right\rangle \\ &=\left\langle\nabla\varphi_{\varepsilon}(x)-\nabla\varphi_{\delta}(y),x-y\right\rangle-\varepsilon|\nabla\varphi_{\varepsilon}(x)|^2-\delta|\nabla\varphi_{\delta}(y)|^2 \\ &\qquad+(\varepsilon+\delta)\left\langle\nabla\varphi_{\varepsilon}(x),\nabla\varphi_{\delta}(y)\right\rangle\end{aligned}$$

and then

$$\begin{aligned} &a) \quad \langle \nabla\varphi_\varepsilon(x) - \nabla\varphi_\varepsilon(y), x-y\rangle \geq a(1-r)|x-y|^2 \\ &b) \quad \langle \nabla\varphi_\varepsilon(x) - \nabla\varphi_\delta(y), x-y\rangle \geq a(1-r)|x-y|^2 \\ &\qquad\qquad - (\varepsilon+\delta)\,|\nabla\varphi_\varepsilon(x)||\nabla\varphi_\delta(y)| \end{aligned} \tag{6.28}$$

for all $x, y \in \mathbb{H}$, $r \in (0,1)$, $\varepsilon, \delta > 0$ such that $0 \leq a(1-r)\varepsilon \leq r$, $0 \leq a(1-r)\delta \leq r$.

Let $u_0 \in \mathbb{H}$ and $r_0 \geq 0$ be such that

$$\{u_0 + r_0 v : |v| \leq 1\} \subset \text{Dom}\varphi.$$

Note that if

$$\varphi^{\#}_{u_0,r_0} \left(\stackrel{def}{=} \sup\{\varphi\,(u_0 + r_0 v) : |v| \leq 1\}\right) < \infty,$$

then we have for all $(x,\hat{x}) \in \partial\varphi$

$$\begin{aligned} &a) \quad r_0|\hat{x}| + \varphi(x) \leq \langle \hat{x}, x-u_0\rangle + \varphi^{\#}_{u_0,r_0}, \quad \forall\ (x,\hat{x}) \in \partial\varphi, \\ &b) \quad r_0|\hat{x}| + |\varphi(x) - \varphi\,(u_0)| \leq \langle \hat{x}, x-u_0\rangle \\ &\qquad\qquad +2\left|(\partial\varphi)^0\,(u_0)\right|\,|x-u_0| + \varphi^{\#}_{u_0,r_0} - \varphi\,(u_0) \end{aligned} \tag{6.29}$$

and in particular for $r_0 = 0$

$$|\varphi(x) - \varphi\,(u_0)| \leq \langle \hat{x}, x-u_0\rangle + 2\left|(\partial\varphi)^0\,(u_0)\right|\,|x-u_0|\,. \tag{6.30}$$

Let us prove (6.29). For $(x,\hat{x}) \in \partial\varphi$ and $|v| \leq 1$ we have

$$\langle \hat{x}, u_0 + r_0 v - x\rangle + \varphi(x) \leq \varphi((u_0 + r_0 v) \leq \varphi^{\#}_{u_0,r_0}$$

and consequently

$$r_0\,\langle \hat{x}, v\rangle + \varphi(x) \leq \langle \hat{x}, x-u_0\rangle + \varphi^{\#}_{u_0,r_0}$$

which yields (6.29-a) taking the $\sup_{|v|\leq 1}$.

On the other hand for all arbitrary $\hat{u}_0 \in \partial\varphi\,(u_0)$,

$$\langle \hat{u}_0, x-u_0\rangle + \varphi\,(u_0) \leq \varphi(x),$$

which yields

$$|\varphi(x) - \varphi\,(u_0)| \leq \varphi(x) - \varphi\,(u_0) + 2\,|\hat{u}_0|\,|x-u_0|\,.$$

Hence for all $|v| \leq 1$:

$$r_0 \langle \hat{x}, v \rangle + |\varphi(x) - \varphi(u_0)| \leq \langle \hat{x}, x - u_0 \rangle + 2|\hat{u}_0| |x - u_0| + \varphi^{\#}_{u_0, r_0} - \varphi(u_0)$$

which yields (6.29-b).

Observing that $\nabla\varphi_\varepsilon(x) \in \partial\varphi(J_\varepsilon x)$, we have

$$r_0 |\nabla\varphi_\varepsilon(x)| + |\varphi(J_\varepsilon x) - \varphi(u_0)| \leq \langle \nabla\varphi_\varepsilon(x), J_\varepsilon x - u_0 \rangle + 2|\hat{u}_0| |J_\varepsilon x - u_0| + \varphi^{\#}_{u_0, r_0} - \varphi(u_0).$$

But

$$\langle \nabla\varphi_\varepsilon(x), J_\varepsilon x - u_0 \rangle = (\nabla\varphi_\varepsilon(x), x - u_0 - \varepsilon|\nabla\varphi_\varepsilon(x)|^2$$

and

$$\begin{aligned} |J_\varepsilon x - u_0| &\leq |J_\varepsilon x - J_\varepsilon u_0| + |J_\varepsilon u_0 - u_0| \\ &\leq |x - u_0| + \varepsilon |\hat{u}_0|. \end{aligned}$$

Hence for all $\varepsilon \in]0, 1]$, $x \in \mathbb{H}$ and $\hat{u}_0 \in \partial\varphi(u_0)$:

$$\begin{aligned} &r_0 |\nabla\varphi_\varepsilon(x)| + |\varphi(J_\varepsilon x) - \varphi(u_0)| + \varepsilon|\nabla\varphi_\varepsilon(x)|^2 \\ &\quad \leq \langle \nabla\varphi_\varepsilon(x), x - u_0 \rangle + 2|\hat{u}_0| |x - u_0| + \left[2|\hat{u}_0|^2 + \varphi^{\#}_{u_0, r_0} - \varphi(u_0)\right]. \end{aligned} \tag{6.31}$$

In particular for $u_0 = 0$ and $\hat{u}_0 = 0$ we obtain

♦ If $\varphi(x) \geq \varphi(0) = 0$, for all $x \in \mathbb{H}$ and

$$\varphi^{\#}_{r_0} = \sup\{\varphi(r_0 v) : |v| \leq 1\} < \infty,$$

then:

$$\begin{array}{ll} a) & r_0 |\hat{x}| + \varphi(x) \leq \langle \hat{x}, x \rangle + \varphi^{\#}_{r_0}, \quad \forall\ (x, \hat{x}) \in \partial\varphi, \\ b) & r_0 |\nabla\varphi_\varepsilon(x)| + \varphi(J_\varepsilon x) + \varepsilon|\nabla\varphi_\varepsilon(x)|^2 \leq \langle \nabla\varphi_\varepsilon(x), x \rangle + \varphi^{\#}_{r_0}, \\ & \qquad \forall\, \varepsilon > 0, \forall\, x \in \mathbb{H}. \end{array} \tag{6.32}$$

6.3.7.3 Convex Functions on $C([0, T]; \mathbb{R}^d)$

Proposition 6.35. *If* $\varphi : \mathbb{R}^d \to]-\infty, +\infty]$ *is a proper convex l.s.c. function and* $\Phi : C([0, T]; \mathbb{R}^d) \to]-\infty, +\infty]$,

$$\Phi(x) = \begin{cases} \int_0^T \varphi(x(t))dt, & \text{if } \varphi(x) \in L^1(0,T), \\ +\infty, & \text{otherwise,} \end{cases} \tag{6.33}$$

then

$c_1)$ *Φ is a proper convex l.s.c. function,*

$c_2)$ $\partial\Phi(x) \overset{def}{=} \Big\{ k \in BV([0,T];\mathbb{R}^d)$:

$$\int_0^T \langle y(r) - x(r), dk(r)\rangle + \Phi(x) \le \Phi(y), \ \forall\, y \in C([0,T];\mathbb{R}^d) \Big\}$$

is a maximal monotone operator.

Proof. We shall prove only the maximal property of the operator $\partial\Phi$, since the other properties are immediate. Let $\mathbb{X} = C([0,T];\mathbb{R}^d)$. Then the dual space is $\mathbb{X}^* = BV([0,T];\mathbb{R}^d)$. Let

$$\langle k - \zeta, x - z\rangle \ge 0, \text{ for all } (z,\zeta) \in \partial\Phi. \tag{6.34}$$

The function $\psi(z) = \Phi(z) + \dfrac{1}{2}\|z - x\|_{\mathbb{X}}^2 - \langle k, z\rangle$ defined on $\mathbb{X}$ is a proper convex l.s.c. function. Furthermore, there exists a $c \in \mathbb{R}$ such that $\Phi(z) \ge c, \forall z \in \mathbb{X}$. By Ekeland's principle there exists a $z_\varepsilon \in \mathbb{X}$ such that

$$\psi(z_\varepsilon) \to \inf\{\psi(z) : z \in \mathbb{X}\},$$

$$\psi(z_\varepsilon) \le \psi(z) + \sqrt{\varepsilon}\,\|z - z_\varepsilon\|_{\mathbb{X}} = \tilde{\psi}(z), \ \forall z \in \mathbb{X}.$$

Then $0 \in \partial\tilde{\psi}(z_\varepsilon)$, which means

$$\partial\Phi(z_\varepsilon) + F(z_\varepsilon - x) - k + \sqrt{\varepsilon}\theta_\varepsilon \ni 0, \tag{6.35}$$

where $F : \mathbb{X} \rightrightarrows \mathbb{X}^*$ is the duality mapping and $\|\theta_\varepsilon\|_{\mathbb{X}^*} \le 1$. Multiplying by $z_\varepsilon - x$ we have $\langle \zeta_\varepsilon - k, z_\varepsilon - x\rangle + \|z_\varepsilon - x\|_{\mathbb{X}}^2 + \sqrt{\varepsilon}\langle\theta_\varepsilon, z_\varepsilon - x\rangle = 0$, for some $\zeta_\varepsilon \in \partial\Phi(z_\varepsilon)$, which implies by (6.34) $\|z_\varepsilon - x\|_{\mathbb{X}} \le \sqrt{\varepsilon}$. Hence $z_\varepsilon \overset{\mathbb{X}}{\to} x$, and by (6.35) $\zeta_\varepsilon \overset{\mathbb{X}^*}{\to} k$, as $\varepsilon \to 0$. From the definition of the subdifferential operator: $\langle\zeta_\varepsilon, y - z_\varepsilon\rangle + \Phi(z_\varepsilon) \le \Phi(y)$, $\forall\, y \in \mathbb{X}$ and passing to the limit as $\varepsilon \to 0$ we obtain $(x,k) \in \partial\Phi$. ∎

Proposition 6.36. *If $\varphi : \mathbb{R}^d \to]-\infty, +\infty]$ is a proper convex l.s.c. function, Φ is defined by (6.33), $x \in C([0,T];\mathbb{R}^d)$ and $k \in C([0,T];\mathbb{R}^d) \cap BV([0,T];\mathbb{R}^d)$, then the following assertions are equivalent:*

$$
\begin{aligned}
&a_1) \quad \int_s^t \langle z - x(r), dk(r)\rangle + \int_s^t \varphi(x(r))dr \le (t-s)\varphi(z), \\
&\qquad\qquad \forall\, z \in \mathbb{R}^d,\ \forall\, 0 \le s \le t \le T, \\
&a_2) \quad \int_s^t \langle y(r) - x(r), dk(r)\rangle + \int_s^t \varphi(x(r))dr \le \int_s^t \varphi(y(r))dr, \\
&\qquad\qquad \forall\, y \in C([0,T];\mathbb{R}^d),\ \forall\, 0 \le s \le t \le T, \\
&a_3) \quad \int_s^t \langle x(r) - z, dk(r) - \hat{z}dr\rangle \ge 0,\ \forall\ (z,\hat{z}) \in \partial\varphi, \\
&\qquad\qquad \forall\, 0 \le s \le t \le T, \\
&a_4) \quad \int_s^t \langle x(r) - y(r), dk(r) - \hat{y}(r)dr\rangle \ge 0,\ \forall\, y, \hat{y} \in C([0,T];\mathbb{R}^d), \\
&\qquad\qquad (y(r), \hat{y}(r)) \in \partial\varphi,\ \forall\, r \in [0,T],\ \forall\, 0 \le s \le t \le T, \\
&a_5) \quad (x,k) \in \partial\Phi, \textit{that is, } \forall y \in C([0,T],\mathbb{R}^d): \\
&\qquad \int_0^T \langle y(r) - x(r), dk(r)\rangle + \int_0^T \varphi(x(r))dr \le \int_0^T \varphi(y(r))dr.
\end{aligned}
\tag{6.36}
$$

Proof. We shall show that $a_1 \Leftrightarrow a_2 \Rightarrow a_3 \Rightarrow a_4 \Rightarrow a_5 \Rightarrow a_2$.
$a_2 \Rightarrow a_1$: is evident.
$a_1 \Rightarrow a_2$: Let $y \in C\left([0,T];\mathbb{R}^d\right)$. We extend $y(t) = y(0)$ for $t \le 0$ and $y(t) = y(T)$ for $t \ge T$. The same extension will be considered for the functions x and k. To prove a_2) it is sufficient to consider the case $0 < s < t < T$.

Since φ is bounded from below by an affine function, from a_1) we deduce that $\varphi(x) \in L^1(0,T)$.

Let $n_0 \in \mathbb{N}^*$ be such that $0 < \frac{1}{n_0} < s < t < t + \frac{1}{n_0} < T$ and $n \ge n_0$. Let $u \in [s,t]$. From (a_1) we have for $z = y(u)$

$$\int_{u-1/n}^{u} \langle y(u) - x(r), dk(r)\rangle + \int_{u-1/n}^{u} \varphi(x(r))\, dr \le \frac{1}{n}\,\varphi(y(u)).$$

Integrating on $[s,t]$ with respect to u we deduce that

$$
\begin{aligned}
\int_s^t \left(n \int_{u-1/n}^u \langle y(u) - x(r), dk(r)\rangle\right) du + \int_s^t \left(n \int_{u-1/n}^u \varphi(x(r))\, dr\right) du \\
\le \int_s^t \varphi(y(u))\, du.
\end{aligned}
\tag{6.37}
$$

By Fatou's Lemma we have

$$\int_s^t \varphi(x(u))\, du \le \liminf_{n\to+\infty} \int_s^t \left(n \int_{u-1/n}^u \varphi(x(r))\, dr\right) du.$$

On the other hand by the Lebesgue dominated convergence theorem

$$\begin{aligned}
&\int_s^t \left(n \int_{u-1/n}^{u} \langle y(u) - x(r), dk(r) \rangle \right) du \\
&= \int_{-\infty}^{+\infty} \left(\int_{-\infty}^{+\infty} n \mathbf{1}_{[s,t]}(u) \mathbf{1}_{[u-1/n,u]}(r) \langle y(u) - x(r), dk(r) \rangle \right) du \\
&= \int_{-\infty}^{+\infty} \left\langle \int_{-\infty}^{+\infty} n \mathbf{1}_{[s,t]}(u) \mathbf{1}_{[r,r+1/n]}(u) [y(u) - x(r)] \, du, dk(r) \right\rangle \\
&= \int_{-\infty}^{+\infty} \left\langle n \int_r^{r+1/n} \mathbf{1}_{[s,t]}(u) [y(u) - x(r)] \, du, dk(r) \right\rangle \\
&\to \int_{-\infty}^{+\infty} \mathbf{1}_{[s,t]}(r) \langle y(r) - x(r), dk(r) \rangle, \quad \text{as } n \to \infty.
\end{aligned}$$

Passing to $\liminf_{n\to+\infty}$ in (6.37) (a_2) follows.
$a_2 \Rightarrow a_3$: is obtained by adding the following inequalities term by term:

$$\int_s^t \langle z - x(r), dk(r) \rangle + \int_s^t \varphi(x(r)) dr \le \int_s^t \varphi(z) dr$$

$$\int_s^t \langle x(r) - z, \hat{z} \rangle \, dr + \int_s^t \varphi(z) dr \le \int_s^t \varphi(x(r)) dr.$$

$a_3 \Rightarrow a_4$: is proved in Proposition 6.17 since $A = \partial\varphi$ is maximal monotone.
$a_4 \Rightarrow a_5$: Let $(\tilde{x}, \tilde{k}) \in \partial\Phi$ be arbitrary. Hence for all y, $\hat{y} \in C([0,T]; \mathbb{R}^d)$, $(y(r), \hat{y}(r)) \in \partial\varphi$ we have: $(y, \int_0^{\cdot} \hat{y} dt) \in \partial\Phi$ and

$$\int_0^T \left\langle \tilde{x}(r) - y(r), d\tilde{k}(r) - \hat{y}(r) \, dr \right\rangle \ge 0 \quad (\partial\Phi \text{ is monotone}),$$
$$\int_0^T \langle x(r) - y(r), dk(r) - \hat{y}(r) \, dr \rangle \ge 0 \ (\text{by } a_4).$$

Since $A = \partial\varphi$ is maximal monotone, by Proposition 6.17 we have

$$\int_0^T \left\langle \tilde{x}(t) - x(t), d\tilde{k}(t) - dk(t) \right\rangle \ge 0,$$

where $(\tilde{x}, \tilde{k}) \in \partial\Phi$ is arbitrary. But by Proposition 6.35, $\partial\Phi$ is a maximal monotone operator. Hence $(x, k) \in \partial\Phi$.
$a_5 \Rightarrow a_2$: Let $a, b \ge 0$ such that $\varphi(y) + a|y| + b \ge 0$. From a_5) it follows that $\varphi(x) \in L^1(0,T)$. Let $\alpha_n \in C([0,T]; \mathbb{R})$, $0 \le \alpha_n \le 1$, and $\alpha_n \nearrow \mathbf{1}_{]s,t[}$. In a_5) we put $y(r) := (1 - \alpha_n(r))x(r) + \alpha_n(r) y(r)$. So we have

$$\int_0^T \langle (y-x)\alpha_n, dk(r)\rangle + \int_0^T \varphi(x)\,dr \le \int_0^T ((1-\alpha_n)\varphi(x) + \alpha_n\varphi(y))\,dr$$

and furthermore

$$\begin{aligned}\int_0^T \langle (y-x)\,\alpha_n, dk(r)\rangle &+ \int_0^T \alpha_n\varphi(x)\,dr \\ &\le \int_0^T \alpha_n\varphi(y)\,dr \\ &\le \int_0^T \alpha_n(\varphi(y) + a|y| + b)\,dr - \int_0^T \alpha_n(a|y|+b)\,dr \\ &\le \int_0^T 1_{[s,t]}(r)(\varphi(y) + a|y| + b)\,dr - \int_0^T \alpha_n(a|y|+b)\,dr \\ &\le \int_s^t \varphi(y)\,dr + \int_0^T \left(1_{[s,t]} - \alpha_n\right)(a|y|+b)\,dr.\end{aligned}$$

Passing to the limit as $n \to \infty$, a_2) follows. ■

Proposition 6.37. *If* $\varphi : \mathbb{R}^d \to]-\infty, +\infty]$ *is a proper convex l.s.c. function and* $\tilde{\Phi}$*:* $L^2(\Omega; C([0,T];\mathbb{R}^d)) \to]-\infty, +\infty]$*,*

$$\tilde{\Phi}(x) = \begin{cases} \mathbb{E}\int_0^T \varphi(x(t)dt, & \text{if } \varphi(x) \in L^1(\Omega\times]0,T[), \\ +\infty, & \text{otherwise} \end{cases} \tag{6.38}$$

then

a) $\tilde{\Phi}$ *is a proper convex l.s.c. function,*

b) $\partial\tilde{\Phi}(x) \overset{def}{=} \Big\{K \in L^2(\Omega;\ BV([0,T];\mathbb{R}^d)) : \mathbb{E}\int_0^T \langle Y_t - X_t, dK_t\rangle$

$$+\mathbb{E}\int_0^T \varphi(X_t)dt \le \mathbb{E}\int_0^T \varphi(Y_t)dt,\ \ \forall\, Y \in L^2(\Omega; C([0,T];\mathbb{R}^d))\Big\}$$

is a maximal monotone operator,

c) $K \in \partial\tilde{\Phi}(x)$ *iff* $K_\cdot(\omega) \in \partial\Phi(X_\cdot(\omega))$*,* $\mathbb{P}$*-a.s.* $\omega \in \Omega$*, with* $\partial\Phi$ *characterized in Proposition 6.36.*

Proof. The assertions a) and b) are obtained in the same manner as c_1) and c_2) from Proposition 6.35. The point c) follows from b) putting $Y := X1_{A^c} + Y1_A$, where $A \in F$ is arbitrary. ■

Proposition 6.38. *Let* $\varphi : \mathbb{R}^d \to]-\infty, +\infty]$ *be a proper convex l.s.c. function such that* int (Dom $(\varphi)) \neq \emptyset$*. Let*

Φ: $C([0,T];\mathbb{R}^d) \to]-\infty,+\infty]$ *be defined by (6.33). Let* $(u_0,\hat{u}_0) \in \partial\varphi$, $r_0 \geq 0$ *and*

$$\varphi^{\#}_{u_0,r_0} \stackrel{def}{=} \sup\{\varphi(u_0+r_0v) : |v| \leq 1\}.$$

Then for all $0 \leq s \leq t \leq T$ *and* $(x,k) \in \partial\Phi$:

$$\begin{aligned} r_0(\updownarrow k\updownarrow_t - \updownarrow k\updownarrow_s) &+ \int_s^t \varphi(x(r))dr \\ &\leq \int_s^t \langle x(r)-u_0, dk(r)\rangle + (t-s)\varphi^{\#}_{u_0,r_0}. \end{aligned} \tag{6.39}$$

Moreover for all $0 \leq s \leq t \leq T$ *and for all* $(x,k) \in \partial\Phi$:

$$\begin{aligned} r_0(\updownarrow k\updownarrow_t - \updownarrow k\updownarrow_s) + \int_s^t |\varphi(x(r)) - \varphi(u_0)|\,dr &\leq \int_s^t \langle x(r)-u_0, dk(r)\rangle \\ +\int_s^t \left(2|\hat{u}_0|\,|x(r)-u_0| + \varphi^{\#}_{u_0,r_0} - \varphi(u_0)\right)dr. \end{aligned} \tag{6.40}$$

Proof. Let $0 \leq s = t_0 < t_1 < \ldots < t_n = t \leq T$, $\max_i(t_{i+1}-t_i) = \delta_n \to 0$. By (6.36-a$_1$) for $z = u_0 + r_0v$. We obtain

$$\begin{aligned} r_0\langle k(t_{i+1}) - k(t_i), v\rangle + \int_{t_i}^{t_{i+1}} \varphi(x(r))dr &\leq \int_{t_i}^{t_{i+1}} \langle x(r)-u_0, dk(r)\rangle \\ &+ (t_{i+1}-t_i)\varphi^{\#}_{u_0,r_0}, \end{aligned}$$

for all $|v| \leq 1$. Hence

$$\begin{aligned} r_0|k(t_{i+1}) - k(t_i)| + \int_{t_i}^{t_{i+1}} \varphi(x(r))dr &\leq \int_{t_i}^{t_{i+1}} \langle x(r)-u_0, dk(r)\rangle \\ &+ (t_{i+1}-t_i)\varphi^{\#}_{u_0,r_0} \end{aligned}$$

and adding term by term for $i = 0$ to $i = n-1$ we have

$$r_0\sum_{i=0}^{n-1}|k(t_{i+1}) - k(t_i)| + \int_s^t \varphi(x(t))dt \leq \int_s^t \langle x(t)-u_0, dk(t)\rangle + (t-s)\varphi^{\#}_{u_0,r_0},$$

which clearly yields (6.39). The second inequality (6.40) now follows, using the fact that

$$|\varphi(x) - \varphi(u_0)| \leq \varphi(x) - \varphi(u_0) + 2|\hat{u}_0|\,|x-u_0|,$$

for all $x \in \mathbb{R}^d$ and $(u_0,\hat{u}_0) \in \partial\varphi$. ∎

Remark 6.39. Since φ is locally bounded on int(Domφ), it follows that for

$$u_0 \in \text{int}(\text{Dom}\varphi)\,[= \text{int}\,(\text{Dom}\,(\partial\varphi))]\,,$$

there exists $r_0 > 0$ and $M_0 \geq 0$, such that

$$\sup\{|\varphi\,(u_0 + r_0 v)| : |v| \leq 1\} \leq M_0.$$

6.3.8 Semiconvex Functions

Let $\varphi : \mathbb{R}^d \to]-\infty, +\infty]$.

Define

$$\text{Dom}\,(\varphi) = \left\{v \in \mathbb{R}^d : \varphi\,(v) < +\infty\right\}.$$

We say that φ is a proper function if Dom $(\varphi) \neq \emptyset$ and Dom (φ) has no isolated points.

Definition 6.40. The (Fréchet) subdifferential of φ at $x \in \mathbb{R}^d$ is defined by

$$\partial^-\varphi\,(x) = \left\{\hat{x} \in \mathbb{R}^d : \quad \liminf_{y \to x} \frac{\varphi\,(y) - \varphi\,(x) - \langle \hat{x}, y - x\rangle}{|y - x|} \geq 0\right\},$$

if $x \in$ Dom (φ), and $\partial^-\varphi\,(x) = \emptyset$, if $x \notin$ Dom (φ).

Example 6.41. If E is a non-empty closed subset of $\mathbb{R}^d$ and

$$\varphi\,(x) = I_E\,(x) = \begin{cases} 0, & \text{if } x \in E, \\ +\infty, & \text{if } x \notin E, \end{cases}$$

then φ is l.s.c. and (by a result of Colombo and Goncharov [17] we have for any closed subset E of a Hilbert space)

$$\begin{aligned} \partial^- I_E\,(x) &= \{\hat{x} \in \mathbb{R}^d : \quad \limsup_{y \to x,\; y \in E} \frac{\langle \hat{x}, y - x\rangle}{|y - x|} \leq 0\} \\ &= \begin{cases} 0, & \text{if } x \in \text{int}\,(E)\,, \\ N_E\,(x)\,, & \text{if } x \in \text{Bd}\,(E)\,, \\ \emptyset, & \text{if } x \notin E, \end{cases} \end{aligned}$$

where $N_E\,(x)$ is the closed normal cone at E in $x \in$ Bd (E)

$$N_E\,(x) \overset{def}{=} \left\{u \in \mathbb{R}^d : \lim_{\varepsilon \searrow 0} \frac{d_E\,(x + \varepsilon u)}{\varepsilon} = |u|\right\}$$

and

$$d_E(z) \overset{def}{=} \inf\{|z-x| : x \in E\}$$

is the distance of a point $z \in \mathbb{R}^d$ to E.

Denote

$$\begin{aligned} a)\quad & \mathrm{Dom}(\partial^-\varphi) = \{x \in \mathbb{R}^d : \partial^-\varphi(x) \neq \emptyset\}, \\ b)\quad & \partial^-\varphi = \{(x,\hat{x}) :\ x \in \mathrm{Dom}(\partial^-\varphi),\ \hat{x} \in \partial^-\varphi(x)\}. \end{aligned}$$

Definition 6.42. A closed set $E \subset \mathbb{R}^d$ is γ–semiconvex, $\gamma \geq 0$, if for all $x \in \mathrm{Bd}(E)$ there exists an $\hat{x} \neq 0$ such that

$$\langle \hat{x}, y-x\rangle \leq \gamma\,|\hat{x}|\,|y-x|^2, \quad \text{for all } y \in E.$$

Note that if E is a semiconvex set, then

$$\partial^- I_E(x) = \{\hat{x} \in \mathbb{R}^d : \langle \hat{x}, y-x\rangle \leq \gamma\,|\hat{x}|\,|y-x|^2, \quad \text{for all } y \in E\}.$$

Definition 6.43. $\varphi : \mathbb{R}^d \to]-\infty,+\infty]$ is a semiconvex function if there exist $\rho, \gamma \geq 0$ such that

(a) $\overline{\mathrm{Dom}(\varphi)}$ is γ–semiconvex;
(b) $\mathrm{Dom}(\partial^-\varphi) \neq \emptyset$;
(c) for all $y \in \mathbb{R}^d$ and for all $(x,\hat{x}) \in \partial^-\varphi$

$$\langle \hat{x}, y-x\rangle + \varphi(x) \leq \varphi(y) + (\rho + \gamma\,|\hat{x}|)\,|y-x|^2.$$

A function φ satisfying the properties of this definition will sometimes be called a (ρ,γ)–semiconvex function, or a γ–semiconvex function (since the second parameter is the most important one).

Note that a convex function is a (ρ,γ)–semiconvex function for all $\rho,\gamma \geq 0$.

A set E is γ–semiconvex iff I_E is a $(0,\gamma)$–semiconvex function.

If we write the definition of semiconvexity for a fixed $(x_0,\hat{x}_0) \in \partial^-\varphi$, then it is clear that we have:

Proposition 6.44. *If $\varphi : \mathbb{R}^d \to]-\infty,+\infty]$ is a semiconvex function, then there exists an $a \geq 0$ such that*

$$\varphi(y) + a\,|y|^2 + a \geq 0, \quad \forall\ y \in \mathbb{R}^d.$$

In particular φ is bounded below on bounded subsets of $\mathbb{R}^d$.

The following properties also hold:

Proposition 6.45. *Let $\varphi : \mathbb{R}^d \to]-\infty, +\infty]$ be a semiconvex function. If there exist $u_0 \in \text{Dom}(\varphi)$, $r_0, M_0 > 0$ such that*

$$\varphi(u_0 + r_0 v) \leq M_0, \quad \forall \ |v| \leq 1,$$

then there exist $\rho_0 > 0$ and $b \geq 0$ such that

$$\rho_0 |\hat{x}| \leq \langle \hat{x}, x - u_0 \rangle + b + b(1 + |\hat{x}|)|x - u_0|^2, \quad \forall \ (x, \hat{x}) \in \partial^- \varphi \tag{6.41}$$

and moreover there exist $M \geq 0$ and $\delta_0 \in]0, r_0]$ such that

$$|\hat{x}| \leq M, \quad \forall \ x \in \bar{B}(u_0, \delta_0) \subset \text{Dom}(\varphi) \ \text{and} \ \hat{x} \in \partial^- \varphi(x). \tag{6.42}$$

Proof. Let $(x, \hat{x}) \in \partial^- \varphi$. Then for all $|v| \leq 1$ and $\lambda \in [0, 1]$:

$$\langle \hat{x}, (u_0 + r_0 \lambda v) - x \rangle + \varphi(x) \leq \varphi(u_0 + r_0 \lambda v) + (\rho + \gamma |\hat{x}|) |(u_0 + r_0 \lambda v) - x|^2,$$

which yields

$$r_0 \lambda \langle \hat{x}, v \rangle \leq \langle \hat{x}, x - u_0 \rangle + \left(a|x|^2 + a\right) + M_0 + 2(\rho + \gamma |\hat{x}|)\left[|x - u_0|^2 + r_0^2 \lambda^2\right].$$

Taking the $\sup_{|v| \leq 1}$, we deduce for $\lambda = 1/(1 + 2\gamma r_0)$:

$$\frac{r_0}{(1 + 2\gamma r_0)^2} |\hat{x}| \leq \langle \hat{x}, x - u_0 \rangle + C + C(1 + |\hat{x}|)|x - u_0|^2$$

that is (6.41).

Moreover if $|x - u_0| \leq \delta_0 = 1 \wedge \frac{\rho_0}{2(1+b)} \wedge r_0$, then

$$\begin{aligned}
\rho_0 |\hat{x}| &\leq \langle \hat{x}, x - u_0 \rangle + b + b(1 + |\hat{x}|)|x - u_0|^2 \\
&\leq \left(\delta_0 + b\delta_0^2\right)|\hat{x}| + b + b\delta_0^2 \\
&\leq \delta_0 (1 + b)|\hat{x}| + 2b \\
&\leq \frac{\rho_0}{2} |\hat{x}| + 2b
\end{aligned}$$

and (6.42) follows. ■

Let E be a non-empty closed subset of $\mathbb{R}^d$ and $\varepsilon > 0$. We denote by

$$U_\varepsilon(E) \stackrel{def}{=} \left\{y \in \mathbb{R}^d : d_E(y) < \varepsilon\right\}$$

the open ε-neighbourhood of E and

$$\overline{U}_{\varepsilon}(E) \stackrel{def}{=} \left\{z \in \mathbb{R}^d : d_E(z) \leq \varepsilon\right\}$$

the closed ε-neighbourhood of E.

Given $z \in \mathbb{R}^d$, we denote by $\Pi_E(z)$ the set of elements $x \in E$ with $|z - x| = d_E(z)$. We remark that $\Pi_E(z)$ is always non-empty since E is non-empty and closed. We also note that if $z \in \mathbb{R}^d$ and $\hat{z} \in \Pi_E(z)$, then $z - \hat{z} \in N_E(\hat{z})$. This follows from the fact that for $0 < \varepsilon < 1$ we have

$$\begin{aligned} d_E(\hat{z} + \varepsilon(z - \hat{z})) &= d_E(z) + d_E(\hat{z} + \varepsilon(z - \hat{z})) - d_E(z) \\ &\geq |z - \hat{z}| - |\hat{z} + \varepsilon(z - \hat{z}) - z| \\ &= \varepsilon|z - \hat{z}| \end{aligned}$$

and

$$\begin{aligned} d_E(\hat{z} + \varepsilon(z - \hat{z})) &= d_E(\hat{z} + \varepsilon(z - \hat{z})) - d_E(\hat{z}) \\ &\leq \varepsilon|z - \hat{z}|. \end{aligned}$$

We recall the notations

$$\begin{aligned} B(y, r) &= \left\{u \in \mathbb{R}^d : |u - y| < r\right\}, \quad \text{and} \\ \overline{B}(y, r) &= \left\{u \in \mathbb{R}^d : |u - y| \leq r\right\}. \end{aligned}$$

Definition 6.46. We say that E satisfies the "uniform exterior ball condition" (abbreviated UEBC) if

- $N_E(x) \neq \{0\}$ for all $x \in \mathrm{Bd}(E)$,
- $\exists\, r_0 > 0$ such that, $\forall\, x \in \mathrm{Bd}(E)$ and $\forall\, u \in N_E(x)$, $|u| = r_0$:

$$d_E(x + u) = r_0 \quad \text{or equivalently} \quad B(x + u, r_0) \cap E = \emptyset,$$

(in this case we say that E satisfies r_0-UEBC).

Note that for all $v \in N_E(x)$, $|v| \leq r_0$, we also have

$$d_E(x + v) = |v|. \tag{6.43}$$

Indeed since

$$0 \leq d_E(x + v) = d_E(x + v) - d_E(x) \leq |v|$$

and

$$\begin{aligned}
|v| &= r_0 + (|v| - r_0) \\
&= d_E\left(x + \frac{r_0}{|v|}v\right) + (|v| - r_0) \\
&\leq \left|d_E\left(x + \frac{r_0}{|v|}v\right) - d_E(x+v)\right| + d_E(x+v) + (|v| - r_0) \\
&\leq \left(\frac{r_0}{|v|} - 1\right)|v| + d_E(x+v) + (|v| - r_0) \\
&= d_E(x+v),
\end{aligned}$$

then (6.43) follows.

It is clear that, under the uniform exterior ball condition with ball radius r_0, for all $z \in \mathbb{R}^d$ with $d_E(z) < r_0$, the set $\Pi_E(z)$ is a singleton. The unique element of $\Pi_E(z)$ is called the projection of z on E, and it is denoted by $\pi_E(z)$.

We have the following characterization of the notion of the uniform exterior ball condition:

Lemma 6.47. *Let E be a non-empty closed subset of $\mathbb{R}^d$. The following assertions are equivalent:*

(i) E satisfies the uniform exterior ball condition;
(ii) E is a semiconvex subset of $\mathbb{R}^d$, that is $\exists \gamma \geq 0$ and for all $x \in \mathrm{Bd}(E)$ there exists an $\hat{x} \neq 0$ such that

$$\langle \hat{x}, y - x \rangle \leq \gamma |\hat{x}| |y - x|^2, \quad \textit{for all } y \in E,$$

(in this case $\hat{x} \in N_E(x)$ follows);
(iii) $\exists \gamma \geq 0, \forall x, y \in \mathrm{Bd}(E), \forall \lambda \in]0,1[$:

$$d_E((1-\lambda)x + \lambda y) \leq 4\lambda(1-\lambda)\gamma |x - y|^2;$$

(iii′) $\exists \gamma \geq 0, \forall x, y \in E, \forall \lambda \in]0,1[$:

$$d_E((1-\lambda)x + \lambda y) \leq 4\lambda(1-\lambda)\gamma |x - y|^2;$$

(iv) $\exists \gamma \geq 0, \forall x, y \in \mathrm{Bd}(E)$:

$$d_E\left(\frac{x+y}{2}\right) \leq \gamma |x - y|^2,$$

(iv′) $\exists \gamma \geq 0, \forall x, y \in E$:

$$d_E\left(\frac{x+y}{2}\right) \leq \gamma |x - y|^2,$$

(v) $\exists\, \delta > 0$ *and* $\mu > 0$ *such that the function*

$$x \longrightarrow \psi_E^{\mu}(x) \stackrel{def}{=} d_E(x) + \mu\,|x|^2 : U_\delta(E) \to \mathbb{R}$$

is convex on each convex subset of $U_\delta(E)$.

Proof. We first remark that the conditions $(ii)\,,(iii)\,,\left(iii'\right),(iv)\,,\left(iv'\right)$ are satisfied for $\gamma = 0$ if and only if E is convex; the convex sets satisfy the r-UEBC for all $r > 0$.

Step I. $(i) \Leftrightarrow (ii)$

$(i) \Rightarrow (ii)$: Let $x \in \mathrm{Bd}(E)$ and $\hat{x} \in N_E(x)$, $\hat{x} \neq 0$. Then there exists an $r_0 > 0$ such that

$$d_E\left(x + \frac{r_0}{|\hat{x}|}\hat{x}\right) = r_0.$$

We have for all $y \in E$ and $\gamma = \frac{1}{2r_0}$

$$\begin{aligned}\gamma\,|\hat{x}|\,|y-x|^2 - \langle \hat{x}, y-x\rangle &= \frac{1}{2r_0}\,|\hat{x}|\left[\left|y - \left(x + \frac{r_0}{|\hat{x}|}\hat{x}\right)\right|^2 - r_0^2\right] \\ &\geq \frac{1}{2r_0}\,|\hat{x}|\left[d_E^2\left(x + \frac{r_0}{|\hat{x}|}\hat{x}\right) - r_0^2\right] \\ &= 0.\end{aligned}$$

$(ii) \Rightarrow (i)$: Let $r_0 > 0$ be such that $2\gamma r_0 \leq 1$. Let $x \in \mathrm{Bd}(E)$ be arbitrary and $u = r_0\dfrac{\hat{x}}{|\hat{x}|}$. Then

$$\begin{aligned}|u|^2 &= r_0^2 \\ &\leq r_0^2 + \frac{2r_0}{|\hat{x}|}\left[\gamma\,|\hat{x}|\,|y-x|^2 - \langle \hat{x}, y-x\rangle\right] \\ &\leq |y - (x+u)|^2, \quad \forall\, y \in E.\end{aligned}$$

Hence

$$|u| = r_0 \leq d_E(x+u) \leq |u|\,,$$

that is E satisfies the r_0-uniform exterior ball condition.

From this equivalence we have that

$$E \text{ is } r_0 - UEBC \Leftrightarrow E \text{ is } \frac{1}{2r_0}\text{–semiconvex.} \tag{6.44}$$

Step II. (*iii*) ⇔ (*iii′*).

We have to prove only (*iii*) ⇒ (*iii′*). Let $x, y \in E$ and $0 < \lambda < 1$. Let $u_\lambda = (1-\lambda)x + \lambda y = x + \lambda(y-x)$. If $u_\lambda \in E$, then

$$d_E((1-\lambda)x + \lambda y) = 0 \le 4\lambda(1-\lambda)\gamma|x-y|^2.$$

If $u_\lambda \notin E$, then there exist $0 < \alpha < \lambda < \beta < 1$ such that

$$u_\rho = x + \rho(y-x) \notin E, \quad \text{for all } \alpha < \rho < \beta$$

and

$$u_\alpha = x + \alpha(y-x) \in E, \quad u_\beta = x + \beta(y-x) \in E.$$

We have

$$u_\lambda = u_\alpha + \frac{\lambda-\alpha}{\beta-\alpha}\left(u_\beta - u_\alpha\right)$$

and consequently

$$\begin{aligned} d_E((1-\lambda)x + \lambda y) &= d_E(u_\lambda) \\ &\le 4\frac{\lambda-\alpha}{\beta-\alpha}\left(1 - \frac{\lambda-\alpha}{\beta-\alpha}\right)\gamma\left|u_\beta - u_\alpha\right|^2 \\ &\le 4(\lambda-\alpha)(\beta-\lambda)\gamma|y-x|^2 \\ &\le 4\lambda(1-\lambda)\gamma|x-y|^2. \end{aligned}$$

Step II. (iii') ⇒ (*iv′*) ⇒ (*iv*) ⇒ (*i*) ⇒ (*iii*).

(iii') ⇒ (*iv′*) ⇒ (*iv*) as particular cases: (*iv′*) for (iii') and (*iv*) for (*iv′*).

(*iv*) ⟹ (*i*): We prove by contradiction. We can assume $\gamma > 0$. We suppose that there is some $z \in \mathbb{R}^d$ in the r_0-neighbourhood of E such that, for two different $x, y \in \mathrm{Bd}(E)$,

$$|z-x| = |z-y| = d_E(z) < r_0 = \frac{1}{2\gamma}.$$

Under this hypothesis the vectors $z - \frac{1}{2}(x+y) = \frac{1}{2}[(z-y)+(z-x)]$ and $2(x-y) = 2[(z-y)-(z-x)]$ are orthogonal and, consequently,

$$d_E^2(z) = |z-x|^2 = \left|z - \frac{1}{2}(x+y)\right|^2 + 4|y-x|^2.$$

Let $u \in \Pi_E\left(\frac{1}{2}(x+y)\right)$. Then, from condition (*iv*) we obtain

$$\begin{aligned}\gamma\,|x-y|^2 &\geq d_E\left(\frac{1}{2}(x+y)\right) = \left|\left(\frac{1}{2}(x+y)\right) - u\right| \\ &\geq |z-u| - \left|z - \left(\frac{1}{2}(x+y)\right)\right| \\ &\geq d_E(z) - \left|z - \left(\frac{1}{2}(x+y)\right)\right| \\ &= \sqrt{\left|z - \frac{1}{2}(x+y)\right|^2 + 4\,|y-x|^2} - \left|z - \left(\frac{1}{2}(x+y)\right)\right|.\end{aligned}$$

Hence, we have

$$\left|z - \frac{1}{2}(x+y)\right|^2 + 4\,|y-x|^2 \leq \left[\gamma\,|x-y|^2 + \left|z - \left(\frac{1}{2}(x+y)\right)\right|\right]^2,$$

from which we easily deduce that

$$\begin{aligned}4 &\leq \gamma^2\,|y-x|^2 + 2\gamma\left|z - \frac{1}{2}(x+y)\right| \\ &\leq \gamma^2\,[|z-x| + |z-y|]^2 + \gamma\,[|z-x| + |z-y|] \\ &< 2,\end{aligned}$$

which is a contradiction. Consequently, condition (*iv*) implies the $\frac{1}{2\gamma}$-uniform exterior ball condition.

$(i) \Longrightarrow (iii)$: Let us now suppose that E satisfies the uniform exterior ball condition with an r_0-ball. Let $x, y \in \mathrm{Bd}(E)$. In a first step we assume that x, y are two different elements such that $0 < |x-y| \leq r_0$. Let $\lambda \in]0,1[$ be such that $x_\lambda = x + \lambda(y-x) \notin E$ (if there is not such a λ, we are done), and let $\overline{x}_\lambda \in \Pi_E(x_\lambda)$. We fix any $u_\lambda \in N_E(\overline{x}_\lambda)$, $|u_\lambda| = r_0$ and put $z_\lambda = \overline{x}_\lambda + u_\lambda$. Then, due to condition (i), $|v - z_\lambda| \geq r_0$, for all $v \in E$. In particular, we have

$$|x - z_\lambda| \geq r_0, \quad \text{and} \quad |y - z_\lambda| \geq r_0.$$

We also observe that

$$|x_\lambda - \overline{x}_\lambda| = d_E(x_\lambda) \leq |x_\lambda - x| = \lambda\,|y-x| \leq r_0 = |z_\lambda - \overline{x}_\lambda|,$$

and

$$\alpha_\lambda = \frac{\langle x - z_\lambda, y - z_\lambda\rangle}{|x - z_\lambda|\,|y - z_\lambda|} \in [0,1].$$

Hence,

$$\begin{aligned}
&|x_\lambda - \overline{x}_\lambda| \\
&= r_0 - |z_\lambda - x_\lambda| \\
&= r_0 - \sqrt{(1-\lambda)^2 |x - z_\lambda|^2 + \lambda^2 |y - z_\lambda|^2 + 2\lambda(1-\lambda)|x - z_\lambda||y - z_\lambda|\alpha} \\
&\le r_0 \left(1 - \sqrt{(1-\lambda)^2 + \lambda^2 + 2\lambda(1-\lambda)\alpha}\right) \\
&= r_0 \left(1 - \sqrt{1 - 2\lambda(1-\lambda)(1-\alpha)}\right) \\
&\le r_0 [1 - (1 - 2\lambda(1-\lambda)(1-\alpha))] = 2r_0\lambda(1-\lambda)(1-\alpha).
\end{aligned}$$

On the other hand, for $\gamma \ge 1/(2r_0)$,

$$\begin{aligned}
&4\lambda(1-\lambda)\gamma|x - y|^2 \\
&= 4\lambda(1-\lambda)\gamma\left(|x - z_\lambda|^2 + |y - z_\lambda|^2 - 2|x - z_\lambda||y - z_\lambda|\alpha\right) \\
&\ge 8\lambda(1-\lambda)\gamma|x - z_\lambda||y - z_\lambda|(1-\alpha) \\
&\ge 8\lambda(1-\lambda)\gamma r_0^2(1-\alpha) \ge 2r_0\lambda(1-\lambda)(1-\alpha).
\end{aligned}$$

Consequently, $d_E(x_\lambda) = |x_\lambda - \overline{x}_\lambda| \le 4\lambda(1-\lambda)\gamma|x - y|^2$, if $\gamma \ge 1/(4r_0)$.

In order to complete the proof, we still have to consider the case of $x, y \in \mathrm{Bd}(E)$ with $|x - y| > r_0$. In this case, for $\gamma \ge 1/(2r_0)$, we have

$$\begin{aligned}
&d_E(x + \lambda(y - x))\,(= d_E(y - (1-\lambda)(y - x))) \\
&\le [\lambda \wedge (1-\lambda)]|x - y| \le 2\lambda(1-\lambda)|x - y| \\
&\le 4\lambda(1-\lambda)\frac{1}{2r_0}|x - y|^2.
\end{aligned}$$

This proves that under the r_0-uniform exterior ball condition the statement (*iii*) holds with $\gamma \ge 1/(2r_0)$.

Step III. (*v*) $\Rightarrow$ (*iii*) $\Rightarrow$ (*v*).

(*v*) $\Longrightarrow$ (*iii*): Let $\lambda \in (0,1)$ and $x, y \in \mathrm{Bd}(E)$ with $|x - y| < \delta$. Then $x, y \in B(x;\delta) = \{z \in \mathbb{R}^d : |z - x| < \delta\} \subset U_\delta(E)$, and, consequently,

$$\begin{aligned}
&d_E(\lambda x + (1-\lambda)y) + \mu|\lambda x + (1-\lambda)y|^2 \\
&= \psi_E^\mu(\lambda x + (1-\lambda)y) \le \lambda\psi_E^\mu(x) + (1-\lambda)\psi_E^\mu(y) \\
&= \lambda\mu|x|^2 + (1-\lambda)\mu|y|^2.
\end{aligned}$$

By subtracting $\mu\,|\lambda x + (1-\lambda)\,y|^2$ on the left-hand and the right-hand sides of this inequality we obtain

$$d_E\,(\lambda x + (1-\lambda)\,y) \le \lambda\,(1-\lambda)\,\mu\,|x-y|^2.$$

On the other hand, if $x, y \in \mathrm{Bd}\,(E)$ are such that $|x-y| \ge \delta$, then

$$d_E\,(\lambda x + (1-\lambda)\,y) \le [\lambda \wedge (1-\lambda)]\,|x-y| \le \frac{2}{\delta}\lambda\,(1-\lambda)\,|x-y|^2.$$

This shows that *(iii)* is fulfilled for $\gamma \ge \frac{1}{2\delta} \vee \frac{\mu}{4}$.

(iii) $\Longrightarrow$ (*v*): We fix any $\delta \in (0, r_0)$, and we recall that $\pi_E : \overline{U}_\delta\,(E) \to E$ is Lipschitz continuous with Lipschitz constant $L_\delta = r_0/\,(r_0 - \delta)$. Let $\lambda \in (0,1)$ and $u, v \in \overline{U}_\delta\,(E)$ be such that $(1-\lambda)\,u + \lambda v \in \overline{U}_\delta\,(E)$. For simplicity of notation we put $x = \pi_E\,(u)$, $y = \pi_E\,(v)$, $z_\lambda = (1-\lambda)\,u + \lambda v$, and $\overline{z}_\lambda = (1-\lambda)\,x + \lambda y$. Then,

$$\begin{aligned} d_E\,((1-\lambda)\,u + \lambda v) &= d_E\,(z_\lambda) \\ &\le |z_\lambda - \pi_E\,(\overline{z}_\lambda)| \\ &\le |z_\lambda - \overline{z}_\lambda| + |\overline{z}_\lambda - \pi_E\,(\overline{z}_\lambda)| \\ &\le (1-\lambda)\,d_E\,(u) + \lambda d_E\,(v) + d_E\,(\overline{z}_\lambda) \\ &\le (1-\lambda)\,d_E\,(u) + \lambda d_E\,(v) + 4\lambda\,(1-\lambda)\,\gamma\,|x-y|^2 \\ &\le (1-\lambda)\,d_E\,(u) + \lambda d_E\,(v) + 4\lambda\,(1-\lambda)\,\gamma L_\delta^2\,|u-v|^2. \end{aligned}$$

Hence, for $\mu \ge 4\gamma L_\delta^2$,

$$\begin{aligned} \psi_E^\mu\,((1-\lambda)\,u + \lambda v) &= d_E\,((1-\lambda)\,u + \lambda v) + \mu\,|(1-\lambda)\,u + \lambda v|^2 \\ &\le (1-\lambda)\left[d_E\,(u) + \mu\,|u|^2\right] + \lambda\left[d_E\,(v) + \mu\,|v|^2\right] \\ &= (1-\lambda)\,\psi_E^\mu\,(u) + \lambda \psi_E^\mu\,(v). \end{aligned}$$

This proves that ψ_E^μ is convex on each convex subset of $\overline{U}_\delta\,(E)$. ∎

Corollary 6.48. *If E is a closed subset of $\mathbb{R}^d$ and satisfies the r_0-uniform exterior ball condition, then for all $x \in E$*

$$N_E\,(x) = \left\{\hat{x} \in \mathbb{R}^d : \langle \hat{x}, y - x\rangle \le \frac{1}{2r_0}\,|\hat{x}|\,|y-x|^2; \quad \forall y \in E\right\}$$

and $\varphi = I_E$ is a $(0, \dfrac{1}{2r_0})$–semiconvex l.s.c. function. Moreover $N_E\,(x) = \partial^- I_E\,(x)$.

Let $r_0 > 0$. The set E satisfies the r_0-uniform exterior ball condition if and only if E is $\frac{1}{2r_0}$–semiconvex.

We recall the following well-known property of the projection.

Lemma 6.49. *Suppose that E satisfies the uniform exterior ball condition with ball radius r_0 and $\varepsilon \in]0, r_0[$. Then the projection π_E restricted to $\overline{U}_\varepsilon(E)$ (the closed ε-neighbourhood of E) is Lipschitz with Lipschitz constant $L_\varepsilon = r_0/(r_0 - \varepsilon)$, and the function d_E^2 is of class C^1 on $\overline{U}_\varepsilon(E)$ with*

$$\frac{1}{2}\nabla d_E^2(z) = z - \pi_E(z), \quad \textit{and} \quad z - \pi_E(z) \in N_E(\pi_E(z)),$$

for all $z \in \overline{U}_\varepsilon(E)$.

Proof. To simplify we denote $\pi = \pi_E$ and $d = d_E$. Let $x, y \in \overline{U}_\varepsilon(E)$. Then we have $x - \pi(x) \in N_E(\pi(x))$, $y - \pi(y) \in N_E(\pi(y))$ and

$$\begin{aligned}|\pi(x) - \pi(y)|^2 &= \langle y - \pi(y), \pi(x) - \pi(y)\rangle + \langle x - \pi(x), \pi(y) - \pi(x)\rangle \\ &\quad + \langle x - y, \pi(x) - \pi(y)\rangle \\ &\le \frac{\varepsilon}{r_0}|\pi(x) - \pi(y)|^2 + |x - y|\,|\pi(x) - \pi(y)|.\end{aligned}$$

Hence

$$|\pi(x) - \pi(y)| \le \frac{r_0}{r_0 - \varepsilon}|x - y|. \tag{6.45}$$

To obtain the second part of lemma it is sufficient to show that there exist a positive constant $C = C_{\varepsilon, r_0}$ such that

$$-C|y - x|^2 < d^2(y) - d^2(x) - 2\langle x - \pi(x), y - x\rangle \le C|y - x|^2. \tag{6.46}$$

We have

$$\begin{aligned}&d^2(y) - d^2(x) - 2\langle x - \pi(x), y - x\rangle \\ &= |(y - x) + (x - \pi(x)) + \pi(x) - \pi(y)|^2 - |x - \pi(x)|^2 \\ &\quad - 2\langle x - \pi(x), y - x\rangle \\ &= |y - x|^2 + |\pi(x) - \pi(y)|^2 + 2\langle y - x, \pi(x) - \pi(y)\rangle \\ &\quad + 2\langle x - \pi(x), \pi(x) - \pi(y)\rangle.\end{aligned}$$

Since

$$\langle x - \pi(x), \pi(x) - \pi(y)\rangle \ge -\frac{\varepsilon}{2r_0}|\pi(y) - \pi(x)|^2$$

and

$$\begin{aligned}&\langle x-\pi(x),\pi(x)-\pi(y)\rangle\\&\leq\langle y-\pi(y),\pi(x)-\pi(y)\rangle+\langle x-y,\pi(x)-\pi(y)\rangle\\&\leq\frac{\varepsilon}{2r_0}|\pi(x)-\pi(y)|^2+|x-y|\,|\pi(x)-\pi(y)|\end{aligned}$$

the inequality (6.46) follows from this and (6.45). ■

6.3.9 Differential Equations

Let $\mathbb{H}$ be a separable real Hilbert space. If $A:\mathbb{H}\rightrightarrows\mathbb{H}$ is a maximal monotone operator, $u_0\in\overline{D(A)}$, $f\in L^1(0,T;\mathbb{H})$, then the strong solution of the Cauchy problem

$$\begin{cases}\dfrac{du(t)}{dt}+Au(t)\ni f(t), & a.e.\ t\in\,]0,T[\,,\\ u(0)=u_0,\end{cases}\tag{6.47}$$

is defined as a function $u\in C([0,T];\mathbb{H})$ satisfying:

i) $u(t)\in D(A)\quad a.e.\ t\in\,]0,T[\,,$

ii) $\exists\, h=h^{(u)}\in L^1(0,T;\mathbb{H})$ such that $h(t)\in Au(t)$, $a.e.\ t\in\,]0,T[\,$, and

$$u(t)+\int_0^t h(s)\,ds=u_0+\int_0^t f(s)\,ds,\ \forall\, t\in[0,T],$$

and we shall write $u=\mathcal{S}(A;u_0,f)$. Note that the strong solution is unique when it exists. Indeed if u, v are two solutions corresponding to (u_0,f), (v_0,g), respectively, then

$$\begin{aligned}&|u(t)-v(t)|^2+2\int_0^t\left\langle h^{(u)}(s)-h^{(v)}(s),u(s)-v(s)\right\rangle ds\\&=|u_0-v_0|^2+2\int_0^t\langle f(s)-g(s),u(s)-v(s)\rangle\,ds\end{aligned}$$

and by the monotonicity of A it follows that

$$|u(t)-v(t)|^2\leq|u_0-v_0|^2+2\int_0^t|f(s)-g(s)|\,|u(s)-v(s)|\,ds.$$

Using Gronwall's inequality (Lemma 6.63, Annex C) we obtain

$$|u(t) - v(t)| \leq |u_0 - v_0| + \int_0^t |f(s) - g(s)|\, ds. \tag{6.48}$$

We recall from Barbu [3], p. 31, that the following proposition holds:

Proposition 6.50. *If A is maximal monotone operator on $\mathbb{H}$, $u_0 \in D(A)$ and $f \in W^{1,1}([0,T];\mathbb{H})$, then the Cauchy problem (6.47) has a unique strong solution $u \in W^{1,\infty}([0,T];\mathbb{H})$. Moreover if A_ε is the Yosida approximation of the operator A and u_ε is the solution of the approximate equation*

$$\frac{du_\varepsilon}{dt} + A_\varepsilon u_\varepsilon = f, \quad u_\varepsilon(0) = u_0,$$

then for all $(x_0, y_0) \in A$ there exists a constant $C = C(\alpha, T, x_0, y_0) > 0$ such that

$c_1)$ $\|u_\varepsilon\|_{C([0,T];\mathbb{H})} \leq C\left(1 + |u_0| + \|f\|_{L^1(0,T;\mathbb{H})}\right)$, *and*
$c_2)$ $\lim_{\varepsilon \searrow 0} u_\varepsilon = u$ *in* $C([0,T];\mathbb{H})$.

We introduce the notation

$$W^{1,p}([0,T];\mathbb{H}) = \Big\{ f : \exists\, a \in \mathbb{H},\ g \in L^p(0,T;\mathbb{H}) \text{ such that } f(t) = a + \int_0^t g(s)\, ds,\ \forall\, t \in [0,T] \Big\}.$$

From Barbu [2] (Chap. IV, p. 197, Theorem 2.5) we recall:

Proposition 6.51. *Let A be a maximal monotone operator on $\mathbb{H}$ such that*

$$\operatorname{int}(D(A)) \neq \emptyset.$$

If $u_0 \in \overline{D(A)}$ and $f \in W^{1,1}([0,T];\mathbb{H})$, then the Cauchy problem (6.47) has a unique strong solution $u \in W^{1,1}([0,T];\mathbb{H})$.

By the continuity property (6.48) one can generalize the notion of the solution of Eq. (6.47) as follows:

♦ u is a generalized solution of the Cauchy problem (6.47) with

$$u_0 \in \overline{D(A)}, \quad f \in L^1(0,T;\mathbb{H}),$$

(and we shall write $u = \mathcal{GS}(A; u_0, f)$) if

◇ $u \in C([0,T];\mathbb{H})$ and
◇ there exist $u_{0n} \in D(A)$, $f_n \in W^{1,1}([0,T];\mathbb{H})$ such that

$$\begin{array}{lll} a) & u_{0n} \to u_0 & \text{in } \mathbb{H}, \\ b) & f_n \to f & \text{in } L^1(0,T;\mathbb{H}), \\ c) & u_n = \mathcal{S}(A;u_{0n},f_n) \to u & \text{in } C([0,T];\mathbb{H}). \end{array}$$

Clearly we have:

Proposition 6.52. *If A is a maximal monotone operator on $\mathbb{H}$, $u_0 \in \overline{D(A)}$ and $f \in L^1(0,T;\mathbb{H})$, then the Cauchy problem (6.47) has a unique generalized solution $u \in C([0,T];\mathbb{H})$. Moreover if $u = \mathcal{GS}(A;u_0,f)$ and $v = \mathcal{GS}(A;v_0,g)$ then*

$$|u(t) - v(t)| \le |u_0 - v_0| + \int_0^t |f(s) - g(s)|\, ds \tag{6.49}$$

and for all $(x_0,\hat{x}_0) \in A$ there exists a constant $C = C(T,x_0,\hat{x}_0) > 0$ such that

$$\|u\|_{C([0,T];\mathbb{H})} \le C\left(1 + |u_0| + \|f\|_{L^1(0,T;\mathbb{H})}\right). \tag{6.50}$$

In the case when $\operatorname{int}(D(A)) \ne \emptyset$ one can give supplementary properties of generalized solutions.

Proposition 6.53. *Let $A \subset \mathbb{H} \times \mathbb{H}$ be a maximal monotone operator such that*

$$\operatorname{int}(D(A)) \ne \emptyset.$$

Let $u_0 \in \overline{D(A)}$ and $f \in L^1(0,T;\mathbb{H})$. Then:

I. there exists a unique pair (u,k) such that

$$(P_A): \begin{cases} a) & u \in C([0,T];\mathbb{H}), \quad u(t) \in \overline{D(A)} \;\; \forall t \in [0,T],\; u(0) = u_0, \\ b) & k \in C([0,T];\mathbb{H}) \cap BV([0,T];\mathbb{H}),\; k(0) = 0, \\ c) & u(t) + k(t) = u_0 + \displaystyle\int_0^t f(s)ds,\; \forall t \in [0,T], \\ d) & \displaystyle\int_s^t \langle u(r) - x, dk(r) - \hat{x}dr\rangle \ge 0, \\ & \qquad\qquad \forall\, 0 \le s \le t \le T,\; \forall\,(x,\hat{x}) \in A; \end{cases}$$

II. $u = \mathcal{GS}(A;u_0,f)$ if and only if u is solution of the problem (P_A);
III. the following estimate holds:

$$\|u\|^2_{C([0,T];\mathbb{H})} + \|k\|_{BV([0,T];\mathbb{H})} \le C\left(1 + |u_0|^2 + \|f\|^2_{L^1(0,T;\mathbb{H})}\right),$$

where C is a positive constant independent of u_0 and f.

Proof. Uniqueness. If (u,k) and (v,ℓ) are two solutions of the problem (P_A) corresponding to (u_0,f), (v_0,g) respectively, then

$$|u(t)-v(t)|^2+2\int_0^t \langle dk(s)-d\ell(s), u(s)-v(s)\rangle\, ds$$
$$=|u_0-v_0|^2+2\int_0^t \langle f(s)-g(s), u(s)-v(s)\rangle\, ds.$$

But by Proposition 6.17, the monotonicity of A and (P_A-d) we have

$$\int_0^t \langle dk(s)-d\ell(s), u(s)-v(s)\rangle\, ds \geq 0.$$

Hence

$$|u(t)-v(t)|^2 \leq |u_0-v_0|^2+2\int_0^t |f(s)-g(s)|\,|u(s)-v(s)|\, ds,$$

which yields (6.49) and, in particular, the uniqueness follows.

Existence. Let $u_{0n}\in D(A)$, $f_n\in W^{1,1}([0,T];\mathbb{H})$ such that

$$u_{0n}\to u_0 \quad \text{in } \mathbb{H} \quad \text{and} \quad f_n\to f \quad \text{in } L^1(0,T;\mathbb{H}).$$

Let $u_n=\mathcal{S}(A;u_{0n},f_n)$ be the strong solution corresponding to $(A;u_{0n},f_n)$. Hence there exists an $h_n\in L^1(0,T;\mathbb{H})$ such that $h_n(t)\in Au_n(t)$, $a.e.\ t\in\,]0,T[$ and denoting $k_n(t)=\displaystyle\int_0^t h_n(s)\,ds$ we have

$$\begin{aligned} &a) \quad u_n(t)+k_n(t)=u_{0n}+\int_0^t f_n(s)\,ds, \quad \forall\, t\in[0,T],\\ &b) \quad \int_s^t \langle u_n(r)-x, dk_n(r)-\hat{x}dr\rangle \geq 0,\\ &\qquad\qquad \forall\, 0\leq s\leq t\leq T,\ \forall\,(x,\hat{x})\in A. \end{aligned} \tag{6.51}$$

Let $x_0\in \operatorname{int}(D(A))$ and $\hat{x}_0\in A(x_0)$. Then

$$|u_n(t)-x_0|^2+2\int_0^t \langle h_n(s), u_n(s)-x_0\rangle\, ds$$
$$=|u_{0n}-x_0|^2+2\int_0^t \langle f_n(s), u_n(s)-x_0\rangle\, ds.$$

Since

$$\langle h_n(s), u_n(s)-x_0\rangle \geq \langle \hat{x}_0, u_n(s)-x_0\rangle,$$

we infer

$$|u_n(t) - x_0|^2 \le |u_{0n} - x_0|^2 + 2\int_0^t [|f_n(s)| + |\hat{x}_0|]\,|u_n(s) - x_0|\,ds.$$

By the Gronwall type inequality from Lemma 6.63, Annex C, we obtain

$$\begin{aligned}|u_n(t) - x_0| &\le |u_{0n} - x_0| + \int_0^T |f_n(s)|\,ds + T\,|\hat{x}_0| \\ &\le C\left[1 + |u_{0n}| + \int_0^T |f_n(s)|\,ds\right],\end{aligned}$$

where $C = C(x_0, \hat{x}_0, T) > 0$.

By Proposition 6.5 we have $a.e.\ t \in]0, T[$:

$$r_0\,|h_n(t)| \le \langle h_n(t), u_n(t) - x_0\rangle + M_0\,|u_n(t) - x_0| + r_0 M_0,$$

and then

$$\begin{aligned}&2r_0 \int_0^t |h_n(s)|\,ds \\ &\quad\le |u_{0n} - x_0|^2 + 2\int_0^t (|f_n(s)| + M_0)\,|u_n(s) - x_0|\,ds + 2r_0 M_0 T \\ &\quad\le C\left[1 + |u_{0n}|^2 + \left(\int_0^T |f_n(s)|\,ds\right)^2\right]\end{aligned}$$

with C a constant depending on $x_0, \hat{x}_0, T, M_0, r_0$.

Hence $k_n(t) = \int_0^t h_n(s)\,ds$ is bounded in $BV([0,T];\mathbb{H})$. Then there exists a $k \in BV([0,T];\mathbb{H})$ such that on a subsequence also denoted by k_n we have

$$k_n \overset{w^*}{\to} k \quad \text{in } BV([0,T];\mathbb{H}).$$

The sequence $(u_n)_{n\in\mathbb{N}^*}$ is a Cauchy sequence in $C([0,T];\mathbb{H})$ since if $u_m = \mathcal{S}(A; u_{0m}, f_m)$ then

$$\sup_{t\in[0,T]} |u_n(t) - u_m(t)| \le |u_{0n} - u_{0m}| + \int_0^T |f_n(s) - f_m(s)|\,ds.$$

Then there exists a $u \in C([0,T];\mathbb{H})$ such that

$$u_n \to u \qquad \text{in} \quad C([0,T];\mathbb{H}).$$

Passing to the limit in (6.51), we obtain that (u,k) satisfies (P_A). The proof is complete. ■

If the assumption $\operatorname{int}(D(A)) \neq \emptyset$ has a smoothing effect as we saw in Proposition 6.51, the maximal monotone $A = \partial\varphi$ also has a smoothing effect.

Consider the differential equation

$$\begin{cases} \dfrac{du(t)}{dt} + \partial\varphi u(t) \ni f(t), & a.e.\ t \in]0,T[, \\ u(0) = u_0, \end{cases} \tag{6.52}$$

where

$$\varphi : \mathbb{H} \to]-\infty,+\infty] \quad \text{is a proper convex l.s.c. function.}$$

Proposition 6.54. *If $u_0 \in \overline{D(\partial\varphi)}\left(= \overline{\operatorname{Dom}(\varphi)}\right)$ and $f \in L^2(0,T;\mathbb{H})$, then the Cauchy problem (6.52) has a unique strong solution. Moreover $u \in W^{1,2}(\delta,T;\mathbb{H})$, $\forall \delta > 0$, $\sqrt{t}\dfrac{du}{dt} \in L^2(0,T;\mathbb{H})$, $\varphi(u) \in L^1(0,T)$ and if $u_0 \in \operatorname{Dom}(\varphi)$, then $\dfrac{du}{dt} \in L^2(0,T;\mathbb{H})$ and $\varphi(u) \in L^\infty(0,T)$.*

Consider now the Cauchy problem

$$\begin{cases} \dfrac{dy(t)}{dt} + \partial^-\varphi(x(t)) \ni g(t), & a.e.\ t \in [0,T] \\ x(0) = x_0, \end{cases} \tag{6.53}$$

where

$$\begin{array}{ll} (i) & \varphi : \mathbb{R}^d \to]-\infty,+\infty] \text{ is a proper l.s.c. } (\rho,\gamma)\text{–semiconvex function,} \\ (ii) & \operatorname{Dom}(\varphi) \text{ is a locally closed subset of } \mathbb{R}^d, \end{array} \tag{6.54}$$

and

$$\begin{array}{ll} (i) & x_0 \in \operatorname{Dom}(\varphi), \\ (ii) & g \in L^2\left(0,T;\mathbb{R}^d\right). \end{array} \tag{6.55}$$

Hence for all $(x,\hat{x}) \in \partial^-\varphi$

$$\langle \hat{x}, z-x\rangle + \varphi(x) \le \varphi(z) + (\rho + \gamma|\hat{x}|)|z-x|^2, \quad \forall z \in \mathbb{R}^d.$$

We denote here by $\partial^-\varphi(x)$ the Fréchet subdifferential given in Definition 6.40. Recall that $E \subset \mathbb{R}^d$ is locally closed if for all $x \in E$, there exists a $\delta > 0$ such that $E \cap \overline{B}(x,\delta)$ is closed.

From Degiovanni–Marino–Tosques [21] and Rossi–Savaré [66] we have:

Proposition 6.55. *Let the assumptions (6.54) and (6.55) be satisfied. Then there exist* $h \in L^2\left(0,T;\mathbb{R}^d\right)$ *and a unique absolutely continuous function* $x:[0,T]\rightarrow$ Dom (φ) *such that:*

$$\begin{array}{ll}(a) & \displaystyle\int_0^T \left[|x'(t)|^2+|\varphi(x(t))|\right]dt<\infty,\\ (b) & x(t)\in \mathrm{Dom}\left(\partial^-\varphi\right), \qquad a.e.\ t\in\,]0,T[\,,\\ (c) & h(t)\in\partial^-\varphi(x(t)), \qquad a.e.\ t\in\,]0,T[\,,\end{array}$$

and

$$\left(P_g\right):\begin{cases} x'(t)+h(t)=g(t), & a.e.\ t\in\,]0,T[\\ x(0)=x_0.\end{cases}$$

Moreover $a.e.\ t,s\in\,]0,T[\,,\ s<t$*:*

$$\int_s^t \left|x'(r)\right|^2 dr=\varphi(x(s))-\varphi(x(t))+\int_s^t \left(g(r),x'(r)\right)dr$$

and there exists a positive constant C_T *(independent of* x_0 *and* g*) such that*

$$\|x\|_T+\|\varphi(x)\|_T+\int_0^T \left|x'(r)\right|^2 dr\le C_T\left(|x_0|^2+\varphi^+(x_0)+\int_0^T |g(r)|^2 dr\right).$$

Remark 6.56. If we put

$$k(t)=\int_0^t h(s)\,ds$$

then

$$(GSP):\begin{cases} j) & k\in BV\left([0,T];\mathbb{R}^d\right), \quad k(0)=0,\\ jj) & x(t)+k(t)=x_0+\displaystyle\int_0^t g(s)\,ds, \quad \forall\, t\in[0,T],\\ jv) & \forall\, 0\le s\le t,\ \forall y:[0,\infty[\,\rightarrow\mathbb{R}^d \text{ continuous:}\\ & \displaystyle\int_s^t \langle y(r)-x(r),dk(r)\rangle+\int_s^t \varphi(x(r))\,dr\\ & \qquad \le \displaystyle\int_s^t \varphi(y(r))\,dr+\int_s^t |y(r)-x(r)|^2\left(\rho dr+\gamma d\updownarrow k\updownarrow_r\right),\end{cases}$$

that is (x,k) is the solution of the generalized Skorohod problem $(x_0,m,\partial^-\varphi)$ with $m(t)=\int_0^t g(s)\,ds$ (see Definition 4.29).

6.3.10 Auxiliary Results

Proposition 6.57. *If $g \in L^1(0,T)$ and*

$$\rho_\lambda(t) = e^{-\lambda t} \int_0^t |g(s)| e^{\lambda s} ds, \quad t \in [0,T], \ \lambda > 0,$$

then

$$\lim_{\lambda \to \infty} \left[\sup_{t \in [0,T]} \rho_\lambda(t) \right] = 0.$$

Proof. Let the continuous function $t \longmapsto G(t) = \int_0^t |g(s)| ds$ and $\mathbf{m}_G(\varepsilon)$ be the modulus of continuity of G on $[0,T]$. We have for all $t \in [0,T]$ and $\lambda > 0$:

$$\begin{aligned} 0 \le \rho_\lambda(t) &= e^{-\lambda t} \left[\int_0^{\left(t-\sqrt{1/\lambda}\right)^+} |g(s)| e^{\lambda s} ds + \int_{\left(t-\sqrt{1/\lambda}\right)^+}^t |g(s)| e^{\lambda s} ds \right] \\ &\le e^{-\lambda t} e^{\lambda \left(t-\sqrt{1/\lambda}\right)^+} \int_0^{\left(t-\sqrt{1/\lambda}\right)^+} |g(s)| ds + e^{-\lambda t} e^{\lambda t} \mathbf{m}_G(\sqrt{1/\lambda}) \\ &\le e^{-\sqrt{\lambda}} G(T) + \mathbf{m}_G(\sqrt{1/\lambda}), \end{aligned}$$

which yields the result. ■

We now give a variant of the Banach fixed point theorem.

Let $\{(\mathbb{V}_a, d_a) : a \ge 0\}$ be a family of complete metric spaces such that for all $0 \le a \le b$:

$$\mathbb{V}_b \subset \mathbb{V}_a$$

with a continuous embedding. Let

$$\mathbb{V} = \bigcap\nolimits_{a \ge 0} \mathbb{V}_a = \bigcap\nolimits_{a \in \mathbb{N}^*} \mathbb{V}_a,$$

and assume $\mathbb{V} \neq \emptyset$. Then $\mathbb{V}$ is a complete metric space with respect to the metric

$$\rho(x,y) = \sum_{a \in \mathbb{N}} \frac{1}{2^a} \frac{d_a(x,y)}{1 + d_a(x,y)},$$

and if $x_n, x \in \mathbb{V}$, $n \in \mathbb{N}^*$, then as $n \to \infty$,

$$x_n \to x \text{ in } \mathbb{V} \iff x_n \to x \text{ in } \mathbb{V}_a,\ \forall\, a \geq 0.$$

Lemma 6.58. *Let $\Gamma : \mathbb{V} \to \mathbb{V}$ be a mapping satisfying: there exists an $a_0 \geq 0$ and for all $a \geq a_0$ there exists a $\delta_a \in]0,1[$ such that*

$$d_a\left(\Gamma(x), \Gamma(y)\right) \leq \delta_a\, d_a(x,y), \text{ for all } x, y \in \mathbb{V}.$$

Then Γ has a unique fixed point, i.e. there exists a unique $x \in \mathbb{V}$ such that

$$x = \Gamma(x).$$

(Banach's fixed point theorem corresponds to the case $(\mathbb{V}_a, d_a) \equiv (\mathbb{V}_0, d_0)$ for all $a \geq 0$.)

Proof. We define

$$x_0 \in \mathbb{V}, \quad x_{n+1} = \Gamma(x_n).$$

Then by recurrence we deduce that

$$x_n \in \mathbb{V}, \quad \text{for all } n \in \mathbb{N},$$

and

$$d_a\left(x_{n+p}, x_n\right) \leq \frac{\delta_a^n}{1-\delta_a} d_a(x_1, x_0),$$

for all $a \geq a_0$, $n, p \in \mathbb{N}^*$. Hence there exists a unique $x^{(a)} \in \mathbb{V}_a$ such that as $n \to \infty$

$$x_n \to x^{(a)} \quad \text{in } \mathbb{V}_a.$$

Moreover by the continuity of the embedding $\mathbb{V}_a \subset \mathbb{V}_b$ for $0 \leq b \leq a$, we infer

$$x_n \to x^{(a)} \quad \text{in } \mathbb{V}_b.$$

Consequently $x^{(a)} = x^{(a_0)}$ for all $a \geq a_0$, $x \overset{def}{=} x^{(a_0)} \in \mathbb{V}$ and for $a \geq a_0$

$$\begin{aligned} d_a(x, \Gamma(x)) &\leq d_a(x, x_{n+1}) + d_a(\Gamma(x_n), \Gamma(x)) \\ &\leq d_a(x, x_{n+1}) + \delta_a d_a(x_n, x) \\ &\to 0, \quad \text{as } n \to \infty, \end{aligned}$$

which yields

$$x = \Gamma(x).$$

The fixed point x is unique, since if $x, y \in \mathbb{V}$ are two fixed points, then for $a \geq a_0$

$$d_a(x, y) = d_a(\Gamma(x), \Gamma(y)) \leq \delta_a d_a(x, y)$$

and $x = y$ follows. ■

6.4 Annex C: Deterministic and Stochastic Inequalities

6.4.1 Deterministic Inequalities

Proposition 6.59 (Stieltjes–Gronwall Inequality). *Let $K : [0, T] \to \mathbb{R}$ be a continuous increasing function, $a : [0, T] \to [0, \infty[$ be an increasing function and $x : [0, T] \to \mathbb{R}$ be a measurable function such that*

$$\int_0^T |x(r)|\, dK(r) < \infty.$$

If

$$x(t) \leq a(t) + \int_0^t x(r)\, dK(r), \quad \forall\, t \in [0, T],$$

then

$$x(t) \leq a(t)\, e^{K(t)-K(0)}, \quad \forall\, t \in [0, T]. \tag{6.56}$$

Proof. I. Note that if $\alpha, \beta_0, \beta_1, \ldots, \beta_n$ and $z_0, z_1, \ldots, z_n \in \mathbb{R}$ satisfy

$$\begin{aligned} &z_0 \leq \alpha, \\ &z_i \leq \alpha + \beta_0 z_0 + \beta_1 z_1 + \cdots + \beta_{i-1} z_{i-1}, \ 1 \leq i \leq n, \end{aligned}$$

then

$$z_i \leq \alpha e^{\beta_0 + \beta_1 + \cdots + \beta_{i-1} + \beta_i}.$$

Indeed, associating the sequence

$$x_0 = \alpha, \quad x_i = \alpha + \beta_0 x_0 + \beta_1 x_1 + \cdots + \beta_{i-1} x_{i-1}, \ 1 \leq i \leq n,$$

by recurrence

$$z_i \leq x_i = \alpha(1 + \beta_0)(1 + \beta_1) \cdots (1 + \beta_{i-1}) \leq \alpha e^{\beta_0 + \beta_1 + \cdots + \beta_{i-1}}$$

follows.

Let

$$g(t) = a(t) + \int_0^t x^+(r)\, dK(r).$$

Clearly g is an increasing function and

$$x(t) \le x^+(t) \le g(t) \le a(t) + \int_0^t g(r)\, dK(r).$$

Let $0 < t_1 < \ldots < t_n = t$ be such that

$$1 > \max\left\{K(t_i) - K(t_{i-1}) : i \in \overline{1,n}\right\} \left(\stackrel{def}{=} \gamma_n\right) \to 0, \quad \text{as } n \to \infty.$$

Let $g_i = g(t_i)$, $c_0 = 0$, $c_i = \int_{t_{i-1}}^{t_i} dK(r) = K(t_i) - K(t_{i-1}) \le \gamma_n$. We have

$$\begin{aligned} g_i &\le a(t_i) + \sum_{j=1}^{i} \int_{t_{j-1}}^{t_j} g(r)\, dK(r) \\ &\le a(t) + \sum_{j=1}^{i} g_j \int_{t_{j-1}}^{t_j} dK(r) \\ &\le a(t) + (c_0 g_0 + c_1 g_1 + \cdots + c_{i-1} g_{i-1}) + \gamma_n g_i, \end{aligned}$$

which yields

$$g_i \le \frac{a(t)}{1-\gamma_n} + \frac{c_0}{1-\gamma_n} g_0 + \frac{c_1}{1-\gamma_n} g_1 + \cdots + \frac{c_{i-1}}{1-\gamma_n} g_{i-1}$$

for all $i \in \{1, 2, \ldots, n\}$. Hence

$$\begin{aligned} x(t) \le g(t) = g_n &\le \frac{a(t)}{1-\gamma_n} \exp\left[\frac{1}{1-\gamma_n} \sum_{j=0}^{n} c_i\right] \\ &= \frac{a(t)}{1-\gamma_n} \exp\left[\frac{1}{1-\gamma_n} [K(t) - K(0)]\right]. \end{aligned}$$

The inequality (6.56) follows by letting $n \to \infty$. ■

For $K(t) = \int_0^t b(r)\, dr$, where $b : [0, \infty[\to [0, \infty[$ is a locally integrable function, the following lemma holds.

Corollary 6.60 (Gronwall Inequality). *Let $a : [0, T] \to [0, \infty[$ be an increasing function and $x, b : [0, T] \to \mathbb{R}$, $b \ge 0$, be integrable functions such that*

$$\int_0^T b(t)\,|x(t)|\,dt < \infty.$$

If

$$x(t) \le a(t) + \int_0^t b(s)\,x(s)\,ds, \quad \forall\, t \in [0,T],$$

then

$$x(t) \le a(t) \exp\left(\int_0^t b(s)\,ds\right), \quad \forall\, t \in [0,T]. \tag{6.57}$$

Corollary 6.61 (Backward Stieltjes–Gronwall Inequality). *Let $\tilde{K} : [0,T] \to \mathbb{R}$ be a continuous increasing function, $\tilde{a} : [0,T] \to [0,\infty[$ be a decreasing function and $y : [0,T] \to \mathbb{R}$ be a measurable function such that*

$$\int_0^T |y(r)|\,d\tilde{K}(r) < \infty.$$

If

$$y(t) \le \tilde{a}(t) + \int_t^T y(r)\,d\tilde{K}(r), \quad \forall\, t \in [0,T],$$

then

$$y(t) \le \tilde{a}(t)\,e^{\tilde{K}(T)-\tilde{K}(t)}, \quad \forall\, t \in [0,T]. \tag{6.58}$$

Proof. Let $x(t) = y(T-t)$, $a(t) = \tilde{a}(T-t)$ and $K(t) = \tilde{K}(T) - \tilde{K}(T-t)$. Then

$$x(t) \le a(t) + \int_0^t x(r)\,dK(r), \quad \forall\, t \in [0,T],$$

and by Proposition 6.59

$$x(t) \le a(t)\,e^{K(t)-K(0)}, \quad \forall\, t \in [0,T],$$

that is (6.58) replacing t by $T-t$. ■

In particular for $K(t) = \int_0^t b(r)\,dr$, we have:

Corollary 6.62 (Backward Gronwall Inequality). *Let $\tilde{a} : [0,T] \to [0,\infty[$ be a decreasing function and $y, b : [0,T] \to \mathbb{R}$, $b \ge 0$, be integrable functions such that*

$$\int_0^T b(t)\,|y(t)|\,dt < \infty.$$

If

$$y(t) \le \tilde{a}(t) + \int_t^T b(s)\,y(s)\,ds, \quad \forall\, t \in [0,T],$$

then

$$y(t) \le \tilde{a}(t) \exp\left(\int_t^T b(s)\,ds\right), \quad \forall\, t \in [0,T]. \tag{6.59}$$

We now give some other deterministic inequalities used in the book.

Lemma 6.63. *Let* $\alpha, \beta \in L^1_{loc}([0,\infty[)$.

I. If $\alpha \ge 0$ *a.e. and* $x : [0,\infty[\to \mathbb{R}^d$ *is an absolutely continuous function such that*

$$\langle x'(t), x(t)\rangle \le \alpha(t)\,|x(t)| + \beta(t)\,|x(t)|^2, \quad a.e.\ t \ge 0,$$

then

$$|x(t)| \le |x(\tau)|\, e^{\int_\tau^t \beta(s)ds} + \int_\tau^t \alpha(s)\, e^{\int_s^t \beta(r)dr} ds \tag{6.60}$$

for all $0 \le \tau \le t$.

II. If $\alpha, \beta \ge 0$ *a.e.,* $a : [0,\infty[\to [0,\infty[$ *is an increasing function and* $\varphi : [0,\infty[\to [0,\infty[$ *is a continuous function such that* $\forall\, t \ge 0$

$$\varphi^2(t) \le a(t) + 2\int_0^t \alpha(s)\,\varphi(s)\,ds + 2\int_0^t \beta(s)\,\varphi^2(s)\,ds,$$

then

$$\varphi(t) \le \sqrt{a(t)}\, e^{\int_0^t \beta(s)ds} + \int_0^t \alpha(s)\, e^{\int_s^t \beta(r)dr} ds, \ \forall\, t \ge 0. \tag{6.61}$$

Proof. I. Let $u_\varepsilon(t) = |x(t)|^2 e^{-2\int_0^t \beta(s)ds} + \varepsilon$, $\varepsilon > 0$. Then

$$\begin{aligned} u_\varepsilon'(t) &= 2\langle x'(t), x(t)\rangle e^{-2\int_0^t \beta(s)ds} - 2\beta(t)\,|x(t)|^2 e^{-2\int_0^t \beta(s)ds} \\ &\le 2\alpha(t)\,|x(t)|\, e^{-2\int_0^t \beta(s)ds} \\ &\le 2\alpha(t)\sqrt{u_\varepsilon(t)}\, e^{-\int_0^t \beta(s)ds}, \end{aligned}$$

which yields

$$\frac{d}{dt}\left(\sqrt{u_\varepsilon(t)}\right) = \frac{u'_\varepsilon(t)}{2\sqrt{u_\varepsilon(t)}}$$
$$\leq \alpha(t)\, e^{-\int_0^t \beta(s)ds}.$$

Hence

$$\sqrt{u_\varepsilon(t)} \leq \sqrt{u_\varepsilon(\tau)} + \int_\tau^t \alpha(s)\, e^{-\int_0^s \beta(r)dr} ds.$$

Passing to the limit as $\varepsilon \searrow 0$ the inequality (6.60) follows.

II. Let $\theta \in [0, T]$ be fixed and

$$x(t) = \left(a(\theta) + 2\int_0^t \alpha(s)\varphi(s)\,ds + 2\int_0^t \beta(s)\varphi^2(s)\,ds\right)^{1/2}.$$

Then for all $t \in [0, \theta]$:

$$\varphi^2(t) \leq a(\theta) + 2\int_0^t \alpha(s)\varphi(s)\,ds + 2\int_0^t \beta(s)\varphi^2(s)\,ds = x^2(t),$$

and

$$x'(t)\,x(t) = \alpha(t)\varphi(t) + \beta(t)\varphi^2(t)$$
$$\leq \alpha(t)\,x(t) + \beta(t)\,x^2(t),$$

which implies, by the first part, that for $t \in [0, \theta]$:

$$\varphi(t) \leq x(t) \leq x(0)\, e^{\int_0^t \beta(s)ds} + \int_0^t \alpha(s)\, e^{\int_s^t \beta(r)dr} ds,$$

which is (6.61) if we choose $t = \theta$. ■

Corollary 6.64. *If* $\alpha, \beta \geq 0$ *a.e.,* $\tilde{a} : [0, T] \to [0, \infty[$ *is a decreasing function and* $\psi : [0, T] \to [0, \infty[$ *is a continuous function such that* $\forall\, t \in [0, T]$*:*

$$\psi^2(t) \leq \tilde{a}(t) + 2\int_t^T \alpha(s)\psi(s)\,ds + 2\int_t^T \beta(s)\psi^2(s)\,ds,$$

then

$$\psi(t) \leq \sqrt{\tilde{a}(t)} e^{\int_t^T \beta(s)ds} + \int_t^T \alpha(s)\, e^{\int_t^s \beta(r)dr} ds, \quad \forall\, t \in [0, T]. \tag{6.62}$$

Proof. Note that $\forall\, t \in [0,T]$:

$$\begin{aligned}
&\psi^2(T-t)\\
&\le \tilde{a}(T-t) + 2\int_{T-t}^{T} \alpha(s)\,\psi(s)\,ds + 2\int_{T-t}^{T} \beta(s)\,\psi^2(s)\,ds\\
&= \tilde{a}(T-t) + 2\int_0^t \alpha(T-s)\,\psi(T-s)\,ds + 2\int_0^t \beta(T-s)\,\psi^2(T-s)\,ds.
\end{aligned}$$

Hence by (6.61)

$$\psi(T-t) \le \tilde{a}(T-t)\,e^{\int_0^t \beta(T-s)ds} + \int_0^t \alpha(T-s)\,e^{\int_s^t \beta(T-r)dr}ds,$$

which clearly yields (6.62) replacing $T-t$ by t. ■

If $f, g \in BV_{loc}([0,\infty[)$ $(= BV_{loc}([0,\infty[;\mathbb{R}))$, we say that $df(s) \le dg(s)$ as signed measures on $[0,\infty[$ if

d1.

$$\int_t^s \varphi(r)\,df(r) \le \int_t^s \varphi(r)\,dg(r),$$

for all $0 \le t \le s$ and for all continuous function $\varphi : [0,\infty[\to [0,\infty[$, or equivalently

d2. $f(s) - f(t) = \int_t^s df(r) \le \int_t^s dg(r) = g(s) - g(t), \quad \forall\, 0 \le t \le s$, or equivalently

d3. $h(s) = f(s) - g(s)$ is a decreasing function on $[0,\infty[$.

Lemma 6.65. *Let $x, N, V \in BV_{loc}([0,\infty[)$. If*

$$x(s) \le x(t) + \int_t^s [dN(r) + x(r)\,dV(r)], \quad \forall\, 0 \le t \le s,$$

or equivalently

$$dx(r) \le dN(r) + x(r)\,dV(r)$$

as signed measures on $[0,\infty[$, then for all $0 \le t \le s$:

$$e^{-V_s}x(s) \le x(t)\,e^{-V_t} + \int_t^s e^{-V_r}dN(r). \tag{6.63}$$

Proof. We have

$$d\left(x(r)\,e^{-V_r}\right) = e^{-V_r}dx(r) - e^{-V_r}x(r)\,dV(r) \\ \leq e^{-V_r}dN(r)$$

and the result follows. ■

Corollary 6.66. *Let $\alpha, \beta \in L^1_{loc}([0,\infty[)$ and $y : [0,\infty[\to \mathbb{R}$ be a continuous function such that*

$$y(t) \leq y(s) + \int_t^s [\alpha(r) + \beta(r)\,y(r)]\,dr, \quad \forall\, 0 \leq t \leq s,$$

then

$$e^{\int_0^t \beta(u)du} y(t) \leq y(s)\,e^{\int_0^s \beta(u)du} + \int_t^s \alpha(r)\,e^{\int_0^r \beta(u)du} dr. \tag{6.64}$$

Proof. By Lemma 6.65 and

$$(-y(s)) \leq (-y(t)) + \int_t^s [\alpha(r) - \beta(r)(-y(r))]\,dr,$$

the result follows. ■

Finally we have:

Proposition 6.67. *Let $x \in BV_{loc}\left([0,\infty[\,,\mathbb{R}^d\right)$ and $V \in BV_{loc}([0,\infty[\,;\mathbb{R})$ be continuous functions. Let $R, N : [0,\infty[\to [0,\infty[$ be continuous increasing functions. If*

$$\langle x(t), dx(t)\rangle \leq dR(t) + |x(t)|\,dN(t) + |x(t)|^2\,dV(t)$$

as signed measures on $[0,\infty[$, then for all $0 \leq t \leq T$:

$$\left\|e^{-V}x\right\|_{[t,T]} \leq 2\left[\left|e^{-V(t)}x(t)\right| + \left(\int_t^T e^{-2V(s)}dR(s)\right)^{1/2} + \int_t^T e^{-V(s)}dN(s)\right]$$

and

$$\|x\|_{[t,T]} \leq 2e^{\updownarrow V\updownarrow_{[t,T]}}\left[|x(t)| + \sqrt{R(T) - R(t)} + (N(T) - N(t))\right].$$

If $R = 0$ then for all $0 \leq t \leq s$:

$$|x(s)| \leq e^{V(s)-V(t)}\,|x(t)| + \int_t^s e^{V(s)-V(r)}dN(r). \tag{6.65}$$

Proof. Let $u_\varepsilon(r) = |x(r)|^2 e^{-2V_r} + \varepsilon$, $\varepsilon > 0$. We have as signed measures on $[0, \infty[$

$$\begin{aligned} du_\varepsilon(r) &= -2e^{-2V(r)} |x(r)|^2 dV(r) + 2e^{-2V(r)} \langle x(r), dx(r)\rangle \\ &\leq 2e^{-2V(r)} dR(r) + 2e^{-2V(r)} |x(r)| \, dN(r) \\ &\leq 2e^{-2V(r)} dR(r) + 2e^{-V(r)} \sqrt{u_\varepsilon(r)} dN(r). \end{aligned}$$

If $R = 0$ then

$$d\left(\sqrt{u_\varepsilon(r)}\right) = \frac{du_\varepsilon(r)}{2\sqrt{u_\varepsilon(r)}} \leq e^{-V(r)} dN(r),$$

and consequently

$$\sqrt{u_\varepsilon(s)} \leq \sqrt{u_\varepsilon(t)} + \int_t^s e^{-V(r)} dN(r),$$

which yields (6.65) passing to the limit as $\varepsilon \to 0$.

If $R \neq 0$ then

$$\begin{aligned} &e^{-2V(s)} |x(s)|^2 \\ &\leq e^{-2V(t)} |x(t)|^2 + 2\int_t^s e^{-2V(r)} dR(r) + 2\int_t^s e^{-2V(r)} |x(r)| \, dN(r) \\ &\leq e^{-2V(t)} |x(t)|^2 + 2\int_t^s e^{-2V(r)} dR(r) + 2 \left\| e^{-V} x \right\|_{[t,T]} \int_t^s e^{-V(r)} dN(r) \\ &\leq \left| e^{-V(t)} x(t) \right|^2 + 2\int_t^T e^{-2V(r)} dR(r) + \frac{1}{2} \left\| e^{-V} x \right\|_{[t,T]}^2 + 2\left(\int_t^T e^{-V(r)} dN(r) \right)^2. \end{aligned}$$

Hence for all $t \leq \tau \leq T$

$$\begin{aligned} e^{-V(\tau)} |x(\tau)| &\leq \left\| e^{-V} x \right\|_{[t,T]}^2 \\ &\leq 2e^{-2V(t)} |x(t)|^2 + 4\int_t^T e^{-2V(s)} dR(s) + 4\left(\int_t^T e^{-V(s)} dN(s) \right)^2 \end{aligned}$$

and the results follow. ■

6.4.2 Stochastic Inequalities

In this subsection $\{B_t : t \geq 0\}$ is a k-dimensional Brownian motion with respect to a given stochastic basis $(\Omega, \mathcal{F}, \mathbb{P}, \{\mathcal{F}_t\}_{t \geq 0})$.

Proposition 6.68 (Stochastic Gronwall Inequality). *Let*

◊ $a, b : [0, \infty[\to [0, \infty[$ *be measurable deterministic functions and*
◊ $H, \alpha, \beta, \gamma, \delta : \Omega \times [0, \infty[\to [0, \infty[$ *be stochastic processes, where* H *is a continuous stochastic processes. If for all* $t \geq 0$

$$|X_t| + |U_t| \leq |H_t| + \int_0^t (\alpha_s + a(s)|X_s|)\, ds + \left|\int_0^t G_s dB_s\right|, \ \mathbb{P}\text{-a.s.}, \quad (6.66)$$

where

$$\begin{array}{ll} i) & X, U \in S_d^0, \quad G \in \Lambda_{d\times k}^0, \\ ii) & |G_t| \leq \beta_t + b(t)|X_t|, \quad d\mathbb{P} \otimes dt\text{-a.e.}, \end{array}$$

then for all $q \geq 1$ *there exists a positive constant* C_q *such that for all* $T \geq 0$*:*

$$\mathbb{E} \sup_{t\in[0,T]} |X_t|^q + \mathbb{E} \sup_{t\in[0,T]} |U_t|^q \leq \left[\mathbb{E}\,\|H\|_T^q + \mathbb{E}\left(\int_0^T \alpha_s ds\right)^q + \mathbb{E}\left(\int_0^T \beta_s^2 ds\right)^{q/2}\right] \times \exp\left\{C_q\left[1 + T^{q-1}\int_0^T \left(a^q(s) + b^{2q}(s)\right) ds\right]\right\}. \quad (6.67)$$

In particular if the right-hand side of the inequality (6.67) is finite then

$$X, U \in S_d^q, \qquad G \in \Lambda_{d\times k}^q.$$

Proof. Clearly we can assume that the right-hand side of the inequality (6.67) is finite. Denote by C_q different constants depending only q and which can be changed from one line to another. For each $n \geq 1$, we define the stopping time

$$\tau_n(\omega) = \inf\{t \geq 0 : |X_t(\omega)| \geq n\} \wedge n.$$

Note that for all positive stochastic processes Z,

$$\int_0^{t\wedge\tau_n} |X_s|^p Z_s ds = \int_0^{t\wedge\tau_n} |X_{s\wedge\tau_n}|^p Z_s ds \leq \int_0^t \sup_{r\in[0,s]} |X_{r\wedge\tau_n}|^p Z_s ds.$$

By the convexity of the function $\varphi(r) = |r|^q$ we have

$$|X_{t\wedge\tau_n}|^q + |U_{t\wedge\tau_n}|^q \leq 2^q \|H\|_{t\wedge\tau_n}^q + 4^q \left|\int_0^{t\wedge\tau_n} (\alpha_s + a_s|X_s|)\, ds\right|^q + 4^q \left|\int_0^{t\wedge\tau_n} G_s dB_s\right|^q.$$

By the Burkholder–Davis–Gundy and Hölder inequalities:

$$4^q \mathbb{E} \sup_{s\in[0,t]} \left| \int_0^{s\wedge\tau_n} G_r dB_r \right|^q \leq C_q \mathbb{E}\left[\left(\int_0^{t\wedge\tau_n} |G_s|^2 ds \right)^{q/2} \right]$$

$$\leq C_q \mathbb{E} \left(2\int_0^t \beta_s^2 ds + 2\int_0^t b^2(s) |X_{s\wedge\tau_n}|^2 ds \right)^{q/2}$$

$$\leq C_q \mathbb{E} \left(\int_0^t \beta_s^2 ds \right)^{q/2} + C_q \mathbb{E}\Big[\sup_{s\in[0,t]} |X_{s\wedge\tau_n}|^{q/2} \left(\int_0^t b^2(s) \sup_{r\in[0,s]} |X_{r\wedge\tau_n}| ds \right)^{q/2} \Big]$$

$$\leq \frac{1}{4} \mathbb{E} \sup_{s\in[0,t]} |X_{s\wedge\tau_n}|^q + C_q \mathbb{E} \left(\int_0^t \beta_s^2 ds \right)^{q/2} + C_q \mathbb{E} \left(\int_0^t b^2(s) \sup_{r\in[0,s]} |X_{r\wedge\tau_n}| ds \right)^q$$

$$\leq \frac{1}{4} \mathbb{E} \sup_{s\in[0,t]} |X_{s\wedge\tau_n}|^q + C_q \mathbb{E} \left(\int_0^t \beta_s^2 ds \right)^{q/2}$$

$$+ C_q t^{q-1} \int_0^t b^{2q}(s) \mathbb{E} \sup_{r\in[0,s]} |X_{r\wedge\tau_n}|^q ds.$$

Also

$$4^q \left| \int_0^{t\wedge\tau_n} (\alpha_s + a_s |X_s|) ds \right|^q$$

$$\leq C_q \left(\int_0^t \alpha_s ds \right)^q + C_q \left(\int_0^t a(s) |X_{s\wedge\tau_n}| ds \right)^q$$

$$\leq C_q \left(\int_0^t \alpha_s ds \right)^q + C_q t^{q-1} \int_0^t a^q(s) \sup_{r\in[0,s]} |X_{r\wedge\tau_n}|^q ds.$$

Hence, defining

$$K_{q,t} = \mathbb{E} \|H\|_t^q + \mathbb{E} \left[\left(\int_0^t \alpha_s ds \right)^q + \left(\int_0^t \beta_s^2 ds \right)^{q/2} \right],$$

we have

$$\mathbb{E} \sup_{s\in[0,t]} |X_{s\wedge\tau_n}|^q + \mathbb{E} \sup_{s\in[0,t]} |U_{s\wedge\tau_n}|^q \leq 2\mathbb{E} \sup_{s\in[0,t]} \left[|X_{s\wedge\tau_n}|^q + |U_{s\wedge\tau_n}|^q \right]$$

$$\leq C_q K_{q,t} + C_q t^{q-1} \int_0^t \left(a^q(s) + b^{2q}(s) \right) \mathbb{E} \sup_{r\in[0,s]} |X_{r\wedge\tau_n}|^q ds.$$

Using Gronwall's inequality (6.57) we obtain

$$\mathbb{E} \sup_{s\in[0,t]} |X_{s\wedge\tau_n}|^q \leq C_q K_{q,t} e^{C_q A_q(t)} < \infty, \tag{6.68}$$

where

$$A_q(t) = t^{q-1}\int_0^t \left(a^q(s) + b^{2q}(s)\right).$$

Since $1 + xe^{ax} \le e^{(a+1)x}$, for all $x \ge 0$, it follows that

$$\begin{aligned}\mathbb{E}\sup_{s\in[0,t]} |U_{s\wedge\tau_n}|^q &\le C_q\left[K_{q,t} + A_q(t)\, C_q K_{q,t} e^{C_q A_q(t)}\right] \\ &\le C_q K_{q,t} e^{2C_q A_q(t)} < \infty.\end{aligned} \tag{6.69}$$

We also have

$$\mathbb{E}\left[\left(\int_0^{t\wedge\tau_n} |G_s|^2\, ds\right)^{q/2}\right] \le \hat{C}_{q,t} < \infty \tag{6.70}$$

for some $\hat{C}_{q,t}$ independent of n. Passing to the limit in (6.68)–(6.70) as $n \to \infty$, we obtain $X, U \in S_d^q[0,T]$, $G \in \Lambda^q_{d\times k}(0,T)$ and (6.67) follows. ■

Proposition 6.69. *Let $\delta \in \{-1, 1\}$. Let $\{B_t : t \ge 0\}$ be a k-dimensional Brownian motion. Let $Y, K, V : \Omega \times \mathbb{R}_+ \to \mathbb{R}$ and $G : \Omega \times \mathbb{R}_+ \to \mathbb{R}^k$ be progressively measurable stochastic processes such that*

i) Y, K, V *are continuous stochastic processes,*
ii) $V_\cdot, K_\cdot \in BV_{loc}([0,\infty[\,;\mathbb{R})$, $V_0 = K_0 = 0$, $\mathbb{P}$-*a.s.*,
iii) $\int_t^s |G_r|^2\, dr < \infty$, $\mathbb{P}$-*a.s.*, $\forall\, 0 \le t \le s$.

If for all $0 \le t \le s$,

$$\delta\left(Y_t - Y_s\right) \le \int_t^s \left(dK_r + Y_r dV_r\right) + \int_t^s \langle G_r, dB_r\rangle, \quad \mathbb{P}\text{-a.s.},$$

then

$$\delta\left(Y_t e^{\delta V_t} - Y_s e^{\delta V_s}\right) \le \int_t^s e^{\delta V_r} dK_r + \int_t^s e^{\delta V_r}\langle G_r, dB_r\rangle, \quad \mathbb{P}\text{-a.s.}$$

Proof. Denoting

$$M_s = \int_0^s \langle G_r, dB_r\rangle, \ \tilde{Y}_s = -M_s - \delta Y_s, \tag{6.71}$$

we obtain

$$\tilde{Y}_s \le \tilde{Y}_t + \int_t^s \left[dK_r + \left(-\delta\tilde{Y}_r - \delta M_r\right) dV_r\right].$$

Hence

$$s \longmapsto L_s \overset{def}{=} \tilde{Y}_s - \int_0^s \left[dK_r + \left(-\delta \tilde{Y}_r - \delta M_r\right) dV_r\right]$$

is a decreasing function and then

$$\begin{aligned} d\left(\tilde{Y}_s e^{\delta V_s}\right) &= \left\{dL_s + \left[dK_s + \left(-\delta \tilde{Y}_s - \delta M_s\right) dV_s\right]\right\} e^{\delta V_s} + \delta\, \tilde{Y}_s e^{\delta V_s} dV_s \\ &\leq -\delta M_s e^{\delta V_s} dV_s + e^{\delta V_s} dK_s \end{aligned}$$

and integrating from t to s

$$\begin{aligned} \tilde{Y}_s e^{\delta V_s} &\leq \tilde{Y}_t e^{\delta V_t} - \int_t^s \delta M_r e^{\delta V_r} dV_r + \int_t^s e^{\delta V_r} dK_r \\ &= \tilde{Y}_t e^{\delta V_t} - M_s e^{\delta V_s} + M_t e^{\delta V_t} + \int_t^s e^{\delta V_r} \langle G_r, dB_r \rangle + \int_t^s e^{\delta V_r} dK_r. \end{aligned}$$

Now by (6.71) we obtain the conclusions. ■

6.4.3 Forward Stochastic Inequalities

In this subsection $\{B_t : t \geq 0\}$ is a k-dimensional Brownian motion with respect to a stochastic basis $(\Omega, \mathcal{F}, \mathbb{P}, \{\mathcal{F}_t\}_{t \geq 0})$.

We shall derive some estimates on the local semimartingale $X \in S_d^0$ of the form

$$X_t = X_0 + K_t + \int_0^t G_s dB_s, \quad t \geq 0, \quad \mathbb{P}\text{-a.s.}, \tag{6.72}$$

where

◊ $K \in S_d^0$; $K_\cdot \in BV_{loc}\left([0, \infty[\,; \mathbb{R}^d\right)$, $K_0 = 0$, $\mathbb{P}$-a.s.;
◊ $G \in \Lambda^0_{d \times k}$.

Notation 6.70. *Let* $p \geq 1$ *and* $m_p \overset{def}{=} 1 \vee (p-1)$.

Proposition 6.71. *Let* $X \in S_d^0$ *be a local semimartingale of the form (6.72). Assume there exist* $p \geq 1$, *a* $\mathcal{P}$*-m.i.c.s.p.* D *and a* $\mathcal{P}$*-m.b-v.c.s.p.* V, $D_0 = V_0 = 0$, *such that as signed measures on* $[0, \infty[$

$$dD_t + \langle X_t, dK_t \rangle + \frac{1}{2} m_p \left|G_t\right|^2 dt \leq |X_t|^2 dV_t, \quad \mathbb{P}\text{-a.s.}, \tag{6.73}$$

then for all $0 \leq t \leq s$*:*

$$\mathbb{E}^{\mathcal{F}_t} \left|e^{-V_s} X_s\right|^p + p\, \mathbb{E}^{\mathcal{F}_t} \int_t^s e^{-pV_r} \left|X_r\right|^{p-2} dD_r \leq \left|e^{-V_t} X_t\right|^p, \ \mathbb{P}\text{-}a.s. \tag{6.74}$$

Moreover for all $\delta \geq 0$, $0 \leq t \leq s$:

$$\begin{aligned} &\mathbb{E}^{\mathcal{F}_t} \frac{\left|e^{-V_s} X_s\right|^p}{\left(1+\delta\left|e^{-V_s} X_s\right|^2\right)^{p/2}} + p\, \mathbb{E}^{\mathcal{F}_t} \int_t^s \frac{e^{-pV_r}\left|X_r\right|^{p-2}}{\left(1+\delta\left|e^{-V_r} X_r\right|^2\right)^{(p+2)/2}} dD_r \\ &\qquad \leq \frac{\left|e^{-V_t} X_t\right|^p}{\left(1+\delta\left|e^{-V_t} X_t\right|^2\right)^{p/2}}, \ \mathbb{P}\text{-}a.s. \end{aligned} \tag{6.75}$$

The proof of this Proposition is contained in the proof of the next Proposition.

Remark 6.72. Since by (2.27)

$$\mathbf{1}_{X_t=0} \left|G_t\right|^2 dt = 0,$$

we see that the condition (6.73) yields

$$\mathbf{1}_{X_t=0} dD_t = 0.$$

We now formulate a more general assumption.

(FB) *There exist*

- $p \geq 1$, $\lambda \geq 0$,
- *three* $\mathcal{P}$-*m.i.c.s.p.* D, R, N, $D_0 = R_0 = N_0 = 0$, *and*
- *a* $\mathcal{P}$-*m.b-v.c.s.p.* V, $V_0 = 0$,

such that, as signed measures on $[0, \infty[$:

$$\begin{aligned} &dD_t + \langle X_t, dK_t\rangle + \left(\frac{1}{2} m_p + 9p\lambda\right) \left|G_t\right|^2 dt \\ &\qquad \leq \mathbf{1}_{p\geq 2} dR_t + \left|X_t\right| dN_t + \left|X_t\right|^2 dV_t. \end{aligned} \tag{6.76}$$

Remark 6.73. From the condition (6.76), we deduce that

$$\begin{aligned} &\mathbf{1}_{X_t=0} dD_t = 0, \quad \text{if } 1 \leq p < 2, \text{ and} \\ &\mathbf{1}_{X_t=0} dD_t \leq \mathbf{1}_{X_t=0} dR_t \leq dR_t, \quad \text{if } p \geq 2. \end{aligned}$$

Proposition 6.74. *Let* $X \in S_d^0$ *be a local semimartingale of the form (6.72). Assume that there exist* $p \geq 1$ *and* $\lambda > 1$ *such that* **(FB)** *is satisfied. Then there exists a positive constant* $C_{p,\lambda}$ *depending only on* (p, λ) *such that for all* $\delta \geq 0$, *and* $0 \leq t \leq s$:

$$
\begin{aligned}
&\mathbb{E}^{\mathcal{F}_t}\frac{\left\|e^{-V}X\right\|^p_{[t,s]}}{\left(1+\delta\left\|e^{-V}X\right\|^2_{[t,s]}\right)^{p/2}}+\mathbb{E}^{\mathcal{F}_t}\int_t^s\frac{e^{-pV_r}|X_r|^{p-2}}{\left(1+\delta\left|e^{-V_r}X_r\right|^2\right)^{(p+2)/2}}dD_r\\
&\quad+\mathbb{E}^{\mathcal{F}_t}\left(\int_t^s\frac{e^{-2V_r}}{\left(1+\delta\left|e^{-V_r}X_r\right|^2\right)^2}dD_r\right)^{p/2}\\
&\quad+\mathbb{E}^{\mathcal{F}_t}\left(\int_t^s\frac{e^{-2V_r}}{\left(1+\delta\left|e^{-V_r}X_r\right|^2\right)^2}|G_r|^2\,dr\right)^{p/2} \qquad (6.77)\\
&\leq C_{p,\lambda}\Bigg[\frac{\left|e^{-V_t}X_t\right|^p}{\left(1+\delta\left|e^{-V_t}X_t\right|^2\right)^{p/2}}+\mathbb{E}^{\mathcal{F}_t}\left(\int_t^s e^{-2V_r}\mathbf{1}_{p\geq 2}dR_r\right)^{p/2}\\
&\quad+\mathbb{E}^{\mathcal{F}_t}\left(\int_t^s e^{-V_r}dN_r\right)^p\Bigg],\quad \mathbb{P}\text{-}a.s.
\end{aligned}
$$

If we set $\delta = 0$ in (6.77), we obtain the following:

Corollary 6.75. *Under the assumption* (**FB**), *for all* $0 \leq t \leq s$*:*

$$
\begin{aligned}
&\mathbb{E}^{\mathcal{F}_t}\left\|e^{-V}X\right\|^p_{[t,s]}+\mathbb{E}^{\mathcal{F}_t}\int_t^s e^{-pV_r}|X_r|^{p-2}\,dD_r\\
&\quad+\mathbb{E}^{\mathcal{F}_t}\left(\int_t^s e^{-2V_r}dD_r\right)^{p/2}\\
&\quad+\mathbb{E}^{\mathcal{F}_t}\left(\int_t^s e^{-2V_r}|G_r|^2\,dr\right)^{p/2} \qquad (6.78)\\
&\leq C_{p,\lambda}\Bigg[\left|e^{-V_t}X_t\right|^p+\mathbb{E}^{\mathcal{F}_t}\left(\int_t^s e^{-2V_r}\mathbf{1}_{p\geq 2}dR_r\right)^{p/2}\\
&\quad+\mathbb{E}^{\mathcal{F}_t}\left(\int_t^s e^{-V_r}dN_r\right)^p\Bigg],\quad \mathbb{P}\text{-}a.s.
\end{aligned}
$$

Proof (of Proposition 6.74). In view of the monotone convergence theorem it suffices to treat the case $\delta > 0$, which we assume from now on.

To simplify, we define

$$
\begin{aligned}
J_r &\overset{def}{=} \frac{\left|e^{-V_r}X_r\right|}{\left(1+\delta\left|e^{-V_r}X_r\right|^2\right)^{1/2}}\\
&\leq \frac{1}{\sqrt{\delta}},
\end{aligned}
$$

and

$$
\begin{aligned}
\hat{J}_r^{(p)} &\overset{def}{=} \frac{\left|e^{-V_r}X_r\right|^{p-2}\mathbf{1}_{X_r\neq 0}}{\left(1+\delta\left|e^{-V_r}X_r\right|^2\right)^{(p+2)/2}}\\
&= J_r^{p-2}\frac{\mathbf{1}_{X_r\neq 0}}{\left(1+\delta\left|e^{-V_r}X_r\right|^2\right)^2}.
\end{aligned}
$$

We remark that

$$\hat{J}_r^{(p)} e^{-V_r} |X_r| \le J_r^{p-1} \mathbf{1}_{X_r \neq 0} \text{ and } \hat{J}_r^{(p)} e^{-2V_r} |X_r|^2 \le J_r^p.$$

Step 1. General calculation.
We begin by assuming a condition which is more general than the assumptions (6.73) and (6.76), namely that there exists a $\gamma \ge 0$ such that

$$\begin{aligned} &dD_r + \langle X_r, dK_r \rangle + \left(\frac{m_p}{2} + \gamma\right) |G_r|^2 \, dr \\ &\le \mathbf{1}_{p \ge 2} dR_r + |X_r| dN_r + |X_r|^2 dV_r. \end{aligned} \tag{6.79}$$

Since by Itô's formula

$$e^{-V_t} X_t = X_0 + \int_0^t \left(e^{-V_r} dK_r - e^{-V_r} X_r dV_r\right) + \int_0^t e^{-V_r} G_r dB_r,$$

it follows from the inequality (2.28) in Corollary 2.28 that for all $0 \le t \le s$ and any stopping time θ

$$\begin{aligned} J_{s\wedge\theta}^p &\le J_{t\wedge\theta}^p + p \int_t^s \mathbf{1}_{r<\theta} \, \hat{J}_r^{(p)} e^{-2V_r} \langle X_r, G_r dB_r \rangle \\ &\quad + p \int_t^s \mathbf{1}_{r<\theta} \, \hat{J}_r^{(p)} e^{-2V_r} \left[\langle X_r, dK_r - X_r dV_r \rangle + \frac{1}{2} m_p |G_r|^2 \right] dr, \ a.s. \end{aligned}$$

But

$$J_s^p \mathbf{1}_{s<\theta} \le J_{s\wedge\theta}^p;$$

hence we deduce that

$$\begin{aligned} &J_s^p \mathbf{1}_{s<\theta} + p \int_t^s \mathbf{1}_{r<\theta} \, \hat{J}_r^{(p)} e^{-2V_r} dD_r + p\gamma \int_t^s \mathbf{1}_{r<\theta} \, \hat{J}_r^{(p)} e^{-2V_r} |G_r|^2 \, dr \\ &\le J_{t\wedge\theta}^p + p \int_t^s \mathbf{1}_{r<\theta} \, \hat{J}_r^{(p)} e^{-2V_r} \langle X_r, G_r dB_r \rangle \\ &\quad + p \int_t^s \mathbf{1}_{r<\theta} \, \hat{J}_r^{(p)} e^{-2V_r} \left[dD_r + \langle X_r, dK_r - X_r dV_r \rangle + \left(\frac{1}{2} m_p + \gamma\right) |G_r|^2 \right] dr, \end{aligned}$$

and using the assumption (6.79) it follows that for any stopping time θ and for all $0 \le t \le s$, $\mathbb{P}$-$a.s.$:

$$\begin{aligned}
&J_s^p \mathbf{1}_{s<\theta} + p\int_t^s \mathbf{1}_{r<\theta}\, \hat{J}_r^{(p)} e^{-2V_r} dD_r + p\gamma \int_t^s \mathbf{1}_{r<\theta}\, \hat{J}_r^{(p)} e^{-2V_r}\, |G_r|^2\, dr \\
&\quad \le J_{t\wedge\theta}^p + p\int_t^s \mathbf{1}_{r<\theta}\, \hat{J}_r^{(p)} e^{-2V_r}\, \langle X_r, G_r dB_r\rangle \\
&\qquad + p\int_t^s \mathbf{1}_{r<\theta}\, J_r^{p-2} \mathbf{1}_{X_r\neq 0} e^{-2V_r} \mathbf{1}_{p\ge 2} dR_r + p\int_t^s \mathbf{1}_{r<\theta}\, J_r^{p-1} \mathbf{1}_{X_r\neq 0} e^{-V_r} dN_r .
\end{aligned} \tag{6.80}$$

Since for all $T>0$:

$$\begin{aligned}
\int_0^T \left|\hat{J}_r^{(p)} e^{-2V_r} X_r^* G_r\right|^2 dr &\le \sup_{r\in[0,T]} \left[e^{-pV_r}\, |X_r|^{p-1}\right] \int_0^T |G_r|^2\, dr \\
&< \infty, \quad \mathbb{P}\text{-a.s.},
\end{aligned}$$

it follows that for all $0\le t\le s$:

$$\int_t^s \hat{J}_r^{(p)} e^{-2V_r} dD_r + \gamma \int_t^s \hat{J}_r^{(p)} e^{-2V_r}\, |G_r|^2\, dr < \infty, \quad a.s.$$

For each $n\in\mathbb{N}^*$ we define the stopping time

$$\begin{aligned}
\theta_n = \inf\Big\{ t\ge 0 : &\int_0^t J_r^{p-2} \mathbf{1}_{X_r\neq 0} e^{-2V_r} \mathbf{1}_{p\ge 2} dR_r + \int_0^t J_r^{p-1} \mathbf{1}_{X_r\neq 0} e^{-V_r} dN_r \\
&+ \int_0^t \left|\hat{J}_r^{(p)} e^{-2V_r} X_r^* G_r\right|^2 dr \ge n \Big\}.
\end{aligned} \tag{6.81}$$

Note that for $\theta=\theta_n$

$$M_t^n = p\int_0^t \mathbf{1}_{r<\theta_n} \hat{J}_r^{(p)} \left\langle e^{-V_r} X_r, e^{-V_r} G_r dB_r\right\rangle$$

is a martingale and consequently, for all $0\le t\le s$:

$$\mathbb{E}^{\mathcal{F}_t} \int_t^s \mathbf{1}_{r<\theta_n}\, \hat{J}_r^{(p)} e^{-2V_r} dD_r + \gamma \mathbb{E}^{\mathcal{F}_t} \int_t^s \mathbf{1}_{r<\theta_n}\, \hat{J}_r^{(p)} e^{-2V_r}\, |G_r|^2\, dr < \infty, \quad a.s.$$

Step 2. Proof of the inequality (6.75).
In view of the first step, the assumption (6.73) yields (6.80) with $\gamma=0$ and $R=N=0$, from which we deduce

$$\mathbb{E}^{\mathcal{F}_t} J_s^p \mathbf{1}_{s<\theta_n} + p\mathbb{E}^{\mathcal{F}_t} \int_t^s \mathbf{1}_{r<\theta_n} \hat{J}_r^{(p)} e^{-2V_r} dD_r \le J_{t\wedge\theta_n}^p, \text{ a.s.,} \tag{6.82}$$

and passing to the limit as $n\to\infty$ (the first two terms converge monotonically and the third one converges $a.s.$) the estimate (6.75) follows in view of Remark 6.73, since $R=0$.

Step 3. Proof of the inequality (6.77).

(**A**) Let $\gamma > 0$. From (6.80) we have

$$\begin{aligned}
&\mathbb{E}^{\mathcal{F}_t} \sup_{r\in[t,s]} \left(J_r^p \mathbf{1}_{r<\theta_n}\right) + p\mathbb{E}^{\mathcal{F}_t} \int_t^s \mathbf{1}_{r<\theta_n} \hat{J}_r^{(p)} e^{-2V_r} dD_r \\
&\quad + p\gamma \mathbb{E}^{\mathcal{F}_t} \int_t^s \mathbf{1}_{r<\theta_n} \hat{J}_r^{(p)} \left|e^{-V_r} G_r\right|^2 dr \\
&\le 2J_{t\wedge\theta_n}^p + 2p\mathbb{E}^{\mathcal{F}_t} \int_t^s \mathbf{1}_{r<\theta_n} J_r^{p-2} \mathbf{1}_{X_r\neq 0} e^{-2V_r} \mathbf{1}_{p\ge 2} dR_r \\
&\quad + 2p\mathbb{E}^{\mathcal{F}_t} \int_t^s \mathbf{1}_{r<\theta_n} J_r^{p-1} \mathbf{1}_{X_r\neq 0} e^{-V_r} dN_r \\
&\quad + 2p\mathbb{E}^{\mathcal{F}_t} \sup_{u\in[t,s]} \left| \int_t^u \mathbf{1}_{r<\theta_n} \hat{J}_r^{(p)} e^{-2V_r} X_r^* G_r dB_r \right| .
\end{aligned}$$

By the Burkholder–Davis–Gundy inequality

$$\begin{aligned}
&2p\, \mathbb{E}^{\mathcal{F}_t} \sup_{u\in[t,s]} \left| \int_t^u \mathbf{1}_{r<\theta_n} \hat{J}_r^{(p)} \left\langle e^{-V_r} X_r, e^{-V_r} G_r dB_r \right\rangle \right| \\
&\le 6p\, \mathbb{E}^{\mathcal{F}_t} \sqrt{\int_t^s \left| \mathbf{1}_{r<\theta_n} \hat{J}_r^{(p)} e^{-2V_r} X_r^* G_r \right|^2 dr} \\
&\le 6p\, \mathbb{E}^{\mathcal{F}_t} \left[\sqrt{\sup_{r\in[t,s]} \hat{J}_r^{(p)} e^{-2V_r} \left|X_r\right|^2 \mathbf{1}_{r<\theta_n}} \sqrt{\int_t^s \mathbf{1}_{r<\theta_n} \hat{J}_r^{(p)} \left|e^{-V_r} G_r\right|^2 dr} \right] \\
&\le \frac{1}{\lambda}\, \mathbb{E}^{\mathcal{F}_t} \sup_{r\in[t,s]} \left(J_r^p \mathbf{1}_{r<\theta_n}\right) + 9p^2\lambda\, \mathbb{E}^{\mathcal{F}_t} \int_t^s \mathbf{1}_{r<\theta_n} \hat{J}_r^{(p)} \left|e^{-V_r} G_r\right|^2 dr,
\end{aligned}$$

for all $\lambda > 0$. Hence

$$\begin{aligned}
&\left(1 - \frac{1}{\lambda}\right) \mathbb{E}^{\mathcal{F}_t} \sup_{r\in[t,s]} \left(J_r^p \mathbf{1}_{r<\theta_n}\right) + p\mathbb{E}^{\mathcal{F}_t} \int_t^s \mathbf{1}_{r<\theta_n} \hat{J}_r^{(p)} e^{-2V_r} dD_r \\
&\qquad + p\left(\gamma - 9p\lambda\right) \mathbb{E}^{\mathcal{F}_t} \int_t^s \mathbf{1}_{r<\theta_n} \hat{J}_r^{(p)} \left|e^{-V_r} G_r\right|^2 dr \\
&\le 2J_{t\wedge\theta_n}^p + 2p\, \mathbb{E}^{\mathcal{F}_t} \int_t^s \mathbf{1}_{r<\theta_n} J_r^{p-2} \mathbf{1}_{X_r\neq 0} e^{-2V_r} \mathbf{1}_{p\ge 2} dR_r \\
&\qquad + 2p\, \mathbb{E}^{\mathcal{F}_t} \int_t^s \mathbf{1}_{r<\theta_n} J_r^{p-1} \mathbf{1}_{X_r\neq 0} e^{-V_r} dN_r .
\end{aligned}$$

Let $\gamma = 9p\lambda$, $\lambda > 1$. By Hölder's inequality

$$2p\mathbb{E}^{\mathcal{F}_t}\int_t^s \mathbf{1}_{r<\theta_n}J_r^{p-2}\mathbf{1}_{X_r\neq 0}e^{-2V_r}\mathbf{1}_{p\geq 2}dR_r + 2p\mathbb{E}^{\mathcal{F}_t}\int_t^s \mathbf{1}_{r<\theta_n}J_r^{p-1}\mathbf{1}_{X_r\neq 0}e^{-V_r}dN_r$$
$$\leq 2p\mathbb{E}^{\mathcal{F}_t}\left[\sup_{r\in[t,s]}\left(J_r^{p-2}\mathbf{1}_{X_r\neq 0}\mathbf{1}_{p\geq 2}\mathbf{1}_{r<\theta_n}\right)\int_t^s e^{-2V_r}\mathbf{1}_{p\geq 2}dR_r\right]$$
$$+ 2p\mathbb{E}^{\mathcal{F}_t}\left[\sup_{r\in[t,s]}\left(J_r^{p-1}\mathbf{1}_{X_r\neq 0}\mathbf{1}_{r<\theta_n}\right)\int_t^s e^{-V_r}dN_r\right]$$
$$\leq \frac{1}{2}\left(1-\frac{1}{\lambda}\right)\mathbb{E}^{\mathcal{F}_t}\sup_{r\in[t,s]}\left(J_r^p\mathbf{1}_{r<\theta_n}\right) + C_{p,\lambda}\mathbb{E}^{\mathcal{F}_t}\left(\int_t^s e^{-2V_r}\mathbf{1}_{p\geq 2}dR_r\right)^{p/2}$$
$$+ C_{p,\lambda}\mathbb{E}^{\mathcal{F}_t}\left(\int_t^s e^{-V_r}dN_r\right)^p.$$

We deduce from the above that

$$\mathbb{E}^{\mathcal{F}_t}\sup_{r\in[t,s]}\left(J_r^p\mathbf{1}_{r<\theta_n}\right) + \mathbb{E}^{\mathcal{F}_t}\int_t^s \mathbf{1}_{r<\theta_n}\hat{J}_r^{(p)}e^{-2V_r}dD_r$$
$$\leq C_{p,\lambda}\left[J_{t\wedge\theta_n}^p + \mathbb{E}^{\mathcal{F}_t}\left(\int_t^s e^{-2V_r}\mathbf{1}_{p\geq 2}dR_r\right)^{p/2} + \mathbb{E}^{\mathcal{F}_t}\left(\int_t^s e^{-V_r}dN_r\right)^p\right]. \tag{6.83}$$

The argument used in order to take the limit in (6.82) yields as $n\to\infty$:

$$\begin{aligned}\mathbb{E}^{\mathcal{F}_t}\sup_{r\in[t,s]}J_r^p + \mathbb{E}^{\mathcal{F}_t}\int_t^s \hat{J}_r^{(p)}e^{-2V_r}dD_r \leq C_{p,\lambda}\Big[J_t^p \\ +\mathbb{E}^{\mathcal{F}_t}\left(\int_t^s e^{-2V_r}\mathbf{1}_{p\geq 2}dR_r\right)^{p/2} + \mathbb{E}^{\mathcal{F}_t}\left(\int_t^s e^{-V_r}dN_r\right)^p\Big].\end{aligned} \tag{6.84}$$

(B) From (6.80) for $p=2$, $\gamma=1$ and $\theta=\theta_n$ we have

$$J_{s\wedge\theta_n}^2 + 2\int_t^s \mathbf{1}_{r<\theta_n}\,\hat{J}_r^{(2)}e^{-2V_r}dD_r + 2\int_t^s \mathbf{1}_{r<\theta_n}\,\hat{J}_r^{(2)}\left|e^{-V_r}G_r\right|^2dr$$
$$\leq J_{t\wedge\theta_n}^2 + 2\int_t^s \mathbf{1}_{r<\theta_n}\,\mathbf{1}_{X_r\neq 0}e^{-2V_r}dR_r + 2\int_t^s \mathbf{1}_{r<\theta_n}\,J_r\mathbf{1}_{X_r\neq 0}e^{-V_r}dN_r$$
$$+2\int_t^s \mathbf{1}_{r<\theta_n}\,\hat{J}_r^{(2)}\left\langle e^{-V_r}X_r, e^{-V_r}G_rdB_r\right\rangle,$$

which yields

$$\mathbb{E}^{\mathcal{F}_t}\left(\int_t^s \mathbf{1}_{r<\theta_n}\hat{J}_r^{(2)}e^{-2V_r}dD_r\right)^{p/2} + \mathbb{E}^{\mathcal{F}_t}\left(\int_t^s \mathbf{1}_{r<\theta_n}\hat{J}_r^{(2)}\left|e^{-V_r}G_r\right|^2dr\right)^{p/2}$$
$$\leq C_p\,J_t^p + C_p\,\mathbb{E}^{\mathcal{F}_t}\left(\int_t^s e^{-2V_r}\mathbf{1}_{p\geq 2}dR_r\right)^{p/2}$$

$$+ C_p\, \mathbb{E}^{\mathcal{F}_t} \sup_{r\in[t,s]} \left(J_r^{p/2}\mathbf{1}_{r<\theta_n}\right)\left(\int_t^s e^{-V_r}dN_r\right)^{p/2}$$

$$+ C_p\, \mathbb{E}^{\mathcal{F}_t} \sup_{u\in[t,s]} \left|\int_t^u \mathbf{1}_{r<\theta_n}\hat{J}_r^{(2)} \left\langle e^{-V_r}X_r, e^{-V_r}G_r dB_r\right\rangle\right|^{p/2}.$$

By the Burkholder–Davis–Gundy inequality (2.8)

$$\begin{aligned}
&C_p\, \mathbb{E}^{\mathcal{F}_t} \sup_{u\in[t,s]} \left|\int_t^u \mathbf{1}_{r<\theta_n}\hat{J}_r^{(2)} \left\langle e^{-V_r}X_r, e^{-V_r}G_r dB_r\right\rangle\right|^{p/2} \\
&\leq C_p'\, \mathbb{E}^{\mathcal{F}_t}\left(\int_t^s \mathbf{1}_{r<\theta_n}\left|\hat{J}_r^{(2)}e^{-2V_r}\left|X_r^* G_r\right|\right|^2 dr\right)^{p/4} \\
&\leq C_p'\mathbb{E}^{\mathcal{F}_t} \sup_{r\in[t,s]} \left(J_r^{p/2}\mathbf{1}_{r<\theta_n}\right)\left(\int_t^s \mathbf{1}_{r<\theta_n}\hat{J}_r^{(2)}\left|e^{-V_r}G_r\right|^2 dr\right)^{p/4} \\
&\leq C_p''\, \mathbb{E}^{\mathcal{F}_t} \sup_{r\in[t,s]} \left(J_r^{p}\mathbf{1}_{r<\theta_n}\right) + \frac{1}{2}\mathbb{E}^{\mathcal{F}_t}\left(\int_t^s \mathbf{1}_{r<\theta_n}\hat{J}_r^{(2)}\left|e^{-V_r}G_r\right|^2 dr\right)^{p/2}.
\end{aligned}$$

Hence

$$\begin{aligned}
&\mathbb{E}^{\mathcal{F}_t}\left(\int_t^s \mathbf{1}_{r<\theta_n}\hat{J}_r^{(2)}e^{-2V_r}dD_r\right)^{p/2} + \frac{1}{2}\mathbb{E}^{\mathcal{F}_t}\left(\int_t^s \mathbf{1}_{r<\theta_n}\hat{J}_r^{(2)}\left|e^{-V_r}G_r\right|^2 dr\right)^{p/2} \\
&\leq C_p\left[\mathbb{E}^{\mathcal{F}_t}\sup_{r\in[t,s]} J_r^p + \mathbb{E}^{\mathcal{F}_t}\left(\int_t^s e^{-2V_r}\mathbf{1}_{p\geq 2}dR_r\right)^{p/2} + \mathbb{E}^{\mathcal{F}_t}\left(\int_t^s e^{-V_r}dN_r\right)^{p}\right].
\end{aligned} \tag{6.85}$$

We take the limit as $n\to\infty$ in the last inequality and the estimate (6.77) follows from (6.84), (6.85), Remark 6.73 and the identity

$$\sup_{r\in[t,s]} J_r^p = \frac{\left\|e^{-V}X\right\|_{[t,s]}^p}{\left(1+\delta\left\|e^{-V}X\right\|_{[t,s]}^2\right)^{p/2}}.$$

This last fact follows from the increasing monotonicity of the function

$$r \mapsto \frac{r^p}{(1+\delta r^2)^{p/2}} : [0,\infty[\,\to [0,\infty[\,.$$

The proof is complete. ■

We shall give a supplementary result in the case when R, N, V are deterministic functions.

Corollary 6.76. *Let $X \in S_d^0$ be a local semimartingale of the form*

$$X_t = X_0 + K_t + \int_0^t G_s dB_s, \quad t \geq 0, \quad \mathbb{P}\text{-a.s.},$$

where

◊ $K \in S_d^0$; $K_\cdot \in BV_{loc}\left([0,\infty[\,;\mathbb{R}^d\right)$, $K_0 = 0$, $\mathbb{P}$-a.s.;
◊ $G \in \Lambda_{d\times k}^0$.

Assume that there exist

- $p \geq 1$, $m_p \stackrel{def}{=} 1 \vee (p-1)$;
- *two continuous increasing deterministic functions* $R, N : [0,\infty[\to [0,\infty[$, $R(0) = N(0) = 0$, *and*
- *a continuous deterministic function with bounded variation* $V : [0,\infty[\to \mathbb{R}$, $V(0) = 0$,

such that as signed measures on $[0,\infty[$:

$$\langle X_t, dK_t\rangle + \frac{1}{2} m_p |G_t|^2 dt \leq \mathbf{1}_{p\geq 2} dR(t) + |X_t| dN(t) + |X_t|^2 dV(t). \quad (6.86)$$

Define

$$\begin{aligned} Q(t) &= 2R(t)\mathbf{1}_{p\geq 2} + N(t), \\ P(t) &= (p-2)R(t)\mathbf{1}_{p\geq 2} + (p-1)N(t) + pV(t) \quad \text{and} \\ M(t) &= \int_0^t e^{-P(r)} dQ(r). \end{aligned}$$

Then for all $\delta \geq 0$ *and* $0 \leq t \leq s$:

$$\mathbb{E}\frac{|X_s|^p e^{-P(s)}}{\left(1+\delta|X_s|^2\right)^{p/2}} \leq \mathbb{E}\frac{|X_t|^p e^{-P(t)}}{\left(1+\delta|X_t|^2\right)^{p/2}} + M(s) - M(t). \quad (6.87)$$

In particular for $\delta \searrow 0$ *and* $0 = t \leq s$:

$$\begin{aligned} &(a) \quad e^{-P(s)}\mathbb{E}\,|X_s|^p \leq \mathbb{E}\,|X_0|^p + M(s), \\ &(b) \quad \int_0^\infty e^{-P(s)-\alpha M(s)-\lambda s}\left(\mathbb{E}\,|X_s|^p\right) ds \leq \frac{1}{\lambda}\left(\mathbb{E}\,|X_0|^p + \frac{1}{\alpha}\right) \end{aligned} \quad (6.88)$$

for all $\alpha, \lambda > 0$.

Proof. We follow from (6.80) the first steps from the proof of Proposition 6.74 but now

$$J_r = \frac{|X_r|}{\left(1+\delta\,|X_r|^2\right)^{1/2}} \quad \text{and} \quad \hat{J}_r^{(p)} = \frac{|X_r|^{p-2}\,\mathbf{1}_{X_r\neq 0}}{\left(1+\delta\,|X_r|^2\right)^{(p+2)/2}}$$

and $\theta = \theta_n$ is defined similarly.

From the inequality (2.28) in Corollary 2.28, we have for all $0 \le t \le s$

$$\begin{aligned}
&\mathbb{E}\,\left(J_s^p \mathbf{1}_{s<\theta_n}\right)\\
&\le \mathbb{E}\,J_{s\wedge\theta_n}^p\\
&\le \mathbb{E}\,J_{t\wedge\theta_n}^p + p\mathbb{E}\int_t^s \mathbf{1}_{r<\theta_n}\,\hat{J}_r^{(p)}\left[\langle X_r, dK_r\rangle + \frac{1}{2}m_p\,|G_r|^2\right]dr\\
&\le \mathbb{E}\,J_{t\wedge\theta_n}^p\\
&\quad + p\mathbb{E}\int_t^s \mathbf{1}_{r<\theta_n}\,\left[J_r^{p-2}\mathbf{1}_{X_r\neq 0},\mathbf{1}_{p\ge 2}dR\left(r\right) + J_r^{p-1}\mathbf{1}_{X_r\neq 0}dN\left(r\right) + J_r^p dV\left(r\right)\right].
\end{aligned}$$

Taking into account that

$$J_r^{p-2}\mathbf{1}_{X_r\neq 0} \le \frac{2}{p} + \frac{p-2}{p}J_r^p \quad \text{and} \quad J_r^{p-1}\mathbf{1}_{X_r\neq 0} \le \frac{1}{p} + \frac{p-1}{p}J_r^p,$$

and passing to the limit as $n \to \infty$ we have for all $0 \le t \le s$:

$$\begin{aligned}
\mathbb{E}\,J_s^p &\le \mathbb{E}\,J_t^p + 2\int_t^s \mathbf{1}_{p\ge 2}dR\left(r\right) + \int_t^s dN\left(r\right)\\
&\quad + \int_t^s \left[\left(p-2\right)\mathbf{1}_{p\ge 2}dR\left(r\right) + \left(p-1\right)dN\left(r\right) + pdV\left(r\right)\right]\mathbb{E}\left(J_r^p\right), \quad \text{a.s.,}
\end{aligned}$$

that is

$$\mathbb{E}\,J_s^p \le \mathbb{E}\,J_t^p + \int_t^s dQ\left(r\right) + \int_t^s \mathbb{E}\left(J_r^p\right)dP\left(r\right).$$

By Gronwall's inequality (Proposition 6.69), we have for all $0 \le t \le s$:

$$e^{-P(s)}\mathbb{E}\,J_s^p \le e^{-P(t)}\mathbb{E}\,J_t^p + \int_t^s e^{-P(r)}dQ\left(r\right),$$

and the inequality (6.87) follows. The inequality (6.88-b) clearly follows from (6.88-a) using the elementary inequality

$$ye^{-x-\alpha y-\lambda s} \le \frac{1}{\alpha}e^{-\lambda s}, \quad \text{for all } x, y, s, \lambda \ge 0 \text{ and } \alpha > 0.$$

■

Let $X, \hat{X} \in S_d^0$ be two semimartingales given by

$$\begin{array}{l} X_t = X_0 + K_t + \int_0^t G_s dB_s, \; t \geq 0, \\ \hat{X}_t = \hat{X}_0 + \hat{K}_t + \int_0^t \hat{G}_s dB_s, \; t \geq 0, \end{array} \tag{6.89}$$

where

◊ $K, \hat{K} \in S_d^0$;
◊ $K_\cdot(\omega), \hat{K}_\cdot(\omega) \in BV_{loc}\left([0,\infty[\,;\mathbb{R}^d\right), \; K_0(\omega) = \hat{K}_0(\omega) = 0, \; \mathbb{P}\text{-a.s. } \omega \in \Omega$;
◊ $G, \hat{G} \in \Lambda^0_{d\times k}$.

(FB'): *Assume there exist $p \geq 1$ and $\lambda \geq 0$ and a $\mathcal{P}$-m.b-v.c.s.p. V, $V_0 = 0$, such that as measures on $[0,\infty[$:*

$$\left\langle X_t - \hat{X}_t, dK_t - d\hat{K}_t \right\rangle + \left(\frac{1}{2}m_p + 9p\lambda\right)\left|G_t - \hat{G}_t\right|^2 dt \leq |X_t - \hat{X}_t|^2 dV_t. \tag{6.90}$$

Corollary 6.77. *Let $p \geq 1$ and A be a $\mathcal{P}$-m.i.c.s.p., $A_0 = 0$.*

(I) If the assumption (6.90) is satisfied with $\lambda = 0$, then for all $\delta \geq 0$, $0 \leq t \leq s$:

$$\begin{aligned} \mathbb{E}^{\mathcal{F}_t} \frac{e^{-p(V_s+A_s)}\left|X_s-\hat{X}_s\right|^p}{\left(1+\delta e^{-2(V_r+A_r)}\left|X_r-\hat{X}_r\right|^2\right)^{p/2}} &+ \mathbb{E}^{\mathcal{F}_t}\int_t^s \frac{e^{-p(V_r+A_r)}\left|X_r-\hat{X}_r\right|^p}{\left(1+\delta e^{-2(V_r+A_r)}\left|X_r-\hat{X}_r\right|^2\right)^{(p+2)/2}} dA_r \\ &\leq \frac{e^{-p(V_t+A_t)}\left|X_t-\hat{X}_t\right|^p}{\left(1+\delta e^{-2(V_t+A_t)}\left|X_t-\hat{X}_t\right|^2\right)^{p/2}}, \quad \mathbb{P}-a.s. \end{aligned}$$

In particular for $\delta = 0$

$$\begin{aligned} \mathbb{E}^{\mathcal{F}_t} e^{-p(V_s+A_s)}\left|X_s - \hat{X}_s\right|^p &+ \mathbb{E}^{\mathcal{F}_t}\int_t^s e^{-p(V_r+A_r)}\left|X_r - \hat{X}_r\right|^p dA_r \\ &\leq e^{-p(V_t+A_t)}\left|X_t - \hat{X}_t\right|^p, \quad \mathbb{P}\text{-a.s.}, \end{aligned}$$

for all $0 \leq t \leq s$.

(II) If the assumption (6.90) is satisfied with $\lambda > 1$, then there exists a positive constant $C_{p,\lambda}$ depending only on (p,λ) such that for all $\delta \geq 0$, $0 \leq t \leq s$:

$$\begin{aligned} \mathbb{E}^{\mathcal{F}_t} \frac{\left\|e^{-V-A}(X-\hat{X})\right\|^p_{[t,s]}}{\left(1+\delta\left\|e^{-V-A}(X-\hat{X})\right\|^2_{[t,s]}\right)^{p/2}} &+ \mathbb{E}^{\mathcal{F}_t}\left(\int_t^s \frac{e^{-2(V_r+A_r)}\left|X_r-\hat{X}_r\right|^2}{\left(1+\delta e^{-2V_r-2A_r}\left|X_r-\hat{X}_r\right|^2\right)^2} dA_r\right)^{p/2} \\ &\leq C_{p,\lambda} \frac{e^{-p(V_t+A_t)}\left|X_t-\hat{X}_t\right|^p}{\left(1+\delta e^{-2(V_t+A_t)}\left|X_t-\hat{X}_t\right|^2\right)^{p/2}}, \quad \mathbb{P}\text{-a.s.} \end{aligned}$$

In particular for $\delta = 0$

$$\mathbb{E}^{\mathcal{F}_t}\left\|e^{-V-A}\left(X-\hat{X}\right)\right\|_{[t,s]}^{p}+\mathbb{E}^{\mathcal{F}_t}\left(\int_t^s e^{-2(V_r+A_r)}\left|X_r-\hat{X}_r\right|^2 dA_r\right)^{p/2}$$
$$\leq C_{p,\lambda}\, e^{-p(V_t+A_t)}\left|X_t-\hat{X}_t\right|^p, \quad \mathbb{P}\text{-a.s.},$$

for all $0 \leq t \leq s$.

Proof. Since the assumption (6.90) is equivalent to

$$dD_t+\left\langle X_t-\hat{X}_t, dK_t-d\hat{K}_t\right\rangle+\left(\frac{1}{2}m_p+9p\lambda\right)\left|G_t-\hat{G}_t\right|^2 dt$$
$$\leq |X_t-\hat{X}_t|^2 d\left(V_t+A_t\right),$$

with

$$D_t=\int_0^t\left|X_r-\hat{X}_r\right|^2 dA_r,$$

the results clearly follow from Propositions 6.71 and 6.74 applied to the identity

$$X_t-\hat{X}_t=X_0-\hat{X}_0+\left(K_t-\hat{K}_t\right)+\int_0^t\left(G_s-\hat{G}_s\right)dB_s.$$

■

Since

$$\frac{1}{2}(r\wedge 1)\leq\frac{r}{\left(1+r^2\right)^{1/2}}\leq r\wedge 1, \quad \forall\, r\geq 0,$$

we have:

Corollary 6.78. *If the assumption (6.90) is satisfied with* $\lambda > 1$ *and* $p \geq 1$*, then there exists a positive constant* $C_{p,\lambda}$ *depending only on* (p,λ) *such that* $\mathbb{P}$*-a.s.*

$$\mathbb{E}^{\mathcal{F}_t}\left[1\wedge\left\|e^{-V}\left(X-\hat{X}\right)\right\|_{[t,s]}^{p}\right]\leq C_{p,\lambda}\left[1\wedge\left|e^{-V_t}\left(X_t-\hat{X}_t\right)\right|^p\right],$$

for all $0 \leq t \leq s$.

6.4.4 Backward Stochastic Inequalities

Let $\{B_t : t\geq 0\}$ be a k-dimensional Brownian motion with respect to a given stochastic basis $\left(\Omega,\mathcal{F},\mathbb{P},\{\mathcal{F}_t^B\}_{t\geq 0}\right)$, where $\mathcal{F}_t^B$ is the natural filtration associated to $\{B_t : t\geq 0\}$.

Notation 6.79. *For $p > 1$ define*

$$n_p \overset{def}{=} 1 \wedge (p-1).$$

In this subsection we shall derive some estimates on $(Y, Z) \in S_m^0 \times \Lambda_{m\times k}^0$ satisfying for all $T \geq 0$ and $t \in [0, T]$:

$$Y_t = Y_T + (K_T - K_t) - \int_t^T Z_s dB_s, \ \mathbb{P}\text{-a.s.}, \tag{6.91}$$

where $K \in S_m^0$ and $K_\cdot(\omega) \in BV_{loc}(\mathbb{R}_+; \mathbb{R}^m)$, $\mathbb{P}$-a.s. $\omega \in \Omega$.

We note that if the interval $[0, T]$ is fixed then the equality (6.91) will be extended to $\mathbb{R}_+$ by $Y_s = Y_T$, $K_s = K_T$ and $Z_s = 0$ for all $s > T$.

Proposition 6.80. *Let $(Y, Z) \in S_m^0 \times \Lambda_{m\times k}^0$ satisfy*

$$Y_t = Y_T + \int_t^T dK_s - \int_t^T Z_s dB_s, \ 0 \leq t \leq T, \quad \mathbb{P}\text{-a.s.},$$

where $K \in S_m^0$ and $K_\cdot(\omega) \in BV_{loc}(\mathbb{R}_+; \mathbb{R}^m)$, $\mathbb{P}$-a.s. $\omega \in \Omega$.
Assume given

- ▲ *three $\mathcal{P}$-m.i.c.s.p. D, R, N, $D_0 = R_0 = N_0 = 0$,*
- ▲ *a $\mathcal{P}$-m.b-v.c.s.p. V, $V_0 = 0$,*
- ▲ *two stopping times τ and σ such that $0 \leq \tau \leq \sigma < \infty$.*

(A) If $\lambda < 1$, $q > 0$ and

$$dD_t + \langle Y_t, dK_t\rangle \leq dR_t + |Y_t| dN_t + |Y_t|^2 dV_t + \frac{\lambda}{2}|Z_t|^2 dt,$$

then there exists a positive constant $C_{q,\lambda}$, depending only on (q, λ), such that

$$\begin{aligned} &\mathbb{E}^{\mathcal{F}_\tau}\left(\int_\tau^\sigma e^{2V_r} dD_r\right)^{q/2} + \mathbb{E}^{\mathcal{F}_\tau}\left(\int_\tau^\sigma e^{2V_r}|Z_r|^2 dr\right)^{q/2} \\ &\leq C_{q,\lambda}\mathbb{E}^{\mathcal{F}_\tau}\left[\sup_{s\in[\tau,\sigma]}\left|e^{V_s}Y_s\right|^q + \left(\int_\tau^\sigma e^{2V_s} dR_s\right)^{q/2} + \left(\int_\tau^\sigma e^{V_s} dN_s\right)^q\right], \\ &\qquad \mathbb{P}\text{-a.s.} \end{aligned} \tag{6.92}$$

(B) If $\lambda < 1 < p$,

$$\begin{aligned} &(i) \quad dD_t + \langle Y_t, dK_t\rangle \leq \left(\mathbf{1}_{p\geq 2} dR_t + |Y_t| dN_t + |Y_t|^2 dV_t\right) + \frac{n_p}{2}\lambda |Z_t|^2 dt, \\ &(ii) \quad \mathbb{E}\sup_{s\in[\tau,\sigma]} e^{pV_s}|Y_s|^p < \infty, \end{aligned} \tag{6.93}$$

then there exists a positive constant $C_{p,\lambda}$, depending only on (p,λ), such that $\mathbb{P}$-a.s.,

$$\begin{aligned}
&\mathbb{E}^{\mathcal{F}_\tau}\left(\sup_{s\in[\tau,\sigma]}\left|e^{V_s}Y_s\right|^p\right)+\mathbb{E}^{\mathcal{F}_\tau}\left(\int_\tau^\sigma e^{2V_s}dD_s\right)^{p/2} \\
&\qquad\qquad\qquad +\mathbb{E}^{\mathcal{F}_\tau}\left(\int_\tau^\sigma e^{2V_s}\left|Z_s\right|^2 ds\right)^{p/2} \\
&+\mathbb{E}^{\mathcal{F}_\tau}\int_\tau^\sigma e^{pV_s}\left|Y_s\right|^{p-2}\mathbf{1}_{Y_s\neq 0}dD_s+\mathbb{E}^{\mathcal{F}_\tau}\int_\tau^\sigma e^{pV_s}\left|Y_s\right|^{p-2}\mathbf{1}_{Y_s\neq 0}\left|Z_s\right|^2 ds \\
&\leq C_{p,\lambda}\,\mathbb{E}^{\mathcal{F}_\tau}\left[\left|e^{V_\sigma}Y_\sigma\right|^p+\left(\int_\tau^\sigma e^{2V_s}\mathbf{1}_{p\geq 2}dR_s\right)^{p/2}+\left(\int_\tau^\sigma e^{V_s}dN_s\right)^p\right].
\end{aligned} \tag{6.94}$$

Proof. Step I.

By the Itô formula, we have for all $0\leq t\leq s$:

$$\begin{aligned}
\left|Y_t\right|^2 e^{2V_t}+\int_t^s e^{2V_r}\left|Z_r\right|^2 dr &= \left|Y_s\right|^2 e^{2V_s}+2\int_t^s e^{2V_r}\left(\langle Y_r,dK_r\rangle-\left|Y_r\right|^2 dV_r\right) \\
&\quad -2\int_t^s e^{2V_r}\langle Y_r,Z_r dB_r\rangle, \quad a.s.
\end{aligned}$$

Since

$$\langle Y_r,dK_r\rangle-\left|Y_r\right|^2 dV_r\leq -dD_r+dR_r+\left|Y_r\right|dN_r+\frac{\lambda}{2}\left|Z_r\right|^2 dr,$$

we get

$$\begin{aligned}
&\left|Y_t\right|^2 e^{2V_t}+2\int_t^s e^{2V_r}dD_r+(1-\lambda)\int_t^s e^{2V_r}\left|Z_r\right|^2 dr \\
&\leq \left|Y_s\right|^2 e^{2V_s}+2\int_t^s e^{2V_r}dR_r+2\int_t^s e^{V_r}\left|Y_r\right|dN_r-2\int_t^s e^{2V_r}\langle Y_r,Z_r dB_r\rangle.
\end{aligned} \tag{6.95}$$

Let the stopping times $0\leq\tau\leq\sigma<\infty$ and

$$\begin{aligned}
\theta_n=\sigma\wedge\inf\Big\{s\geq\tau:\left\|e^V Y-e^{V_\tau}Y_\tau\right\|_s+\int_\tau^{s\vee\tau} e^{2V_r}dD_r+\textstyle\int_\tau^s e^{2V_r}\left|Z_r\right|^2 dr \\
+\int_\tau^{s\vee\tau} e^{2V_r}dR_r+\int_\tau^{s\vee\tau} e^{V_r}dN_r\geq n\Big\}.
\end{aligned}$$

We have $\tau\leq\theta_n\leq\sigma$ and $\theta_n\nearrow\sigma$ $\mathbb{P}$-a.s. Replacing in (6.95) t by τ and s by θ_n we obtain

$$
\begin{aligned}
&2\int_\tau^{\theta_n} e^{2V_r}dD_r + (1-\lambda)\int_\tau^{\theta_n} e^{2V_r}|Z_r|^2 dr\\
&\quad\le |Y_{\theta_n}|^2 e^{2V_{\theta_n}} + 2\int_\tau^{\theta_n} e^{2V_r}(dR_r + |Y_r|dN_r)\\
&\qquad\qquad - 2\int_\tau^{\theta_n} e^{2V_r}\langle Y_r, Z_r dB_r\rangle\\
&\quad\le |Y_{\theta_n}|^2 e^{2V_{\theta_n}} + \sup_{r\in[\tau,\sigma]}\mathbf{1}_{[\tau,\theta_n]}(r)\left|e^{V_r}Y_r\right|^2 + 2\int_\tau^{\theta_n} e^{2V_r}dR_r\\
&\qquad\qquad + \left(\int_\tau^{\theta_n} e^{V_r}dN_r\right)^2 - 2\int_\tau^{\sigma}\mathbf{1}_{[\tau,\theta_n]}(r)\, e^{2V_r}\langle Y_r, Z_r dB_r\rangle .
\end{aligned}
$$

Moreover, by Minkowski's inequality we infer for all $q>0$

$$
\begin{aligned}
&\mathbb{E}^{\mathcal{F}_\tau}\left(\int_\tau^{\theta_n} e^{2V_r}dD_r\right)^{q/2} + \mathbb{E}^{\mathcal{F}_\tau}\left(\int_\tau^{\theta_n} e^{2V_r}|Z_r|^2 dr\right)^{q/2}\\
&\le C_{q,\lambda}\mathbb{E}^{\mathcal{F}_\tau}\sup_{r\in[\tau,\sigma]}\left|e^{V_r}Y_r\right|^q + C_{q,\lambda}\mathbb{E}^{\mathcal{F}_\tau}\left(\int_\tau^{\sigma} e^{2V_r}dR_r\right)^{q/2}\\
&+C_{q,\lambda}\mathbb{E}^{\mathcal{F}_\tau}\left(\int_\tau^{\sigma} e^{V_r}dN_r\right)^{q} + C_{q,\lambda}\mathbb{E}^{\mathcal{F}_\tau}\left|\int_\tau^{\sigma}\mathbf{1}_{[\tau,\theta_n]}(r)\, e^{2V_r}\langle Y_r, Z_r dB_r\rangle\right|^{q/2}.
\end{aligned}
\tag{6.96}
$$

But by the Burkholder–Davis–Gundy and Cauchy–Schwarz inequalities, we get

$$
\begin{aligned}
&C_{q,\lambda}\mathbb{E}^{\mathcal{F}_\tau}\left|\int_\tau^{\sigma}\mathbf{1}_{[\tau,\theta_n]}(r)\, e^{2V_r}\langle Y_r, Z_r dB_r\rangle\right|^{q/2}\\
&\le C_{q,\lambda}\mathbb{E}^{\mathcal{F}_\tau}\left(\int_\tau^{\sigma}\mathbf{1}_{[\tau,\theta_n]}(r)\, e^{4V_r}|Y_r|^2|Z_r|^2 dr\right)^{q/4}\\
&\le C_{q,\lambda}\mathbb{E}^{\mathcal{F}_\tau}\left[\sup_{r\in[\tau,\sigma]}\left(\mathbf{1}_{[\tau,\theta_n]}(r)\left|e^{V_r}Y_r\right|^{q/2}\right)\left(\int_\tau^{\sigma}\mathbf{1}_{[\tau,\theta_n]}(r)\, e^{2V_r}|Z_r|^2 dr\right)^{q/4}\right]\\
&\le C'_{q,\lambda}\mathbb{E}^{\mathcal{F}_\tau}\sup_{r\in[\tau,\theta_n]}\left|e^{V_r}Y_r\right|^q + \frac{1}{2}\mathbb{E}^{\mathcal{F}_\tau}\left(\int_\tau^{\theta_n} e^{2V_r}|Z_r|^2 dr\right)^{q/2}.
\end{aligned}
$$

Since $\int_\tau^{\theta_n} e^{2V_r}|Z_r|^2 dr$ is finite, from (6.96) we infer

$$\begin{aligned}&\mathbb{E}^{\mathcal{F}_\tau}\left(\int_\tau^{\theta_n} e^{2V_r}dD_r\right)^{q/2}+\frac{1}{2}\mathbb{E}^{\mathcal{F}_\tau}\left(\int_\tau^{\theta_n} e^{2V_r}\left|Z_r\right|^2dr\right)^{q/2}\\&\le C_{q,\lambda}\left[\mathbb{E}^{\mathcal{F}_\tau}\sup_{r\in[\tau,\sigma]}\left|e^{V_r}Y_r\right|^q+\mathbb{E}^{\mathcal{F}_\tau}\left(\int_\tau^{\sigma} e^{2V_r}dR_r\right)^{q/2}+\mathbb{E}^{\mathcal{F}_\tau}\left(\int_\tau^{\sigma} e^{V_r}dN_r\right)^{q}\right].\end{aligned}\tag{6.97}$$

By the monotone convergence theorem as $n\to\infty$ the inequality (6.92) follows. *Step II.* Let us first assume that $p\ge 1$.

Noting that

$$e^{V_t}Y_t=Y_0-\int_0^t e^{V_r}\left(dK_r-Y_rdV_r\right)+\int_0^t e^{V_r}Z_rdB_r,$$

then by the inequality (2.30) from Corollary (2.30) we get, for $p\ge 1$ and for all stopping times $t\in[\tau,\sigma]$

$$\begin{aligned}&e^{pV_t}\left|Y_t\right|^p+\frac{p}{2}n_p\int_t^{\theta_n} e^{pV_r}\left|Y_r\right|^{p-2}\mathbf{1}_{Y_r\neq 0}\left|Z_r\right|^2dr\le e^{pV_{\theta_n}}\left|Y_{\theta_n}\right|^p\\&\quad+p\int_t^{\theta_n} e^{pV_r}\left|Y_r\right|^{p-2}\mathbf{1}_{Y_r\neq 0}\Big(\langle Y_r,dK_r\rangle-\left|Y_r\right|^2dV_r\Big)\\&\quad-p\int_t^{t\vee\theta_n} e^{pV_r}\left|Y_r\right|^{p-2}\mathbf{1}_{Y_r\neq 0}\langle Y_r,Z_rdB_r\rangle.\end{aligned}\tag{6.98}$$

We note that the right-hand side of (6.98) is finite $\mathbb{P}$-$a.s.$ and consequently

$$0\le n_p\int_\tau^{\theta_n} e^{pV_r}\left|Y_r\right|^{p-2}\mathbf{1}_{Y_r\neq 0}\left|Z_r\right|^2dr<\infty,\quad \mathbb{P}\text{-a.s.}$$

By the assumption (6.93)

$$\langle Y_r,dK_r\rangle-\left|Y_r\right|^2dV_r\le -dD_r+\big(\mathbf{1}_{p\ge 2}dR_r+\left|Y_r\right|dN_r\big)+\frac{n_p}{2}\lambda\left|Z_r\right|^2dr.$$

It follows that

$$\begin{aligned}&e^{pV_t}\left|Y_t\right|^p+p\int_t^{\sigma}\mathbf{1}_{[\tau,\theta_n]}(r)\,e^{pV_r}\left|Y_r\right|^{p-2}\mathbf{1}_{Y_r\neq 0}dD_r\\&\quad+\frac{p}{2}n_p(1-\lambda)\int_t^{\sigma}\mathbf{1}_{[\tau,\theta_n]}(r)\,e^{pV_r}\left|Y_r\right|^{p-2}\mathbf{1}_{Y_r\neq 0}\left|Z_r\right|^2dr\\&\quad\le e^{pV_{\theta_n}}\left|Y_{\theta_n}\right|^p+\left(U_{\theta_n}-U_t\right)-\left(M_{\theta_n}-M_t\right),\end{aligned}\tag{6.99}$$

where

$$U_s=p\int_0^s\mathbf{1}_{[\tau,\theta_n]}(r)\,e^{pV_r}\left|Y_r\right|^{p-2}\mathbf{1}_{Y_r\neq 0}\big(\mathbf{1}_{p\ge 2}dR_r+\left|Y_r\right|dN_r\big)\tag{6.100}$$

and

$$M_s = p\int_0^s \mathbf{1}_{[\tau,\theta_n]}(r)\, e^{pV_r}\,|Y_r|^{p-2}\,\mathbf{1}_{Y_r\neq 0}\,\langle Y_r, Z_r dB_r\rangle\,.$$

Note that $\{M_s : s\in[0,T]\}$ is a martingale since

$$\begin{aligned}\mathbb{E}\sqrt{<M>_T} &\leq p\mathbb{E}\,\left(\int_\tau^{\theta_n} e^{2pV_r}\,|Y_r|^{2p-4}\,\mathbf{1}_{Y_r\neq 0}\,|Y_r|^2\,|Z_r|^2\,dr\right)^{1/2}\\ &\leq p\mathbb{E}\left[\left(\left|e^{V_\tau}Y_\tau\right|+n\right)^{p-1}\left(\int_\tau^{\theta_n} e^{2V_r}\,|Z_r|^2\,dr\right)^{1/2}\right]\\ &\leq C_p\left(\mathbb{E}\left|e^{V_\tau}Y_\tau\right|^{p-1}+n^{p-1}\right)\sqrt{n}.\end{aligned}$$

Therefore from (6.99),

$$e^{pV_\tau}\,|Y_\tau|^p \leq \mathbb{E}^{\mathcal{F}_\tau}e^{pV_{\theta_n}}\,|Y_{\theta_n}|^p + \mathbb{E}^{\mathcal{F}_\tau}\left(U_{\theta_n}-U_\tau\right). \tag{6.101}$$

From here we assume that $p>1$. From (6.99) we also get

$$\begin{aligned}&p\mathbb{E}^{\mathcal{F}_\tau}\int_\tau^{\theta_n} e^{pV_r}\,|Y_r|^{p-2}\,\mathbf{1}_{Y_r\neq 0}dD_r\\ &\qquad+\frac{p}{2}n_p\,(1-\lambda)\,\mathbb{E}^{\mathcal{F}_\tau}\int_\tau^{\theta_n} e^{pV_r}\,|Y_r|^{p-2}\,\mathbf{1}_{Y_r\neq 0}\,|Z_r|^2\,dr\\ &\leq \mathbb{E}^{\mathcal{F}_\tau}\,e^{pV_{\theta_n}}\,|Y_{\theta_n}|^p+\mathbb{E}^{\mathcal{F}_\tau}\,U_{\theta_n}.\end{aligned} \tag{6.102}$$

Since

$$\sup_{t\in[\tau,\sigma]}|M_\sigma - M_t| \leq 2\sup_{t\in[\tau,\sigma]}|M_t - M_\tau| = 2\sup_{t\in[\tau,\sigma]}|M_t|\,,$$

we obtain from (6.99) that

$$\begin{aligned}&\mathbb{E}^{\mathcal{F}_\tau}\sup_{t\in[\tau,\theta_n]}\left(e^{pV_t}\,|Y_t|^p\right)\\ &\quad\leq \mathbb{E}^{\mathcal{F}_\tau}\,e^{pV_{\theta_n}}\,|Y_{\theta_n}|^p+\mathbb{E}^{\mathcal{F}_\tau}\left(U_{\theta_n}-U_\tau\right)+2\mathbb{E}^{\mathcal{F}_\tau}\sup_{t\in[\tau,\theta_n]}|M_t|\,.\end{aligned} \tag{6.103}$$

By the Burkholder–Davis–Gundy inequality (2.8) and (6.102):

$$\begin{aligned}&\mathbb{E}^{\mathcal{F}_\tau}\sup_{t\in[\tau,\sigma]}|M_t|\\ &\leq 3p\,\mathbb{E}^{\mathcal{F}_\tau}\left(\int_\tau^\sigma \mathbf{1}_{[\tau,\theta_n]}(r)\,e^{2pV_r}\,|Y_r|^{2p-4}\,\mathbf{1}_{Y_r\neq 0}\,|Y_r|^2\,|Z_r|^2\,dr\right)^{1/2}\end{aligned}$$

$$\leq 3p\,\mathbb{E}^{\mathcal{F}_\tau}\left[\sup_{r\in[\tau,\theta_n]} e^{(p/2)V_r}\,|Y_r|^{p/2}\left(\int_\tau^{\theta_n} e^{pV_r}\,|Y_r|^{p-2}\,\mathbf{1}_{Y_r\neq 0}\,|Z_r|^2\,dr\right)^{1/2}\right]$$

$$\leq \frac{1}{4}\mathbb{E}^{\mathcal{F}_\tau}\sup_{r\in[\tau,\theta_n]}\left(e^{pV_r}\,|Y_r|^p\right)+C_p\,\mathbb{E}^{\mathcal{F}_\tau}\int_\tau^{\theta_n} e^{pV_r}\,|Y_r|^{p-2}\,\mathbf{1}_{Y_r\neq 0}\,|Z_r|^2\,dr$$

$$\leq \frac{1}{4}\mathbb{E}^{\mathcal{F}_\tau}\sup_{r\in[\tau,\theta_n]} e^{pV_r}\,|Y_r|^p+C_{p,\lambda}\mathbb{E}^{\mathcal{F}_\tau}\,e^{pV_{\theta_n}}\,|Y_{\theta_n}|^p+C_{p,\lambda}\mathbb{E}^{\mathcal{F}_\tau}\,\left(U_{\theta_n}-U_\tau\right).$$

Plugging this last estimate into (6.103) we obtain with another constant $C_{p,\lambda}$

$$\mathbb{E}^{\mathcal{F}_\tau}\sup_{r\in[\tau,\theta_n]} e^{pV_r}\,|Y_r|^p\leq C_{p,\lambda}\mathbb{E}^{\mathcal{F}_\tau}\,e^{pV_{\theta_n}}\,|Y_{\theta_n}|^p+C_{p,\lambda}\mathbb{E}^{\mathcal{F}_\tau}\,\left(U_{\theta_n}-U_\tau\right). \quad (6.104)$$

We deduce from (6.102) and (6.104)

$$\begin{aligned}\mathbb{E}^{\mathcal{F}_\tau}\sup_{r\in[\tau,\theta_n]} e^{pV_r}\,|Y_r|^p+\mathbb{E}^{\mathcal{F}_\tau}\int_\tau^{\theta_n} e^{pV_r}\,|Y_r|^{p-2}\,\mathbf{1}_{Y_r\neq 0}dD_r&\\ +\mathbb{E}^{\mathcal{F}_\tau}\int_\tau^{\theta_n} e^{pV_r}\,|Y_r|^{p-2}\,\mathbf{1}_{Y_r\neq 0}\,|Z_r|^2\,dr&\\ \leq C_{p,\lambda}\mathbb{E}^{\mathcal{F}_\tau}\,e^{pV_{\theta_n}}\,|Y_{\theta_n}|^p+C_{p,\lambda}\mathbb{E}^{\mathcal{F}_\tau}\,U_{\theta_n}.&\end{aligned}$$

But

$$\begin{aligned}&C_{p,\lambda}\mathbb{E}^{\mathcal{F}_\tau}\left(U_{\theta_n}-U_\tau\right)\\ &\leq C_{p,\lambda}\mathbb{E}^{\mathcal{F}_\tau}\left[\sup_{r\in[\tau,\theta_n]}\left[e^{(p-2)V_r}\,|Y_r|^{p-2}\,\mathbf{1}_{Y_r\neq 0}\mathbf{1}_{p\geq 2}\right]\int_\tau^{\theta_n} e^{2V_r}\mathbf{1}_{p\geq 2}dR_r\right]\\ &\qquad+\mathbb{E}^{\mathcal{F}_\tau}\left[\sup_{r\in[\tau,\theta_n]}\left[e^{(p-1)V_r}\,|Y_r|^{p-1}\,\mathbf{1}_{Y_r\neq 0}\right]\int_\tau^{\theta_n} e^{V_r}\,dN_r\right]\\ &\qquad\leq \frac{1}{2}\mathbb{E}^{\mathcal{F}_\tau}\sup_{r\in[\tau,\theta_n]} e^{pV_r}\,|Y_r|^p+C'_{p,\lambda}\mathbb{E}^{\mathcal{F}_\tau}\left(\int_\tau^{\sigma} e^{2V_r}\mathbf{1}_{p\geq 2}dR_r\right)^{p/2}\\ &\qquad\qquad+C'_{p,\lambda}\mathbb{E}^{\mathcal{F}_\tau}\left(\int_\tau^{\sigma} e^{V_r}\,dN_r\right)^p.\end{aligned}$$

Hence

$$\begin{aligned}&\mathbb{E}^{\mathcal{F}_\tau}\sup_{r\in[\tau,\theta_n]} e^{pV_r}\,|Y_r|^p+\mathbb{E}^{\mathcal{F}_\tau}\int_\tau^{\theta_n} e^{pV_r}\,|Y_r|^{p-2}\,\mathbf{1}_{Y_r\neq 0}dD_r\\ &\quad+\mathbb{E}^{\mathcal{F}_\tau}\int_\tau^{\theta_n} e^{pV_r}\,|Y_r|^{p-2}\,\mathbf{1}_{Y_r\neq 0}\,|Z_r|^2\,dr\end{aligned}$$

$$\leq C_{p,\lambda}\mathbb{E}^{\mathcal{F}_\tau}\, e^{pV_{\theta_n}}\left|Y_{\theta_n}\right|^p + C_{p,\lambda}\mathbb{E}^{\mathcal{F}_\tau}\left(\int_\tau^\sigma e^{2V_r}\mathbf{1}_{p\geq 2}dR_r\right)^{p/2}$$

$$+ C_{p,\lambda}\mathbb{E}^{\mathcal{F}_\tau}\left(\int_\tau^\sigma e^{V_r}dN_r\right)^p .$$

Now letting $n \to \infty$, by the Beppo Levi monotone convergence theorem for the first member and by the Lebesgue dominated convergence theorem for the right-hand side of the inequality, we conclude (6.94) (using of course the first step: inequality (6.92)).

The proof is complete. ■

Corollary 6.81. *Let* $(Y, Z) \in S_m^0 \times \Lambda_{m\times k}^0$ *satisfy*

$$Y_t = Y_T + \int_t^T dK_s - \int_t^T Z_s dB_s,\ 0 \leq t \leq T, \quad \mathbb{P}\text{-a.s.},$$

where $K \in S_m^0$ *and* $K_\cdot(\omega) \in BV_{loc}(\mathbb{R}_+;\mathbb{R}^m)$, $\mathbb{P}$*-a.s.* $\omega \in \Omega$.
Assume given

- ▲ D *and* N *are* $\mathcal{P}$*-m.i.c.s.p.,* $N_0 = 0$,
- ▲ V *a* $\mathcal{P}$*-m.b-v.c.s.p.,* $V_0 = 0$,
- ▲ τ, θ *and* σ *are three stopping times such that* $0 \leq \tau \leq \theta \leq \sigma < \infty$.

If

$$\begin{array}{ll} (a) & dD_t + \langle Y_t, dK_t\rangle \leq |Y_t|dN_t + |Y_t|^2 dV_t, \\ (b) & \mathbb{E}\sup\limits_{s\in[\tau,\sigma]}\left|e^{V_s}Y_s\right| < \infty, \end{array}$$

then

$$e^{V_\tau}\left|Y_\tau\right| \leq \mathbb{E}^{\mathcal{F}_\tau}e^{V_\sigma}\left|Y_\sigma\right| + \mathbb{E}^{\mathcal{F}_\tau}\int_\tau^\sigma e^{V_r}dN_r \tag{6.105}$$

and for all $0 < \alpha < 1$

$$\begin{aligned} \sup_{\theta\in[\tau,\sigma]}\left[\mathbb{E}\left(e^{V_\theta}\left|Y_\theta\right|\right)\right]^\alpha + \mathbb{E}\left(\sup_{s\in[\tau,\sigma]}\left|e^{V_s}Y_s\right|^\alpha\right) + \mathbb{E}\left(\int_\tau^\sigma e^{2V_r}\left|Z_r\right|^2 dr\right)^{\alpha/2} \\ +\mathbb{E}\left(\int_\tau^\sigma e^{2V_r}\left|D_r\right|^2 dr\right)^{\alpha/2} \\ \leq C_\alpha\left[\left(\mathbb{E}\left(e^{V_\sigma}\left|Y_\sigma\right|\right)\right)^\alpha + \left(\mathbb{E}\int_\tau^\sigma e^{V_r}dN_r\right)^\alpha\right]. \end{aligned} \tag{6.106}$$

Proof. From (6.101) for $p = 1$ we deduce, using the definition (6.100) of U_s, that

$$\begin{aligned} e^{V_\tau} |Y_\tau| &\leq \mathbb{E}^{\mathcal{F}_\tau} e^{V_{\theta_n}} |Y_{\theta_n}| + \mathbb{E}^{\mathcal{F}_\tau} U_{\theta_n} \\ &\leq \mathbb{E}^{\mathcal{F}_\tau} e^{V_{\theta_n}} |Y_{\theta_n}| + \mathbb{E}^{\mathcal{F}_\tau} \int_\tau^\sigma e^{V_r} dN_r \end{aligned}$$

and the inequality (6.105) follows as $n \to \infty$. Moreover

$$\sup_{\theta \in [\tau,\sigma]} \mathbb{E}\left(e^{V_\theta} |Y_\theta|\right) \leq \mathbb{E}\left(e^{V_\sigma} |Y_\sigma|\right) + \mathbb{E}\int_\tau^\sigma e^{V_r} dN_r$$

and by the martingale inequality (1.11-A_3) from Theorem 1.60 we infer

$$\mathbb{E}\left(\sup_{s \in [\tau,\sigma]} \left|e^{V_s} Y_s\right|^\alpha\right) \leq \frac{1}{1-\alpha}\left[\mathbb{E}\left(e^{V_\sigma} |Y_\sigma| + \int_\tau^\sigma e^{V_r} dN_r\right)\right]^\alpha.$$

The inequality (6.106) is now a consequence of (6.92). ∎

Corollary 6.82. *Let* $(Y, Z) \in S_m^0 \times \Lambda_{m \times k}^0$ *satisfy*

$$Y_t = Y_T + \int_t^T dK_s - \int_t^T Z_s dB_s, \ 0 \leq t \leq T, \quad \mathbb{P}\text{-a.s.},$$

where $K \in S_m^0$ *and* $K_\cdot(\omega) \in BV_{loc}\left(\mathbb{R}_+; \mathbb{R}^m\right)$, $\mathbb{P}$*-a.s.* $\omega \in \Omega$.
Assume given

- ▲ *a* $\mathcal{P}$*-m.b-v.c.s.p.* V, $V_0 = 0$,
- ▲ τ, θ *and* σ *are three stopping times such that* $0 \leq \tau \leq \theta \leq \sigma < \infty$.

If $\lambda < 1 \leq p$, $n_p = 1 \wedge (p-1)$ *and*

$$\begin{aligned} &(a) \qquad \langle Y_t, dK_t \rangle \leq |Y_t|^2 dV_t + \frac{n_p}{2}\lambda |Z_t|^2 dt, \\ &(b) \qquad \mathbb{E} \sup_{s \in [\tau,\sigma]} e^{pV_s} |Y_s|^p < \infty, \end{aligned}$$

then for all $1 \leq q \leq p$,

$$e^{qV_\tau} |Y_\tau|^q \leq \mathbb{E}^{\mathcal{F}_\tau} e^{qV_\sigma} |Y_\sigma|^q, \quad \mathbb{P}\text{-a.s.} \tag{6.107}$$

If $p > 1$ *then*

$$\mathbb{E} \sup_{s \in [\tau,\sigma]} e^{pV_s} |Y_s|^p + \mathbb{E}\left(\int_\tau^\sigma e^{2V_r} |Z_r|^2 dr\right)^{p/2} \leq C_{p,\lambda} \mathbb{E}\left(e^{pV_\sigma} |Y_\sigma|^p\right),$$

and if $p = 1$ *(and* $n_p = 0$*) then for all* $0 < \alpha < 1$,

$$\sup_{\theta\in[\tau,\sigma]} \left(\mathbb{E}e^{V_\theta} |Y_\theta|\right)^{\alpha} + \mathbb{E} \sup_{s\in[\tau,\sigma]} e^{\alpha V_s} |Y_s|^{\alpha} + \mathbb{E} \left(\int_\tau^\sigma e^{2V_r} |Z_r|^2 dr\right)^{\alpha/2} \leq C_\alpha \left(\mathbb{E}\, e^{V_\sigma} |Y_\sigma|\right)^{\alpha}. \tag{6.108}$$

Proof. Since

$$\mathbb{E} \sup_{s\in[\tau,\sigma]} e^{qV_s} |Y_s|^q \leq \left(\mathbb{E} \sup_{s\in[\tau,\sigma]} e^{pV_s} |Y_s|^p\right)^{q/p} < \infty,$$

the inequality (6.101) with p replaced by q yields (6.107). The next two inequalities follow from Proposition 6.80 and Corollary 6.81, respectively. ■

Corollary 6.83. *Let* $p \geq 1$ *and* $(V_t)_{t\geq 0}$ *be a bounded variation continuous progressively measurable stochastic process with* $V_0 = 0$*. Let* $T > 0$ *and* $\eta : \Omega \to \mathbb{R}^m$ *be a random variable such that* $\mathbb{E}\left(\sup_{r\in[0,T]} e^{pV_r} |\eta|^p\right) < \infty$*. If* $(\xi, \zeta) \in S_m^p [0, T] \times \Lambda_{m\times k}^p [0, T]$ *satisfies*

$$\xi_s = \mathbb{E}^{\mathcal{F}_T} \eta - \int_s^T \zeta_r dB_r, \ s \in [0, T], \ a.s.$$

(the pair (ξ, ζ) *exists and it is unique by the martingale representation: Corollary 2.44), then there exists a* $C = C(p) > 0$ *such that for all* $t \in [0, T]$*, for* $p > 1$

$$\mathbb{E}^{\mathcal{F}_t} \sup_{s\in[t,T]} e^{pV_s} |\xi_s|^p + \mathbb{E}^{\mathcal{F}_t} \left(\int_t^T e^{2V_s} |\zeta_s|^2 ds\right)^{p/2} \leq C_p\, \mathbb{E}^{\mathcal{F}_t} \left(\sup_{r\in[0,T]} e^{pV_r} |\eta|^p\right) \tag{6.109}$$

and for $p = 1$

$$\sup_{s\in[0,T]} \left(\mathbb{E}e^{V_s} |\xi_s|\right)^{\alpha} + \mathbb{E} \sup\nolimits_{t\in[0,T]} \left|e^{V_t} \xi_t\right|^{\alpha} + \mathbb{E} \left(\int_0^T e^{2V_s} |\zeta_s|^2 ds\right)^{\alpha/2} \leq C_\alpha \left(\mathbb{E}\left(\sup\nolimits_{t\in[0,T]} e^{V_r} |\eta|\right)\right)^{\alpha}, \quad \text{for all } 0 < \alpha < 1. \tag{6.110}$$

Proof. We see at once that the stochastic pair (ξ, ζ) satisfy the equation

$$\xi_t = \xi_T - \int_t^T \zeta_s dB_s, \ t \in [0, T], \ a.s.$$

The stochastic process $\tilde{V}_t = \sup_{s\in[0,t]} V_s$; $\tilde{V}$ is increasing continuous progressively measurable and $\tilde{V}_0 = 0$. Since for all $t \in [0,T]$

$$\mathbb{E}^{\mathcal{F}_t}\left|e^{\tilde{V}_t}\xi_t\right| = \left|e^{\tilde{V}_t}\xi_t\right| = e^{\tilde{V}_t}\left|\mathbb{E}^{\mathcal{F}_t}\eta\right| \leq \mathbb{E}^{\mathcal{F}_t}e^{\tilde{V}_t}\left|\eta\right| \leq \mathbb{E}^{\mathcal{F}_t}e^{\tilde{V}_T}\left|\eta\right| \tag{6.111}$$

by Proposition 1.56 we infer for all $p > 1$

$$\mathbb{E}\left\|\xi e^{\tilde{V}}\right\|_{[0,T]}^{p} \leq \left(\frac{p}{p-1}\right)^{p}\mathbb{E}\left(e^{p\tilde{V}_T}\left|\eta\right|^{p}\right) < \infty$$

and consequently by Proposition 6.80-B (for $(Y,Z) = (\xi,\zeta)$ with $\lambda = 0$, $K = R = N = 0$, $dD_t = |\xi_t|^2\, d\tilde{V}$) the inequality (6.109) follows; we also use that $V \leq \tilde{V}$ and

$$\mathbb{E}^{\mathcal{F}_t}\left|e^{\tilde{V}_T}\xi_T\right|^{p} = \mathbb{E}^{\mathcal{F}_t}\left[\left|e^{\tilde{V}_T}\mathbb{E}^{\mathcal{F}_T}\eta\right|^{p}\right] \leq \mathbb{E}^{\mathcal{F}_t}\left|e^{\tilde{V}_T}\eta\right|^{p}.$$

In the case $p = 1$ we have for all $0 < \alpha < 1$, by Proposition 1.56

$$\mathbb{E}\sup_{t\in[0,T]} e^{\alpha V_t}\left|\xi_t\right|^{\alpha} \leq \mathbb{E}\sup_{t\in[0,T]}\left|e^{\tilde{V}_t}\xi_t\right|^{\alpha} \leq \frac{1}{1-\alpha}\left(\mathbb{E}e^{\tilde{V}_T}\left|\eta\right|\right)^{\alpha} \tag{6.112}$$

and by Proposition 6.80-A

$$\mathbb{E}\left(\int_0^T e^{2\tilde{V}_s}\left|\zeta_s\right|^2 ds\right)^{\alpha/2} \leq C_1\mathbb{E}\sup_{t\in[0,T]}\left|e^{\tilde{V}_t}\xi_t\right|^{\alpha}. \tag{6.113}$$

Also we can see that from (6.111) $\mathbb{E}\left|e^{\tilde{V}_t}\xi_t\right| \leq \mathbb{E}\left(e^{\tilde{V}_T}\left|\eta\right|\right)$ and therefore

$$\sup_{t\in[0,T]}\mathbb{E}\left|e^{\tilde{V}_t}\xi_t\right| \leq \mathbb{E}\left(e^{\tilde{V}_T}\left|\eta\right|\right). \tag{6.114}$$

From (6.112)–(6.114) the inequality (6.110) follows. ■

Let $(Y,Z), (\hat{Y},\hat{Z}) \in S_m^0[0,T] \times \Lambda_{m\times k}^0(0,T)$ satisfying for all $t \in [0,T]$:

$$Y_t = Y_T + \int_t^T dK_s - \int_t^T Z_s dB_s, \quad \mathbb{P}\text{-a.s.},$$

and respectively

$$\hat{Y}_t = \hat{Y}_T + \int_t^T d\hat{K}_s - \int_t^T \hat{Z}_s dB_s, \quad \mathbb{P}\text{-a.s.},$$

where

◊ $K, \hat{K} \in S_m^0$,
◊ $K_\cdot(\omega), \hat{K}_\cdot(\omega) \in BV_{loc}([0,\infty[;\mathbb{R}^m)$, $\mathbb{P}$-a.s. $\omega \in \Omega$.

Assume there exist $\lambda < 1 \le p$ and V a $\mathcal{P}$-m.b-v.c.s.p., $V_0 = 0$, such that as signed measures on $[0,T]$:

$$\langle Y_t - \hat{Y}_t, dK_t - d\hat{K}_t\rangle \le |Y_t - \hat{Y}_t|^2 dV_t + \frac{n_p}{2}\lambda \left|Z_t - \hat{Z}_t\right|^2 dt, \tag{6.115}$$

where $n_p = 1 \wedge (p-1)$.

Corollary 6.84. *Let $\lambda < 1 \le p$ be given. Let the assumption (6.115) be satisfied and $\{A_t : t \ge 0\}$ be a $\mathcal{P}$-m.i.c.s.p., $A_0 = 0$, such that*

$$\mathbb{E} \sup_{t\in[0,T]} \left(e^{p(A_t+V_t)}\left|Y_t - \hat{Y}_t\right|^p\right) < \infty.$$

Then for all $0 \le t \le T$,

$$e^{pV_t}\left|Y_t - \hat{Y}_t\right|^p \le \mathbb{E}^{\mathcal{F}_t}\left(e^{pV_T}\left|Y_T - \hat{Y}_T\right|^p\right), \quad \mathbb{P}\text{-a.s.}$$

Moreover if $p > 1$, then

$$\begin{aligned}
&\mathbb{E}^{\mathcal{F}_t}\left(\sup_{s\in[t,T]} e^{p(A_s+V_s)}\left|Y_s - \hat{Y}_s\right|^p\right) + \mathbb{E}^{\mathcal{F}_t}\int_t^T e^{p(A_s+V_s)}\left|Y_s - \hat{Y}_s\right|^p dA_s \\
&+\mathbb{E}^{\mathcal{F}_t}\left(\int_t^T e^{2(A_s+V_s)}\left|Y_s - \hat{Y}_s\right|^2 dA_s\right)^{p/2} + \mathbb{E}^{\mathcal{F}_t}\left(\int_t^T e^{2(A_s+V_s)}\left|Z_s - \hat{Z}_s\right|^2 ds\right)^{p/2} \\
&\qquad\le C_{p,\lambda}\mathbb{E}^{\mathcal{F}_t} e^{p(A_T+V_T)}\left|Y_T - \hat{Y}_T\right|^p, \quad \mathbb{P}\text{-a.s.},
\end{aligned}$$

where $C_{p,\lambda}$ is a positive constant depending only on (p,λ)..

Proof. The results clearly follow from Corollary 6.82 and the inequality (6.94) from Proposition 6.80, applied to

$$Y_t - \hat{Y}_t = Y_T - \hat{Y}_T + \int_t^T d\left(K_s - \hat{K}_s\right) - \int_t^T \left(Z_s - \hat{Z}_s\right) dB_s,$$

satisfying

$$dD_t + \langle Y_t - \hat{Y}_t, dK_t - d\hat{K}_t\rangle \le |Y_t - \hat{Y}_t|^2 d\left(A_t + V_t\right) + \frac{n_p}{2}\lambda\left|Z_t - \hat{Z}_t\right|^2 dt$$

with

$$dD_t = |Y_t - \hat{Y}_t|^2 dA_t.$$

■

6.5 Annex D: Viscosity Solutions

The aim of this section is to introduce the notion of viscosity solutions to second order elliptic and parabolic PDEs and give uniqueness results for such solutions. This notion, which was invented by Crandall and Lions, allows us to state that a continuous function satisfies a PDE, without any differentiability requirement on that function. This notion has been invented specifically for nonlinear equations, for which the notion of weak solutions in the sense of distributions is not convenient. We use this notion here for linear and semilinear equations.

This section is divided into four parts. In the first part, we state the main definitions of viscosity solutions to elliptic and parabolic PDEs (or systems of PDEs). We prove three uniqueness results in the next three parts. We do not prove any existence results, since such results for the equations considered in this book are provided by our probabilistic formulas. Concerning uniqueness, it would be too long and repetitive to give a uniqueness result for each PDE considered in this book. The last three parts of this section give uniqueness results, corresponding to three large classes of semilinear PDEs or systems of PDEs. All other relevant results can be proved by combining the arguments given in those three proofs.

The first uniqueness result concerns an elliptic PDE with Dirichlet boundary condition at the boundary of a bounded set. We shall also explain how the proof can be adapted to the parabolic case. The second result treats the case of a system of parabolic PDEs in the whole space. Finally the third result concerns a parabolic PDE with subdifferential operators and nonlinear Neumann boundary condition.

We refer to the well-known "user's guide" of Crandall et al. [18] for more details, which complements the material presented here.

6.5.1 Definitions

Let $\mathcal{O}$ be a locally closed subset of $\mathbb{R}^d$, that is for all $x \in \mathcal{O}$ there exists a $\delta > 0$ such that $\mathcal{O} \cap \overline{B}(x, \delta)$ is closed.

A function $h : \mathcal{O} \subset \mathbb{R}^d \to \mathbb{R}$ is lower semicontinuous and we write $h \in LSC(\mathcal{O})$ if there exist $\{h_n,\ n \geq 1\} \subset C(\mathcal{O})$ such that

$$h_1(x) \leq \cdots \leq h_n(x) \leq \cdots \leq h(x) \quad \text{and} \quad \lim_{n\to\infty} h_n(x) = h(x),\ \forall\, x \in \mathcal{O}.$$

The function $h : \mathcal{O} \subset \mathbb{R}^d \to \mathbb{R}$ is upper semicontinuous and we write $h \in USC(\mathcal{O})$ if $-h$ is lower semicontinuous.

In particular for all $R > 0$ we have

$$\begin{aligned} &(i) \quad \inf_{x\in\mathcal{O},\, |x|\leq R} h(x) > -\infty, \quad \text{if } h \in LSC(\mathcal{O}), \\ &(ii) \quad \sup_{x\in\mathcal{O},\, |x|\leq R} h(x) < \infty, \quad \text{if } h \in USC(\mathcal{O}). \end{aligned}$$

6.5.1.1 Elliptic PDE

Consider the PDE

$$\Phi(x, u(x), Du(x), D^2u(x)) = 0, \quad x \in \mathcal{O}, \tag{6.116}$$

where

$$\Phi : \mathcal{O} \times \mathbb{R} \times \mathbb{R}^d \times \mathbb{S}^d \to \mathbb{R},$$

and $\mathbb{S}^d$ denotes the set of symmetric $d \times d$ matrices.

Definition 6.85. (i) $u \in USC(\mathcal{O})$ is a *viscosity sub-solution* of (6.116) if for any $\varphi \in C^2(\mathcal{O})$ and $\hat{x} \in \mathcal{O}$ a local maximum of $u - \varphi$:

$$\Phi(\hat{x}, u(\hat{x}), D\varphi(\hat{x}), D^2\varphi(\hat{x})) \leq 0.$$

(ii) $u \in LSC(\mathcal{O})$ is a *viscosity super-solution* of (6.116) if for any $\varphi \in C^2(\mathcal{O})$ and $\hat{x} \in \mathcal{O}$ a local minimum of $u - \varphi$:

$$\Phi(\hat{x}, u(\hat{x}), D\varphi(\hat{x}), D^2\varphi(\hat{x})) \geq 0.$$

(iii) $u \in C(\mathcal{O})$ is *viscosity solution* if it is both a viscosity sub- and super-solution.

In these definitions we can also assume that $u(\hat{x}) = \varphi(\hat{x})$ since we can translate φ.

Note that the class of PDEs for which probabilistic formulas are given in this book is the class of semilinear equations, where the function Φ has the following particular form

$$\Phi(x, r, p, X) = -\frac{1}{2}\mathbf{Tr}\left[g(x)g^*(x)X\right] - \langle f(x), p\rangle - F(x, r, p). \tag{6.117}$$

In the Definition 6.85 we can replace local maximum (minimum) by strict global maximum (minimum).

Remark 6.86. Let $\mathcal{O}$ be an open subset of $\mathbb{R}^d$ and $u \in C^2(\mathcal{O})$.

(i) If u is a *viscosity solution of* (6.116), then u is a classical solution.
(ii) If u is a classical solution *of* (6.116) and Φ satisfies the degenerate ellipticity condition

$$X \leq Y \Rightarrow \Phi(x, r, p, X) \geq \Phi(x, r, p, Y), \ \forall\, x, r, p,$$

then u is a *viscosity solution.*

Definition 6.87. A function $u \in USC(\mathcal{O})$ satisfies the maximum principle if for all $\varphi \in C^2(\mathcal{O})$ and all open subsets $D \subseteq \mathcal{O}$ the inequality

$$\Phi(x, \varphi(x), D\varphi(x), D^2\varphi(x)) > 0, \ \forall x \in D$$

implies that at every $\hat{x} \in D$ which is a local maximum of $u - \varphi$:

$$u\,(\hat{x}) < \varphi\,(\hat{x})\,.$$

Proposition 6.88. *Let $\mathcal{O}$ be an open subset of $\mathbb{R}^d$ and*

$$r \leq s \ \Rightarrow \ \Phi(x, r, p, X) \leq \Phi(x, s, p, X), \ \forall\, x, p, X.$$

Then each viscosity sub-solution u satisfies the maximum principle.

Proof. If we assume that there exist $\varphi \in C^2(\mathcal{O})$, an open subset $D \subseteq \mathcal{O}$ such that

$$\Phi(x, u(x), D\varphi(x), D^2\varphi(x)) > 0, \ \forall x \in D,$$

and $\hat{x} \in D$ a local maximum of $u - \varphi$ such that $u\,(\hat{x}) \geq \varphi\,(\hat{x})$ then

$$\Phi(\hat{x}, \varphi(\hat{x}), D\varphi(\hat{x}), D^2\varphi(\hat{x})) \leq \Phi(\hat{x}, u(\hat{x}), D\varphi(\hat{x}), D^2\varphi(\hat{x})) \leq 0,$$

since u is a sub-solution. Hence necessarily $u(\hat{x}) < \varphi(\hat{x})$. ∎

We next introduce the notion of a proper function (in the sense of the theory of viscosity solutions, which should not be confused with the notion of proper convex function), for which the notion of a viscosity solution makes sense.

Definition 6.89. A continuous function

$$\Phi : \mathcal{O} \times \mathbb{R} \times \mathbb{R}^d \times \mathbb{S}^d \to \mathbb{R}$$

is said to be proper, if Φ satisfies:

(1) Monotonicity condition

$$r \le s \;\Rightarrow\; \Phi(x, r, p, X) \le \Phi(x, s, p, X),\ \forall\, x, p, X,$$

and

(2) Degenerate ellipticity condition

$$X \le Y \Rightarrow \Phi(x, r, p, X) \ge \Phi(x, r, p, Y),\ \forall\, x, r, p. \tag{6.118}$$

Definition 6.85 of a viscosity solution can be reformulated in terms of subjets and superjets of u.

Definition 6.90. Let $\mathcal{O}$ be a locally closed subset of $\mathbb{R}^d$, $u : \mathcal{O} \to \mathbb{R}$ and $x \in \mathcal{O}$.

(i) $(p, X) \in \mathbb{R}^d \times \mathbb{S}^d$ is a superjet to u at x if

$$\limsup_{\mathcal{O} \ni y \to x} \frac{u(y) - u(x) - \langle p, y - x\rangle - \frac{1}{2}\langle X(y-x), y-x\rangle}{|y-x|^2} \le 0.$$

The set of superjets to u at x will be denoted $J^{2,+}_{\mathcal{O}} u(x)$.

(ii) $(p, X) \in \mathbb{R}^d \times \mathbb{S}^d$ is a subjet to u at x if

$$\liminf_{\mathcal{O} \ni y \to x} \frac{u(y) - u(x) - \langle p, y - x\rangle - \frac{1}{2}\langle X(y-x), y-x\rangle}{|y-x|^2} \ge 0.$$

The set of subjets to u at x will be denoted $J^{2,-}_{\mathcal{O}} u(x)$.

If $\mathcal{O} = \mathbb{R}^d$, then the index $\mathcal{O}$ will be omitted.

Proposition 6.91. *Let $\mathcal{O}$ be a locally closed subset of $\mathbb{R}^d$.*

(i) Let $u \in USC(\mathcal{O})$ and $\tilde{x} \in \mathcal{O}$.

(a) If $(p, X) \in J^{2,+}_{\mathcal{O}} u(\tilde{x})$, then there exists a $\varphi \in C^2(\mathcal{O})$ such that $u(\tilde{x}) = \varphi(\tilde{x})$,

$$(p, X) = \left(\varphi'_x(\tilde{x}), \varphi''_{xx}(\tilde{x})\right)$$

and $\tilde{x}$ is a strict global maximum of $u - \varphi$ in $\mathcal{O}$.

(b) If $\varphi \in C^2(\mathcal{O})$ and $\tilde{x}$ is a local maximum of $u - \varphi$ in $\mathcal{O}$, then

$$\left(\varphi'_x(\tilde{x}), \varphi''_{xx}(\tilde{x})\right) \in J^{2,+}_{\mathcal{O}} u(\tilde{x}).$$

(ii) Let $u \in LSC(\mathcal{O})$ and $\tilde{x} \in \mathcal{O}$.

(a) If $(p, X) \in J^{2,-}_{\mathcal{O}} u(\tilde{x})$, then there exists a $\varphi \in C^2(\mathcal{O})$ such that $u(\tilde{x}) = \varphi(\tilde{x})$,

$$(p, X) = \left(\varphi'_x(\tilde{x}), \varphi''_{xx}(\tilde{x})\right)$$

and $\tilde{x}$ is a strict global minimum of $u - \varphi$ in $\mathcal{O}$.

(b) *If $\varphi \in C^2(\mathcal{O})$ and $\tilde{x}$ is a local minimum of $u - \varphi$ in $\mathcal{O}$, then*

$$\left(\varphi'_x(\tilde{x}), \varphi''_{xx}(\tilde{x})\right) \in J^{2,-}_{\mathcal{O}} u(\tilde{x}).$$

Proof. It is sufficient to prove (i) since $J^{2,-}_{\mathcal{O}} u(\tilde{x}) = -J^{2,+}_{\mathcal{O}} (-u)(\tilde{x})$. Also the equivalence is clear if $\tilde{x}$ is an isolated point of $\mathcal{O}$.

Let $\tilde{x}$ be a non-isolated point of $\mathcal{O}$.

($\Rightarrow$): Let $(p, X) \in J^{2,+}_{\mathcal{O}} u(\tilde{x})$. Then there exists a strictly increasing function $\rho = \rho^{(\tilde{x})} : [0, +\infty[\to [0, +\infty[$, $\rho(0+) = 0$ such that $\forall y \in \mathcal{O}$

$$u(y) \leq u(\tilde{x}) + \langle p, y - \tilde{x}\rangle + \frac{1}{2}\langle X(y - \tilde{x}), y - \tilde{x}\rangle + \rho(|y - \tilde{x}|)|y - \tilde{x}|^2. \quad (6.119)$$

One can define ρ by

$$\rho(r) = r + \sup_{y \in \mathcal{O},\, |y - \tilde{x}| \leq r} \frac{\left(u(y) - u(\tilde{x}) - \langle p, y - \tilde{x}\rangle - \frac{1}{2}\langle X(y - \tilde{x}), y - \tilde{x}\rangle\right)^+}{|y - \tilde{x}|^2}.$$

Let

$$\beta(r) = \frac{1}{r^2}\int_r^{2r}\int_{r_2}^{2r_2}\int_{r_1}^{2r_1} \rho\left(\sqrt{\tau}\right) d\tau dr_1 dr_2, \quad \text{for } r > 0,$$

and $\beta(r) = 0$ if $r \leq 0$. Then $r\rho(\sqrt{r}) < \beta(r) < 8r\rho(8\sqrt{r})$ for all $r > 0$, $\beta \in C^2(]0, \infty[)$, $\beta(0+) = \beta'(0+) = 0$ and

$$\lim_{r \searrow 0} r\beta''(r) = 0.$$

Define $\varphi \in C^2(\mathbb{R}^d)$ by

$$\varphi(y) \stackrel{def}{=} u(\tilde{x}) + \langle p, y - \tilde{x}\rangle + \frac{1}{2}\langle X(y - \tilde{x}), y - \tilde{x}\rangle + \beta(|y - \tilde{x}|^2).$$

Then

$$\varphi'_x(\tilde{x}) = p \quad \text{and} \quad \varphi''_{xx}(\tilde{x}) = X$$

and $\tilde{x}$ is a strict global maximum of $u - \varphi$ since for $y \in \mathcal{O} \setminus \{\tilde{x}\}$:

$$\begin{aligned} u(y) - \varphi(y) &\leq \rho(|y - \tilde{x}|)|y - \tilde{x}|^2 - \beta(|y - \tilde{x}|^2) \\ &< 0 = \varphi(\tilde{x}) - u(\tilde{x}). \end{aligned}$$

($\Leftarrow$): Let $\varphi \in C^2(\mathcal{O})$ and $\tilde{x}$ be a local maximum of $u - \varphi$. Let

$$\psi(y) = \varphi(y) - \varphi(\tilde{x}) + u(\tilde{x}).$$

By Taylor's formula

$$\begin{aligned} 0 = & \lim_{y \to \tilde{x}} \frac{\psi(y)-\psi(\tilde{x})-\langle \psi_x'(\tilde{x}), y-\tilde{x}\rangle - \frac{1}{2}\langle \psi_{xx}''(\tilde{x})(y-\tilde{x}), y-\tilde{x}\rangle}{|y-\tilde{x}|^2} \\ \geq & \limsup_{y \to \tilde{x}, y \in \mathcal{O}} \frac{u(y)-u(\tilde{x})-\langle \varphi_x'(\tilde{x}), y-\tilde{x}\rangle - \frac{1}{2}\langle \varphi_{xx}''(\tilde{x})(y-\tilde{x}), y-\tilde{x}\rangle}{|y-\tilde{x}|^2}. \end{aligned}$$

■

Corollary 6.92. *Let $\mathcal{O}$ be a locally closed subset of $\mathbb{R}^d$.*

(i) $u \in USC(\mathcal{O})$ is a viscosity sub-solution of (6.116) iff for any $x \in \mathcal{O}$ and $(p, X) \in J_{\mathcal{O}}^{2,+}u(x)$

$$\Phi(x, u(x), p, X) \leq 0.$$

(ii) $u \in LSC(\mathcal{O})$ is a viscosity super-solution of (6.116) iff for any $x \in \mathcal{O}$ and $(p, X) \in J_{\mathcal{O}}^{2,-}u(x)$

$$\Phi(x, u(x), p, X) \geq 0.$$

Definition 6.93. Let $u : \mathcal{O} \to \mathbb{R}$ and $x \in \mathcal{O}$.
$\overline{J}_{\mathcal{O}}^{2,+}u(x)$ (respect. $\overline{J}_{\mathcal{O}}^{2,-}u(x)$) is the set of $(p, X) \in \mathbb{R}^d \times \mathbb{S}^d$ such that there exists a sequence $(x_n, p_n, X_n) \in \mathcal{O} \times \mathbb{R}^d \times \mathbb{S}^d$, $n \in \mathbb{N}^*$, with the properties

$$(p_n, X_n) \in J_{\mathcal{O}}^{2,+}u(x_n), \text{ (respect. } (p_n, X_n) \in J_{\mathcal{O}}^{2,-}u(x_n)), \ \forall\, n \in \mathbb{N}^*,$$

and

$$(x_n, u(x_n), p_n, X_n) \to (x, u(x), p, X), \quad \text{as } n \to \infty.$$

6.5.1.2 Systems of PDEs

Backward stochastic differential equations naturally give probabilistic formulas for systems of PDEs, not just for single PDEs.

Let $\mathcal{O}$ be an open subset of $\mathbb{R}^d$, $\Phi \in C(\overline{\mathcal{O}} \times \mathbb{R}^m \times \mathbb{R}^d \times \mathbb{S}^d; \mathbb{R}^m)$. We want to explain what we mean by the fact that $u \in C(\mathcal{O}, \mathbb{R}^m)$ solves in the viscosity sense the following systems of PDEs

$$\Phi_i(x, u(x), Du_i(x), D^2u_i(x)) = 0, \ 1 \leq i \leq m, \ x \in \mathcal{O}. \tag{6.120}$$

Note that the various equations are coupled only through the vector $u(x)$. The i-th equation depends upon all coordinates of $u(x)$, but only on the i-th coordinate of $Du(x)$ and $D^2u(x)$. This is essential for the following definition to make sense.

Definition 6.94. Let $\mathcal{O}$ be a locally closed subset of $\mathbb{R}^d$.

- *(i)* $u \in USC(\mathcal{O})$ is a viscosity sub-solution of (6.120) if

$$\Phi_i(x, u(x), p, X) \leq 0 \text{ for } x \in \mathcal{O},\ 1 \leq i \leq m,\ (p, X) \in \overline{J}_{\mathcal{O}}^{2,+} u_i(x).$$

- *(ii)* $u \in LSC(\mathcal{O})$ is a viscosity super-solution of (6.120) if

$$\Phi_i(x, u(x), p, X) \geq 0 \text{ for } x \in \mathcal{O},\ 1 \leq i \leq m,\ (p, X) \in \overline{J}_{\mathcal{O}}^{2,-} u_i(x).$$

- *(iii)* $u \in C(\mathcal{O})$ is a viscosity solution of (6.120) if it is both a viscosity sub- and super-solution.

6.5.1.3 Boundary Conditions

We now discuss the formulation of the boundary condition in the framework of viscosity solutions. Suppose for simplicity that the boundary $\partial\mathcal{O}$ of the open set $\mathcal{O}$ is of class C^1 and that $\mathcal{O}$ satisfies the uniform exterior ball condition. We shall consider two types of boundary conditions, namely:

- Dirichlet boundary conditions, of the form

$$u(x) - \kappa(x) = 0, \quad x \in \partial\mathcal{O};$$

- Nonlinear Neumann boundary conditions, of the form

$$\langle n(x), Du(x)\rangle + G(x, u(x)) = 0, \quad x \in \partial\mathcal{O},$$

  where $n(x)$ denotes the outward normal vector to the boundary $\partial\mathcal{O}$ at x.

  Consider the function

$$\Gamma : \partial\mathcal{O} \times \mathbb{R} \times \mathbb{R}^d \to \mathbb{R}$$

defined in the case of the Dirichlet boundary condition by

$$\Gamma(x, r, p) = r - \kappa(x),$$

and in the case of the Neumann boundary condition by

$$\Gamma(x, r, p) = \langle n(x), p\rangle - G(x, r),$$

where $G \in C(\partial\mathcal{O} \times \mathbb{R})$ and $r \to G(x,r)$ is assumed to be nonincreasing for all $x \in \partial\mathcal{O}$. The correct formulation of the boundary value problem

$$\begin{cases} \Phi(x, u(x), Du(x), D^2u(x)) = 0, \ x \in \mathcal{O}, \\ \Gamma(x, u(x), Du(x)) = 0, \ x \in \partial\mathcal{O}, \end{cases} \tag{6.121}$$

is as follows.

Definition 6.95. Let $\mathcal{O}$ be an open subset of $\mathbb{R}^d$, $\Phi \in C(\overline{\mathcal{O}} \times \mathbb{R} \times \mathbb{R}^d \times \mathbb{S}^d)$ be proper and $\Gamma \in C(\mathcal{O} \times \mathbb{R} \times \mathbb{R}^d)$ be as defined above.

- *(i)* $u \in USC(\overline{\mathcal{O}})$ is a viscosity sub-solution of (6.121) if

$$\begin{cases} \Phi(x, u(x), p, X) \le 0 \ \text{ for } x \in \mathcal{O}, \ (p, X) \in \overline{J}^{2,+}_{\overline{\mathcal{O}}} u(x), \\ \Phi(x, u(x), p, X) \wedge \Gamma(x, u(x), p) \le 0 \ \text{ for } x \in \partial\mathcal{O}, \ (p, X) \in \overline{J}^{2,+}_{\overline{\mathcal{O}}} u(x). \end{cases}$$

- *(ii)* $u \in LSC(\overline{\mathcal{O}})$ is a viscosity super-solution of (6.121) if

$$\begin{cases} \Phi(x, u(x), p, X) \ge 0 \ \text{ for } x \in \mathcal{O}, \ (p, X) \in \overline{J}^{2,-}_{\overline{\mathcal{O}}} u(x), \\ \Phi(x, u(x), p, X) \vee \Gamma(x, u(x), p) \ge 0 \ \text{ for } x \in \partial\mathcal{O}, \ (p, X) \in \overline{J}^{2,-}_{\overline{\mathcal{O}}} u(x). \end{cases}$$

- *(iii)* $u \in C(\overline{\mathcal{O}})$ is a viscosity solution of (6.121) if it is both a viscosity sub- and super-solution.

6.5.1.4 Parabolic PDEs

One might think that a parabolic PDE is an elliptic PDE with one more variable, namely time t. However, because we are considering equations with first derivatives in t only, the variable t plays a specific role. In particular, there will be a boundary condition either at the initial point or at the final point of the time interval, not at both.

Given $\mathcal{O} \subset \mathbb{R}^d$ and $\Phi \in C([0,T] \times \mathcal{O} \times \mathbb{R} \times \mathbb{R}^d \times \mathbb{S}^d)$, we consider the parabolic equation

$$\begin{cases} \dfrac{\partial u}{\partial t}(t,x) + \Phi(t, x, u(t,x), Du(t,x), D^2u(t,x)), \ 0 < t < T, \ x \in \mathcal{O}, \\ u(0,x) = \kappa(x), \ x \in \mathcal{O}, \end{cases} \tag{6.122}$$

where as previously, Du stands for the vector of first order partial derivatives with respect to the x_i's, and D^2u for the matrix of second order derivatives with respect to x_i and x_j, $1 \le i,j \le d$. Only in the case $\mathcal{O} = \mathbb{R}^d$ can we hope that the above parabolic PDE is well posed. If $\mathcal{O} \ne \mathbb{R}^d$, some boundary condition is needed. This will be discussed later.

We denote by $\mathcal{P}^{2,+}_{\mathcal{O}}$ and $\mathcal{P}^{2,-}_{\mathcal{O}}$ the parabolic analogs of $J^{2,+}_{\mathcal{O}}$ and $J^{2,-}_{\mathcal{O}}$. More specifically, for $\mathcal{O}$ a locally compact subset of $\mathbb{R}^d$, $T > 0$, denoting $\mathcal{O}_T = (0,T) \times \mathcal{O}$, if $u : \mathcal{O}_T \to \mathcal{R}$, $0 < s,t < T$, $x, y \in \mathcal{O}$, $(p,q,X) \in \mathbb{R} \times \mathbb{R}^d \times \mathbb{S}^d$, we say that $(p,q,X) \in \mathcal{P}^{2,+}_{\mathcal{O}} u(t,x)$, whenever

$$u(s,y) \leq u(t,x) + p(s-t) + \langle q, y-x \rangle + \frac{1}{2}\langle X(y-x), y-x \rangle$$
$$+o(|s-t| + |y-x|^2).$$

Moreover $\mathcal{P}^{2,-}_{\mathcal{O}} u = -\mathcal{P}^{2,+}_{\mathcal{O}}(-u)$. The corresponding definitions of $\overline{\mathcal{P}}^{2,+}_{\mathcal{O}} u(t,x)$ and $\overline{\mathcal{P}}^{2,-}_{\mathcal{O}} u(t,x)$ are now clear.

We now give a definition of the notion of a viscosity solution of equation (6.122).

Definition 6.96. With the above notation:

- *(i)* $u \in USC([0,T) \times \mathcal{O})$ is a viscosity sub-solution of Eq. (6.122) if $u(0,x) \leq \kappa(x)$, $x \in \mathcal{O}$ and

$$p + \Phi(t,x,u(t,x),q,X) \leq 0, \text{ for } (t,x) \in \mathcal{O}_T, (p,q,X) \in \mathcal{P}^{2,+}_{\mathcal{O}} u(t,x).$$

- *(ii)* $u \in LSC([0,T) \times \mathcal{O})$ is a viscosity super-solution of Eq. (6.122) if $u(0,x) \geq \kappa(x)$, $x \in \mathcal{O}$ and

$$p + \Phi(t,x,u(t,x),q,X) \geq 0, \text{ for } (t,x) \in \mathcal{O}_T, (p,q,X) \in \mathcal{P}^{2,-}_{\mathcal{O}} u(t,x).$$

- *(iii)* $u \in C([0,T) \times \mathcal{O})$ is a viscosity solution of (6.122) if it is both a sub- and a super-solution.

We remark that $u(t,x)$ solves the parabolic PDE (6.122) if and only if $v(t,x) = e^{\lambda t} u(t,x)$ solves the same equation with Φ replaced by $\Phi + \lambda r$, which in the case where Φ has the form (6.117) is proper iff $r \to \lambda r - F(t,x,r,q)$ is increasing for any (t,x,q). The fact that this is true for some λ is one of our standing assumptions on F for existence and uniqueness of the solution to the associated BSDE.

Note that we also consider parabolic PDEs with a final condition (at time $t = T$) rather than an initial condition (at time $t = 0$). In that case, the equation becomes

$$-\frac{\partial u}{\partial t}(t,x) + \Phi(t,x,u(t,x),Du(t,x),D^2u(t,x)) = 0,$$

and the condition $u(0,x) \leq \kappa(x)$ (resp. $u(0,x) \geq \kappa(x)$) becomes $u(T,x) \leq \kappa(x)$ (resp. $u(T,x) \geq \kappa(x)$).

Finally we explain what we mean by a viscosity solution of the parabolic PDE

$$\frac{\partial u}{\partial t}(t,x) + \Phi(t,x,u(t,x),Du(t,x),D^2u(t,x)) + \partial\varphi(u(t,x)) \ni 0,$$

where $\partial\varphi$ is the subdifferential of the convex lower semicontinuous function $\varphi : \mathbb{R} \to (-\infty, +\infty]$.

A sub-solution is a function $u \in USC(\mathcal{O}_T)$ which is such that for any $(t,x) \in \mathcal{O}_T$, $u(t,x) \in \mathrm{Dom}(\varphi)$ and whenever $(p,q,X) \in \mathcal{P}_{\mathcal{O}}^{2,+}u(t,x)$,

$$p + \Phi(t,x,u(t,x),q,X) + \varphi'_-(u(t,x)) \leq 0,$$

where $\varphi'_-(r)$ is the left derivative of φ at the point r. A super-solution is defined similarly with the usual changes, the left derivative of φ being replaced by its right derivative.

6.5.2 A First Uniqueness Result

Let $\mathcal{O}$ be an open subset of $\mathbb{R}^d$ and $\Phi \in C(\mathcal{O} \times \mathbb{R} \times \mathbb{R}^d \times \mathbb{S}^d)$.

The basic assumptions of this subsection are:

(A_1) *Super-monotonicity*: there exists a $\delta > 0$ such that for all $x \in \mathcal{O}$, $p \in \mathbb{R}^d$, $X \in \mathbb{S}^d$, $r, s \in \mathbb{R}$:

$$r_1 \leq r_2 \Rightarrow \Phi(x,r_2,p,X) - \Phi(x,r_1,p,X) \geq (r_2 - r_1)\,\delta,$$

and

(A_2) *Super-degenerate-ellipticity*: for all $R > 0$ there exists an increasing function $\mathbf{m}_R : \mathbb{R}_+ \to \mathbb{R}_+$, $\mathbf{m}_R(0+) = 0$ such that if $\alpha > 0$, $X, Y \in \mathbb{S}^d$ and

$$\begin{pmatrix} X & 0 \\ 0 & -Y \end{pmatrix} \leq 3\alpha \begin{pmatrix} I & -I \\ -I & I \end{pmatrix}, \tag{6.123}$$

or equivalently

$$\langle Xz, z\rangle - \langle Yw, w\rangle \leq 3\alpha\,|z-w|^2, \quad \forall\, z, w \in \mathbb{R}^d,$$

then for all $x, y \in \mathcal{O} \cap \overline{B(0,R)}$, $r \in \mathbb{R}$:

$$\Phi(y,r,\alpha(x-y),Y) - \Phi(x,r,\alpha(x-y),X) \leq \mathbf{m}_R\left(|x-y| + \alpha\,|x-y|^2\right). \tag{6.124}$$

Note that if X and Y satisfy (6.123) then $Y \leq X$ (setting $z = w$).

In the particular case of the function Φ given by (6.117), the super-monotonicity of Φ is a consequence of the same property for $-F$. As for the super degenerate ellipticity, we have the following:

Lemma 6.97. *If g is globally Lipschitz, f is globally monotone, and $-F$ satisfies (6.124), then Φ is super-degenerate-elliptic.*

Proof. The global monotonicity of f implies that

$$-\langle f(y) - f(x), \alpha(x - y)\rangle \leq \mu\alpha|x - y|^2.$$

Now consider the term involving g. We take advantage of (6.123) and the Lipschitz continuity of g:

$$\begin{aligned}
\mathbf{Tr}\left[gg^*(x)X\right] - \mathbf{Tr}\left[gg^*(y)Y\right] &= \mathbf{Tr}\left[g^*(x)Xg(x) - g^*(y)Yg(y)\right] \\
&= \sum_{i=1}^{d} \left[\langle Xg(x)e_i, g(x)e_i\rangle - \langle Yg(y)e_i, g(y)e_i\rangle\right] \\
&\leq 3\alpha \sum_{i=1}^{d} |g(x)e_i - g(y)e_i|^2 \\
&\leq C|x - y|^2.
\end{aligned}$$

■

Theorem 6.98 (Comparison Principle). *Let $\mathcal{O}$ be a bounded open subset of $\mathbb{R}^d$ and assume that $\Phi : \mathcal{O} \times \mathbb{R} \times \mathbb{R}^d \times \mathbb{S}^d \to \mathbb{R}$ satisfies (A_1) and (A_2). If*

$$\begin{aligned}
&(j) \quad u \in USC\left(\overline{\mathcal{O}}\right) \text{ is a sub-solution of } \Phi = 0 \text{ in } \mathcal{O}, \\
&(jj) \quad v \in LSC(\overline{\mathcal{O}}) \text{ is a super-solution of } \Phi = 0 \text{ in } \mathcal{O}, \\
&(jjj) \quad u(x) \leq v(x), \quad \forall\, x \in \partial\mathcal{O},
\end{aligned}$$

then

$$u(x) \leq v(x) \quad \forall\, x \in \overline{\mathcal{O}}.$$

We first prove auxiliary results.

Lemma 6.99. *Given $u, v \in C(\bar{\mathcal{O}})$, $\alpha > 0$, we define*

$$\psi_\alpha(x, y) = u(x) - v(y) - \frac{\alpha}{2}|x - y|^2.$$

Let $(\hat{x}, \hat{y})$ be a local maximum in $\mathcal{O} \times \mathcal{O}$ of ψ_α. Then there exist $X, Y \in \mathbb{S}^d$ such that

(j) $(\alpha(\hat{x} - \hat{y}), X) \in \bar{J}^{2,+}_{\mathcal{O}} u(\hat{x})$,

(jj) $(\alpha(\hat{x} - \hat{y}), Y) \in \bar{J}^{2,-}_{\mathcal{O}} v(\hat{y})$,

(jjj) $\begin{pmatrix} X & 0 \\ 0 & -Y \end{pmatrix} \leq 3\alpha \begin{pmatrix} I & -I \\ -I & I \end{pmatrix}$.

Proof. We shall use the notation

$$A = \alpha \begin{pmatrix} I & -I \\ -I & I \end{pmatrix}.$$

It is sufficient to prove the proposition in case $\mathcal{O} = \mathbb{R}^d$, $\hat{x} = \hat{y} = 0$, $u(0) = v(0) = 0$, $(0,0)$ is a global maximum of ψ_α, u and $-v$ are bounded from above. Hence we may assume that for all $x, y \in \mathbb{R}^d$,

$$u(x) - v(y) \leq \frac{1}{2}\left\langle A\begin{pmatrix} x \\ y \end{pmatrix}, \begin{pmatrix} x \\ y \end{pmatrix}\right\rangle, \tag{6.125}$$

and we need to show that there exist $X, Y \in S_d$ such that

(j') $(0, X) \in \bar{J}^{2,+}u(0)$,
(jj') $(0, Y) \in \bar{J}^{2,+}v(0)$,
(jjj') $\begin{pmatrix} X & 0 \\ 0 & -Y \end{pmatrix} \leq 3A.$

With the notations $\bar{x} = \begin{pmatrix} x \\ y \end{pmatrix}$, $\bar{\xi} = \begin{pmatrix} \xi \\ \eta \end{pmatrix}$, we deduce from Schwarz's inequality that (with the notation $\|A\| \stackrel{def}{=} \sup\{|\langle A\bar{\xi}, \bar{\xi}\rangle|; |\bar{\xi}| \leq 1\}$):

$$\begin{aligned} \langle A\bar{x}, \bar{x}\rangle &= \langle A\bar{\xi}, \bar{\xi}\rangle + \langle A(\bar{x} - \bar{\xi}), \bar{x} - \bar{\xi}\rangle + 2\langle \bar{x} - \bar{\xi}, A\bar{\xi}\rangle \\ &\leq \langle A\bar{\xi}, \bar{\xi}\rangle + \frac{1}{\alpha}|A\bar{\xi}|^2 + (\alpha + \|A\|)|\bar{x} - \bar{\xi}|^2 \\ &\leq \left\langle (A + \frac{1}{\alpha}A^2)\bar{\xi}, \bar{\xi}\right\rangle + (\alpha + \|A\|)|\bar{x} - \bar{\xi}|^2. \end{aligned}$$

Hence if $B \stackrel{def}{=} 3A = A + \frac{1}{\alpha}A^2$, $\lambda \stackrel{def}{=} \alpha + \|A\|$, and $w(\bar{x}) \stackrel{def}{=} u(x) - v(y)$, (6.125) implies

$$w(\bar{x}) - \frac{\lambda}{2}|\bar{x} - \bar{\xi}|^2 \leq \frac{1}{2}\langle B\bar{\xi}, \bar{\xi}\rangle. \tag{6.126}$$

We now introduce inf- and sup-convolutions. Let

$$\begin{aligned} \hat{w}(\bar{\xi}) &\stackrel{def}{=} \sup_{\bar{x}}(w(\bar{x}) - \frac{\lambda}{2}|\bar{x} - \bar{\xi}|^2) \\ &= \hat{u}(\xi) - \hat{v}(\eta), \end{aligned}$$

where

$$\hat{u}(\xi) = \sup_x (u(x) - \frac{\lambda}{2}|x - \xi|^2),$$

$$\hat{v}(\eta) = \inf_y (v(y) + \frac{\lambda}{2}|y - \eta|^2).$$

Since a supremum (resp. an infimum) of convex (resp. concave) functions is convex (resp. concave), the mappings

$$\bar{\xi} \to \hat{w}(\bar{\xi}) + \frac{\lambda}{2}|\bar{\xi}|^2, \quad \text{and} \quad \xi \to \hat{u}(\xi) + \frac{\lambda}{2}|\xi|^2$$

are convex, while

$$\eta \to \hat{v}(\eta) - \frac{\lambda}{2}|\eta|^2$$

is concave. Hence $\hat{w}$, $\hat{u}$ and $-\hat{v}$ are "semiconvex", i.e. they are the sum of a convex function and a function of class C^2. Note that the hyphen is here on purpose, in order to distinguish this notion from the notion of semiconvex functions, as introduced in Chap. 4.3.

Moreover:

$$\hat{w}(0) \leq w(0) = 0,$$

and from (6.126)

$$\hat{w}(\bar{\xi}) \leq \frac{1}{2}\left\langle B\bar{\xi}, \bar{\xi}\right\rangle,$$

hence

$$\hat{w}(0) \leq 0,$$

and consequently

$$\hat{w}(0) = \max_{\bar{\xi}}\left(\hat{w}(\bar{\xi}) - \frac{1}{2}\left\langle B\bar{\xi}, \bar{\xi}\right\rangle\right).$$

If $\hat{w}$ is smooth, we could deduce that there exists an $\mathcal{X} \in S_{2d}$ such that $(0, \mathcal{X}) \in J^2\hat{w}(0)$, and $\mathcal{X} \leq B$. Since $\hat{w}$ is semiconvex, it is possible to show, using Alexandrov's theorem (which says that a semiconvex function is a.e. twice differentiable), and a lemma due to R. Jensen, which states that the above is essentially true in the sense that it is true provided the first condition is changed to $(0, \mathcal{X}) \in \bar{J}^2\hat{w}(0)$. We refer to the user's guide [18] for more details. Now, since $\hat{w}(\bar{\xi}) = \hat{u}(\xi) - \hat{v}(\eta)$, it is not hard to deduce that $\mathcal{X} = \begin{pmatrix} X & 0 \\ 0 & -Y \end{pmatrix}$, and $(0, X) \in \bar{J}^2\hat{u}(0)$, $(0, Y) \in \bar{J}^2\hat{v}(0)$.

The magical property of sup-convolution is that this is enough to conclude that $(0, X) \in \bar{J}^{2,+}u(0)$ and $(0, Y) \in \bar{J}^{2,-}v(0)$, which is a consequence of the next Lemma. ■

Lemma 6.100. *Let $\lambda > 0$, $u \in C(\mathbb{R}^d)$ be bounded from above, and*

$$\hat{u}(\zeta) = \sup_{x \in \mathbb{R}^d} (u(x) - \frac{\lambda}{2}|x - \zeta|^2).$$

If $\eta, q \in \mathbb{R}^d$, $X \in S_d$ and $(\eta, X) \in J^{2,+}\hat{u}(\eta)$, then $(q, X) \in J^{2,+}u(\eta + q/\lambda)$.

Proof. We assume that $(q, X) \in J^{2,+}\hat{u}(\eta)$. Let $y \in \mathbb{R}^d$ be such that

$$\hat{u}(\eta) = u(y) - \frac{\lambda}{2}|y - \eta|^2.$$

Then for any $x, \zeta \in \mathbb{R}^d$,

$$\begin{aligned} u(x) - \frac{\lambda}{2}|x - \zeta|^2 &\leq \hat{u}(\zeta) \\ &\leq \hat{u}(\eta) + \langle q, \zeta - \eta \rangle + \frac{1}{2} \langle X(\zeta - \eta), \zeta - \eta \rangle + o(|\zeta - \eta|^2) \\ &= u(y) - \frac{\lambda}{2}|y - \eta|^2 + \langle q, \zeta - \eta \rangle \\ &+ \frac{1}{2} \langle X(\zeta - \eta), \zeta - \eta \rangle + o(|\zeta - \eta|^2) \\ &= u(y) - \frac{\lambda}{2}|y - \eta|^2 + \langle q, \zeta - \eta \rangle + O(|\zeta - \eta|^2). \end{aligned}$$

If we choose $\zeta = x - y + \eta$, then we deduce from the above that

$$u(x) \leq u(y) + \langle q, x - y \rangle + \frac{1}{2} \langle X(x - y), x - y \rangle + o(|x - y|^2).$$

On the other hand, choosing $x = y$ and $\zeta = \eta + \alpha(\lambda(\eta - y) + q)$, we obtain that

$$0 \leq \alpha|\lambda(\eta - y) + q|^2 + O(\alpha^2).$$

The first inequality says that $(q, X) \in J^{2,+}u(y)$, while the second, with $\alpha < 0$ small enough in absolute value, implies that $y = \eta + \frac{q}{\lambda}$. The result is proved. ■

We shall also need the following:

Lemma 6.101. *Let $\mathcal{O}$ be locally closed subset of $\mathbb{R}^d$, $\Phi \in USC(\mathcal{O})$, $\Psi \in LSC(\mathcal{O})$, $\Psi \geq 0$, $\varepsilon > 0$ and*

$$M_\varepsilon = \sup_{x \in \mathcal{O}} \left\{ \Phi(x) - \frac{1}{\varepsilon} \Psi(x) \right\}.$$

If $\lim_{\varepsilon \to 0} M_\varepsilon$ exists in $\mathbb{R}$ and $x_\varepsilon \in \mathcal{O}$ satisfies

$$\lim_{\varepsilon \to 0} \left[M_\varepsilon - \Phi(x_\varepsilon) + \frac{1}{\varepsilon} \Psi(x_\varepsilon) \right] = 0,$$

then

$$\lim_{\varepsilon \to 0} \frac{\Psi(x_\varepsilon)}{\varepsilon} = 0. \tag{6.127}$$

Moreover if $\hat{x} \in \mathcal{O}$ and there exists an $\varepsilon_n \to 0$ such that $x_{\varepsilon_n} \to \hat{x}$, then

$$\Psi(\hat{x}) = 0, \text{ and } \lim_{\varepsilon \to 0} M_\varepsilon = \Phi(\hat{x}) = \sup \{ \Phi(x) : x \in \mathcal{O}, \Psi(x) = 0 \}. \tag{6.128}$$

Proof. Let $\alpha_\varepsilon = M_\varepsilon - \Phi(x_\varepsilon) + \frac{1}{\varepsilon} \Psi(x_\varepsilon)$. Note that for $0 < \varepsilon < \delta$ we have $M_\varepsilon \leq M_\delta$ and

$$M_{2\varepsilon} \geq \Phi(x_\varepsilon) - \frac{1}{2\varepsilon} \Psi(x_\varepsilon) = M_\varepsilon - \alpha_\varepsilon + \frac{1}{2\varepsilon} \Psi(x_\varepsilon).$$

Then

$$\frac{\Psi(x_\varepsilon)}{\varepsilon} \leq 2 (M_{2\varepsilon} - M_\varepsilon + \alpha_\varepsilon)$$

and (6.127) follows. Moreover by the lower semicontinuity of Ψ

$$0 \leq \Psi(\hat{x}) \leq \liminf_{\varepsilon_n \to 0} \Psi(x_{\varepsilon_n}) = 0.$$

Using now the upper semicontinuity of Φ we have

$$\begin{aligned} \Phi(\hat{x}) &\geq \limsup_{\varepsilon_n \to 0} \Phi(x_{\varepsilon_n}) \\ &= \limsup_{\varepsilon_n \to 0} \left[M_{\varepsilon_n} - \alpha_{\varepsilon_n} + \frac{1}{\varepsilon_n} \Psi(x_{\varepsilon_n}) \right] \\ &= \lim_{\varepsilon \to 0} M_\varepsilon \end{aligned}$$

$$\geq \sup \{\Phi(x) : x \in \mathcal{O}, \Psi(x) = 0\}$$
$$\geq \Phi(\hat{x}).$$

The result follows. ■

Proof of the comparison principle. Assume that

$$M \stackrel{def}{=} \sup_{x \in \overline{\mathcal{O}}} \{u(x) - v(x)\} > 0.$$

Let $\varepsilon > 0$ and

$$M_\varepsilon \stackrel{def}{=} \sup_{(x,y) \in \overline{\mathcal{O}} \times \overline{\mathcal{O}}} \left[u(x) - v(y) - \frac{1}{2\varepsilon} |x - y|^2 \right].$$

Clearly for $\delta > \varepsilon$, $M_\delta \geq M_\varepsilon \geq u(x) - v(x)$, $\forall\, x \in \overline{\mathcal{O}}$ and consequently M_ε converges in $\mathbb{R}$ as $\varepsilon \to 0$,

$$M_\varepsilon \geq M > 0 \text{ and } \lim_{\varepsilon \to 0} M_\varepsilon \geq M.$$

Since $\overline{\mathcal{O}}$ is compact and $(x, y) \mapsto u(x) - v(y)$ is upper semicontinuous on $\overline{\mathcal{O}} \times \overline{\mathcal{O}}$, there exists $(x_\varepsilon, y_\varepsilon) \in \overline{\mathcal{O}} \times \overline{\mathcal{O}}$ such that

$$u(x_\varepsilon) - v(y_\varepsilon) - \frac{1}{2\varepsilon} |x_\varepsilon - y_\varepsilon|^2 = M_\varepsilon.$$

By Lemma 6.101, with $\Phi(x, y) = u(x) - v(y)$ and $\Psi(x, y) = \frac{1}{2} |x - y|^2$ we obtain

$$\lim_{\varepsilon \to \infty} \frac{1}{\varepsilon} |x_\varepsilon - y_\varepsilon|^2 = 0.$$

We now conclude that there exists an $\varepsilon_0 > 0$ such that

$$x_\varepsilon, y_\varepsilon \in \mathcal{O}, \text{ for all } 0 < \varepsilon \leq \varepsilon_0.$$

Since $u(x) \leq v(x)$, $\forall\, x \in \partial\mathcal{O}$ and whenever $\varepsilon_n \to 0$ and $x_{\varepsilon_n}, y_{\varepsilon_n} \to \hat{x}$, it follows that

$$\begin{aligned}
\lim_{\varepsilon \to 0} M_\varepsilon &= u(\hat{x}) - v(\hat{x}) \\
&= \sup \left\{ \Phi(x, y) : (x, y) \in \overline{\mathcal{O}} \times \overline{\mathcal{O}}, \Psi(x, y) = 0 \right\} \\
&= \sup_{x \in \overline{\mathcal{O}}} \{u(x) - v(x)\} \\
&> 0.
\end{aligned}$$

By Lemma 6.99, for $0 < \varepsilon \leq \varepsilon_0$ there exist $X_\varepsilon, Y_\varepsilon \in \mathbb{S}^d$ such that

$$\begin{aligned} &\left(\frac{1}{\varepsilon}\Psi'_x(x_\varepsilon, y_\varepsilon), X_\varepsilon\right) \in \overline{J}^{2,+}_{\mathcal{O}} u(x_\varepsilon), \quad \text{and} \\ &\left(-\frac{1}{\varepsilon}\Psi'_y(x_\varepsilon, y_\varepsilon), Y_\varepsilon\right) \in \overline{J}^{2,-}_{\mathcal{O}} v(y_\varepsilon) \end{aligned} \tag{6.129}$$

and the inequality (jjj) in Lemma 6.99 reads here

$$\begin{pmatrix} X_\varepsilon & 0 \\ 0 & -Y_\varepsilon \end{pmatrix} \leq \frac{3}{\varepsilon}\begin{pmatrix} I & -I \\ -I & I \end{pmatrix}.$$

Let $R > 0$ such that $\mathcal{O} \subset B(0, R)$. From (A_2) with $\alpha = \varepsilon^{-1}$, we deduce that

$$\begin{aligned} &\Phi\left(y_\varepsilon, v(y_\varepsilon), \frac{x_\varepsilon - y_\varepsilon}{\varepsilon}, Y_\varepsilon\right) - \Phi\left(x_\varepsilon, v(y_\varepsilon), \frac{x_\varepsilon - y_\varepsilon}{\varepsilon}, X_\varepsilon\right) \\ &\leq \mathbf{m}_R\left(|x_\varepsilon - y_\varepsilon| + \frac{1}{\varepsilon}|x_\varepsilon - y_\varepsilon|^2\right), \end{aligned}$$

and since $u(x_\varepsilon) > v(y_\varepsilon)$ for ε small enough, we deduce from (A_1) that

$$\begin{aligned} &\Phi\left(x_\varepsilon, v(y_\varepsilon), \frac{x_\varepsilon - y_\varepsilon}{\varepsilon}, X_\varepsilon\right) - \Phi\left(x_\varepsilon, u(x_\varepsilon), \frac{x_\varepsilon - y_\varepsilon}{\varepsilon}, X_\varepsilon\right) \\ &\leq \delta\left[v(y_\varepsilon) - u(x_\varepsilon)\right] \\ &= -\delta\left[\frac{1}{2\varepsilon}|x_\varepsilon - y_\varepsilon|^2 + M_\varepsilon\right]. \end{aligned}$$

It follows that

$$\begin{aligned} &\Phi\left(y_\varepsilon, v(y_\varepsilon), \frac{x_\varepsilon - y_\varepsilon}{\varepsilon}, Y_\varepsilon\right) - \Phi\left(x_\varepsilon, u(x_\varepsilon), \frac{x_\varepsilon - y_\varepsilon}{\varepsilon}, X_\varepsilon\right) \\ &\leq \mathbf{m}_R\left(|x_\varepsilon - y_\varepsilon| + \frac{1}{\varepsilon}|x_\varepsilon - y_\varepsilon|^2\right) - \delta\left[\frac{1}{2\varepsilon}|x_\varepsilon - y_\varepsilon|^2 + M_\varepsilon\right]. \end{aligned}$$

Since u is a viscosity sub-solution and v is a viscosity super-solution of the equation $\Phi = 0$, we deduce from (6.129) that

$$\Phi\left(x_\varepsilon, u(x_\varepsilon), \frac{x_\varepsilon - y_\varepsilon}{\varepsilon}, X_\varepsilon\right) \leq 0 \leq \Phi\left(y_\varepsilon, v(y_\varepsilon), \frac{x_\varepsilon - y_\varepsilon}{\varepsilon}, Y_\varepsilon\right).$$

Hence

$$0 \leq \mathbf{m}_R\left(|x_\varepsilon - y_\varepsilon| + \frac{1}{\varepsilon}|x_\varepsilon - y_\varepsilon|^2\right) - \delta\left[\frac{1}{2\varepsilon}|x_\varepsilon - y_\varepsilon|^2 + M_\varepsilon\right],$$

then also

$$0 < \delta\, M \leq \delta\, M_\varepsilon \leq \mathbf{m}_R \left(|x_\varepsilon - y_\varepsilon| + \frac{1}{\varepsilon} |x_\varepsilon - y_\varepsilon|^2 \right) - \frac{\delta}{2\varepsilon} |x_\varepsilon - y_\varepsilon|^2$$

and letting $\varepsilon \to 0$, we infer the contradiction

$$0 < \delta\, M \leq 0.$$

The Theorem is established. ■

We deduce from this theorem the uniqueness of the viscosity solution for the Dirichlet problem.

Corollary 6.102. *Under the assumptions of Theorem 6.98, if $u, v \in C\left(\overline{\mathcal{O}}\right)$ are two viscosity solutions of $\Phi = 0$ on $\mathcal{O}$ then*

$$u(x) = v(x)\,,\ \forall\, x \in \partial\mathcal{O} \quad \Longrightarrow \quad u(x) = v(x)\,,\ \forall\, x \in \overline{\mathcal{O}}.$$

This Corollary proves that our probabilistic formula provides the unique solution of the corresponding elliptic PDE, satisfying the Dirichlet boundary condition in the classical sense. However it follows from Theorem 7.9 in [18] that it is also the unique solution in the larger class of those solutions satisfying the Dirichlet boundary condition in the (relaxed) viscosity sense.

Let us now indicate how the above proof can be modified, in order to treat the case of a parabolic PDE with Dirichlet condition at the boundary of a bounded set.

Let $\mathcal{O}$ be a bounded open subset of $\mathbb{R}^d$. Consider the Cauchy–Dirichlet problem

$$\begin{cases} \dfrac{\partial u}{\partial t} + \Phi(t, x, u, u'_x, u''_{xx}) = 0 \ \text{ in }]0, T[\times \mathcal{O}, \\ u(t, x) = \kappa(t, x), \quad (t, x) \in]0, T[\times \partial\mathcal{O}, \\ u(0, x) = \kappa(0, x) \quad x \in \overline{\mathcal{O}}, \end{cases} \tag{6.130}$$

where $\kappa \in C([0, T[\times\overline{\mathcal{O}})$.

The notion of a viscosity solution to (6.130) is expressed as in Definition 6.96, adding the requirement $u(t, x) \leq \kappa(t, x)$ (resp. $\geq$) for $(t, x) \in (0, T) \times \partial\mathcal{O}$ for u to be a sub-solution (resp. a super-solution).

We have the comparison principle:

Theorem 6.103. *Let $\Phi \in C([0, T] \times \overline{\mathcal{O}} \times \mathbb{R} \times \mathbb{R}^d \times \mathbb{S}^d)$ be a proper function satisfying (A_1) and (A_2) for each fixed $t \in [0, T[$, with the same δ and $\mathbf{m}_R$. If $u \in USC([0, T) \times \overline{\mathcal{O}})$ is a viscosity sub-solution of (6.130) and $v \in LSC([0, T) \times \overline{\mathcal{O}})$ is a viscosity super-solution of (6.130) then*

$$u(t, x) \leq v(t, x)\,, \ \textit{for all } (t, x) \in [0, T) \times \overline{\mathcal{O}}.$$

An essential tool for the proof of this Theorem is the parabolic analog of Lemma 6.99, which is as follows:

Lemma 6.104. *Given $u, v \in C(\mathcal{O}_T)$, $\alpha > 0$, let*

$$\psi_\alpha(t,x,y) = u(t,x) - v(t,y) - \frac{\alpha}{2}|x-y|^2.$$

Let $(\hat{t},\hat{x},\hat{y})$ be a local maximum of ψ_α in $(0,T)\times\mathcal{O}\times\mathcal{O}$. Suppose moreover that there is an $r > 0$ such that for every $M > 0$ there is a C with the property that whenever $(p,q,X) \in \mathcal{P}^{2,+}_{\mathcal{O}}u(t,x)$, $|x-\hat{x}|+|t-\hat{t}| \le r$ and $|u(t,x)|+|q|+|X| \le M$, then $p \le C$, and the same is true if we replace $\mathcal{P}^{2,+}_{\mathcal{O}}u(t,x)$ by $-\mathcal{P}^{2,-}_{\mathcal{O}}v(t,x)$. Then there exist $p \in \mathbb{R}$, $X, Y \in \mathbb{S}^d$ such that

(j) $(p, \alpha(\hat{x}-\hat{y}), X) \in \overline{\mathcal{P}}^{2,+}_{\mathcal{O}}u(\hat{t},\hat{x})$,

(jj) $(-p, \alpha(\hat{x}-\hat{y}), Y) \in \overline{\mathcal{P}}^{2,-}_{\mathcal{O}}v(\hat{t},\hat{x})$,

(jjj) $\begin{pmatrix} X & 0 \\ 0 & -Y \end{pmatrix} \le 3\alpha \begin{pmatrix} I & -I \\ -I & I \end{pmatrix}.$

Proof of the Theorem. We only sketch the proof. We first observe that it suffices to prove that $\tilde{u}(t,x) = u(t,x) - \varepsilon/(T-t) \le v(t,x)$ for all $(t,x) \in (0,T)\times\mathcal{O}$ and all $\varepsilon > 0$. Now $\tilde{u}$ satisfies

$$\begin{cases} \dfrac{\partial \tilde{u}}{\partial t}(t,x) + \Phi(t,x,\tilde{u}(t,x), D\tilde{u}(t,x), D^2\tilde{u}(t,x)) \le -\dfrac{\varepsilon}{(T-t)^2}, \\ \lim\limits_{t\to T} \tilde{u}(t,x) = -\infty. \end{cases}$$

From now on we write u instead of $\tilde{u}$. We want to contradict the assumption that $\max_{(0,T)\times\mathcal{O}}[u-v] = \delta > 0$. Let $(\hat{t},\hat{x},\hat{y})$ be a local maximum of $\psi_\alpha(t,x,y)$ from Lemma 6.104, and write

$$M_\alpha - u(\hat{t},\hat{x}) - v(\hat{t},\hat{y}) - \frac{\alpha}{2}|\hat{x}-\hat{y}|^2.$$

From our standing assumption, $M_\alpha \ge \delta > 0$. It is not hard to show that for α large enough, $0 < \hat{t} < T$, $\hat{x}, \hat{y} \in \mathcal{O}$. Arguing as in the proof of Theorem 6.98 with the help this time of Lemma 6.104, we conclude that there exist $p \in \mathbb{R}$, $X, Y \in \mathbb{S}^d$ $c > 0$ such that

$$p + \Phi(\hat{t},\hat{x},u(\hat{t},\hat{x}),\alpha(\hat{x}-\hat{y},X) \le -c,$$

$$-p + \Phi(\hat{t},\hat{y},v(\hat{t},\hat{y}),\alpha(\hat{x}-\hat{y},Y) \ge 0,$$

while

$$\begin{pmatrix} X & 0 \\ 0 & -Y \end{pmatrix} \le 3\alpha \begin{pmatrix} I & -I \\ -I & I \end{pmatrix}.$$

We deduce that

$$\begin{aligned} c &\le \Phi(\hat{t}, \hat{y}, v(\hat{t}, \hat{y}), \alpha(\hat{x} - \hat{y}, Y) - \Phi(\hat{t}, \hat{x}, u(\hat{t}, \hat{x}), \alpha(\hat{x} - \hat{y}, X) \\ &\le m(\alpha|\hat{x} - \hat{y}|^2 + |\hat{x} - \hat{y}|), \end{aligned}$$

from which a contradiction follows. ■

6.5.3 A Second Uniqueness Result

We are given a continuous and globally monotone $f : \mathbb{R}^d \to \mathbb{R}^d$ and a globally Lipschitz $g : \mathbb{R}^d \to \mathbb{R}^{d \times d}$ together with

$$\kappa \in C(\mathbb{R}^d; \mathbb{R}^m), \qquad \text{and } F \in C([0, T] \times \mathbb{R}^d \times \mathbb{R}^m \times \mathbb{R}^{m \times d}; \mathbb{R}^m)$$

such that, for each $1 \le i \le k$, $F_i(t, x, y, z)$ depends on the matrix z only through its i-th column z_i. As already explained, this assumption is essential for the notion of a viscosity solution of the system of partial differential equations to be considered below to make sense. We assume specifically that for some constants $C, p > 0$:

(A.2i) $|F(t, x, 0, 0, 0)| \le C(1 + |x|^p), \quad |\kappa(x)| \le C(1 + |x|^p)$,
(A.2ii) $F = F(t, x, y, z)$ is globally Lipschitz in (y, z), uniformly in (t, x).

Remark 6.105. In the case of systems of equations, it does not seem possible to weaken the Lipschitz continuity of F in y to a monotonicity condition as we do in the case $m = 1$.

Under the assumptions (A.2i) and (A.2ii), for each $t \in [0, T]$ and $x \in \mathbb{R}^d$, we consider the system of PDEs

$$\begin{cases} -\dfrac{\partial u_i}{\partial t}(t, x) + \Phi_i(t, x, u(t, x), Du_i(t, x), D^2 u_i(t, x)) = 0, \\ \qquad\qquad\qquad\qquad (t, x) \in [0, T] \times \mathbb{R}^d, \quad 1 \le i \le k, \\ u_i(T, x) = \kappa_i(x), \ x \in \mathbb{R}^d, \quad 1 \le i \le m, \end{cases} \tag{6.131}$$

where

$$\Phi_i(t, x, r, q, X) = -\frac{1}{2} Tr[gg^*(x) X] - \langle f(x), q \rangle - F_i(t, x, r, q).$$

The notion of a viscosity solution for such a system is easily deduced from a combination of Definitions 6.94 and 6.96.

We can replace "global maximum point" or "global minimum point" by "strict global maximum point" or "strict global minimum point". The proof of this claim is very simple and we leave it as an exercise for the reader.

Now we give a uniqueness result for (6.131). This result is obtained under the following additional assumption:

(A.2 iii) $|F(t,x,r,p) - F(t,y,r,p)| \le \mathbf{m}_R(|x-y|(1+|p|))$,

for all $x, y \in \mathbb{R}^d$ such that $|x| \le R$, $|y| \le R$, $r \in \mathbb{R}^m$, $p \in \mathbb{R}^d$, where for each $R > 0$, $\mathbf{m}_R \in C(\mathbb{R}_+)$ is increasing and $\mathbf{m}_R(0) = 0$.

Our result is the following:

Theorem 6.106. *Assume that f, g satisfy (A2). Then there exists at most one viscosity solution u of (6.131) such that*

$$\lim_{|x|\to+\infty} |u(t,x)|e^{-\delta[\log(|x|)]^2} = 0, \tag{6.132}$$

uniformly for $t \in [0,T]$, for some $\delta > 0$.

Remark 6.107. Notice that any function which has at most a polynomial growth at infinity satisfies (6.132).

The growth condition (6.132) is optimal to obtain such a uniqueness result for (6.131). Indeed, consider the equation

$$\frac{\partial u}{\partial t} - \frac{x^2}{2}\frac{\partial^2 u}{\partial x^2} - \frac{x}{2}\frac{\partial u}{\partial x} = 0 \quad \text{in } (0,T)\times(0,+\infty), \tag{6.133}$$

then u is a solution of (6.133) if and only if the function $v(t,y) = u(t,e^y)$ is a solution of the Heat Equation

$$\frac{\partial v}{\partial t} - \frac{1}{2}\frac{\partial^2 v}{\partial x^2} = 0 \quad \text{in } (0,T)\times\mathbb{R}. \tag{6.134}$$

But it is well-known that, for the Heat Equation, the uniqueness holds in the class of solutions v satisfying

$$\lim_{|y|\to+\infty} |v(t,y)|e^{-\delta|y|^2} = 0, \tag{6.135}$$

uniformly for $t \in [0,T]$, for some $\delta > 0$. And (6.135) gives back (6.132) for (6.133) since $y = \log(x)$.

Let us finally mention that, in our case, the growth condition (6.132) is mainly a consequence of the assumptions on the coefficients of the differential operator and in particular on $a = gg^*$; under the assumptions of Theorem 6.106, the matrix a has, a priori, a quadratic growth at infinity. If a is assumed to have a linear growth at infinity, an easy adaptation of the proof of Theorem 6.106 shows that the uniqueness holds in the class of solutions satisfying

$$\lim_{|x|\to+\infty} |u(t,x)|e^{-\delta|x|} = 0,$$

uniformly for $t \in [0,T]$, for some $\delta > 0$.

Proof of Theorem 6.106. Let u and v be two viscosity solutions of (6.131). The proof consists of two steps. We first show that $u - v$ and $v - u$ are viscosity sub-solutions of an integral partial differential system; then we build a suitable sequence of smooth super-solutions of this system to show that $|u - v| = 0$ in $[0, T] \times \mathbb{R}^d$. Here and below, we denote by $| \cdot |$ the sup norm in $\mathbb{R}^m$.

Lemma 6.108. *Let u be a sub-solution and v a super-solution of (6.131). Then the function $\omega := u - v$ is a viscosity sub-solution of the system*

$$-\frac{\partial \omega_i}{\partial t} - \mathcal{A}\omega_i - \tilde{K}\left[|\omega| + |\nabla \omega_i g|\right] = 0 \text{ in } [0, T] \times \mathbb{R}^d, \tag{6.136}$$

for $1 \le i \le k$, where $\tilde{K}$ is the Lipschitz constant of F in (r, p).

Proof. Let $\varphi \in C^2([0, T] \times \mathbb{R}^d)$ and let $(t_0, x_0) \in (0, T) \times \mathbb{R}^d$ be a strict global maximum point of $\omega_i - \varphi$ for some $1 \le i \le k$.

We introduce the function

$$\psi_n(t, x, y) = u_i(t, x) - v_i(t, y) - n|x - y|^2 - \varphi(t, x),$$

where n is devoted to tend to infinity.

Since (t_0, x_0) is a strict global maximum point of $u_i - v_i - \varphi$, by a classical argument in the theory of viscosity solutions, there exists a sequence (t_n, x_n, y_n) such that:

(i) (t_n, x_n, y_n) is a global maximum point of ψ_n in $[0, T] \times (\overline{B}_R)^2$, where B_R is a ball with a large radius R;
(ii) $(t_n, x_n), (t_n, y_n) \to (t_0, x_0)$ as $n \to \infty$;
(iii) $n|x_n - y_n|^2$ is bounded and tends to zero as $n \to \infty$.

It follows from a variant of Lemma 6.104, see also Theorem 8.3 in the user's guide [18], that there exist $X, Y \in \mathbb{S}^d$ such that

$$\left(\frac{\partial \varphi}{\partial t}(t_n, x_n),\, q_n + D\varphi(t_n, x_n),\, X\right) \in \mathcal{P}^{2,+} u_i(t_n, x_n)$$

$$(0, q_n, Y) \in \mathcal{P}^{2,-} v_i(t_n, y_n)$$

$$\begin{pmatrix} X & 0 \\ 0 & -Y \end{pmatrix} \le 4n \begin{pmatrix} I & -I \\ -I & I \end{pmatrix} + \begin{pmatrix} D^2\varphi(t_n, x_n) & 0 \\ 0 & 0 \end{pmatrix},$$

where

$$q_n = 2n(x_n - y_n).$$

Modifying if necessary ψ_n by adding terms of the form $\chi(x)$ and $\chi(y)$ with supports in $B^c_{R/2}$, we may assume that (t_n, x_n, y_n) is a global maximum point of ψ_n in $([0,T] \times \mathbb{R}^d)^2$. Since u and v are respectively sub and super-solutions of (6.131), we have

$$-\frac{\partial \varphi}{\partial t}(t_n, x_n) - \frac{1}{2} Tr(a(x_n)X) - \langle f(x_n),\, q_n + D\varphi(t_n, x_n)\rangle$$
$$-F_i(t_n, x_n, u(t_n, x_n),\, (q_n + D\varphi(t_n, x_n))g(x_n)) \le 0$$

and

$$-\frac{1}{2} Tr(a(y_n)Y) - \langle f(y_n), q_n\rangle - F_i(t_n, y_n, v(t_n, y_n), p_n g(y_n)) \ge 0.$$

The computation of Lemma 6.97 yields

$$\frac{1}{2} Tr[gg^*(x_n)X] - \frac{1}{2} Tr[gg^*(y_n)Y] + \langle f(x_n) - f(y_n), q_n\rangle$$
$$\le n|x_n - y_n|^2 + Tr[gg^*(x_n) D^2\varphi(t_n, x_n)].$$

Finally, we consider the difference between the nonlinear terms

$$F_i(t_n, x_n,\, u(t_n, x_n),\, (q_n + D\varphi(t_n, x_n))g(x_n)) - F_i(t_n, y_n,\, v(t_n, y_n),\, q_n g(y_n))$$
$$\le \mathbf{m}(|x_n - y_n|(1 + |p_n g(y_n)|)) + \tilde{K}|u(t_n, x_n) - v(t_n, y_n)|$$
$$+\tilde{K}|q_n(g(x_n) - g(y_n)) + D\varphi(t_n, x_n)g(x_n)|.$$

The first term on the right-hand side comes from (A.2 iii): we have denoted by $\mathbf{m}$ the modulus $\mathbf{m}_R$ which appears in this assumption for R large enough. The two last terms come from the Lipschitz continuity of F_i with respect to the two last variables

We notice that

$$|q_n(g(x_n) - g(y_n))| \le Cn|x_n - y_n|^2,$$

because of the Lipschitz continuity of g and that

$$|x_n - y_n| \cdot |q_n g(y_n)| \le Cn|x_n - y_n|^2.$$

Now we subtract the viscosity inequalities for u and v: thanks to the above estimates, we can write the obtained inequality in the following way

$$-\frac{\partial \varphi}{\partial t}(t_n, x_n) - \mathcal{A}\varphi(t_n, x_n) - \tilde{K}|u(t_n, x_n) - v(t_n, y_n)| \le \omega_1(n),$$

where we have gathered in $\omega_1(n)$ all the terms of the form $n|x_n - y_n|^2$ and $|x_n - y_n|$; $\omega_1(n) \to 0$ when n tends to ∞. To conclude we let $n \to \infty$. Since (t_n, x_n), $(t_n, y_n) \to (t_0, x_0)$, we obtain:

$$-\frac{\partial \varphi}{\partial t}(t_0, x_0) - \mathcal{A}\varphi(t_0, x_0) - \tilde{K}|\omega(t_0, x_0)| - \tilde{K}|D\varphi(t_0, x_0)g(x_0)| \leq 0,$$

and therefore ω is a sub-solution of the desired equation. ∎

Now we are going to build suitable smooth super-solutions for the equation (6.136).

Lemma 6.109. *For any $\delta > 0$, there exists a $C_1 > 0$ such that the function*

$$\chi(t, x) = \exp\left[(C_1(T - t) + \delta)\psi(x)\right]$$

where

$$\psi(x) = \left[\log\left((|x|^2 + 1)^{1/2}\right) + 1\right]^2 ,$$

satisfies

$$-\frac{\partial \chi}{\partial t} - \mathcal{A}\chi - \tilde{K}\chi - \tilde{K}|D\chi g| > 0 \text{ in } [t_1, T] \times \mathbb{R}^d$$

for $1 \leq i \leq k$ where $t_1 = T - \delta/C_1$.

Proof. We first estimate the term $K\chi$, the main point being its dependence in x. For the sake of simplicity of notation, we denote below by C all the positive constants which enter into these estimates. These constants depend only on δ and on the bounds on the coefficients of the equations.

We first give estimates on the first and second derivatives of ψ: easy computations yield

$$|D\psi(x)| \leq \frac{2[\psi(x)]^{1/2}}{(|x|^2 + 1)^{1/2}} \leq 4 \quad \text{in } \mathbb{R}^d ,$$

and

$$|D^2\psi(x)| \leq \frac{C(1 + [\psi(x)]^{1/2})}{|x|^2 + 1} \quad \text{in } \mathbb{R}^d .$$

These estimates imply that, if $t \in [t_1, T]$

$$\begin{aligned}|D\chi(t, x)| &\leq (C_1(T - t) + \delta)\chi(t, x)|D\psi(x)| \\ &\leq C\chi(t, x)\frac{[\psi(x)]^{1/2}}{(|x|^2 + 1)^{1/2}} ,\end{aligned}$$

and, in the same way

$$|D^2\chi(t,x)| \le C\chi(t,x)\frac{\psi(x)}{|x|^2+1}\ .$$

It is worth noticing that, because of our choice of t_1, the above estimates do not depend on C_1.

Since gg^* and $\langle f(x), x\rangle$ grow at most quadratically at infinity, we have

$$-\frac{\partial\chi}{\partial t}(t,x) - \mathcal{A}\chi(t,x) - \tilde{K}\chi(t,x) - \tilde{K}|D\chi(t,x)g(x)|$$
$$\ge \chi\Big[C_1\psi(x) - C\psi(x) - C\frac{\psi(x)}{|x|^2+1} - \tilde{K} - C\tilde{K}[\psi(x)]^{1/2} - C\tilde{K}\frac{[\psi(x)]^{1/2}}{(|x|^2+1)^{1/2}}\Big].$$

Since $\psi(x) \ge 1$ in $\mathbb{R}^d$, by using the Cauchy–Schwartz inequality, it is clear enough that for C_1 large enough the quantity in the brackets is positive and the proof is complete. ■

To conclude the proof, we are going to show that $\omega = u - v$ satisfies

$$|\omega(t,x)| \le \alpha\chi(t,x) \ \text{ in } [0,T]\times\mathbb{R}^d$$

for any $\alpha > 0$. Then we will let α tend to zero.

To prove this inequality, we first remark that because of (6.132)

$$\lim_{|x|\to+\infty} |\omega(t,x)|e^{-\delta[\log((|x|^2+1)^{1/2})]^2} = -\infty$$

uniformly for $t \in [0,T]$, for some $\delta > 0$. From now on we choose δ in the definition of χ such that this holds. Then $|\omega_i| - \alpha\chi$ is bounded from above in $[t_1,T]\times\mathbb{R}^d$ for any $1 \le i \le k$ and

$$M = \max_{1\le i\le m}\ \max_{[t_1,T]\times\mathbb{R}^d} (|\omega_i| - \alpha\chi)(t,x)e^{-\tilde{K}(T-t)}$$

is attained at some point (t_0, x_0) and for some i_0.

We first remark that, since $|\cdot|$ is the sup norm in $\mathbb{R}^m$, we have

$$M = \max_{[t_1,T]\times\mathbb{R}^d} (|\omega| - \alpha\chi)(t,x)e^{-\tilde{K}(T-t)}$$

and $|\omega_{i_0}(t_0,x_0)| = |\omega(t_0,x_0)|$. We may assume without loss of generality that $|\omega_{i_0}(t_0,x_0)| > 0$, otherwise we are done.

There are two cases: either $\omega_{i_0}(t_0,x_0) > 0$ or $\omega_{i_0}(t_0,x_0) < 0$. We treat the first case, the second one is treated in a similar way since the roles of u and v are symmetric.

From the maximum point property, we deduce that

$$\omega_{i_0}(t,x) - \alpha\chi(t,x) \leq (\omega_{i_0} - \alpha\chi)(t_0,x_0)e^{-\tilde{K}(t-t_0)}$$

and this inequality can be interpreted as the property for the function $\omega_{i_0} - \phi$ to have a global maximum point at (t_0, x_0), where

$$\phi(t,x) = \alpha\chi(t,x) + (\omega_{i_0} - \alpha\chi)(t_0,x_0)e^{-\tilde{K}(t-t_0)}.$$

Since ω is a viscosity sub-solution of (6.136), if $t_0 \in [t_1, T[$, we have

$$-\frac{\partial\phi}{\partial t}(t_0,x_0) - \mathcal{A}\phi(t_0,x_0) - \tilde{K}|\omega(t_0,x_0)| - \tilde{K}|D\phi(t_0,x_0)g(x_0)| \leq 0.$$

But the left-hand side of this inequality is nothing but

$$\alpha\left[-\frac{\partial\chi}{\partial t}(t_0,x_0) - \mathcal{A}\chi(t_0,x_0) - \tilde{K}\chi(t_0,x_0) - \tilde{K}|D\chi(t_0,x_0)g(x_0)|\right],$$

since $\omega_{i_0}(t_0,x_0) = |\omega(t_0,x_0)|$; so, by Lemma 6.109, we have a contradiction. Therefore $t_0 = T$ and since $|\omega(T,x)| = 0$, we have

$$|\omega(t,x)| - \alpha\chi(t,x) \leq 0 \text{ in } [t_1,T] \times \mathbb{R}^d.$$

Letting α tend to zero, we obtain

$$|\omega(t,x)| = 0 \text{ in } [t_1,T] \times \mathbb{R}^d.$$

Applying successively the same argument on the intervals $[t_2, t_1]$ where $t_2 = (t_1 - \delta/C_1)^+$ and then, if $t_2 > 0$, on $[t_3, t_2]$ where $t_3 = (t_2 - \delta/C_1)^+ \ldots$ etc, we finally obtain that

$$|\omega(t,x)| = 0 \quad \text{in } [0,T] \times \mathbb{R}^d$$

and the proof is complete.

6.5.4 A Third Uniqueness Result

Let D be an open connected bounded subset of $\mathbb{R}^d$ of the form

$$D = \left\{x \in \mathbb{R}^d : \phi(x) < 0\right\}, \quad \mathrm{Bd}(D) = \left\{x \in \mathbb{R}^d : \phi(x) = 0\right\},$$

where $\phi \in C_b^3\left(\mathbb{R}^d\right)$, $|\nabla\phi(x)| = 1$, for all $x \in \mathrm{Bd}(D)$.

We define the outward normal derivative of v at the point $x \in \text{Bd}(D)$ by

$$\frac{\partial v(x)}{\partial n} = \sum_{j=1}^{d} \frac{\partial \phi(x)}{\partial x_j} \frac{\partial v(x)}{\partial x_j} = \langle \nabla \phi(x), \nabla v(x) \rangle .$$

The aim of this section is to prove uniqueness of a viscosity solution for the following parabolic variational inequality (PVI) with a mixed nonlinear multivalued Neumann–Dirichlet boundary condition:

$$\begin{cases} \dfrac{\partial u(t,x)}{\partial t} - \mathcal{A}_t u(t,x) + \partial \varphi(u(t,x)) \ni F(t,x,u(t,x),(\nabla u g)(t,x)), \\ \hfill t>0,\ x \in D, \\ \dfrac{\partial u(t,x)}{\partial n} + \partial \psi(u(t,x)) \ni G(t,x,u(t,x)), \quad t>0,\ x \in \text{Bd}(D), \\ u(0,x) = \kappa(x), \quad x \in \overline{D}, \end{cases} \tag{6.137}$$

where the operator $\mathcal{A}_t$ is given by

$$\mathcal{A}_t v(x) = \frac{1}{2} \text{Tr}\big[g(t,x) g^*(t,x) D^2 v(x)\big] + \big\langle f(t,x), \nabla v(x) \big\rangle.$$

We will make the following assumptions:

(I) The functions

$$\begin{array}{l} f : [0,\infty) \times \mathbb{R}^d \to \mathbb{R}^d, \\ g : [0,\infty) \times \mathbb{R}^d \to \mathbb{R}^{d\times d}, \\ F : [0,\infty) \times \overline{D} \times \mathbb{R} \times \mathbb{R}^d \to \mathbb{R}, \\ G : [0,\infty) \times \text{Bd}(D) \times \mathbb{R} \to \mathbb{R}, \\ \kappa : \overline{D} \to \mathbb{R} \end{array} \tag{6.138}$$

are continuous.

We assume that for all $T > 0$, there exist $\alpha \in \mathbb{R}$ and $L, \beta, \gamma \geq 0$ (which can depend on T) such that $\forall t \in [0,T]$, $\forall x, \tilde{x} \in \mathbb{R}^d$:

$$\langle f(t,x) - f(t,\tilde{x}), \frac{x-\tilde{x}}{|x-\tilde{x}|} \rangle^+ + \big| g(t,x) - g(t,\tilde{x}) \big| \leq L|x-\tilde{x}|, \tag{6.139}$$

and $\forall t \in [0,T]$, $\forall x \in \overline{D}$, $x' \in \text{Bd}(D)$, $y, \tilde{y} \in \mathbb{R}$, $z, \tilde{z} \in \mathbb{R}^d$:

$$\begin{array}{rl} (i) & (y-\tilde{y})(F(t,x,y,z) - F(t,x,\tilde{y},z)) \leq \alpha |y-\tilde{y}|^2, \\ (ii) & \big|F(t,x,y,z) - F(t,x,y,\tilde{z})\big| \leq \beta |z-\tilde{z}|, \\ (iii) & \big|F(t,x,y,0)\big| \leq \gamma\big(1+|y|\big), \\ (iv) & (y-\tilde{y})(G(t,x',y) - G(t,x',\tilde{y})) \leq \alpha |y-\tilde{y}|^2, \\ (v) & \big|G(t,x',y)\big| \leq \gamma(1+|y|). \end{array} \tag{6.140}$$

In fact, the conditions (6.140-i) and (6.140-iv) mean that, for all $t \in [0,T]$, $x \in \overline{D}$, $x' \in \text{Bd}(D)$, $z \in \mathbb{R}^d$,

$$r \mapsto \alpha y - F(t,x,ry,z) \quad \text{and} \quad r \mapsto \alpha r - G\left(t,x',r\right)$$

are increasing functions.

(II) We assume that

$$\begin{array}{ll} (i) & \varphi, \psi : \mathbb{R} \to (-\infty,+\infty] \text{ are proper convex l.s.c. functions,} \\ (ii) & \varphi(y) \geq \varphi(0) = 0 \text{ and } \psi(y) \geq \psi(0) = 0, \ \forall\ y \in \mathbb{R}, \end{array} \tag{6.141}$$

and there exists a positive constant M such that

$$\begin{array}{ll} (i) & \left|\varphi\left(\kappa(x)\right)\right| \leq M, \ \ \forall x \in \overline{D}, \\ (ii) & \left|\psi\left(\kappa(x)\right)\right| \leq M, \ \ \forall x \in \text{Bd}(D). \end{array} \tag{6.142}$$

Remark 6.110. Condition (6.141-ii) is generally satisfied after a translation of both the functions φ, ψ and their arguments.

We define

$$\begin{array}{l} \text{Dom}(\varphi) = \{u \in \mathbb{R} : \varphi(u) < \infty\}, \\ \partial\varphi(u) = \{\hat{u} \in \mathbb{R} : \hat{u}(v-u) + \varphi(u) \leq \varphi(v), \forall v \in \mathbb{R}\}, \\ \text{Dom}(\partial\varphi) = \{u \in \mathbb{R} : \partial\varphi(u) \neq \emptyset\}, \\ (u,\hat{u}) \in \partial\varphi \Leftrightarrow u \in \ \text{Dom}\partial\varphi, \ \ \hat{u} \in \partial\varphi(u) \end{array}$$

and we will use the same notions with φ replaced by ψ.

At every point $y \in \text{Dom}(\varphi)$ we have

$$\partial\varphi(y) = \mathbb{R} \cap \left[\varphi'_{-}(y), \varphi'_{+}(y)\right],$$

where $\varphi'_{-}(y)$ and $\varphi'_{+}(y)$ are resp. the left and right derivatives of φ at y.

For the reader's convenience we recall here from Sect. 5.8 the definition of a viscosity solution of the parabolic variational inequality (6.137). We define

$$\Phi(t,x,r,q,X) := -\frac{1}{2}\text{Tr}\left((gg^*)(t,x)X\right) - \langle f(t,x),q\rangle - F(t,x,r,qg(t,x)),$$

$$\Gamma(t,x,r,q) := \langle \nabla\phi(x),q\rangle - G(t,x,r).$$

Definition 6.111. Let $u : [0,\infty)\times\overline{D} \to \mathbb{R}$ be a continuous function, which satisfies $u(0,x) = \kappa(x)$, $\forall\ x \in \overline{D}$.

(a) u is a viscosity sub-solution of (6.137) if:

$$\left| \begin{array}{l} u(t,x) \in \mathrm{Dom}\,(\varphi)\,, \quad \forall (t,x) \in (0,\infty) \times \overline{D}, \\ u(t,x) \in \mathrm{Dom}\,(\psi)\,, \quad \forall (t,x) \in (0,\infty) \times \mathrm{Bd}\,(D)\,, \end{array} \right.$$

and for any $(t,x) \in (0,\infty) \times \overline{D}$, any $(p,q,X) \in \mathcal{P}^{2,+}u(t,x)$:

$$\begin{cases} p + \Phi\,(t,x,u(t,x),q,X) + \varphi'_{-}\,(u(t,x)) \leq 0 \;\; if \;\; x \in D, \\ \min \Big\{ p + \Phi\,(t,x,u(t,x),q,X) + \varphi'_{-}\,(u(t,x))\,, \\ \qquad\qquad \Gamma(t,x,u(t,x),q) + \psi'_{-}\,(u(t,x)) \Big\} \leq 0 \;\; if \;\; x \in \mathrm{Bd}\,(D)\,. \end{cases} \tag{6.143}$$

(b) The viscosity super-solution of (6.137) is defined in a similar manner as above, with $\mathcal{P}^{2,+}$ replaced by $\mathcal{P}^{2,-}$, the left derivative replaced by the right derivative, min by max, and the inequalities $\leq$ by $\geq$.
(c) A continuous function $u : [0,\infty) \times \overline{D}$ is a viscosity solution of (6.137) if it is both a viscosity sub- and super-solution.

We now present the main result of this section.

Theorem 6.112. *Let the assumptions* (6.138)–(6.142) *be satisfied. If moreover the function*

$$r \to G(t,x,r) \text{ is decreasing for all } t \geq 0,\; x \in \mathrm{Bd}\,(D)\,, \tag{6.144}$$

and there exists a continuous function $\mathbf{m} : [0,\infty) \to [0,\infty)$, $\mathbf{m}\,(0) = 0$, *such that*

$$\begin{array}{l} \left| F(t,x,r,q) - F(t,y,r,q) \right| \leq \mathbf{m}\,(|x-y|\,(1+|q|))\,, \\ \forall\, t \geq 0,\; x,y \in \overline{D},\; q \in \mathbb{R}^d, \end{array} \tag{6.145}$$

then the parabolic variational inequality (6.137) has at most one viscosity solution.

Proof. It is sufficient to prove uniqueness on a fixed arbitrary interval $[0,T]$.

Also, it suffices to prove that if u is a sub-solution and v is a super-solution such that $u(0,x) = v(0,x) = \kappa\,(x)$, $x \in \overline{D}$, then $u \leq v$.

Clearly by adding a constant we may assume that $\phi(x) \geq 0$ on $\overline{D}$.

For $\lambda = \alpha^{+} + 1$ and $\delta, \varepsilon, c > 0$ let

$$\begin{array}{l} \bar{u}\,(t,x) = e^{-\lambda t}u\,(t,x) - \delta\phi(x) - c \\ \bar{v}\,(t,x) = e^{-\lambda t}v\,(t,x) + \delta\phi(x) + c + \dfrac{\varepsilon}{T-t}. \end{array}$$

Let

$$\begin{aligned}\tilde{\Phi}(t,x,r,q,X) &= \lambda r - \frac{1}{2}\mathrm{Tr}\big[(gg^*)(t,x)X\big] - \langle f(t,x),q\rangle \\ &\quad - e^{-\lambda t}F\big(t,x,e^{\lambda t}r,e^{\lambda t}qg(t,x)\big), \\ \tilde{\Gamma}(t,x,r,q) &= \langle\nabla\phi(x),q\rangle - e^{-\lambda t}G(t,x,e^{\lambda t}r).\end{aligned} \tag{6.146}$$

Clearly $r \to \tilde{\Phi}(t,x,r,q,X)$ is an increasing function for all $(t,x,q,X) \in [0,T] \times \mathbb{R}^d \times \mathbb{R}^d \times \mathbb{S}^d$. Moreover, since

$$\sup_{(t,x)\in[0,T]\times\overline{D}} \big\{|\phi(x)| + |D\phi(x)| + |D^2\phi(x)| + |f(t,x)| + |g(t,x)|\big\} < \infty,$$

then for any $\delta > 0$, we can choose $c = c(\delta) > 0$ such that $c(\delta) \to 0$ as $\delta \to 0$ and for all $\delta, \varepsilon > 0$,

$$\begin{aligned}\tilde{\Phi}(t,x,r,q,X) &\le \tilde{\Phi}(t,x,r+\delta\phi+c,q+\delta D\phi,X+\delta D^2\phi), \\ \tilde{\Phi}(t,x,r-\delta\phi-c-\frac{\varepsilon}{T-t},q-\delta D\phi,X-\delta D^2\phi) &\le \tilde{\Phi}(t,x,r,q,X).\end{aligned}$$

We will prove that $\bar{u} \le \bar{v}$ for all $\delta > 0$, $\varepsilon > 0$, $c = c(\delta)$. This will imply $u \le v$ on $[0,T) \times \overline{D}$ by letting $\delta, \varepsilon \to 0$. The result will follow, since T is arbitrary.

Using the two last properties, assumption (6.144) and the fact that the left and right derivative of φ and ψ are increasing we infer that $\bar{u}$ satisfies in the viscosity sense:

$$\begin{cases}\dfrac{\partial\bar{u}}{\partial t}(t,x) + \tilde{\Phi}\big(t,x,\bar{u}(t,x),D\bar{u}(t,x),D^2\bar{u}(t,x)\big) + e^{-\lambda t}\varphi'_-\big(e^{\lambda t}\bar{u}(t,x)\big) \le 0 \\ \hfill \text{if } x \in D, t > 0 \\ \min\Big\{\dfrac{\partial\bar{u}}{\partial t}(t,x) + \tilde{\Phi}\big(t,x,\bar{u}(t,x),D\bar{u}(t,x),D^2\bar{u}(t,x)\big) + e^{-\lambda t}\varphi'_-\big(e^{\lambda t}\bar{u}(t,x)\big), \\ \tilde{\Gamma}(t,x,\bar{u}(t,x),D\bar{u}(t,x)) + \delta + e^{-\lambda t}\psi'_-\big(e^{\lambda t}\bar{u}(t,x)\big)\Big\} \le 0 \\ \hfill \text{if } x \in \mathrm{Bd}(D), t > 0.\end{cases} \tag{6.147}$$

Analogously we see that $\bar{v}$ satisfies in the viscosity sense:

$$\begin{cases}\dfrac{\partial\bar{v}}{\partial t}(t,x) + \tilde{\Phi}\big(t,x,\bar{v}(t,x),D\bar{v}(t,x),D^2\bar{v}(t,x)\big) \\ +e^{-\lambda t}\varphi'_+\big(e^{\lambda t}\bar{v}(t,x)\big) - \dfrac{\varepsilon}{(T-t)^2} \ge 0, \text{ if } x \in D,\ t > 0, \\ \max\Big\{\dfrac{\partial\bar{v}}{\partial t}(t,x) + \tilde{\Phi}\big(t,x,\bar{v}(t,x),D\bar{v}(t,x),D^2\bar{v}(t,x)\big) + e^{-\lambda t}\varphi'_+\big(e^{\lambda t}\bar{v}(t,x)\big) \\ \quad -\dfrac{\varepsilon}{(T-t)^2}, \tilde{\Gamma}(t,x,\bar{v}(t,x),D\bar{v}(t,x)) - \delta + e^{-\lambda t}\psi'_+\big(e^{\lambda t}\bar{v}(t,x)\big)\Big\} \ge 0 \\ \hfill \text{if } x \in \mathrm{Bd}(D), t > 0.\end{cases} \tag{6.148}$$

For simplicity of notation we write from now on u, v instead of $\bar{u}, \bar{v}$ respectively.

We now assume that

$$\max_{[0,T]\times\overline{D}} (u-v)^{+} > 0. \tag{6.149}$$

By an argument similar to that of Theorem 6.103, see Theorem 4.2 in [56] for more details, there exists $(\hat{t}, \hat{x}) \in (0, T] \times \mathrm{Bd}\,(D)$ such that

$$u(\hat{t}, \hat{x}) - v(\hat{t}, \hat{x}) = \max_{[0,T]\times\overline{D}} (u-v)^{+} > 0.$$

We now let

$$\psi_n\,(t,x,y) = u\,(t,x) - v\,(t,y) - \rho_n\,(t,x,y)\,,\ \text{with}\ (t,x,y) \in [0,T]\times\overline{D}\times\overline{D},$$

where

$$\begin{aligned}\rho_n\,(t,x,y) &= \frac{n}{2}\,|x-y|^2 + e^{-\lambda\hat{t}}G\big(\hat{t},\hat{x},e^{\lambda\hat{t}}u(\hat{t},\hat{x})\big)\big\langle\nabla\phi\,(\hat{x})\,,x-y\big\rangle + |x-\hat{x}|^4\\ &+|t-\hat{t}|^4 - e^{\lambda\hat{t}}\psi'_{-}\big(e^{-\lambda\hat{t}}u(\hat{t},\hat{x})\big)\big\langle\nabla\phi\,(\hat{x})\,,x-y\big\rangle.\end{aligned} \tag{6.150}$$

Let (t_n, x_n, y_n) be a maximum point of ψ_n.

We observe that $u\,(t,x) - v\,(t,x) - |x-\hat{x}|^4 - |t-\hat{t}|^4$ has $(\hat{t}, \hat{x})$ as its unique maximum point. Then, by Lemma 6.101, we have that as $n \to \infty$

$$\begin{aligned}&t_n \to \hat{t},\ x_n \to \hat{x},\ y_n \to \hat{x},\ n\,|x_n - y_n|^2 \to 0,\\ &u\,(t_n, x_n) \to u(\hat{t},\hat{x}),\ v\,(t_n, x_n) \to v(\hat{t},\hat{x}).\end{aligned} \tag{6.151}$$

But the domain D satisfies the uniform exterior sphere condition:

$$\exists\, r_0 > 0 \text{ such that } S\big(x + r_0\nabla\phi\,(x)\,, r_0\big) \cap D = \emptyset\,,\quad \text{for all } x \in \mathrm{Bd}\,(D)\,,$$

where $S\,(x, r_0)$ denotes the closed ball of radius r_0 centered at x.

Then

$$\big|y - x - r_0\nabla\phi\,(x)\,\big|^2 > r_0^2\,,\ \text{for } x \in \mathrm{Bd}\,(D)\,,\ y \in \overline{D},$$

or equivalently

$$\big\langle\nabla\phi\,(x)\,, y-x\big\rangle < \frac{1}{2r_0}\,|y-x|^2\ \text{for } x \in \mathrm{Bd}\,(D)\,,\ y \in \overline{D}. \tag{6.152}$$

If $x_n \in \mathrm{Bd}(D)$, we have, using the form of ρ_n given by (6.150) and (6.152), that

$$\begin{aligned}
\tilde{\Gamma}\big(t_n, x_n, u(t_n, x_n), D_x\rho_n(t_n, x_n, y_n)\big) &= \tilde{\Gamma}\Big(t_n, x_n, u(t_n, x_n), n(x_n - y_n) \\
&\quad + e^{-\lambda\hat{t}} G\big(\hat{t}, \hat{x}, e^{\lambda\hat{t}} u(\hat{t}, \hat{x})\big) \nabla\phi(\hat{x}) + 4|x_n - \hat{x}|^2 (x_n - \hat{x}) \\
&\quad - e^{-\lambda\hat{t}} \psi'_-\big(e^{\lambda\hat{t}} u(\hat{t}, \hat{x})\big) \nabla\phi(\hat{x})\Big) \\
&\geq -\frac{n}{2r_0}|x_n - y_n|^2 + e^{-\lambda\hat{t}} G\big(\hat{t}, \hat{x}, e^{\lambda\hat{t}} u(\hat{t}, \hat{x})\big)\big\langle \nabla\phi(\hat{x}), \nabla\phi(x_n)\big\rangle \\
&\quad - e^{-\lambda t_n} G\big(t_n, x_n, e^{\lambda t_n} u(t_n, x_n)\big) + 4|x_n - \hat{x}|^2 \big\langle \nabla\phi(x_n), x_n - \hat{x}\big\rangle \\
&\quad - e^{-\lambda\hat{t}} \psi'_-\big(e^{\lambda\hat{t}} u(\hat{t}, \hat{x})\big)\big\langle \nabla\phi(\hat{x}), \nabla\phi(x_n)\big\rangle.
\end{aligned}$$

Then (6.151) and the lower semicontinuity property of ψ'_- implies that along a subsequence $\{x_n\}$ which belongs to ∂D:

$$\liminf_{n\to\infty}\Big[\tilde{\Gamma}(t_n, x_n, u(t_n, x_n), D_x\rho_n(t_n, x_n, y_n)) + \delta + e^{-\lambda t_n}\psi'_-\left(e^{\lambda t_n} u(t_n, x_n)\right)\Big] > 0. \tag{6.153}$$

Analogously if $y_n \in \partial D$ we infer

$$\limsup_{n\to\infty}\Big[\tilde{\Gamma}\left(t_n, y_n, v(t_n, y_n), -D_y\rho_n(t_n, x_n, y_n)\right) - \delta + e^{-\lambda t_n}\psi'_+\left(e^{\lambda t_n} v(t_n, x_n)\right)\Big] < 0. \tag{6.154}$$

From Lemma 6.104 we deduce that there exists

$$(p, X, Y) \in \mathbb{R} \times \mathbb{S}^d \times \mathbb{S}^d,$$

such that

$$\begin{aligned}
&\big(p, D_x\rho_n(t_n, x_n, y_n), X\big) \in \overline{\mathcal{P}}^{2,+} u(t_n, x_n), \\
&\big(p, -D_y\rho_n(t_n, x_n, y_n), Y\big) \in \overline{\mathcal{P}}^{2,-} v(t_n, y_n),
\end{aligned}$$

and

$$\begin{pmatrix} X & 0 \\ 0 & -Y \end{pmatrix} \leq A + \frac{1}{n}A^2, \tag{6.155}$$

where $A = D^2_{x,y}\rho_n(t_n, x_n, y_n)$. From (6.150) we have

$$A = n\begin{pmatrix} I & -I \\ -I & I \end{pmatrix} + O\big(|x_n - \hat{x}|^2\big),$$
$$A^2 = 2n^2\begin{pmatrix} I & -I \\ -I & I \end{pmatrix} + O\big(n\,|x_n - \hat{x}|^2 + |x_n - \hat{x}|^4\big).$$

Then (6.155) becomes

$$\begin{pmatrix} X & 0 \\ 0 & -Y \end{pmatrix} \leq 3n\begin{pmatrix} I & -I \\ -I & I \end{pmatrix} + \delta_n\begin{pmatrix} I & 0 \\ 0 & I \end{pmatrix}, \tag{6.156}$$

where $\delta_n \to 0$.

Then from (6.147), (6.148) together with (6.154) and (6.153), we deduce that for n large enough

$$p + \tilde{\Phi}\big(t_n, x_n, u(t_n, x_n), D_x\rho_n(t_n, x_n, y_n), X\big) + e^{-\lambda t_n}\varphi'_-\big(e^{\lambda t_n}u(t_n, x_n)\big) \leq 0,$$

and

$$p + \tilde{\Phi}\big(t_n, y_n, v(t_n, y_n), -D_y\rho_n(t_n, x_n, y_n), Y\big) + e^{-\lambda t_n}\varphi'_+\big(e^{\lambda t_n}v(t_n, y_n)\big)$$
$$\geq \frac{\varepsilon}{(T - t_n)^2}.$$

Subtracting the last two inequalities, we deduce that

$$\begin{aligned} &\frac{\varepsilon}{(T - t_n)^2} \\ &\leq \tilde{\Phi}\big(t_n, y_n, v(t_n, y_n), -D_y\rho_n(t_n, x_n, y_n), Y\big) + e^{-\lambda t_n}\varphi'_+\big(e^{\lambda t_n}v(t_n, y_n)\big) \\ &\quad -\tilde{\Phi}\big(t_n, x_n, u(t_n, x_n), D_x\rho_n(t_n, x_n, y_n), X\big) - e^{-\lambda t_n}\varphi'_-\big(e^{\lambda t_n}u(t_n, x_n)\big). \end{aligned} \tag{6.157}$$

By (6.149) and (6.151) there exists an $N \geq 1$ such that for all $n \geq N$, the above holds together with

$$u(t_n, x_n) > v(t_n, y_n), \tag{6.158}$$

and consequently

$$e^{-\lambda t_n}\varphi'_-\big(e^{\lambda t_n}u(t_n, x_n)\big) \geq e^{-\lambda t_n}\varphi'_+\big(e^{\lambda t_n}v(t_n, y_n)\big).$$

Combining this with (6.157), we deduce that

$$\begin{aligned}\frac{\varepsilon}{(T-t_n)^2} &\le \tilde{\Phi}\left(t_n, y_n, v(t_n, y_n), -D_y\rho_n(t_n, x_n, y_n), Y\right)\\ &\quad -\tilde{\Phi}(t_n, x_n, u(t_n, x_n), D_x\rho_n(t_n, x_n, y_n), X)\\ &\le \frac{1}{2}\mathrm{Tr}\big[(gg^*)(t_n, x_n)X - (gg^*)(t_n, y_n)Y\big] + Cn|x_n - y_n|^2 + \omega_n,\end{aligned}$$

where $\omega_n \to 0$ as $n \to \infty$. Note that we have used the assumption (6.145), (6.151), (6.158), the fact that $r \to \lambda r - F(t,x,r,z)$ is increasing, and the Lipschitz continuity of F with respect to its last variable.

From (6.156), $\forall\ q, \tilde{q} \in \mathbb{R}^d$,

$$\langle Xq, q\rangle - \langle Y\tilde{q}, \tilde{q}\rangle \le 3n\,|q-\tilde{q}|^2 + \left(|q|^2 + |\tilde{q}|^2\right)\delta_n.$$

Hence by the same computation as in Lemma 6.97 we obtain

$$\mathrm{Tr}\big[\left(gg^*\right)(t_n, x_n)X - \left(gg^*\right)(t_n, y_n)Y\big]$$

$$\le 3C\ n\,|x_n - y_n|^2 + \left(|g(t_n, x_n)|^2 + |g(t_n, y_n)|^2\right)\delta_n,$$

and consequently taking the limit in the above set of inequalities yields

$$\frac{\varepsilon}{(T-\hat{t})^2} \le 0,$$

which is a contradiction.

Then

$$u(t,x) \le v(t,x),\ \forall\,(t,x) \in [0,T] \times \overline{D}.$$

■

6.6 Annex E: Hints for Some Exercises

Chapter 1

Exercise 1.7

By Proposition 1.34 we have

$$\begin{aligned}\mathbb{E}\left(g\left(B_T\right)|\mathcal{F}_t\right) &= \mathbb{E}\left(g\left(\frac{B_T - B_t}{\sqrt{T-t}}\sqrt{T-t} + B_t\right)|\mathcal{F}_t\right)\\ &= \int_{\mathbb{R}} g\left(x\sqrt{T-t} + B_t\right)\rho(x)\,dx.\end{aligned}$$

Setting here $g(u) = \mathbf{1}_{(-\infty,a]}(u)$, the second assertion follows.

Exercise 1.15: Let $N > 0$ and a sequence $\varepsilon_n \searrow 0$ as $n \to \infty$. Then

$$\begin{aligned}
\mathbb{P}\left(\limsup_{n\to+\infty} \frac{|B_{t+\varepsilon_n} - B_t|}{\varepsilon_n} > N\right) &= \mathbb{P}\left(\bigcap\nolimits_{n\geq 1} \downarrow \left(\bigcup\nolimits_{k\geq n} \left(|B_{t+\varepsilon_k} - B_t| > N\varepsilon_k\right)\right)\right) \\
&= \lim_{n\to\infty} \mathbb{P}\left(\bigcup\nolimits_{k\geq n} \left(|B_{t+\varepsilon_k} - B_t| > N\varepsilon_k\right)\right) \\
&\geq \liminf_{n\to+\infty} \mathbb{P}\left(|B_{t+\varepsilon_n} - B_t| > N\varepsilon_n\right) \\
&= \liminf_{n\to+\infty} \mathbb{P}\left(|B_1| > N\sqrt{\varepsilon_n}\right) \\
&= \mathbb{P}\left(|B_1| > 0\right) \\
&= 1.
\end{aligned}$$

Exercise 1.16: Let us write $S_n^{(p)} = S_{\Delta_n}^{(p)}(B.; [s,t])$. The results are consequences of the following inequalities (see Proposition 1.86 for the first one) combined with Proposition 1.14 and Proposition 1.7:

$$\mathbb{E}\left[S_n^{(2)} - (t-s)\right]^2 = \mathrm{Var}\left(S_n^{(2)}\right) \leq 2\,\|\Delta_n\|\,(t-s)$$

and

$$\begin{aligned}
S_{\Delta_n}^{(p)} &\leq S_{\Delta_n}^{(2)} \times \left(\mathbf{m}_B\left(\|\Delta_n\|\right)\right)^{p-2}, \quad \text{for } p > 2, \quad \text{and} \\
S_{\Delta_n}^{(2)} &\leq S_{\Delta_n}^{(p)} \times \left(\mathbf{m}_B\left(\|\Delta_n\|\right)\right)^{2-p}, \quad \text{for } 1 \leq p < 2,
\end{aligned}$$

where

$$\mathbf{m}_B(\delta) = \sup\left\{|B_u - B_v| : u, v \in [s,t],\ |u-v| \leq \delta\right\},$$

is the modulus of continuity of $\{B_u : u \in [s,t]\}$.

Exercise 1.17: Applying the inequality (1.25) with $\alpha = \frac{1}{2} - \frac{\varepsilon}{2}$ and $p = \dfrac{2}{\varepsilon}$, we deduce that for all $s, t \in [0,T]$

$$|X_t(\omega) - X_s(\omega)| \leq \xi(\omega)\, T^{\varepsilon}\, |t-s|^{\frac{1}{2}-\varepsilon},$$

where

$$\xi(\omega) = \xi_{\varepsilon,T}(\omega) = \begin{cases} 0, & \text{if } T = 0, \\ \dfrac{C_\varepsilon}{T^\varepsilon}\left(\displaystyle\int_0^T\int_0^T \frac{|B_u(\omega) - B_r(\omega)|^{\frac{2}{\varepsilon}}}{|u-r|^{\frac{1}{\varepsilon}}} dudr\right)^{\frac{\varepsilon}{2}}, & \text{if } T > 0. \end{cases}$$

Let $1 \le q \le \frac{2}{\varepsilon} \le p$. By Lyapunov's inequality and Minkowski's inequality (1.24) from Exercise 1.2 we obtain

$$\begin{aligned}
\|\xi\|_{L^q(\Omega,\mathcal{F},\mathbb{P})} &\le \|\xi\|_{L^p(\Omega,\mathcal{F},\mathbb{P})} \\
&= \frac{C_\varepsilon}{T^\varepsilon} \left\| \int_0^T \int_0^T \frac{|B_u - B_r|^{\frac{2}{\varepsilon}}}{|u-r|^{\frac{1}{\varepsilon}}} dudr \right\|_{L^{\varepsilon p/2}(\Omega,\mathcal{F},\mathbb{P})}^{\frac{\varepsilon}{2}} \\
&\le \frac{C_\varepsilon}{T^\varepsilon} \left(\int_0^T \int_0^T \frac{\left\| |B_u - B_r|^{\frac{2}{\varepsilon}} \right\|_{L^{\varepsilon p/2}(\Omega,\mathcal{F},\mathbb{P})}}{|u-r|^{\frac{1}{\varepsilon}}} dudr \right)^{\frac{\varepsilon}{2}} \\
&= C_{\varepsilon,p},
\end{aligned}$$

since

$$\begin{aligned}
\left\| |B_u - B_r|^{\frac{2}{\varepsilon}} \right\|_{L^{\varepsilon p/2}(\Omega,\mathcal{F},\mathbb{P})} &= \left(\mathbb{E}\,|B_u - B_r|^p\right)^{\frac{2}{\varepsilon p}} \\
&= \left(C_p\,|u-r|^{p/2}\right)^{\frac{2}{\varepsilon p}}.
\end{aligned}$$

Exercise 1.19: Deduce from the proof of Theorem 1.40 that for any $0 < \delta < b/a$, there exists a constant $K = K(M, T, a, b, \delta)$ such that for all $\varepsilon, \lambda > 0$,

$$\mathbb{P}\left(\mathbf{m}_{X^n}(\varepsilon;[0,T]) \ge \lambda\right) \le \frac{1}{\lambda^a}\mathbb{E}\left(\mathbf{m}^a_{X^n}(\varepsilon;[0,T])\right) \le \frac{K}{\lambda^a}\varepsilon^{b-a\delta}$$

and conclude that (ii) in Theorem 1.46 is satisfied.

Exercise 1.20 (2) By Lemma 1.73 and Proposition 1.65, we infer that $(U_t^{(\lambda)})_{t\in[0,T]}$ and $(Z_t^{(\lambda)})_{t\in[0,T]}$ are continuous martingales.

(3) Let the stopping time $\tau_n = \inf\{t \ge 0 : |M_t| + \langle M\rangle_t \ge n\}$. Then $\left\{Z^{(\lambda)}_{t\wedge\tau_n}; t \ge 0\right\}$ is a martingale and for all $0 \le s \le t$

$$\mathbb{E}^{\mathcal{F}_s} Z_t^{(\lambda)} \le \liminf_{n\to+\infty} \mathbb{E}^{\mathcal{F}_s} Z^{(\lambda)}_{t\wedge\tau_n} = \liminf_{n\to+\infty} Z^{(\lambda)}_{s\wedge\tau_n} = Z_s^{(\lambda)}.$$

(5) By Proposition 1.59 with $\varphi(x) = e^{ax}$, $\left\{e^{aM_{t\wedge\theta}}; t \ge 0\right\}$ is a sub-martingale and the inequality follows by Doob's inequality (Theorem 1.60) and Hölder's inequality.

(6) The inequality yields that $\left\{Z^{(\lambda)}_{t\wedge\theta_n}; n \in \mathbb{N}^*\right\}$ is uniformly integrable and consequently $\mathbb{E}\,Z_t^{(\lambda)} = \lim_{n\to\infty}\mathbb{E}Z^{(\lambda)}_{t\wedge\theta_n} = 1$.

(7) In the inequality from (6) with $A = \Omega$, one passes to the limit as $n \to \infty$ and then $\lambda \nearrow 1$.

(8) We have $\mathbb{E}\left(e^{\frac{1}{2}M_T}\right) = \mathbb{E}\left(\sqrt{Z}_T e^{\frac{1}{4}<M>_T}\right) \le (\mathbb{E}\, Z_T)^2\, \mathbb{E}\left(e^{\frac{1}{2}<M>_T}\right) \le \mathbb{E}\left(e^{\frac{1}{2}<M>_T}\right) < \infty$.

Chapter 2

Exercise 2.1: ($\Leftarrow$): From the theory of the Riemann–Stieltjes integral we know that if $g \in BV[0,T]$, then $S_n(f)$ converges (to the Riemann–Stieltjes integral $\int_0^T f(t)\, dg(t)$).

($\Rightarrow$): Let $S_n(f)$ be convergent for all $f \in C[0,T]$. Then $S_n : C[0,T] \to R$ is a bounded linear operator such that

$$\sup_{n\ge 1} |S_n(f)| < \infty,$$

and by the Banach–Steinhauss Theorem

$$\sup_{n\ge 1} \|S_n\| = M < \infty,$$

where $\|S_n\| = \sup\{|S_n(f)| : \|\!|f|\!\|_T \le 1\}$. For a fixed n we can construct $h_n \in C[0,T]$ such that $h_n(t_i^n) = \text{sign}\left\{g\left(t_{i+1}^n\right) - g\left(t_i^n\right)\right\}$ and $\|\!|h_n|\!\|_T = 1$. Hence

$$\sum_{i=0}^{n-1} \left|g\left(t_{i+1}^n\right) - g\left(t_i^n\right)\right| = S_n(h_n) \le \|S_n\| \le M,$$

and as a consequence g is of finite variation.

Note (Banach–Steinhauss Theorem). Let X be a Banach space and let Y be a normed linear space. Let $S_i : X \to Y$, $i \in I$, be a family of bounded linear operators. If for each $x \in X$ the set $\{S_i(x) : i \in I\}$ is bounded then the set $\{\|S_i\| : i \in I\}$ is bounded.

Remark: This is not a contradiction since the subsequence $\{n_k\}$ depends on f.

Exercise 2.3: If $\mathcal{E}$ is the linear subspace of $L^2(\mathbb{R}_+)$ consisting of those functions f of the form:

$$f = \sum_{i=0}^{n-1} a_i \mathbf{1}_{[t_i, t_{i+1}[}, \quad n \in \mathbb{N}^*; 0 = t_0 < t_1 < \cdots < t_n;\ a_i \in \mathbb{R},\ i \le n,$$

then $H[B]$ is the closure of $\{\mathbb{B}(f),\ f \in \mathcal{E}\}$, which coincides with $\{\mathbb{B}(f),\ f \in L^2(\mathbb{R}_+)\}$. Moreover the set $\{B_t = \mathbb{B}(\mathbf{1}_{[0,t]}),\ t > 0\}$ is total in $H[B]$.

Exercise 2.4: Let $s \in [0,T]$. We have

$$\mathbb{E}\left[\left(\int_0^T f(t)\, dB_t + \int_0^T f'(t) B_t\, dt\right) B_s\right] = \int_0^s f(t)\, dt + \int_0^T f'(t)(s \wedge t)\, dt$$
$$= f(T)s.$$

Since $\mathbb{E}|B_t| = \sqrt{\frac{2t}{\pi}}$, it follows that $\int_0^\infty |f'(t)||B_t|\,dt < \infty \quad a.s.$, and

$$\begin{aligned}\int_0^\infty f^2(t)\,dt &= \int_0^\infty \left(\int_0^\infty \mathbf{1}_{[t,\infty[}(u)\, f'(u)\,du \int_0^\infty \mathbf{1}_{[t,\infty[}(v)\, f'(v)\,dv\right) dt \\ &\le \int_0^\infty \int_0^\infty (u \wedge v)|f'(u)||f'(v)|\,du\,dv \\ &\le \int_0^\infty \int_0^\infty \sqrt{uv}|f'(u)||f'(v)|\,du\,dv \\ &= \left(\int_0^\infty \sqrt{u}|f'(u)|\,du\right)^2 < \infty.\end{aligned}$$

Exercise 2.5: Note that

$$g'(x) = 30(x-1)^2(2-x)^2 \ge 0 \quad \text{and} \quad g''(x) = 60(x-1)(2-x)(3-2x).$$

and for $x \in [1,2]$

$$0 \le (x-1)(2-x) \le \left(\frac{x-1+2-x}{2}\right)^2 = \frac{1}{4},$$

and therefore for all $x \in [1,2]$,

$$0 \le g'(x) \le 2, \quad |g''(x)| \le 15.$$

The relation (2.67) follows by taking the limit as $\varepsilon \to 0$ in Itô's formula for $\varphi_\varepsilon(X_t)$.

Chapter 3

Exercise 3.1: Consider the equation

$$\begin{aligned}X_t &= \xi + \textstyle\int_0^t F(s,X_s)\,ds \\ &\quad + \int_0^t \left(-\mu(s) - \frac{m_p}{2}\ell^2(s) - \frac{a}{p}\right) X_s ds + \textstyle\int_0^t G(s,X_s)\,dB_s.\end{aligned} \tag{6.159}$$

By Theorem 3.27, it has a unique solution $X \in S_d^0$ and from the inequality (3.18) we clearly have (3.131)

$$U_t = \left(-\mu(t) - \frac{m_p}{2}\ell^2(t) - \frac{a}{p}\right) X_t, \tag{6.160}$$

where $X \in S_d^0$ is the solution of the Eq. (6.159). The inequality (3.132) shows us that $Y_t = e^{at}\frac{|X_t|^p}{(1+\delta|X_t|^2)^{p/2}}$ is a super-martingale and then (3.134).

Exercise 3.3: First deduce the following from the stochastic Gronwall inequalities (Annex C)

$$\mathbb{E}\sup_{t\in[0,T]}|X_t^\varepsilon - X_t|^p \le C_p\mathbb{E}\left(\int_0^T |F_\varepsilon(r,X_r) - F(r,X_r)|\,dr\right)^p e^{C_p\int_0^t[\mu^+(r)+\ell^2(r)]dr}.$$

Exercise 3.9: $(1i)$ We clearly have

$$\mathbb{E}\exp\left(C|x+B_t|^b\right) = \frac{1}{(2\pi)^{d/2}}\int_{\mathbb{R}^d}\exp\left(C\left|x+\sqrt{t}u\right|^b - \frac{|u|^2}{2}\right)du < \infty,$$

for all $C, t \ge 0$ if and only if $0 \le b < 2$.

$(1ii)$ If $0 \le a < 2$, then by Jensen's inequality

$$\mathbb{E}\exp\left(C\int_0^t |x+B_s|^a\,ds\right) \le \frac{1}{t}\int_0^t \mathbb{E}\exp\left(Ct|x+B_s|^a\right)ds < \infty.$$

If $-1 < a < 0$, then by Corollary 2.30 we have

$$|x+B_t|^{a+2} = (a+2)\int_0^t |x+B_s|^a\langle x+B_s, dB_s\rangle + \frac{(a+2)(a+1)}{2}\int_0^t |x+B_s|^a\,ds.$$

Hence by (2.62-b)

$$\begin{aligned}
&\mathbb{E}\exp\left(C\int_0^t |x+B_s|^a\,ds\right)\\
&\le \left[\mathbb{E}\exp\left(C_1|x+B_t|^{a+2}\right)\right]^{1/2}\left[\mathbb{E}\exp\left(C_2\int_0^t |x+B_s|^a\langle x+B_s, dB_s\rangle\right)\right]^{1/2}\\
&\le \left[\mathbb{E}\exp\left(C_1|x+B_t|^{a+2}\right)\right]^{1/2}\left[\mathbb{E}\exp\left(2C_2\int_0^t |x+B_s|^{2a+2}\,ds\right)\right]^{1/2}\\
&< \infty.
\end{aligned}$$

$(1iii)$

$$\begin{aligned}
&\mathbb{E}\left\{\exp\left[C\log^2(|x+B_t|)\right]\right\}\\
&\ge \int_{[0,1]^k} e^{C\log^2|u|}\frac{1}{(2\pi t)^{k/2}}e^{-|u-x|^2/2t}du\\
&\ge \frac{1}{(2\pi t)^{k/2}}e^{-(k+|x|)^2/2t}\int_{[0,1]^k} e^{\frac{C}{4}\log^2\left(u_1^2+\cdots+u_k^2\right)}du_1\ldots du_k
\end{aligned}$$

$$\geq C_{k,t,x}\int_0^1 e^{C\log^2 u_1}du_1$$
$$= C_{k,t,x}\int_0^\infty e^{Cy^2}e^{-y}dy$$
$$= \infty.$$

$(1iv)$ Observe that for every $\alpha \in]0,1[$, there exists a $C_\alpha > 0$ such that

$$\log^2|x| \leq C_\alpha + |x| + |x|^{-\alpha}$$

and consequently (iii) follows from (ii).

(2). Existence follows in both cases from Lemma 2.49 and Girsanov's Theorem 2.51. Uniqueness in law on $(\Omega,\mathcal{F}_{n\wedge\tilde{T}_n})$ (resp. $\left(\Omega,\mathcal{F}_{n\wedge\hat{T}_n}\right)$) follows again from Girsanov's Theorem, where

$$\tilde{T}_n = \inf\left\{t>0:\int_0^t |g(X_s)|^2\log^2(|X_s|)ds > n\right\},$$
$$\hat{T}_n = \inf\left\{t>0:\int_0^t |g(X_s)|^2|X_s|^a ds > n\right\}.$$

It remains to note that $\tilde{T}_n\to\infty$, $\hat{T}_n\to\infty$, as $n\to\infty$.

Exercise 3.10 The function $F:\mathbb{R}\to\mathbb{R}$, $F(x) = f(x)\sqrt{|x|}\operatorname{sign}(x)$ is locally monotone and $xF(x)\leq 0$, but it is not locally Lipschitz.

Chapter 4

Exercise 4.1

1. The existence and the uniqueness of the solution $X^n \in S^2[0,T]$ follows from Theorem 3.17; by the comparison result from Proposition 3.12 we have $X_t^{n+1} \geq X_t^n$, for all $t\in[0,T]$, $\mathbb{P}$-$a.s.$
2. Let L and ℓ be the Lipschitz constants of f and, respectively, g. We have

$$X_t^n - 1 = (x-1) + \int_0^t dK_s^n + \int_0^t g(X_s^n)dB_s,$$

with $dK_s^n = \left[f(X_s^n) + n(X_s^n)^-\right]ds$ and $G_s^n = g(X_s^n)$. Since

$$dD_t^n + (X_t^n - 1)\,dK_t^n + |G_t^n|^2\,dt \leq dR_t + |X_t^n - 1|^2\,dV_t,$$

where $D_t^n = n\int_0^t\left[(X_s^n)^-\right]^2 ds + n\int_0^t (X_s^n)^- ds$, $R_t = \left(\frac{1}{2}|f(1)|^2 + 2|g(1)|^2\right)t$ and $V_t = \left(L + \frac{1}{2} + 2\ell^2\right)t$, it follows by (6.78) (with $p=2$ and $\lambda = 1/18$) that

$$\mathbb{E} \sup_{t\in[0,T]} |X_t^n - 1|^2 + \mathbb{E}\left(\int_0^T n\left[(X_t^n)^-\right]^2 dt\right) + \mathbb{E}\left(\int_0^T n(X_t^n)^- dt\right) \le C_2.$$

3. Since, moreover, $\left(X_t^n\right)^- \ge \left(X_t^{n+1}\right)^-$ for all $t \in [0,T]$, $\mathbb{P}$-a.s., it follows that $\lim_{n\to\infty}\left(X_t^n\right)^- = 0$, $d\mathbb{P}\otimes dt - a.e.$ By Itô's formula for $\left[\left(X_t^n\right)^-\right]^2$ (see Proposition 2.35), we deduce $\mathbb{E} \sup_{0\le t\le T} |\left(X_t^n\right)^-|^2 \to 0$, as $n\to\infty$.
4. Since

$$\begin{aligned}
&\left(X_t^n - X_t^m\right)\left[f\left(X_t^n\right) - f\left(X_t^m\right) + n(X_t^n)^- - m(X_t^m)^-\right] dt \\
&\qquad + \left|g\left(X_t^n\right) - g\left(X_t^m\right)\right|^2 dt \\
&\quad \le (n+m)\left[(X_t^n)^-(X_t^m)^-\right] dt + \left(L+\ell^2\right)|X_t^n - X_t^m|\, dt,
\end{aligned}$$

we see, by (3.138), that

$$\begin{aligned}
\mathbb{E} \sup_{t\in[0,T]} |X_t^n - X_t^m|^2 &\le C\mathbb{E}\int_0^T (n+m)\left[(X_t^n)^-(X_t^m)^-\right] dt \\
&\le C\mathbb{E}\left(\mathbb{E} \sup_{t\in[0,T]} \left[(X_t^m)^-\right]^2\right)^{1/2}\left[\mathbb{E}\left(\int_0^T n(X_t^n)^- dt\right)^2\right]^{1/2} \\
&\quad + C\mathbb{E}\left(\mathbb{E} \sup_{t\in[0,T]} \left[(X_t^n)^-\right]^2\right)^{1/2}\left[\mathbb{E}\left(\int_0^T m(X_t^m)^- dt\right)^2\right]^{1/2} \\
&\to 0, \quad \text{as } n,m\to\infty.
\end{aligned}$$

9. It is sufficient to prove that the SDE

$$X_t = x + \int_0^t \left(f\left(X_s\right) + [f(0)]^- \mathbf{1}_{X_s=0}\right) ds + \int_0^t g\left(X_s\right) dB_s$$

has a unique positive solution $X \in S^2[0,T]$. The uniqueness of positive solutions follows from

$$\begin{aligned}
&\left(X_s - \hat{X}_s\right)\left[f\left(X_s\right) + [f(0)]^- \mathbf{1}_{X_s=0} - f\left(\hat{X}_s\right) - [f(0)]^- \mathbf{1}_{\hat{X}_s=0}\right] \\
&\qquad\qquad + \left|g\left(X_s\right) - g\left(\hat{X}_s\right)\right|^2 \\
&\le \left(L+\ell^2\right)\left|X_s - \hat{X}_s\right|^2
\end{aligned}$$

and Corollary 6.77. The existence of a positive solution follows from the approximating equation

$$X_t^\varepsilon = x + \int_0^t \left[f\left(X_s^\varepsilon\right) + [f(0)]^- \left(1 - \frac{|X_s^\varepsilon|}{\varepsilon}\right)^+ \right] ds + \int_0^t g\left(X_s^\varepsilon\right) dB_s.$$

Note that $\tilde{X}^\varepsilon = 0$ is the unique solution of the SDE

$$\tilde{X}_t^\varepsilon = 0 + \int_0^t \left[f\left(\tilde{X}_s^\varepsilon\right) - f(0) \left(1 - \frac{|\tilde{X}_s^\varepsilon|}{\varepsilon}\right)^+ \right] ds + \int_0^t g\left(\tilde{X}_s^\varepsilon\right) dB_s,$$

and $f(0) + [f(0)]^- \left(1 - \frac{|0|}{\varepsilon}\right)^+ \geq 0$, which yields (by Proposition 3.12) $X_t^\varepsilon \geq 0$.

10. By Remark 2.27 we have for all $t \geq 0$,

$$0 = \int_0^t \mathbf{1}_{X_s = y} g^2(X_s)\, ds = g^2(y) \int_0^t \mathbf{1}_{X_s = y} ds.$$

Exercise 4.2: On each interval $\mathbf{I}_i^n$ the equations from the schema (4.149) have unique adapted solutions U^n, V^n and Y^n, respectively; U^n is absolutely continuous; $H_\cdot^n = F_1(\cdot, U_\cdot^n) - \frac{d}{dt} U_\cdot \in L^1(\Omega \times]0,T[)$. Let $K_t^n = \int_0^t H_s^n ds$. To prove (4.150) the steps are:

1. $\mathbb{E}\left(|U_t^n|^4 + |V_t^n|^4 + |Y_t^n|^4 + |X_t^n|^4 + \updownarrow K^n \updownarrow_t^2\right) \leq C\left(1 + \mathbb{E}|H_0|^4\right)$;
2. $\mathbb{E} \sup_{t\in[0,T]} |V_t^n - U_t^n|^4 \leq \frac{C}{n^3}\left(1 + \mathbb{E}|H_0|^4\right)$;
3. $\mathbb{E} \sup_{t\in[0,T]} |Y_t^n - U_t^n|^4 + \mathbb{E} \sup_{t\in[0,T]} |X_t^n - U_t^n|^4 \leq \frac{C}{n}\left(1 + \mathbb{E}|H_0|^4\right)$;
4. $\mathbb{E} \sup_{t\in[0,T]} |Y_t^n - U_t^n|^2 \leq \frac{C}{\sqrt{n}}\left(1 + \mathbb{E}|H_0|^4\right)$;
5. Let $t \in \mathbf{I}_i^n$. By Itô's formula for $|Y_t^n - X_t|^2$ and the above estimates we obtain (4.150).

Exercise 4.3: In the same manner as the estimate from Proposition 4.8 is obtained, we derive using Proposition 6.74 the boundedness of approximating quantities. Then estimating, via the same Proposition 4.8, $X^\varepsilon - X$and $\hat{X}^\varepsilon - X$and using Proposition 6.9 the convergence results follow.

Exercise 4.4: For the first four questions, choose the control in feedback form as follows:

$$U_s = -\left(\mu(s) + \frac{1}{2} m_p \ell^2(s) + \frac{a}{p}\right)(X_s - x_0).$$

For the last question, choose

$$\tilde{U}_t = -\left[\mu(s) + \left(\frac{1}{2}m_p + 9p\lambda\right)\ell^2(s) + \frac{a}{p}\right](\tilde{X}_s - x_0).$$

Exercise 4.5: The equivalence follows easily from Example 4.79.
Exercise 4.6: 1&2 Let $\hat{x} \in \Pi_E(x)$ and $\hat{y} \in \Pi_E(y)$. Then

$$\begin{aligned} d_K^2(x) - d_K^2(y) &\leq |x - \hat{y}|^2 - |y - \hat{y}|^2 \\ &= |x - y|^2 + 2\langle x - y, y - \hat{y}\rangle \\ &\leq |x - y|(|x - y| + 2|y - a|). \end{aligned}$$

3. Let $0 < \lambda < 1$ and $x, y \in \mathbb{R}^d$. Put $z = \lambda x + (1-\lambda)y$. Then there exists a $\hat{z} \in E$ such that $d_K(z) = \|z - \hat{z}\|$. Hence

$$\begin{aligned} |z|^2 - d_K^2(z) &= |z|^2 - |z - \hat{z}|^2 \\ &= 2\langle z, \hat{z}\rangle - |\hat{z}|^2 \\ &= \lambda\left(2\langle x, \hat{z}\rangle - |\hat{z}|^2\right) + (1-\lambda)\left(2\langle y, \hat{z}\rangle - |\hat{z}|^2\right) \\ &= \lambda(|x|^2 - |x - \hat{z}|^2) + (1-\lambda)(|y|^2 - |y - \hat{z}|^2) \\ &\leq \lambda(|x|^2 - d_E^2(x)) + (1-\lambda)(|y|^2 - d_E^2(y)). \end{aligned}$$

4. According to Alexandrov's Theorem (1939),[1] the function $x \mapsto |x|^2 - d_K^2(x)$ is almost everywhere twice differentiable, consequently so is $x \mapsto d_K^2(x)$.

Chapter 5

Exercise 5.1

Let $p \geq 2$, $\delta \geq 0$ and the Banach space

$$\mathbb{V}_{m,k}^{\delta,p}(0,T) \stackrel{def}{=} \left\{(Y,Z) \in S_m^0[0,T] \times \Lambda_{m\times k}^0(0,T) : \|(Y,Z)\|_{\delta V} < \infty\right\},$$

where

$$\begin{aligned} \|(Y,Z)\|_{\delta V}^p \stackrel{def}{=} \mathbb{E}\sup_{s\in[0,T]} e^{\delta p V_s}|Y_s|^p + \mathbb{E}\left(\int_0^T e^{2\delta V_s}|Y_s|^2 L_s dQ_s\right)^{p/2} \\ + \mathbb{E}\left(\int_0^T e^{2\delta V_s}|Z_s|^2 ds\right)^{p/2}, \end{aligned}$$

and the complete metric space $\mathbb{V}_{m,k}^p(0,T) = \bigcap_{\delta\geq 0}\mathbb{V}_{m,k}^{\delta,p}(0,T)$.

[1] Alexandrov, Alexandr Danilovich (1939) *The existence almost everywhere of the second differential of a convex function and some associated properties of convex surfaces.* (in Russian), Ucenye Zapiski Leningrad. Gos. Univ. Ser. Math. Vol. 37, N. 6, pp. 3–35.

Using Lemma 6.58 we show that the mapping $\Gamma : \mathbb{V}^p_{m,k}(0,T) \to \mathbb{V}^p_{m,k}(0,T)$ given by

$$\begin{cases} (Y,Z) = \Gamma(X,U) \\ Y_t = \eta + \int_t^T \Phi(s, X_s, U_s)\, dQ_s - \int_t^T Z_s dB_s \end{cases}$$

has a unique fixed point in $\mathbb{V}^p_{m,k}(0,T)$. First Γ is well defined because by Corollary 2.45 $\mathbb{E}\ \sup_{t\in[0,T]} e^{p\delta V_t}|Y_t|^p < \infty$ and by the inequality

$$\begin{aligned} &|Y_s|^2 L_s dQ_s + \langle Y_s, \Phi(s, X_s, U_s)\rangle dQ_s \\ &\leq \tfrac{1}{4(\delta-1)}|X_s|^2 L_s dQ_s + \tfrac{1}{2\delta}|U_s|^2 ds + |Y_s|\,|\Phi(s,0,0)|\, dQ_s + |Y_s|^2 \delta dV_s,\ \forall \delta > 1 \end{aligned}$$

and Proposition 5.2 we get $\|(Y,Z)\|^p_{\delta V} < \infty$ for all $\delta > 1$.

From the inequality

$$\begin{aligned} &|Y_s - Y_s'|^2 L_s dQ_s + \langle Y_s - Y_s', \Phi(s, X_s, U_s) - \Phi(s, X_s', U_s')\rangle dQ_s \\ &\leq \tfrac{1}{2\delta}|U_s - U_s'|^2 ds + \tfrac{1}{4(\delta-1)}|X_s - X_s'|^2 L_s dQ_s + |Y_s - Y_s'|^2 \delta dV_s,\ \forall \delta > 1 \end{aligned}$$

and Proposition 5.2 we obtain

$$\|(Y - Y', Z - Z')\|^p_{\delta V} \leq \frac{C_p}{(\delta-1)^{p/2}} \|(X,U) - (X',U')\|^p_{\delta V},\ \forall \delta > 1$$

which tells us there exists a $\delta_0 > 1$ such that Γ is a *strict contraction* on $\left(\mathbb{V}^p_{m,k}(0,T), \|\cdot\|_{\delta V}\right)$, for all $\delta \geq \delta_0$, and consequently, by Lemma 6.58, Γ has a unique fixed point in $\mathbb{V}^p_{m,k}(0,T)$.

Exercise 5.3: Since

$$\begin{aligned} &\left(Y_t^\varepsilon - Y_t^\delta\right)\left(G_\varepsilon\left(t, Y_t^\varepsilon, Z_t^\varepsilon\right) - G_\delta\left(t, Y_t^\delta, Z_t^\delta\right)\right) \\ &\leq L\left|Y_t^\varepsilon - Y_t^\delta\right|\left(2 + |Y_t^\varepsilon| + \left|Y_t^\delta\right| + |Z_t^\varepsilon| + \left|Z_t^\delta\right|\right) \end{aligned}$$

we obtain, by Proposition 5.2, with $N = 0$, $V = 0$, $\lambda = 0$, that

$$\begin{aligned} &\mathbb{E}\left(\sup_{s\in[0,T]}\left|Y_s^\varepsilon - Y_s^\delta\right|^p\right) + \mathbb{E}\left(\int_0^T \left|Z_s^\varepsilon - Z_s^\delta\right|^2 ds\right)^{p/2} \\ &\leq C_p \mathbb{E}\left(\int_0^T L\left|Y_s^\varepsilon - Y_s^\delta\right|\left(2 + |Y_s^\varepsilon| + \left|Y_s^\delta\right| + |Z_s^\varepsilon| + \left|Z_s^\delta\right|\right) ds\right)^{p/2} \end{aligned}$$

$$\leq C_p \left[\mathbb{E} \left(\sup_{s\in[0,T]} |Y_s^\varepsilon - Y_s^\delta|^p \right) \right]^{1/2}$$

$$\times \left[\mathbb{E} \left(\int_0^T L \left(2 + |Y_s^\varepsilon| + \left|Y_s^\delta\right| + |Z_s^\varepsilon| + \left|Z_s^\delta\right|\right) ds \right)^p \right]^{1/2} .$$

Exercise 5.5: We apply the existence and uniqueness result from Theorem 5.27 and the comparison result from Theorem 5.33 for the BSDE

$$Y_t = \eta + \int_t^T Y_s \left(1 - Y_s^+\right) ds - \int_t^T \langle Z_s, dB_s \rangle$$

with $0 \leq \eta \leq 1$.

Exercise 5.7: Assume that E is not convex. We shall show there exists a bounded continuous function $g : \mathbb{R}^k \to E$ such that

$$P\left(\{Y_t \notin E\}\right) > 0, \text{ for some } t \in [0, T].$$

If E is not convex, we can find $a, b \in \operatorname{Bd}(E)$ such that $a \neq b$ and $a + \lambda (b - a) \notin E$ for all $\lambda \in]0, 1[$. Let $\delta = \frac{1}{4} d_E(\frac{a+b}{2}) > 0$. Define $g : \mathbb{R}^k \to E$ by $g\left(x^{(1)}, x^{(2)}, \ldots, x^{(k)}\right) = a + (b - a)\, \mathbf{1}_{(-\infty,1]}\left(x^{(1)}\right)$. By Exercise 1.7 we have

$$\mathbb{E}\left(g\left(B_T^{(1)}\right)|\mathcal{F}_t\right) = a + (b - a)\, \Phi\left(\frac{1 - B_t^{(1)}}{\sqrt{T - t}}\right),$$

where

$$\Phi(r) := \frac{1}{\sqrt{2\pi}} \int_{-\infty}^r e^{-\frac{x^2}{2}} dx, \; r \in \mathbb{R}.$$

We also have

$$\left|Y_t - \mathbb{E}[g\left(B_T^{(1)}\right)|\mathcal{F}_t]\right| \leq \mathbb{E}\left|\int_t^T F_s ds|\mathcal{F}_t\right| \leq M\,(T - t) \leq \delta,$$

if $t \in [T - \frac{\delta}{M}, T]$, where $M > 0$ denotes the bound of F.

Then for all $t \in [T - \frac{\delta}{M}, T]$,

$$\left|Y_t - \frac{a+b}{2}\right| \leq \left|Y_t - \mathbb{E}[g\left(B_T^{(1)}\right)|\mathcal{F}_t]\right| + \left|\mathbb{E}[g\left(B_T^{(1)}\right)|\mathcal{F}_t] - \frac{a+b}{2}\right|$$

$$\leq \delta + |b - a| \left|\Phi\left(\frac{1 - B_t^{(1)}}{\sqrt{T - t}}\right) - \frac{1}{2}\right|.$$

Therefore

$$
\begin{aligned}
0 < \mathbb{P}&\left[\left|\Phi\left(\frac{1 - B_t^{(1)}}{\sqrt{T-t}}\right) - \frac{1}{2}\right| \leq \frac{\delta}{|b-a|}\right] \\
\leq& \mathbb{P}\left(\left|Y_t - \frac{a+b}{2}\right| \leq 2\delta\right) \\
\leq& \mathbb{P}\left(Y_t \notin E\right).
\end{aligned}
$$

References

1. Asiminoaei, Ioan; Răşcanu, Aurel: are not cited in the text. Please provide the citation or delete them from the list. *Approximation and simulation of stochastic variational inequalities—splitting up method*, Numer. Funct. Anal. Optim. **18** (1997), no. 3-4, 251–282.
2. Barbu, Viorel: *Nonlinear semigroups and differential equations in Banach spaces*, Editura Academiei Republicii Socialiste România, Bucharest, 1976, Translated from the Romanian.
3. Barbu, Viorel: *Optimal Control of Variational Inequalities*, Pitman Research Notes in Mathematics, 100, London - Boston, 1984.
4. Barbu, Viorel: *Mathematical Methods in Optimization of Differential Systems*, Kluwer Academic Publishers, 1994.
5. Barbu, Viorel; Răşcanu, Aurel: *Parabolic variational inequalities with singular inputs*, Differential Integral Equations **10** (1997), no. 1, 67–83.
6. Barles, Guy; Buckdahn, Rainer; Pardoux, Etienne: *Backward stochastic differential equations and integral-partial differential equations*, Stochastics and Stochastic Reports **60** (1997), 57–83.
7. Bensoussan, Alain: *Some existence results for stochastic partial differential equations*, Stochastic partial differential equations and applications (Trento, 1990), pp.37–53, Pitman Res. Notes Math. Ser., 268, Lagman Sci. Tech., Harlow, 1992.
8. Bensoussan, Alain; Răşcanu, Aurel: *Parabolic variational inequalities with random inputs*, Les grands systèmes des sciences et de la technologie, RMA Res. Notes Appl. Math., vol. 28, Masson, Paris, 1994, pp. 77–94.
9. Bensoussan, Alain; Răşcanu, Aurel: *Stochastic variational inequalities in infinite-dimensional spaces*, Numer. Funct. Anal. Optim. **18** (1997), no. 1-2, 19–54.
10. Billingsley, Patrick: *Probability and measure*, third ed., Wiley Series in Probability and Mathematical Statistics, John Wiley & Sons Inc., New York, 1995, A Wiley-Interscience Publication.
11. Billingsley, Patrick: *Convergence of Probability Measures*, second ed., Wiley Series in Probability and Mathematical Statistics, John Wiley & Sons Inc., New York, 1999.
12. Brezis, Haïm: *Opérateurs maximaux monotones et semi-groupes de contractions dans les espaces de Hilbert*, Math. studies 5, North Holland (1973).
13. Briand, Philippe; Deylon, Bernard; Hu, Ying; Pardoux, Etienne; Stoica, Lucretiu: *L^p solutions of backward stochastic differential equations*, Stochastic Process. Appl., **108** (2003), 109–129.
14. Buckdahn, Rainer; Quincampoix, Marc; Rainer, Catherine; Răşcanu, Aurel: *Viability of moving sets for stochastic differential equation*, Adv. Differential Equations **7** (2002), no. 9, 1045–1072.

E. Pardoux and A. Răşcanu, *Stochastic Differential Equations, Backward SDEs, Partial Differential Equations*, Stochastic Modelling and Applied Probability 69, DOI 10.1007/978-3-319-05714-9,

15. Buckdahn, Rainer; Quincampoix, Marc; Răşcanu, Aurel: *Viability property for a backward stochastic differential equation and applications to partial differential equations*, Probab. Theory Related Fields **116** (2000), no. 4, 485–504.
16. Buckdahn, Rainer; Răşcanu, Aurel: *On the existence of stochastic optimal control of distributed state system*, Nonlinear Anal. **52** (2003), no. 4, 1153–1184.
17. Colombo, Giovanni; Goncharov, Vladimir: *Variational Inequalities and Regularity Properties of Closed Sets in Hilbert Spaces,* Journal of Convex Analysis, Volume 8 (2001), No. 1, 197–221.
18. Crandall, Michael G.; Ishii, Hitoshi; Lions, Pierre-Louis: *User's guide to viscosity solutions of second order partial differential equations*, Bull. Amer. Soc. **27** (1992), 1–67.
19. Darling, Richard W. R.: Constructing gamma-martingales with prescribed limit, using backwards SDE. Ann. Probab. **23** (1995), no. 3, 1234–1261.
20. Darling, Richard W. R.; Pardoux, Etienne: *Backwards SDE with random terminal time and applications to semilinear elliptic PDE*, Ann. Probab. **25** (1997), no. 3, 1135–1159.
21. Degiovanni, Marco; Marino, Antonio; Tosques, Mario: *Evolution equations with lack of convexity*, Nonlinear Anal. Th. Meth. Appl. 9(2) (1985), 1401–1443.
22. Delarue, François: *On the existence and uniqueness of solutions to FBSDEs in a non-degenerate case*, Stochastic Process. Appl. **99** (2002), no. 2, 209–286.
23. Delarue, François; Guatteri, Giuseppina.: *Weak existence and uniqueness for forward-backward SDEs*, Stochastic Process. Appl. **116** (2006), no. 12, 1712–1742.
24. Dupuis, Paul; Ishii, Hitoshi: *SDEs with oblique reflection on nonsmooth domains*, Ann. Probab. **21** (1993), no. 1, 554–580.
25. Dynkin, Eugene B.: *Superprocesses and partial differential equations*, Annals of Prob. **21**, 1185–1262, 1993.
26. Ekeland, Ivar: *On the Variational Principle*, J. Math. Anal. Appl., 47, p. 324–353, 1974.
27. El Karoui, Nicole; Kapoudjian, Christophe; Pardoux, Étienne; Peng, Shige; Quenez, Marie Claire: *Reflected solutions of backward SDE's, and related obstacle problems for PDE's*, Ann. Probab. **25** (1997), no. 2, 702–737.
28. El Karoui, Nicole; Mazliak, Laurent (eds.): *Backward stochastic differential equations*, Pitman Research Notes in Mathematics Series, vol. 364, Longman, Harlow, 1997, Papers from the study group held at the University of Paris VI, Paris, 1995–1996.
29. El Karoui, Nicole; Pardoux, Étienne; Quenez, Marie Claire: *Reflected backward SDEs and American options*, Numerical methods in finance, Publ. Newton Inst., Cambridge Univ. Press, Cambridge, 1997, pp. 215–231.
30. El Karoui, Nicole; Peng, Shige; Quenez, Marie Claire: *Backward stochastic differential equations in finance*, Math. Finance **7** (1997), no. 1, 1–71.
31. Fleming, Wendell H.; Soner, Halil Mete: *Controlled Markov Processes and Viscosity Solutions*, Springer-Verlag, New York, 1993.
32. Friedman, Avner: *Stochastic differential equations and applications. Vol. 1*, Academic Press, New York, Vol. 1 - 1975, Vol. 2 - 1976, Probability and Mathematical Statistics, Vol. 28.
33. Gassous, Anouar; Răşcanu, Aurel; Rotenstein, Eduard: *Stochastic variational inequalities with oblique subgradients*, Stochastic Process. Appl., **122** (2012), 2668–2700.
34. Gégout-Petit, Anne; Pardoux, Étienne: *Équations différentielles stochastiques rétrogrades réfléchies dans un convexe*, Stochastics and Stochastics Rep. **57**, no. 1–2, 111–128, 1996.
35. Gikhman, Iosif Ilich; Skorokhod, Anatoliy Volodymyrovych: *Stochastic differential equations*, Springer-Verlag, New York, 1972, Translated from the Russian by Kenneth Wickwire, Ergebnisse der Mathematik und ihrer Grenzgebiete, Band 72.
36. Gyöngy, Istvan; Krylov, Nikolai Vladimirovich: *On stochastic equations with respect to semimartingales. I,* Stochastics 4, no. 1, 1–21., 1980/81.
37. Hamadene, Said; Lepeltier, Jean-Pierre; Peng, Shige: *BSDEs with continuous coefficients and stochastic differential games*, Backward stochastic differential equations (Harlow) (N. El Karoui and L. Mazliak, eds.), Longman, 1997, pp. 115–128.
38. Ikeda, Nobuyuki; Watanabe, Shinzo: *Stochastic differential equations and diffusion processes,* North–Holland/Kodansha, 1981.

39. Itô, Kiyosi: *Stochastic integral*, Proc. Imp. Acad. Tokyo 20 (1944), 519–524.
40. Jacod, Jean: *Une condition d'existence et d'unicité pour les solutions fortes d'équations différentielles stochastiques,* (French) Stochastics 4, no. 1, 23–38, 1980/81.
41. Jakubowski, Adam: *A non-Skorohod Topology on the Skorohod space*, Electronic Journal of Probability 2 (1997), 1–21.
42. Karatzas, Ioannis; Shreve, Steven E.: *Brownian motion and stochastic calculus*, second ed., Graduate Texts in Mathematics, vol. 113, Springer-Verlag, New York, 1991.
43. Lions, Pierre-Louis; Sznitman, Alain-Sol: *Stochastic differential equations with reflecting boundary conditions,* Comm. Pure Appl. Math. 37 (1984), no. 4, 511–537.
44. Ma, Jin; Yong, Jiongmin: *Forward-backward stochastic differential equations and their applications*, Lecture Notes in Mathematics, vol. 1702, Springer-Verlag, Berlin, 1999.
45. MacKean, Henry P.: *A class of Markov processes associated with nonlinear parabolic equations*, Proc. Nat. Acad. Sci. **56**, 1907–1911, 1966.
46. Maticiuc, Lucian; Răşcanu, Aurel: *On the continuity of the probabilistic representation of a semi linear Neumann–Dirichlet problem*, Bernoulli (2014), to appear.
47. Maticiuc, Lucian; Răşcanu, Aurel: *Backward Stochastic Variational Inequalities on Random Intervals*, submitted.
48. Maticiuc, Lucian; Pardoux, Étienne; Răşcanu, Aurel; Zalinescu, Adrian: *Viscosity solutions for systems of parabolic variational inequalities*, Bernoulli 16 (2010), no. 1, 258–273.
49. Øksendal, Bernt: *Stochastic differential equations*, sixth ed., Universitext, Springer-Verlag, Berlin, 2003.
50. Paley, Raymond; Wiener, Norbert; Zygmund, Antoni: *Notes on random functions*, Mathematische Zeitschrift, **3**, (1933), 647–668.
51. Pardoux, Étienne; Peng, Shige: *Adapted solution of a backward stochastic differential equation*, Syst. Control Lett. **14** (1990), 55–61.
52. Pardoux, Étienne: *Backward stochastic differential equations and viscosity solutions of systems of semilinear parabolic and elliptic PDEs of second order*, Stochastic analysis and related topics, VI (Geilo, 1996), Progr. Probab., vol. 42, Birkhäuser Boston, Boston, MA, 1998, pp. 79–127.
53. Pardoux, Étienne: *BSDEs, weak convergence and homogenization of semilinear PDEs*, Nonlinear analysis, differential equations and control (Montreal, QC, 1998), NATO Sci. Ser. C Math. Phys. Sci., vol. 528, Kluwer Acad. Publ., Dordrecht, 1999, pp. 503–549.
54. Pardoux, Étienne; Peng, Shige: *Backward stochastic differential equations and quasilinear parabolic partial differential equations*, Lecture Notes in Control and Inform. Sci., vol. 176, Springer, 1992, pp. 200–217.
55. Pardoux, Étienne; Pradeilles, Frédéric; Rao, Zusheng: *Probabilistic interpretation of a system of semi-linear parabolic partial differential equations*, Ann. Inst. H. Poincaré Probab. Statist. **33** (1997), no. 4, 467–490.
56. Pardoux, Étienne; Răşcanu, Aurel: *Backward stochastic differential equations with subdifferential operator and related variational inequalities*, Stochastic Process. Appl. **76** (1998), no. 2, 191–215.
57. Pardoux, Étienne; Răşcanu, Aurel: *Backward stochastic variational inequalities*, Stochastics and Stochastics Rep. **67** (1999), no. 3-4, 159–167.
58. Pardoux, Étienne; Sow, Ahmadou Bamba: *Probabilistic interpretation of a system of quasilinear parabolic PDEs*, Stochastics and Stochastic Rep. **76** (2004), 429–477.
59. Pardoux, Étienne; Zhang, Shuguang: *Generalized BSDEs and nonlinear Neumann boundary value problems*, Probab. Theory Related Fields **110** (1998), no. 4, 535–558.
60. Pham, Huyên: *Continuous-time Stochastic Control and Optimization with Financial Applications*, Stochastic Modelling and Applied Probability, Vol. **61**, Springer 2009 (Original French edition published as volume 61 in the series Mathématiques & Applications).
61. Rao, Malempati Mahusudana; Swift, Randall: *Probability Theory with Applications*, New York, Springer, 2006.
62. Răşcanu, Aurel: *Existence for a class of stochastic parabolic variational inequalities*, Stochastics **5** (1981), no. 3, 201–239.

63. Răşcanu, Aurel: *Deterministic and stochastic differential equations in Hilbert spaces involving multivalued maximal monotone operators*, Panamer. Math. J. **6** (1996), no. 3, 83–119.
64. Revuz, Daniel; Yor, Marc: *Continuous martingales and Brownian motion*, third ed., Grundlehren der Mathematischen Wissenschaften [Fundamental Principles of Mathematical Sciences], vol. 293, Springer-Verlag, Berlin, 1999.
65. Rockafellar, R. T.: *Convex analysis*, Princeton University Press, Princeton, N.J. Princeton Mathematical Series, No. 28, 1970.
66. Rossi, Riccarda; Savaré, Giuseppe: *Gradient flows of non convex functionals in Hilbert spaces,* ESAIM Control Optim. Calc. Var. 12 (2006) 564–614.
67. Saisho, Yasumasa: *Stochastic differential equations for multidimensional domain with reflecting boundary,* Probab. Theory Related Fields 74 (1987), no. 3, 455–477.
68. Stroock Daniel W.: *Lectures on stochastic analysis: diffusion theory*, London Mathematical Society Student Texts, vol. 6, Cambridge University Press, Cambridge, 1987.
69. Stroock, Daniel W.; Varadhan, S.R.S.: *On the support of diffusion processes with applications to the strong maximum principle*, Proc. of the sixth Berkeley Symposium on Mathematical Statistics and Probability (Univ. of California Press) 3: 333–359, 1972.
70. Vrabie, Ioan: *C_0-semigroups and applications,* North-Holland, Elsevier, Amsterdam, 2003.
71. Zălinescu, Constantin: *Convex analysis in general vector spaces*, World Scientific, Singapore, 2002.

Index

E. Pardoux and A. Răşcanu, *Stochastic Differential Equations, Backward SDEs, Partial Differential Equations*, Stochastic Modelling and Applied Probability 69,
DOI 10.1007/978-3-319-05714-9,

GPSR Compliance
The European Union's (EU) General Product Safety Regulation (GPSR) is a set of rules that requires consumer products to be safe and our obligations to ensure this.

If you have any concerns about our products, you can contact us on

ProductSafety@springernature.com

In case Publisher is established outside the EU, the EU authorized representative is:

Springer Nature Customer Service Center GmbH
Europaplatz 3
69115 Heidelberg, Germany

www.ingramcontent.com/pod-product-compliance
Ingram Content Group UK Ltd.
Pitfield, Milton Keynes, MK11 3LW, UK
UKHW021435280726

14060UKWH00001BA/95

9783319057132